The Ecological Bases for Lake and Reservoir Management

Developments in Hydrobiology 136

Series editor
H. J. Dumont

The Ecological Bases for Lake and Reservoir Management

Proceedings of the Ecological Bases for Management of Lakes and Reservoirs Symposium, held 19–22 March 1996, Leicester, United Kingdom

Edited by

David M. Harper, Bill Brierley, Alastair J.D. Ferguson and Geoff Phillips

Reprinted from Hydrobiologia, volumes 395/396 (1999)

Kluwer Academic Publishers
Dordrecht / Boston / London

Library of Congress Cataloging-in-Publication Data

Harper, David M.
The ecological bases for lake and reservoir management / edited by David M. Harper ... [et al.]
p. cm. -- (Developments in hydrobiology ; v. 136)
ISBN 0-7923-5785-X (alk. paper)
1. Freshwater ecology. 2. Ecosystem management. 3. Sustainable development. I. Title. II. Series: Developments in hydrobiology ; 136.
QH541.5.F7H37 1999
577.6--dc21 99-15358

ISBN 0-7923-5785-X

Published by Kluwer Academic Publishers,
P.O. Box 17, 3300 AA Dordrecht, The Netherlands

Sold and distributed in North, Central and South America
by Kluwer Academic Publishers,
101 Philip Drive, Norwell, MA 02061, U.S.A.

In all other countries, sold and distributed
by Kluwer Academic Publishers Group,
P.O. Box 322, 3300 AH Dordrecht, The Netherlands

Printed on acid-free paper

Printed in the Netherlands

Hydrobiologia **395/396:** v–vii, 1999.
D.M. Harper, B. Brierley, A.J.D. Ferguson & G. Phillips (eds), The Ecological Bases for Lake and Reservoir Management.

Contents

Section 4: Management Strategies for Different Lake Types

Section 5: Regional and National Strategies for Lake and Catchment Management

Hydrobiologia **395/396:** ix–x, 1999.
D.M. Harper, B. Brierley, A.J.D. Ferguson & G. Phillips (eds), The Ecological Bases for Lake and Reservoir Management.

Preface

The papers in this volume were all presented at a conference of the same name held in March 1996 at the University of Leicester. This followed a successful conference on the ecological basis for river management three years earlier (Harper & Ferguson, 1995). The 1993 conference was supported by the National Rivers Authority and the later one by its successor, the Environment Agency. The philosophy behind both was that there was a significant gap between our scientific understanding of processes in rivers and lakes, and our application of this understanding in practical management.

In lakes and reservoirs, such a gap might not at first seem very obvious. For thirty years scientists have known about the relationship between phosphorus and algal growth as first revealed by Vollenweider (Vollenweider, 1968) and a great deal of successful management has been based upon this seemingly simple relationship. However, the complexities of phosphorus dynamics and sediment–water relationships, together with the difficulties in many circumstances (such as lowland catchments) of reducing phosphorus concentrations low enough have meant that the majority of lakes need a multi-faceted approach to manage or restore them. Hence the development over the past twenty years of 'biomanipulation' as well as physical management techniques such as artificial mixing within lake basins coupled with many ways of dealing with phosphorus in transit from its sources in the catchment to the lake downstream, techniques well represented in this volume.

One reason that there is a gap between what we know about freshwater catchments and how we manage them, is that eutrophication has 'crept back' as a problem of the 1990s. Only a decade or so ago it was possible to say that eutrophication was not a problem for the United Kingdom because of advanced water purification technology. Almost exactly a decade ago, Harper wrote in a book preface that eutrophication had been 'eclipsed' by newer problems of aquatic systems such as acidification (Harper, 1992). He couldn't have been more wrong – that book was written just before the magnitude of the problems of diffuse sources of phosphorus from intensive agriculture became widely appreciated. This volume now has important contributions which deal with the protection of oligotrophic water bodies from relatively recent agricultural intensification, as well as those which deal with the integration of several methods of catchment and in-lake methods for nutrient management.

This is not to imply that problems of acidification, and of lake management other than for nutrients, have gone away. On the contrary, lake managers are faced with a range of problems of lakes. One response to this has been to develop methods for classifying lakes which recognise not just their present state and problems, but their past 'natural' state. Methods of palaeolimnology and hindcasting from catchment land use information are both valuable approaches which feature in this volume.

Modern technology has an increasing role to play in classification and management, in particularly remote sensing and GIS techiques. Although relatively under-developed compared with oceans and terrestrial vegetation, new methodologies for evaluation of catchment processes and lake state are also described here, together with computer-based systems for including the general public in decision-making. Whilst this latter development is illustrated for urban lakes, it may well prove to be a central feature of lake management in the 21st century.

Sustainability is the key word in water management. It is easy to define in words, but hard to implement. There is much debate on what the words actually mean in practice but one of the few areas where all practitioners are likley to agree is that they will involve far more integration of ecological techniques into catchment management decisions. The editors and the authors of the papers in this volume hope that it will make a significant contribution to that integration.

Leicester, March 1999

DAVID HARPER
BILL BRIERLEY
ALASTAIR FERGUSON
GEOFF PHILLIPS

References

Harper, D.M., 1982. Eutrophication of Freshwaters. Chapman & Hall, London.

Harper, D.M. & A.J.D. Ferguson, 1995. The Ecological Basis for River Management. John Wiley & Sons, Chichester.

Section One

Twenty-first Century Challenges for Lake Management

Hydrobiologia **395/396:** 3–11, 1999.
D.M. Harper, B. Brierley, A.J.D. Ferguson & G. Phillips (eds), The Ecological Bases for Lake and Reservoir Management.

Ecological challenges for lake management

Brian Moss
School of Biological Sciences, Derby Building, University of Liverpool, Liverpool L69 3BX, U.K.

Key words: ecosystem approach, restoration, philosophy of science

Abstract

The need for management, in natural systems, and possibly in all systems, reflects an inability of the systems to operate in self-sustaining ways, due to interference, or damage to an extent that is beyond the capacities of the system for self-repair. Management need is thus a symptom of failure of the system. The more management needed, the greater the failure and management is always costly. Most freshwater systems have been seriously altered by human activities. We may wish to restore them to self-sustaining systems which provide conservation or amenity values, or products such as potable water or fish, with minimal management, or preferably none at all. This may not be possible in many cases; it is completely impossible without profound understanding of their functioning. The ecological 'challenges' in lake management are thus to remove blocks which impede this understanding. There are at least three groups of blocks: fundamental; those of approach and method; and political. The latter is no less ecological than the others; human beings and their behaviour are ecological phenomena like those of any other animal. Fundamental blocks include the complexity and stochasticity of natural systems, which may change faster than they can be dissected and understood. They include also the scarcity of complete systems for investigation. Lakes are parts of greater systems in the landscape and if these systems are no longer intact, through insensitive land use, such as drainage, it is not possible to understand interactions and influences which were not confined to the lake basin. Examples are given concerning fish and bird movements. Methodological blocks concern the scale of investigations. The need for statistical rigour confines many investigations to small scale, easily replicable phenomena. Lake systems, however, are influenced by larger scale, lakewide or more extensive effects. An example is given concerning the effects of piscivorous birds. Unless whole lakes and their catchments are subject to experimentation, these influences will remain obscure, though they may be at least as important as within-lake effects. Replication is difficult and areas with large numbers of broadly similar lakes are needed. Such areas exist, but the costs of such large scale and long-term experimentation are rarely met. Political blocks are the ultimate blocks. A society which chooses to continue exploiting natural resources in an unsustainable way is not particularly interested in the fundamental understanding which underlies sustainability. There are many symptoms that we are choosing to follow the imperatives of our selfish genes, rather than make the ethical choices of which we are alternatively capable. They include the greatly reduced support for fundamental research in the U.K., the failure to use, in the spirit in which it was intended, existing European legislation for environmental protection, and the decreased strength of imperatives for conservation in the Environment Act 1995, compared with the Water Act 1989 and the Water Resources Act of 1991.

Words and language

Words, no less than numbers, are precision tools. They can be used as fine probes to convey significant shades of meaning but, like all fine tools, they can also be easily blunted and made into bludgeons. William Shakespeare was one of our finest wordsmiths. His invectives alone are unsurpassed (Hill & Ottchen, 1991). You might disdain from being called the 'fount from which small brooks flow', a 'triton of the minnows', 'caddis garter', 'foolless gudgeon', or 'lily-livered boy', despite their connection with freshwater ecology.

Words can also be used to convey messages of subtler content, though equally forcibly intended. W.

H. Auden's poem seems to me to make an important statement, to which I will later return:

"We would rather be ruined than changed,
we would rather die in our dread
than climb the cross of the moment
and let our illusions die."

Those who use words well are keenly sensitive to the horrors of verbal debasement. One of Shakespeare's epithets was 'corruptor of words', and the Book of Job talks of 'he who multiplieth words without knowledge' (Job, 35,16). The odious Humpty Dumpty, in Lewis Carroll's 'Through the Looking Glass', used words to mean 'just what I choose it to mean – neither more nor less' and George Orwell articulated the current problem in his novel, '1984'. 'New speak' was the deliberately ambiguous and misleading language of bureaucrats and politicians. *Challenge, vision, relevant, target and management* are part of the new speak of today. And so I reach two of the key words in the title I was given, and accepted.

I suppose the intended agendum of the title is that lakes have to be managed, that they are complex, that there is inadequate understanding, that therefore there needs to be more research, that understanding is just a matter of time as a beneficent government and industry supports the relevant research to attain this target. The challenge, I presume, is to gain this understanding. And the word *challenge*, though it can have many shades of meaning, is intended in the context of aggressively meeting the problem, a throwing down of a gauntlet. It seems to me that such aggressive new speech is part of the problem, not of the solution.

Blocks to understanding

A more accurate term than *challenge* would be *block*. The word challenge is currently used in the hope of injecting zest into situations where, due to ignorance or dogma, support has been reduced, and motivation has been eroded. 'Management' is also a weasel word. The need for management, in natural systems, and perhaps in others, is a reflection that all is not well. It is a measure of a failing to conserve systems that could be self-sustaining. The more management that is needed, the greater the failure, and management is costly. Our aim should therefore be to restore systems to conditions where only minimal management, or none, is needed. To achieve this requires profound knowledge of how the systems function.

The title, in plain English, is thus translated to *What blocks are there to increasing our understanding of how lake systems function*? Unless function, with all its antecedents of evolution and biogeography, is understood, there cannot be sensible management nor predictable restoration. I see blocks to understanding that are fundamental to the systems themselves, those that are concerned with method and approach, and those that are political and imposed by government and its agencies.

The fundamental blocks concern the complexity of the systems themselves, and the current fragmentation of these systems in a landscape greatly influenced by human activities. The methodological blocks concern the scale on which we are able and willing to work. The governmental blocks concern the motives and aspirations of society as reflected, perhaps, in its government. We need to ask if it is really intended that the *challenges* be met?

They are all, even the latter, ecological blocks for, notwithstanding the real concerns of many non-biologists, human beings are biologically driven to a significant extent and their behaviour reflects the operation of the selfish gene no less than the beaver as dam builder (Diamond, 1991). The dominant themes of politics, as those of art and literature, are those of human behaviour, those largely of the selfish gene. There are differences from other animals, of course. The powers our particular genes have given for modifying the Earth's environment are stupendous on the one hand, but, on the other, we may choose not to use these powers. This is a choice given to no other organism.

Fundamental blocks – determinism, complexity and stochasticity

Criteria of a full understanding of an ecosystem might be that we know exactly what species are in it, and how many of each there are, why they are there, what they are doing, and why they are doing it. In some of the simpler Antarctic fellfield communities of a few species of bacteria, algae, mosses and mites, considerable understanding might be possible. In a lake system there are several thousands of bacterial (Torsvik et al., 1990a, b) and algal genotypes, several hundreds of those of invertebrates and several tens of vertebrates and higher plants. Detailed understanding will clearly be unlikely, if not impossible, even were the community completely deterministic – fixed by the conditions extant at that place and time.

A popular approach of British-trained plant ecologists in the middle of this century assumed that the ecology of plants depended on the plant's physiology and the physical and chemical properties of the soil, all seen in the context of the local microclimate. Given particular combinations of these environmental factors, the plant would be present and able to grow and reproduce. It was an approach I implicitly adopted in carrying out culture-based studies on the distribution of eutrophic and oligotrophic algae in the late 1960s (Moss, 1973).

Hutchinson (1957) began to upset this view by his definition of the fundamental and realised niches of organisms. The fundamental niche included, in a multidimensional hyperspace, all the possible combinations of circumstances in the habitat that an organism could tolerate. But, because the organisms themselves and their interactions with one another were also included as dimensions, the possibility of precise determinism receded. When it was realised that particular combinations of circumstances could be occupied by many different organisms and that competition for niche space was usual, a stochastic element was introduced.

In 1965, Hutchinson introduced the idea of the 'ecological theatre' and the 'evolutionary play'. The organisms evolve; they continually change in response to changes in the environment and the environment, being partly determined by the organisms, reciprocally responds. Environmental determinism became only a small part of ecological understanding. The realities of chance colonisation and chance extinction confound even further a simple predictability of community composition. For example, organisms can persist in habitats which they have not reached by natural means but to which they have been introduced. Their presence or absence may be randomly determined. Reynoldson (1966) provided a good example for flatworms. Four species of carnivorous flatworm are common in U.K. standing waters – *Polycelis tenuis*, *P. nigra*, *Dugesia polychroa* and *Dendrocoelum lacteum*. There was a degree of determinism in that *Dugesia* and *Dendrocoelum* declined in numbers with decreasing calcium concentrations and were usually absent from waters with less than 5 mg l^{-1} Ca. Northern Scottish waters were thus devoid of all but *Polycelis* species. However, all four species would persist indefinitely in waters of even the lowest calcium concentrations, if supplied with suitable food. The absence of all species from lakes on some Scottish islands, and of two of the four species from others, seems simply to reflect that in the comparatively short period of 10 000 years or so since deglaciation, there has simply not been time for natural colonisation to have occurred. British waters are similarly devoid of many fish species that were probably interglacial residents, and could be current ones, for similar reasons of prevention of colonisation by sea level rise following deglaciation, and isolation from sources in mainland Europe.

It is also well established that many alternative communities can occupy a given set of conditions and that there is a high degree of chance as to which particular collection is extant at a given time (McIntosh, 1995; Drake, 1989). The stochastic element in communities thus blocks a full understanding.

There are, for example, at least 30 submerged aquatic plant species that can persist in moderately fertile shallow lakes in the U.K. An average lake, however, might have almost any permutation of five, dependent on accidents of colonisation and local extinction over many years. The number of combinations of five from 30 is a little over 17 million. The number of possible interaction pairs in 5 is 20. At its simplest (ignoring within-habitat clonal differences among each species or any complications arising from the heterogeneous distribution of associated organisms such as invertebrates), understanding the ecology of submerged plant communities in shallow, fertile lakes at the species level thus draws on consideration of about 340 million potential interactions. With the additional thousands of bacterial and algal genotypes, and the invertebrates and vertebrates, this approach – building up the functioning of the community in a strictly reductionist way, clearly has no future. There is a fundamental block in the size and stochasticity of the problem.

To a large extent this has been recognised in the use of functional groups or guilds of various kinds. A good example has been the concept of shredders, scrapers, collectors and predators among stream invertebrates (Cummins, 1974). This approach has also been particularly useful in understanding phytoplankton dynamics, where the behaviour of small, easily grazeable cells can be contrasted with that of large rapidly sinking ones, such as diatoms or of buoyant cyanophytes. Specific studies on particular predominant species, such as the classic work of Lund (1964) on *Asterionella formosa* and species of *Melosira* (*Aulacoseira*) have also contributed much understanding of how the phytoplankton communities function.

There will always be limits to understanding through these approaches because there will always be deviation from the general models, but functional groups limit the number of interactions that have to be considered and define an attainable ceiling for understanding. The problem arises in that important subtleties will be missed if we select particular groups or species on the basis of our existing, inevitably prejudiced, concepts. Thus an unsuspected keystone predator or parasite, present at low biomass, may be of far greater importance than the much larger biomass of some organism influenced by it.

Fundamental blocks – incomplete systems

The second fundamental block to ecological understanding is that we rarely have systems on which to work that are not severely altered by human activity. It is thus difficult to reconstruct the workings of a pristine system so that a damaged system can be restored and managed. The original blueprints have been destroyed. I am arguing, however, that human beings, whatever they do, are still ecological agents, to be understood as part of the current systems rather than as something separate, so this may seem an untenable argument. The difficulty comes because we are unwilling to accept this. We still want lakes that have features recorded in the past when our activities were at much lower levels and we assume that we can, by judicious restoration and management still have this. It may not be possible.

For example, many lakes in lowland regions were parts of complex floodplain wetland systems. To understand the lake ecosystem it is, among other things, necessary to understand its interactions with the surrounding wetlands, and the rise and fall of the river flood. There are obvious chemical and hydrological implications for this, but perhaps less obvious biological ones, particularly where vertebrates are concerned. This is important because birds and fish, though traditionally conceived as separate or dependent on the functioning of freshwater systems, may be crucial in structuring such systems (Leah et al., 1978, 1980; Andersson, 1981; Moss & Leah, 1982; Hurlbert et al., 1986; Andersson et al., 1988).

Piscivorous birds, for example, may take as much fish prey as piscivorous fish (Winfield, 1990); pike, through feeding on chicks, may influence duck populations (Hill et al., 1987); ducks and fish may compete for invertebrate prey (Winfield et al., 1989); and piscivorous and zooplanktivorous fish are key players in determining the clarity of water and the ultimate growth of plants in shallow lakes (Irvine et al., 1989; Moss, 1989, 1990, 1991, 1994; Scheffer et al., 1993). Grazing birds, such as coot (*Fulica atra*), may devastate aquatic plant communities if the birds gather at high densities (van Donk et al., 1994; van Donk & Gulati, 1995). Many of these vertebrates would have moved freely over the various sub-habitats in an intact floodplain system.

Most such floodplain systems in Europe have been drained and the rivers canalised. The annual flood is not permitted and the complexity of the former floodplain system has been lost. Movements of fish are severely restricted and although birds may move, there may be little alternative habitat to which they may move. They too, are thus confined. Klinge et al. (1995) suggested that in the original system in Holland, the pike (*Esox lucius*) populations built up in the extended habitat of the floodplain at high spring water levels and then concentrated in the residual lakes of the floodplain in summer. There, at highly concentrated population densities, they may have controlled zooplanktivorous fish and helped maintain clear water and dominance by aquatic plants, rather than phytoplankton.

The residual lakes alone, now cannot support such populations because of lack of spawning and lurking habitat, and indeed the former high summer populations of pike could only be sustained for a short time – but a crucial one for the establishment of plants. If this is correct, we may not be able to maintain a stable, diverse, plant-dominated community in many isolated lakes, without continual intervention, such as the stocking of predators or the removal of zooplanktivores.

Isolation of lakes through removal of surrounding natural habitat may also result in concentration of migratory birds. This may occur if they provide suitable habitat and food and there are few, or no, neighbourhood alternatives. The birds, especially grazers like coot and swan, may then destroy the plant communities through overgrazing. This appears to have happened at Zwemlust, in the Netherlands, where restored plant communities in an isolated lake have been devastated by invasions of coot (van Donk & Gulati, 1995), which have no alternative habitat in the immediate area.

The implications of these findings are that sustainable restoration, with minimal continued management, of such lakes, is not possible without restoration of the greater system of which they were formerly a part. Countries such as Denmark (Madsen, 1995) have already taken considerable steps towards the restoration of floodplain ecosystems but other European countries have yet to do so.

Methodological blocks – the ecosystem concept and scale

Sir Arthur Tansley provided a great service to ecology in his articulation (Tansley, 1935) of the ecosystem concept – that there exist more or less self-contained collections of organisms and their detritus in a particular set of physical and chemical conditions. The concept had the idea of self regulation and balance; the ecosystem was something around which a boundary line, if a hatched one, could be drawn. It was a concept very much conditioned by the compartmentalisation of the English countryside, with which Tansley, an inveterate field worker, was deeply familiar. It still forms the base of much management of habitats in the U.K., where remaining fragments of natural and semi-natural systems do appear as apparently self-contained islands in a sea of agricultural and urban land. It might also seem highly appropriate to the distinctive boundary created by the lake water's edge.

The ecosystem concept, however, has undergone evolution since Tansley's time, though his original idea of more or less self-containment strongly persists among landscape and conservation ecologists, if not among population and evolutionary ecologists. Where freshwater systems are concerned, the original concept can now form a block to understanding. Lakes do not only have hydrological catchments, from which water and substances are drawn, and without consideration of which, any real understanding is impossible. They also are influenced by processes outside their hydrological catchment due to the movements of atmospheric gases and migratory animals. The example of acid precipitation is obvious; the linkages of remote systems by migratory species have been more recent insights. Any idea of self-containment of a freshwater system is illusory.

Thus the headwater lakes of many river systems depend for their nutrient supplies, their periphyton growth, their invertebrate production and their salmonid (or other anadromous fish) recruitment on the carcases of adult fish that have made much of their growth in the sea and died after spawning (Krokhin, 1975; Richey et al., 1975; Durbin et al., 1979; Kline et al., 1990). The retention of the carcases, and hence the nutrients, in a flowing system depends on obstructions in the streams provided by naturally collapsed timber. A further link with the terrestrial systems, comes with the feeding on the migrating fish by bears. The lake is thus ultimately influenced by downstream forested areas and the ocean, as well as its immediate catchment and riparian vegetation. Ultimately, the only system around which a boundary of sorts can be drawn is the entire biosphere.

This point has not been taken widely enough by limnologists. There is a block to understanding when too reductionist an approach confines work to what is going on only in the lake itself, and sometimes to laboratory abstractions of these processes. Few have been able or willing to carry out the more desirable whole lake experiments. This is partly because of limited opportunities, at least in the U.K., which, except for Scotland, is not so well endowed with lakes as in Scandinavia or Canada, where such experiments have been more common. Partly it is also because of the statistical difficulties of handling unreplicated manipulations, and an unwillingness sometimes on the part of peers to accept data from such situations.

Primarily, however, it is a problem of obtaining replicated, manipulable systems on which to work, unhampered by other contemporary users, and with sufficient resources to carry out the extensive sampling and measurement such approaches demand. It is regrettable that the Canadian government proposes to withdraw much of its support from one of the few such facilities, the Experimental lakes Area in Ontario, where much classic work has already been carried out.

It is also a function of a culture that is dominated by reductionist science, in which experiments that give precise answers to simple questions are valued over those that give more equivocal answers to more complex and perhaps more appropriate questions about large and complex systems. We are wont to assume that the results of a laboratory experiment can be applied to a larger system, but this is most unlikely to be the case. For example, in one lake, Little Mere in Cheshire, laboratory release rates of phosphorus from sediments have been found to be much higher than those calculated from mass balance studies in the lake itself. Concentrations found in the lake were consistent with the whole lake calculations not the laboratory experiments. Results from mesocosm experiments carried out in the lake, on the other hand, were consistent (Beklioglu & Moss, 1996). This raises questions about how large experimental mesocosms should be, for the smaller they are, the greater the possibilities of replication.

The results of laboratory experiments, for example with algal cultures, have been valuable in elucidating whole lake phenomena, such as the succession of phytoplankton species. But it is only when such data are considered, together with larger scale phenomena, such as washout rates, that they really explain what is happening in the lake. The case of the cyanobacteria is particularly revealing. There is an immense literature on the cyanobacteria because of their particular interest as prokaryotes containing chlorophyll *a*, their great antiquity, their production of toxins and blooms and their migratory behaviour in the water column. The results of many observations and experiments suggest that, given a suitable inoculum, their growth is favoured, *inter alia*, by high pH, low free carbon dioxide concentrations, low nitrogen to phosphorus ratios, the presence of grazers on other, potentially competing, algae, stratified conditions, microaerophilic habitats, high temperatures and high retention time of the water mass (Shapiro, 1990). In a given lake, however, with many of these conditions present, cyanobacteria populations may still fail to develop, despite abundant inocula from sources upstream (Beklioglu & Moss, 1995).

The need to realise the greater dimensions of lake processes, than those of the limits of the basin, or a subsample of water from it, and the availability of properly replicated experimental systems on which to investigate some of these dimensions, are thus blocks to understanding.

Political blocks

Political blocks may seem far removed from ecological challenges. Yet they constitute the greatest of these. In the past, there was considerable support for fundamental science and a growing official acknowledgement of the importance of environmental issues. Particularly since the mid-1980s, there has been a continual erosion of that support. Now, much of the funding that is described as research funding in the U.K., and increasingly elsewhere, is in fact funding for technological development work, descriptive survey and monitoring or literature review. None of this is likely to increase understanding. It may proliferate knowledge, but that is a different matter.

The opportunities for fundamental research have greatly diminished and the application of market principles of competitive tendering for research contracts, has dissolved the former efficient cooperativeness that existed between different research institutions and individuals. It has also greatly increased the time spent, and often wasted, by active scientists in the raising of funds and the production of large numbers of ephemeral reports. These are often not available to the community at large and have not been screened by peer review. The value of this system has not been proved. The absence of publication of material in support of it indeed suggests that such proof is not to be found; it must certainly have been sought.

It should not be necessary to have to make a case for fundamental research, yet it appears that such is needed. The case given by Victor Weisskopf in 1965, when the American government realised that its scientific endeavours were being eclipsed by those of the Soviet Union, is still valid:

> "The value of fundamental research does not lie only in the ideas it produces. There is more to it. It affects the whole intellectual life of a nation by determining its way of thinking and the standards by which actions and intellectual production are judged. If science is highly regarded and if the importance of being concerned with the most up-to-date problems of fundamental research is recognized, then a spiritual climate is created which influences the other activities. An atmosphere of creativity is established which penetrates every cultural frontier. Applied sciences and technology are forced to adjust themselves to the highest intellectual standards which are developed in the basic sciences. This influence works in many ways; some fundamental students go into industry; the techniques which are applied to meet the stringent requirements of fundamental research serve to create new technological methods. The style, the scale and the level of scientific and technical work are determined in pure research; that is what attracts productive people and what brings scientists to those countries where science is at the highest level. Fundamental research sets the standards of modern scientific thought; it creates the intellectual climate in which our modern civilisation flourishes. It pumps the lifeblood of idea and inventiveness not only into the technological laboratories and factories, but into every cultural activity of our time. The case for generous support for pure and fundamental science is as simple as that".

But the generous support is no longer there. There has been a systematic plundering of research institutes and universities by which the background support for fundamental research has been removed. The ability to update instrumentation through a disinterested public sector has been lost and there has ensued a dependency on government agencies and worse, private industry, to maintain most scientific work. It is not possible to sustain impartial investigation under these conditions.

Indeed the fundamental contribution to freshwater ecology by British scientists has dwindled, absolutely and proportionately. For five major freshwater ecological journals published between 1980 and 1991, Hildrew (1993) found a fall in contribution from the U.K. from about 10% to 6% of papers, whilst North American contributions stayed steady at about 60% and mainland Europe increased its contribution from 14 to 25%. In 1992, Britain was overtaken by Australasia. It is not a consolation that despite maintaining its paper contribution, the North American limnological community feels equally beleagured (Jumars, 1990; Kalff, 1991; Wetzel, 1991).

The funds directed at responsive mode (that in which choice of research topic is determined only by quality through extensive peer review) by the research councils have effectively dwindled (Motluk, 1996). More of the proportionately reduced total 'pot' has been diverted into directed programmes, referred to as relevant, but hardly innovative, and usually directed at wealth creation and therefore at exploitation.

It is tempting to attribute this to confusion and ignorance rather than conspiracy, but the signs are there that it represents a deliberate move to hamper progress in fundamental understanding, especially in the environmental arena, lest it conflicts with particular economic and political aspirations. The interesting aspect is that these aspirations can be predicted quite closely from the operation of selfish genes.

The evidence of deliberate undermining of fundamental inquiry can be seen in budget patterns, in the nature and weakening importance of ministerial support for science, in the undermining of environmental agencies like the former U.K. Nature Conservancy Council by reorganisation, and in the phraseology of legislation. Thus the U.K.'s Water Act of 1989 and the Water Resources Act of 1991, which respectively set up and consolidated the powers of the National Rivers Authority, spoke, in Sections 8 and 16, respectively, of the duty of Ministers and the Authority to further the conservation and enhancement of natural beauty and the conservation of flora, fauna and geological or physiographical features of special interest.

The Environment Act of 1995, ostensibly strengthening the powers of environmental regulation of a new body, the Environmental Agency, formed from an amalgamation of the National Rivers Authority with other bodies, speaks only (Sections 8 and 9) of the recognition of codes of practice, for which the Agency must have regard and puts the onus for notifying issues of conservation importance on other bodies. This harks back to the Water Act of 1984, which set up the Water Authorities, where again there was no duty to conservation, but simply a requirement to have regard for such issues. *Having regard* is much weaker phraseology than *duty*. The same weakening can be seen also in the ways in which the U.K. Department of the Environment, Transport and the Regions is failing to use environmental safeguards provided through the EU Urban Waste Waters Treatment Directive (Carvalho & Moss, 1995).

In a general disregard for environmental issues, recent U.K. governments essentially behaved like the members of any other species, produced by natural selection, and subject to the prerogatives of its genetic endowment. Its strategies are the maximising of resource acquisition (wealth creation) so as to promulgate its particular strategies as prolifically as possible. In doing this, it takes no future view. Natural selection responds to the current environment, not even the subsequent one in which the selected generation will live.

It is an interpretation of human behaviour which has been denied by sociologists, but for which there is increasing evidence, as the strategies of human behaviour are teased out and compared with those of other animals. On this biological interpretation, our present environmental problems – of lakes and everything else – are understandable. No more than the beaver, creating his lodge, cares for the upstream impacts of damming a stream and flooding a woodland community, does the human dam-builder worry about the ultimate morality of his actions. He may predict some consequences, and seek to mitigate them, but there is no real question that the structure is not in the human interest. To the dam builder, the creation of hydroelectric power or the storage of irrigation or drinking water clearly is.

The alternative side of this issue is the essence of being human. It is to decide against responding to the strategy of the genes, to make decisions on ethical grounds, to regard the biosphere as a collaborator rather than a slave; to understand first and control only after much reflection. To look ahead is to prolong the span of the human species and to create a more equitable world, where the selfish gene is not allowed its ruthless consequences. Part of such an approach is a return to more enlightened attitudes towards fundamental understanding of natural systems, and to the removal of the political blocks now being increasingly placed on our abilities to understand our world. The lesson of history is that suppression of new ideas ultimately always fails, but it is a lesson that has had continually to be re-learned.

The immediate, the proximate and the ultimate

The reasons for any phenomenon, boil down to the immediate, the proximate and the ultimate. With respect to understanding ecological issues in lake systems, fundamental issues might seem to be the ultimate blocks and political ones merely immediate. However, because the political ones are, in the final analysis, powerful products of biological phenomena, I believe that they are the truly ultimate ones. The immediate ones are the fundamental ones – those of the nature of the systems themselves and the material with which we have to work. Some of these can never be under our control and must be accepted, but the realisation of the importance of stochasticity is, in itself, a contribution to understanding. The proximate reasons are the blocks of methodology and scientific approach. They are dissoluble given sufficient support. Ecological science, no less than nuclear physics and biotechnology, has to be 'big science'. The pond net has now to be supplemented by the mass spectrometer, the large-scale controlled-environment chamber, and the system of replicated, instrumented experimental lakes or ponds.

Finally, the ultimate political blocks would be insuperable if we were completely governed by selfish genes, by the inherent ecology of one species, ourselves. But we are not. We can make choices. Frequently the behaviour of an individual as an individual, is much less selfish than that of the same individual competing in a group. Different choices are made. It is these choices to which Auden's poem, introduced at the beginning of this paper, refers.

The illusions of which he speaks are that we can continue to use resources as we do, and continue our present lifestyle, yet still expect to maintain a sustainable biosphere. We need to change in ways that are not palatable when it is felt that others in the group may steal a march, but which are otherwise reasonable and acceptable. The transition from being group-driven to being an individual is the cross to be climbed. It is a cross that many vested interests in the consumption of resources might wish were not there, whose existence they conspire to conceal. Tudge (1993) put it well:

> "Only when we are straight in our own heads, and have structured societies that are able to override their own innate tendency to be overtaken by hawks and hawkishness, can we hope to create the kind of world that can be sustained, for only the meek *can* inherit the Earth".

In the final analysis it is a very personal cross. I have not yet met any columbine professors.

References

Andersson, G., 1981. Influence of fish on waterfowl in lakes. (In Swedish, English summary). Anser 20: 21–34.

Andersson, G., W. Graneli & J. Stenson, 1988. The influence of animals on phosphorus cycling in lake ecosystems. Hydrobiologia 170: 267–284.

Beklioglu, M. & B. Moss, 1995. The impact of pH on interactions among phytoplankton algae, zooplankton and perch (*Perca fluviatilis*) in a shallow, fertile lake. Freshwat. Biol. 33: 497–509.

Beklioglu, M. & B. Moss, 1996. Mesocosm experiments on the interaction of sediment influence, fish predation and aquatic plants on the structure of phytoplankton and zooplankton communities. Freshwat. Biol.

Carvalho, L. & B. Moss, 1995. The current status of a sample of English Sites of Special Scientific Interest subject to eutrophication. Aquatic Conservation: Marine and Freshwater Ecosystems 5: 191–204.

Cummins, K., 1974. Structure and function of stream ecosystems. Bioscience 24: 631–641.

Diamond, J., 1991. The Rise and Fall of the Third Chimpanzee. Vintage Press, London.

Drake, J. A., 1989. Communities as assembled structures: do rules govern pattern? Trends in Ecology and Evolution 5: 159–163.

Durbin, A. G., S. W. Nixon & C. A. Oviatt, 1979. Effects of the spawning migration of the alewife, *Alosa pseudoharengus*, on freshwater ecosystems. Ecology 60: 8–17.

Hildrew, A. G., 1993. Freshwater Ecology in Britain – a case of decline? Freshwater Forum 3: 237–242.

Hill, D., R. Wright & M. Street, 1987. Survival of mallard ducklings *Anas platyrhynchos* and competition with fish for invertebrates on a flooded gravel quarry in England. Ibis 129: 159–167.

Hill, W. F. & C. J. Öttchen, 1991. Shakespeare's Insults. Vermillion Press, London.

Hurlbert, S. H., W. Loayza & T. Moreno, 1986. Fish-flamingo-plankton interactions in the Peruvian Andes. Limnol. Oceanogr. 31: 457–468.

Hutchinson, G. E., 1957. Concluding remarks. Cold Spring Harbor Symposium on Quantitative Biology 22: 415–427.

Hutchinson, G. E., 1965. The Ecological Theatre and the Evolutionary Play. Yale University Press, New Haven.

Irvine, K. A., B. Moss & H. Balls, 1989. The loss of submerged plants with eutrophication II Relationships between fish and zooplankton in a set of experimental ponds and conclusions. Freshwat. Biol. 22: 89–107.

Jumars, P. A., 1990. W(h)ither limnology? Limnol. Oceanogr. 35: 1216–1218.

Kalff, J., 1991. On the teaching and funding of limnology. Limnol. Oceanogr. 36: 1499–1501.

Kline, T. C., J. J. Goering, O. A. Mathisen & P. H. Hoe, 1990. Recycling of elements transported upstream by runs of Pacific salmon. 1. ^{15}N and ^{13}C evidence in Sashin Creek, Southeastern Alaska. Can. J. Fish. aquat. Sci. 47: 136–144.

Klinge, M., M. P. Grimm & S. H. Hosper, 1995. Eutrophication and ecological rehabilitation of Dutch lakes: Explanation and prediction by a new conceptual framework. Water Science and Technology 31: 207–218.

Krokhin, E. M., 1975. Transport of nutrients by salmon migrating from the sea into lakes. In A. D Hasler (ed.), Coupling of Land and Water Ecosytems. Springer Verlag, New York: 153–156.

Leah, R. T., B. Moss & D. E. Forrest, 1978. Experiments with large enclosures in a fertile, shallow, brackish lake, Hickling Broad, United Kingdom. Int. Rev. ges. Hydrobiol. 63: 291–310.

Leah, R. T., B. Moss & D. E. Forrest, 1980. The role of predation in causing major changes in the limnology of a hyper-eutrophic lake. Int. Rev. ges. Hydrobiol. 65: 223–247.

Lund, J. W. G., 1964. Primary production and periodicity of phytoplankton. Verh. int. Ver. Limnol. 15: 37–56.

Madsen, B., 1995. Danish Watercourses – Ten years with the New watercourse Act. Danish Environmental Protection Agency, Copenhagen.

McIntosh, R. P., 1995. H. A. Gleason's 'Individualistic Concept' and theory of animal communities: a continuing controversy. Biological Reviews 70: 317–357.

Moss, B., 1973. The influence of environmental factors on the distribution of freshwater algae: an experimental study. IV. Growth of test species in natural lake waters and conclusion. J. Ecol. 61: 193–211.

Moss, B., 1989. Water pollution and the management of ecosystems: a case study of science and scientist. In P. J. Grubb & J. B. Whittaker (eds), Toward a More Exact Ecology. Blackwell Scientific Publications, Oxford: 401–422

Moss, B., 1990. Engineering and biological approaches to the restoration from eutrophication of shallow lakes in which aquatic plant communities are important components. Hydrobiologia 200/201: 367–377.

Moss, B., 1991. The role of nutrients in determining the structure of lake ecosystems and implications for the restoration of submerged plant communities to lakes which have lost them. International Conference on N, P and Organic matter. Contributions by invited experts. Danish Environmental Protection Agency, Copenhagen.

Moss, B., 1994. Brackish and freshwater shallow lakes – different systems or variations on the same theme? Hydrobiologia 275/276: 1–14.

Moss, B. & R. T. Leah, 1982. Changes in the ecosystem of a guanotrophic and brackish shallow lake in eastern England: potential problems in its restoration. Int. Rev ges. Hydrobiol. 67: 625–659.

Motluk, A., 1996. And then there was one. New Scientist 149 (No 2015): 12–13.

Reynoldson, T. B., 1966. The distribution and abundance of lake dwelling triclads – towards a hypothesis. Adv. ecol. Res. 3: 1–71.

Richey, J. E., M. A. Perkins & C. R. Goldman, 1975. Effects of Kokanee salmon (*Onchorhynkus nerka*) decomposition on the ecology of a subalpine stream. J. Fish. Res. Bd Can. 32: 817–820.

Scheffer, M., S. H. Hosper, M-L. Meier, B. Moss & E. Jeppesen, 1993. Alternative equilibria in shallow lakes. Trends in Ecology and Evolution 8: 275–279.

Shapiro, J., 1990. Current beliefs regarding dominance by bluegreens: the case for the importance of carbon dioxide and pH. Verh. int. Ver. Limnolog. 24: 38–54.

Tansley, A. G., 1935. The use and abuse of vegetational concepts and terms. Ecology 16: 284–307.

Torsvik, V., J. Goksoyr & F. L. Daae, 1990a. High diversity in DNA of soil bacteria. Apl. envir. Microbiol. 56: 782–787.

Torsvik, V., K. Salte, R. Sorheim & J. Goksoyr, 1990b. Comparison of phenotypic diversity and DNA heterogeneity in a population of soil bacteria. Apl. envir. Microbiol. 56: 776–781.

Tudge, C. 1993. The Engineer in the Garden. Pimlico, London.

van Donk, E., E. De Deckere, J. G. P. Klein Breteler, & J. T. Meulemans, 1994. Herbivory by waterfowl and fish on macrophytes in a biomanipulated lake: effects on long-term recovery. Verh. int. Ver. Limnol. 25: 2139–2143.

Van Donk, E. & R. D. Gulati, 1995. Transition of a lake to turbid state six years after biomanipulation: mechanisms and pathways. Water Science & Technology 32: 197–206.

Weisskopf, V. F., 1965. Why pure science? Bulletin of the Atomic Scientists 21: 4–8.

Wetzel, R. G., 1991. On the teaching of limnology: need for a national initiative. Limnol. Oceanogr. 36: 213–215.

Winfield, I. J., 1990. Predation pressure from above: observations on the activities of piscivorous birds at a shallow eutrophic lake. Hydrobiologia 191: 223–231.

Winfield, D. K., R. D. Davidson & I. J. Winfield, 1989. Long-term trends (1965–1988) in the numbers of waterfowl overwintering on Lough Neagh and Lough Beg, Northern Ireland. Irish Birds 4: 19–42.

Hydrobiologia **395/396:** 13–18, 1999.
D.M. Harper, B. Brierley, A.J.D. Ferguson & G. Phillips (eds), The Ecological Bases for Lake and Reservoir Management.

Economic challenges for lake management

Timothy O'Riordan
School of Environmental Sciences University of East Anglia, Norwich NR4 7TJ, U.K.

Key words: interdisciplinary science, environmental politics, eutrophication, broads, environmental valuation

Abstract

Restoring lakes from a degraded state is a costly and risky enterprise. It is costly partly because 'rewinding' the cycles of degradation involves great scientific uncertainties. Therefore, the only way forward is by careful, expensive monitoring, and much adaptation of treatment as the evidence unfolds. This process requires much patience and political commitment. Therefore, the economic challenge is to find a relationship between the ecological exploration of restoration techniques and the economic justification of the outcome. Unfortunately, the basis of that justification is an economy that creates environmental degradation. So there is a cruel circularity in the economic appraisal: the process of benefit calculation is predicted on an economy that is, at present, non sustainable. This suggests that there should be another approach to economic valuation, based more on participatory processes of involving interests with a stake in the outcome and in the gains and losses that will inevitably be incurred in reaching the outcome of lake restoration. In that process, ecological science plays a vital role of explaining possible future pathways for restoration, and the dilemma of uncertainty is handled by creating various scenarios rather than models. The real challenge, therefore, is to devise a fair and full process of bargaining over lake futures, within which both ecology and economics play important, but subsidiary roles.

The economics of lake restoration

Despite their common root, economics and ecology do not readily mix. Indeed there are only two academic programmes of ecological economics in the U.K., in York and in East Anglia, and neither has managed to attract sustained research funding. The lively International Association of Ecological Economics is filled far more with economists than with ecologists, and a scrutiny of its conference proceedings reveals that integrated research in this important area is still in its infancy. Why is it proving so difficult to integrate economics and ecology?

I propose four reasons, all of which make for uncomfortable reading. In summary form, these reasons are:

1. Economists rely on willingness to pay as a basis for valuation of, say, restoring a damaged lake, but willingness to pay is an expression of an economic system that damages the lake in the first instance. Taken more generally, any economics-based technique of environmental valuation cannot be justified until an economy operates on a much more environmentally sustainable basis. Only at that point, can willingness to pay be used as a true basis for valuation.
2. Ecologists are trapped by an imperfect understanding of cause, effect and synergies in the process of lake restoration. This point is made repeatedly throughout this volume, introduced by Brian Moss, who calls for much more dedicated scientific funding and for more properly designed experimental conditions so as to be sure that ecological science can be pursued in the most rigorous manner possible. Even then, he admits, there will be much uncertainty as to what precisely makes lakes 'tick' and hence how they can either restore themselves or be consciously helped to do so by sympathetic and knowledgeable humans.
3. Maybe this is the wrong way of going forward. Is there nowadays any such thing as an ecologically isolatable lake, free from the vagaries of human interference? If one looks at the Norfolk and Suffolk Broads, (see Phillips et al., 1999; Madgwick, 1999) one is struck by an ecosystem that is continually interfered with by human activ-

ity. Nutrient-rich rain is falling from a polluted atmosphere and through groundwater migration, toxic chemicals are also falling from the air and seeping through springs. Salt intrusion is a constant threat, created in part by a rising tide and falling fresh water flows due to abstraction and unusually dry conditions, themselves a product of human intervention. The science of lake restoration has evolved considerably as a result of the focused effort of the EC LIFE programme funding backed by the Broads Authority and the Environment Agency. But it is clear that, at various points in the process, there is real chaos (ie. unpredictable uncertainties) in the interrelationships between water quality, surface sediment, fish feeding habits and invertebrate distributions that defy modelling with any reliability. So maybe we have to think of an ecological science that is coupled to human intent and intervention, and is not in the strict sense, a separable physical-biological field of enquiry. This is a very challenging statement to make in a text on ecological restoration, but one that at least should be aired. The conclusion that follows is that ecological science on its own is not a basis for assessing the pathways to restoring a lake.

4. This conclusion suggests that economic valuation and ecological restoration may not be analysable on the basis of conventional economic and ecological theory. Maybe it is only possible to contemplate a process of valuation as an outcome of debate and bargaining over various scenarios of future lake states, where those with an interest in the outcome, or whose welfare may be affected by the commitment of funds to get to those possible outcomes, meet to find common positions through consensus and compensation (where necessary). Thus the true economic valuation is not a surrogate calculation of a willingness to pay in a market that is imperfectly designed to generate that signal. It is a process of creating an agreed set of outcomes amongst groups who might not otherwise realise they had a common purpose and a shared commitment to a more sustainable future.

Economic valuation techniques

Let us look first at the state of economic theory. This is aptly summarised by Bateman (1995) and reproduced in Figure 1. The basic issue is that any benefit arising from lakes' restoration cannot be captured by a market price through direct willingness to pay. Consumers do not pay at an entrance gate to see a rejuvenated lake, unless it is purely a private enterprise. In the process, an improved lake benefits those who visit, normally without directly paying, as well as those who never visit but who enjoy the satisfaction of knowing that the watercourse is more healthy. So, the economist has to simulate a demand curve, to guess at what might be the measure of willingness to pay for more or less lake improvement.

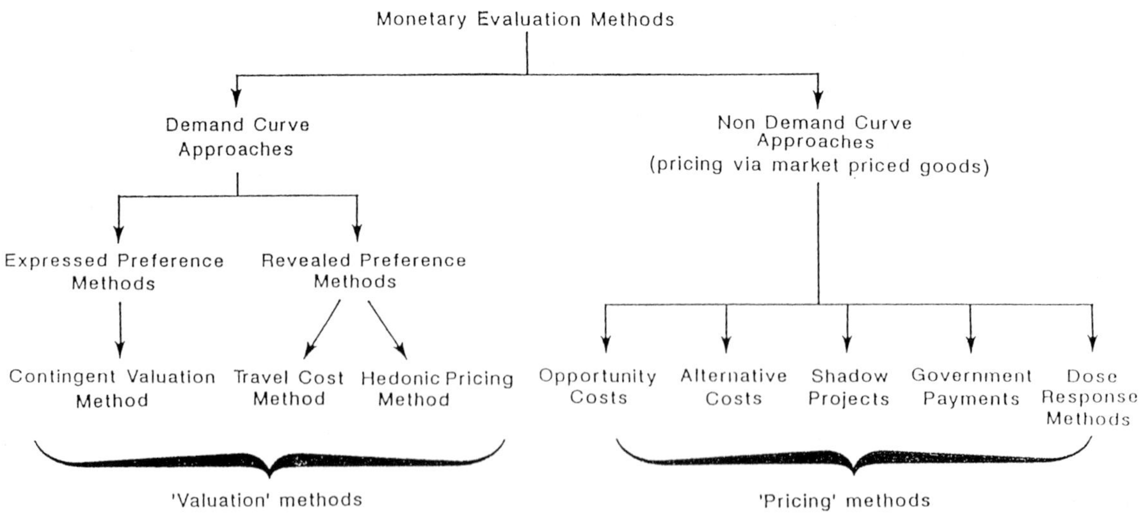

Figure 1. Methods for the valuation and pricing of environmental goods and services (from Bateman, 1995).

Figure 1 shows that there are two basic approaches. One is through assumptions of value based on politically acceptable expenditures. That political acceptability becomes the democratic surrogate for willingness to pay. There are four approaches here:

1. The cost of what is lost directing expenditure in one way and not in another is the 'opportunity' lost. Thus an investment of, say £150 000 per year in the restoration of the Norfolk and Suffolk Broads means that that money cannot be spent on fen management or information centres. Ideally the level of investment should be equal to the point where any incremental benefit of further restoration is just not met by the incremental cost of reaching that restoration effort. In practice, the level of ecological science is too imperfect to provide a reliable indicator of this point of equalising marginal benefits and costs, so the outcome is more a matter of political horsetrading amongst competing interests. In that sense, the opportunity cost approach is more a function of political power and skilled lobbying, together with the prevailing ecological morality, than it is based on economic theory.
2. The cost of providing a fully restored lake elsewhere is another technique, namely estimating the expenditure necessary to finance an equivalent ecological and amenity asset. Again this is, in reality, only a benchmark calculation. But it often is salutary to show how costly it is to create a purpose built piece of uncontaminated water habitat. Increasingly, now that the EC Habitats Directive is calling for no net loss of high profile conservation sites, this so-called 'alternative cost' approach will have greater significance in economic valuation. This is because a threat to such a site needs to be evaluated in terms of the cost of creating an equivalent site elsewhere. At least such a legally backed policy might help the cause of restoration ecologists!
3. Social values attached to landscapes and lakes can be imputed from policy measures that reflect political preferences. An example is the investments in Environmentally Sensitive Area and Countryside Stewardship schemes that are aimed at maintaining and enhancing particular habitats or landscapes. For example, the Broads grazing marshes are protected by an expenditure of £2 million annually through the ESA scheme. This is not necessarily the true value of those marshes, but it can be regarded as an indicator of social preferences, and hence a base value on which to assess a fuller valuation.
4. Polluting emissions, such as nutrient inputs, are doses, and their removal creates a response. One way of assessing the value of restoration is to try to cost the consequences of removing the dose and measuring the response. In practice, for the reasons of imperfect ecological understanding, such a calculation is rarely precise. But if the precautionary principle is invoked, then it is possible to estimate the costs of precautionary removal of particular nutrient inputs, for example from sewage works discharges, and calculating the likely benefits in, say visitor pleasure, or freer navigation movements.

All these techniques are surrogates of demand curves, or patterns of willingness to pay for greater or lesser amounts of restoration, where the valuation is achieved by means of ingenious variants of estimating the equivalent of a fully or partially restored lake. One additional method, as yet still in its infancy, is to calculate the economic gains of the various ecological functions a healthy lake provides – say in buffering toxics and nutrients, in sequestering pollutants in muds or in biomass, or in protecting floodwalls from erosion by maintaining the physical barrier of a fringing reed bed. Ideally the value of these services, in terms of what it would cost to provide the equivalent package of ecological functions if the lake did not exist, or at least was not so healthy, would form the basis of the valuation. This is a difficult area, but it is one where ecologists and economists have the greatest scope for constructive cooperation.

Nevertheless, none of these approaches is ideal, for the four reasons stated at the outset. The real dilemma is whether the state of ecological and economic theory is ready for the degree of reformulation necessary to answer these challenges. Proponents of both sets of theories probably will argue that no fundamental change is necessary. The contention in this chapter is that inevitably, though slowly, these modulations will occur.

Simulating demand curves

The left-hand side of Figure 1 lists the four conventional economic techniques aimed at creating a demand curve based on the total economic value of a restored lake. Bateman (1995) provides a summary of that approach as depicted in Figure 2. The pack-

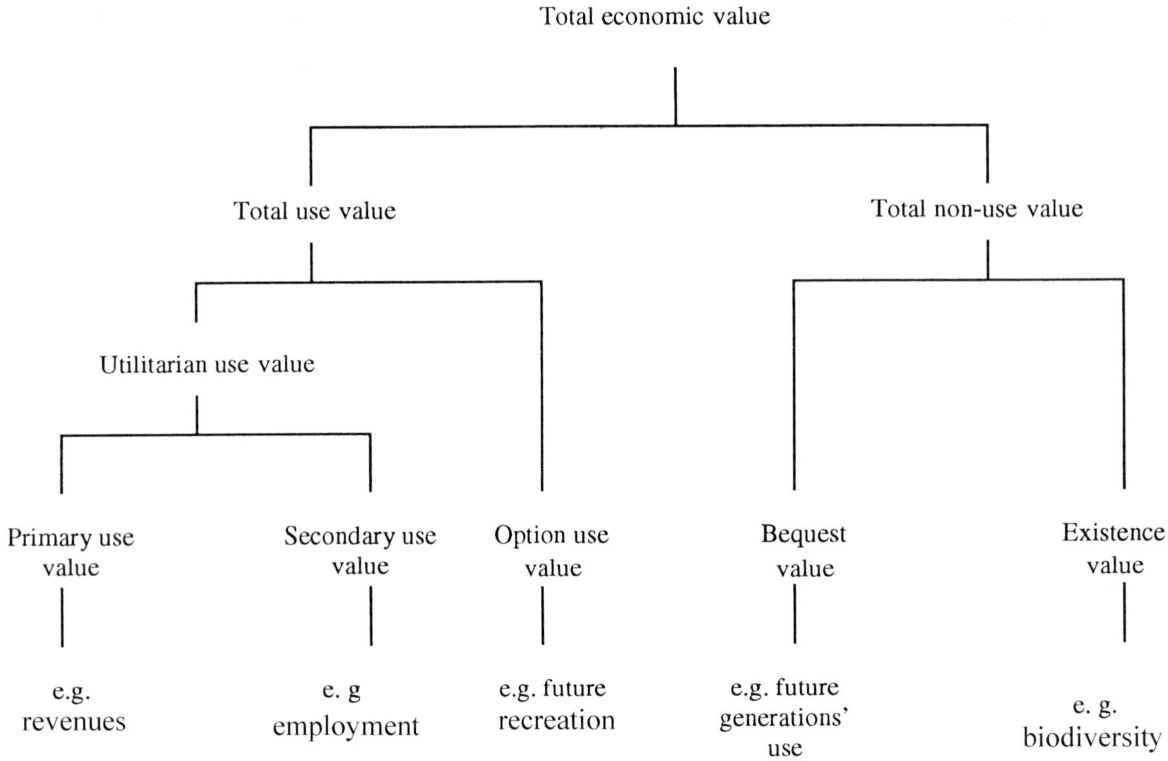

Figure 2. Total economic value (from Bateman, 1995).

age of values divides into those associated with actual use, such as navigation or angling, together with the associated economic gains of employment from servicing these pastimes or providing wardens. There are also non-use values. These are more psychological or 'feel good' factors linked to simply knowing that a once-damaged waterbody is more healthy, and that its ecological and aesthetic functions can be appreciated by others. Obviously the amounts that people will be prepared to pay for such feelings depends on the state of the economy, their income, their knowledge and preferences over environmental degradation and restoration generally, and the number of other lakes either requiring restoration or in a healthy condition. Capturing all these circumstantial factors is beyond economic theory at present, and lies mostly in the realm of political biasses and interest group lobbying. This is why economic valuation tools are only preliminary, and often distortionary devices for valuing the benefits of lake restoration. This is because the tools are rooted in a paradigm that accepts a degree of environmental damage as an unavoidable element of economic and social progress, even though the political rhetoric is that of advocating sustainable development.

Within this broad pattern of valuation, three techniques for simulating willingness to pay stand out. These are:

1. The travel cost method assumes that visitors' valuations are broadly the same, so that the costs of travel incurred by the longest travelling visitor provide the basis for calculating how visitor numbers might change should a 'shadow' entrance price be placed at the restored lake. The gap between the cost incurred by the longest distance traveller and shorter distance travellers is the 'consumer surplus' of the trip. That surplus can be cut back by a shadow entrance price set in stages. Hence the demand curve is set up.
2. The restored lake will add to the amenity of surrounding property. This should encourage people to buy or sell property at prices higher than would be the case if the lake remained polluted. Obviously such prices depend on the vagaries of the property markets, but that 'noise' can be reduced. This approach is called the 'hedonic price method', but would be only used for urban lakes restoration. Like all the techniques mentioned in this section, it depends on the level of awareness

of lake clean up, the degree to which restoration can be guaranteed to be long lasting, and the cultural significance of resuscitating the natural world. None of these items have their roots only in an economic paradigm, hence the limited role of this technique.

3. The third approach is simply to ask visitors how much they would be prepared to pay for more lake restoration or to be compensated for less restoration. This is called the contingent valuation technique, and is hotly disputed amongst economists and non-economists alike (see Norton, 1994 for a non-economist critique). Bateman (1995) cites a U.S. study (Desvousges et al., 1987) of willingness to pay for three levels of improvement of a river in Pennsylvania, on the basis of maintaining boatable quality, upgrading to fishing quality and upgrading again to swimming quality. This is the kind of analysis that is reasonable to do because respondents have a relatively clear idea of what is involved. The results show that the willingness to pay for progressively higher standards of water quality diminishes. Nevertheless, despite the plausibility for economic theory for such a study, the full range of benefits, as indicated in Figure 2, would not be captured because there is no guarantee that respondents would know of them.

Contingent valuation techniques are flawed because they are based on a number of false assumptions about knowledge, awareness of options, the actual process of bidding different levels of restoration, and the difficulty of what to do about extreme answers that may be motivated more by a dislike of having to place money on an environmental good, or by a belief that the actual price would never have to be paid but the higher the bid more others would have to pay to see the investment actually take place. Mitchell & Carson (1989) debate these issues in detail. So motivation behind the entry bid and successive bids is again more a political matter than an economic one. If there is no clear ecological case for how different lake outcomes might look with varying degrees of clean-up, then the contingent valuation method is effectively useless.

This conclusion is reinforced by a study by Clark & Burgess (1996). They interviewed two groups of people who had participated in a CVM survey of the value of protecting the Pevensey Levels in Sussex. Clark & Burgess found that none of the people who offered a price had a particularly clear vision in their mind of what the Levels were, nor was there any agreement as to that vision. Furthermore, the bid depended very much on the respondent's belief in the reliability and competence of the managers of the Levels (the local authorities and the wildlife agencies). The lower the trust, the more unwilling they were to bid a high price.

This finding suggests that all of these techniques are of limited value insofar as they rely on the degree of knowledge and rationality on the part of users and visitors. Since neither assumption is valid, this suggests that other approaches are required if a more politically tenable approach to restoration valuation is to be followed.

Valuation by consultative negotiation

In Australia, public option and social behaviour have been altered by imaginative publicity in order to improve the water quality of the Murray-Darling basin (Rogers, 1998). This shows that consumers and citizens can be educated into a more supportive environmental awareness of a restoration programmeand hence give it political backing. The study shows that economic valuation of lake restoration can be enhanced and indeed made politically more acceptable if the consequences of not complying with certain patterns of water use and disposal are more widely appreciated. The key to this programme lay in its creation of a sense of collective self-interest in the restoration scheme, and a recognition of how others might become hurt if that socially and environmentally sound behaviour was not followed.

This is an approach that requires further experimentation. One way forward would be to bring various interested parties together in the form of regular, but sequential, discussion groups to examine both the mutual interest of co-operation and the basis for compensation or give and take in the process of reconciliation (Georgiou et al., 1998). Ideally this kind of stakeholder mediation process requires good facilitation by experienced negotiators, plus means of imaging how different investment programmes would result in varying degrees of ecological resilience in a restored lake. This scenario approach, which could be imaged on computers, avoids the problems of imperfect ecological knowledge, and takes into account the coupling effect of natural processes and human interference. It would also get around the difficulties of person-based visualisation that plagues many valuation techniques.

Such an approach would be multi-disciplinary, educative, community bonding and democratic. It would obviously be costly. But in terms of reaching consensus, in clarifying outcomes that generate widespread support in reaching outcomes that are fairer because losers are actually recognised and may be compensated by gainers – for all these reasons, the interactive participatory approach to valuation surely must become the scientifically and politically acceptable face for lake restoration for the 21st century. Maybe in the process, ecologists and economists can work in a common endeavour. Let us hope so.

References

Bateman, I., 1995. Environmental and economic appraisal, In T. O'Riordan (ed.), Environmental Science for Environmental Management. Longman, Harlow: 45–65.

Clark, J. & J. Burgess, 1996. Asking questions about answering questions: a case study of public understanding of a contingent valuation survey. In P. Lowe (ed.), Environmental Valuation and Public Policy. CAB International, Wallingford.

Desvousges, W. H., V. K. Smith & A. Fisher, 1987. Option price estimates for water quality improvement: a contingent valuation study of the Mongolla River. Journal of Environmental Economics and Management 14: 248–267.

Georgiou, S., I. H. Langford, I. J. Bateman & R. K. Turner, 1998. Determinants of individual willingness to pay for perceived reductions in environnmental health risks: a case study of bathing water quality. Environment and Planning A 30: 577–594.

Madgwick, J., 1999. Strategies for conservation management of lakes. Hydrobiologia 395/396 (Dev. Hydrobiol. 136): 309–323.

Mitchell, R. C. & R. T. Carson, 1985. Using Surveys to Value Public Goods: The Contingent Valuation Method. Resources for the Future, Washington DC.

Norton, B. C., 1994. Economists preferences and preferences of economists. Environmental Values 3: 311–332.

Phillips, G., A. Bramwell, J. Pitt, J. Stansfield & M. Perrow, 1999. Practical application of 25 years' research into the management of shallow lakes. Hydrobiologia 395/396 (Dev. Hydrobiol. 136): 61–76.

Rodgers, M., 1998. Applying the Precautionary Principle to River Management: Unpublished PhD Thesis, Charles Stuart University, Australia.

Hydrobiologia **395/396:** 19–28, 1999.
D.M. Harper, B. Brierley, A.J.D. Ferguson & G. Phillips (eds), The Ecological Bases for Lake and Reservoir Management.

Lake restoration: capabilities and needs

Stephen R. Carpenter[1] & Richard C. Lathrop[1,2]
[1]*Center for Limnology, 680 North Park Street, University of Wisconsin, Madison, WI 53706, U.S.A.*
[2]*Bureau of Integrated Science Services, Wisconsin Department of Natural Resources, 1350 Femrite Drive, Madison, WI 53716, U.S.A.*

Key words: eutrophication, lake, management, restoration, Wisconsin

Abstract

Lake degradation results from excessive nutrient inputs, toxic substances, habitat loss, overfishing, species invasions and extirpations. The scientific basis of lake degradation is generally well understood, although each restoration project requires some level of new site-specific research. Remediation may require management actions which are difficult to implement for social or institutional reasons. Even where large-scale remediations are attempted, it is difficult to sustain scientific assessments for long enough to evaluate success. Collaborations of scientists and managers have sometimes succeeded in overcoming limitations to lake restoration, and produced important advances in our capability to restore lakes.

Introduction

Lakes provide humankind with many services: esthetic enjoyment, recreation, fish, transportation, water for irrigation, drinking and dilution of pollutants (Postel & Carpenter, 1997). These services are impaired by exploitation of lakes and the lands of their catchments. The goal of management is to balance the uses of lakes with conservation measures to sustain ecosystem services over time. Research can provide understanding of lakes, their catchments and the mechanisms that sustain ecosystem services; the causes of lake degradation; and methods and technologies for lake restoration. This brief paper summarizes the state of scientific knowledge relevant to lake restoration, and discusses some of the linkages between research and management.

Complex problems with multiple causes

In relatively undisturbed lakes, water quality is maintained by several mechanisms, each acting at different scales of space and time (Carpenter & Cottingham, 1997). In these lakes, impacts of climate fluctuations or terrestrial ecosystem changes are damped in many ways. Riparian vegetation and wetlands delay or prevent the transport of nutrients to lakes from eroding upland soils. Wetlands release humic substances which reduce the response of algae to nutrients. Zooplankton prevent the buildup of algal biomass and efficiently transfer nutrients to higher trophic levels. The water remains clear enough for macrophyte growth, and considerable amounts of nutrients are stored in macrophytes. Food web structure is regulated by large piscivorous fishes, which use the habitat provided by macrophyte beds, wetlands and trees fallen from riparian forests. The resilience of lake water quality thus depends on mechanisms which have similar effects at different scales.

Eutrophication is a syndrome that develops when several resilience mechanisms are broken down (Harper, 1992; National Research Council, 1992). Agriculture and urban development increase nutrient inputs. Loss of riparian vegetation and wetlands increases the efficiency of nutrient transport to lakes. Humic inputs decline, and humic constraints on algal growth are less effective. Piscivorous fish abundance is reduced by overfishing, so planktivorous and benthivorous fishes become more abundant and the large zooplanktonic grazers are reduced. Consequently, incoming nutrients accumulate in phytoplankton biomass, especially blue-green algae which are no longer controlled by grazing. Macrophyte beds decline with losses of water clarity. Loss of crucial habitat – macrophytes, wetlands, fallen trees – leads to further breakdown of the food web. The result is a lake with few piscivorous fishes, abundant planktivorous and benthivorous fishes, few large herbivorous zooplankton, few macrophytes, dense algal blooms, and risks of anoxia and algal toxins. Species invasions, losses of biodiversity, and toxic pollutants also degrade lakes, and may interact with effects of nutrient enrichment, habitat loss and overfishing.

Status of lake restoration

Lake degradation is a syndrome with multiple causes. For some restoration problems, proven solutions exist (National Research Council, 1992; Cooke et al., 1993, chapters in this volume). For other restoration problems, methods are insufficient. Gaps occur in scientific knowledge, institutional mechanisms, or both. Here we briefly note some major needs in lake restoration.

Excess nutrient input has received more attention than the other causes of lake degradation. Point sources of nutrient input can often be controlled by sewage treatment plants. Nonpoint sources of nutrients result from erosion, livestock waste, overfertilization of crops and urban wastes (Carpenter et al., 1998). These sources account for most of the water pollution in the United States (National Research Council, 1992). Control of nonpoint pollution requires improvements in land use that are difficult to implement for economic and political reasons (National Research Council, 1992). The best treatment for excess nutrient problems is reduction of nutrient input. However, a number of methods exist for removing or inactivating nutrients after they have entered lakes (Cooke et al., 1993). These methods can succeed where input reductions are slow or insufficient, and where recycling of nutrients from bottom sediment maintains eutrophication, even after inputs are reduced (National Research Council, 1992).

Prediction of phosphorus inputs from land use data is a complex interdisciplinary problem, for which many solutions have been proposed (Poiani & Bedford, 1995; Soranno et al., 1996). Existing approaches leave considerable room for improvement, and research toward better non-point pollution assessments deserves greater effort. Also, management practices designed to control non-point pollution are rarely tested at the large scale of actual applications. Non-point pollution control projects should be operated as large-scale experiments to determine whether the management actions have any effect on nonpoint pollution (Carpenter et al., 1997).

The fundamental policy problem in lake restoration is that those who cause non-point pollution do not benefit from reduced pollution, especially in large agricultural catchments. Conversely, the beneficiaries of non-point pollution control are not those who cause the pollution, except in some urban lakes. This mismatch between polluters and beneficiaries is the root of institutional shortcomings that prevent the success of non-point pollution control programs.

Toxic substances, such as metals, organochlorine compounds and acid, have been managed most successfully by reducing inputs at their sources. However, non-point inputs of toxins can be large. Examples are airborne inputs of mercury and PCBs to lakes (Swackhamer & Armstrong, 1986; Driscoll et al., 1994). Once they are added to lakes, persistent toxins like mercury and organochlorine compounds are difficult to remediate (National Research Council, 1992). Clean-ups can be successful when pollution is concentrated in a restricted area. In some cases, sediment-bound pollutants can be removed by localized dredging. In other cases, however, resuspension of these sediments would exacerbate problems and it is better to leave the sediment undisturbed. Lakewide pollution may dissipate very slowly even after point-source inputs have ceased (Stow et al., 1995). Cleanup of persistent toxic pollution at large spatial scales poses enormous technological and fiscal challenges, and may be impossible in many cases (National Research Council, 1992).

Restoration of habitats – wetlands, macrophytes and fallen trees – is a crucial aspect of self-sustaining lake restorations. Wetland restoration is the subject of substantial and ongoing research effort (National Research Council, 1992; Zedler 1996). Restoration of macrophytes in shallow lakes has yielded some notable successes (Scheffer et al., 1992; chapters in this volume). Restorations that target particularly desirable species of macrophytes are an important priority for further research.

Fallen trees are a critical, but vanishing, habitat element in both lakes and streams (Maser & Sedell, 1994; Christensen et al., 1996). Losses of woody habitat are caused by declining riparian forests and deliberate removal of fallen trees from the littoral zone. In some Wisconsin lakes, it will take centuries to restore wood habitat to its former abundance (Christensen et al., 1996). Effects of these habitat losses on fish production in lakes are potentially large but not yet well quantified (Heck & Crowder, 1991). Although artificial structures have long been used in fish management (Johnson & Stein, 1979), the extent to which these structures mimic the invertebrate production and shelter afforded by fallen trees is not known. Further research is needed on the effects of littoral woody habitat on fishes. Managers need to consider new mechanisms for preserving riparian vegetation and littoral woody habitat in lakes.

Overfishing affects water quality by reducing populations of large piscivorous fishes, thereby increasing populations of zooplanktivores (which consume grazers) and benthivores (which resuspend sediments and recycle nutrients). Angling has powerful effects on fish populations (Johnson & Carpenter, 1994; Carpenter et al., 1994). Kitchell & Carpenter (1993) argued that fishing increases the temporal variability of lakes by increasing the probability of recruitment events, stock collapses, and their reverberations through the food web. Although anglers have strong effects on sport fish populations and lake food webs, relatively few studies have examined the interactions of sport fishers and their prey. Anglers can respond rapidly and creatively to changes in the resource (Johnson & Carpenter, 1994). If anglers do not agree with the need for regulation, then regulations are not likely to work. Hilborn et al. (1995) argued that management has largely failed to preserve commercial fisheries, and suggest that sustainable management of commercial fisheries is impossible. Sport fish management may face similar institutional limitations.

Species invasions and extirpations have been addressed only occasionally by lake restoration projects (National Research Council, 1992). Our abilities to anticipate or predict invasions are limited (Lodge, 1993). Removal of exotic species has succeeded only rarely, although exotic species can sometimes be reduced to tolerable levels. There have been few successful re-introductions of extirpated species to lakes.

A case study: Lake Mendota, Wisconsin

Because lake degradation has multiple causes, lake restoration usually requires multiple interventions. Here we summarize past and continuing efforts to restore Lake Mendota, Wisconsin. Lake Mendota has all of the problems described above, with the exception of toxic pollution. The lake has been the subject of substantial research and management programs conducted by agency and university staff over the past century. Efforts to restore Lake Mendota exemplify the current state of the art.

Lake and catchment description

Bordered by Wisconsin's capital city and leading university, Lake Mendota is symbolic of the quality of life in a state with more than 15 000 lakes and a multi-billion dollar per year recreation industry centered on lakes (Figure 1). Management of the lake is highly visible to government officials.

The lake (A_0=40 km^2, Z_{max}=25.3 m) is located in a rich agricultural region, and its eutrophic state derives largely from agricultural runoff (Lathrop, 1992b; Lathrop et al., 1998). At present, the 601 km^2 catchment is 80% agriculture, 9% urban, and 11% wetlands, forest and lake area. The urban area is expected to increase to 15% by the year 2020 (Soranno et al., 1996). Although most of the catchment is prime agricultural land, the lake's urban setting attracts a heavy use of the lake. Fishing, boating and swimming are important, but scenic enjoyment undoubtedly is the lake's most popular use. The amenities of Lake Mendota contribute to the region's high property values and burgeoning population growth.

Nutrient enrichment

Most of Madison's sewage effluents were discharged downstream from Lake Mendota after a sanitary sewer collection system and treatment facility were first built in the early 1900s (Lathrop 1992b; Lathrop et al., 1992). However, sewage from small communities in the lake's catchment began entering the lake via inflowing streams in the 1920s. Dissolved phosphorus (P) concentrations in these streams increased substantially after the end of World War II (Figure 2). The sewage effluents continued entering the lake until the sewage was diverted in late 1971. This diversion reduced biologically available phosphorus loading to the lake by about 30% (Lathrop, 1990).

Over the past century and a half, however, nutrients from nonpoint pollution have probably been more important than sewage inputs (Lathrop, 1992b; Lathrop et al., 1992). Urban runoff has been adding increasing amounts of nutrients to the lake, especially from construction site erosion, as urban population growth has increased steeply in recent decades. Agriculture, however, accounts for most of the non-point inputs. While the arable land of the catchment was farmed by 1870, an increase in corn (maize) production occurred soon after World War II. Because corn

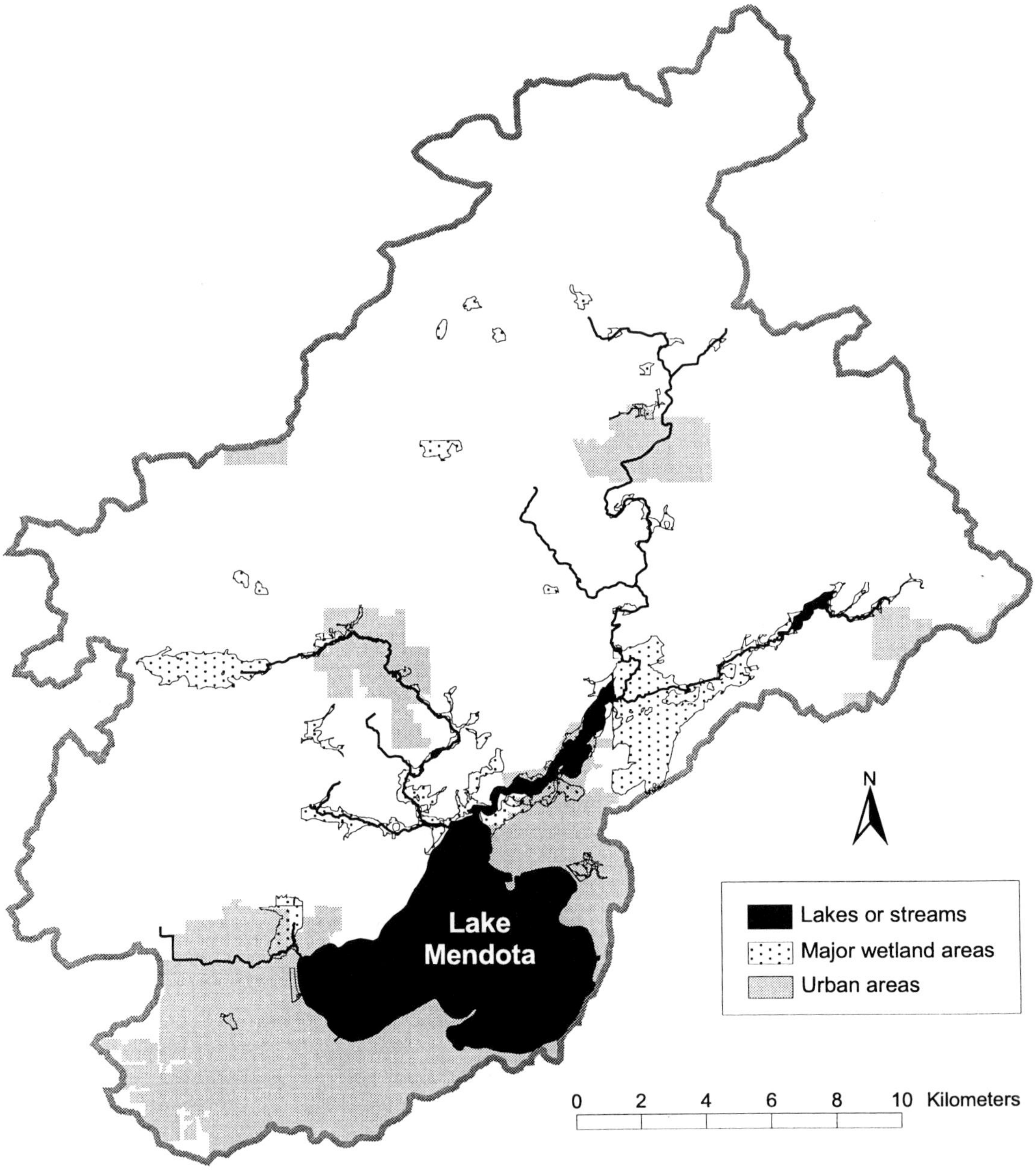

Figure 1. Lake Mendota Watershed, Dane County, Wisconsin, U.S.A.

exposes soil to erosion for longer periods of time than other grains, nutrient loadings also increased at that time. In addition, the use of artificial fertilizers increased after the war. As a consequence, phosphorus concentrations of soils susceptible to runoff have increased in recent years. The numbers of cattle and hogs (pigs) on farms have also increased. Nutrients leached from manure enter the lake in greater amounts, especially during times when manure is being spread on frozen ground. These changes in agriculture have significantly increased nutrient inputs from non-point sources.

Water quality problems apparently first became an issue in Lake Mendota by the mid-1940s. While summer cyanobacterial blooms were common since at least the late 1800s (Brock, 1985; Lathrop & Carpenter, 1992a), algal blooms became severe enough during the mid-1940s that a special study was com-

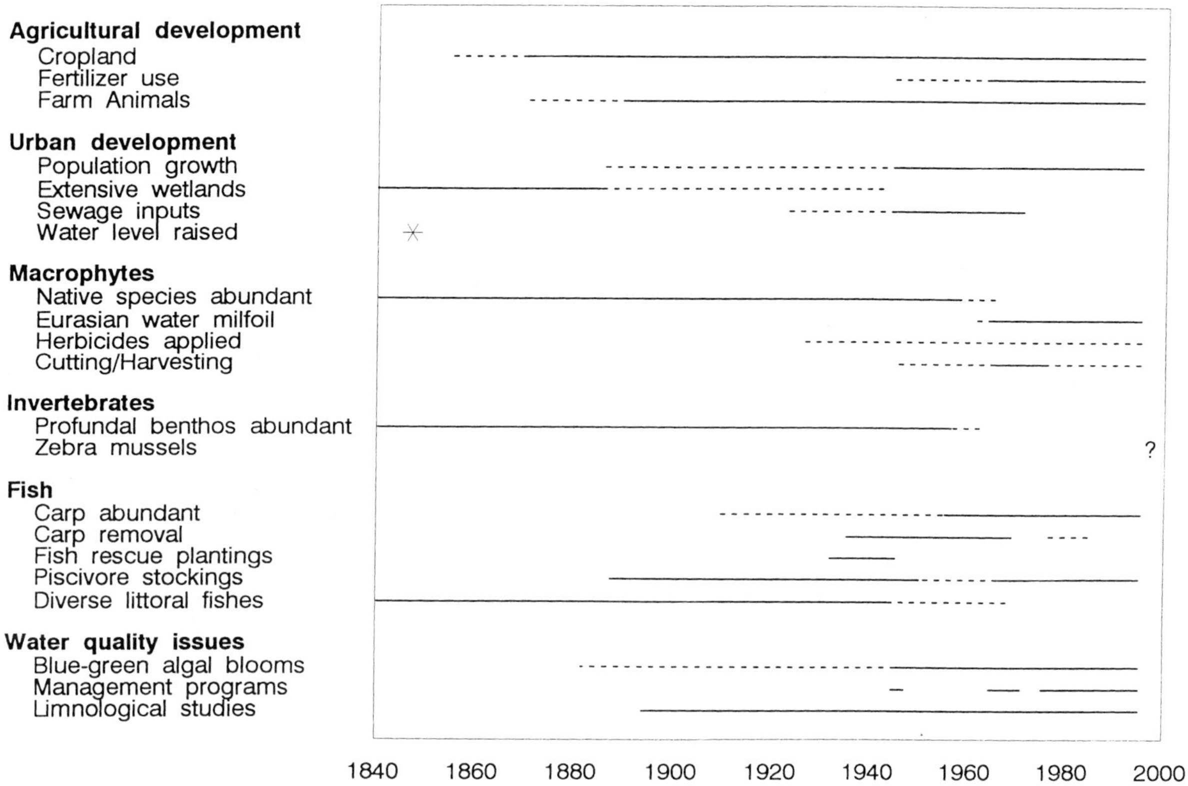

Figure 2. Time line of major human disturbances and management programs for Lake Mendota, Wisconsin from 1840 to present. Solid lines indicate major activity, dashed lines indicate minor activity. Asterisk denotes the time of water level increase. Question mark indicates that zebra mussel invasion is deemed likely to occur in the near future.

missioned to address the causes for these blooms. However, little was done to control the nutrient inputs to Lake Mendota. Rather, attention was focused on diverting Madison's sewage effluents out of the downstream lakes, which had much worse water quality problems. This diversion was extremely controversial, both economically and politically. Because of lawsuits related to the new discharge channel that carried the sewage around all the lakes, the diversion was not completed until 1958.

Concerns about the sewage going into Lake Mendota was not a major concern until the mid-1960s, when serious water quality problems coincided with the invasion of Eurasian water milfoil (*Myriophyllum spicatum*). In 1965, the Lake Mendota Problems Committee (LMPC) was formed, bringing together university and governmental scientists and managers to work on various aspects of lake management. As a result of that work, the sewage was diverted out of the lake in 1971. The LMPC also addressed the importance of reducing non-point pollution to Lake Mendota from agricultural sources, but no action was taken.

Beginning in 1975, the Dane County Regional Planning Commission began preparing a plan with the assistance of local and state government agencies to reduce nonpoint pollution to Lake Mendota and other area waters. This plan was prepared under the provisions of the federal government's 1972 Water Pollution Control Act Amendments and the 1977 Clean Water Act. While many sources of nonpoint pollution were identified, corrective measures were voluntary with some limited cost-sharing money available for implementation.

Further efforts at reducing non-point pollution from drainage basins in the western half of the lake's catchment were conducted during the early to mid-1980s, when the Wisconsin Department of Natural Resources (WDNR) designated the area a 'Priority Watershed Project'. However, because the project was the first to be conducted in the state, and because corrective measures were voluntary, there was little participation in the program by area farmers. Also, many of the 'best management practices' implemented were untested and later proved ineffective at reducing phosphorus loads to the lake.

Recognizing that only an aggressive non-point pollution abatement program would improve water quality in Lake Mendota, the WDNR designated the entire watershed as a Priority Watershed Project in 1993 (Betz et al., 1997). This designation allows state funds to identify significant sources of pollution and to cost-share with both rural landowners and local municipalities, to apply management practices that will reduce phosphorus loads to the lake. Approximately 9 million U.S.$ of state monies have been committed for implementation of the management practices. Local sources (mostly municipalities) are asked to spend an equal amount for cost-sharing practices and other pollution control activities. To date, the project has produced inventories of pollution sources (Betz et al., 1997), and determined the phosphorus load reductions that will be needed to reduce the frequency and severity of cyanobacterial blooms in the lake (Lathrop et al., 1998). The implementation phase of the project began in 1998 and will continue for 10 years. Because much experience has been gained since the early nonpoint source priority watershed projects in Wisconsin, and because mechanisms are now in place for regulating severe pollution sources, expected reductions of agricultural and urban nonpoint pollution to Lake Mendota are significant. This project will be monitored closely by scientists and managers. It is viewed as a long-term experiment by both university and agency staff, who plan to monitor the effects of the program for up to 20 years.

Overfishing

A biomanipulation project begun in 1987 established a strong connection between Lake Mendota's water quality and the fishery (Kitchell, 1992). The experiment employed massive stockings of walleye (*Stizostedion vitreum*) and northern pike (*Esox lucius*) to decrease planktivorous fish populations and increase zooplankton grazing on phytoplankton in Lake Mendota. These piscivores were protected by size limits and bag limits that (at the time) were the most restrictive in Wisconsin (Johnson & Staggs, 1992). However, angling effort increased substantially in response to publicity about the project. Increased angler effort resulted in a four-fold increase in exploitation on walleyes, even with the restrictive fishing regulations (Johnson & Staggs, 1992; Johnson & Carpenter, 1994).

Angler response to the biomanipulation has heightened managers' concerns about overfishing. Technologically advanced fishing gear, the proximity of large urban centers with effective media dissemination of fishing hot spots and the increase in power boat usage, have made anglers rapidly responsive to changes in fishing conditions (Johnson & Staggs, 1992). Overfishing has depleted populations of panfish such as yellow perch (*Perca flavescens*) and bluegill (*Lepomis macrochirus*), as well as predators such as walleye, northern pike and largemouth and smallmouth bass (*Micropterus salmoides* and *M. dolomieu*). Maintenance of these species at high densities may require special fishing regulations – shorter open seasons, larger size limits, and smaller bag limits – to offset increasing fishing effort (Johnson & Staggs, 1992). Additional options include restoring habitat for piscivores and educating anglers about reasonable exploitation rates to sustain viable fisheries.

Habitat loss

Significant losses of wetland habitat from Lake Mendota's catchment have occurred since the area was first settled by Europeans. In 1847, the water level was raised approximately 1.5 m by a dam constructed at the outlet (Lathrop et al., 1992). The increase altered littoral habitat, flooded the upstream wetlands and created new wet areas up the tributary streams. In later years, approximately 50% of the original wetlands in the lake's catchment were either filled for urbanization or ditched and drained for agricultural use. The remaining wetlands are now protected by state laws and local zoning ordinances, but urbanization near wetlands continues to have negative impacts, particularly from sediments eroded from construction sites. Local ordinances for construction site erosion standards are now in place county-wide, but enforcement has been lacking. One of the primary objectives of the Mendota Priority Lake Project is to provide money to enforce these erosion standards.

The decline in area, density and species richness of aquatic macrophytes is another important loss of habitat (Lathrop et al., 1992; Nichols et al., 1992; Nichols & Lathrop, 1994). Surveys and other accounts of macrophytes from the late 1800s into the 1950s indicate that as much as 25% of the total lake area was covered with a diverse community of macrophytes, with plants growing to water depths of 5 m and more.

Major plant species included wild celery (*Vallisneria americana*), various pondweeds (*Potamogeton* spp.), coontail (*Ceratophyllum demersum*), and native milfoil (*Myriophyllum* spp.). By the early 1960s, the submersed macrophyte community changed drastically. At the same time that native species were declining, Eurasian water milfoil invaded and spread rapidly in dense stands throughout Lake Mendota. However, plant growth became restricted to water depths less than 3 m.

Before and during the early years of the milfoil invasion, primitive weed cutters were used to manage nuisance macrophytes. The cut plants were removed by work crews on barges after the plants had floated to shore. This practice undoubtedly aided in the rapid spread of Eurasian milfoil throughout the area's lakes, because milfoil can grow readily from vegetative fragments. To solve this problem, university engineers developed and tested prototype mechanical harvesters that removed the cut plants from the lakes (Livermore & Wunderlich, 1969). A program of harvesting nuisance aquatic macrophytes continues to this day, although Eurasian milfoil no longer attains the same biomass levels as in former years (Carpenter 1980; Nichols & Lathrop 1994). In more recent years, native aquatic macrophytes have been slowly increasing, but the littoral habitat is still not as extensive as it was before the milfoil invasion (Deppe & Lathrop 1993; Nichols & Lathrop 1994).

Invading species

A number of other species, in addition to Eurasian milfoil, have successfully invaded Lake Mendota. The invasion of European common carp (*Cyprinus carpio*) also had significant impacts. This species was introduced intentionally to lakes throughout southern Wisconsin in the late 1800s (Lathrop et al., 1992). It proliferated rapidly and was soon recognized as a cause of poor water clarity. While commercial fishing of carp began in the early 1900s, carp became so numerous in the state that in 1934 Wisconsin established a major carp removal program that was active until 1969. Since 1969, commercial fishing has been minimal due to low demand and poor market prices for carp.

Although studies conducted during the 1940s and 1950s indicated that carp were harming aquatic macrophytes and associated fish communities in many southern Wisconsin lakes, Lake Mendota was perceived to be relatively unaffected by carp through those years (Lathrop et al., 1992; Magnuson & Lathrop 1992). However, after nutrients increased in the mid-1940s (Lathrop, 1992b), carp densities also increased. Increased carp populations may have caused the hundred-fold decline in profundal zoobenthos densities that occurred between the mid-1950s and the mid-1960s (Lathrop, 1992a). The depauperate profundal zoobenthos may have had negative impacts on native fish species, such as yellow perch.

Other species have invaded Lake Mendota, although impacts have been less pronounced than for Eurasian water milfoil and carp. Invading fish species include the freshwater drum (*Aplodinotus grunniens*) and the yellow bass (*Morone mississippiensis*) (Lathrop et al., 1992; Magunson & Lathrop, 1992). Yellow bass were inadvertently introduced to local waters from fish 'rescued' from shallow sloughs of the Mississippi River during the 1930s–1940s. They first were recorded in Lake Mendota in 1957, and were common until a major die-off occurred in 1976 when they became rare. Freshwater drum were not present in Lake Mendota and other area lakes in the early 1900s. Drum may have been introduced from fish rescue operations or they may have migrated into the lakes from the Mississippi River. Freshwater drum were uncommon in Lake Mendota during the mid-1900s, but have reached moderate densities since the mid-1970s. A few other fish invasions have occurred, but their impact on the lake apparently has been minimal. Other exotic or non-native fish currently found in the Great Lakes or the Mississippi River are likely to spread to Lake Mendota.

Records of pelagic zooplankton species do not show any significant change in dominant species between the early 1900s and more recent years (Lathrop & Carpenter, 1992b). *Eubosmina coregoni*, a European exotic that invaded the Great Lakes in the early 1960s, was found in the recent sediment record of Lake Mendota (Kitchell & Sanford, 1992). It achieved modest abundances during the fall months of the mid-1980s (Lathrop & Carpenter, 1992b), but has virtually disappeared from the plankton in more recent years. Other exotic species such as the zooplankton *Bythotrephes* sp. and the benthic zebra mussel (*Dreissena polymorpha*) have colonized nearby Lake Michigan, but have yet to invade Lake Mendota. Impacts of zebra mussel on the lake's food web will likely be substantial, but at present no policy exists that can prevent their introduction to Lake Mendota from other lakes.

Extirpations and loss of genetic diversity

Extirpations of many different plants and animals have occurred in Lake Mendota during this past century. The species diversity of aquatic macrophytes declined after the mid-1950s. Many species of the genus *Potamogeton*, including the once abundant *P. amplifolius*, are no longer present (Nichols & Lathrop, 1994). Declining numbers of migratory waterfowl may be related to declines of favored forage such as wild celery (*Vallisneria americana*) and sago pondweed (*Potamogeton pectinatus*). While some plant species are slowly returning to the lake, it is unknown whether the macrophyte community structure in Lake Mendota will return to its former diversity, now that Eurasian milfoil is no longer growing in dense monotypic stands.

A less noticeable extirpation has been the fingernail clam (*Pisidium* sp.) that once was found in moderate densities throughout the profundal sediments until the mid-1960s when other zoobenthos densities declined (Lathrop, 1992a). Its ecological importance in reprocessing detrital organic matter in Lake Mendota is unknown.

Finally, extirpations of many small littoral-zone fishes have occurred in Lake Mendota since the early 1900s (Lyons, 1989). Most of these fish species are intolerant of environmental degradation. Some species may have declined because they prefer habitat with extensive, diverse macrophytes – a condition that changed around the time of the invasion of Eurasian milfoil in the early 1960s. Another factor that may have contributed to their decline is greater predation due to an increase in piscivore stocking in the 1960s (Magnuson & Lathrop, 1992). These fish may have been an important food source for piscivorous fish, particularly in the spring before the young-of-the-year panfish are available. Now a single species, the brook silverside (*Labidesthes sicculus*) dominates the littoral fish community. Abundance of silversides fluctuates considerably, and effects of these fluctuations on recruitment of the larger fish species is unknown.

While the loss of species diversity has occurred through extirpations, the loss of genetic diversity in some piscivorous fish has occurred through the massive stockings in Lake Mendota (Lathrop et al., 1992). Concerns have been raised about hatchery strains of fish being planted in the state's waters. Low survival of walleye fry and fingerlings stocked during the biomanipulation project on Lake Mendota was attributed to the unsuitability of stocking northern Wisconsin strains of these fish (Johnson & Staggs, 1992). Northern strains were stocked because most of the state's hatchery production of walleyes was from that area. However, future stockings will use local strains more suited for southern Wisconsin's warmer, more eutrophic waters.

Summary

Lake Mendota exemplifies the status of research and management for many eutrophic lakes of agricultural and urban regions. Non-point phosphorus pollution is the keystone cause of the lake's problems. Non-point pollution has been difficult to control. The principal reason is that farmers lack the incentives to manage phosphorus appropriately. At present, hopes are high that the Priority Lake Project possesses the regulatory capabilities and the financial incentives needed to solve the problem. Biomanipulation can substantially improve the lake's water clarity (Lathrop et al., 1996), but intensive angling has made it difficult to sustain the large piscivore populations needed for biomanipulation (Johnson & Staggs, 1992; Johnson & Carpenter, 1994). More restrictive harvest regulations could potentially sustain adequate levels of piscivory in the lake. Restorations of wetlands, native macrophytes, riparian forests and woody habitat of the littoral zone are needed to improve fish habitat. Riparian forests and wetlands also contribute to nonpoint pollution control. Zebra mussels are likely to invade Lake Mendota in the next few years. Their effect on the food web is likely to be strong, but more specific predictions are highly uncertain.

Collaboration of scientists and managers

Lake restoration involves unprecedented manipulations of large, complex systems. Consequently, restoration projects often yield surprising results. While the particular surprises cannot be forecast, we should expect the need to change tactics as restorations develop and new scientific information becomes available.

Successful lake restoration projects have often involved collaborations of scientists and managers (Gulati, 1990; Kitchell, 1992; National Research Council, 1992; Moss et al., 1996; chapters in this volume). In our experience, successful collaborations have certain characteristics. These are:

1. Scientists and managers are involved in the design, implementation and assessment of the project. There is clear communication of goals and expect-

ations among participants. Public outreach is an effective component of the program.

2. The partners agree on the scope of the project. Generally, this means that scientists must work on spatial units larger than those studied by many ecologists. Managers must work at the temporal scale of ecosystem responses, which is long in comparison to typical agency processes.
3. The restoration is conducted as an experiment (Carpenter et al., 1995). Both premanipulation and postmanipulation studies are performed. If possible, a reference ecosystem is studied in parallel with the manipulated ecosystem. Replicate reference and manipulated ecosystems are included, if possible. The manipulation is strong, sustained and consistent. Confounding factors and alternative explanations are carefully evaluated.

Several advantages of collaboration are discussed in Kitchell (1992). These include the new and promising ideas that come from interactions of scientists and managers; opportunities for funding and career advancement that derive from collaboration; academics and students who are better informed about the challenges of environmental management; and managers who are better informed about current scientific understanding (Rudstam & Johnson, 1992; Staggs, 1992). Such collaborations are very successful at overcoming scientific limitations to lake restoration, and have produced important advances in our ability to restore lakes. They have been less successful at overcoming political, institutional and economic constraints to the success of restoration projects.

Acknowledgements

We are grateful for the support of the North Temperate Lakes Long-Term Ecological Research Program, the Pew Foundation and the Wisconsin Department of Natural Resources. We thank Michelle Richardson, Dane County Land Conservation Department, for preparation of the map.

References

Betz, C. R., M. L. Lowndes & S. Porter, 1997. Nonpoint source control plan for the Lake Mendota Priority Watershed Project. Wisconsin Department of Natural Resources, Madison, Wisconsin, U.S.A., Publication number WT-481-97.

Brock, T. D., 1985. A Eutrophic Lake: Lake Mendota, Wisconsin. Springer-Verlag, New York.

Carpenter, S. R., 1980. The decline of *Myriophyllum spicatum* in a eutrophic Wisconsin lake. Can. J. Bot. 58: 527–535.

Carpenter, S. R., D. Bolgrien, R. C. Lathrop, C. A. Stow, T. Reed & M. A. Wilson, 1998. Ecological and economic analysis of lake eutrophication by nonpoint pollution. Australian Journal of Ecology 23: 68–79.

Carpenter, S. R., N. F. Caraco, D. L. Correll, R. W. Howarth, A. N. Sharpley & V. H. Smith, 1998. Nonpoint pollution of surface waters with phosphorus and nitrogen. Ecological Applications 8: 559–568.

Carpenter, S. R., S. W. Chisholm, C. J. Krebs, D. W. Schindler & R. F. Wright, 1995. Ecosystem experiments. Science 269: 324–327.

Carpenter, S. R. & K. L. Cottingham, 1997. Resilience and restoration of lakes. Conservation Ecology 1. Internet at http://www.consecol.org/vol1/iss1/art2.

Carpenter, S. R., A. Muñoz del Rio, S. Newman, P. Rasmussen & B. M. Johnson, 1994. Interactions of anglers and walleyes in Escanaba Lake, Wisconsin. Ecological Applications 4: 822–832.

Christensen, D. L., B. J. Herwig, D. E. Schindler & S. R. Carpenter, 1996. Impacts of lakeshore residential development on coarse woody debris in north temperate lakes. Ecological Applications 6: 1143–1149.

Cooke, G. D., E. B. Welch, S. A. Peterson & P. R. Newroth, 1993. Restoration and Management of Lakes and Reservoirs. Lewis Publishers, Boca Raton, Florida, U.S.A.

Deppe, E. R. & R. C. Lathrop, 1993. Recent changes in the aquatic macrophyte community of Lake Mendota. Trans. Wis. Acad. Sci. Arts Lett. 81: 47–58.

Driscoll, C. T., C. Yan, C. L. Schofield, R. Munson & J. Holsapple, 1994. The mercury cycle and fish in Adirondack lakes. Envir. Sci. Technol. 28: 137–143.

Gulati, R. D., E. H. R. R. Lammens, M.-L. Meijer & E. van Donk (eds), 1990. Biomanipulation – Tool for Water Management. Developments in Hydrobiology 61. Kluwer Academic Publishers, Dordrecht, 628 pp. Reprinted from Hydrobiologia 200/201.

Gunderson, L. H., C. S. Holling & S. S. Light (eds), 1995. Barriers and Bridges to the Renewal of Ecosystems and Institutions. Columbia University Press, New York, U.S.A.

Harper, D. M, 1992. Eutrophication of Freshwaters. Chapman & Hall, London, England.

Heck, K. L. & L. B. Crowder, 1991. Habitat structure and predator-prey interactions in vegetated aquatic systems. In S. S. Ball, E. D. McCoy & H.R. Mushinsky (eds), Habitat Structure: The Physical Arrangement of Objects in Space. Chapman & Hall, London: 281–299.

Hilborn, R., 1992. Can fisheries agencies learn from experience? Fisheries 17: 6–14.

Hilborn, R., C. J. Walters & D. Ludwig, 1995. Sustainable exploitation of renewable resources. Ann. Rev. Ecol. Syst. 26: 25–44.

Johnson, B. M. & S. R. Carpenter, 1994. Functional and numerical responses: A framework for fish-angler interactions? Ecol. Applic. 4: 808–821.

Johnson, B. M. & M. D. Staggs. 1992. The fishery. In J. F. Kitchell (ed.), Food Web Management: a Case Study of Lake Mendota. Springer-Verlag, New York: 353–375.

Johnson, D. L. & R. A. Stein (eds), 1979. Response of Fish to Habitat Structure in Standing Water. American Fisheries Society Special Publication Number 6: Columbus, Ohio, U.S.A.

Kitchell, J. F. (ed.), 1992. Food Web Management: a Case Study of Lake Mendota. Springer-Verlag, New York.

Kitchell, J. F. & S. R. Carpenter, 1993. Variability in lake ecosystems: Complex responses by the apical predator. In M. J.

McDonnell & S. T. A. Pickett (eds), Humans as Components of Ecosystems. Springer-Verlag, N.Y., U.S.A.: 111–124.

Kitchell, J. F. & P. R. Sanford, 1992. Paleolimnological evidence of food web dynamics in Lake Mendota. In J. F. Kitchell (ed.), Food Web Management: a Case Study of Lake Mendota. Springer-Verlag, New York: 31–47.

Lathrop, R. C., 1990. Response of Lake Mendota (Wisconsin, U.S.A.) to decreased phosphorus loadings and the effect on downstream lakes. Verh. int. Ver. Limnol. 24: 457–463.

Lathrop, R. C., 1992a. Decline in zoobenthos densities in the profundal sediments of Lake Mendota (Wisconsin, U.S.A.). Hydrobiologia 235/236: 353–361.

Lathrop, R. C., 1992b. Nutrient loadings, lake nutrients and water clarity. In J. F. Kitchell (ed.), Food Web Management: a Case Study of Lake Mendota. Springer-Verlag, New York: 71–98.

Lathrop, R. C. & S. R. Carpenter, 1992a. Phytoplankton and their relationship to nutrients. In J. F. Kitchell (ed.), Food Web Management: a Case Study of Lake Mendota. Springer-Verlag, New York: 99–128.

Lathrop, R. C. & S. R. Carpenter, 1992b. Zooplankton and their relationship to phytoplankton. In J. F. Kitchell (ed.), Food Web Management: a Case Study of Lake Mendota. Springer-Verlag, New York: 129–152.

Lathrop, R. C., S. B. Nehls, C. L. Brynildson & K. R. Plass, 1992. The fishery of the Yahara lakes. Tech. Bull. No. 181. Wis. Dep. Nat. Resour., Madison, Wis.

Lathrop, R. C., S. R. Carpenter & L. G. Rudstam, 1996. Water clarity in Lake Mendota since 1900: Responses to differing levels of nutrients and herbivory. Can. J. Fish. aquat. Sci. 53: 2250–2261.

Lathrop, R. C., S. R. Carpenter, C. A. Stow, P. A. Soranno & J. C. Panuska, 1998. Phosphorus loading reductions needed to control blue-green algal blooms in Lake Mendota. Can. J. Fish. aquat. Sci. 55: 1169–1178.

Livermore, D. F. & W. E., Wunderlich, 1969. Mechanical removal of organic production from waterways. In G. A. Rohlich (ed.) Eutrophication: Causes, Consequences, Correctives. National Academy of Science, Washington, D.C.: 494–520.

Lodge, D. M., 1993. Biological invasions: lessons for ecology. Trends in Ecology & Evolution 8: 133–137.

Lyons, J., 1989. Changes in the abundance of small littoral-zone fishes in Lake Mendota, Wisconsin. Can. J. Zool. 67: 2910–2916.

Magnuson, J. J. & R. C. Lathrop, 1992. Historical changes in the fish community. In J. F. Kitchell (ed.), Food Web Management: a Case Study of Lake Mendota. Springer-Verlag, New York: 195–231.

Maser, C. & J. R. Sedell, 1994. From the Forest to the Sea: The Ecology of Wood in Streams, Rivers, Estuaries and Oceans. St. Lucie Press, Delray Beach, Florida, U.S.A.

Moss, B., J. Stansfield, K. Irvine, M. Perrow & G. Phillips, 1996. Progressive restoration of a shallow lake: a 12-year experiment in isolation, sediment removal and biomanipulation. J. appl. Ecol. 33: 71–86.

National Research Council, 1992. Restoration of Aquatic Ecosystems. National Academy Press, Washington, D.C.

Nichols, S. A. & R. C. Lathrop, 1994. Cultural impacts on macrophytes in the Yahara lakes since the late 1800s. Aquat. Bot. 47: 225–247.

Nichols, S. A., R. C. Lathrop & S. R. Carpenter, 1992. Long-term vegetation trends – a history. In J. F. Kitchell (ed.), Food Web Management: a Case Study of Lake Mendota. Springer-Verlag, New York: 153–173.

Poiani, K. A. & B. L. Bedford, 1995. GIS-based nonpoint source pollution modeling: considerations for wetlands. Journal of Soil & Water Conservation 50: 613–619.

Postel, S. & S. R. Carpenter, 1997. Freshwater ecosystem services. In G. Daily (ed.), Ecosystem Services. Island Press, Washington, D.C.

Rudstam, L. G. & B. M. Johnson, 1992. Development, evaluation, and transfer of new technology. In J. F. Kitchell (ed.), Food Web Management: A Case Study of Lake Mendota. Springer-Verlag, New York, USA: 507–524.

Scheffer, M., S. H. Hosper, M.-L. Meijer, B. Moss & E. Jeppesen, 1993. Alternative equilibria in shallow lakes. Trends in Ecology & Evolution 8: 275–279.

Soranno, P. A., S. L. Hubler, S. R. Carpenter & R. C. Lathrop, 1996. Phosphorus loads to surface waters: A simple model to account for the spatial pattern of land use. Ecological Applications. 6: 865–878.

Staggs, M. D., 1992. Benefits on a larger scale. In J. F. Kitchell (ed.), Food Web Management: A Case Study of Lake Mendota. Springer-Verlag, New York, USA: 525–538.

Stow, C. A., S. R. Carpenter, C. P. Madenjian, L. A. Eby & L. J. Jackson, 1995. Fisheries management to reduce contaminant consumption. BioScience 45: 752–758.

Swackhamer, D. L. & D. E. Armstrong, 1986. Estimation of the atmospheric and nonatmospheric contributions and losses of polychlorinated biphenyls for Lake Michigan on the basis of sediment records of remote lakes. Envir. Sci. Technol. 20: 879–883.

Zedler, J. B., 1996. Ecological issues in wetland mitigation: an introduction to the Forum. Ecol. Applic. 6: 33–38.

Hydrobiologia **395/396:** 29–38, 1999.
D.M. Harper, B. Brierley, A.J.D. Ferguson & G. Phillips (eds), The Ecological Bases for Lake and Reservoir Management.

Towards sustainable development of still water resources

Mark Everard
The Natural Step U.K., 9 Imperial Square, Cheltenham, Gloucestershire GL50 1QB, U.K.

Key words: sustainable management, freshwater

Abstract

Evidence suggests a net decline in the quality and quantity of still waters over most of the developed world. A strategy predicated on merely 'sustaining' them as isolated resources is ultimately unsustainable. True sustainable development takes account of the ecological, social and economic values of still waters within the context of living catchments. Catchment management has begun to move thinking about water management 'upstream', towards causes and integrated solutions, rather than local amelioration of 'downstream' effects. Application of *systems thinking*, which focuses on holistic dynamic systems and the factors that influence them, offers a powerful means to extend this 'upstream' thinking. This includes, as an important component, a movement towards cyclical resource utilisation and away from linear flows based on a *mine-use-dispose* ethos. 'The Natural Step' presents a generic yet science-based framework to support decision-makers across the diversity of sectors of society that influence the water cycle, and which ultimately affect still waters. It is rooted in systems thinking, taking the biospheric system of matter and energy cycling as its frame of reference, yet presents these as simple and readily-comprehensible "rules" that are powerful through their very simplicity and generic applicability. Application of 'The Natural Step' systems thinking to complex and diverse problems affecting still waters offers high-level insight, comprehensible to decision-makers in all sectors of society, which may potentially lead to cost-effective and sustainable solutions.

Introduction

The papers in this volume cover aspects of lakes and reservoirs ranging from basic ecology and conservation, through amenity and other social values, to practical implications for their management. One of the overall goals of this book, and of the conference upon which it was based, is to achieve a better linkage between the fundamental and applied sciences and the formulation of wise management decisions that affect practical, day-to-day activities. Ultimately, these management decisions should be consistent with the goal of achieving sustainable development.

The quality and quantity of various types of still fresh waters in the British Isles are in decline (Pond Action, 1993; Sansom, 1993; Biggs et al., 1994; Collinson et al., 1995; Duigan et al., 1996; Moss et al., 1996; Bennion et al., 1997; Everard, 1997a, 1998a and 1998b). The situation is little different in other developed countries and probably all other countries in the world. It is a trend which is clearly inconsistent with the overarching goal of achieving sustainable development; formulation and delivery of new policies is therefore essential for the achievement of this goal. This paper questions what sustainable development means in practice in the context of still waters, and explores new approaches to decision-making that are founded in sound science yet incorporate environmental needs across the broad range of activities that impact upon the quantity and quality of still waters.

Distinguishing 'sustaining' from 'sustainable development'

It is valuable at this point to define what is meant by sustainable development and, in particular, to draw a clear distinction between sustainable development *per se* and the process of purely *sustaining* a still water body.

Sustaining

The Concise Oxford Dictionary defines 'sustain', as '*Keep going continuously*' (Sykes, 1982). The national trends indicated in this volume demonstrate that we are failing our still water resource even in this basic respect.

Largely due to necessity, the historic focus of U.K. still water conservation effort has been upon the protection of remaining bodies of high conservation importance through a series of designations such as Local Nature Reserves (LNRs), National Nature Reserves (NNRs), Sites of Special Scientific Interest (SSSIs), Sites of Particular Value to Wildfowl identified under the Ramsar Convention (Ramsar, 1971), Special Protection Areas for conservation under the EC Directive on Birds (CEC, 1979) and Special Areas for Conservation under the EC Habitats Directive (CEC, 1992). The fragmented pattern of contemporary land ownership and use generally necessitates site management agreements entailing some degree of subsidy. Fixes to perceived hydrological and water quality problems are often undertaken in isolation from other factors affecting the overall water cycle within which the designated site lies.

Everard (1998) identified this spatial approach to conservation as being economically unsustainable in the longer-term. By treating still waters as isolated resources, out of context of the living landscapes of which they are a component, protective policies tend towards the preservative and the exclusive. The resultant public perception is that protection of the aquatic environment is a net and continuing financial burden on the nation for the conservation of relatively small fixed locations, often fenced-off from general access and therefore of little perceived value to the public. This process externalises the wider social and economic benefits arising from healthy aquatic ecosystems. Thus, merely *sustaining* still fresh waters in isolation – just 'keeping them going' on a largely *ad hoc* basis – is itself an unsustainable strategy.

Sustainable development

The modern concept of sustainable development stems from a long heritage of concern about the many ways in which man affects the environment that sustains his health, quality of life, material needs and business interests. Significant milestones within this long heritage of environmental consciousness include:

- The *land ethic* proposed as long ago as the 1930s by Aldo Leopold (Leopold, 1949).
- The seminal *Limits to Growth* concept (Meadows et al., 1972), which introduced the notion that the earth's ecosystems are limited and that population growth and economic activity should therefore be managed to take account of these limits.
- The Ramsar Convention, recognising the global importance of the dwindling wetland resources (Ramsar, 1971), later developing into the concept of *Wise Use* of global wetland resources (Ramsar, 1988).
- The *Global 2000 Report* to the US President, which contained the words "*given the urgency, scope, and complexity of the challenges before us, the efforts now underway around the world fall far short of what is needed. An era of unprecedented global co-operation and commitment is essential*" (US Council on Environmental Quality and US Department of State, 1980).
- The *World Conservation Strategy* (IUCN/UNEP/WWF, 1980) which sought to address the integration of society within the earth's biospheric processes. The *World Conservation Strategy* contributed significantly to an understanding of the global nature of environmental problems, as well as the importance of adopting a 'systems thinking' approach to take account of biospheric processes. Systems thinking offers a means for cutting through the mass of political and economic issues that surround the essential processes impacted by human activities. The lines of thought and proposals for practical actions contained in the 1980 report were taken further forwards in *Caring for the Earth: A Strategy for Sustainable Living* (IUCN/UNEP/WWF, 1991).
- The World Commission on Environment and Development (WCED) was established in 1983 to formulate long-term global strategies for changing the ways in which society interacts with the environment, with the goal of attaining sustainable development. WCED's final report, perhaps better known as the 'Brundtland' report (WCED, 1987), contained a detailed and holistic analysis of the environmental, social and economic issues to be addressed if we are to achieve sustainability. One of the key successes of the report was the production of a common definition of sustainable development. This definition, more commonly known as the 'Brundtland definition', is: 'Development that meets the needs of the present without compromising the ability of future generations to meet their own needs'.
- Reinforcement of prior commitments by the 150 signatories to the *Declaration of Principles for Sustainable Development* at the United Nations

Conference on Environment and Development (UNCED), or 'Earth Summit', held in Rio de Janiero in 1992. Over 150 nation states and the European Union also signed the *Convention on Biological Diversity* at the Earth Summit, recognising the crucial importance of maintaining diversity and productivity in nature for the long-term interests of global and local communities (DoE, 1994).

Sustainable development is often defined as addressing the 'triple bottom line'; the integrated and interdependent aspects relating to economy, ecology and society. In this model, all externalities – both net costs and attendant benefits – are taken fully into account in the marketplace, and in decision-making, to inform wise use decisions.

The sustainable development of aquatic ecosystems

The implication for the sustainable development of aquatic ecosystems is that they should be considered within the context of the living landscapes of which they are a part, and of the processes that form and sustain them (Everard, 1998b). Protective policies should therefore be addressed not at prohibition of local activities nor application of 'end of pipe' technology to ameliorate local effects; rather it should be proactive and relate to all social and economic activities within the catchment.

In the aquatic environment, considerable progress has been made in developing an understanding of the value of wetlands beyond merely their importance to wildlife, taking account also of the hydrological, physico-chemical and ecological functions that they perform (Dugan, 1990; Maltby, 1991a, 1991b; Denny, 1995). Acknowledgement of the ecological and societal values arising from wetlands is helpful in identifying the ways in which man benefits from the natural services, regenerative capacity, resources, assimilative capabilities and so forth of the biospheric processes. However, in practice, they are currently largely unrecognised in contemporary planning and land use decision systems (Everard, 1997a,1998c) even though a recent study has estimated the value of the global ecosystem 'services' at 33 trillion U.S. dollars, or just over twice the global annual GDP (Costanza et al., 1997).

Sustainable development goes substantially beyond mere 'greening', extending concerns about the effects of human activities way beyond local considerations and into the whole social, economic and environmental 'footprint' of that activity (Wackernagel & Rees, 1996). There is therefore a two-fold need: for those concerned with the protection of still waters to influence policy-making decisions across all sectors of society; and for these decision-makers to use sound scientific advice as regards the biospheric processes with which they interact, and which ultimately influence still waters.

There is, therefore, an overwhelming need to extend our understanding and appreciation of these biospheric services in two corresponding ways: to develop our understanding to enable wise and sustainable decisions to be undertaken at the catchment and water cycle scales; and to translate the body of science into a practical form usable by those responsible for decisions about land use and development planning. We must bear in mind that the transfer of sound science to decision-makers must be achieved in generic terms, readily comprehensible and useable by people in all sectors – business, government, regulatory, general public, etc. – who generally possess little specific knowledge about issues of scientific detail.

A 'systems thinking' approach to sustainable water management

A key obstacle to the sustainable use of the water environment has been the historic perception that local problems need to be addressed at a local level (Everard, 1997a, 1998b; Everard & Porritt, 1997). In common with traditional approaches to nature conservation, national strategies for pollution control, flood defence and water resource management have also been planned on a relatively parochial basis. In England and Wales, for instance, the regulatory framework in place to govern investment by the water industry is tied strongly to traditional engineering solutions to overcome perceived local problems (Everard & Porritt, 1997).

In contrast to this essentially piecemeal approach, a 'systems thinking' approach looks at problems from an holistic level, examining systems as a whole and applying the first-order principles that govern the system. In the present instance, systems thinking therefore necessarily focuses upon the water cycle, and indeed the wider cycling of energy and matter within the biosphere, rather than taking as its starting point various fragmented aspects of local human interest. To a significant extent, catchment management has been successful in working to find solutions to problems 'upstream' (in the process sense) and has therefore taken us at least some way along the path of systems thinking. However, it is a widely-shared perception that, in its practical application by overworked regulatory practitioners, application of the catchment management concept often fails to achieve this holistic goal but is rather used as a 'basket' in which to place essentially non-integrated plans.

The holistic systems thinking approach to sustainable development of the water cycle is consistent with The Natural Step (TNS) approach, the development and application of which is documented by Robért (1991), Holmberg et al. (1996), Everard (1997b, 1997c, 1998b), Holmberg & Robért (in press), Slavin (1998) and Henderson (1998). TNS systems thinking takes as its scientific reference point the thermodynamic 'rules' governing the cycling of matter and energy within the larger system of the planet's biosphere, developing these into generic first-order principles to guide wise decisions within business, government, regulatory agencies, etc. It addresses the need to take account of environmental matters as a matter not merely of the continued survival and health of the human species and the ecosystems that support it, but as crucially important in maintaining longer-term economic performance and social wellbeing. These economic and social factors are an expression of the material quality of life that a society can support, since economic activity, social factors and quality of life, as well as ecological health, ultimately depend upon the cycles of energy and matter that govern the global biosphere (Georgescu-Roegen, 1971). Thus, The Natural Step framework integrates the 'triple bottom line' – economy, ecology and society – in a pragmatic yet scientifically-robust expression of direct use to decision-makers who may have little in the way of detailed scientific knowledge.

Sustainable development of still water resources

This ecologically-founded systems thinking (or 'ecocyclic') approach to sustainable development demands that we consider the first order principles governing the water cycle, including both the water itself and the substances cycling within it. Still fresh waters must therefore be considered not merely as individual units, for protection merely at the parochial scale through local pollution control and hydrological schemes, but as components within the larger context of the overall water cycle. Solutions to today's problems do not therefore lie merely in the construction of new treatment plant and water level management schemes to alleviate local problems, but by the extension of our view to the wider context of wise use of the entire water cycle. Since it is, for example, ultimately unsustainable to seek merely to sustain the current quality of a single designated still fresh water body against a rising tide of urban growth, unsympathetic land use, and loss of habitat in the upstream catchment, sustainable use of the entire water cycle offers the only long-term solutions. This clearly challenges existing land use, pollution control and land right assumptions. Nevertheless, if critical conditions governing the overall water cycle system are breached, the consequences for adverse "downstream" (in both a process and a geographic sense) impacts are significant and diverse.

TNS thinking embodies a transition from linear to cyclic resource use, and the four first-order principles at the heart of TNS, known as the 'system conditions', provide a comprehensive yet common-sense framework to guide wise decisions. A brief consideration of cyclic versus linear systems, and several examples based on the four system conditions, will help to illustrate this point.

Linear versus cyclic resource usage

Through approximately 4.5 billion years of evolution, nature has evolved in a highly efficient manner, producing no net waste. Waste products from isolated species or processes are the essential resources for other components of a highly complex ecosystem, containing diverse pathways for efficient and adaptive processing. All matter within the system is used and reused cyclically, powered by net input of solar energy, and all consequent societal activities and economic processes are dependent upon these solar-powered energetic processes (Georgescu-Roegen, 1971; Odum, 1982, 1983).

Cyclic use of natural resources by society – and water is perhaps the ultimate example of such an infinitely renewable resource – is ultimately integrated within the biosphere's carrying capacity and therefore, by systems thinking definition, a sustainable practice. In contrast, production of substances alien to nature and therefore which cannot be reintegrated into natural cycles is by definition unsustainable, as is the over-harvesting of natural resources or destruction of productive surfaces, or the loss of diversity of ecosystems that equip them with efficient and adaptive metabolic pathways. Utilisation of recycled resources, or procedures that ensure that the waste products from non-renewable resources do not accumulate in the biosphere, are consistent with this cyclic model provided all externalities, such as energy inputs etc., are also taken fully into account.

The sad history of human society since the Industrial Revolution is one of increasing reliance upon linear resource flows (Jackson, 1996). Abundant cheap energy has been unlocked by mining of fossil fuels, the wastes of which do not disappear but tend to accumulate in the biosphere. Furthermore, the industrial revolution and modern manufacturing practices have been founded on the linear extraction, use and disposal of resources, resulting in a diversity of adverse effects including depletion of resources, excessive land-filling, and build-up of synthetic as well as naturally-occurring pollutants in nature and in human tissues. Modern high-input, high-output agricultural practices are no less polluting or resource intensive. A generic framework for decision-making is therefore essential to help those charged with policy development, in all sectors of society, to make wise decisions leading to resource use patterns that are cyclic and which will not lead to an unnecessary accumulation of wastes in the biosphere.

System condition 1

System condition 1 of The Natural Step states that '*Substances from the Earth's crust must not systematically increase in nature*'. It is in fact a simplification of a large body of science exploring environmental capacity. In essence, in a sustainable society, fossil fuels, metals and other materials must not be extracted at a faster pace than their slow redeposit and reintegration into the Earth's crust. The reasoning behind this rests in the fact that, throughout billions of years of cycling and evolution in the biosphere, substances have become immobilised in the earth's crust and, once reintroduced into the biosphere at rates considerably exceeding natural flows, tend to overload nature's ability to 'tidy them away' by biomineralisation, sedimentation, etc. They therefore tend systematically to increase in concentration with potentially adverse and unpredictable impacts upon climate, ecology and society, and the economic interests that ultimately derive from them. When system condition 1 is breached, it leads to such widespread phenomena as eutrophication, increasing carbon dioxide levels in the upper atmosphere, metal accumulation in soils and waters, and so forth.

A practical example of system condition 1 is the trend towards increasing nutrient loadings, predominantly of phosphorus and nitrogen, in the still fresh waters of the UK (Moss et al., 1996; Bennion et al., 1997). Local solutions to these problems might include installation of narrow buffer zones to reduce nutrient and sediment flux from agricultural land (recently reviewed by Haycock et al., 1997), the installation of tertiary nutrient-stripping plant at sewage treatment plants, and other subsidised land use controls. However, it is the continued reliance upon a linear input of nutrient substances, mined from the Earth's crust, refined and transported into catchments as artificial fertilisers, detergents, etc., that contributes directly or indirectly to elevating nutrient loadings within the catchments of the U.K. Indeed, the extraction, refinement and transportation of these fertilisers also directly breaches system condition 1 due to a reliance on energy derived predominantly from fossil fuels, with their resultant loads of waste carbon dioxide, NO_x and SO_x gases, VOCs, etc. Everard & Porritt (1997) commend schemes in Eastern England that seek to close this nutrient loop, reducing reliance upon the mined nutrients by spreading nutrient-rich sludge back onto land local to the treatment works, providing much of the fertiliser needs for crop production by the local population. (Obviously, sludge application must conform with 'best practice' recommended by regulatory authorities for land spreading.) Further 'best practice' in closing the loop is provided by Wessex Water, in the English West Country, who offer economic incentives to encourage farmers in sensitive Wessex Water catchment areas to convert to more natural low-input farming methods and to control the uses of artificial fertilisers (Wessex Water, 1998), directly benefiting the company through improved water quality. These approaches address problems 'upstream' (in the process, as well as fluvial, senses) consistent with the ecocyclic principles advocated by The Natural Step, and as such go some way towards reducing dependence upon linear usage of resources otherwise immobilised in the Earth's crust. Furthermore, by returning substances formerly treated as waste back into the catchment as a valuable resource, widespread and diverse ecological problems are averted and significant cost savings accrue directly to the water companies, farmers, and other water users. By application of systems thinking, sustainable solutions to eutrophication of still waters may be achieved not by local 'end of pipe' technological responses to perceived local problems, at net cost and with their own environmental impacts, but by basing wise decisions upon the dynamics of the wider water cycle and eliminating dependence upon such substances 'upstream'. A business or local government enterprise that choses incrementally to reduce dependence on these substances is achieving a long-term competitive advantage in the face of increasing resource scarcity and cost (including in transport and processing energy), capacity to dispose of wastes, likelihood of more stringent legislation to control the adverse environmental effects of the substances, and adverse public opinion, etc.

A further example of system condition 1 in action is provided by Everard & Porritt (1997). Water pollution control schemes in England and Wales have frequently been developed on too parochial a basis, in which the identified 'environmental improvement' benefits are not adequately offset by their, often significant, environmental costs. Seriously to address the criteria for sustainable development, the investment planning framework for the privatised water industry (the Asset Management Planning process, or AMP) needs to be adjusted to take account of implications for the environment far wider in scale than merely local environmental aspirations. It must necessarily cover, for example, the climatic impacts of increased greenhouse gas emissions resultant from increased energy demand for pumping or sludge-drying plant, dependence upon mined or synthetic substances, habitat loss resulting from extended effluent treatment plant, increased flood risk lower in a catchment resulting from wetland loss or other hydrological impacts upstream, etc. (Everard & Porritt, 1997). The scale of consideration must also take in not only local environmental aspirations, but the consequences of activities on the environment taken as a whole, and their interaction with economic and social factors.

System condition 2

System condition 2 states that '*Substances produced by society must not systematically increase in nature*'. In essence, in a sustainable society, substances new to nature, and to which nature is therefore not equipped throughout evolution to recycle and reintegrate within its cycles, must not be permitted to be released at a faster pace than they can be broken down or deposited into the Earth's crust. Concerns about synthetic chemicals are far from new. Carson (1962) was perhaps the first to publicise and popularise such concerns in the seminal *Silent Spring*. In practice, it is difficult, or perhaps even impossible, to predict tolerance limits for substances to which nature has not been adapted, and which are not therefore recycled efficiently by biospheric processes, particularly within the complexity of natural ecosystems. Given the wide range of synthetic chemicals currently in use, and the largely incomplete state of knowledge about their acute and/or chronic toxicity, application of the precautionary principle should counsel us that, in a sustainable world, humankind must cease to depend upon the production and usage of artificial substances which are chemically stable and persistent, and which tend to accumulate in organisms.

Application of systems thinking is helpful in the face of this complexity, suggesting that decisions should ultimately be based on broad principles relating to the behaviour of substances in the biosphere, rather than in the confusion of fine detail about each of the estimated 100,000 artificial substances in current production (Hawken, 1993). Further political impetus has been given to this debate by the Chemicals Policy Committee, an official committee of the Swedish Government, which proposes several targets for the eventual phasing out of persistent and bioaccumulative substances regardless of current toxicological data on the basis that "*...new, unexpected forms of toxicity may be uncovered in the future*" (reported in ENDS, 1997). Examples of such new, and previously unforeseen, toxicological effects include the relatively recent discovery of endocrine disruption, a widespread but largely unquantified problem in U.K. waters (as reviewed recently by Environment Agency, 1998). It is from this systems thinking level of generic and broadly-agreed principle that decision-makers can formulate far-sighted and sustainable policies about chemicals policy, yet today most public environmental debate still focuses upon details for which little consensus exists. Were system condition 2 already to be observed, endocrine disruptors, pesticides and other substances would not be so widespread in U.K. waters.

System condition 3

System condition 3 states that '*The physical basis for the productivity and diversity of nature must not be systematically diminished*'. Put simply, in a sustainable society, the productive surfaces of nature, including their biological diversity and the processes that ecosystems perform, must not be further diminished in quality or quantity, thus ensuring that the 'engine' of biological productivity, and the complexity of pathways provided by an adaptable and complex ecosystem, are not depleted. Also, we must not harvest more from nature than can be recreated and renewed, nor seriously compromise the processes performed by natural systems. As both form part of the ecocyclic processes that sustain nature and mankind, sustainable development and biological diversity are in practice indivisible as it is only through a shift to sustainable modes of resource use that habitat, soils and genetic resources can be sustained and equitably shared (World Resources Institute, 1992).

A local-scale example of this cited by Everard & Porritt (1997) is of a habitat improvement scheme intended as mitigation for fish stocks to offset perceived problems of poor fish recruitment resultant from over-abstraction in a river catchment. One of the most significant 'spin-off' benefits of the cited scheme was the solution of a long-standing *Cryptosporidium* pollution problem, believed to have been mediated by both the 'buffer zone' effect of the improved riparian habitat and the increased metabolism of microorganisms within the river system. The functional analysis work of Dugan (1990), Maltby (1991a, 1991b) and Denny (1995) would clearly suggest that commensurate benefits would also accrue in terms of enhanced metabolism of nutrients, organic matter and other substances, in addition to the partial restoration of hydrological and ecological processes within the catchment. This example confirms in practice that wetland habitats do in fact perform valuable functions of direct benefit to society, and that their restoration may avert the need for expensive and potentially polluting technology 'downstream' (in both a process and a literal sense). The protection or restoration of natural water cycle processes throughout catchments therefore has direct impacts upon their capacity to store water, general hydrology, chemistry, sediment flows, fish recruitment and ecological health, resulting in a broad range of societal and economic benefits (NRA, 1995; Everard, 1997a, 1998c).

System condition 3 is also a restatement, in simple but applied terms, both of the three key objectives of the World Conservation Strategy (IUCN/UNEP/WWF, 1980) and of the 'wise use' principles proposed and developed by the Ramsar Convention (Ramsar, 1971; 1988). A large body of science has been established with this objective in mind, including studies in such diverse fields as sustainable use criteria for fishery yields and other ecosystems, critical natural capital, etc. The Natural Step framework brings this concept into the context of decision-making in business, government, regulation, and other sectors of society.

System condition 4

System condition 4 states that "*There needs to be fair and efficient use of resources with respect to meeting human needs*". This system condition differs in character from the previous three in that it relates more to the societal conditions that permit conformance with the broad principles of biospheric cycling, and includes social equity and the support of basic human needs by the most resource-efficient means. It therefore relates less to the underlying science and more to the practicality of 'making it happen' at the social and economic level.

This includes an implied commitment to the development of cleaner and more efficient technologies, and also the transfer of such technologies to developing countries (for example, as supported by the U.K.'s Darwin Initiative (DoE, 1992)). It also counsels against increasing dependence upon the often short-term benefits derived from unsympathetic development of land and water resources, which may be offset by long-lasting dis-benefits for water bodies lower in the catchment. At present, these remote dis-benefits are not routinely screened in development consents. The rising per capita water consumption in the U.K. has to be viewed in this context, establishing demand management as an urgent priority. Equity issues also exist at differing scales, including both the temporal factor of earlier exploitation of aquatic habitats for agriculture and other forms of land use, which deprive future generations of these same benefits, and the geographic factor of upstream pollution inputs and loss of hydrological processes disadvantaging those downstream (Everard, 1998). There is also a significant international inequity issue, both economic and social, resulting from reliance upon less-developed countries for the production of essential resources to compensate for declining productivity at home, or for the cheap production of luxury items such as 'out of season' vegetables and fruit, cut flowers, etc.

Climate change, an issue of significant concern for the U.K. (DoE, 1996), is also a result of activities occurring on a worldwide stage. Consequently, it is in the interest of every nation to put in place policies and practices that reduce dependence upon exploitation of the developing world, where much of the future demand for energy is likely to arise as a result both of increasing population and expected material 'quality of life'. The importance of addressing ethical, cultural, social and economic needs in addition to the merely technological is supported by many international studies (WCED, 1987; IIED, 1994), and so the importance of this system condition must not be underestimated. As such, there is clear opportunity to use The Natural Step's decision-making framework to help in, for example, identifying domestic critical natural capital as well as prioritising the apportionment of overseas development aid.

The generic nature of the four system conditions

One of the other key strengths of The Natural Step's four system conditions is that they comprise a comprehensive 'shared mental model', rooted in sound science. Their generic nature offers a common language for dialogue about complex environmental issues between decision-makers in various sectors. It is essential that large-scale problems are addressed in common terms by all sectors of society – small, medium and large businesses, governments and local communities, etc – without recourse to detailed scientific analysis of the fine details of every chemical, process or decision.

Utilising a systems thinking approach to apply protection of the overall water cycle would rely upon a more holistic and concerted approach to planning and land use decisions that take account of the interaction between water bodies, water courses, and their catchments. The inextricable linkage between environmental, economic, and social aspects of our water resources lies at the heart of sustainable development. The Natural Step offers a frame of reference for integrating these three factors, and thereby basing decision-making in the biospheric processes that ultimately support all human economic, social and physical wellbeing.

Conclusions

The challenge of sustainable development is to move beyond peripheral 'greening', that merely softens the environmental impact of human activities. Truly sustainable development is about reappraising core business and social activities, and adapting them to fit within the carrying capacity of the biospheric processes that sustain them. This calls for decisions, and decision-support mechanisms, that have their roots in robust science yet are generic across business sectors.

Any serious address to the challenge of sustainable development must necessarily cause us to reassess what would constitute a sustainable water cycle. At its 1996 Brisbane meeting, the Ramsar Commission recognised that water will be a limiting factor worldwide to future human development (Ramsar, 1996), raising the profile of an already pressing issue. However, fundamental though the development of a robust scientific base might be for future wise water management decisions, it is equally important that appropriate technology transfer techniques are developed in parallel to ensure that this science is made available in a useable form for decision-makers in industry, agriculture, development planning and land use (Everard, 1997a, 1998b). Everard (1997a) has already argued for the establishment of a National Wetland Forum to bring into open debate the conflicting ecological, social and economic issues associated with the development of a wise use strategy for the U.K.'s wetlands.

The longer-term goal of achieving a sustainable water cycle seems daunting. However, this book serves to illustrate that considerable progress is being made in furthering our understanding of the ecological processes that govern still waters. Considerable further work is clearly required to unravel the details, the wider-scale implications, and the generic principles upon which wise management decisions will be founded. The sustainable water cycle is therefore a field ripe for further research by ecologists, economists and social scientists and, crucially, for those charged with making the best available science available to decision-makers on the ground.

The Natural Step framework offers an appropriate and useful decision-support framework, relevant at a sufficiently large ecological scale, for addressing the implications of moving society towards to a truly sustainable water cycle. It further offers a robust yet comprehensive basis for the development of future applied research that takes account of the place of the limiting resource of the water environment within biospheric cycles, of its ecological health, and the implications for human health, wealth creation and wellbeing that depend upon it. As outlined previously in this paper, the innovations of The Natural Step carry forward a long-established tradition of science-based models for addressing the world's more serious problems. Challenging though it might be to traditional linear industrial and commercial procedures, and within the current constraints of population, land use and historic rights upon water and land use, the 'systems thinking' approach provides a bedrock of sound science and a generic approach to problem-solving from which to reconsider holistic solutions to the development of water resources in a way that is sustainable in the longer term.

It is widely acknowledged that a commitment to sustainable development must necessarily break down barriers between scientific, business, government, educational and social sectors of society. Limnologists too must recognise the need to go beyond the barriers represented by traditional approaches to conservation, moving from local-scale preservation to catchment-scale influence on the activities affecting water resources. Systems thinking reinforces this need to think 'upstream', addressing problems at source through the broad principles governing interactions with the water cycle, and supporting wise decision-making across the broad range of economic and social activities occurring within the catchment.

References

Bennion, H., R. Harriman, & R. Battarbee, 1997. A chemical survey of standing waters in south-east England with reference to acidification and eutrophication. Freshwater Forum 8: 28–44.

Biggs, J., A Corfield, D. Walker, M. Whitfield & P. J. Williams, 1994. New approaches to the management of ponds. British Wildlife 5: 273–287.

Carson, R., 1962. Silent Spring. Hamish Hamilton, London.

CEC, 1979. Directive on the conservation of wild birds. 79/409/EEC.

CEC, 1992. EC directive on the conservation of natural habitats and of wild fauna and flora, 92/43/EEC, CEC Brussels.

Collinson, N. H., J. Biggs, A. Corfield, M. J. Hodson, D. Walker, M. Whitfield & P. J. Williams, 1995. Temporary and permanent ponds: an assessment of the effects of drying out on the conservation value of aquatic macroinvertebrate communities. Biological Conservation 74: 125–133.

Costanza, R., R. d'Arge, R. de Groot, S. Farber, M. Grasso, B. Hannon, K. Limburg, S. Haeem, R. V. O'Neill, J. Paruelo, R. G. Raskin, P. Sutton, & M. van den Belt, 1997. The Value of the World's Ecolosystem Services and Natural Capital. Nature 387: 253–260.

Denny, P., 1995. Benefits and priorities for wetland conservation: the case for national wetland conservation strategies. In M. Cox, V. Straker & D. Taylor (eds), Proceedings of the International Conference on Wetlands Archaeology and Nature Conservation. HMSO, London.

DoE, 1992. This Common Inheritance. HMSO, London. DoE, 1994. Biodiversity: The U.K. Action Plan. HMSO, London.

DoE, 1996. Review of the Potential Effects of Climate Change in the United Kingdom. United Kingdom Climate Change Impacts Review Goup, Second Report. HMSO, London.

Dugan, P. J., 1990. Wetland Conservation: A Review of Current Issues and Required Action. IUCN, Gland, Switzerland, 96 pp.

Duigan, C. A., T. E. H. Allott, H. Bennion, J. Lancaster, D. T. Monteith, S. T. Patrick, J. Ratclifffe & J. M. Seda, 1996. The Anglesey Lakes, Wales, UK – A conservation resource. Aquatic Conservation: Marine and Freshwater Ecosystems 6: 31–55.

ENDS, 1997. Sweden sets the agenda for tomorrow's chemicals policy. ENDS Report 269: 21–25.

Environment Agency, 1998. Endocrine-disrupting Substances in the Environment: What Should We Do? Environmental Issues Series, 1. Enviroment Agency, Bristol.

Everard, M., 1997a. Development of a British wetland strategy. Aquatic Conservation 7: 239-254.

Everard, M., 1997b. Stepping Stones. The Natural Step, Cheltenham, U.K.

Everard, M., 1997c. TNS: From the Basic Science to the System Conditions. Stepping Stones, 6, p1-7. The Natural Step, Cheltenham, U.K.

Everard, M., 1998. Floodplain protection: challenges for the next millennium. In R. G. Bailey, P. V. José & B. R. Sherwood (eds), United Kingdom Floodplains. Westbury Academic and Scientific Publishing, West Yorkshire: 477–485.

Everard, M. & J. Porritt, 1997. Concerns about OFWAT's position on AMP3. Utility Week.

Hawken, P., 1993. The Ecology of Commerce: A Declaration of Sustainability. HarperCollins, New York.

Haycock, N. E., T. P. Burt, K. W. T. Goulding & G. Pinay, 1997. Introduction. In N. E. Haycock, T. P. Burt, K. W. T. Goulding & G. Pinay, (eds), Buffer Zones: Their Processes and Potential in Water Protection. Quest Environmental, Harpenden, U.K.

Georgescu-Roegen, N., 1971. The Entropy Law and the Economic Process. Harvard University Press, Cambridge, MA.

Henderson, C., 1998. Natural Step to Sustainability. Financial Times, 7th Jan 1998. Financial Times, London.

Holmberg, J., K-H. Robért, & K-E. Eriksson, 1996. Socio-ecological principles for sustainability. In R. Costanza, S. Olman & J. Martinez-Alier (eds), Getting Down to Earth – Practical Applications of Ecological Economics. International Society of Ecological Economics, Island Press, Washington.

Holmberg, J. & K-H. Robért, *in press*. The system conditions for sustainability: a tool for strategic planning. Journal of Industrial Ecology.

IIED, 1994. Whose Eden? Empowering local communities to manage their wildlife resources. International Institute for Environment and Development Perspectives 13: 3–5.

IUCN/UNEP/WWF, 1980. World Conservation Strategy: Living Resource Conservation for Sustainable Development. IUCN, Gland, Switzerland.

IUCN/UNEP/WWF, 1991. Caring for the Earth: A Strategy for Sustainable Living. IUCN, Gland, Switzerland.

Jackson, T., 1996. Material Concerns. Routledge, London.

Leopold, A., 1949. A Sand County Almanac: and Essays on Conservation from Round River. Oxford University Press, New York.

Maltby, E., 1991a. Wetlands and their Values. In M. Finlayson & M. Moser (eds), Wetlands. Facts on File, Oxford: 8–26.

Maltby, E., 1991b. Wetland goods and services – real values amid unreal economics? In P. Driver, (ed.), Harmonising Environmental Conservation and Economic Development. IUCN Special Publication, Gland, Switzerland.

Meadows, D. H., D. L. Meadows, J. Randers & W. W. Behrens, 1972. The Limits to Growth: A Report for the Club of Rome's Project on the Predicament of Mankind. Earth Island Limited, London.

Moss, B., P. J. Johnes & G. L. Phillips, 1996. The monitoring of ecological quality and the classification of standing waters in temperate regions: a review and proposal based on a worked scheme for British waters. Biol. Rev. 71: 301–339.

National Rivers Authority, 1995. Understanding Riverbank Erosion from a Conservation Perspective. National Rivers Authority, Northumbria and Yorkshire Region, Leeds.

Odum, E. P., 1983. Basic Ecology. Saunders College Publishing, Philadelphia.

Odum, H. T., 1982. Systems Ecology. John Wiley, New York.

Pond Action, 1993. The Oxfordshire Pond Survey: A Report to the World Wide Fund for Nature. Pond Action, Oxford Brookes University.

Ramsar, 1971. Convention on Wetlands of International Importance Especially as Waterfowl Habitat 2/2/1971 Ramsar, Iran. The Ramsar Bureau, Gland, Switzerland.

Ramsar, 1988. Proceedings of the Third Meeting of the Conference of the Contracting Parties. Ramsar Convention Bureau, IUCN, Gland, Switzerland: 588 pp.

Ramsar, 1996. The Ramsar 25th Anniversary Statement. Resolution VI.14, 6th Meeting of the Conference of the Contracting Parties, Brisbane, March 1996.

Robért, K-H., 1991. Educating a nation: The Natural Step. In Context 28: 10.

Sansom, A., 1993. Ponds and Conservation. NRA Northumbria and Yorkshire Region, York.

Slavin, T., 1998. Sustainable argument takes a step forward. The Observer, 22nd March 1998.

Sykes, J. B. (ed.), 1982. The Concise Oxford Dictionary. 7th edn. Oxford University Press, Oxford.

US Council on Environmental Quality and US Department of State, 1980. The Global 2000 Report to the President, 1(27), Washington DC.

Wackernagel, M. & W. Rees, 1996. Our Ecological Footprint. New Society Publishers, New York.

WCED, 1987. Our Common Future. Oxford University Press, Oxford, England.

Wessex Water, 1998. Wessex Water to pay farmers to protect water resources. Wessex Water Press Release, 7th April 1998. Wessex Water, Bristol.

World Resources Institute, The World Conservation Union (IUCN) and United Nations Environment Programme, 1992. Global Biodiversity Strategy: Guidelines for Action to Save, Study, and Use Earth's Biotic Wealth Sustainably and Equitably. World Resources Institute, Washington.

Section Two

Holistic Approaches to Successful Lake Restoration and Management

Hydrobiologia **395/396:** 41–60, 1999.
D.M. Harper, B. Brierley, A.J.D. Ferguson & G. Phillips (eds), The Ecological Bases for Lake and Reservoir Management.

Understanding lake and catchment history as a tool for integrated lake management

P. J. Johnes
Department of Geography, University of Reading, Reading, U.K.

Key words: nitrogen, phosphorus, non-point source, export, catchment modelling

Abstract

Sustainable lake management for nutrient-enriched lakes must be underpinned by an understanding of both the functioning of the lake, and the origins of changes in nutrient loading from the catchment. To date, limnologists have tended to focus on studying the impact of nutrient enrichment on the lake biota, and the dynamics of nutrient cycling between the water column, biota and sediments within the lake. Relatively less attention has been paid to understanding the specific origins of nutrient loading from the catchment and nutrient transport pathways linking the lake to its catchment. As such, when devising catchment management strategies to reduce nutrient loading on enriched lakes, assumptions have been made regarding the relative significance of non-point versus point sources in the catchment. These are not always supported by research conducted on catchment nutrient dynamics in other fields of freshwater science. Studies on nutrient enrichment in lakes need to take account of the history of catchment use and management specific to each lake in order to devise targeted and sustainable management strategies to reduce nutrient loading to enriched lakes. Here a modelling approach which allows quantification of the relative contribution of nutrients from each specific point and non-point catchment source over the course of catchment history is presented. The approach has been applied to three contrasting catchments in the U.K. for the period 1931 to present. These are the catchment of Slapton Ley in south Devon, the River Esk in Cumbria and the Deben Estuary in Suffolk. Each catchment showed marked variations in the nature and intensity of land use and management. The model output quantifies the relative importance of point source versus non-point livestock and land use sources in each of the catchments, and demonstrates the necessity for an understanding of site-specific catchment history in devising suitable management strategies for the reduction of nutrient loading on enriched lakes.

Introduction

Limnology has a long history of research which has considered the lake and its catchment as a functional whole. However, limnological research to date has been strongly weighted toward the study of in-lake processes, with far less emphasis placed on the functioning of the catchment and its linkage to the lake ecosystem.

In the wide range of studies published to date on nutrient enrichment in lakes, the focus of research has been on nutrient cycling dynamics within the lake ecosystem, and the impact of increased nutrient loading on the biotic communities, taking account of only the gross nutrient loading from catchment sources. There have been more detailed nutrient budgeting studies conducted on a number of long-term research sites, examples of which include Loch Leven in Scotland (Holden & Caines, 1974; Bailey-Watts & Kirika, 1987; Bailey-Watts et al., 1990), Esthwaite Water in England (Lund, 1972; Talling & Heaney, 1988; George et al., 1990) and Lake Washington, North America (Edmondson, 1970; 1985; Edmondson & Lehman, 1981). However, the majority of studies have focused on the more easily identified point source discharges from sewage treatment works and livestock production units as the sources of nutrient enrichment. Rarely has the same degree of scrutiny been applied to the sources and pathways of nutrient export from catchment sources, with the result that the delivery of nutrients from non-point sources in the catchment is poorly understood, and may thus be underestimated in limnological studies.

Understanding catchment history for sustainable lake management

There are a number of notable examples of limnological research where nutrient export from catchment sources has been studied in greater detail, including sites in both the U.K. (for example studies by Foy et al. (1982) on Lough Neagh, N. Ireland, and by Moss et al. (1988) and Phillips et al. (1994) on a number of the Norfolk Broads and their river catchments) and further afield (see for example Knuuttila et al. (1994) on Lake Kotojärvi and Lake Villikkatanjärvi, Finland). However, there has been a propensity in limnological studies to take the findings from these few studies and assume that they are replicated in other lake catchments with very different environmental characteristics. As with any model, whether mathematical or conceptual, the model is only applicable within the range of variation encompassed by the original study. For example, in the studies conducted on Lake Washington, North America, the increase in the discharge of phosphorus-rich effluent from sewage treatment works in the catchment was identified as the primary source of nutrient enrichment and subsequent eutrophication of the lake (Edmondson, 1970, 1985; Edmondson & Lehman, 1981). The management strategy adopted for this site involved reduction of the contribution from these point sources. The assumption that point source contributions were the primary cause of the eutrophication of Lake Washington was supported by the success of the management strategy in reducing nutrient loading on the lake. However, Lake Washington has a very unintensively farmed catchment compared to most U.K. catchments, resulting in a much lower contribution from non-point agricultural sources than would be expected in the intensively farmed land of lowland U.K. The findings from Lake Washington are not therefore directly applicable to all U.K. lakes and their catchments, where the non-point sources of nutrients in the catchments are more extensive and abundant.

The hydrology and topography of a catchment must also be taken into account in assessing the relative contributions of non-point and point sources to nutrient enrichment in water bodies. For example, in studies on the nutrient budget of the River Bure in Norfolk, which flows into many of the Norfolk Broads, Moss et al. (1988) concluded that the phosphorus load was largely derived from point sources discharges from sewage treatment works in the catchment, with the majority of the load in the form of soluble reactive phosphorus. Nitrogen loading was more directly attributed to non-point sources in the catchment. These findings are supported by recent research on nutrient loading in the River Bure (Johnes, 1996a) where the low total phosphorus loading within the Bure is attributed to two causes. First, the Broadland region receives a low rate of rainfall and has a high evapotranspiration rate compared to the central and western regions of the U.K. resulting in low rates of catchment runoff. The Bure catchment also has a high proportion of flat arable land with no vegetation cover for much of the year, and fewer hedgerows to reduce wind velocities, compared to the wetter, colder west of the U.K. As a result, sediment-bound phosphorus applied to the land from fertilisers and livestock wastes is exported *via* wind- rather than water-erosion pathways, delivered to the adjacent marine environment rather than the adjacent freshwaters of the Norfolk Broads. The rate of phosphorus export from non-point sources along the limited hydrological pathways is therefore a proportion of the net phosphorus load remaining in the soil after wind erosion. Furthermore, once arrived in the River Bure, the low gradient of the river results in a high trapping efficiency for sediments and sediment-associated nutrients, further reducing the non-point source contribution to the total phosphorus load in the river. As a result, point source contributions are proportionally much more significant in this catchment than might be expected given the intensity of land use and management in its catchment. These findings are not, however, applicable in the majority of lowland Britain, where rates of rainfall and the slope of the land promotes much higher rates of nutrient export from non-point sources along hydrological flow pathways. This is demonstrated by considering the water balance of three catchments in Britain, the catchment of Slapton Ley in south Devon, the River Esk in Cumbria and the River Bure in Norfolk (Figure 1), employing the water balance equation, and using 30 year mean annual average rainfall and actual evapotranspiration rate data from the Meteorological Office (1989, 1992). Thus:

$$Q = \mathrm{Rf} - \mathrm{AE} \pm \Delta S,$$

where Q = mean annual discharge (mm); Rf = mean annual rainfall (mm); AE= mean annual actual evapotranspiration (mm); and ΔS = storage (mm).

In the catchment of the Slapton Ley in the south west of the U.K., mean annual rainfall is 1280 mm, mean annual actual evapotranspiration is 529 mm, giving a mean annual discharge of 751 mm, assuming

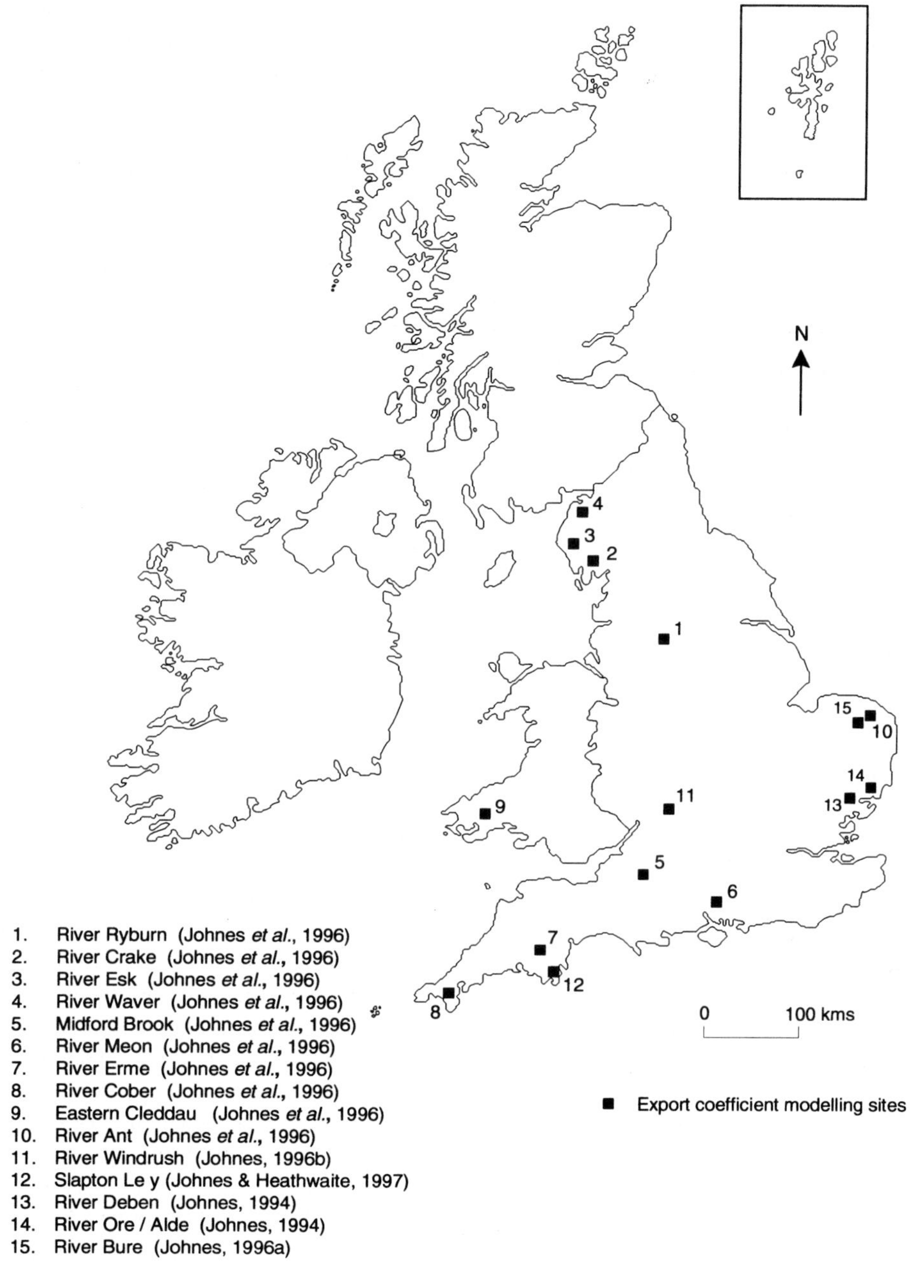

Figure 1. Export coefficient modelling sites in the U.K.

storage to be constant over the 30 year period. In the colder, wetter Esk catchment in the north west of the U.K., mean annual rainfall is 2264 mm, mean annual actual evapotranspiration is 475 mm, giving a mean annual discharge of 1789 mm, more than twice that of the Slapton catchment. In contrast, in the warmer, drier catchment of the River Bure, mean annual rainfall is 636 mm, mean annual actual evapotranspiration is 478 mm, giving a mean annual discharge of 158 mm, less than one quarter of that produced in the Slapton catchment, and less than a tenth of that produced in the Esk catchment. The potential for nutrient export along hydrological pathways can be argued, on this basis, to be ten times higher in the Esk catchment, and almost five times higher in the Slapton catchment than in the Bure catchment, with the realisation of this intrinsic nutrient export capacity being determined by the actual land use and management in each catchment. These patterns are further exaggerated if the slope of the land, the nature of the soils and underlying geology, and therefore the speed and efficacy of the nutrient transport pathways are taken into account, with the steepest sloping land in the Esk catchment, followed by the Slapton catchment, and a negligible degree of slope in the catchment of the River Bure. Clearly then, quite aside from land use and management in a catchment, the physical characteristics of the catchments must be taken into account if the non-point source contribution to nutrient loading in lakes is to be fully and accurately quantified.

Despite the absence of this detail in most limnological studies to date, information is available to the limnologist on the dynamics of nutrient export along non-point source transport pathways from other fields of freshwater research, most notably from hydrological studies of river catchments. This has been a major research area in hydrological science research conducted in the U.K., mainland Europe and North America for the past 25 years. For a full discussion of this research, reference can be made to Johnes & Burt (1993), Heathwaite & Johnes (1996), Heathwaite et al. (1996) and Johnes (1996b). Figure 2, which outlines the hydrological pathways linking non-point catchment nutrient sources to surface freshwaters, summarises this.

Sediment-associated nutrient fractions are transported along surface runoff or overland flow pathways, with the efficacy of this pathway determined by the permeability of the soil surface, and by the surface roughness and therefore the sediment trapping efficiency of land downslope. The dominant nutrient transport pathway for soluble nutrient fractions is subsurface throughflow, with soluble nutrient fractions leached downward by infiltration through the soil profile, and then percolation vertically to groundwater in catchments with permeable bedrock, and horizontal flow parallel to the hillslope along shallow and deep throughflow pathways. The efficacy of this transport pathway is dependent on soil physical properties, the depth of flow through or below the root zone, and the nature of downslope vegetation cover, particularly in the riverside or riparian zone where the subsurface flow may return to, or near the land surface as return flow, particularly during storm events.

Hydrological research has not yet answered all of the questions which arise when attempting to quantify nutrient export from non-point catchment sources, because the emphasis has been on nitrogen rather than phosphorus. This is in marked contrast to the preoccupations of limnologists, with the dynamics of phosphorus cycling in lakes as the major nutrient limiting biotic production. The assumption in hydrological sciences has been that phosphorus is so strongly bound to sediment particles in the soil profile, and that it is in such short supply for biological production, that it rarely leaves the soil. The relatively high phosphorus concentrations in many lowland rivers in the U.K. with intensively farmed agricultural land and low human population density, however, suggest that this assumption is not valid. The focus of these studies to date has therefore been on the movement of bioavailable, inorganic nitrogen forms, particularly nitrate (NO_3-N), with leaching of nitrate along subsurface flow pathways being the dominant route of non-point source nutrient export to surface freshwaters and groundwaters. Less is understood about the dynamics of phosphorus movement in catchments, and of the movement of particulate and dissolved organic nitrogen fractions along these pathways. This is despite evidence for these fractions comprising a substantial proportion of the total nitrogen load in the few river systems in which the dynamics of nitrogen speciation and phosphorus fractionation dynamics has been studied (see for example Ryding & Forsberg, 1979; Stevens & Stewart, 1982; Johnes & Burt, 1991; Heathwaite & Johnes, 1996). Recent research on phosphorus fractionation dynamics along hydrological flow pathways has, however, suggested that it is effectively leached along subsurface flow pathways in intensively fertilised or grazed soils, in much the same way as nitrate (see for example Duffy et al., 1978; Withers, 1994; Dils & Heathwaite, 1996; Heckrath et al., 1996). It has also been demonstrated that sediment-associated nutrients including bioavailable ammonium (NH_4-N), particulate nitrogen and particulate phosphorus are transported to surface freshwaters along overland flow pathways during storm events, with the efficacy of this transport route determined by surface roughness and sediment trapping efficiency of downslope land (Sharpley & Syers, 1976; Johnes & Burt, 1991; Heathwaite & Johnes, 1996). There is also some suggestion from these studies that, once arrived in a turbulent, oxygen rich freshwater environment, the sediment-bound phosphorus and nitrogen loads rapidly dissociate from the sediment particles to form bioavailable ammonium and soluble reactive phosphorus, although the breakdown of particulate organic nitrogen and insoluble unreactive phosphorus is

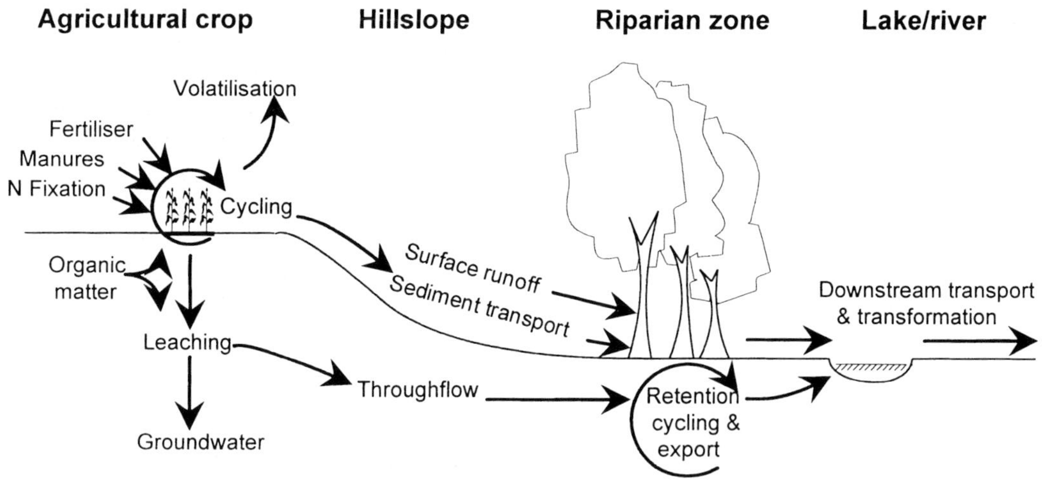

Figure 2. Non-point source nutrient transport pathways in agricultural catchments.

much slower. Further research is needed to quantify each of these routes, if the full potential nutrient loading delivered to freshwaters from non-point sources in their catchments is to be quantified.

Nevertheless, the findings from research in this field to date have indicated the close linkage of non-point catchment nutrient sources to surface freshwater systems. They have also demonstrated for the intensively farmed rural lowlands and the sparsely populated uplands of the U.K., the dominance of non-point, rather than point sources, of nutrient loading on freshwater systems. The wealth of field experimental studies on the processes controlling nutrient export rates have identified marked spatial variations in the rates of nitrogen and phosphorus export from different non-point agricultural sources in these catchments. In addition, the rates of export of both nitrogen and phosphorus to surface waters in the U.K. have increased substantially over the past century, with the majority of the increases in agricultural catchments derived from non-point, rather than point sources. For nutrient-enriched lakes in lowland U.K., this has important implications for the design of lake restoration programmes where a reduction in nutrient loading on the lake is considered important. Where the majority of the nutrient load derives from non-point sources, focusing management on the point source discharges is unlikely to be successful in the long term. What is needed is fully integrated lake and catchment management, informed by an understanding of both lake processes and catchment history.

Successful reduction of nutrient loading on a water body requires targeted catchment management, underpinned by an awareness of the origins of any increases in nutrient loading. This can be achieved through the use of catchment based models to determine the relative contribution of each nutrient source in a catchment to the total nutrient load delivered to a water body.

The role of modelling in integrated lake and catchment management

Lake- and catchment-based process models help to explain the mechanisms controlling the cycling and bioavailability of nutrients in transit between land and water. In terms of management, there are two types of model that might help understand different aspects of the nutrient enrichment of lakes.

The first group are catchment-scale, distributed, mechanistic models of nutrient export from the catchment under different land-use scenarios, which can be used to determine the origins of increased nutrient loading on a lake. They do not explain nutrient transport mechanisms, and cannot predict short-term, extreme variations in the concentrations of individual nutrient species or fractions in receiving waters.

The second group are process-based models of in-lake nutrient cycling dynamics which can be used to understand the cycling and transport of individual nutrient species within lake ecosystems, and the mechanisms controlling bioavailability. The same type of modelling approach can be applied at the catchment scale, but is limited in its suitability for answering management related questions because of the extremely high data requirements of such models.

Export coefficient modelling

Export coefficient modelling is a very simple approach, which calculates the total load of nitrogen or phosphorus arriving in a water body as the sum of the individual nutrient loads exported from each separate nutrient source in the catchment as a function of the rate of nutrient input to that source, and the nutrient export potential of each land use type, livestock variety or people. Thus:

$$\overset{n}{\underset{i=1}{L}} = \_E_i(A_i(I_i)) + p,$$

where: L = Loss of nutrients; E = Export coefficient for nutrient source i; A = Area of catchment occupied by land use type i, or number of livestock type i, or of people; I = Input of nutrients to source i; p = Input of nutrients from precipitation.

The scale of modelling uses the field as the spatial unit, and produces predictions of total nitrogen and total phosphorus load delivered to a water body annually. The rate of nutrient export for each source is selected on the basis of the processes controlling nutrient cycling and export for each source type under different environmental conditions, such that the rate of nutrient export will vary between different catchments for each source type. As such, this approach can be described as a semi-distributed, mechanistic model.

The export coefficient modelling procedure is outlined in Figure 3 (Johnes, 1996b). The model is initially constructed through the collection of data on land use and management in the catchment, and nutrient loading in the receiving waters determined in a field monitoring programme and questionnaire survey. Further information may be extracted from databases on the number of people living in the catchment in the modelling year, the rate of nutrient input from atmospheric sources, and nutrient inputs from fertiliser and livestock wastes.

A set of export coefficients is initially selected from the literature, from case studies with broadly similar land use, topography, climate and geology. The model is then constructed for an initial run, and the model predictions of nutrient load exported to the water body are compared with observed total nitrogen and total phosphorus loads from the field monitoring programme. The model is run through a sensitivity analysis step to determine which model parameters have the greatest influence on model output. These are then adjusted within the published range of coefficients for North Temperate Regions, and the model re-run to compare with observed water quality in the calibration step.

This is necessary, as a number of answers might be generated by placing a different weighting on each component of the equation. Whether the right set of coefficients has been selected, which places the correct weighting on each nutrient source, is established in the validation step. Here the value of each component in the equation changes, as the number of livestock and people, and the areas of each land use type in the catchment change over the course of the catchment history, and the outcome of the equation varies reflecting changes in observed nutrient concentrations in the receiving water over time. Thus, if the model hindcasts for this period of catchment history agree closely with observed water quality for the validation period, then the model can be considered to provide an accurate representation of the relative importance of different catchment nutrient sources to the nutrient enrichment of the water body. The model may then be used to evaluate the origins of nutrient enrichment in a lake, and to determine the likely future response of the catchment to a range of management strategies designed to reduce nutrient loading on the lake.

Application of the export coefficient modelling approach to U.K. catchments

To date, this approach has been successfully applied to 16 British catchments (Figure 1), in which the models were able to explain >95% of the variance in observed data for each catchment (see Johnes, 1996b; Johnes et al., 1996; Johnes & Heathwaite, 1997). The modelling approach was initially developed in a study on the catchment of the River Windrush, in the Cotswold region (NGR SP 404014), outlined in full in Johnes (1996b). It has also been applied to the catchment of Slapton Ley, south Devon (NGR SX 825440), where detailed modelling studies have been conducted, linked to field experimental studies of nutrient export rates in these catchments. The modelling approach was adapted for 10 U.K. catchments in the first phase of the development of the Lake Classification and Monitoring scheme for the National Rivers Authority (now the Environment Agency) where it was used to determine an approach for establishing the past nutrient loadings on lakes in the baseline period of the 1930s (Johnes et al., 1996). Finally it has been applied to a group of three Suffolk Estuaries, including the Deben and Ore/Alde, in order to determine the rel-

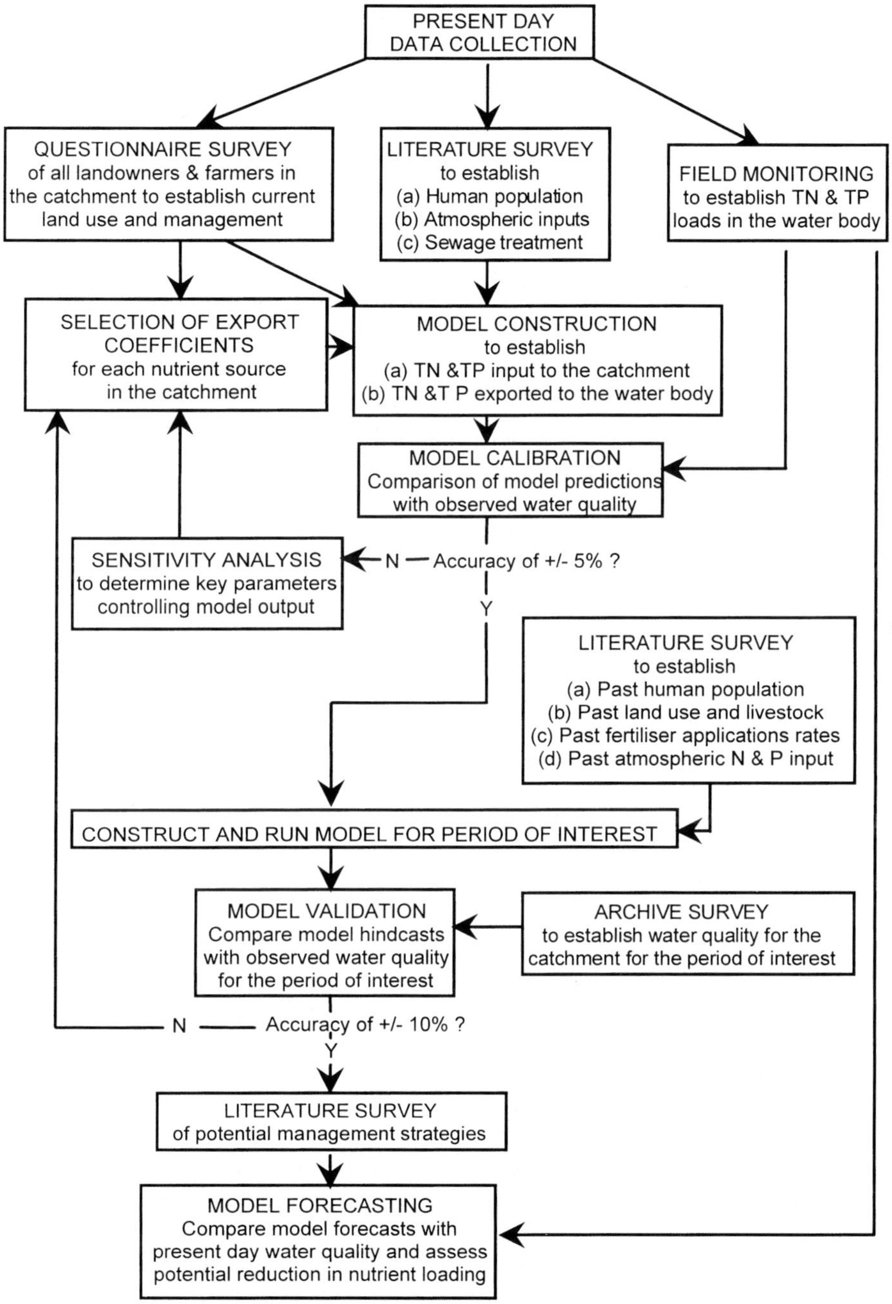

Figure 3. Export coefficient modelling procedure (after Johnes, 1996b).

ative contribution of non-point and point sources of nutrient enrichment in the freshwater catchment of the Deben estuary, to try to explain the occurrence of algal blooms where no such problems exist in the seemingly similar Ore and Alde estuaries (Johnes, 1994). In applying the export coefficient modelling approach to a range of different catchments, the model was successfully run using a limited range of coefficients, and it has begun to emerge that the same set of coefficients may be equally successfully applied to different catchments with broadly similar land use, management, soils, geology and climate (see Johnes et al., 1996).

Here the application of this approach for quantifying the relative contribution of nitrogen and phosphorus to water bodies from non-point and point sources, as these vary over the course of catchment history and is illustrated for these three contrasting U.K. catchments.

The study sites

Slapton Ley, Devon

Slapton Ley is a shallow, freshwater coastal lake, which has undergone substantial nutrient enrichment, to the point where it is commonly considered to be in a hypertrophic state (O'Sullivan, 1994; Heathwaite, 1994). It has a mean depth of about 1 m, a retention time of 2–3 weeks and a rapidly disappearing plant community (Johnes & Wilson, 1996). The catchment is steeply sloping and is intensively farmed for dairy and beef cattle production. There are no major urban developments and the human population is located in isolated farms and small villages throughout the catchment. The Slapton village sewage treatment works discharges directly into Slapton Ley, with a second STW discharging into the River Gara in the upper catchment.

The history of land use and management in the catchment is illustrated in Figure 4 (Johnes & Heathwaite, 1997). The trends do not mirror the widely held view of arable expansion in the U.K. The area of permanent grass in the catchment has declined, replaced by temporary grass which is more intensively fertilised and supports higher stocking densities. The number of people in the catchment has changed very little over this time period, and they are clearly not a major source of the nutrient enrichment which has occurred in Slapton Ley. There have, however, been substantial increases in the numbers of sheep and cattle, leading to increased stocking densities on the grassland. In addition, there have been marked increases in fertiliser applications to both crops and grass in this catchment over the past 60 years (Johnes & Heathwaite, 1997), increasing the non-point source nutrient pool potentially available for export to Slapton Ley.

The River Esk, Cumbria

The River Esk drains west from the English Lake District, and is predominantly an upland catchment, with steep slopes, a high rate of both rainfall and runoff, and a sparse human population. There are no major urban areas, and the human population is scattered amongst isolated farms and villages. The farming system is extensive livestock production in the upper east of the catchment, with a small area of more intensive livestock farming in the lower west of the catchment near the basin outlet. Changes in catchment land use and management since 1930 are shown in Figure 5. Over the past 60 years, there has been a marked increase in the area of fertilised permanent grass at the expense of the unfertilised rough grazing land in the catchment. Fertiliser application rates to crops and grass have also increased substantially for both nitrogen and phosphorus, particularly since 1940 (Johnes et al., 1996). In addition, whilst there are only some 700 people in this catchment, not an important nutrient source in this instance, there has been a substantial increase in the numbers of livestock, particularly sheep of which there are now some 55 000 in the catchment. As with the Slapton catchment, changes in both the numbers of livestock and fertiliser applications to an increasing area of agricultural land have increased the quantity of nutrients potentially available for export to the River Esk. Other upland catchments in the U.K. have probably seen similar changes in nutrient loadings in their receiving waters.

The Deben Estuary, Suffolk

The Deben Estuary drains into the North Sea near Ipswich. The catchment has the flattest topography and lowest rainfall of the three, and is very intensively used for mixed arable farming, and for pig and poultry production in intensive rearing units. There are a number of major urban settlements in the catchment, including the towns of Woodbridge, Kelsale and Martlesham. The estuary itself has recently experienced algal blooms, in marked contrast to the neighbouring Ore/Alde Estuary, which has a very similar pattern of land use and management in its catchment, but fewer large urban centres.

Changes in land use and management in the Deben catchment since 1930 are presented in Figure 6. The trends shown for the Deben catchment are very different from those for the Slapton and Esk catchments, and reflect the changes in land use and management characteristic of many areas of south eastern England. There has been a conversion of permanent grass to cereal cultivation, with an increase in fertiliser applications to all crops and grass over the 60-year period (Johnes, 1994). The most striking change in this catchment is the increase in the number of people and pigs in the post-war period, with over 15 000 people and

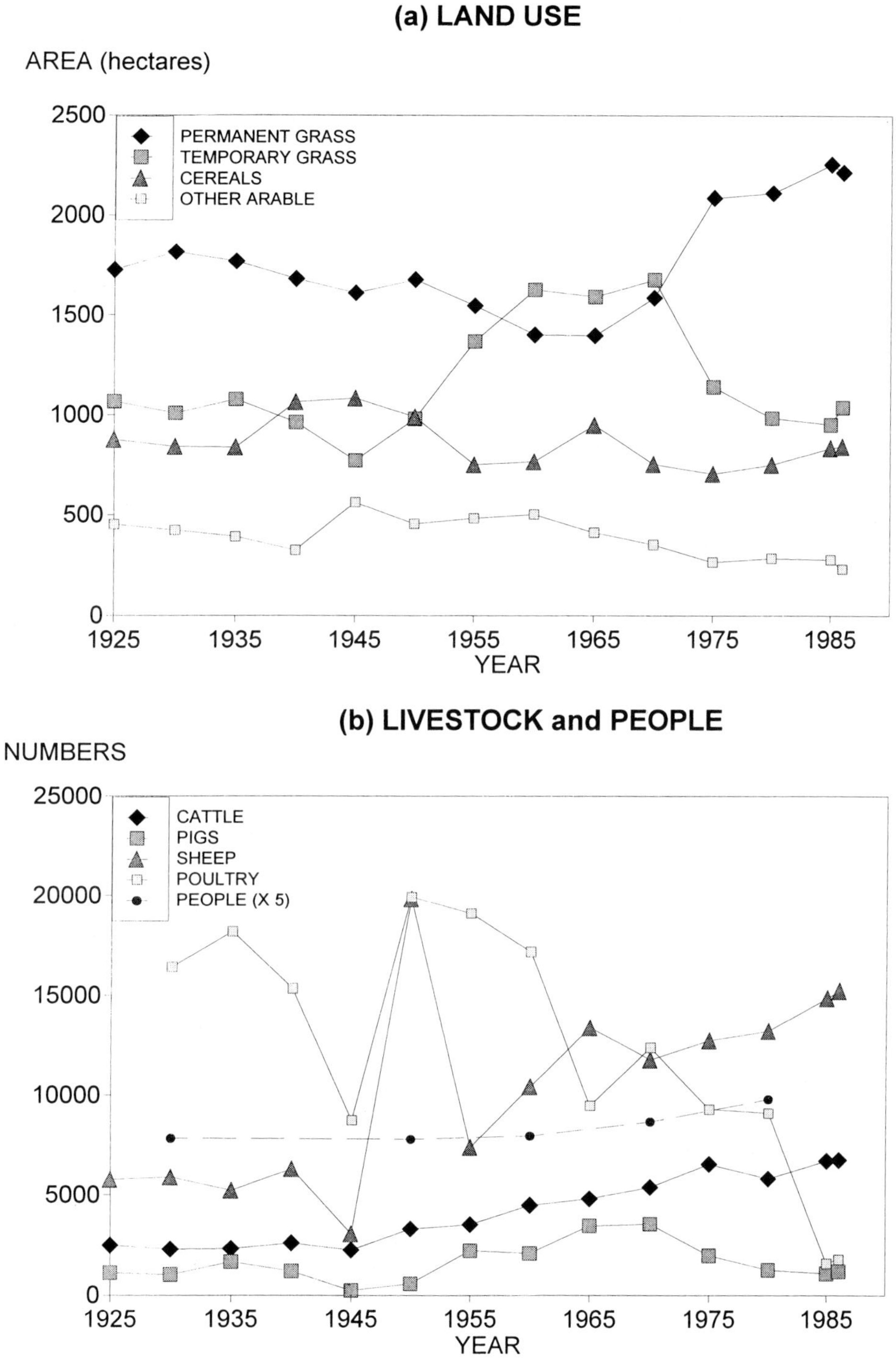

Figure 4. Land use, livestock and people in the Slapton catchment, 1925–1986 (after Johnes & Heathwaite, 1997).

14 000 pigs at present. This reflects an expansion of the area of urban land in the catchment, particularly around the estuary and periphery of major towns, and change from extensive to intensive rearing of pigs and poultry in factory units.

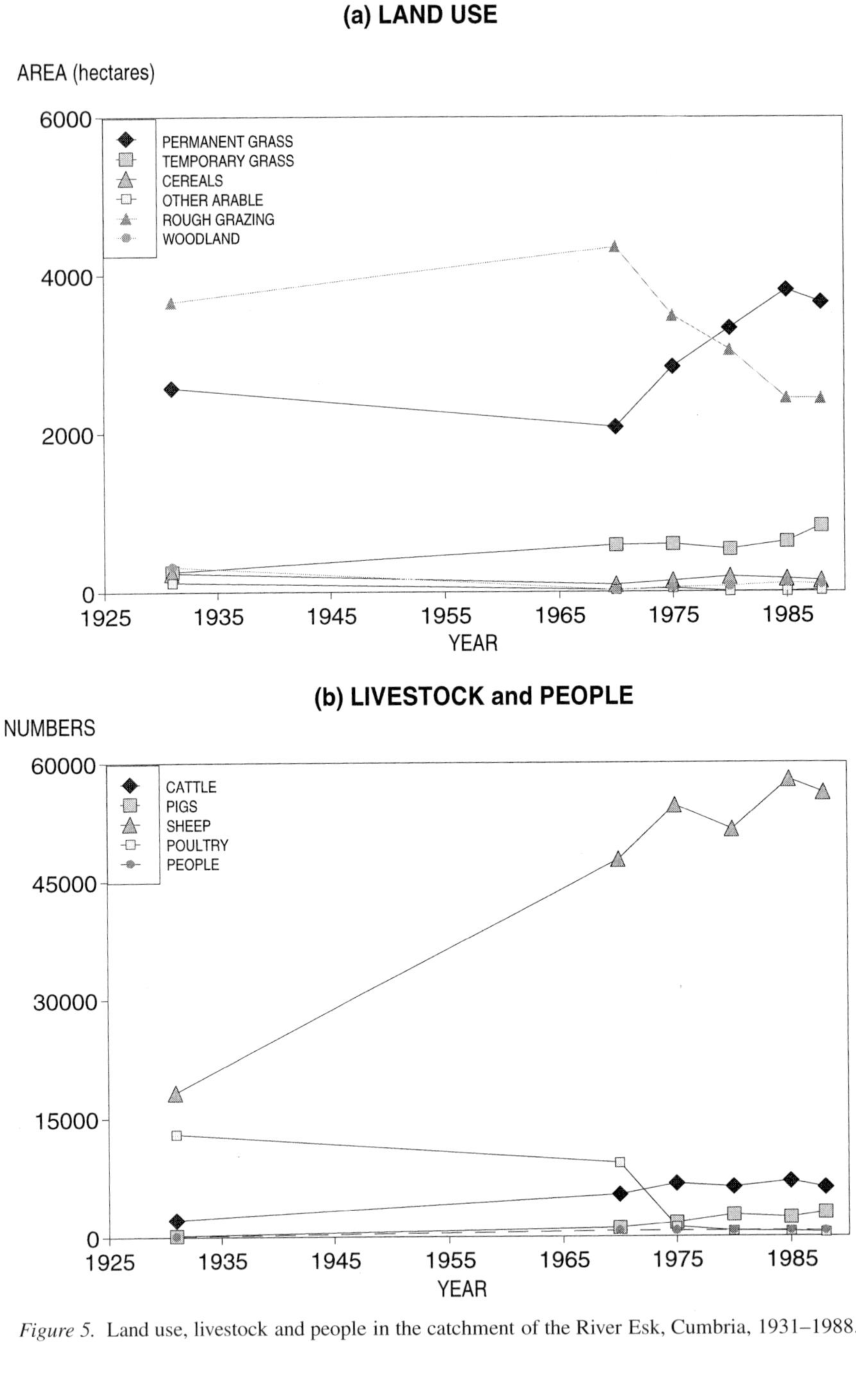

Figure 5. Land use, livestock and people in the catchment of the River Esk, Cumbria, 1931–1988.

Application of export coefficient modelling to the study sites

Slapton Ley

The results of application of the export coefficient modelling approach to the Slapton catchment are presented in Figure 7 for the period 1930–1990, together with long-term trends in observed nutrient concentrations in the streams flowing into Slapton Ley from its catchment (see Johnes & Heathwaite, 1997). There is a very close agreement between model hind-

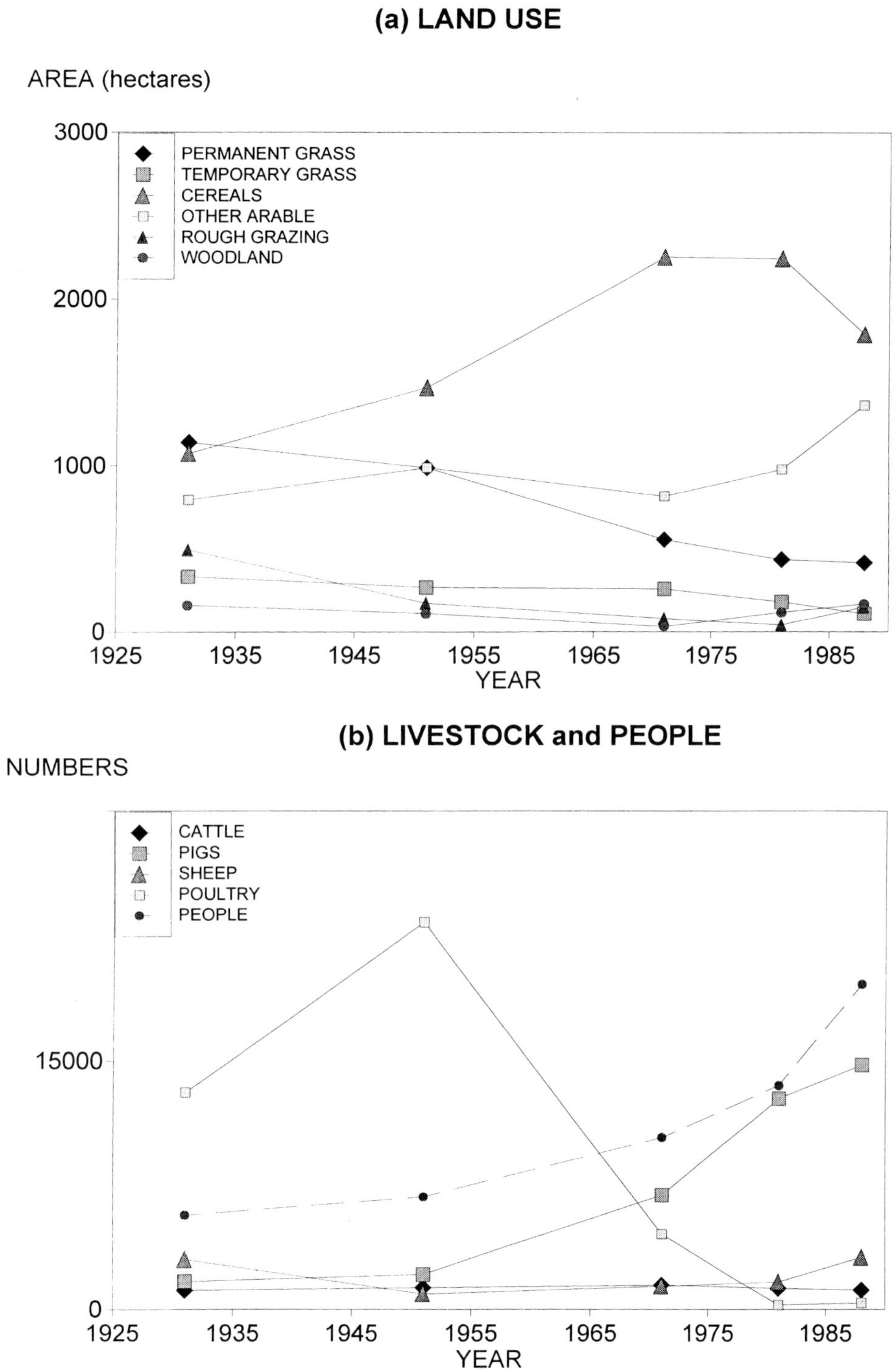

Figure 6. Land use, livestock and people in the catchment of the Deben Estuary, Suffolk, 1931–1988.

casts and observed water quality for the period 1970–1990, with a less than 3% error for both nitrogen and phosphorus over the 20 year validation period.

The major sources of the increase in nitrogen loading on Slapton Ley over the past 60 years have been an increase in exports from non-point sources in

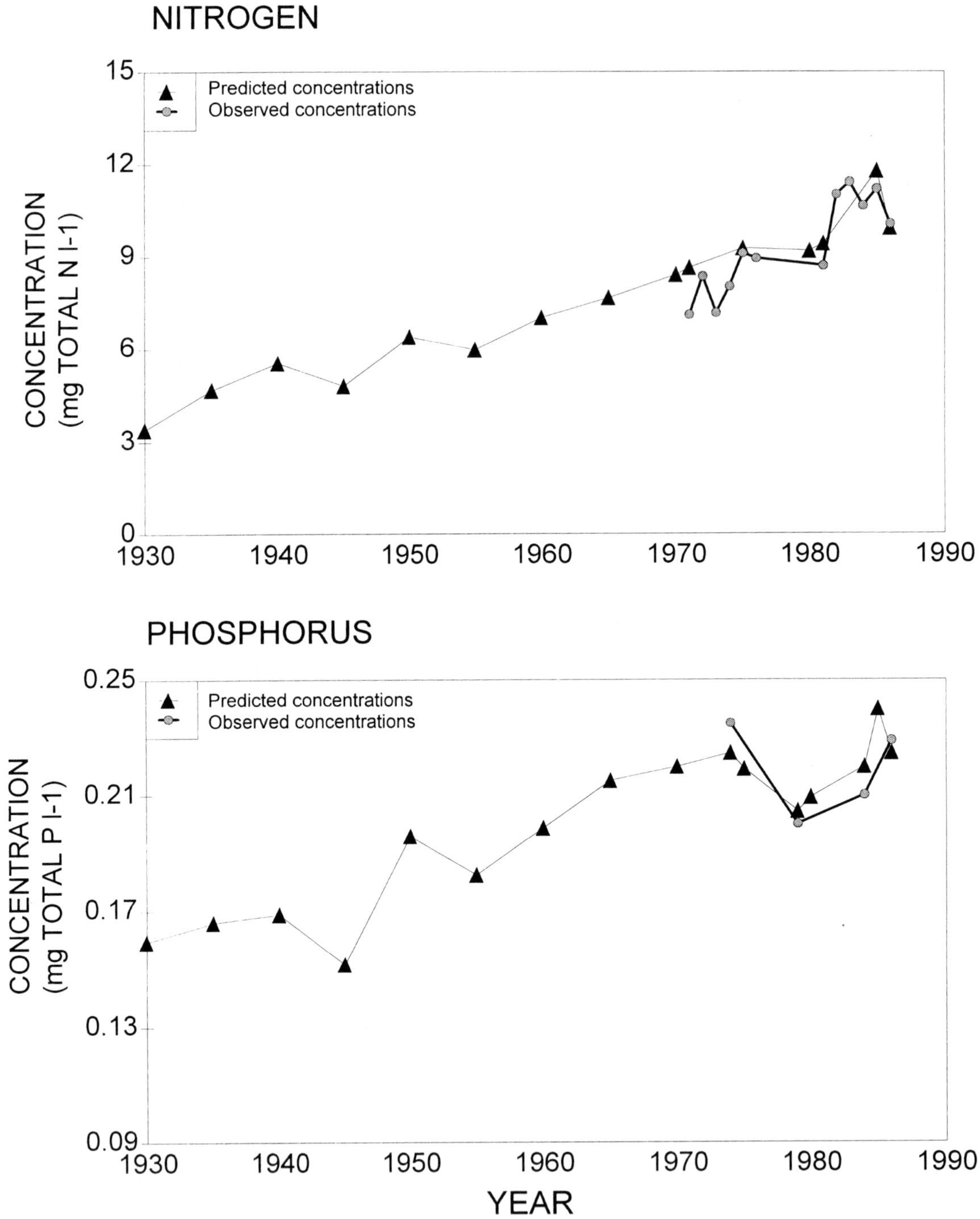

Figure 7. Model hindcasts and observed nutrient concentrations in inflowing streams in the Slapton catchment, 1930–1986 (after Johnes & Heathwaite, 1997).

the catchment, including increases in the number of livestock, particularly cattle, and in fertiliser applications to crops and grass in the catchment (Johnes & Heathwaite, 1997). Contributions from point source discharges from the human population contributed less than 2% of the total nitrogen loading on Slapton Ley in 1986, and should clearly not be a focus in the design of a management strategy to reduce nutrient loading on Slapton Ley. Increases in phosphorus loading on Slapton Ley have largely resulted from the same factors described for nitrogen, although the contribution of phosphorus from sewage treatment works comprises approximately 15% of the total load delivered to Slapton Ley in 1986. Nevertheless, the model suggests that more than 85% of the total external phosphorus loading on Slapton Ley derives from non-point sources, and these must clearly be addressed in developing a suitable catchment management strategy to reduce future nutrient loading on Slapton Ley from its catchment.

There is also further external verification of the phosphorus hindcasts for the Slapton catchment from the lake sedimentary phosphorus record. Figure 8 presents some preliminary results of comparisons between model hindcasts, observed nutrient chemistry and the record derived from a core from Slapton Ley (Foster et al., 1996; Johnes & Wilson, 1996). There is remarkably good agreement between the trends, except for the last 10 years or so, reflecting the top 10 cm of the core. Some researchers have interpreted this as indicating a switch to hypertrophy for this part of Slapton Ley, following changes in catchment hydrology during the 1976 drought, when the lower part of Slapton Ley became hydrologically cut off from the upper part of the system, known as the Higher Ley, and thus isolated from 70% of its agricultural drainage waters (O'Sullivan, 1994). However, we do know that the Ley has undergone periods of anoxia in the hypolimnion in the past 15 years, which were not recorded prior to this, and this mis-fit in the upper layers may simply reflect phosphorus diffusion up the sediment profile. Either way, despite scepticism about sedimentary phosphorus records, and in particular the value of the sediment record in Slapton Ley, where the sediments are so frequently disturbed by wind and wave action, it would seem that both the model hindcasts and the sedimentary phosphorus record give a good representation of changes in the nutrient chemistry of Slapton Ley, and more importantly a verification of each approach. The sedimentary record is not as good over the past 20 years, however, which supports the hypothesis that sediment diagenesis in shallow, eutrophic lakes undergoing periods of anoxia makes the sediment geochemical record for phosphorus unreliable. Comparison of the model hindcasts and observed water quality with a diatom-inferred total phosphorus record (Anderson et al., 1993; Bennion, 1994; Bennion et al., 1995) for Slapton Ley in Phase 2 of the Environment Agency Lake Classification and Monitoring scheme, may provide a more reliable comparison with model output. This type of verification is very valuable in helping us to determine the accuracy of different approaches to interpreting the history of nutrient loading in nutrient enriched lakes. It may be that the sedimentary record is more valuable than previously thought for these shallow eutrophic systems, but it is the model which indicates the origins of these trends, providing an indication of the relative contribution of point and non-point sources to the nutrient loading on this system, giving a strong indication of the necessary focus for a successful management strategy to reduce nutrient loading on Slapton Ley.

The River Esk

The model output for the River Esk is presented in Figure 9, and again there is a close agreement between model hindcasts and observed concentrations of both nitrogen and phosphorus since 1980. There has been a substantial increase in the concentrations of each in the Esk over the 60 year period, approximately a 100% increase in total phosphorus and 150% increase in total nitrogen concentrations in the river. These changes are largely attributable to increases in the export of nutrients from non-point sources in the catchment. In 1988, the human population contributed 1.2% of the total nitrogen loading on the River Esk, and less than 6% of the total phosphorus loading. In the same year, 66% of the nitrogen loading and 34% of the phosphorus loading derived from non-point source export of livestock wastes, particularly from sheep and cattle, with 27% of the phosphorus load derived from fertiliser applications to land in the catchment. Nitrogen input from atmospheric sources also contributed over 27% to the total nitrogen load delivered to the Esk in 1988, reflecting the importance of atmospheric deposition to the nutrient enrichment of freshwaters in upland areas of the U.K. The export coefficient model is not very sophisticated in dealing with inputs from atmospheric sources at present, since there are limited long-term historical data available for nitrogen deposition in the U.K. Work currently in progress in Phase 2 of the Lake Classification and Monitoring scheme is addressing this issue. However, these trends for the Esk catchment are not isolated, and are mirrored in published case studies outside the U.K., particularly from data produced for some of the long-term lake monitoring sites in the Adirondack Mountains, North America (see for example, Driscoll & Van Dreason, 1993; Stoddard, 1994). In Constable

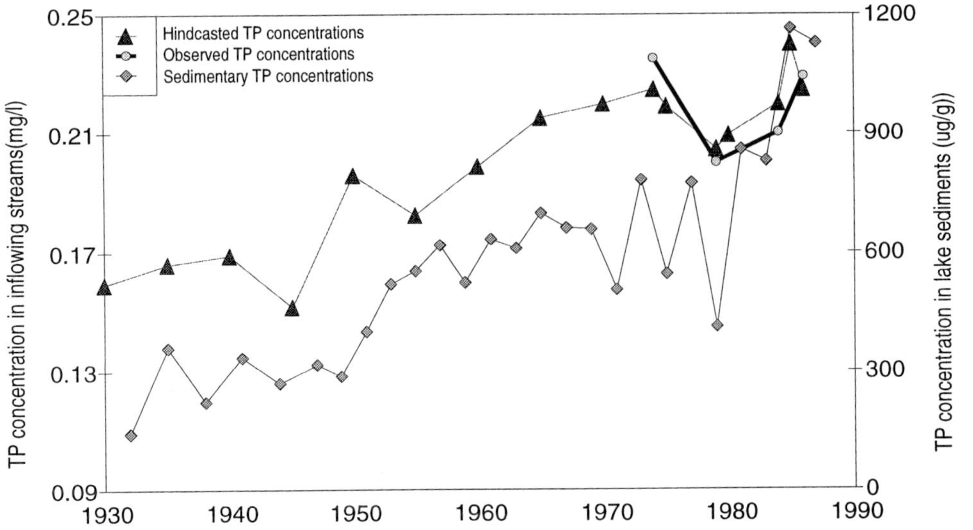

Figure 8. Comparison of model hindcasts and observed total phosphorus concentrations in inflowing streams in the Slapton catchment with a sedimentary reconstruction of total phosphorus in the Slapton Ley for the period 1930–1986 (based on data presented in Foster et al., 1996; Johnes & Wilson, 1996).

Pond, these authors observed an increase of 50–75% in nitrate concentrations from 1982–1990 at a rate of 1.65 mequiv. l^{-1} ann^{-1}, with a parallel decrease in Acid Neutralising Capacity, suggesting that the trend was attributable to nitrogen deposition from atmospheric sources. Whilst the nutrient concentrations in these upland sites are very low, compared to the data reported for lowland agricultural catchments, the rate and extent of the observed and modelled changes may have significant ecological implications in terms of loss of species diversity. This type of information may make us rethink our traditional views of such systems as being largely unpolluted, and to look at potential ecological responses to such changes at the bottom end of the nutrient scale.

The Deben and Ore/Alde estuaries

Model output for the Deben estuary for the period 1930–1990 was validated against observed nutrient concentrations using a different approach, since long-term historical water quality records were not available for this catchment. The model was therefore calibrated for a headwater catchment of the River Deben above Naunton Hall, above the confluence of the Deben with its major tributary the River Fynn. The predictions for nutrient loading in 1988 were within 1% of observed nitrogen and phosphorus loadings in the Deben at Naunton Hall (Johnes, 1994). The calibrated model was then applied to the headwater catchment of the River Alde at Farnham in the adjacent Ore/Alde catchment. The model developed for the Deben above Naunton Hall, predicted within 3% of the observed nitrogen loading, and 5% of the observed phosphorus loading in the River Alde at Farnham.

The purpose in modelling the two systems was to determine the relative contributions of nitrogen and phosphorus from point and non-point sources in the two freshwater catchments, to establish whether there were differences which might account for the occurrences of algal blooms in the Deben estuary. The model was therefore applied to each major subcatchment within the Deben and Ore/Alde catchments. The model hindcasts of nutrient loading on each estuary from their subcatchments in the period 1930–1990 are presented in Figure 10 as annual nitrogen and phosphorus export rate in kg N ha^{-1} and kg P ha^{-1} respectively. For nitrogen, the patterns of increase over time are broadly similar in both the Ore/Alde and Deben catchments, with a slow increase from 1931–1951, followed by a greater rate of increase between 1951–1971, and then 1971–1988, essentially reflecting increases in non-point source nitrogen export from the intensification of agricultural production in the headwater catchments. However, the rate of increase in nitrogen loading in the Deben catchment is much greater than in the Ore/Alde, with an overall increase of 150–200% in nitrogen loading on the rivers in this catchment since 1931, compared to 100–120% in-

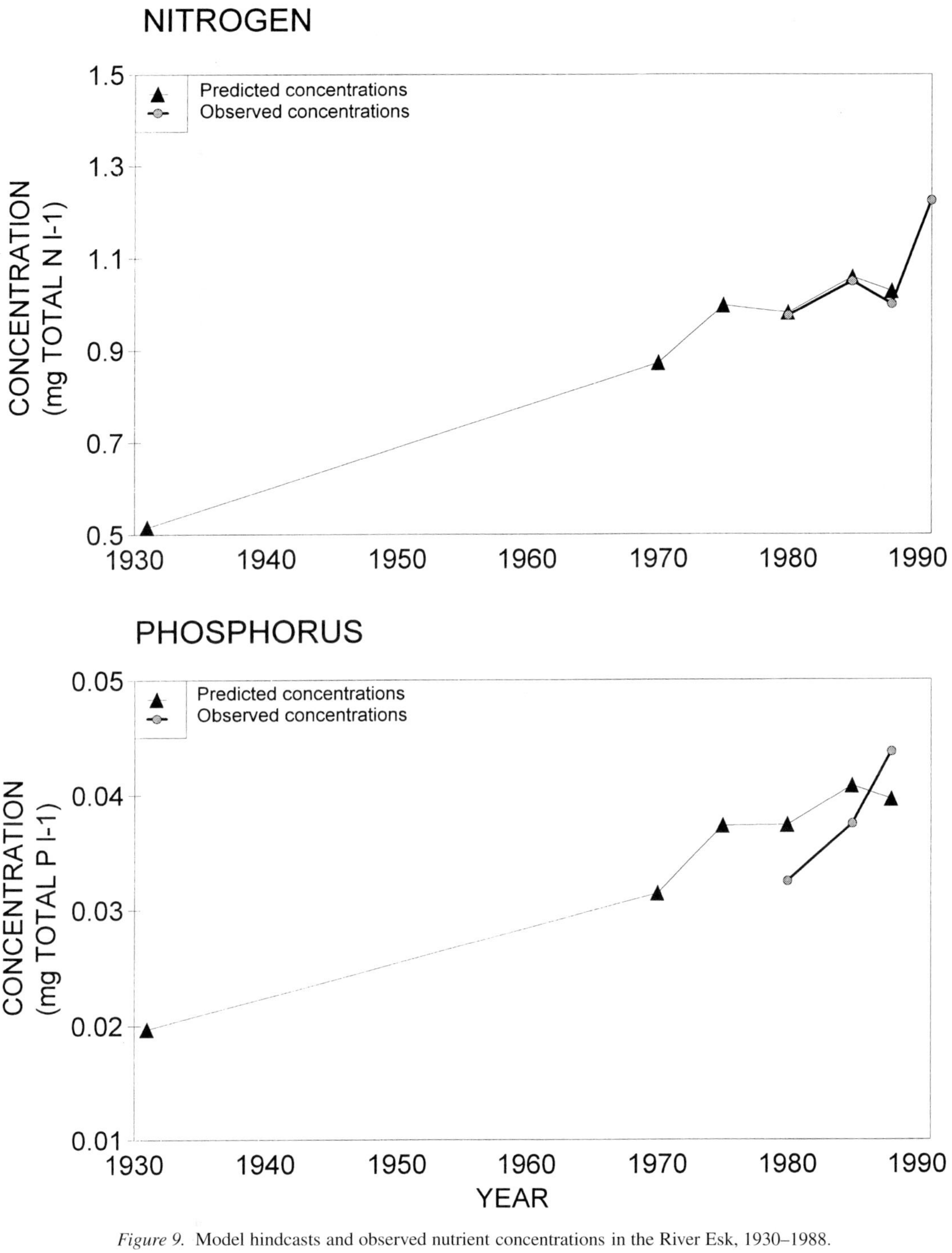

Figure 9. Model hindcasts and observed nutrient concentrations in the River Esk, 1930–1988.

crease in loading on rivers in the Ore/Alde catchment. This trend is most marked in the Fynn subcatchment and the downstream parishes with direct estuarine input, where increases in human population density have been most marked. For phosphorus, there are greater differences, both between the Ore/Alde and Deben catchment, and between their individual subcatchment areas. In the Ore/Alde, phosphorus loadings have in-

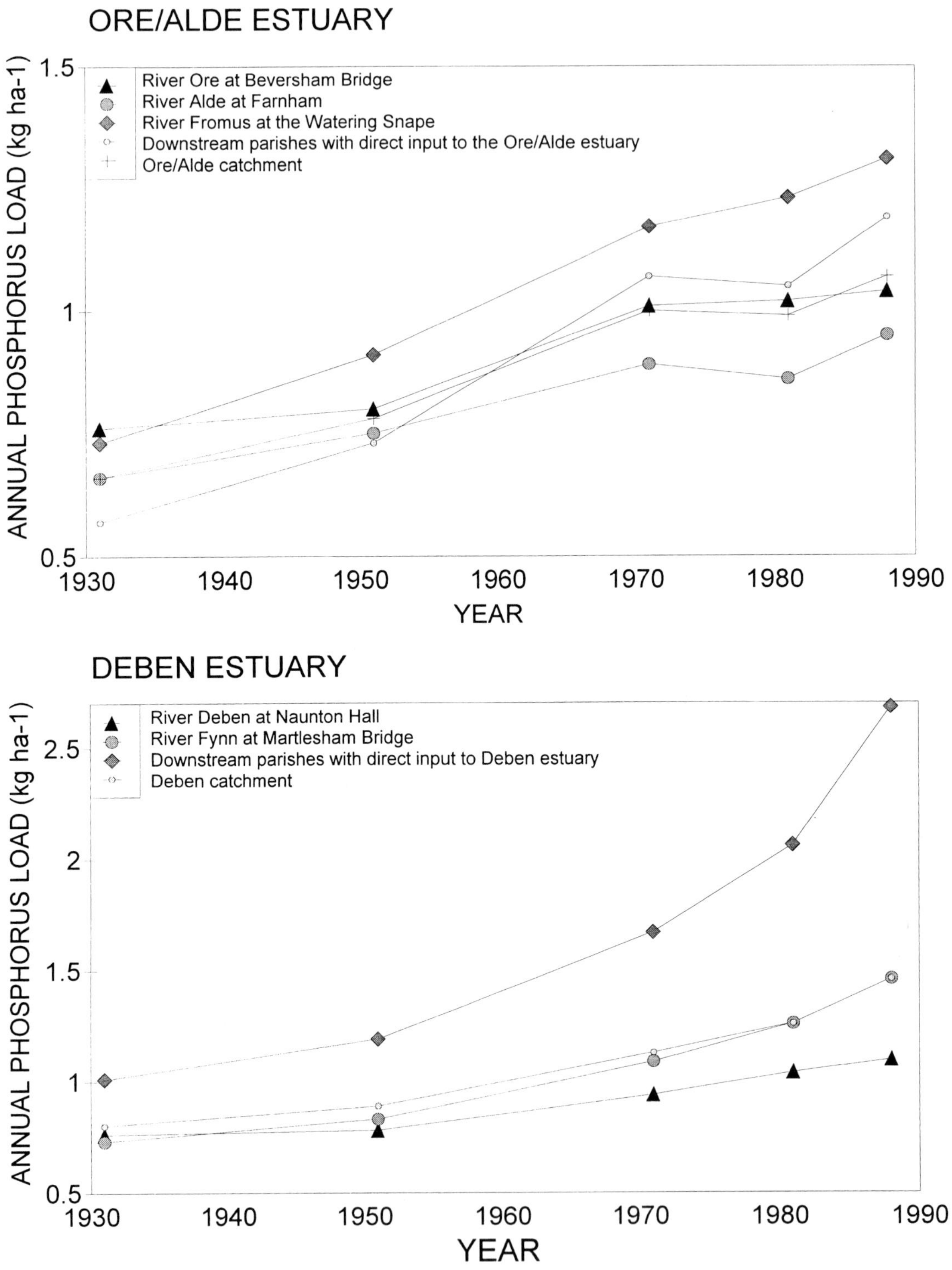

Figure 10. Model hindcasts of the nitrogen and phosphorus loadings delivered to the Ore/Alde and Deben estuaries from their catchments, 1930–1988.

creased by approximately 75–125% since 1931, with the greatest increase occurring in the downstream parishes and the Fromus subcatchment, where human population density has increased more rapidly. However, the rate of increase has declined over the past 20 years, perhaps reflecting a reduction in the rates of agricultural intensification, and human population expansion in this catchment. Both the Ore and Alde subcatchments have undergone some depopulation over the past 60 years through mechanisation of agriculture in the catchment, and the loss of the agricultural labour force. In the Deben catchment, the pattern more closely replicates the patterns in increase in nitrogen loading since 1930, with a slower rate of increase in phosphorus loading from 1931–1951 than in the ensuing period. The most notable pattern here, however, is the much greater increase in phosphorus loading on the Deben estuary from the parishes with direct estuary input, reflecting the sharp increases in human population density in this region of the Deben over the past 60 years, and particularly in the last 15–20 years, especially around the towns of Woodbridge, Kelsale and Martlesham. Clearly, in considering a management strategy for this system, effluents from major STWs in these near-estuary urban centres should be focused on as a key source of nutrient enrichment in this estuary.

Catchment management for the reduction of nutrient loading on the study sites

A range of catchment management strategies are available for the reduction of nutrient loading on freshwaters. The catchment manager might aim to reduce nutrient export at source. This would require either point source control, involving phosphorus precipitation at major sewage treatment works, or modification of discharge consents. Alternatively he/she might seekdiffuse source control, requiring catchment zoning for nutrient export risk, allowing spatial relocation of high risk land uses to areas of the catchment with the greatest intrinsic nutrient retention capacity, or modification of land use and management in the more vulnerable zones. U.K. examples include Nitrate Vulnerable Zones or Environmentally Sensitive Areas.

If reduction of nutrient export at source is considered impracticable, or too expensive, the catchment manager can implement measures to interrupt the nutrient transport pathway linking non-point sources in the catchment to the lake. This might involve land use zonation, to place uses such as rough grazing and woodland in the path of nutrients exported from land upslope, or the creation of semi-natural vegetated strips in the riparian zones with a high nutrient retention capacity.

Finally, if reduction of nutrient export at source, and interruption of the nutrient transport pathway are not adopted, the catchment manager can focus on mitigating the impact of nutrient loading on the receiving waters through a number of measures. These include reduction of bioavailable phosphorus in the water column through ferric dosing, reducing the sedimentary phosphorus store in shallow lakes by sediment dredging or biomanipulation.

Importantly, however, the catchment manager must note that not all strategies will be equally suitable in different water bodies, and for this reason, it is important that the selection of a management strategy is informed by an understanding of the following:

1. The origins of the increases in nutrient loading.
2. The nutrient transport pathways linking the point and non-point sources in the catchment to the water body.
3. The likely reduction in nutrient loading which might be associated with different management strategies.

The application of the export coefficient modelling approach to the three study sites has produced a very good fit with observed data for these three catchments, indicating that each water body has undergone nutrient enrichment over the past 60 years. The models for each catchment also give an indication of the origins of this enrichment, and the relative contributions of nitrogen and phosphorus from each non-point and point source in each catchment. As a final step, the models can, therefore, be used to predict the likely reductions in nutrient loading on each water body which might be effected through changes in catchment management.

A range of potential catchment management options have been evaluated here for the River Esk, Slapton Ley and the Deben estuary. The options selected include the introduction of phosphorus removal at all major sewage treatment works, a 20% reduction of fertiliser application rates to all crops and grass, conversion of all agricultural land to rough grazing receiving no fertiliser applications, a 20% reduction of stocking densities on grassland, conversion of 20% of all arable land to fertilised permanent grass and a 20% reduction of atmospheric nitrogen deposition rates. The model forecasts for each option in the three catchments are presented as % change from 1990

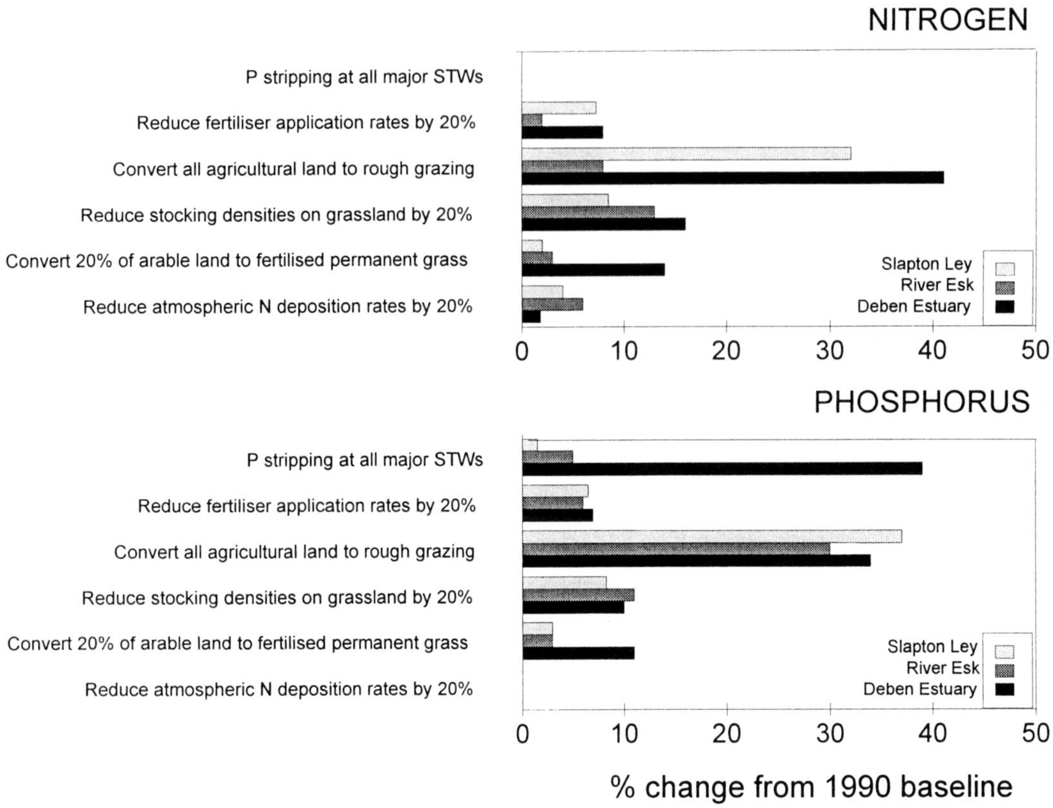

Figure 11. Predicted changes in nutrient loading on Slapton Ley, the River Esk and the Deben estuary in relation to selected catchment management strategies.

baseline in Figure 11. The information contained here is very revealing. Phosphorus removal would give no reduction in nitrogen loading, but there will also be very little reduction in phosphorus loading in either the Slapton or Esk catchment for what is a relatively expensive management strategy. However, for the Deben system, the model predicts a very substantial reduction in phosphorus loading on the estuary, the highest of all options evaluated here, and as such, it might be considered the Best Practicable Environmental Option (BPEO) for the Deben Estuary. If a reduction in nitrogen loading is also sought, then focusing on a reduction in the number of pigs in this catchment might have very successful results.

For the Slapton catchment, the story is very different, and if a reduction in both nitrogen and phosphorus loadings on the system is required, the most successful option would be to convert all agricultural land in the catchment to unfertilised rough grazing. However desirable this might be, it is unlikely to be practicable on a wider scale. Instead, the Slapton model predicts that a 20% reduction in stocking densities to 1970s levels would achieve a reduction of approximately 10% in both nitrogen and phosphorus concentrations in waters draining into Slapton Ley. This is unlikely in the long-term to lead to a recovery of the macrophyte community, and might be best linked to in-lake management through biomanipulation.

For the Esk catchment, the model output is interesting in that no one technique, other than converting all agricultural land to rough grazing is likely to achieve a reduction in both nitrogen and phosphorus loading on the river. Instead, the Esk model would suggest a combination of reduction of stocking densities, and atmospheric deposition to reduce the nitrogen load, and a reduction of fertiliser applications to all crops and grass to achieve a reduction in phosphorus loading.

Conclusions

An understanding of catchment history is fundamental to the development of sustainable management strategies for nutrient enriched lakes. This paper has demonstrated the use of catchment scale models of nutrient export as a tool for the integrated management of lakes, and in particular the use of export coefficient modelling as a means of interpreting the role of changes in catchment history which have led to increases in nutrient loading on lakes.

This type of approach is very powerful as one of a number of tools available to the catchment manager, and is perhaps most powerful when linked to models of in-lake nutrient dynamics and studies of the ecological mechanisms controlling lake state under nutrient enrichment. What this approach allows us to do is to understand the role of catchment history in the nutrient enrichment of lakes, and to make informed judgements as to the origins of nutrient enrichment and the best means of managing the system to reduce these loadings as part of an integrated catchment and lake management programme

Recently, this approach has been used in Phase 2 of the Environment Agency's Lake Classification and Monitoring scheme to hindcast nutrient export rates for all land in England and Wales in the 1930s to provide a baseline, prior to the main phase of agricultural intensification and expansion in Britain. Subsequent increases in nutrient loading on lakes may then be determined (Moss et al., 1996). This allows determination of present trophic status for any lake in England and Wales as a function of its baseline state, and provides a valuable resource for the management of nutrient enriched water bodies in England and Wales. The database produced from this research is equally applicable to any surface freshwater system in England and Wales, including rivers, streams and estuaries.

References

Anderson, N. J., B. Rippey & C. E. Gibson, 1993. A comparison of sedimentary and diatom-inferred phosphorus profiles: implication for defining pre-disturbance nutrient conditions. Hydrobiologia 253: 357–366.

Bennion, H., 1994. A diatom-phosphorus transfer function for shallow, eutrophic ponds in southeast England. Hydrobiologia 275/276: 391–410.

Bennion, H., S. Wunsam & R. Schmidt, 1995. The validation of diatom-phosphorus transfer functions: an example from Mondsee, Austria. Freshwat. Biol. 34: 271–283.

Bailey-Watts, A. E. & A. Kirika, 1987. A re-assessment of the phosphorus inputs to Loch Leven (Kinross, Scotland): rationale and an overview of results in instantaneous loadings with special reference to run-off. Trans. r. Soc. Edin. Earth Sciences 78: 351–367.

Bailey-Watts, A. E., A. Kirika, L. May & D. H. Jones, 1990. Changes in phytoplankton over various time scales in a shallow, eutrophic lake: the Loch Leven experience with special reference to the influence of flushing rate. Freshwat. Biol. 23: 85–111.

Dils, R. M. & A. L. Heathwaite, 1996. Phosphorus transport in agricultural runoff, In S. Brooks & M. Anderson (eds), Advances in Hillslope Processes, Wiley, Chichester.

Driscoll, C. T. & R. Van Dreason, 1993. Seasonal and long-term temporal patterns in the chemistry of Adirondack lakes. Wat. Air Soil Pollut. 67: 319–344.

Duffy, P. D., J. D. Schreiber, D. C. McClurkin & L. L. McDowell, 1978. Aqueous- and sediment-phase phosphorus yields from five southern pine watersheds. J. Envir. Qual. 7: 45–50.

Edmondson, W. T., 1970. Phosphorus, nitrogen and algae in Lake Washington after diversion of sewage. Science 169: 690–691.

Edmondson, W. T., 1985. Recovery of Lake Washington from eutrophication, In Proceedings of the International Congress on Lake Pollution and Recovery, European Water Pollution Control Association, Rome, April 15–18: 228–234.

Edmondson, W. T. & J. T. Lehman, 1981. The effect of changes in the nutrient income on the condition of Lake Washington, Limnol. Oceanogr. 26: 1–29.

Foster, I. D. L., P. N. Owens & D. E. Walling, 1996. Sediment yields and sediment delivery in the catchments of Slapton Lower Ley, South Devon, U.K. Field Studies 8: 629–661.

Foy, R. H., Smith, R. V., Stevens, R. J. & D. A. Stewart, 1982. Identification of factors affecting nitrogen and phosphorus loadings to Lough Neagh. J. Environmental Management 15: 109–129.

George, D. G., D. P. Hewitt, J. G. W. Lund & W. J. P. Smyly, 1990. The relative effects of enrichment and climate change on the long-term dynamics of *Daphnia* in Esthwaite Water, Cumbria. Freshwat. Biol. 23: 55–70.

Heathwaite, A. L. & P. J. Johnes, 1996. The contribution of nitrogen species and phosphorus fractions to stream water quality in agricultural catchments. Hydrological Processes 10: 971–983.

Heathwaite, A. L., P. J. Johnes & N. E. Peters, 1996. Trends in water quality 6: trends in nutrients, Hydrological Processes 10: 263–293.

Heckrath, G., P. C. Brooks, P. R. Poulton & K. W. T. Goulding, 1996. Phosphorus losses in drainage waters from arable silty clay loam soil. In H. Tunney & O. Caton (eds), Phosphorus Loss to Water from Agriculture. C.A.B. International: 367–9.

Holden, A. V. & L. A. Caines, 1974. Nutrient chemistry of Loch Leven, Kinross. Proc. Royal. Society Edin. B 74: 101–122.

Johnes, P. J., 1994. Modelling nutrient loading on the Ore/Alde and Deben estuaries, 1930-1990. Unpublished report for National Rivers Authority, Anglian Region, Peterborough.

Johnes, P. J., 1996a. Nutrient export to the River Bure from its catchment, 1930–1995. Unpublished report for the Environment Agency, Anglian Region, Ipswich.

Johnes, P. J., 1996b. Evaluation and management of the impact of land use change on the nitrogen and phosphorus load delivered to surface waters: the export coefficient modelling approach. Journal of Hydrology 183: 323–349.

Johnes, P. J. & T. P. Burt, 1991. Water quality trends in the Windrush catchment: nitrogen speciation and sediment interactions. In N. E. Peters & D. E. Walling (eds), Sediment and Stream Water Quality in a Changing Environment: Trends and Explanation. IAHS Publication 203, Wallingford 349–357.

Johnes, P. J. & T. P. Burt, 1993. Nitrate in surface waters. In T. P. Burt, A. L. Heathwaite & S. T. Trudgill (eds), Nitrate: Processes, Patterns and Control. John Wiley, Chichester: 269–320.

Johnes, P. J. & A. L. Heathwaite, 1997. Modelling the impact of land use change on water quality in agricultural catchments. Hydrological Processes 11: 269–286.

Johnes, P. J., B. Moss & G. L. Phillips, 1996. The determination of water quality by land use, livestock numbers and population date – testing of a model for use in conservation and water quality management, Freshwat. Biol. 36: 451–473.

Johnes, P. J. & H. M. Wilson, 1996. The limnology of Slapton Ley. Field Studies 8: 585–612.

Knuuttila, S., O-P. Pietiläinen & L. Kauppi, 1994. Nutrient balances and phytoplankton dynamics in two agriculturally loaded shallow lakes. Hydrobiologia 275/276: 359–369.

Lund, J. W. G., 1972. Changes in the biomass of blue-green and other algae in an English lake from 1945-69. In T. V. Desikachary (ed.), Taxonomy and Biology of Blue-Green Algae, Madras Symposium, 1970: 305–327.

Meteorological Office, 1989. Climatological Data for Agricultural Land Classification. The Meteorological Office, Bracknell.

Meteorological Office, 1992. MORECS – the Meteorological Office Rainfall and Evaporation Calculation System database. The Meteorological Office, Bracknell.

Moss, B., H. Balls, I. Booker, K. Manson & M. Timms, 1988. Problems in the construction of a nutrient budget for the River Bure and its Broads (Norfolk) prior to its restoration from eutrophication. In F. E. Round (ed.), Algae and The Aquatic Environment Biopress Ltd., Bristol: 326–353.

Moss, B., P. J. Johnes & G. L. Phillips, 1996. The monitoring of ecological quality and the classification of standing waters in temperate regions: a review and proposal based on a worked scheme for British waters. Biol. Rev. 71: 301–339.

O'Sullivan, P. E., 1994. The Natural History of Slapton Ley National Nature Reserve XXI: the palaeolimnology of the uppermost sediments of the Lower Ley, with interpretations based on ^{210}Pb dating the historical record. Field Studies 8: 403–449.

Phillips, G. L., R. Jackson, C. Bennett & A. Chilvers, 1994. The importance of sediment phosphorus release in the restoration of very shallow lakes (The Norfolk Broads, England) and implications for biomanipulation. Hydrobiologia 275/276: 445–456.

Ryding, S-O. & C. Forsberg, 1979. Nitrogen, phosphorus and organic matter in running waters: studies from six drainage basins. Vatten 1: 46–58.

Sharpley, A. N., 1996. Identification of critical source areas for phosphorus export from agricultural catchments. In S. Brooks & M. Anderson (eds), Advances in Hillslope Processes, Wiley, Chichester.

Sharpley, A. N. & J. K. Syers, 1976. Phosphorus transport in surface runoff as influenced by fertiliser and grazing cattle. New Zealand Journal of Science 19: 277–282.

Stevens, R. J. & B. M. Stewart, 1982. Concentration, fractionation and characterisation of soluble organic phosphorus in river water entering Lough Neagh. Wat. Res. 16: 1507–1519.

Stoddard, J. L., 1994. Long-term changes in watershed retention of nitrogen, In L. A. Baker (ed.), Environmental Chemistry of Lakes and Reservoirs. Advances in Chemistry Series 237, American Chemical Society, Washington, D.C.

Talling, J. F. & S. I. Heaney, 1988. Long term changes in some English (Cumbrian) lakes subjected to increased nutrient inputs, In F. E. Round (ed.), Algae and the Aquatic Environment, Biopress, Bristol 1–29.

Withers, P. J. A., 1994. The significance of agriculture as a source of phosphorus pollution to inland and coastal waters in the U.K. Unpublished report to MAFF, London, 90 pp.

Hydrobiologia **395/396:** 61–76, 1999.
D.M. Harper, B. Brierley, A.J.D. Ferguson & G. Phillips (eds), The Ecological Bases for Lake and Reservoir Management.

Practical application of 25 years' research into the management of shallow lakes

Geoff Phillips[1], Alison Bramwell[1], Jo Pitt[1], Julia Stansfield[1] & Martin Perrow[2]
[1]*Environment Agency, Ipswich, U.K.*
[2]*ECON, Biological Sciences, University of East Anglia, Norwich, U.K.*

Abstract

The Norfolk Broads are a series of shallow, man-made lakes dug in medieval times for peat extraction, in Eastern England. Their eutrophic state has been well-documented and, since the early 1980s, their restoration has been attempted using a variety of techniques. The restoration began with the removal of point sources of phosphorus from sewage treatment works, which then revealed the role of sediment release when lake phosphorus levels failed to decline following inflow phosphorus levels. Small-scale removal of sediment layers in isolated broads demonstrated the feasibility, both technical and economic, of this technique, but experience then showed that sediment removal alone could not provide long-term restoration. Biomanipulation following sediment removal now offers the most reliable route to restoration, but the mechanisms by which a stable submerged plant community can be maintained after biomanipulation are still not clear.

Introduction

Lakes are subject to many influences, but the one that has received the most attention is the impact of eutrophication. The causes and effects are well known, indeed the concept was recognised early in the century, (Thieneman, 1918; Naumann, 1919; Pearsall, 1921) and much has subsequently been written describing the effects and attempts to reverse them (e.g. Harper, 1992). While the central paradigm linking the supply of catchment-derived nutrients, particularly phosphorus, to enhanced production of algae and higher plants remains valid, the last decade has seen a much broader understanding of the way aquatic ecosystems respond to eutrophication. Firstly, it has become clear that lakes do not always respond as rapidly or as predictably to a reduction in phosphorus supply, as simple correlation models might suggest (Reynolds, 1992) and secondly, the effects of food webs on the expression of the enhanced productivity can have profound implications for both the ecology of lakes and their perceived quality (Gulati et al., 1990; Carpenter & Kitchell, 1993).

This is not a happy state of affairs for those charged with maintaining and improving the quality of our lakes, for they can no longer be assured that a given investment directed simply at an improvement to water quality will result in a predictable ecological change. However, the recognition that reversing a decline in ecological quality may be neither a simple, nor a cheap, option is perhaps the first step in establishing a viable eutrophication control strategy. This chapter provides a review of the progress and lessons that have been learned in the restoration of a group of very shallow eutrophic lakes in eastern England, known as the Norfolk Broads. It will highlight the important issues, and illustrate how practical management derived from the best available ecological knowledge has been used to both improve the environment and an understanding of the science on which it is based. Details of methods used can be found in Phillips (1984), Phillips & Kerrison (1991) and Phillips et al. (1994).

Background to eutrophication in the Norfolk Broads and details of study area

The Norfolk Broads are a group of small, very shallow lakes, interconnected by a tidal river system in eastern England (Figure 1). Created in medieval times by peat digging, the area has been influenced by man from its creation to the present day. Until the turn of this century, the rivers were important arteries for local transport, but with the development of railways their value as a recreational resource began to be recognised. Fortunately, this meant that many Victorian naturalists provided documentary evidence of the wealth and diversity of the biota found in these shallow lakes, but a growing local population, coupled with agricultural development and demands for recreation led to increasing pressure on the area. By the late 1960s, surveys revealed that the now highly eutrophic Broads had changed considerably (Morgan, 1972; Mason & Bryant, 1975) from the state described at the turn of the century. The once-clear water, dominated by a variety of submerged aquatic plants, including a diverse array of charophytes, had almost completely disappeared. Associated with this loss was a dramatic reduction in invertebrate diversity and an impoverished fish community of low biomass dominated by small roach (*Rutilus rutilus*) and bream (*Abramis abramis*) (Moss et al., 1979; Phillips et al., 1996).

The loss of submerged aquatic plants was the most obvious change and this was linked directly to increased phosphorus entering the aquatic system from the now rapidly-growing population living and using the area (Phillips et al., 1978). Unfortunately the biggest changes to the ecosystem took place prior to the onset of recent ecological research and monitoring in the area. However in one lake, Hickling Broad, records of the amount of weed removed from the lake to maintain navigation revealed that the change from a plant- to an algal-dominated community took place over a relatively short period (approximately 3 years) in the early 1970s (Phillips & Moss, 1978). In all probability, the lake was already severely stressed by eutrophication by this time, but these records, taken in the context of our current understanding of shallow lake systems, reveal the vulnerability of stressed aquatic ecosystems to management actions and the potential speed of ecosystem change (Moss et al., 1997).

Extensive limnological surveys carried out during the 1970s (reviewed by Moss, 1983) identified that the key source of phosphorus in the Broads was from treated sewage effluent. This, together with the apparent success of phosphorus reduction schemes, such as at Lake Washington (Edmondson, 1977), resulted in the initiation of a restoration programme for the Broads based on the control of phosphorus discharged from key sewage treatment works. At the time it was assumed that results would be relatively quick to achieve, although it is interesting to note that an early paper by Mason (1976) correctly identified many of the additional interactions, that have now been shown to be critical in explaining the changes that have taken place over the last 25 years.

Impact of phosphorus reduction from STWs discharging to the River Ant

The experimental restoration programme initially focused on the River Ant, which flows through Barton Broad, the second largest lake in the Broads system (Figure 1). Phosphorus removal began in 1977 with the installation of tertiary chemical dosing at Stalham sewage treatment works which discharges into the River Ant, approximately 2 km upstream of Barton Broad (Phillips, 1984; Kerrison et al., 1989). In 1980, the effluent from another substantial sewage treatment works further upstream at North Walsham was diverted from the catchment to the coast. This was followed by the installation of phosphorus removal at an industrial source and at all of the remaining small sewage treatment works with direct discharges to the river. For the majority of this period, the removal of phosphorus has been undertaken voluntarily by the dischargers, but is now controlled by statutory discharge consents issued under the U.K.'s Water Resources Act 1989.

Details of the River Ant catchment and its nutrient status, prior to phosphorus removal, may be found in Osborne (1981). Since 1977, monitoring of the River Ant and significant effluents discharging to it have been undertaken by the statutory agencies (Anglian Water Authority 1977–1989, National Rivers Authority 1989–1996, Environment Agency post-1996) using a similar strategy to Osborne (1981). Sample sites and frequency have varied slightly throughout this time, but generally water samples have been analysed at fortnightly intervals for a number of determinands, including total and soluble reactive phosphorus at a minimum of three key river sites upstream of Barton Broad and at a site in the broad itself (Figure 1). Similar weekly analysis of effluent samples have also been undertaken. These phosphorus data have been multiplied by daily river and effluent flows to derive estimates of total phosphorus load. Chlorophyll *a* concentration has also been measured in Barton Broad and from the two river sites to assess the impact of these changes on phytoplankton development.

The uppermost river sample site in the catchment (presented here) is Honing Lock, a disused lock struc-

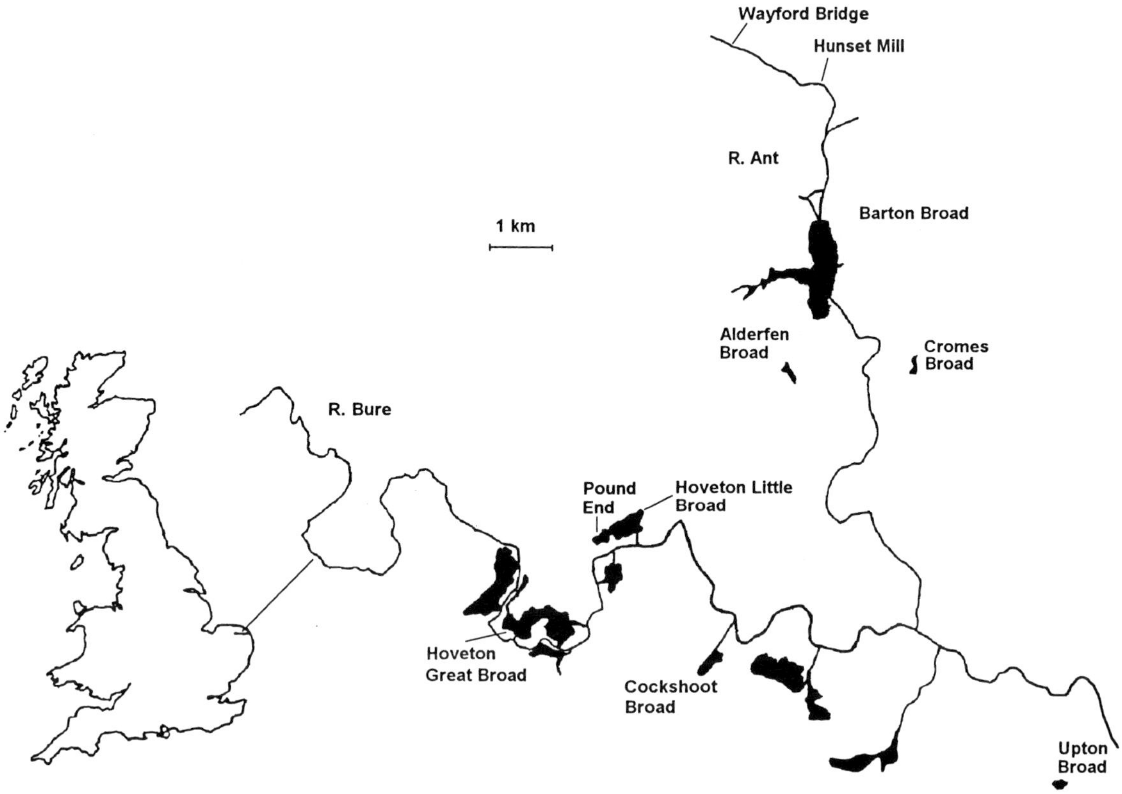

Figure 1. Map showing the location of the Broads and position of sample sites and experimental locations.

ture separating the upper faster flowing river from the lower sluggish tidal system, positioned 6km downstream of the discharge from North Walsham sewage treatment works. The site contains a flow gauging station and represents a nodal point in the river at which the load of phosphorus discharging from the upper catchment into the tidal river can be easily estimated. Further downstream, Wayford Bridge is located 1km upstream of the discharge from Stalham sewage treatment works. Here the river is large enough to be used by pleasure craft, and tidal action is sufficient to halt fluvial flow during the flood tide. A final sample point at Hunsett Mill is below Stalham sewage treatment works and is a convenient point to determine the quality of water entering Barton Broad.

Phosphorus loads from 1978–1995 are presented for these three key river sites, together with the load discharged from sewage treatment works or other point sources upstream of them (Figure 2). Phosphorus removal at Stalham sewage treatment works was operational prior to the start of this monitoring programme, and in 1976 this discharge would have contributed an additional total phosphorus load of approximately 16 kg d^{-1} (Osborne, 1981). At this time, Osborne measured a total phosphorus load entering Barton Broad of 21 and 17 kg d^{-1} (1975, 1976), values very similar to those estimated in this study (Figure 2c), suggesting that the impact of the Stalham discharge on the loading to the broad was relatively small. However, the diversion of the North Walsham sewage treatment works effluent in 1980 resulted in a further substantial (c 90%) reduction in the discharged phosphorus load. River total phosphorus load clearly responded to this change, although the 90% reduction in discharged load was only matched by a 50% reduction in river load at the point where it enters Barton Broad (Hunset Mill) (Figure 2c).

Prior to 1980, the measured total phosphorus load in the river could only account for about half of that discharged from point sources upstream, and it is clear that there must have been a considerable loss of phosphorus from the water column during its passage downstream. However, since 1980 the river load at Honing Lock has been higher than that discharged from remaining upstream point sources, revealing the presence of other diffuse inputs. This difference was

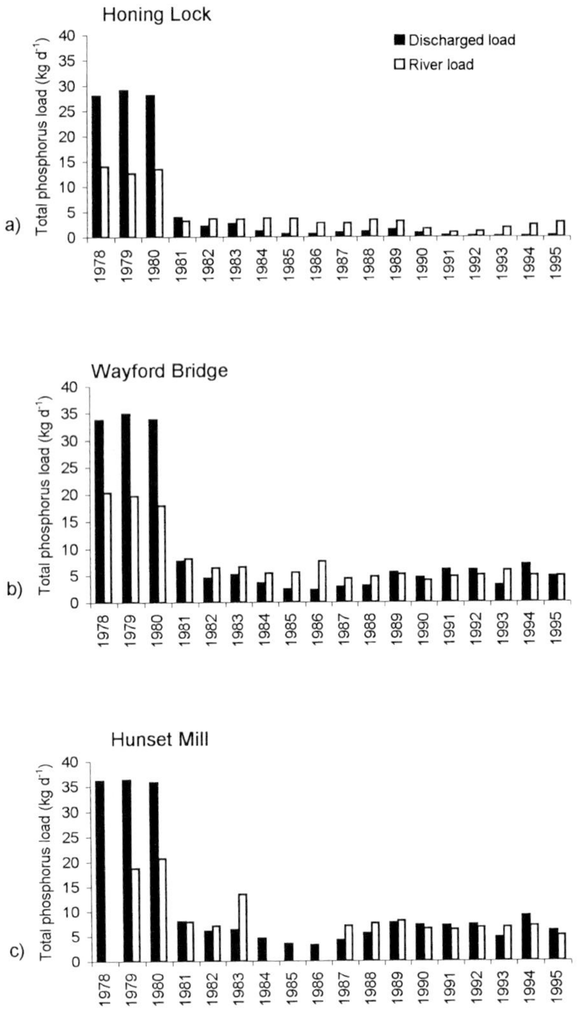

Figure 2. Total phosphorus load discharged upstream of key points in the River Ant (dark bars) and measured load in the river (open bars) at (a) Honing Lock, (b) Wayford Bridge, (c) Hunset Mill (No data for Honing Lock 1984–1986).

found to be 0.12 kg P ha^{-1} ann^{-1} which is lower than the estimates of 'agricultural loading' made by Osborne (1981), but similar to those of Phillips (1977) for Alderfen Broad and typical of diffuse sources in the area estimated by export modelling (Johnes, 1996). Further downstream, the discrepancy between river load and upstream point sources are less apparent, but if the diffuse load estimated above is added to that from point sources, the measured load in the river can no longer account for all of the phosphorus discharged (Table 1).

It is clear from these data that there was a substantial loss of phosphorus from the water column prior to 1980 and that some loss still occurs. A number of other

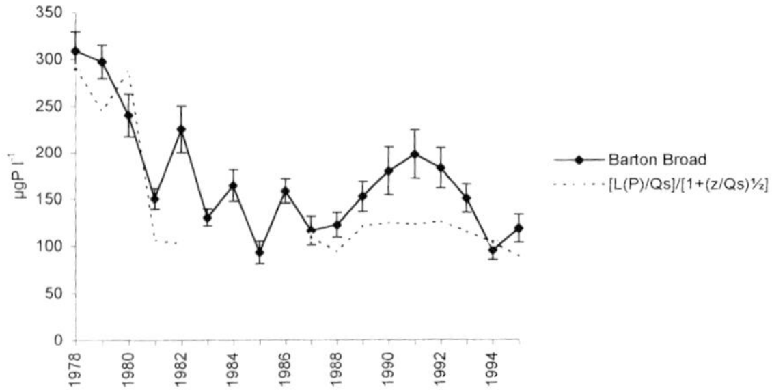

Figure 3. Comparison between annual total phosphorus concentration in Bartin Broad and that predicted from river load corrected for hydraulic load and depth (Vollenweider & Kerekes, 1982).

nutrient budget studies in the Broads (Osborne, 1981; Moss et al., 1988; Johnes, 1996) have identified similar losses and suggest that uptake of phosphorus by the sediment is the most likely explanation. This sediment can move downstream under flood conditions, undetected by spot-sampling regimes, ultimately to be deposited in Barton Broad where subsequent chemical and biological transformation can make it available to the overlying water (Phillips et al., 1994).

Influent phosphorus load can be related to lake concentration using the OECD regression model of Vollenweider & Kerekes (1982):

$$P_l^y = [L(P)/q_s]/[1 + (z/q_s)],$$

where: $P_l{}^y$ is annual mean lake phosphorus concentration (mg m^{-2}); $L(P)$ is the areal loading (g m^{-2} ann^{-1}); z is mean depth (m); and q_s is the hydraulic load (m a^{-1}).

In Barton Broad, the model provides a reasonable fit to the observed data, although after 1980 the broad generally had higher than predicted total phosphorus

Table 1. Comparison between measured total phosphorus load in River Ant and estimated inputs, mean data 1987–1995

	Honing Lock	Wayford Bridge (kg/d)	Hunsett Mill
P controlled STW	0.46	0.17	1.83
Other point sources	0.09	4.05	0.00
Diffuse	1.66	1.73	0.11
Upstream		2.21	8.16
Total	2.21	8.16	10.10
Measured	2.21	4.88	6.81
Difference	0.00	−3.28	−3.29

concentrations (Figure 3). This elevated lake phosphorus concentration can be attributed to the release of phosphorus from the sediments, which is known to provide a significant internal loading during the summer months (Phillips & Jackson, 1990; Phillips et al., 1994). The most marked increase in 1982, and from 1989, coincided with periods of abnormally low river flow (Figure 4), caused by low rainfall, when the reduced flushing of the broad resulted in increased impact from this internal phosphorus supply.

These observations demonstrate the difficulty of relating estimates of source loadings of phosphorus to that transported along a river and ultimately entering a lake without a knowledge of loss rates, diffuse inputs and potential sediment release rates. Simple comparisons of source loads to river inflow loads may underestimate the total potential load and thus provide an unrealistic assessment of load reduction. River sediments may act as a net sink for phosphorus, but if these sediments can enter the lake, it may be released into the overlying water. Thus this 'bed load' can be an important and often an invisible component of the phosphorus load on the lake. It is thus essential when planning a phosphorus reduction programme to obtain a good estimate of all potential sources, before predicting the possible impact of planned reduction measures from individual point sources.

Impact on phytoplankton populations

The river upstream of Honing lock is relatively fast flowing and phytoplankton only develops below this in the tidal river system. Typical growth patterns can be seen from 16-year averages of monthly chlorophyll *a* concentrations (Figure 5). The river from Wayford Bridge can contain substantial phytoplankton populations during the spring and early summer. Barton Broad shows a similar spring growth, but this is followed by a more substantial summer population, which is sustained well into the early winter. Since 1980, chlorophyll *a* concentration in both the river and the broad have been reduced, and there is a suggestion of a long term trend of reduced phytoplankton growth (Figure 6). However, the impact of the low flow period from 1990–1992 is again apparent, demonstrating the importance of flow as an additional factor influencing phytoplankton growth.

The annual mean chlorophyll *a* concentration in Barton Broad can be fitted to a power function of the mean total phosphorus concentration, $\mathrm{Chl}a_{\mathrm{y}} = 2.24\ \mathrm{TP}_{\mathrm{y}}^{0.8\pm0.29}$ with an R^2 value of 0.68 ($p < 0.001$) (Figure 7). The exponent of this function is very similar to that obtained in the OECD study (0.91) suggesting a similar rate of response to phosphorus control, although the relative amount of chlorophyll produced is

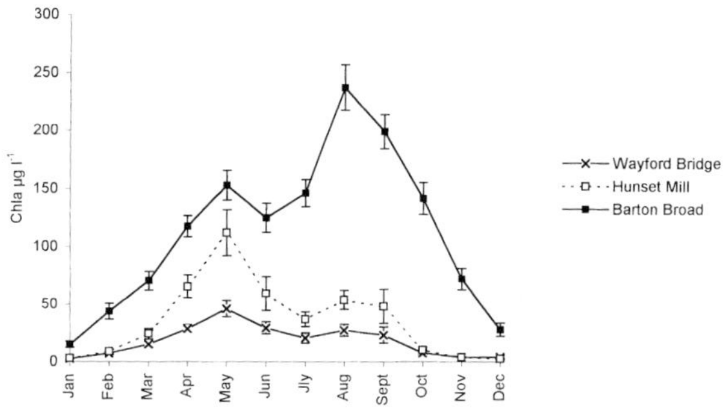

Figure 5. Sixteen year monthly mean chlorophyll *a* concentrations (1980–1995) in the River Ant and Barton Broad (± SE).

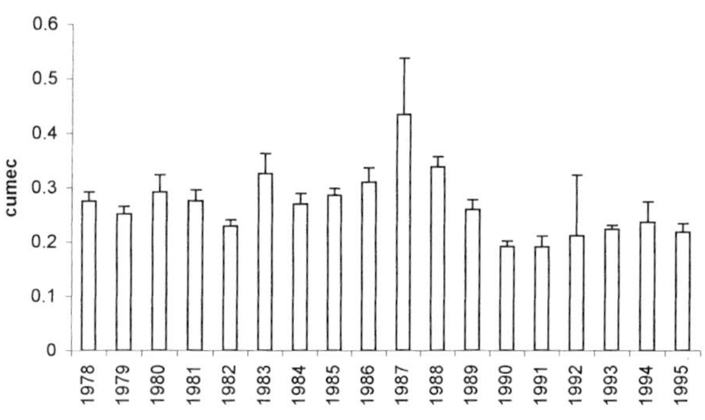

Figure 4. Summer (April–September inclusive) mean daily river discharge in the River Ant at Honing Lock (± SE).

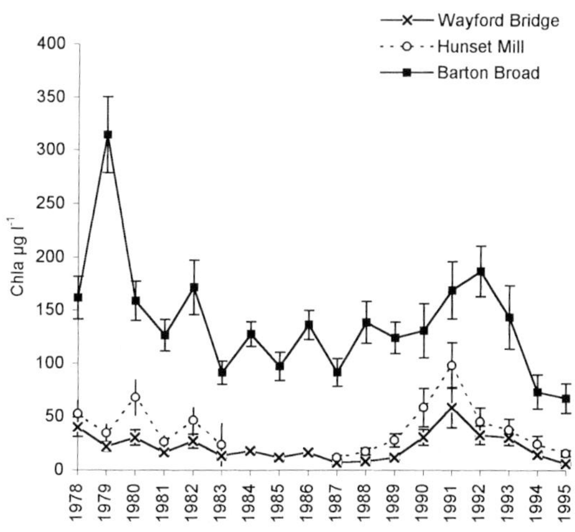

Figure 6. Annual mean chlorophyll *a* concentration in the River Ant and Barton Broad (± SE).

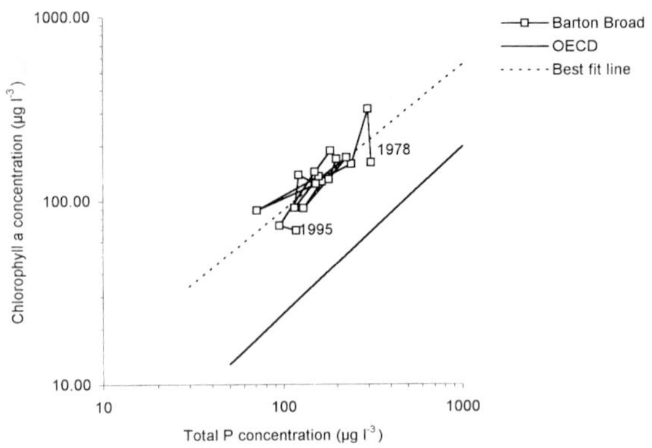

Figure 7. Annual mean chlorophyll *a* concentration in Barton Broad as a function of annual mean lake total phosphorus concentration. Best fit line shown and for comparison the OECD regression (Vollenweider & Kerekes, 1982).

clearly greater. As predicted for shallow lakes, where phosphorus has become growth-limiting, there is no evidence of a lag response in the curve, where mean phosphorus is reduced, but chlorophyll *a* fails to respond (Sas, 1989; Reynolds, 1992). Osborne (1978) found little evidence of phosphorus limitation of the phytoplankton of Barton Broad in his studies prior to phosphorus control (1974–1976), but, since 1981, soluble reactive phosphorus in Barton Broad has been at or less than 10 mg l^{-1} for more than 50% of the growing season, a definition of phosphorus-limitation used by Sas (1989). As a result of phosphorus control, the lake appears to be capable of responding directly to the reduced phosphorus concentrations. The much higher chlorophyll *a* concentrations in the lake relative to those predicted by the OECD regression are probably a result of the lake being shallow and extremely well mixed, the combined effects of which will maximise phytoplankton growth.

Thus, despite 16 years of relatively consistent and substantial reduction in the amount of phosphorus discharged to the catchment of Barton Broad, the lake remains dominated by phytoplankton. There is evidence for a reduction in algal biomass but the desired change to a macrophyte-dominated lake community has not occurred. Inter-annual variation remains high (Figure 7) and this appears to be linked to the continuing internal supply of phosphorus from the sediment. If the supply of phosphorus from the sediment can be reduced, then algal biomass should respond and this should improve the chances of re-establishing aquatic plants in the lake.

Controlling sediment phosphorus release

There is now a wealth of evidence for the release of phosphorus from the sediment of shallow lakes (Marsden, 1989) and the magnitude of release in the Broads can be an order of magnitude greater than that from catchment sources (Phillips & Jackson, 1990). The mechanism is still not clearly understood, but in the Broads it is clear that the iron content of the sediment is an important factor. According to the classic model of phosphorus exchange (Mortimer, 1941, 1942), iron (III) will bind phosphorus and thus under oxidising conditions, no phosphorus release should occur. As a result, aerobic sediments high in iron should retain phosphorus. Sediments in the broads have a high iron content (10–30 mg Fe g^{-1}) (Pitt et al., 1997) and, given the well stirred aerobic water column, they should be capable of retaining phosphorus. Even under reducing conditions, created below the sediment water surface by decomposition, iron (II) will form in the pore water, and provided this is in a molar excess in relation to the soluble phosphorus in the pore water, release of phosphorus into the overlying lake water is unlikely due to subsequent binding with particulate iron (III) in the oxidised microzone at the sediment water interface (Holdren & Armstrong, 1986).

In the broads, as sediment temperature increases, microbial sulphate reduction generates hydrogen sulphide, with the result that iron (II) is removed from the pore water as iron sulphides. These adhere to the particulate sediment matrix leaving the phosphorus in the pore water free to 'leak' from the sediment to the overlying water. Phosphorus release measurements, made in the laboratory from undisturbed sediment cores into aerobic water, reveal very variable rates, but values in excess of 30 mg P m^{-2} d^{-1} are generally only found when there is an excess of soluble phosphorus over iron (II) in the pore water (Phillips et al., 1994). These rates cannot be accounted for by diffusion and other factors, including bioturbation, may be important (Phillips et al., 1994). The broads are very shallow (1–2 m average depth), the sediment is extremely unstable and, in the field, wind stirring may also be a very important mechanism introducing phosphorus into the overlying water.

Whatever the mechanism, it is clear that the sediment pore water can contain substantially higher soluble phosphorus than the overlying water. Typical values are 500–1500 mg P l^{-1}, but levels as high as 9.0 mg P l^{-1} were measured in Alderfen Broad in

1993 (Pitt et al., in press), and soluble phosphorus in the sediment remains a significant internal source of nutrient that continues to provide the phytoplankton with nutrients.

The importance of iron in controlling phosphorus release, and the observation that phosphorus concentration is very low in the overlying water of broads, where deep drainage of surrounding land (with acid sulphate soils) has enriched the sediment with very high iron concentrations (94.7 mg Fe kg^{-1}), generated interest in adding iron to the sediment artificially using a technique developed in the Netherlands (Quaak, 1993). Although pilot laboratory experiments provided encouraging results (Jackson, 1991), subsequent field trials at various scales other than a whole lake, demonstrated that a significant proportion of the added iron was subsequently released to the water column (Pitt et al., 1997). As the released iron (II) should re-precipitate in the aerobic lake water column, it is possible that, on a whole lake scale, loss of iron from the sediment might be replaced by sedimenting iron (III). However, in experimental systems phosphorus control could not be demonstrated, and the technique has, at least for the time being, been abandoned in favour of sediment removal.

Sediment removal

The removal of sediment from shallow eutrophic lakes has been used as a management tool in a number of countries, usually in combination with reduction in external nutrient load (Bengtsson et al., 1975; Peterson, 1981; Pokorny & Hauser, 1994). It was first used in the Norfolk Broads in 1981 in an attempt to restore Cockshoot Broad (5.5 ha), a lake that had lost much of its open water by silt accumulation. In 1981, sufficient sediment was pumped out to give a water depth of about 1m, and the broad was then isolated from the River Bure by a dam. As this project resulted in subsequent water quality improvements it was generally assumed that the removal of the sediment resulted in a reduction of the phosphorus released from the sediments (Moss et al., 1986). Few specific investigations have been carried out on the sediment of Cockshoot Broad, although sediment cores collected from the site in 1990 and 1995, after sediment removal, showed that it contained relatively high phosphorus concentrations (Table 2) in comparison with other broads. Total phosphorus concentration in the overlying water decreased following isolation and dredging, but since 1990, mean summer concentration has risen again (Figure 8).

In 1992/93 Alderfen Broad, another small (4.7 ha) isolated broad was dredged using a suction dredger to re-create open water conditions in a lake, where water depth was frequently reduced to less than half a metre. A nutrient budget for this lake had demonstrated the release of phosphorus from its sediment (Phillips, 1977) and, since the lake's small inflow stream was diverted around the lake in 1979, the sediment has almost certainly been the source of its very high phosphorus concentration (Moss et al., 1986). To assess the impact of suction dredging on the release of phosphorus from Alderfen Broad, either three (1993/1994) or five replicate (1994/1995) sediment cores were collected from the lake, monthly until September 1995, and used to determine depth profiles (top 15 cm) of soluble reactive phosphorus in the pore water. Similar cores were used to measure phosphorus release rates during the summer of 1994 and 1995 using batch incubations at ambient temperature in the laboratory (Pitt et al., 1997). Some additional cores were collected at intervals during 1993 & 1994 to determine the total phosphorus content of the sediment.

Table 2. Changes in total phosphorus in sediment of Alderfen Broad following mud pumping. Mean of 5 cores over depth of 12 cm except where stated ± standard error. Comparative values from a range of other broads (Pitt et al., in press)

Date	Total phosphorus (mg P g^{-1})
12/4/93	2.13[1]
12/4/93	0.68[2]
4/10/93	0.87[3]
21/3/94	1.73 ± 0.23
18/4/94	1.51 ± 0.15
6/6/94	1.10 ± 0.11
5/7/94	1.29 ± 0.12
31/8/94	1.40 ± 0.14
17/8/94	1.75 ± 0.40
Other sites	**Total phosphorus (mg P g^{-1})**
Barton	1.2 ± 0.21
Hoveton Gt	1.02 ± 0.07
Cockshoot	2.13 ± 0.32
Ranwoth	1.11 ± 0.05
Upton	1.89 ± 0.20

[1] Prior to dredging (mean of 2 cores)
[2] After dredging (mean of 2 cores)
[3] (mean of 2 cores)

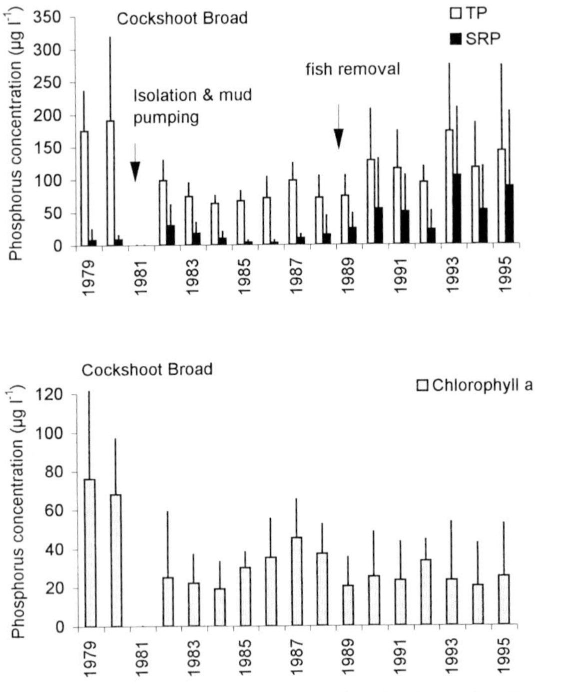

Figure 8. Mean growing season (March–October inc.) phosphorus and chlorophyll *a* concentrations in Cockshoot Broad, following isolation from the River Bure (± SE).

Dredging removed the upper half metre of sediment and the newly exposed sediment, which was more peaty in nature than the original material, had a substantially lower phosphorus content, although within a year the total phosphorus content had increased to about 80% of its original value and is typical of the area (Table 2). Interstitial phosphorus formed a typical profile ranging in concentration from around 0.1 mg P l^{-1} at the surface, to values in excess of 2.0 mg P l^{-1} at 15 cm. The mean concentration of interstitial phosphorus in the upper 15 cm generally increased in May reaching a maximum by July or August, coinciding with the period when phosphorus release could be detected in the laboratory cores (Pitt et al., in press). During 1994, cores were also collected from an area of original undisturbed sediment that had been left around the margin of the lake. Phosphorus release rates and overall mean interstitial phosphorus concentration from the two areas were not significantly different (Table 3), and release rates remain similar to those measured in other broads (Phillips et al., 1994).

Prior to the removal of sediment, the total phosphorus concentration in the overlying water of Alderfen Broad had become extremely high during the summer months. Maximum values approached 3.0 mg l^{-1}, considerably greater than previously observed maximum in 1984 (c 1.0 mg l^{-1}) (Moss et al., 1986) and there was a strong smell of hydrogen sulphide when the sediment was disturbed. Following the removal of sediment, phosphorus concentrations in the overlying water were substantially reduced, but they still remain higher than they were in 1987 and 1988 (Figure 9). This may be a short-term response to sediment removal, but the lake has previously gone through periodic fluctuations in total phosphorus concentration. Periods of high phosphorus were reported in 1974/75, 1979, 1984/85 and 1991 (Phillips 1977; Moss et al., 1986, 1990; Perrow et al., 1994) and the increase in 1991/93 fits the pattern described by Moss et al. (1990). They suggested that the release of phosphorus was dependant on the organic load imposed on the sediment by the periodic development of aquatic vegetation in the lake. which peaked in 1982 (Moss et al., 1990) and this is supported by the observations, reported in Perrow et al. (1994), of a further peak of plant growth in 1990 just prior to the onset of high phosphorus concentrations. This cyclical nature of plant growth remains to be satisfactorily ex-

Table 3. Changes in interstitial soluble reactive phosphorus in upper 15 cm of sediment from Alderfen Broad and phosphorus release rates from intact cores

Year	Interstitial SRP		Phosphorus release rate			
	Pumped	Un-pumped	Pumped		Un-pumped	
	mean (mg/l)	mean (mg/l)	mean (mg/m²/day)	max	mean (mg/m²/day)	max
1993	1.34 ± 0.04	–	–	–	–	–
1994	1.38 ± 0.05	1.64 ± 0.12	13.8 ± 4.4	41.4	11.1 ± 4.3	31.3
1995	1.35 ± 0.04	–	12.9 ± 4.3	22.9	–	–

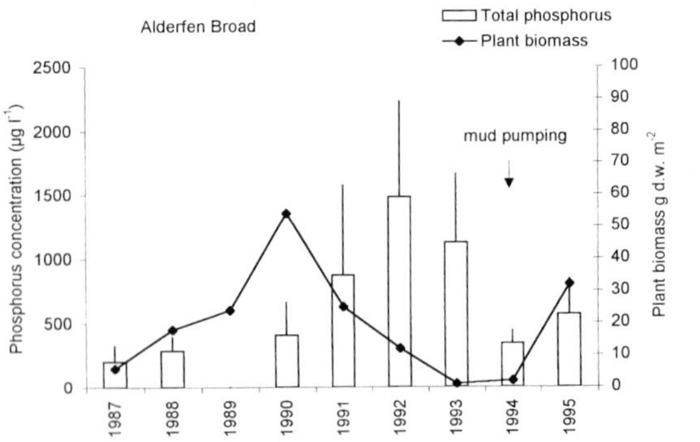

Figure 9. Mean growing season (April–September inc.) phosphorus concentration and plant biomass in Alderfen Broad. (Data 1987 & 1988 from Moss et al., 1990).

plained, but its presence confuses the interpretation of the impact of sediment removal on phosphorus release.

The mechanism for phosphorus release is still not fully understood in these shallow lake systems, but when interstitial phosphorus concentrations are substantially higher than those of the overlying water, the potential for phosphorus release remains. Sediment-pumping removes a considerable reservoir of phosphorus which can be potentially released, and when applied to lakes that have received enriched water for many decades, this may reduce the length of time to establish new equilibria following phosphorus reduction programmes. However, in isolated lakes, such as Alderfen and Cockshoot Broads, the small local catchments do not provide sufficient water to adequately flush the lake. The low rainfall of recent summers may have exaggerated this, but in these situations it seems that recycling and subsequent deposition of particulate phosphorus leads to an enrichment of the upper layers of the sediment and an eventual decline in water quality. Aquatic plants may exaggerate this process, although insufficient information is available to clearly determine this, or to decide if they can act as temporary seasonal sinks for critical nutrients, such as phosphorus. Sediment removal can thus be seen as an additional tool for the restoration of lakes. It is most likely to be successful in well flushed systems, and there is currently large-scale sediment removal being undertaken at Barton Broad by the Broads Authority, which will provide valuable insights into its effectiveness.

Role of the food web in lake restoration

Over the last decade, increasing evidence has been gathered about the role of zooplankton grazing in the functioning of lakes, both in the Broads (Timms & Moss, 1984; Moss, 1990; Phillips & Kerrison, 1990; Moss et al., 1996) and elsewhere (Edmondson & Abella, 1988; Gulati et al., 1990; Carpenter & Kitchell, 1993). These studies have demonstrated the potential for large-bodied cladoceran grazers to reduce phytoplankton crops, and have also highlighted the importance of aquatic vegetation in providing a habitat in which these grazers can coexist with fish communities (Phillips et al., 1996; Stansfield et al., 1997).

In the Broads, cladoceran zooplankton communities in the open water are generally dominated numerically by the small-bodied cladoceran *Bosmina longirostris* Müller, although larger-bodied species such as *Daphnia hyalina* Leydig typically occur during May and June (Phillips et al., 1996). The impact of fish predation on the zooplankton community has been well documented in the Broads (Cryer et al., 1986; Irvine et al., 1989; Perrow et al., 1994) and the persistence of *Daphnia* spp into the summer months in broads devoid of aquatic vegetation depends on the strengths of recruitment of 0+ fish, particularly roach. Peak populations of *Daphnia* spp. often coincide with short periods of clear water and, using small mesh enclosures, Phillips & Kerrison (1990) were able to demonstrate that the removal of fish predation could maintain these *Daphnia* populations throughout the summer. The resulting grazing pressure on the phytoplankton created clear water conditions, in which submerged aquatic vegetation was able to flourish.

Further evidence for the potential impact of grazing by zooplankton came from the management work undertaken at Cockshoot Broad. Immediately after the isolation and sediment removal by mud pumping was completed, the phytoplankton chlorophyll *a* concentration declined and it was initially concluded that the lake had responded to nutrient reduction (Moss et al., 1986). However, after 1985 the chlorophyll *a* concentration began to increase again (Figure 8) and a subsequent, more detailed, examination revealed changes in the zooplankton community, which could explain the changes in phytoplankton chlorophyll *a* (Moss et al., 1996). For the three years following isolation, *Daphnia* populations were high and contained animals of large size, including bigger species such as *D. pulex* (Perrow & Irvine, 1992). It appears that the combined effect of isolating the broad and extensive engineering operations depleted the planktivorous fish stock allowing the larger bodied *Daphnia* to dominate the zooplankton. From 1985 until 1989 this fish stock recovered, with a consequent reduction in both the body size and abundance of *Daphnia*; a reduction in grazing and hence increase in phytoplankton. To reverse these changes zooplanktivorous fish have been removed from the broad each winter or spring since 1988/89, with the result that the chlorophyll *a* concentration decreased again (Figure 8) as both the *Daphnia* spp. population and individual body size increased. Taking all the data from 1979 to 1993, Moss et al. (1996) were able to demonstrate a significant negative correlation between *Daphnia* spp. numbers and chlorophyll *a*, and no correlation between total phosphorus and chlorophyll concentration, except in the years when the fish population was high. Thus there seems little doubt that grazing by large cladocerans can substantially reduce phytoplankton, and that the manipulation of the fish population has enabled clear water conditions to be maintained in Cockshoot Broad since 1989. The result is that a steady increase in the amount of submerged vegetation has occurred (Stansfield et al., 1997).

This removal of fish to elicit a top-down effect on the trophic cascade, leading to a reduction in phytoplankton, or biomanipulation, has been used in many lakes to improve water quality (Phillips & Moss, 1994). It can easily be used in isolated lakes, but in the Broads many lakes connected to the tidal river have navigation rights, and in these situations isolation would be difficult or inappropriate. However, an alternative restoration strategy might be to use biomanipulation in large temporary embayments, to reduce phytoplankton and establish aquatic vegetation. This approach has been tested in two broads linked to the river Bure; Hoveton Great Broad and Hoveton Little Broad. In Hoveton Little Broad, a large bay (5.5 ha), known as Pound End was partially isolated by building a simple water-permeable dam, where a spit of land (a peat baulk) almost separates it from the main lake basin. At Hoveton Great Broad, a smaller (1ha) fish exclosure was constructed using wooden posts and sheets of 5mm board made from recycled plastic. In both barriers, steel mesh (1mm) 'windows' were incorporated to enable tidally-induced water levels to equilibrate without allowing fish to enter. Fish were removed from both sites in the winter or early spring using a combination of netting and electro-fishing. The technique is highly effective and at least 90% of the fish are believed to have been removed although, to ensure fish populations remained low, further fish removals have been carried out each subsequent winter/spring.

Fish removal in Hoveton Great Broad was delayed until the spring of 1991, and an undetected breach in the barrier just below the sediment allowed small fish to enter the exclosure during 1991. These fish were removed during the winter of 1992/93 and as only a very small number of planktivorous fish were found in the exclosure in the spring of 1993, the barrier was clearly effective throughout 1992. Fish must have invaded the exclosure during 1993 as over 4600 roach (0+) and 53 bream (mean length 46.3 cm, which clearly could not have recruited from the remaining stock) were removed in the spring of 1994.

At Hoveton Little Broad, fish were removed in the spring of 1990. Unfortunately, the following winter (1990/91) the barrier failed, allowing fish to re-enter the exclosure and the need to reconstruct the barrier meant that these could not be removed until the spring of 1992. The following spring (1993), a considerable number of fish (175 kg ha^{-1}), mainly small roach and bream, were removed. It is not known when these fish re-entered the site, but the presence of *Daphnia* throughout the summer of 1992 (Stansfield et al., 1993) suggests that they probably entered during the winter when water flooded the low lying land separating the two water bodies. A follow-up survey in the spring of 1994 confirmed that planktivorous fish stocks during 1993 were low (c. 13 kg ha^{-1}) (Phillips et al., 1996), but subsequent spring and autumn surveys suggest that planktivorous fish populations increased during 1994 and 1995 (30 kg ha^{-1} and 72 kg ha^{-1} respectively).

Concentrations of phytoplankton chlorophyll *a*, total phosphorus and population density of *Daphnia* in the fish exclosures and open water of these sites are summarised in Table 4. No significant differences in total phosphorus concentrations were found between the fish exclosures and the open water at either site, despite obvious differences in water clarity. At Hoveton Little Broad, significantly larger numbers of *Daphnia* and lower chlorophyll *a* concentrations were found in Pound End in 1990, 1992 and 1993. These years corresponded to those when fish populations were thought to be low and can be attributed to the lack of fish predation.

During 1992, a similar result was obtained in the fish exclosure at Hoveton Great Broad. In 1993 and 1994, mean *Daphnia* numbers in the exclosure were greater than in the open water, although this did not significantly reduce the mean chlorophyll *a* concentration, as the populations only occurred during the early summer (Stansfield et al., 1997).

From these experiments, and observations in Alderfen Broad, where recent summer fish kills and poor recruitment allowed large bodied cladocera to develop (Perrow et al., 1994), it is clear that phytoplankton can be substantially reduced by zooplankton grazing. The difficulties of maintaining these grazers by managing the fish populations are all too clear from the above exmples, but the importance of these conclusions to lake restoration comes from a recognition of the key role of grazers in shallow lake ecosystems, particularly when submerged aquatic plants are found in the lake.

Table 4. Summer (April–September inclusive) mean values (± SD) for total phosphorus, chlorophyll *a* and total *Daphnia* in the open water and fish exclosures in Hoveton Great and Little Broads. Comparison between fish exclosures and open water was made by paired *t*-tests

	Total P (μg l^{-1})		Chlorophyll *a* (μg l^{-1})		Total *Daphnia* (animals l^{-1})	
	Exclosure	Open Water	Exclosure	Open Water	Exclosure	Open Water
a) Hoveton Great Broad						
1990	193 ± 98	215 ± 69	184 ± 181	181 ± 93	78 ± 103	8 ± 12
1991	135 ± 70	154 ± 76	102 ± 69	110 ± 72	11 ± 19	18 ± 34
1992	87 ± 28**	132 ± 41**	20 ± 13***	102 ± 50***	152 ± 138**	18 ± 34**
1993	103 ± 28**	123 ± 40	61 ± 29	94 ± 35	77 ± 148*	4 ± 10*
1994	85 ± 44	111 ± 51	43 ± 35	62 ± 53	114 ± 146*	19 ± 46*
1995	nd	87 ± 45	nd	68 ± 53	nd ±	53 ± 54
b) Hoveton Little Broad						
1990	161 ± 74	165 ± 90	52 ± 26	220 ± 57***	99 ± 46***	1 ± 0***
1991	124 ± 44	121 ± 34	90 ± 29	96 ± 23	19 ± 33	12 ± 24
1992	95 ± 39	83 ± 53	18 ± 19	107 ± 55***	106 ± 93**	17 ± 51**
1993	101 ± 43	73 ± 34	12 ± 9	83 ± 42***	109 ± 84***	28 ± 45***
1994	74 ± 34	50 ± 17	13 ± 8	58 ± 51	75 ± 75	50 ± 88
1995	81 ± 46	49 ± 19	19 ± 14	53 ± 50	48 ± 39	53 ± 71

* p<0.05
** p<0.01
*** p<0.001

In the majority of the broads, large cladoceran grazers are restricted to the spring and their population size appears to be largely controlled by fish predation. This contrasts with the zooplankton community found in Upton Broad which, unlike the other freshwater lakes in the Broads, has a mean summer chlorophyll *a* concentration below 25 mg l^{-1} and is still dominated by submerged vegetation. Here the zooplankton community is dominated by a succession of large bodied grazers. *Daphnia hyalina* forms the typical spring population but, *Ceriodaphnia quadrangula* and *Simocephalus vetulus* replace it as plant biomass begins to increase (Phillips et al., 1996). The plants appear to provide an important habitat for these alternative grazers and may offer a refuge from fish predation (Timms & Moss, 1984). However, the plants may also play a critical role in structuring the fish community in favour of the larger bodied cladoceran grazers, by providing habitat for piscivorous fish species. These directly reduce the population of planktivorous fish (Grimm & Backx, 1990) or their distribution (Persson & Eklov, 1995), and thus indirectly provide refugia for grazing cladocera. These effects may be occurring in Upton Broad as the planktivorous fish population is very low and the lake is dominated by the piscivorous species, pike *Esox lucius* and perch *Perca fluviatilis* (Phillips et al., 1996).

The interactions between zooplankton, piscivorous and planktivorous fish in the context of plant beds are complex and well reviewed in the context of the Broads by Perrow et al., (1997) and in this volume. Studies in Cromes Broad during 1993 (Phillips et al., 1996) demonstrated that planktivorous fish were attracted to plant beds, and that in less-dense beds the predation pressure on the zooplankton remained high. In denser beds, few planktivorous fish were found and, as cladoceran grazers were positively correlated with plant biomass in Cromes Broad (Phillips et al., 1996), it seems that dense plant beds provided a refuge from predation for these cladocera.

The association of cladoceran grazers with plant beds appears to be closely influenced by predation pressure. Where fish are moderate (< 0.45 m^{-2}), as at Pound End in 1994 and in Upton Broad, *Ceriodaphnia* are found in both the open water and plant beds. At the higher densities (c 0.8 m^{-2}) found in Cromes Broad during 1993 and 1994, *Ceriodaphnia* and *Simocephalus* were positively correlated with plant density (Phillips et al., 1996; Stansfield et al., 1997), but in the following autumn when fish densities increased (1.0

m^{-2}), cladoceran populations became almost extinct (Perrow et al., 1997).

From these observation in the Broads, it is clear that, when large cladoceran populations occur, phytoplankton populations can be significantly reduced. For these cladocerans to persist through the summer, the recruitment of 0+ fish needs to be poor. This can occur due to variation in weather, (e.g. spring temperature) and this may account for the variation in the annual mean population density of *Daphnia* in Barton Broad which explained part of the variation in chlorophyll *a* concentration in the broad (Phillips & Kerrison, 1990). Alternatively if aquatic plants produce sufficiently dense stands they may provide a refuge for *Daphnia* or other grazers, provided the density of planktivorous fish is not excessively large. The interaction between plants and piscivores is a further twist to this complex series of relationships, but what becomes clear is that in shallow lakes plants have a key role in stabilising the community structure of the lake.

This has been widely recognised, although the reason has not always been linked to refuges for zooplankton grazers, as other factors such as competition for nutrients (Blindlow et al., 1993; Van Donk et al., 1993) or de-nitrification (Meijer et al., 1990) may also be important. The concept of multiple stable states (Scheffer et al., 1993) where either phytoplankton or macrophyte dominated systems may exist over a range of nutrient conditions (Jeppesen et al., 1990) may be relevant. Figure 10 shows the relationship between total phosphorus and chlorophyll *a* for a number of sites in the Broads. In lakes where fish populations are dominated by small planktivorous species and there is no aquatic vegetation, such as Barton Broad, Hoveton Great and Hoveton Little Broads and in Cockshoot Broad prior to biomanipulation, the data can be fitted to a power function (Table 5). In these situations, phosphorus is apparently limiting the growth of phytoplankton as the exponent of each of the functions are close to unity. In contrast, there is either no relationship between chlorophyll *a* and total phosphorus (Alderfen, Upton Broad) or the exponent is low (Cockshoot when fish are low), and in these sites grazing rather than phosphorus is believed to be controlling phytoplankton growth (Perrow et al., 1994; Moss et al., 1996; Phillips et al., 1996). In all of these sites the chlorophyll *a* concentration, relative to the total phosphorus, is lower, and a similar result can be seen in the years when biomanipulation in Pound End was successful (Figure 10).

General conclusions

From the work carried out in the Broads over the last 20 years, we can draw a number of general conclusions which are important for the management of these and other shallow lakes.

Firstly, the reduction of phosphorus discharged to Barton Broad has clearly reduced phytoplankton chlorophyll *a*. However, the impact of phosphorus released from the sediment reduces the effectiveness of controls applied in the catchment. The impact of sediment release is magnified when river flow is low, as in droughts or at sites where isolation has reduced the lake's natural catchment. In the latter situation, even small amounts of phosphorus can be annually recycled, and although sediment removal can remove a substantial proportion of this phosphorus, the lake will probably still experience a substantial internal phosphorus load.

Secondly, the concept of multiple stable states in lakes (Scheffer et al., 1993) can clearly be seen operating in the Broads. Phytoplankton chlorophyll *a* can be substantially reduced if planktivorous fish populations are reduced, either artificially by biomanipulation (Pound End, Cockshoot and Alderfen post-1994), as a result of mortality (Alderfen), or by piscivorous predation (Upton). In clear water, submerged aquatic plants appear to play a key stabilising role by providing refugia for grazing cladocera and by creating the habitat required to sustain a fish community less dominated by small planktivorous species. The exact conditions that these plants require is still uncertain. At Upton Broad, where phosphorus concentrations are low, the plant population is relatively stable. Sustained clear water in Cockshoot Broad appears to have enabled a diverse community to re-establish. However, at Alderfen Broad, where phosphorus concentrations are much higher, the dominant plant species *Ceratophyllum demersum* appears to undergo long term cyclical fluctuations. Zooplankton grazing can clearly still limit algal growth in Alderfen Broad, but the lake appears to show that the instability predicted by Jeppesen et al., (1991) occurs at high phosphorus concentrations.

Thus the restoration of these shallow lakes can no longer be seen as a simple process of reducing phosphorus inputs from the catchment, but as a series of steps.

1. Reduce the phosphorus input from the catchment to minimise the addition of new phosphorus to the system.

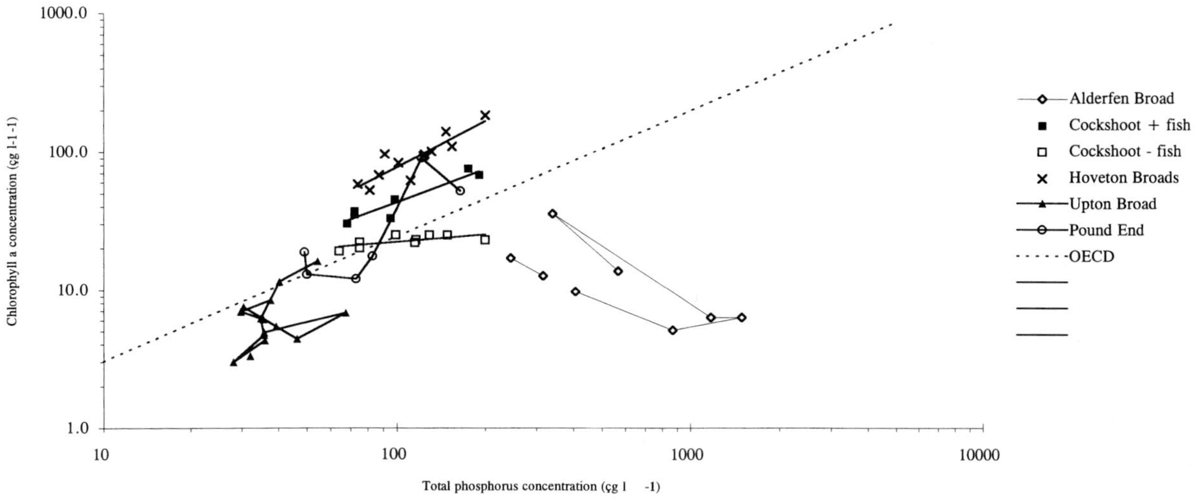

Figure 10. Mean growing season (April–September inc.) phosphorus and chlorophyll *a* concentrations in a variety of broads. Alderfen and Cockshoot Broads (1979–1995), Hoveton Little Broad (1983–1995), Hoveton Great Broad (1990–1995) and Pound End (1990–1995). OECD regression shown for comparison (Vollenweider & Kerekes, 1982).

Table 5. Regression analysis results relating mean chlorophyll *a* and total phosphorus concentrations in sites where planktivorous fish are high or low

Site	Function	R^2	Planktivorous fish
Barton Broad	$\mathrm{CHl}a_y = 2.24\mathrm{TP}_y^{0.8\pm0.13}$	0.68***	high
Hoveton Gt & Lt Broads	$\mathrm{CHl}a_s = 0.37\mathrm{TP}_s^{1.1\pm0.18}$	0.80***	high
Cockshoot Broad when fish population known to be high (cf Moss et al., 1996)	$\mathrm{CHl}a_s = 1.06\mathrm{TP}_s^{0.8\pm0.13}$	0.89***	high
Cockshoot Broad when fish population known to be low (cf Moss et al., 1996)	$\mathrm{CHl}a_s = 9.54\mathrm{TP}_s^{0.2\pm0.07}$	0.47*	low (bio-manipulation)
Upton Broad	not significant	0.21	variable
OECD (Vollenweider & Kerekes, 1982)	$\mathrm{CHl}a_s = 0.37\mathrm{TP}_s^{0.91}$	0.76	

2. Reduce the internal sediment phosphorus load. This step remains relatively uncertain, but in well-flushed systems it is likely to significantly increase the rate that phosphorus in the lake can be reduced.
3. Biomanipulate to create clear water conditions to enable submerged vegetation to become established.
4. Develop a diverse plant and animal community which stabilises the fish community. This final step is the least well understood. General principles have been established, but further research and experimental management will be required before firm conclusions can be drawn.

The work carried out in the Broads over the last 25 years, has involved many people and considerable expenditure. It has reaped many benefits, both in the dramatic improvement of sites such as Cockshoot Broad, and in the steady improvement of larger lakes such as Barton Broad. It has revealed information about the way ecosystems respond to external influences. While they can be very robust, their collapse can be sudden and catastrophic. Once the complexity of the shallow lake ecosystem is lost, it is very difficult to re-create and as a result, the difficulties of shallow lake restoration should not be underestimated. Prevention is better than cure, and it is essential that remaining pristine lakes are given adequate protection.

Acknowledgements

This work has been carried out over the last 25 years as part of an on-going project to restore the Norfolk Broads. It has been supported by range of funding bodies, including Anglian Water, the former Nature Conservancy Council, the Soap and Detergent Industry Association. Throughout this period there has been close cooperation with many organisations, particularly the Broads Authority who have played a major role in the development of the experimental techniques. Much of the work described in this paper results from a joint project carried out by the Broads Authority and the National Rivers Authority (now the Environment Agency) with assistance from the European Commission (LIFE 92-3/UK/031). The views expressed in this paper are those of the authors and not necessarily those of any of the above organisations.

References

Bengtsson, L., G. Fleischer, G. Lindmark & W. Ripl, 1975. Lake Trummen restoration project I. Water and sediment chemistry. Verh. int. Ver. Limnol. 19: 1080–1087.

Blindlow, I., G. Andersson, A. Hargeby & S. Johansson, 1993. Long-term pattern of alternative stable states in two shallow eutrophic lakes. Freshwat. Biol. 30: 159–167.

Carpenter, S. R. & J. F. Kitchell (eds), 1993. The Trophic Cascade in Lakes. Cambridge University Press, Cambridge.

Cryer, M., G. Peirson & C. Townsend, 1986. Reciprocal interactions between roach, *Rutilus rutilus*, and zooplankton in a small lake: prey dynamics and fish growth and recruitment. Limnol. Oceanogr. 31: 1022–1038.

Edmondson, W. T., 1977. Recovery of Lake Washington from eutrophication. In J. Cairns, K. L. Dickson & E. E. Henricks (eds), Recovery and Restoration of Damaged Ecosystems. University Press of Virginia, Charlottesville.

Edmondson, W. T. & S. E. B. Abella, 1988. Unplanned biomanipulation of Lake Washington, Limnologica (Berlin) 19: 73–79.

Grimm, M. P. & J. J. Backx, 1990. The restoration of shallow eutrophic lakes, and the role of norther pike, aquatic vegetation and nutrient concentration. Hydrobiologia 200/201: 557–566.

Gulati, R. D., E. H. R. R. Lammens, M.-L. Meijer & E. van Donk (eds), 1990. Biomanipulation – Tool for Water Management. Developments in Hydrobiology 61. Kluwer Academic Publishers, Dordrecht, 628 pp. Reprinted from Hydrobiologia 200/201.

Harper, D. M., 1992. Eutrophication of Freshwaters: Principles, Problems and Restoration. Chapman & Hall, London.

Holdren, G. C. & D. E. Armstrong, 1986. Interstitial iron concentration as an indicator of phosphorus release and mineral formation in lake sediments. In P. G. Sly (ed.), Sediments and Water Interactions. Springer, New York: 133–147.

Irvine, K., M. T. Bales, B. Moss, J. H. Stansfield & D. Snook, 1993. The changing ecosystem of a shallow brackish lake, Hickling Broad, Norfolk, U.K. I Trophic relationships with special reference to the role of *Neomysis integer* (Leach). Freshwat. Biol. 29: 119–139.

Jackson, R., 1991. The extent of phosphorus release in the Norfolk Broads. Progress report for period 1989–1990, Research Project 518, National Rivers Authority, Ipswich.

Jeppesen, E., J. P. Jensen, P. Kristensen, M. Sondeergaard, E. S. Mortensen & K. Olrik, 1990. Fish manipulations as a lake restoration tool in shallow, eutrophic temperate lakes 2: threshold levels, long-term stability and conclusions. Hydrobiologia 200/201: 219–227.

Jeppesen, E., P. Kristensen, J. P. Jensen, M. Sondergaard, E. Mortensen & T. Lauridsen, 1991. Recovery following a reduction in external phosphorus loading of shallow eutrophic Danish lakes: duration, regulating factors and methods of overcoming resilience. Mem. Inst. ital. Idrobiol. 48: 127–148.

Johnes, P. J., 1996. Nutrient export modelling – River Bure, Norfolk. Anglian Region Operational Investigation Report 581, Environment Agency, Peterborough.

Kerrison, P., B. McEwan, B. Crook & G. L. Phillips, 1989. Redox contolled chemical dosing efficiently removes phosphorus from effluents discharged to the Norfolk Broads. Journal of the Instiitiute of Water and Environmental Management 3: 397–403.

Marsden, M. W., 1989. Lake restoration by reducing external phosphorus loading: the influence of sediment phosphorus release. Freshwat. Biol. 21: 139–162.

Mason, C. F., 1976. Broadland. In Nature in Norfolk, a Heritage in Trust. Jarrold and Sons Ltd, Norwich.

Mason, C. & R. J. Bryant, 1975. Changes in the ecology of the Norfolk Broads. Freshwat. Biol. 5: 257–270.

Meijer, M-L., M. W. de Haan, A. W. Breukelaarand & H. Buiteveld, 1990. Is reduction of the benthivorous fish an important cause of high transparency following biomanipulation in shallow lakes? Hydrobiologia 200/201: 303–315.

Morgan, N. C., 1972. Problems of the conservation of freshwater ecosystems. Symposium of the Zoological Society of London 29: 135–154.

Mortimer, C. H., 1941. The exchange of dissolved substances between mud and water in lakes I. J. Ecol. 29: 280–239.

Mortimer, C. H., 1942. The exchange of dissolved substances between mud and water in lakes II. J. Ecol. 30: 147–201.

Moss, B., 1990. Engineering and biological approaches to the restortion from eutrophication of shallow lakes in which aquatic plant communities are important components. Hydrobiologia 120/121: 367–378.

Moss, B. 1983. The Norfolk Broadland: Experiments in the restoration of a complex wetland. Biol. Rev. 58: 521–561.

Moss, B., H. Balls, I. Booker, K. Manson & M. Timms, 1988. Problems in the construction of a nutrient budget for the R. Bure and its Broads (Norfolk) prior to its restoration from eutrophication. In F. E. Round (ed.), Algae and the Aquatic Environment. Biopress, Bristol.

Moss B., R. T. Leah & B. Clough, 1979. Problems of the Norfolk Broads and their impact on freshwater fisheries. Proceedings of the First British Freshwater Fisheries Conference, University of Liverpool, Liverpool: 67–85.

Moss, B., H. Balls, K. Irvine & J. Stansfield, 1986. Restoration of two lowland lakes by isolation from nutrient-rich water sources with and without the removal of sediment. J. Appl. Ecol. 23: 391–414.

Moss, B., J. Stansfield & K. Irvine, 1990. Problems in the restoration of a hypertrophic lake by diversion of a nutrient- rich inflow. Verh. int. Ver. Limnol. 24: 568–572.

Moss, B., J. Stansfield, K. Irvine, M. Perrow & G. L. Phillips, 1996. Progressive restoration of a shallow lake; a 12-year experiment in isolation, sediment removal and biomanipulation. J. Appl. Ecol. 33: 71–86.

Moss, B., J. Madgwick & G. L. Phillips, 1997. A guide to the restoration of nutrient-enriched shallow lakes. Broads Authority and Environment Agency, Norwich.

Naumann, E., 1919. Några Synpunkter Anagående limnoplanktons ökologi med särskild hänsyn till fytoplankton. Svensk Botanisk Tidskrift. 13: 129–163.

Osborne, P. L., 1978. Relationships between the phytoplankton and nutrients in the River Ant and Barton, Sutton and Stalham Broads, Norfolk. Ph.D. Thesis, University of East Anglia, Norwich.

Osborne, P. L., 1981. Phosphorus and nitrogen budgets of Barton Broad and predicted effects of a reduction in nutrient loading on phytoplankton biomass in Barton, Sutton and Stalham Broads, Norfolk, United Kingdom. Int. Rev. ges. Hydrobiol. 66: 171–202.

Pearsall, W. H., 1921. The development of vegetation in English lakes, considered in relation to the general evolution of glacial lakes and rock basins. Proceedings of the Royal Society of London B 92: 259–284.

Perrow, M. R. & K. Irvine, 1992. The relationship between cladoceran body size and the growth of underyearling roach (Rutilus rutilus) (L.) in two shallow lakes: a mechanism for density-dependent reductions in growth. Hydrobiologia 241: 155–161.

Perrow, M. R., 1994. Practical aspects of the biomanipulation of fish populations. In J. Pitt & G. L. Phillips (eds), The Development of Biomanipulation Techniques and Control of Phosphorus Release from Sediments. Interim Report 475/ 1/A LIFE Project (92-3/UK/031) - Restoration of the Norfolk Broads. National Rivers Authority, Bristol.

Perrow, M. R., B. Moss & J. Stansfield, 1994. Trophic interactions in a shallow lake following a reduction in nutrient loading: a long-term study. Hydrobiologia 275/276: 43–52.

Perrow, M. R., J. Stansfield, A. J. D. Jowitt & L. D. Tench, (1997). Mechanisms Controlling the Stability of Restored Broads. In F. J. Madgwick & G. L. Phillips (eds), Restoration of the Norfolk Broads – Final Report BARS 14b LIFE Project 92-93/UK/031. Broads Authority and Environment Agency, Norwich.

Persson, L. & P. Eklov, 1995. Prey refuges affecting interactions between piscivorous perch and juvenile perch and roach. Ecology 76: 70–81.

Peterson, S. A., 1981. Sediment Removal as a Lake Restoration Technique. EPA 600/3-81-013, Corvallis.

Phillips, G. L., 1977. The mineral nutrient levels in three Norfolk Broads differing in trophic status, and an annual mineral content budget for one of them. J. Ecol. 65: 447–474.

Phillips, G. L., 1984. A large scale field experiment in the control of eutrophication in the Norfolk Broads. Journal of Institute of Water Pollution Control 83: 400–408.

Phillips, G. L. & B. Moss, 1978. The distribution, biomass and productivity of submerged aquatic macrophytes in the Thurne Broads, Norfolk, 1975–1977. Final report to the Nature Conservancy Council. University of East Anglia, Norwich.

Phillips, G. L., D. Eminson & B. Moss, 1978. A mechanism to account for macrophyte decline in progressively eutrophicated freshwaters. Aquat. Bot. 4: 103–126.

Phillips, G. L. & R. Jackson, 1990. The control of eutrophication in very shallow lakes, the Norfolk Broads. Verh. int. Ver. Limnol. 24: 573–575.

Phillips, G. L. & P. Kerrison, 1991. The restoration of the Norfolk Broads: The role of biomanipulation. Mem. Ist. ital. Idrobiol. 48: 75–97.

Phillips, G. L. & B. Moss, 1994. Is biomanipulation a useful technique in lake management? R & D Note 276, National Rives Authority, Bristol.

Phillips, G. L., R. Jackson, C. Bennett & A. Chilvers, 1994. The importance of sediment phosphorus release in the restoration of very shallow lakes (The Norfolk Broads, England) and implications for biomanipulation. Hydrobiologia 275/276: 445–456.

Phillips. G. L., M. R. Perrow & J. Stansfield, 1996. Manipulating the fish-zooplankton ineraction in shallow lakes; a tool for restoration. In S. P. R. Greenstreet & M. L. Tasker (eds), Aquatic Predators and their Prey. Fishing News Books, Oxford.

Pitt, J., A. Kelly & G. L. Phillips, (1997). Control of nutrient release from sediments. In F. J. Madgwick & G. L. Phillips (eds), Restoration of the Norfolk Broads – Final Report BARS 14b LIFE Project 92-93/UK/031. Broads Authority and Environment Agency, Norwich.

Pokorny, J. & V. Hauser, 1994. Restoration of lakes through sediment removal – Vajgar fish pond, Czech Republic, in Restoration of Lake Ecosystems: a Holistic Approach. In M. Eiseltova (ed.), Restoration of Lakes. Publication No 32, IWRB, Slimbridge, Gloucestershire: 141–153.

Quaak, M., J. Van der Does, P. Boers & J. Van der Vlugt, 1993. A new technique to reduce internal phosphorus loading by in-lake phosphate fixation in shallow lakes. Hydrobiologia 253: 337–344.

Reynolds, C. S., 1992. Eutrophication and the management of planktonic algae: what Vollenweider couldn't tell us. In D. W.

Sutcliffe & J. G. Jones (eds), Eutrophication: Research and Application to Water Supply. Freshwater Biological Association, Ambleside: 4–29.

Sas, H., 1989. Lake restoration by reduction of nutrient loadings: expectations, experiences, extrapolations. Academia Verlag Richarz, Sankt Augustin.

Scheffer, M., S. H. Hosper, M.-L. Meijer, B. Moss & E. Jeppesen, 1993. Alternative equilibria in shallow lakes. Trends in Ecology and Evolution 8: 275–279.

Stansfield, J., S. Caswell & M. Perrow, 1997. Biomanipulation as a restoration tool. In F. J. Madgwick & G. L. Phillips (eds), Restoration of the Norfolk Broads – Final Report BARS 14b LIFE Project 92-93/UK/031. Broads Authority and Environment Agency, Norwich.

Stansfield, J., B. Moss, C. Bennett & G. L. Phillips, 1993. Restoration of Broadland – The role of biomanipulation. Anglian Region Operational Investigation 528:, National Rivers Authority, Bristol.

Stansfield, J. H., M. R. Perrow, L. D. Tench, A. J. D Jowitt & A. L. L. Taylor, 1997. Submerged macrophytes as refuges for grazing Cladocera against fish predation: observations on seasonal changes in relation to macrophyte cover and predation pressure. Hydrobiologia 342/343 (Dev. Hydrobiol. 119): 229–240.

Thienemann, A., 1918. Untersuchungen über dir Beziehungen zwischen der Sauerstoffgehält des Wassers und der Zusammensetzung der Fauna in norddeutschen Seen. Arch. Hydrobiol. 12: 1–24.

Timms, R. M. & B. Moss, 1984. Prevention of growth of potentially dense phytoplankton populations by zooplankton grazing in the prescence of zooplanktivorous fish in a shallow wetland ecosystem. Limnol. Oceanogr. 29: 472–486.

Van Donk, E., R. D. Gulati, A. Iedema & J. T. Meulemans, 1993. Macrophyte-related shifts in the nitrogen and phosphorus contents of the different trophic levels in a biomanipulated shallow lake. Hydrobiologia 251: 19–26.

Vollenweider, R. A. & J. Kerekes, 1982. The loading concept as basis for controlling eutrophication philosophy and preliminary results of the OECD programme on eutrophication. Progress in Water Technology 12: 5–38.

Hydrobiologia **395/396:** 77–85, 1999.
D.M. Harper, B. Brierley, A.J.D. Ferguson & G. Phillips (eds), The Ecological Bases for Lake and Reservoir Management.

Multiple techniques for lake restoration

Heléne Annadotter[1], Gertrud Cronberg[1], Richard Aagren[2], Bejron Lundstedt[2], Per-Åke Nilsson[2] & Sven Ströbeck[2]
[1]*Institute of Ecology, Department of Limnology, Lund University, Ecology Building, S-223 62 Lund, Sweden*
[2]*Municipal Services Department, Municipality of Hässleholm, Stadshuset, S-281 80 Hässleholm, Sweden*

Key words: shallow lakes, restoration, artificial wetland, biomanipulation

Abstract

Lake Finjasjön is a shallow, eutrophic lake (area 1100 ha, mean depth 3 m, maximum depth 13 m) in southern Sweden. In the 1920s, the lake was clear, with a summer Secci depth of about 2 m. During the first half of the 20th century, untreated sewage from the town polluted the lake. In the 1930s, the lake began to show eutrophic characteristics, and in the 1940s, the cyanobacterium *Gloetrichia echinulata* dominated in summer. In 1949, the first municipal sewage treatment plant was built. The treatment was, however, insufficient, since the lake continued to be the recipient of the effluent with the result that the occurrence of cyanobacteria became more frequent. Species such as *Microcystis* and *Anabaena* caused skin rash and allergic symptoms among swimmers. The phosphorus load on Lake Finjasjön increased as the population of Hässleholm grew and reached a peak value of 65 tons annum^{-1} in 1965. In 1977, the sewage plant was rebuilt to include chemical flocculation, reducing the total external phosphorus load to about 5 tons annum^{-1}. Despite this improvement the lake did not recover from its chronic and toxic cyanobacterial blooms. Phosphorus-leaking black sediments were identified as the cause of the lake's failure to recover. Some 60% of lakebed area is covered with sediments on average 3 m thick. Dredging the sediments was started on a large scale in 1987. Five years later, 25% of the sediment area had been removed but the dredging was stopped since phosphorus continued to be released into the water from these areas. In 1992, a new restoration policy, a combination of further reduced external nutrient loading and food-web manipulation was initiated. A constructed wetland (30 ha) to reduce phosphorus and nitrogen was created in connection to the effluent from the sewage treatment plant. Protection zones along the feeder streams into Lake Finjasjön were also established. A cyprinid reduction programme by trawling was carried out between 1992 and 1994. When it started, the fish community was composed of 90–95% bream and roach. After two years of trawling, the ratio between piscivorous and planktivorous fish was 1:1. In 1994 and 1995, the transparency increased due to a considerably reduced biomass of phytoplankton and a radically altered phytoplankton community. The monoculture of *Microcystis* was replaced by a diverse phytoplankton community. The increased transparency made possible the development of submerged macrophytes such as *Elodea, Myriophyllum* and *Potamogeton*. The internal loading of phosphorus decreased dramatically in 1994 and 1995, possibly as a result of reduced sedimentation of phytoplankton.

Introduction

Lake Finjasjön (latitude 56° 08′, longitude 13° 42′, 43.2 m above sea level) is a eutrophic, shallow lake in southern Sweden, situated near the town of Hässleholm. It has a mean depth of 3 m, a maximum depth of 12 m and an area of 1110 ha. The retention time is three months. The lake has five feeder streams and one outflow (Figure 1). The catchment area of 260 km^2 is covered by mostly coniferous and deciduous forests with about 10–15% used for agriculture.

In the 1920s, the lake was clear, with a summer transparency of about 2 m. During the first half of the 20th century, untreated sewage from the town of Hässleholm started to pollute the lake. In the 1940s, the cyanobacterium *Gloeotrichia echinulata* domin-

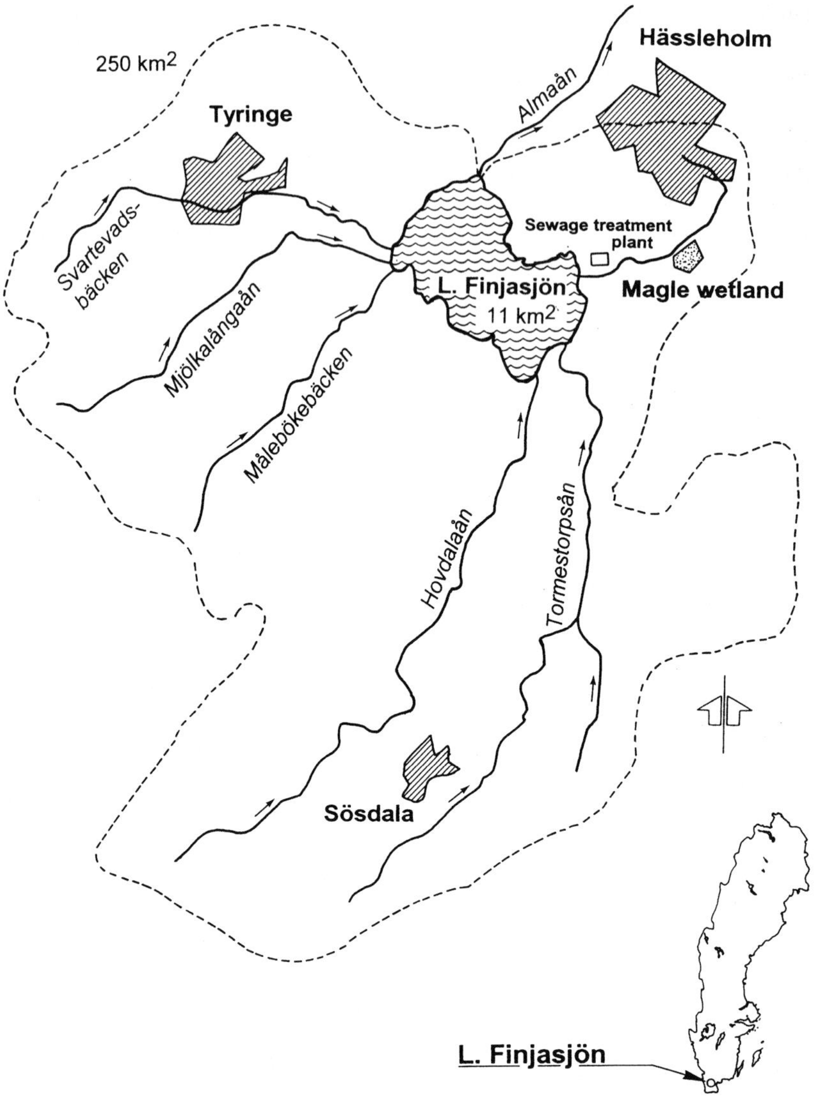

Figure 1. The catchment area of Lake Finjasjön, Sweden. The lake is the recipient for treated sewage water from the sewage treatment plant of Hässleholm. Before entering the lake, the treated sewage water is further treated by flowing through a constructed wetland.

ated during the summer. The bloom was occasionally heavy, making it necessary to prohibit swimming.

In 1949, the first municipal sewage treatment plant, equipped with a mechanical treatment step, was built (Figure 1). However, the blooms did not cease. The treatment was apparently insufficient and the lake was still the recipient of its effluent. Instead of disappearing, the occurrence of cyanobacteria became more frequent. Swimmers suffered from skin rashes and allergic symptoms caused by the increased biomass of the potentially toxic species *Microcystis* and *Anabaena* (Annadotter, 1993).

Despite the fact that the sewage treatment plant was improved in 1963 with an activated sludge step, the phosphorus load on Lake Finjasjön continued to increase due to growth of the population of Hässleholm, and reached a peak value of 65 tons annum^{-1} in 1965 (Figure 2).

At the beginning of the 1970s, the municipality wanted to relocate the outlet of the sewage treatment plant from the lake to its outflow, in order to improve the water quality of the lake. However, the county board of Kristianstad did not permit the municipality to procede with this plan, beacuse, among other things, it was considered unethical to discharge sewage downstream of the lake, and thus encumber the villages situated there. A further argument for not shifting the location of the effluent from the lake was that the lake water dilutes the sewage and a certain amount of purification was assumed to take place within the lake.

In 1977, the sewage plant was rebuilt with chemical flocculation and a three-media filter, reducing the phosphorus load to about 5 tons annum^{-1} (Figure 2). Despite this improvement, the lake did not recover from its chronic cyanobacterial blooms and, quite contrary to expectation, as the phosphorus load decreased dramatically the content of chlorophyll a increased. Diving ducks – goldeneye (*Bucephala clangula*) and pochard (*Aythya Ferina*) – disappeared from the lake when the transparency decreased towards the end of the seventies (Figure 3). During the 1980s, the lake became more hypertrophic and the politicians of the Hässleholm municipality became desperate about its condition, because it also served as the drinking water supply to 25 000 consumers.

A political consensus that the lake had to be restored existed and several limnological investigations were initiated. The municipality of Hässleholm started the lake's restoration in the middle of the 1980s, using the slogan: *'Lake Finjasjön – possible to swim in, in our time, for our children'*. Phosphorus release from anaerobic sediments was identified as the cause of the failure of the lake to recover from reduced loading and so dredging of the sediments was one recommended measure (Löfgren & Forsberg, 1984).

Another method of decreasing the internal phosphorus release was suggested by Ripl & Leonardsson (1980). They investigated the denitrification capacity of the sediments. By adding calcium nitrate to the sediments, they found that the speed of the denitrification process at 12 m depth was double that of 7 m. Efficient phosphorus retention occurred as a result of the denitrification. In systems without nitrate addition, high phosphorus release occurred, especially from the sediments at 12 m depth. Based on their results, they suggested that the municipality should add nitrified sewage effluent to the lake at depths greater than 8 m. They assumed that this step would cause denitrification at the sediment surface, thereby considerably

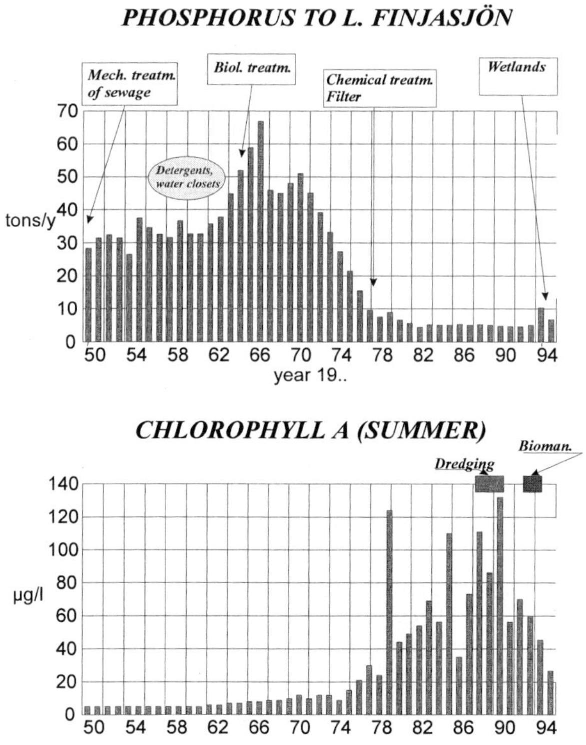

Figure 2. The external phosphorus load to Lake Finjasjön 1950–1995 and the summer values of chlorophyll *a* in Lake Finjasjön 1950–1994.

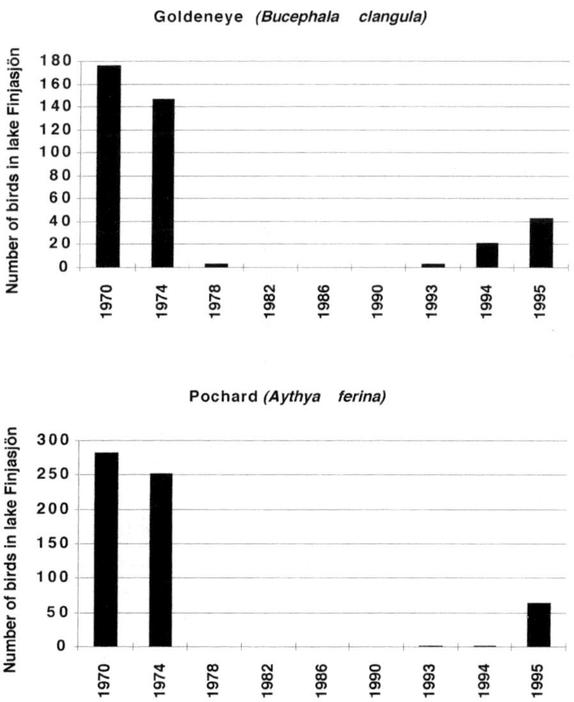

Figure 3. Numbers of goldeneye (*Bucephala clangula*) and pochard (*Aythya ferina*) in Lake Finjasjön 1970–1995.

enhancing sediment phosphorus. Injection of nitrate solution into the sediments ('Riplox'-treatment) was recommended as a complementary measure.

Faced with the choice of these two suggested methods, the Municipality decided to restore the lake by dredging. The politicians did not accept the idea of adding sewage water and nitrate to an already sewage-loaded and over-fertilised lake, whereas dredging the mud sounded more logical.

Sediments cover 60% of the bottom of the lake, on average to a depth of 3 m. Investigations of the sediments showed that the top layer, about 0.5 m thick, had a 30% higher phosphorus concentration and a slightly higher phosphorus/iron ratio than the rest of the sediment profile. The municipality was advised to remove this top layer (Löfgren, 1991) although, after the dredging, a 2 m thick layer of dark, phosphorus-rich sediment still remained.

Suction-dredging of the sediments was started on a large scale in 1987, with the purpose of dredging the whole area of 6.6 square kilometres of sediment. The cost was estimated as at least 100 million Swedish crowns, (SEK), approximately £10 million sterling. Ten sedimentation dams were constructed adjacent to the lake for the purpose of separating the water from the sediment. Three years later, several aspects of the restoration strategy were strongly criticised (Annadotter, 1990), mainly because of the small difference in sediment chemistry between dredged and non-dredged sediments and a lack of awareness of the catchment nutrient load on the lake.

With the purpose of evaluating the success of suction dredging, an investigation of the sediments was carried out in 1990 and 1991. Contents of total phosphorus and iron in dredged and non-dredged sediments were compared. Dredged areas contained 30% less total phosphorus (Nilsson, 1991) and 15% less phosphorus in the pore water, differences which were not considered to be sufficiently great. The sediments remaining after dredging were still typical of sewage-loaded, dark sediments with a low phosphorus retention capacity. There is no explicit relationship between the trophic level of a lake and the phosphorus content of sediments, while the phosphorus concentration in the pore water is believed to reflect the trophic state of a lake (Håkansson & Jansson, 1983). This showed that continued dredging would not produce a clear water lake. By 1991, the dredging had cost about 50 million SEK, (£5 million sterling). Despite the removal of 134 tons of phosphorus from the bottom of Lake Finjasjön, the dredging had not diminished the extent of cyanobacterial blooms.

Between 1987 and 1993, the cyanobacterial toxicity of Lake Finjasjön was found to vary strongly from year to year, as well as over the course of a bloom season (Annadotter, 1993; Annadotter, unpubl.). Because of the occurrence of these toxic species, swimming was still prohibited in Lake Finjasjön, although the lake was still used for drinking water. Cows that drank water from the outflow of the lake suffered liver damage, associated with hepatotoxic blooms, which cows drinking well water did not develop. (Annadotter et al., 1990). The cyanobacteria in the photic zone were toxic, while those sampled below this zone were not (Annadotter et al., 1993). It was unfortunate that the intake to the water works was situated at the outflow of the lake, thus pumping surface water with the highest algal concentration and the highest toxicity to the water works. Apart from the fear of distributing toxic drinking water to consumers, algae caused technical problems, including blocked filters in the water treatment plant.

Thus, several reasons for continuing the restoration of Lake Finjasjön existed even though it was evident that dredging was an inappropriate method. However, an alternative strategy had been suggested in 1990, when the external nutrient load on Lake Finjasjön was evaluated and found to be substantial using Vollenweiders phosphorus loading model (Vollenweider, 1976; Nilsson, 1990). To investigate whether the lake would be suitable for attempted restoration by biomanipulation, a standardized test fishery was carried out, which showed that the fish community was dominated by cyprinid fish (72% cyprinids, 28% piscivorous fish) (Collvin & Månsson, 1990) (Figure 4).

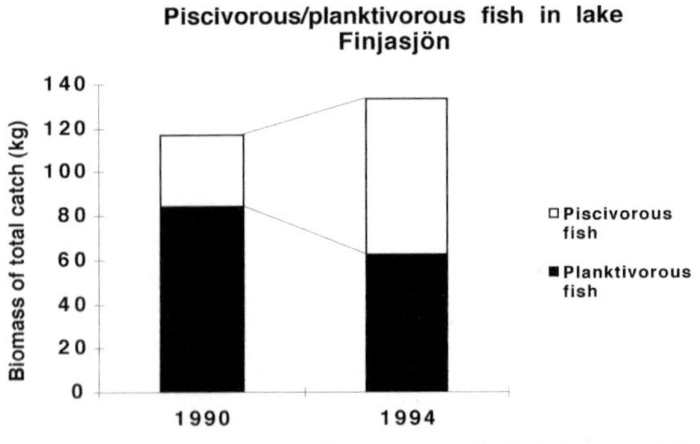

Figure 4. The ratio between piscivorous and planktivorous fish in Lake Finjasjön 1990 and 1994 (before and after the fish reduction).

The new restoration strategy

A new strategy was formulated, based on the idea that it was possible to reduce the internal phosphorus loading, which was still the main problem of Lake Finjasjön, without removing the sediments (Annadotter et al., 1992). This was based upon the following two strategies:

1. The creation of a 'top-down' effect on the phytoplankton by manipulating the food web through cyprinid reduction.
2. Further reduction of external nutrient loading, both from the catchment and from the sewage treatment works.

Different scientists had various opinions about the exact cause of the internal phosphorus loading (e.g. low redox conditions, high pH, sedimentation of algae on the sediment surface). Since all these conditions were caused by a high phytoplankton biomass, the main focus of the new restoration strategy was to minimize this biomass in spring and early summer in order to prevent the subsequent internal phosphorus loading.

In 1992, suction dredging was discontinued. The new restoration strategy was a political minority decision. The former political consensus was broken and the new restoration strategy was strongly questioned by sceptics from the very beginning. Despite the fact that no scientist recommended further dredging of Lake Finjasjön, several politicians and civil servants in the municipality wanted to continue the sediment pumping.

The cyprinid reduction was carried out from 1992 until 1994 by trawling. Two boats were specially designed for trawling in Lake Finjasjön and equipped with a sorting table. The boats worked together using a trawl of 7 × 30 m in the pelagic zone. After one year, it was evident that their capacity was not sufficient. Therefore, in the autumn of 1993, two smaller boats were set to trawl in the littoral zone concurrently with the larger, pelagic trawlers. All the predatory fish caught were released alive into the lake as soon as they had been sorted.

The theoretical catchment phosphorus load to Lake Finjasjön results in a lake total phosphorus concentration of approximately 40 $\mu g\ l^{-1}$. In order to decrease the external load, the most important streams were identified by Krug & Walker (1991) and Vought & Lacoursiére (1992, 1994), and agreements with the landowners drawn up, to give 5 m wide buffer zones which will not be fertilised, ploughed or used for agriculture. The cost of these was 3000 SEK (£300) ha ann^{-1}.

A further reduction of the phosphorus from the sewage treatment works was necessary. The yearly external phosphorus supply to Lake Finjasjön was 5 tons, of which 1 ton came from the sewage treatment plant. While the phosphorus coming from the feeder streams, to a large extent appeared as particulate phosphorus, the phosphorus from the sewage treatment plant appeared as PO_4-P, easily available to phytoplankton. In the summer, when several of the feeder streams had low flow rates, the effluent from the sewage treatment plant became an important and constant inflow with a PO_4-P concentration of about 150 $\mu g\ L^{-1}$, amounting to about 40% of total load.

A 30 ha wetland was constructed to further reduce the phosphorus and nitrogen in the effluent from the sewage treatment plant (Figures 1 and 5).

The success of these measures was evaluated through monitoring of water quality, plankton and fish. Samples for physico-chemical and plankton analyses were taken weekly from the deepest part of the lake. Plankton was sampled once a month in 1992 and 1993, and weekly in 1994 and 1995. Test fishing was carried out according to a standardised test fishing method with survey nets, developed at the Institute of Freshwater Research of the Swedish National Board of Fisheries (Nyberg & Degerman, 1988).

Results of the strategy

By the autumn of 1994, when fish catch weights had been reduced to 20% of initial ones catches, 400 tons of cyprinids had been removed. The ratio between piscivorous and planktivorous fish in test fishing with benthic survey nets was approximately 1:1 (Figure 4), which implies a balance in the fish community (Jeppesen pers. comm.). At this point the fish reduction was stopped.

Evident signs that the fish reduction had had an effect on the plankton community were observed in the summer of 1994. High peaks of daphnids, *Daphnia cucullata* and *Daphnia galeata*, appeared in July (Figure 6), a new phenomenon interpreted as the consequence of reduced predation pressure on large cladocerans. The change in zooplankton community during the season was accompanied by changes in the phytoplankton community. The toxic and chronic *Microcystis* blooms were replaced by a diverse phyto-

Figure 5. Aerial view of the constructed wetland in Hässleholm. Photo: Heléne Annadotter.

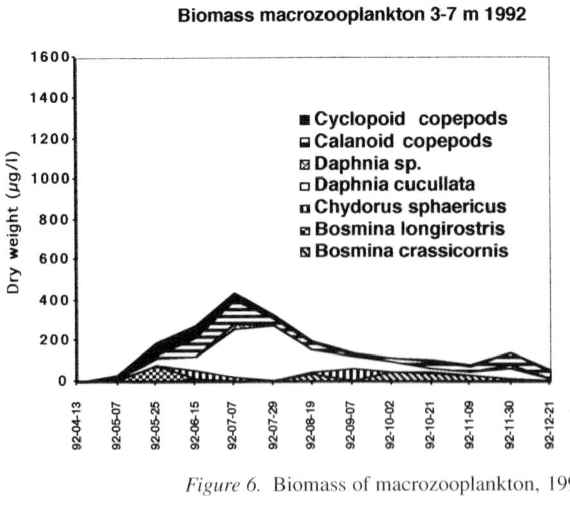

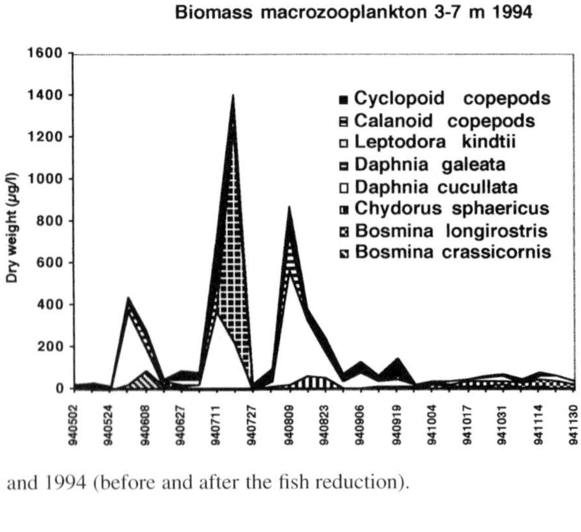

Figure 6. Biomass of macrozooplankton, 1992 and 1994 (before and after the fish reduction).

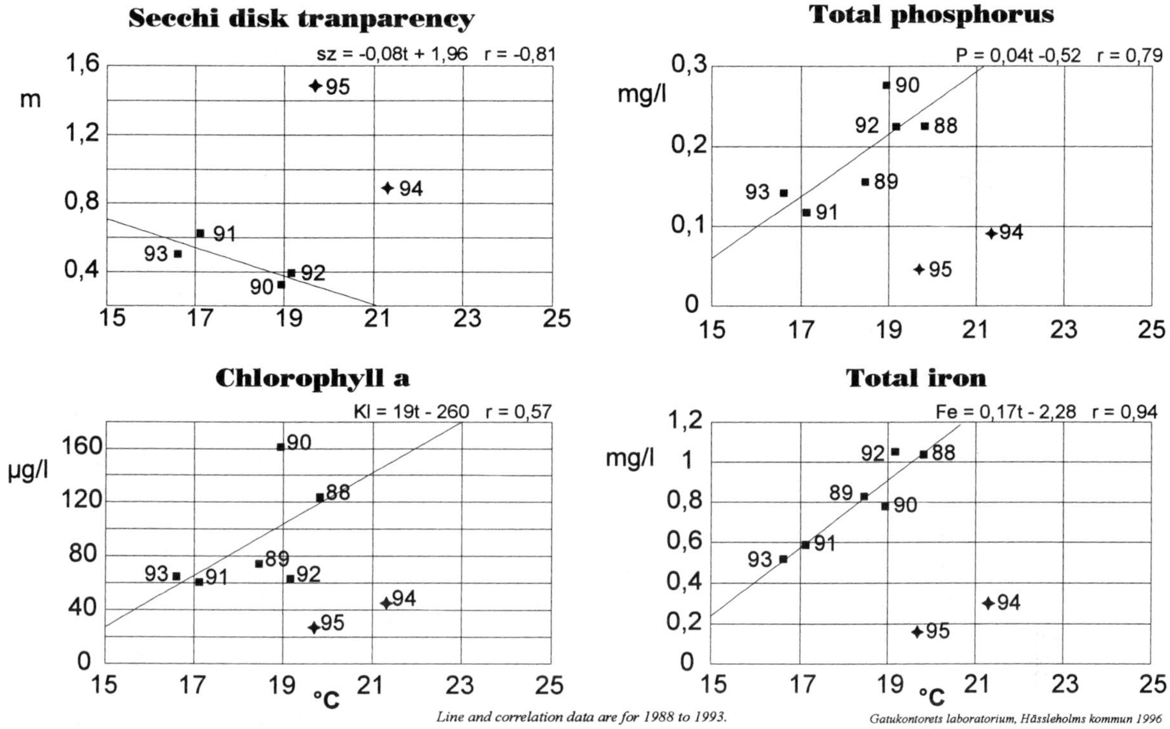

Figure 7. Summer average variations with temperature for Secchi disc transparency, chlorophyll *a*, total phosphorus and total iron in Lake Finjasjön 1988–1995.

plankton community of diatoms, cryptomonads, cyanobacteria, chrysophytes and dinoflagellates.

The high peaks of *Daphnia galeata* in July, coincided with high occurrence of cryptomonads, probably an effect of heavy grazing pressure on phytoplankton.

In addition to the pronounced changes in the phytoplankton community, total phytoplankton biomass and the chlorophyll *a* values were considerably lower than normal and the average summer Secchi disc transparency was twice that of previous years (Figure 7).

The constructed wetland started to function in February 1995. For the rest of that year, the average total phosphorus (mostly phosphate) values were reduced on average by 25% and the average total nitrogen (mostly nitrate) values by 32%. The nitrogen reduction was, as expected, highest during the summer and varied between 31 and 57% during the period May–October.

In 1995, the phytoplankton biomass and chlorophyll *a* levels were considerably lower than in 1994, and the species composition showed less eutrophic characteristics than in 1994. The average summer Secchi disc transparency was 1.5 m in 1995 compared to 0.9 m in 1994.

Before restoration, the Secchi disc transparency and the concentrations of chlorophyll *a*, total-phosphorus and total-iron used to depend on the temperature in the summer (Figure 7). However, in 1994 and 1995, the values deviated from the normal pattern. The internal loading decreased dramatically in 1994 and 1995. In 1995, the values of phosphorus and iron in the lake were the same as the average value of the inflow and the internal phosphorus loading had practically ceased (Figure 8). These results imply that, in certain lakes, it is possible to reduce an internal phosphorus loading without dredging or adding chemicals to the sediments.

When the food web manipulation started in 1992, only about 1% of the area of the lake was covered with submerged macrophytes. In 1995, 20% of the area was colonised by submerged macrophytes such as *Myriophyllum spicatum*, *Elodea canadensis* and *Potamogeton* spp.

As a consequence of the increased transparency and recolonisation of submerged macrophytes, the

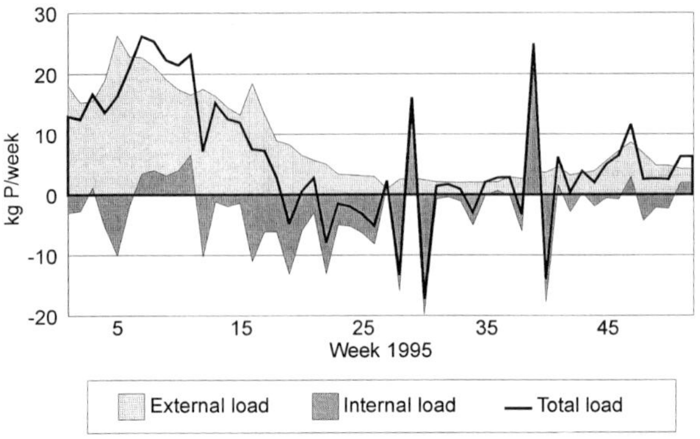

Figure 8. The internal, external and total phosphorus load on Lake Finjasjön, 1995.

birds that had disappeared when the transparency became very low in the middle of the seventies, returned to Lake Finjasjön in 1994 and 1995 (Figure 3).

The cost of the fish reduction was approximately 5 million SEK (£0.5 million), and of the constructed wetland, 6 million SEK (£0.6 million). This is equivalent to £100 000 km^{-2}.

The fact that a single municipality has been able to raise 70 million Swedish crowns including the dredging costs, (£7 million) in order to restore a lake, mainly for recreation purposes, is remarkable. Two factors were crucial in this achievement. The first was the fact that the town of Hässleholm did not have any alterative lake for recreation and not even an outdoor swimming pool.

The second, probably even more important, was that the lake and almost all the catchment are situated in a single municipality. In Sweden, restoration of water is normally financed by the municipalities according to the polluter pays principle. The municipality of Hässleholm polluted Lake Finjasjön, and consequently, the municipality – not the state – had to pay for the restoration.

If several municipalites have to share the costs for a restoration, intrinsic difficulties exist; how to split the costs between the different municipalities? Whether according to the number of inhabitants, according to the area of the lake belonging to each municipality, or according to how much each municipality contributed to the pollution of the lake, could be the source of long and complex argument.

The future water quality, plankton and fish studies in Lake Finjasjön will clarify the question of whether the present situation with clear water, a diverse phytoplankton community and almost no internal phosphorus loading will remain stable or not.

References

Aagren, R., 1994. Test fishing with benthic survey nets in Lake Finjasjön, Autumn 1994. Report. The Municipal Services Department, Municipality of Hässleholm.

Annadotter, H., 1990. Finjasjön – Badbar i vår tid för våra barn? (Lake Finjasjön – able to swim in, in our time, for our children?). Report, The Municipal Services Department, Municipality of Hässleholm (in Swedish).

Annadotter, H., 1993. Algtoxiner i dricksvatten. VA-Forsk rapport 1993-03. (In Swedish with an English abstract): 110 pp.

Annadotter, H., A. Larsson, E. Mattsson, L. Thunberg & T. Willén, 1990. Algtoxisk påverkan på nötkreatur. Rapport, Hässleholms gatukontor.

Annadotter, H., B. Lundstedt, & S. Ströbeck, 1992. Deklaration angående Finjasjöns fortsatta restaurering. Press release 1992-03-27, (in Swedish).

Annadotter, H., G. Cronberg, & L. Jönsson, 1993. The toxicity of blue-green algae (cyanobacteria) in Lake Finjasjön, Sweden in relation to Environmental Factors – A five-year study. In G. Giussani & C. Callieri (eds), Proceedings from the 5th International Conference on the Conservation and Management of Lakes, "Strategies for lake ecosystems beyond 2000".

Annadotter, H., 1995. Toxic blue-green algae in drinking water treatment. Vatten 51: 5–11. (In Swedish with an English abstract).

Annadotter, H., P-Å. Nilsson, B. Lundstedt, & S. Ströbäck, 1995. A new constructed wetland for combined reduction of phosphorus and nitrogen. Vatten 51: 103–106. (In Swedish with an English abstract).

Collvin, L. & J-I., Månsson, 1990. Provfiske i Finjasjön 1990. Rapport, Hässleholms gatukontor.

Håkansson, L. & Jansson, M. 1983. Principles of Lake Sedimentology. Springer Verlag, Berlin & Heidelberg.

Krug, A. & C. Walker, 1991. Tormestorpsån, Inventering och Åtgärdsförslag. Lunds universitet, Limnologiska avdelningen. (In Swedish).

Löfgren, S. & C. Forsberg, 1984. Finjasjöprojektet Report. Institute of Limnology in Uppsala. (In Swedish).

Löfgren, S. 1991. Finjasjön – miljösituation samt förslag till miljömål och åtgärder. Utredning åt Gatukontoret, Hässleholms kommun. (In Swedish).

Nilsson, P-Å., 1990. Restaureringen av Finjasjön – Läge och strategi 1991. Report, Hässleholms gatukontor. (In Swedish).

Nilsson, P-Å., 1991. Finjasjön. Sedimentundersökning 1991. Rapport. Hässleholms gatukontor.

Nyberg, P. & E. Degerman, 1988. Information från Sötvattenslaboratoriet Drottningholm. Standariserat provfiske med översiktsnät. Nr 7. Institute of Freshwater Research of the Swedish National Board of Fisheries.

Ripl, W. & L. Leonardsson, 1980. Förstudie av fosforfastläggning genom inducerad denitrifikation vid sedimentytan i Finjasjön. Rapport Limnologiska institutionen, Lunds universitet & Hässleholms kommun, Gatukontoret. (In Swedish).

Vollenweider, R. A., 1976. Advances in defining critical loading levels for phosphorus in lake eutrophication. Mem. Ist. ital. Idrobiol. 33: 53–83.

Vought, L & J. Lacoursiére, 1992. Hovdalaån, Inventering och Åtgärdsförslag. Lunds universitet, Limnologiska avdelningen. (In Swedish).

Vought, L & J. Lacoursiére, 1994. Svartevadsbäcken, Inventering och Åtgärdsförslag. Lunds universitet, Limnologiska avdelningen. (In Swedish).

Hydrobiologia **395/396:** 87–97, 1999.
D.M. Harper, B. Brierley, A.J.D. Ferguson & G. Phillips (eds), The Ecological Bases for Lake and Reservoir Management.

Integrated management to restore macrophyte domination

Karl Donabaum[1], Michael Schagerl[1] & Martin T. Dokulil[2]
[1]*Institute of Plant Physiology, Department of Hydrobotany, University of Vienna, Austria*
[2]*Institute of Limnology, Dpt. Mondsee, Austrian Academy of Sciences, Mondsee, Austria*

Key words: alternative stable states, macrophytes, integrated restoration

Abstract

Recent changes which have been observed at Alte Donau, a shallow urban lake within the city of Vienna, have been interpreted as a shift to a new stable state. The former macrophyte-dominated state changed to a turbid state dominated by high biomass of filamentous cyanobacteria, associated with a significant reduction in Secchi-depth. Phytoplankton was dominated by the filamentous cyanobacterial species *Cylindrospermopsis raciborskii* (Wolosz.), Seenayya et Subba Raju and *Limnothrix redekei* (Van Goor) Meffert. Integrated restoration plans included internal and external measures. Improvements in the catchment aim to minimize the input of nutrients from contaminated groundwater and from storm water and to reduce large numbers of water fowl. Internal restoration measures included water exchange, chemical flocculation and nitrate oxidation of the sediments. Additionally, macrophyte re-colonisation was enhanced through planting. A pelagic predator (*Aspius aspius* L.) was stocked to reduce bleak (*Alburnus alburnus* L.), the dominant cyprinid planktivore. Results from the period after water exchange and chemical treatment, showed significant reduction of nutrient and chlorophyll *a* concentrations. A shift in the phytoplankton species from cyanobacteria towards diatoms and greens was observed. Secchi depth greatly increased. Macrophyte growth became apparent both through re-colonisation, as well as from the planting.

Introduction

Scheffer et al. (1993) showed that at intermediate nutrient concentrations shallow lakes are either dominated by macrophytes and characterized by clear water, or abundant phytoplankton biomass and low visibility. These two situations may be seen as alternative stable equilibria. This implies that the system, when in equilibrium, is resistant to disturbance and tends to return to equilibrium when disturbed (Blindow et al., 1993). The switch from one to the other stable state can occur as a catastrophic feature, or an ecosystem can oscillate between the two equilibria over longer time periods.

The recent changes in a shallow lake in Vienna, from a macrophyte-dominated state to a turbid state, dominated by high biomass of filamentous cyanobacteria, have been interpreted as a shift to a new equilibrium, and restoration measures have been instigated to reverse the process.

Site description

The Alte Donau, a backwater of the river Danube, was formed during the regulation of the river in the last century (1870–1875), when the former main course was cut off. The water body is subdivided into two main basins of almost equal size, connected by a narrow passage (Figure 1). Hydrological conditions of the Alte Donau are entirely dependent on groundwater exchange and precipitation, because there is no natural surface inflow or outflow. The nearby river Danube and especially the impoundment Neue Donau, which was built to protect the city from flood events, both influence the direction and dynamics of the groundwater. The Alte Donau nowadays is a shallow urban lake within the city of Vienna and a very popular recreation area.

The altitude of the lake is 157 m.a.s.l. It has an area of 1.6 km^2 and an average depth of 2.3 m. The maximum depth is 6.8 m. A theoretial retention time of

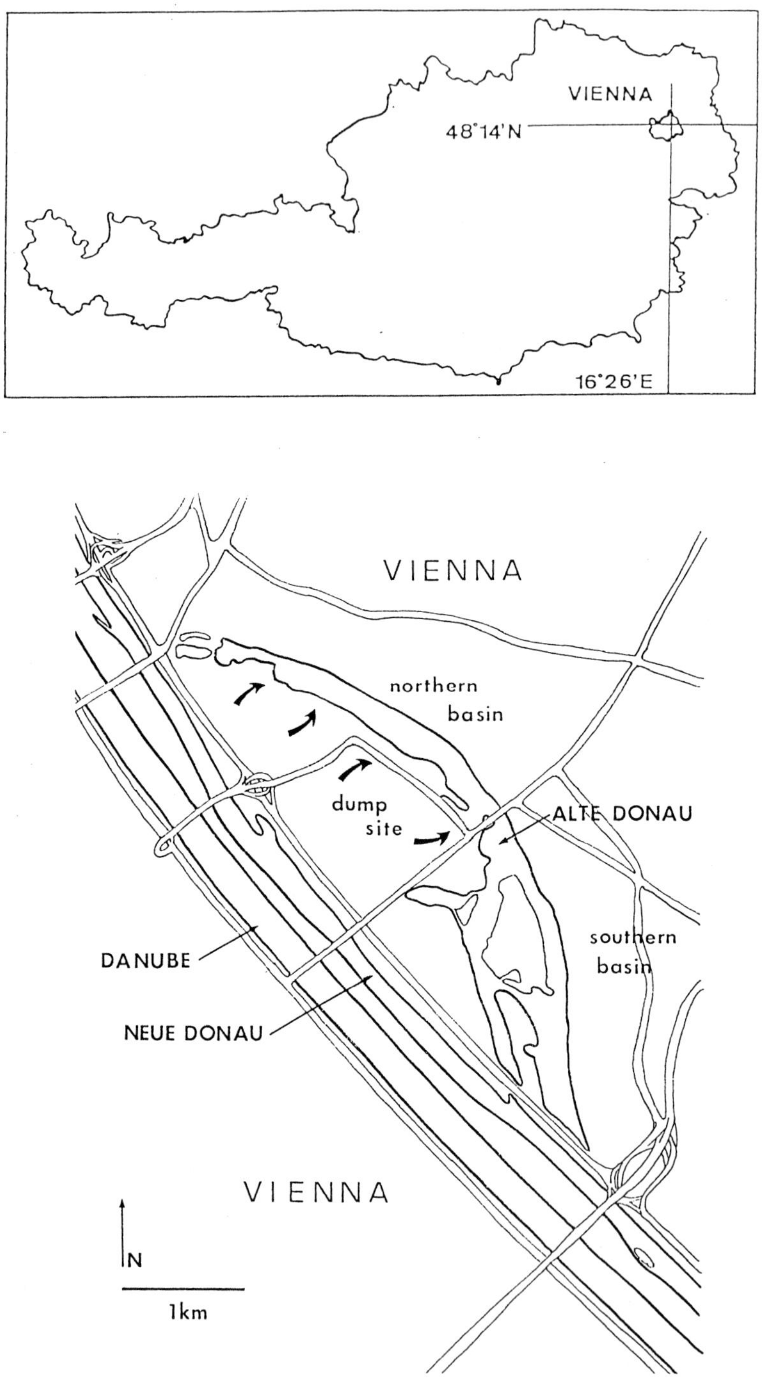

Figure 1. Location of the study site. Arrows indicate direction of groundwater flow in the northern basin.

Table 1. Morphological and hydrological data of Alte Donau according to Löffler (1988) and Ripl (1995).

Parameter	Dimension
Height above sea level	157 m
Area of northern basin	0.62 km^2
Area of southern basin	0.97 km^2
Volume of northern basin	1.43×10^6 m^3
Volume of southern basin	2.227×10^6 m^3
Mean retention time	180 days
Maximum depth	6.8 m
Mean depth	2.33 m

150–210 days (mean value 180 d) was deduced from chloride dilution rates (Ripl, 1995). The Alte Donau is polymictic (Löffler et al., 1988), sometimes stratifies, but never for long periods. Morphological and hydrological data are summarized in Table 1.

Conditions in the past and present changes

Evidence from the literature suggests that phytoplankton dominated in Alte Donau in the beginning of this century (Brunnthaler, 1907; Schiller, 1929). In the past decades, the water body was characterized by high transparency, low chlorophyll *a* concentrations and extensive macrophyte growth. The dominant species in the lake were *Myriophyllum spicatum* L. and *Potamogeton pectinatus* L. Charophytes like *Nitellopsis obtusa* (Desv. in Lois) J. Groves, *Chara tomentosa* L. and *Chara hispida* Wood were also abundant (Löffler, 1988). A similar macrophyte composition was reported from the swedish lakes – Lake Takern and Lake Krankesjon – which have shifted several times between a clear-water state and a turbid state (Andersson et al., 1990; Blindow et al., 1993).

Significant numbers of the blue-green algae *Limnothrix redekei* (Van Goor) Meffert were first recorded in water quality samples in the year 1992. Microscopical observations in the following year revealed a substantial domination of *Cylindrospermopsis raciborskii* (Wolosz.) Seenayya et Subba Raju. Average annual chlorophyll *a* concentrations increased substantially and secchi depth decreased (Dokulil, 1994; Mayer et al., 1997). Associated with these changes was a remarkable decline of submerged macrophytes. Within two years, only remnants of macrophyte and charophyte stands were left (Dokulil & Janauer, 1995).

What has caused the shift?

Several reasons were believed to be responsible for the shift from a clear water state to a turbid state in Alte Donau. These were:

1. Changes in water level dynamics.
2. Increased nutrient input.
3. Insufficient lining of the sewage network.
4. High stocks of water fowl.
5. Extensive recreational activities.
6. High stocks of benthivorous and planktivorous fish.

Water level fluctuations for the period from 1958 to 1993 are shown in Figure 2. Reduced water level dynamics, in combination with a higher water level, can be seen from the late 1970s onwards. The mean water level calculated for the period from 1978 to 1993 was about 37 cm higher than in the period before. Similarly the range of water level fluctuations decreased by 1 m. The sharp transition in hydrological dynamics in 1978 are basically due to the nearby construction of a highway and the impoundment Neue Donau.

Nutrient input from non point sources is a major problem in the Alte Donau. Associated with the rise of the water level in the 1980s was a change in the direction of flow in the northern basin, such that nutrients were eluated from a former dump site (Figure 1). Leakage of septic tanks and insufficient sewerage in the catchment area also contributed to nutrient loading. Additional nutrient input originated from the excretion of water fowl and from recreation activities. Phosphorus input by water fowl was estimated to be 120 kg P (or 53 $\mu g\ l^{-1}$) in a winter period (Steiner, 1986). Estimation of phosphorus addition by

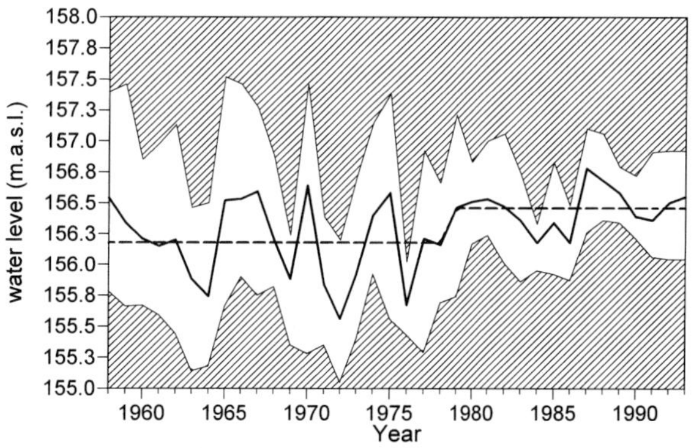

Figure 2. Water level fluctuations from 1958 to 1993. Upper and lower limits of the water level (unfilled area). Mean water level (thick line). Mean water level calculated for different periods (dashed line).

the 400 000 visitors in the summer period amount to approximately 38 kg P or 17 μg l^{-1} (Dokulil, 1993).

These changes seem to be correlated to the severe decline of the macrophytes similar to observations elsewhere (Scheffer et al., 1993; Blindow et al., 1993). Here, several factors may have combined to cause the loss. Macrophytes were extensively removed by the local authorities as a management practice for many years. Grazing by high densities of water fowl may also have contributed to the disappearence of submerged macrophytes in Alte Donau. According to Steiner (1986), water fowl grazed about 194 tons (wet weight) of macrophytes per year. The disappearence was accelerated through light limitation when phytoplankton began to dominate. The decisive environmental variable for macrophyte decrease is an altered underwater light climate (Chambers & Kalff, 1985).

Recent investigations of the fish community (Waidbacher et al., 1996) revealed a dominance of bleak (*Alburnus alburnus* L.). The cyprinids roach (*Rutilus rutilus* L.), rudd (*Scardinius erythrophthalmus* L.), bream (*Abramis brama* L.), white bream (*Blicka bjoerkna L.)* and tench *(Tinca tinca* L.) were abundant. Common carp (*Cyprinus carpio* L.) is the main fish stocked. Perch (*Perca fluviatilis* L.) is also abundant in Alte Donau and the piscivores pike (*Esox lucius* L.) as well as pike-perch (*Stizostedion lucioperca* L.) are stocked. An extensive grazing pressure on zooplankton was assumed by the fish community, especially from bleak. Confirmation came from zooplankton studies in 1994 (Dokulil, 1994). Abundances and species composition were typical for eutrophic conditions. Among the crustaceans, small species like *Eubosmina coregoni* Baird showed higher abundances than larger species such as *Daphnia cucculata* Sars. These results confirm investigations in 1987 (Löffler, 1988), but are in contradiction to earlier reports, where high species diversity amongst the cladocerans was reported (Pesta, 1928).

As a consequence of increased nutrient input, macrophyte decrease and grazing pressure on zooplankton, cyanobacteria became abundant in 1993 and 1994. The water showed a striking yellow colour and transparency decreased to 30 cm.

Investigations on the benthic communities (Waidbacher et al., 1996) showed a loss of characteristic species and poor diversity of macroinvertebrates in comparison to earlier studies (Löffler, 1988) and to similar backwaters. Lack of habitats in the littoral zone and loss of macrophytes were recognized to be responsible.

Restoration measures

Integrated lake management strategies included external and internal measures. Rapid expansion of the sewerage network was initiated when the problem became apparent. To keep contaminated groundwater from the lake, four pumping wells were constructed between the former dump site and the lake. Recent investigations of groundwater fluxes showed that they have been successfully applied.

As a first step, artificial aeration combined with vertical mixing was performed in 1993 in areas where thermal stratification, and hence oxygen depletion near the sediment were expected to create problems. Substantial nutrient release from the sediment under anoxic conditions, shown by laboratory experiments, could thus be avoided (Dokulil, 1994).

Half of the water volume of the lake was exchanged by water low in nutrient concentrations from the impoundment Neue Donau in December 1993 (Dokulil & Gasser, 1994).

Attempts were made to concentrate most of the water fowl in a small appendix of the north-eastern basin which has been cut off from the rest of the lake. Collecting eggs from the ducks was proposed as a measure to reduce the high stock of water fowl (Steiner, 1986). In addition, people were asked not to feed the birds.

Great efforts were made by the local authorities to plant a reed belt to give more structure to the littoral zone in the southern basin and to enhance sedimentation of organic material. Approximately 10% of the shore line is now covered by a narrow reed belt in this basin. Structuring and revitalisation of the shore line is still in progress. Further investigations will show if restructuring of the shore line can supplement recovery of the benthic community.

Re-colonisation of submerged macrophytes was assisted by planting experiments in 1995. Macrophytes were taken from the impoundment Neue Donau and transferred into experimental areas in Alte Donau. The first results obtained for *Myriophyllum spicatum* L. and *Ceratophyllum demersum* L. are promising.

A pelagic predator (*Aspius aspius* L.) was introduced to the ecosystem to reduce bleak (Waidbacher et al., 1996). In the past three years, Alte Donau was stocked with juvenile as well as with adult fish, approximately 500 kg ann^{-1}. It is expected that this biomanipulation will enhance the development of larger cladocerans.

After careful consideration of most of the restoration techniques commonly used, the Riplox-method was selected as the main restoration measure (Ripl 1976, 1978, 1986; Ripl & Lindmark 1978; Ripl & Wolter 1993a, b). It was implemented in spring 1995 in cooperation with the University of Berlin. The method consists of two steps which are performed in chronological sequence (Figure 3). In the first step, $FeCl_3$ buffered with limestone is added to remove phosphorus and suspended material by chemical and mechanical flocculation. The $FeCl_3$ acts as a barrier against phosphorus release from sediments. In a second step, $Ca(NO_3)_2$ is added to the sediments to enhance nitrate oxidation. Nitrate is reduced to elementary nitrogen by anaerobic denitrification. Organic mud is oxidized to carbon-dioxide and water. Consequently, the oxygen deficit caused by heterotrophic metabolism decreases. Thus prolonged anoxic conditions near the sediment surface can be avoided and internal loading reduces.

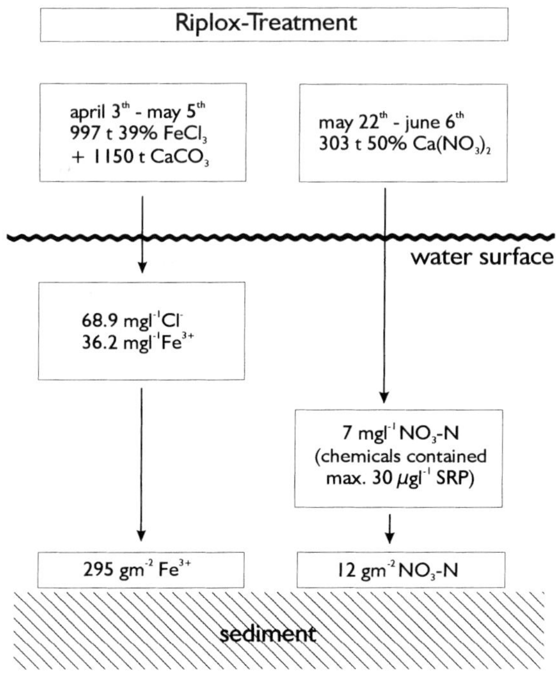

Figure 3. Scheme of the Riplox-treatment in Alte Donau in 1995.

Monitoring of recovery

Detailed descriptions of sampling intervals, sampling stations and methods for the investigation period from 1993 to 1995 are given in Dokulil (1993, 1994) and Mayer et al. (1997). Therefore only a brief description of the methods used is given here. Sampling was at fortnight intervals at ten sampling stations. At each of the sampling stations, water samples were collected near the surface and near the bottom. If not stated otherwise, all values in the figures and in the text are mean values of the different sampling stations and of depth.

Depth (m), temperature (°C), pH (-log H^+), oxygen (mg l^{-1} and % O_2) and conductivity (μS cm^{-1}) were measured with a multiple measuring probe (Hydrolog 2001, Grabner Instruments). Depth profiles were recorded at each sampling station in 50 cm intervals. Whatman GF/F filters were used for filtration of water samples. SRP (soluble reactive phosphorus), DP (total dissolved phosphorus) and TP (total phosphorus) were determined according to Strickland & Parsons (1968). After digestion with potassium persulphate, total phosphorus and DP were determined. Particulate phosphorus was calculated as the difference between total phosphorus and DP.

Chloride (Cl^-), sulphate (SO_4-S) and nitrate-concentrations, higher than 2 mg l^{-1} (NO_3-N), were analysed by means of Ion-chromatography (Merck-Hitachi-HPLC equipment; column Merck polyspher IC AN-1; elution with phthalic acid and Tris-boric acid buffer). Nitrate concentrations below 2 mg l^{-1} were analysed according the method with sodium-salicylate (Legler, 1988). Nitrite was analysed with sulphanylic acid and N-(1-naphthyl) ethylen-diamine-dihydrochloride (Legler, 1988). TN (total nitrogen) and DN (dissolved nitrogen) were determined as NH_4-N following digestion with sulphuric acid and hydrogen peroxide.

Chlorophyll *a* was extracted in 90% acetone after sample homogenisation and measured according the method of Lorenzen (1967). Phytoplankton counts were done from Lugol-fixed samples according to Utermöhl (1958). Cell dimensions were measured by means of an image analysing system (Lucia-M, Nikon) and volumes were calculated using simple geometric forms. For conversion of phytoplankton biovolumes into fresh weight, a specific gravity of 1 was assumed.

Results

The chloride concentrations for 1995 are shown in Figure 4 (bars indicate the treatment period). No immediate response of chloride concentrations to the addition of $FeCl_3$ was observed. In fact, the $FeCl_3$ showed almost no dissociation in the water column. A significant release of chloride ions took place after the deposition of the $FeCl_3$ in the sediments. Maximum chloride values reached almost 100 mg l^{-1}. As chloride is hardly used by organisms, the rapid decline of the chloride concentrations after the treatment indicate high dilution rates by the groundwater.

Changes in the nitrate concentrations are shown in Figure 5. Before treatment, the nitrate content was only about a few hundred $\mu g\ l^{-1}$. In contrast to the chloride values, an immediate reaction to the addition of $Ca(NO_3)_2$ was observed. The nitrate content reached its highest amounts immediately after the addition. After the application was finished, a rapid decline occurred, mainly due to denitrification and water exchange. At the end of the year, nitrate content was as low as in the years before. The degree of denitrification can be deduced from high nitrite concentrations, obtained after the addition of $Ca(NO_3)_2$ (Figure 6). In general, nitrite is an intermediate product of bacterial metabolism. The high values in Alte Donau, up to 120 $\mu g\ l^{-1}$, originated from the pathway of denitrification, where nitrate is reduced to nitrite, nitric oxides (NO, N_2O) and finally to molecular nitrogen (Rheinheimer, 1991). With decreasing nitrate concentrations, decreasing nitrite values were also measured.

Average total phosphorus concentration was 54 $\mu g\ l^{-1}$ in 1993 (Table 2, Figure 7). Cyanobacterial growth took place in late summer and autumn, associated with maximum total phosphorus concentrations of 70 $\mu g\ l^{-1}$. No marked seasonal changes of total phosphorus concentrations were observed. Flushing of the ecosystem in winter 1993 was not accompanied by a significant reduction of total phosphorus. Contrary to

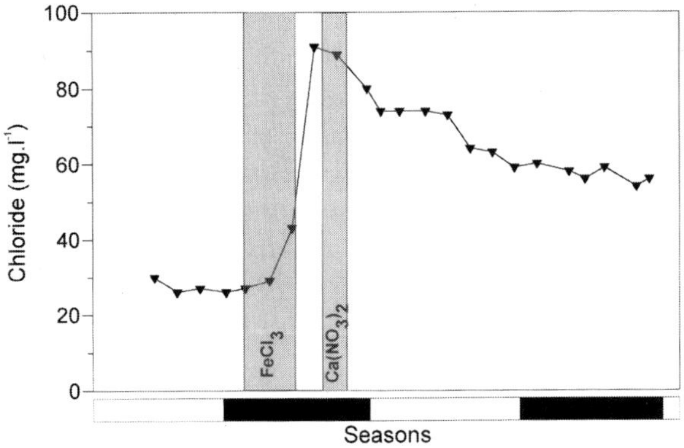

Figure 4. Chloride concentrations in 1995 (bars indicate the treatment period).

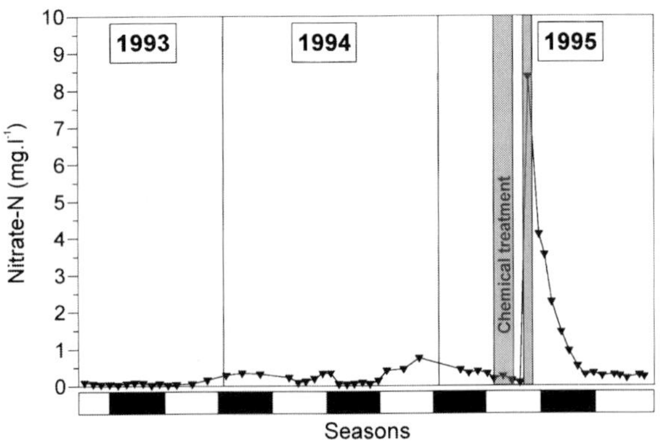

Figure 5. Nitrate-N concentrations from 1993 to 1995 (bars indicate the treatment period).

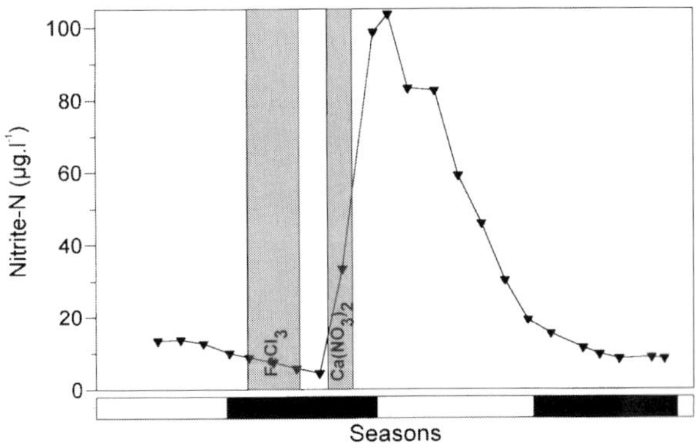

Figure 6. Nitrite-N concentrations in 1995 (bars indicate the treatment period).

Table 2. Changes in mean values of chlorophyll *a*, total phosphorus (both in $\mu g\ l^{-1}$) and percentage chlorophyll *a* to total phosphorus from 1987 to 1995.

Year	Total phosphorus	Chlorophyll *a*	Chl *a*/total P as%
1987	35.1	5.2	15.0
1988	66.1	16.6	25.1
1989	110.0	15.2	13.8
1990	42.4	15.0	35.4
1991	47.1	20.3	43.0
1992	41.4	22.4	54.2
1993	54.2	43.1	79.6
1994	70.0	41.0	58.6
1995	27.3	12.0	43.9

Table 3. Mean values of phosphorus ($\mu g\ l^{-1}$) and nitrogen ($mg\ l^{-1}$) fractions from 1993 to 1995

Parameter	Mean value 1995	Mean value 1994	Mean value 1993
Total phosphorus	0.027	0.07	0.055
Soluble phosphorus	0.007	0.021	0.013
Particulate phosphorus	0.020	0.049	0.042
Soluble reactive phosphorus	0.002	0.003	–
Total nitrogen	1.88	1.75	1.69
Soluble nitrogen	1.63	0.93	–
Particulate nitrogen	0.25	0.83	–
Total N: Total P ratio	70	25	31

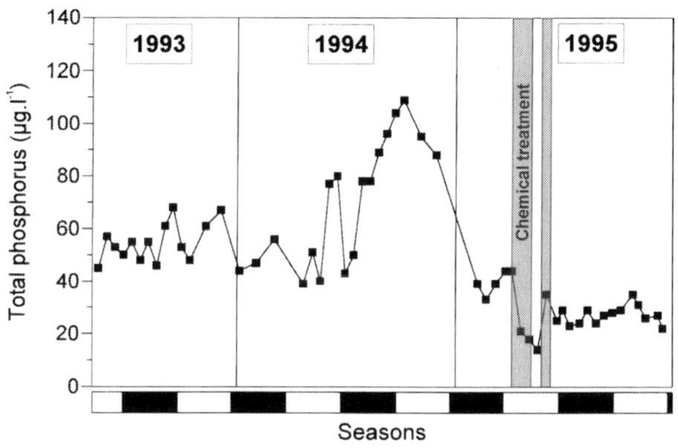

Figure 7. Total phosphorus from 1993 to 1995 (bars indicate the treatment period).

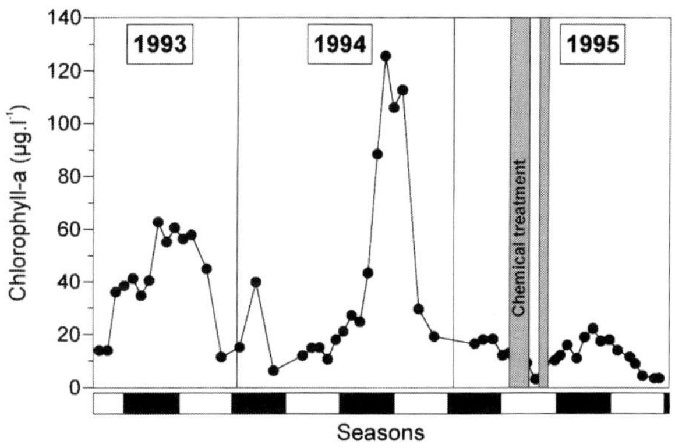

Figure 8. Chlorophyll *a* values from 1993 to 1995 (bars indicate the treatment period).

expectations, highest values were reached in summer 1994 (Figure 7), above 100 $\mu g\ l^{-1}$ in late summer, when cyanobacterial abundance was highest. Internal loading, loss of macrophytes and suppression of zooplankton by planktivorous fish resulted in intensive phytoplankton growth, thus causing these high total phosphorus loadings. The chemical treatment resulted in a considerable decline of total phosphorus concentrations in 1995. The slight increase which occurred after the addition of $Ca(NO_3)_2$ can be explained by phosphorus contamination of the used chemicals (Figure 3). In the remaining investigation period, values were about 25 $\mu g\ l^{-1}$. Concentrations of soluble reactive phosphorus were mostly depleted to trace levels by phytoplankton (data not shown).

As a consequence of cyanobacterial growth high chlorophyll *a* concentrations were observed in late summer 1993 (up to 60 $\mu g\ l^{-1}$) and 1994 (up to 125 $\mu g\ l^{-1}$; Figure 8). Chemical flocculation with $FeCl_3$ resulted in a decline of chlorophyll-*a* concentrations with a minimum value of 3 $\mu g\ l^{-1}$. The maximum value of 22 $\mu g\ l^{-1}$ was observed in the summer period, but the typical chlorophyll *a* peak caused by cyanobacteria in late summer was lacking. The annual average of chlorophyll *a* for 1995 was 12 $\mu g\ l^{-1}$. Chlorophyll *a* measurements for the different sampling stations in 1995 are given in Figure 9. In contrast to the untreated parts of Alte Donau, the majority of the chlorophyll *a* values were clustered near to the lower border of the eutrophic range with single values hypertrophic.

A very similar curve can be obtained from phytoplankton biomass values (Figure 10). Cyanobacteria, mainly *Cylindrospermopsis raciborskii* and *Limnothrix redekei*, were dominant throughout the year in 1993 and 1994, contributing 70% to 90% to total phytoplankton biomass. *Cylindrospermopsis raciborskii*, difficult to distinguish from closly related genera because of morphological variants (Singh, 1962; Cronberg, 1977; Horecka & Komarek, 1979; Hindak, 1988), was the main component of the phytoplankton in the years 1993 and 1994. *C. raciborskii* has been registered as the dominant species in a

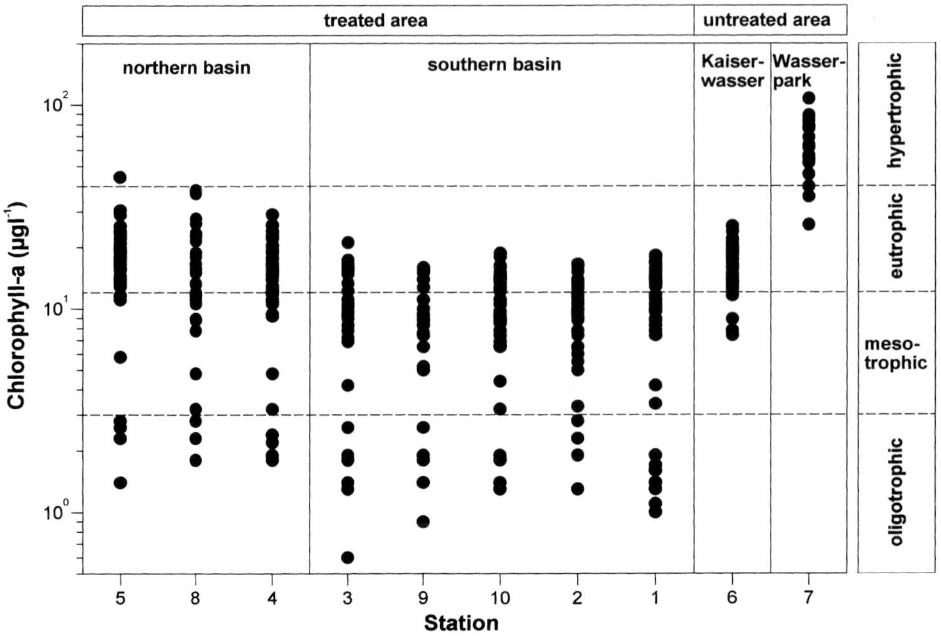

Figure 9. Chlorophyll *a* concentrations of the different sampling stations in 1995. Trophic classification according to Forsberg & Ryding (1980).

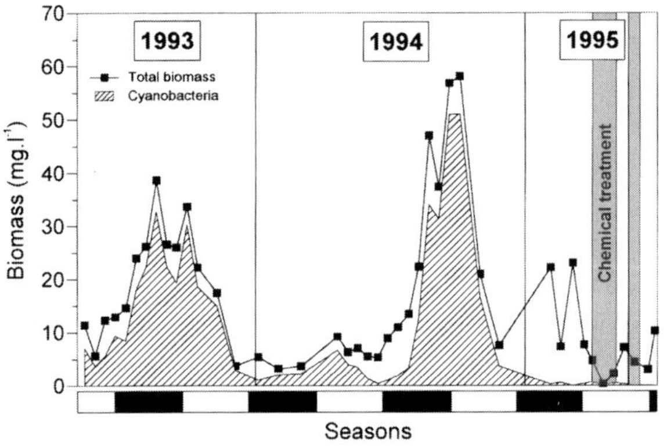

Figure 10. Phytoplankton biomass from 1993 to 1995 (bars indicate the treatment period).

Brazilean reservoir by Cronberg (1977) and Branco & Senna (1994). This species has also been found as a common component in other water bodies in tropical and temperate regions forming blooms at high temperatures and elevated nutrient levels (Hill, 1970; Lind, 1984; Padisak et al., 1984; Lewis, 1986; Kalff & Watson, 1986). The dominance of *C. raciborskii* is favoured by a set of factors such as the ability to absorb rapidly phosphates at even low concentrations, the possibility for nitrogen-fixation by heterocyst formation, high water temperature, long retention time and small predation by zooplankton (Cronberg, 1977; Branco & Senna, 1994).

After the chemical treatment, cyanobacteria were of minor importance. A dramatical shift in phytoplankton biomass composition towards diatoms and coccal chlorophyceans occurred. For short periods cryptophyceans, chrysophyceans, dinoflagellates and coccal cyanobacteria were also abundant. The shift in the phytoplankton composition may be due to the artificially altered N/P ratio. Among other factors, summarized by Varis (1993), low N/P ratios seem to favour cyanobacterial growth (Smith, 1986; Varis, 1993).

Transparency considerably improved in 1995 (Figure 11), re-establishing optimum light-conditions for new colonisation by submerged macrophytes.

Discussion

Eutrophication of freshwaters is a well known and intensively studied problem, creating serious problems for water supply or recreational use (Harper, 1992). Numerous strategies and techniques have been developed in the past to cope with deterioration of fresh waters (e.g. Vollenweider & Kerekes, 1982; Shapiro & Wright, 1984; Sas, 1989). However, restoration of

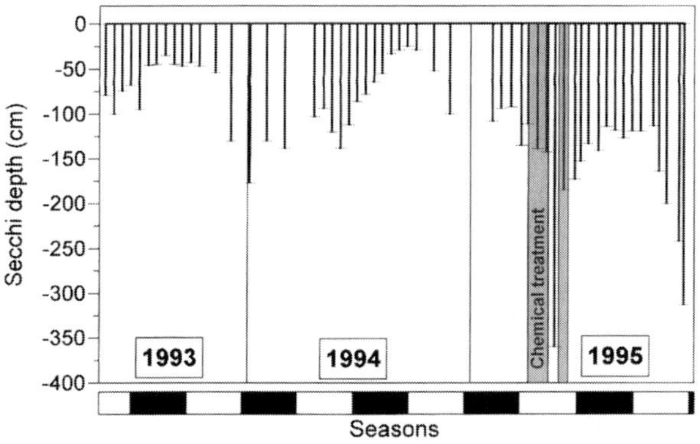

Figure 11. Secchi depth from 1993 to 1995 (bars indicate the treatment period).

eutrophic or hypertrophic shallow lakes is often more complex than those of deep stratifying lakes (Sas, 1989; Moss, 1990). Integrated lake management practices, and detailed concepts, based on investigations of the whole ecosystem, are necessary to cope with these systems. Additionally mathematical modelling may serve as a tool for management in eutrophication control of shallow lakes (Van der Molen et al., 1994). The concept of alternative stable states in shallow lakes (Blindow et al., 1993), implying that the system, when in equilibrium, is resistant to disturbance and tends to return to equilibrium when disturbed, has strong implications for lake management. The impact of any applied restoration measures to an ecosystem must be greater than its resistence to disturbance. Knowledge of the lake's history, and detailed investigations of the ecosystem, are necessary to address problems and to set the objectives in a special case. Nutrient reduction serves as a prerequisite, but without considering trophic cascade effects of the food web (Carpenter & Kitchell, 1993), long term stabilisation of the ecosystem cannot be expected. Many case studies show that nutrient reduction itself is not sufficient to change conditions (e.g. Foy & Fitzsimons, 1987; Annadotter et al., 1999), especially when internal loading from the sediments is high. The importance of the structure of the food web is known from numerous investigations and biomanipulation experiments (Shapiro & Wright, 1984; Kasprzak et al., 1988; Benndorf, 1990; Hanson & Butler, 1990; van Donk et al., 1990a, b; Giussani et al., 1990; Jeppesen et al., 1990a, b; Lyche et al., 1990; Meijer et al., 1990; Vanni et al., 1990; Mazumder, 1994). However, appearence of filamentous cyanobacteria has been reported from different study sites, in response to biomanipulation (e.g. Kasprzak et al., 1993). Due to different political and public interests, removal of fish on a large scale is impossible in many cases. Therefore, specific solutions have to be adopted for each case.

The management practices in Alte Donau are still in progress. From the results obtained so far, a success for restoration of a macrophyte dominated state can be expected (Figure 12, Table 1). Nevertheless, the conflicting interests of conservation, angling, swimming and other recreational activities will necessitate further monitoring and management measures to ensure sustainability. Although integrated management and continuous management practices need high funds, recovery of lakes may justify the high costs, especially in cases where public interest is high.

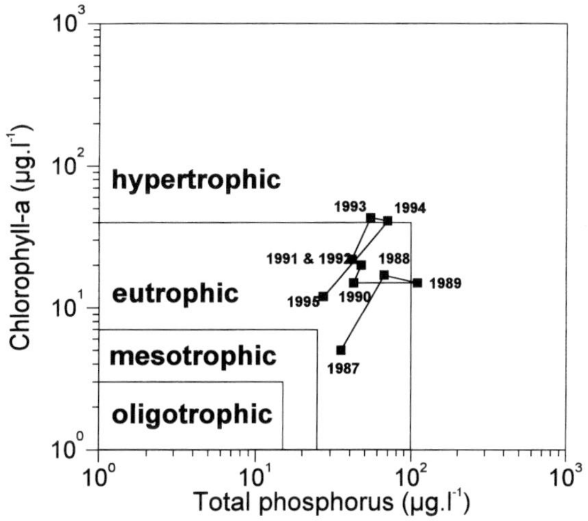

Figure 12. Log–log plot of mean chlorophyll *a* and mean total phosphorus from 1987 to 1995. Trophic classification according to Forsberg & Ryding (1980).

Conclusions

From the results and observations obtained in this study, the following general conclusions can be drawn:

1. Changes in the hydrological conditions, nutrient status and in the food-web structure are responsible for the catastrophic shift from the macrophyte dominated state to one characterized by abundant cyanobacterial appearance.
2. Recovery of the clear state can be achieved by adopting both external and internal restoration strategies.
3. Phosphorus precipitation combined with sediment conditioning, proves to be a reliable technique to

reduce nutrient concentrations in the lake and internal loading from the sediment in groundwater seepage lakes.

4. Flocculation of seston improves the under-water light climate allowing re-colonization by submerged macrophytes.
5. Technical restoration measures should be supplemented by biomanipulation to ensure long-term stabilization of the desired ecosystem condition.
6. Integrated lake management practices, based on detailed investigations, must consider conflicting interests and the carrying capacity of the ecosystem.

Acknowledgements

This study was supported by grants from the MA 45 - Wasserbau, Vienna. We would like to thank all of the numerous collaborators.

References

Andersson, G., I. Blindow, A. Hargeby & S. Johannson. 1990. Det varas för Krankesjön (The recovery of Lake Krankesjön). Anser 29: 53–62.

Annadotter, H., G. Cronberg, R. Aagren, B. Lundstedt, P.-Å. Nilsson & Sv. Ströbeck, 1999. Multiple techniques for Lake Restoration. Hydrobiologia 395/396 (Dev. Hydrobiol. 136): 77–85.

Benndorf, J., 1990. Conditions for effective biomanipulation; conclusions derived from whole-lake experiments in Europe. Hydrobiologia 200/201: 187–203.

Blindow, I., G. Andersson, A. Hargeby & S. Johannson, 1993. Long-term pattern of alternative stable states in two shallow eutrophic lakes. Freshwat. Biol. 30: 159–167.

Branco, C. W. C. & P. A. C. Senna, 1994. Factors influencing the developement of *Cylindrospermopsis raciborskii* and *Microcystis aeruginosa* in the Paranoá Reservoir, Brasília, Brazil. Algological Studies 75: 85–96.

Brunnthaler, J., 1907. Die Algen und Schizophyceen der Altwässer der Donau bei Wien. Verh. k. u. k. Zool.-Bot. Ges. Wien 57: 170–223.

Carpenter, S. R. & J. F. Kitchell (eds), 1993. The trophic cascade in lakes. Cambridge University Press: 385 pp.

Chambers, P. A. & J. Kalff, 1985. Depth distribution and biomass of submerged aquatic macrophyte communities in relation to Secchi depth. Can. J. Fish. aquat. Sci. 42: 701–709.

Cronberg, G., 1977. The Lago do Paranoá project. Phytoplankton ecology and taxonomy. Final report. Project PAHO/WHO 77/WT/BRA/2341/04, 71 pp.

Dokulil, M. T., 1993. Limnologische Untersuchungen zur Sanierung der Alten Donau. Jahresbericht an die Wiener Magistratsabteilung 45 - Wasserbau. 104 pp.

Dokulil, M. T., 1994. Limnologische Untersuchungen zur Sanierung der Alten Donau. Jahresbericht an die Wiener Magistratsabteilung 45 - Wasserbau. 227 pp.

Dokulil, M. T. & R. Gasser, 1994. Begleituntersuchungen zum Wasseraustausch in der Alten Donau 1993. Hydrologische und limnologische Begleitstudie zur Beweissicherung während des Wassertauschs Ende 1993. Bericht an das Magistrat der Stadt Wien - MA 45 - Wasserbau. 42 pp.

Dokulil, M. T. & G. A. Janauer, 1995. Alternative stable states during eutrophication of a shallow urban lake in Vienna, Austria. 6th International Conference on the Conservation and Management of Lakes – Kasumigaura '95. Proceedings Vol. 2: 730–733..

van Donk, E., M. P. Grimm, R. D. Gulati & J. P. G. Klein-Breteler, 1990a. Whole-lake food web manipulation as a means to study community interactions in a small ecosystem. Hydrobiologia 200/201: 275–289.

van Donk, E., M. P. Grimm, R. D. Gulati, P. G. M. Heuts, W. A. Kloet & L. van Liere, 1990b. Whole-lake food web manipulation on a large scale in the Netherlands. Hydrobiologia 200/201: 291–301.

Forsberg, C. & S. O. Ryding, 1980. Eutrophication parameters and trophic state indices in 30 Swedish waste-receiving lakes. Arch. Hydrobiol. 89: 189–207.

Foy, R. H. & A. G. Fitzsimons, 1987. Phosphorus inactivation in a eutrophic lake by the direct addition of ferric aluminium sulphate: changes in phytoplankton populations. Freshwat. Biol. 17: 1–13.

Giussani, G., R. de Bernardi & T. Ruffoni, 1990. Three years of experience in biomanipulating a small eutrophic lake: Lago di Candia (Northern Italy). Hydrobiologia 200/201: 357–366.

Harper, D. M., 1992. Eutrophication of freshwaters – principles, problems and restoration. Chapman & Hall. 327 pp.

Hanson, M. A. & M. G. Butler, 1990. Early responses of plankton and turbidity to biomanipulation in a shallow prairie lake. Hydrobiologia 200/201: 317–329.

Hill, H., 1970. A new form of *Raphidiopsis mediterranea* Skuja found in Minnesota lakes. Phycologia 9: 73–77.

Hindák, F., 1988. Planctic species of two related genera *Cylindrospermopsis* and *Anabaenopsis* from western Slovakia. Algological Studies 50–53: 283–302.

Horecká, M. & J. Komárek, 1979. Taxonomic position of three planktonic blue-green algae from the genera *Aphanizomenon* and *Cylindrospermopsis*. Preslia, Praha 51: 289–312.

Jeppesen, E., J. P. Jensen, P. Kristensen, M. Søndergaard, E. Mortensen, O. Sortkjaer & K. Olrik, 1990a. Fish manipulation as a lake restoration tool in shallow, eutrophic, temperate lakes. 2: Threshold levels, longterm stability and conclusions. Hydrobiologia 200/201: 219–227.

Jeppesen, E., M. Sondergaard, E. Mortensen, P. Kristensen, B. Riemann, H. J. Jensen, J. P. Müller, O. Sortkjaer, J. P. Jensen, K. Christoffersen, S. Bosselmann & E. Dall, 1990a. Fish manipulation as a lake restoration tool in shallow, eutrophic, temperate lakes. 1: cross-analysis of three Danish case studies. Hydrobiologia 200/201: 205–218.

Kalff, J. & S. Watson, 1986. Phytoplankton and its dynamics in two tropical lakes: a tropical and temperate zone comparison. Hydrobiologia 138: 161–176.

Kasprzak, P., J. Benndorf, R. Koschel & F. Recknagel, 1988. Applicability of the food web manipulation in the restoration programme of a hypertrophic stratified lake: Model studies for Lake Haussee (Feldberg, GDR). Limnologica (Berlin) 19: 87–95.

Kasprzak, P., L. Krienitz & R. Koschel, 1993. Changes in the plankton of Lake Feldberger Haussee (Germany, Mecklenburg-Vorpommern) in response to biomanipulation. Arch. Hydrobiol. 128/2: 149–168.

Legler, C., 1988. Ausgewählte Methoden der Wasseruntersuchung. Bd. 1: Chemische, physikalisch-chemische und physikalische Methoden. G. Fischer Verlag, Jena: 517 pp.

Lewis, W. M. Jr., 1986. Phytoplankton succession in Lake Valencia, Venezuela. Hydrobiologia 138: 189–203.

Lind, O. T., 1984. Patterns of phytoplankton populations and their relationship to trophic state in an elongate reservoir. Verh. int. Ver. Limnol. 22: 1465–1469.

Löffler, H. (ed.) 1988. Limnologische Projektstudie - Ökosystem Alte Donau. Bericht im Auftrag der Wasserstraßendirektion – Wien: 272 pp.

Lorenzen, C. J., 1967. Determination of chlorophyll and phaeopigments: Spectrophotometric equations. Limnol. Oceanogr. 12: 343–346.

Lyche, A., B. A. Faafeng & A. Braband, 1990. Predictability and possible mechanisms of plankton response to reduction of planktivorous fish. Hydrobiologia 200/201: 251–261.

Mayer, J., M. T. Dokulil, M. Salbrechter, M. Berger, T. Posch, G. Pfister, A. K. T. Kirschner, B. Velimirov, A. Steitz & T. Ulbricht. 1997. Seasonal successions and trophic relations between phytoplankton, zooplankton, ciliates and bacteria in a hypertrophic shallow lake in Vienna, Austria. Hydrobiologia 342/343: 165–174.

Mazumder, A., 1994. Phosphorus-chlorophyll relationships under contrasting herbivory and thermal stratification: Predictions and patterns. Can. J. Fish. aquat. Sci. 51: 390–400.

Meijer, M.-L., M. W. de Haan, A. W. Breukelaar & H. Buiteveld, 1990. Is reduction of benthivorous fish an important cause of high transparency following biomanipulation in shallow lakes? Hydrobiologia 200/201: 303–315.

Moss, B., 1990. Engineering and biological approaches to the restoration from eutrophication of shallow lakes in which aquatic plant communities are important components. Hydrobiologia 200/201: 367–377.

Padisák, J. L., G. Toth & L. Vörös, 1984. *Anabaenopsis raciborskii* Wolonsz. bloom in Lake Balaton in the summer and autumn of 1982. Biologisches Forschungsinstitut Burgenland 51: 77–81.

Pesta, O., 1928. Berichte zur Limnologie der Alten Donau bei Wien. Arch. Hydrobiol. 19: 301–317.

Rheinheimer, G., 1991. Mikrobiologie der Gewässer. 5. Auflage. Gustav Fischer Verlag. Jena Stuttgart. 294 pp.

Ripl, W., 1976. Biochemical oxidation of polluted lake sediment with nitrate. A new restoration method. Ambio 5: 112–135.

Ripl, W., 1978. Oxidation of lake sediments with nitrate. A restoration method for former recipients. Institute of Limnology, University of Lund. Coden Lunbds I (NBLI-100l) /1-151, ISSN 0348-0798.

Ripl, W., 1986. Restaurierung der Schlei. Bericht über ein Forschungsvorhaben (ed.), Landesamt für Wasserhaushalt und Küsten Schleswig-Holstein, D 5. - Kiel.

Ripl, W., 1995. Sanierung Alte Donau (Wien) – Begleituntersuchung zur kombinierten Eisen und Nitratbehandlung. Bericht an Magistratsabteilung 45 - Wasserbau. 76 pp.

Ripl, W. & G. Lindmark, 1978. Ecosystem control by nitrogen sediment metabolism. Vatten 34: 135–144.

Ripl, W. & K. D. W. Wolter, 1993a. Sanierungsmaßnahmen am Hambutten und Karutschenpfuhl. Berlin: TU-Berlin, Gesellschaft fur Gewässerbewirtschaftung mbH, im Eigenverlag.

Ripl, W. & K. D. W Wolter, 1993b. Sanierungsmaßnahmen am Gross-Glienicker See. Berlin: Tif-Berlin, Bericht an die Senatsverwaltung für Staatsentwicklung und Umweltschutz: 33 pp.

Sas, H., 1989. Lake Restoration by Reduction of Nutrient Loading. Expectations, Experiences, Extrapolations. Academia Verlag Richarz, St. Augustin.

Scheffer, M., S. H. Hosper, M.-L. Meijer, B. Moss & F. Jeppesen, 1993. Alternative equilibria in shallow lakes. TREE 8: 275–279.

Schiller, I., 1929. Neue Chryso- und Cryptomonaden in den AItwässern der Donau bei Wien. Arch. Protistenk. 66: 436–458.

Shapiro, J. & D. I. Wright, 1984. Lake restoration by biomanipulation: Round Lake, Minnesota, the first two years. Freshwat. Biol. 14: 371–383.

Singh, R. N., 1962. Seasonal variants of *Anabaenopsis raciborskii* Wolosz. Hydrobiologia 20: 87–91.

Smith, V. H., 1986. Light and nutrient effects on the relative biomass of blue-green algae in lake phytoplankton. Can. J. Fish. aquat. Sci. 43: 148–153.

Steiner, E., 1986. Untersuchungen zur Dynamik des Wasservogelbestandes an der Alten Donau und dessen Bedeutung für die ökologie dieses Gewässers. Bericht an die Wiener Magistratsabteilung 22: 42 pp.

Strickland, J. D. H. & T. R. Parsons, 1968. A Practical Handbook of Seawater Analysis. Bull. Fish. Res. Bd. Can. 167: 311 pp.

Utermöhl, H., 1958. Zur Verrollkommnung der quantitativen Phytoplankton Methodik. Mitt. Int. Ver. Limnol. 9: 1–38.

Van der Molen, D. T., F. J. Los, L. van Ballegooijen & M. P. van der Vat, 1994. Mathematical modelling as a tool for management in eutrophication control of shallow lakes. Hydrobiologia 275/276: 479–492.

Vanni, M. J., C. Luecke, J. F. Kitchell & J. J. Magnuson, 1990. Effects of planktivorous fish mortality on the plankton community of Lake Mendota, Wisconsin: implications for biomanipulation. Hydrobiologia 200/201: 329–336.

Varis, O., 1993. Cyanobacteria dynamics in a restored Finnish lake: a long term simulation study. Hydrobiologia 268: 129–145.

Vollenweider, R. & Kerekes, J., 1982. Eutrophication of Waters, Monitoring, Assessment and Control. OECD, Paris.

Waidbacher, H., O. Moog, B. Janecek, H. Kummer, H. Mann, A. Schmidt-Kloiber, H. Nesemann & W. Graf, 1996. Endbericht zu fischökologischen und benthosbiozönotischen Aufnahmen im Rahmen der 'Fe$C1_3$ Behandung' in der Alten Donau. Jahresbericht an die Wiener Magistratsabteilung 45 - Wasserbau. 102 pp.

Hydrobiologia **395/396**: 99–106, 1999.
D.M. Harper, B. Brierley, A.J.D. Ferguson & G. Phillips (eds), The Ecological Bases for Lake and Reservoir Management.

Direct and indirect mechanisms behind successful biomanipulation

Timo Kairesalo[1], Sirpa Laine[1], Eira Luokkanen, Tommi Malinen & Juha Keto[2]
[1]*Department of Ecological and Environmental Sciences, University of Helsinki, Niemenkatu 73, FIN-15140 Lahti, Finland*
[2]*Environmental Centre of Lahti, Niemenkatu 73, FIN-15140, Lahti, Finland*

Key words: biomanipulation, cyanobacteria, lake restoration, fishery management, participatory action

Abstract

Lake Vesijärvi is a relatively large (length 25 km; total area 110 km^2), shallow (mean depth 6 m), but stratified lake in southern Finland. The Enonselkä basin (26 km^2), surrounded by the city of Lahti, received its sewage effluent, and changed from a clear water basin with flourishing fisheries from the 1940–50s to one of the most eutrophic lake systems in Finland thereafter. In 1976, the sewage effluent was diverted, resulting in a temporary recovery of water quality. However, in the 1980s, massive surface scums of cyanobacteria degraded the water quality and arrested the recovery of the lake. A restoration strategy providing an ecologically sound basis for the management of the lake was initiated in 1987. This strategy involved biomanipulation (mass removal of coarse fish) together with conventional pollution control measures on discharges to the lake. Biomanipulation was chosen instead of much more expensive chemical and/or technical methods, such as chemical treatment or dredging of the profundal sediment. The large-scale biomanipulation trial was carried out in the Enonselkä basin during 1989–93. Following the mass removal of coarse fish (1000 metric tons of fish; mainly roach and smelt), the biomass of cyanobacteria collapsed concomitantly with a decline of total phosphorus concentration from 45 to 35 mg P m^{-3}, and with an increase of Secchi depth from 1 m to 3.5 m. These observed improvements in the water quality were matched with a large decline in roach-mediated phosphorus movement from littoral to pelagial, from 100 mg P m^{-2} in 1989 to 15 mg P m^{-2} in 1993. Year-to-year variation within the littoral communities, and in the recruitment of fish, could in this way cause large oscillations in the whole ecosystem. The involvement of local people (fishermen, farmers etc.) in controlling non-point nutrient loading and fish stock development, is of prime importance for the long term success of lake restoration.

Introduction

Lake Vesijärvi (Figure 1) is part of the River Kymijoki catchment in southern Finland, between the extensive Salpausselkä Eskers. The lake is characterized by a long retention time and a low number of islands. The hydrological characteristics of L. Vesijärvi are shown in Table 1. Due to the shallowness of the lake, two-thirds of the sediments in the Enonselkä basin are in direct contact with epilimnetic waters (depth <10 m) during stratification periods (May–September).

Lake Vesijärvi ('vesi' means 'water') was originally a clear-water lake, but its eutrophication started during the first decades of this century as a consequence of increased industrial and sewage effluent discharges from its surrounding urban community, the city of Lahti (Keto & Sammalkorpi, 1988). For instance, the first incidence of cattle poisoning in Finland, by cyanobacteria-rich lakewater, was documented in L. Vesijärvi as early as 1928 (Hindersson, 1933). In the 1960s and early 1970s, the lake became known as one of the most eutrophic large lakes in Finland. *Anabaena* and *Microcystis* blooms were common, and taste and odour problems in water and in fish occurred frequently, especially in the Enonselkä basin (Figure 1). In addition, during stratification periods, oxygen deficiency below 10 m depth and a complete depletion of oxygen in the deeper hypolimnion (20–30 m), were a regular phenomena. As a result, the Enonselkä basin was unsuitable for recreation and fishing (Keto, 1982).

During the 1960s and the 1970s, the annual phosphorus loading of sewage effluent to the Enonselkä basin was 50–60 metric tons, equivalent to 1.9–2.3 g P

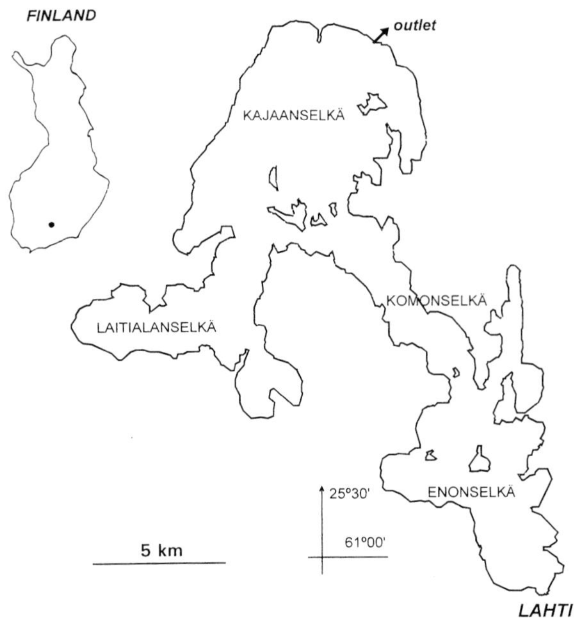

Figure 1. Map of Lake Vesijärvi representing the different lake basins, the location of the City of Lahti and the outlet of the lake.

Table 1. Hydrological properties of Lake Vesijärvi and the Enonselkä basin

Parameter	Whole lake	Enonselkä
Drainage area, km^2	515	84
Surface area, km^2	110	26
Volume at MSL, 10^6 m^3	663	176
Theoretical retention time, yr	5.4	5.6
Maximum depth, m	42	33
Mean depth, m	6.0	6.8

m^{-2}. This nutrient loading was drastically reduced in 1976 when a new biological–chemical sewage treatment plant was opened and the effluent waters were discharged to the River Porvoonjoki instead of to the lake. After the sewage diversion, the loading of phosphorus and nitrogen dropped to 8% of pre-diversion values and the Enonselkä basin showed obvious recovery (Keto, 1982). During winter stagnation periods, artificial aeration was applied in order to prevent development of oxygen deficiency in the hypolimnion. This improved oxygen balance in winter, and consequently, within the next two years the average winter concentration of phosphorus under ice decreased from 175 μg P l^{-1} in 1975, to 50 μg P l^{-1} in 1977 (Figure 2). The aeration, however, had no influence on hypolimnetic oxygen concentrations in summer, and due to high internal loading, phosphorus concentrations regularly doubled in the course of the summer. Hence, despite the gradual improvement of the winter and spring state, the water quality in the Enonselkä basin tended to degrade as a result of massive blooms of *Microcystis* and *Aphanizomenon* cyanobacteria over the longer-term (Keto & Sammalkorpi, 1988).

Overall, the Enonselkä basin provides a typical example of the problems encountered in the restoration of large, eutrophic lakes. Cyanobacterial blooms continued throughout the late 1970s and early 1980s despite traditional restoration measures which, for example, reduced the external phosphorus loading from 2.1 g P m^{-2} ann^{-1} in 1975 to 0.15 g P m^{-2} ann^{-1} in 1978 (Keto & Sammalkorpi, 1988), clearly below the critical level of 0.3 g P m^{-2} ann^{-1} of Vollenweider (1976). Traditional lake management measures were evidently not adequate to restore the lake.

Mesocosm experiments showed that roach (*Rutilus rutilus* (L.)), which in the course of the lake's eutrophication had developed a very dense population (Keto & Sammalkorpi, 1988), had a key role in maintaining high phytoplankton biomass and productivity in the basin (Horppila & Kairesalo, 1990, 1992). Consequently, biomanipulation was started in 1989 as an additional measure in the management program, i.e. removal of roach by trawling. Between the years 1989 and 1993, more than 1000 metric tonnes of fish (nearly 400 kg ha^{-1}) were removed; roach comprised 52% and smelt (*Osmerus eperlanus* (L.)) comprised 28% of the catch. This large-scale manipulation of the fish stock was followed by collapse of cyanobac-

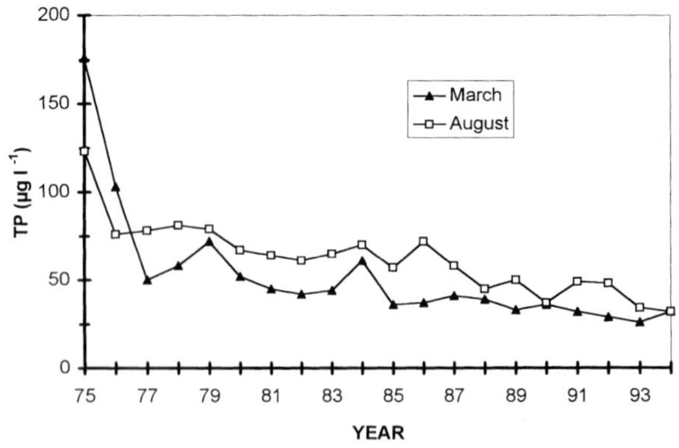

Figure 2. Volumetric mean concentrations of total phosphorus (μg P l^{-1}) during winter stagnation (under ice cover) in March, and during summer stagnation in August, in 1975–94.

terial blooms and then increased water transparency (Kairesalo et al., 1998a).

The mass removal of fish from the Enonselkä basin can be regarded as a large-scale disturbance to the ecosystem. Its execution, and subsequent developments, provide a good opportunity for monitoring and testing the dynamics of the lacustrine ecosystem, in terms of its resilience and resistance to disturbance (cf. Holling, 1973). It was clear that this disturbance exceeded the resistance of the system: the future development of water quality, however, will be a better measure of the resilience of the ecosystem. It is still not known, for instance, whether the roach stock can become dominant again to return the lake system to the 'old equilibrium' where cyanobacteria blooms can re-appear (Horppila & Peltonen, 1994) or whether the foodweb processes involved in the pelagic fish-zooplankton interactions (Horppila, 1994a; Luokkanen, 1995), and in the enlarged submerged macrophyte beds (Scheffer et al., 1993; Kairesalo et al., 1998b), can maintain the system in its new state. The human population of the surrounding urban and rural areas with all its socio-economic feedbacks, however, will be the most prominent but also the most uncertain factor in the lake's management. Here, local people and organizations have become more active as the rehabilitation of the lake has proceeded.

In order to evaluate the limnological changes, samples for physico-chemical analyses were taken monthly from the Enonselkä basin and those for phytoplankton 6–8 times during the growing season. Planktonic primary productivity was measured with the ^{14}C-method using illuminated (5000 lux) constant-temperature (20 °C) baths, with an exposure time of 24 hours. External nutrient loading to the Enonselkä basin was estimated from water samples taken from the major inflowing rivers and brooks during winter, spring, summer and autumn periods, and loading calculated according to nutrient concentrations and flow rates. The average drainage values of the different watersheds were then used for determining the seasonal and annual loadings. A description of the methods for zooplankton sampling and analysis is given by Luokkanen (1995). Algal biomasses given are weighted with volume.

More detailed descriptions of the eutrophication history of Lake Vesijärvi, as well as methods for analyses used, are given by Keto (1982), Keto & Sammalkorpi (1988), Liukkonen et al. (1993) and Kairesalo et al. (1998a). This paper presents the ecological basis for the present and future management of Lake Vesijärvi. Integration of ecologically sustainable measures with socio-economic activities based on regional, functional and administrational co-operation will form a central element of the holistic management plan and program.

Ecological basis for the management of Lake Vesijärvi

Generally, a reduction in the external nutrient load is always the primary restoration measure to provide a sound basis for the recovery of any eutrophic lake. Since phosphorus is most often, as in Lake Vesijärvi, the main factor controlling growth of cyanobacteria, the input reduction of this element has usually been regarded as the principal restoration measure (Sas, 1989). In the Enonselkä basin, different phases in the development of the summertime phosphorus concentration can be seen (Figure 2). At first, the phosphorus concentration responded promptly to the diversion of sewage waters but then remained at a high level, 60–70 μg l^{-1} until 1986, by which time the water mass of the Enonselkä basin had been replaced twice with inflowing waters (retention time 5.4 years). Thereafter, phosphorus concentration seemed to achieve a new 'steady state'-level of 40–50 μg P l^{-1} between 1988–92. From 1993 onwards, there seemed to be a slight decreasing trend in phosphorus concentrations. The observed difference between the August and March concentrations indicate the contribution of summertime loading to the phosphorous concentration (Figure 2).

The collapse of cyanobacteria resulted in a new 'equilibrium' level of the planktonic system, as revealed by the phosphorus/chlorophyll *a* ratio (Figure 3; Kairesalo et al., 1998a). Although cyanobacteria tended to disappear as phosphorus concentrations decreased, their disappearance was not continuous, but occurred rather in stepwise fashion as observed elsewhere (Sas, 1989). In the Enonselkä basin, the collapse of cyanobacteria (roughly, from 50% to 10% of the total algal biomass) occurred with the decline of the average epilimnetic phosphorus concentration from 45 to 35 μg P l^{-1} (Figure 3), which is lower than found for shallow, unstratified *Oscillatoria*-lakes (100–50 μg P l^{-1}), but distinctly higher than that (20–10 μg P l^{-1}) found in deep, stratified European lakes (Sas, 1989). The average annual concentration, however, does not illustrate the seasonal dynamics of phosphorus, which during the stratification may be

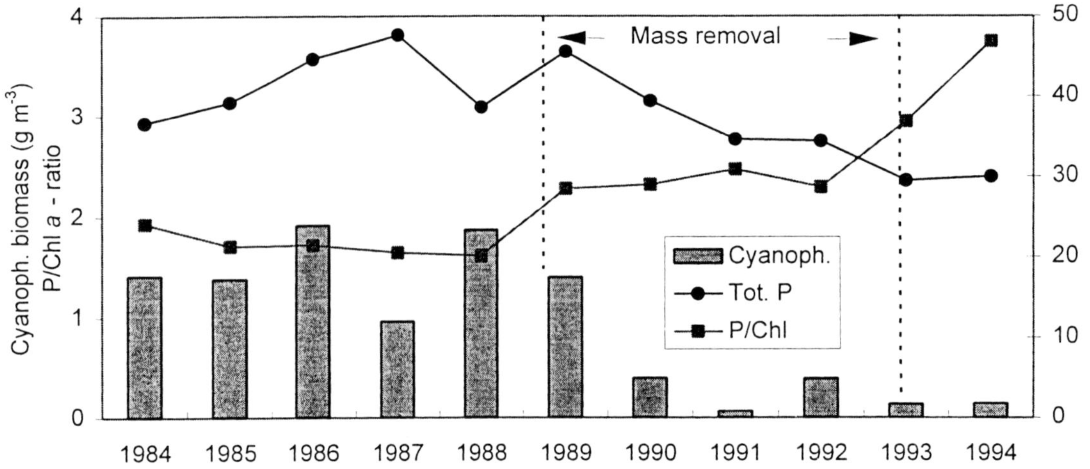

Figure 3. Summer mean concentrations of phosphorus (μg P l^{-1}, right axis); phosphorus/chlorophyll *a* ratio (left axis); and summer means of cyanobacteria biomass (g m^{-3}, left axis) in the epilimnetic water (0–5 m) in the Enonselkä basin in 1984–94.

largely steered by biotic processes and communities. In the Enonselkä basin, a stepwise increase in the epilimnetic phosphorus concentration was regularly recorded in early June during the 1980s (Figure 4), which could not be explained by the traditional means, i.e. neither by external- nor sediment-derived phosphorus inputs (Kairesalo, et al., 1995; Hartikainen et al., 1996). This phosphorus increase, however, coincided with the early-summer migration of adult roach populations from the littoral areas to the pelagic zone (Peltonen & Horppila, 1992; Horppila 1994b). During 1989–93, these migrating fish, in particular, were the target of the trawlings which led to a dramatic reduction in their biomass; from 172 kg ha^{-1} (i.e. about 100 mg P m^{-2}) in 1989 to less than 30 kg ha^{-1} (about 15 mg P m^{-2}) in 1993 (Horppila & Peltonen, 1994). Concomitantly, smaller amounts of sediment and phosphorus were re-suspended in lake water by the smaller roach populations feeding in the littoral environment (Horppila & Kairesalo, 1990, 1992). The fish phosphorus compartment in the Enonselkä basin had previously been of significant magnitude, that half of the limnetic phosphorus pool was bound to, and transported with, the fish biomass. This influenced the pelagic cycling of phosphorus and retention in algal biomass (Figure 3). Hence, the substantial decrease in the fish-mediated phosphorus input caused by the fish removal provided a reasonable explanation for the observed change and decline in the epilimnetic phosphorus concentration, which, consequently, resulted in the displacement of cyanobacteria (Kairesalo et al., 1998a). Temporal and spatial variation in the

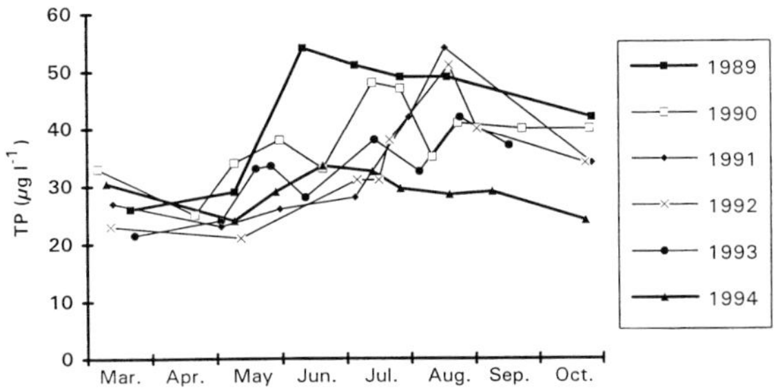

Figure 4. Seasonal variation in the total phosphorus concentration (μg P l^{-1}) in the epilimnion of the Enonselkä basin in 1989–1994.

fish-P compartment may therefore be of crucial, but still poorly quantified, importance in lake phosphorus budgets.

Planktonic primary productivity declined gradually after 1989 when it reached 1100 mg C m^{-3} d^{-1}, both in early July and in September (Figure 5). In 1993, the highest measured primary productivity was 460 mg C m^{-3} d^{-1} in late July, and in 1994, only 280 mg C m^{-3} d^{-1} in late June, very low values compared to the earlier years.

Despite the decline in the density of planktivorous fish there were no significant changes in the pelagic cladoceran community in the years 1991–1994 (Luokkanen, 1995; unpublished). The summer maximum biomass declined from more than 150 μg C l^{-1} to less than 100 μg C l^{-1} in 1994, but the change is largely due to the decline in the epilimnetic *Bosmina crassicornis* population. The biomass peak of this bosminid alone has declined from over 80 μg C l^{-1} in 1991 to less than 30 μg C l^{-1} in 1994. The

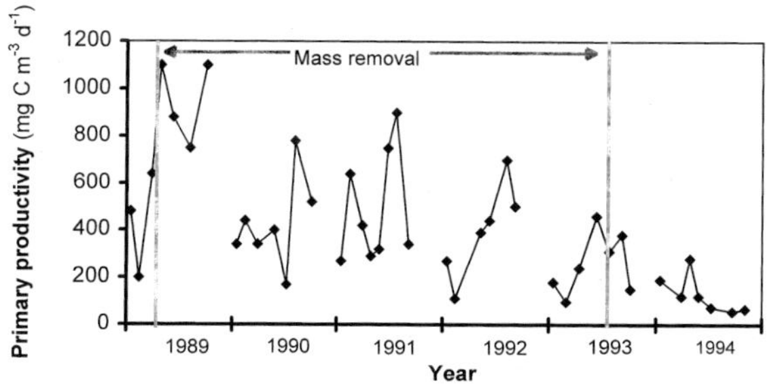

Figure 5. Summertime fluctuation and development of planktonic primary productivity in the Enonselkä basin in 1989–94.

daphnids, *Daphnia cucullata, D. longiremis* and *D. cristata* remained at the same level both in numbers and in biomass, and no increase in their average size was recorded. Thus the disappearance of cyanobacteria cannot be explained by any increased grazing by cladocerans.

The management imperative is now to maintain the improvements in water quality. In practice, this requires the mobilization and action of local people and organizations for the rehabilitation of the lake to achieve two ends – control of external nutrient loading, and maintenance of the fish stock with a balanced prey-predator ratio.

Present management of Lake Vesijärvi

Since the diversion of point-source sewage effluent in the late 1970s, measures for reducing the external nutrient loading into the lake have been focused on diffuse sources of nutrients from the surrounding agricultural areas. Accordingly, during 1991–1994, management plans for reducing the use of fertilizers have been made individually for each of the 321 farms on the drainage area. These management programs have been supported financially from several regional and local funding sources.

It has been estimated that without an active exploitation policy the roach stock would double within 3 years (Horppila & Peltonen, 1994). Former manifestations of eutrophication, driven by roach-mediated nutrient reallocation, could then return. In order to prevent this, the roach population was modelled at different levels of simulated natural and fishing mortalities (Horppila & Peltonen, 1994). This has led to a policy of removal by local fishermen (and other people) of about 50 metric tonnes of roach each year from the Enonselkä basin, using large pound nets and fykes (made by the fishermen).

An alternative method of controlling the 'unwanted' fish stocks is to increase the stocks of predatory fish, formerly weak in this lake. Before biomanipulation, the only predatory fish inhabiting the pelagic zone was perch (*Perca fluviatilis* L.). Over 1 million fingerling pikeperch (*Stizostedion lucioperca* (L.)) were stocked during years 1984–1991, with the intention of increasing predation pressure on roach and smelt. The minor goal was to increase the value of the fishery. The restocking proved successful, as in 1992 natural spawning took place. The growth of stocked pikeperch was also rapid compared to other lakes in Finland (Ruuhijärvi, unpublished), indicating that a dense pikeperch stock can consume a large fraction of the planktivorous fish production. Diet studies showed that small pikeperch fed mainly on smelt, whereas the diet of large ones also comprised perch and roach (Peltonen et al., 1996).

The stocking was a victim of its own success however, because the species rapidly became the most valuable in the gill-net fishery, particularly in the Enonselkä basin. The fishing regulations were very liberal, neither mesh size limits, nor a limit to the number of fishermen, were applied. The minimum size limit for catchable pikeperch (370–400 mm) had no effect in practice because more than 95% of catch was captured with gill nets (Ruuhijärvi, unpublished), providing no chance for released fish to survive. Intense competition between fishermen forced the mesh sizes used, down to 80 mm (stretched mesh) and the catch then consisted mostly of pikeperch in the age groups 3–4 years. This led to a serious risk of overfishing because most female pikeperch in Lake Vesijärvi mature at 5 years old (Ruuhijärvi, unpublished). It was also very likely that, under such pressure, the food consumption by pikeperch was only a fraction of its maximum value (Peltonen, unpublished data). The situation was also undesirable for the fishery itself, as according to other studies in Finland, the recruitment age producing the maximum total yield is 5–7 years (Lehtonen, 1987; Lehtonen & Miina, 1988).

In Finland, water areas are mostly privately owned and decison-making process in fishery management is very complicated, and hence often ineffective. Thus, although the need for an increase in legal minimum mesh size was obvious, no restrictions were applied before 1996, when the classical Y/R analysis was performed with the growth data from Lake Vesijärvi, and the problem was presented in the form of a simple model. Such analyses were not done earlier because data were available only on the growth of pikeperch. However, the Y/R model showed that the conclusions were valid on a wide range of natural mortalities, selection curves and growth speeds. On the basis of this very simple modelling (Figure 6), a decision to set the minimum legal mesh size at 100 mm (stretched mesh) was finally made by the local fishery authorities in April 1996. This is perhaps only a small step towards the optimum mesh size but it is hoped that it will at least protect the stock from extinction.

This mesh size limit will be applied only in the Enonselkä basin. In other areas, especially in the Kajaanselkä basin (Figure 1), the gill-net fishery of whitefish (*Coregonus lavaretus* (L.)) and perch is also important, and the increase of mesh size would perhaps lead to large declines in the yield of these species. Finding an appropriate management strategy for mixed gill-net fisheries is a general problem in Finland where gill nets are also widely used in recreational and 'household' fishing, in contrast to many other countries. The objectives of these fisheries are totally different from those of a commercial fishery.

Future management of Lake Vesijärvi

The future management of Lake Vesijärvi will be based on the integration of ecologically sustainable measures with the participation of local people. The Department of Ecological and Environmental Sciences of University of Helsinki has started a project to study and develop this kind of integration. The project is co-financed by the European Union Life Programme and the City of Lahti. People in the areas surrounding Lake Vesijärvi had already been educated about the problems of lake rehabilitation during the period of mass removal of fish in 1989–93. For the present and future management of the lake, it is even more important to inform the public and thus gain essential support for efforts to maintain the lake in its rehabilitated state. However, mere education is not enough. For efficient and sustainable lake management, increased ecological awareness must be combined with a more intensive participation of the local population.

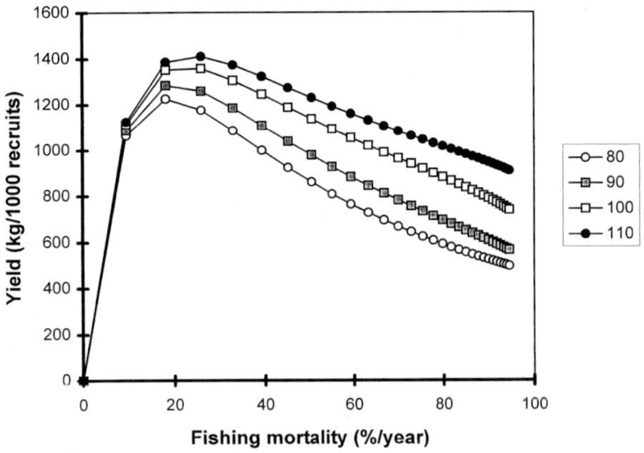

Figure 6. The yield per 1000 recruits as a function of fishing mortality with four minimum mesh sizes. The current mesh size is 80–90 mm (stretched mesh) and fishing mortality very high (more than 80%), new allowable minimum mesh size is 100 mm.

Since the rehabilitation of the lake has improved opportunities for recreation and fishing, the future of Lake Vesijärvi is strongly dependent on the commitment of lake users to the ecologically sustainable management. Commitment of people makes the regulatory measures of authorities less necessary and the rehabilitation thus becomes a continuous process.

Local people will be involved in lake management by means of participatory planning. Mere consultation of the people is no longer considered sufficient. People and their organizations need to be involved throughout the project cycle, from design stage to monitoring and evaluation. The opportunity to participate effectively in the different phases presupposes flexibility in planning and organization of the project. The mode of action most suitable for this is that of action research, conducted as a learning process for all involved (Schneider & Libercier, 1995).

Participatory action research is a cyclical inquiry process that includes a diagnosis of the problem, plans for action, implementation and evaluation of outcomes. The approach aims to help people develop a higher degree of self-determination and self-development capability, so that learning continues after researchers leave the system (Elden & Chisholm, 1993). A careful and often lengthy process of observation, analysis and consultation forms the initial phase of the project. It offers opportunities for the population to put forward its views of problems and priorities. Of importance are differences in the positions and power constellations of the various groups, or categories of the population and the possible conflicts of interest between them (Schneider & Libercier, 1995). Participatory action research emerges over time as a process; it does not appear fully at the outset in most situations. Each case starts as an attempt to solve a particular kind of problem and gradually opens up into a much broader and deeper research process (Greenwood et al., 1993).

The holistic management project of Lake Vesijärvi will start with ascertaining the goals and preferences of fishermen. Fishermen are considered to have a central position in the process of encouraging local people to take responsibility for ecological lake management. Lake Vesijärvi serves an urban population base of 150 000 and there is thus a large pool of potential fishermen who can choose to fish on the lake when fishing conditions are good. Representatives of fishermen will participate in personal computer-aided interviews. The interviews are performed by using the HIPRE 3+ software (Hämäläinen & Lauri, 1993). The decision analyst conducts the interviews, operates the computer and familiarises the stakeholder with the decision analysis methods used. When the objectives of various groups of fishermen – such as anglers and net fishermen – have been defined, the utility outcomes of different management strategies can be computed by ecological modelling. However, because of the stochastic nature of many processes in the ecosystem, these outcomes are uncertain. The uncertainty is modelled with Bayesian decision analysis, which, especially the approach using influence diagrams, has been shown to be a useful tool for analyzing environmental management problems (Varis et al., 1990; Varis & Kuikka, 1990; Varis, 1995). In an influence diagram, controllable (decisions) and uncontrollable (probabilistic or deterministic) variables are connected with one-directional links (Figure 7). The approach is based on the Bayes' theorem, which makes it possible to calculate the probable distribution of the outcomes of different management options.

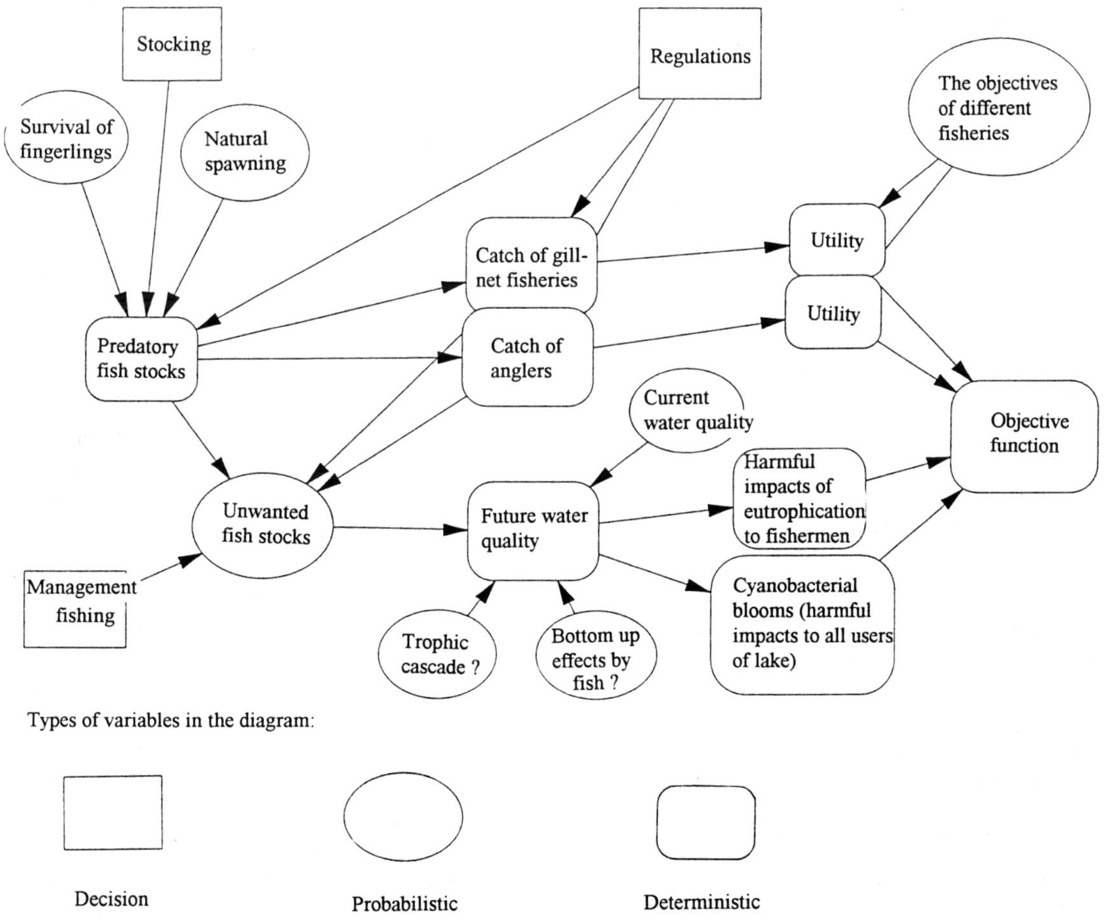

Figure 7. A prior Bayesian influence diagram about the management problem of Lake Vesijärvi. The costs of management options are not presented here.

An important advantage of the interactive method is that by active participation the commitment of stakeholders to the chosen alternative can be increased. The computer-supported interactive decision analysis is a practical and an efficient way of dealing with conflicting opinions and improving citizens' participation in environmental project planning and evaluation (Marttunen & Hämäläinen, 1995).

Acknowledgements

This ecosystem-scale manipulation and management program was financially supported by the City of Lahti, the Academy of Finland and the European Union Life Programme.

References

Elden, M. & R. F. Chisholm, 1993. Emerging Varieties of Action Research: Introduction to the Special Issue. Human Relations 46: 121–142.

Greenwood, D. J., W. F. Whyte & I. Harkavy, 1993. Participatory Action Research as a Process and as a Goal. Human Relations 46: 175–192.

Hartikainen, H., M. Pitkänen, T. Kairesalo & L. Tuominen, 1996. Co-occurrence and potential chemical competition of phosphorus and silicon in lake sediment. Wat. Res. 30: 2472–78.

Hindersson, R., 1933. Förgiftning av nötkreatur genom sötvattensplankton. Finsk veterinär Tidskrift 39: 179–189.

Holling, C. S., 1973. Resilience and stability in ecological systems. Ann. Rev. Ecol. Syst. 4: 1–23.

Horppila, J. 1994a. Interactions between roach (*Rutilus rutilus* L.) stock and water quality in Lake Vesijärvi (southern Finland), Ph.D. Thesis, Helsingin yliopiston Lahden tutkimus- ja koulutuskeskuksen tutkimuksia 5/1994.

Horppila, J. 1994b. The diet and growth of roach (*Rutilus rutilus* L.) in Lake Vesijärvi and possible changes in the course of biomanipulation. Hydrobiologia 294: 35–41.

Horppila, J. & T. Kairesalo, 1990. A fading recovery: the role of roach (*Rutilus rutilus* L.) in maintaining high algal productivity and biomass in Lake Vesijärvi, southern Finland. Hydrobiologia 200/201: 153–165.

Horppila, J. & T. Kairesalo, 1992. Impacts of bleak (*Alburnus alburnus*) and roach (*Rutilus rutilus*) on water quality, sedimentation and internal nutrient loading. Hydrobiologia 243/244: 323–331.

Horppila, J. & H. Peltonen, 1994. The fate of a roach (*Rutilus rutilus*) stock under an extremely strong fishing pressure and its predicted development after the cessation of mass removal. J. Fish Biol. 45: 777–786.

Hämäläinen, R. P. & H. Lauri, 1993. HIPRE 3+ User's guide, Systems Analysis Laboratory. Helsinki University of Technology.

Kairesalo. T., S. Laine, T. Malinen & M. Suoraniemi, 1998a. Comprehensive approach to the rehabilitation and management of lake Vesijärvi (Southern Finland). Boreal Environment Research (in press).

Kairesalo, T., I. Tatrai & E. Luokkanen, 1998b. Impacts of *Elodea canadensis* on fish-plankton interactions in lake littoral. Verh. int. Ver. Limnol. 26: 1846–1851.

Kairesalo, T., L. Tuominen, H. Hartikainen & K. Rankinen, 1995. The role of bacteria in the nutrient exchange between sediment and water in a flow-through system. Microb. Ecol. 29: 129–144.

Keto, J., 1982. The recovery of Lake Vesijärvi after sewage diversion. Hydrobiologia 86: 195–199.

Keto, J. & I. Sammalkorpi, 1988. A fading recovery: A conceptual model for Lake Vesijärvi management and research. Aqua Fennica 18: 193–204.

Lehtonen, H., 1987. Selection of minimum size limit for pike-perch (*Stizostedion lucioperca*) in the coastal waters of Finland. Proc. V Congr. europ. Ichthyol., Stockholm 1985: 351–355.

Lehtonen, H. & T. Miina, 1988. Minimum size of pike-perch (*Stizostedion lucioperca* (L.)) for exploitation in Lake Lohjanjärvi, Southern Finland. Aqua Fennica 18: 157–164.

Liukkonen, M., T. Kairesalo & J. Keto, 1993. Eutrophication and recovery of Lake Vesijärvi (South Finland): Diatom frustules in varved sediments over a 30-year period. Hydrobiologia 269/270: 415–426.

Luokkanen, E., 1995. The species composition, biomass and production of the pelagic cladoceran community in the Enonselkä basin of Lake Vesijärvi. M.Sc. Thesis, Helsingin yliopiston Lahden tutkimus- ja koulutuskeskuksen raportteja ja selvityksiä 25/1995 (in Finnish; summary in English).

Marttunen, M. & R. P. Hämäläinen, 1995. Decision analysis interviews in environmental impact assessment. European Journal of Operational Research 87: 551–563.

Peltonen, H. & J. Horppila, 1992. Effects of mass removal on the roach stock of Lake Vesijärvi estimated with VPA within one season. J. Fish Biol. 40: 293–301.

Peltonen, H., H. Rita & J. Ruuhijärvi, 1996. Diet and prey selection of pikeperch (*Stizostedion lucioperca* (L.)) in Lake Vesijärvi analysed with a logit model. Ann. Zool. Fenn. 33: 481–87.

Sas, H. (ed.), 1989. Lake Restoration by Reduction of Nutrient Loading: Expectations, Experiences, Extrapolations. Academia Verlag, Richarz: XXI + 497 pp.

Scheffer, M., S. H. Hosper, M-L. Meijer, B. Moss & E. Jeppesen, 1993. Alternative equilibria in shallow lakes. Trends in Ecology & Evolution 8: 275–279.

Schneider, H. & M. H. Libercier (eds), 1995. Participatory Development from Advocacy to Action, Development Centre Seminars. Organisation for Economic Co-operation and Development, Paris.

Varis, O., 1995. Resource Management and Bayesian Decision Analysis. MODSIM 95, International Congress on Modelling and Simulation. Newcastle, NSW, Australia, 27–30 November.

Varis, O., J. Kettunen & H. Sirviö, 1990. Bayesian influence diagram approach to complex environmental management including observational design. Computational Statistics & Data Analysis 9: 77–91.

Varis, O. & S. Kuikka. 1990. Analysis of Sardine Fisheries Management of Lake Kariba, Zimbabwe and Zambia - Structuring a Bayesian Influence Diagram Model. International Institute for Applied Systems Analysis WP-90-48.

Vollenweider, R., 1976. Advances in defining critical loading levels for phosphorus in lake eutrophication. Mem. Ist. ital. Idrobiol. 33: 53–83.

Hydrobiologia **395/396:** 107–114, 1999.
D.M. Harper, B. Brierley, A.J.D. Ferguson & G. Phillips (eds), The Ecological Bases for Lake and Reservoir Management.

Minimising the risk and amplifying the opportunities for restoration of shallow reservoirs

Maciej Zalewski
University of Lodz, Poland

Key words: shallow reservoir, eutrophication, restoration, ecohydrology

Abstract

The ecohydrological approach is based upon understanding and integration of processes which regulate water and biota dynamics at catchment scale. Such an approach has recently been applied to the restoration programme of a eutrophic reservoir, by integrating classical methods with those of landscape ecology and recent advances in biogeochemistry at catchment scale. Restoration measures include rehabilitating buffer zones, renaturalising river channel morphology, regulating the flow regime of the incoming river as a means of modifying the nutrient supply, creating wetland systems and applying a variety of biomanipulation techniques.

Introduction

The most important problem for freshwater management is eutrophication (Harper, 1992). The most vulnerable systems to its impacts are reservoirs, which operate as traps for nutrients along the river continuum. The 25 km^2 Sulejow Reservoir in Poland, supplies drinking water for 1 million people, and is such an ecosystem undergoing eutrophication. Its low average depth (3 m) and short retention time (30 days) are the reasons for its high vulnerability to nutrient loading. On the other hand, the forested land use and low human population density in its catchment (the Pilica River; the largest western tributary of the Vistula River) indicate that serious opportunities for successful restoration exist (Zalewski, 1995). The implementation of a catchment-scale project using the new 'ecohydrological' approach is being undertaken, despite our limited ecological and hydrological knowledge (Zalewski et al., 1997). To avoid large scale failure through inadequate experience, the way chosen is the use of 'Adaptative Assessment and Management of Environment' (Gunderson et al., 1995). The basic assumption of this approach is that each stage of the restoration plan should be under permanent monitoring and assesment. The efficiency of restoration measures undertaken should lead to incorporation of corrections as necessary.

The use of ecohydrology in restoration

The ecohydrological approach involves the integration of three components – catchment, water and biota – into an holistic model of river systems at basin scale, analogous to the 'Platonian superorganism', thus implicitly identifying the management objectives as the maintenance of sustainability, biodiversity, water quality and quantity. The general concept of ecohydrology has been based on the two fundamental components of successful strategic action – elimination of threats and amplification of chances. If the action is focused only on one of these components, it never leads to long-term success (Kolodziejski, 1995).

To date, water management, and consequently the spending of societies' taxes, has concentrated on the elimination of threats. The major threats and their elimination have been:

1. Catastrophic floods and droughts, by building dams and levees and canalisation of rivers.
2. Pollution, by construction of sewage treatment plants.
3. Erosion, by planting trees or reduction of slope of rivers valleys by terraces.

However, most of these measures often do not provide the improvements expected, despite heavy expenditure. The lack of positive results may be attributed to neglect of the second component of successful strategic action – amplification of chances.

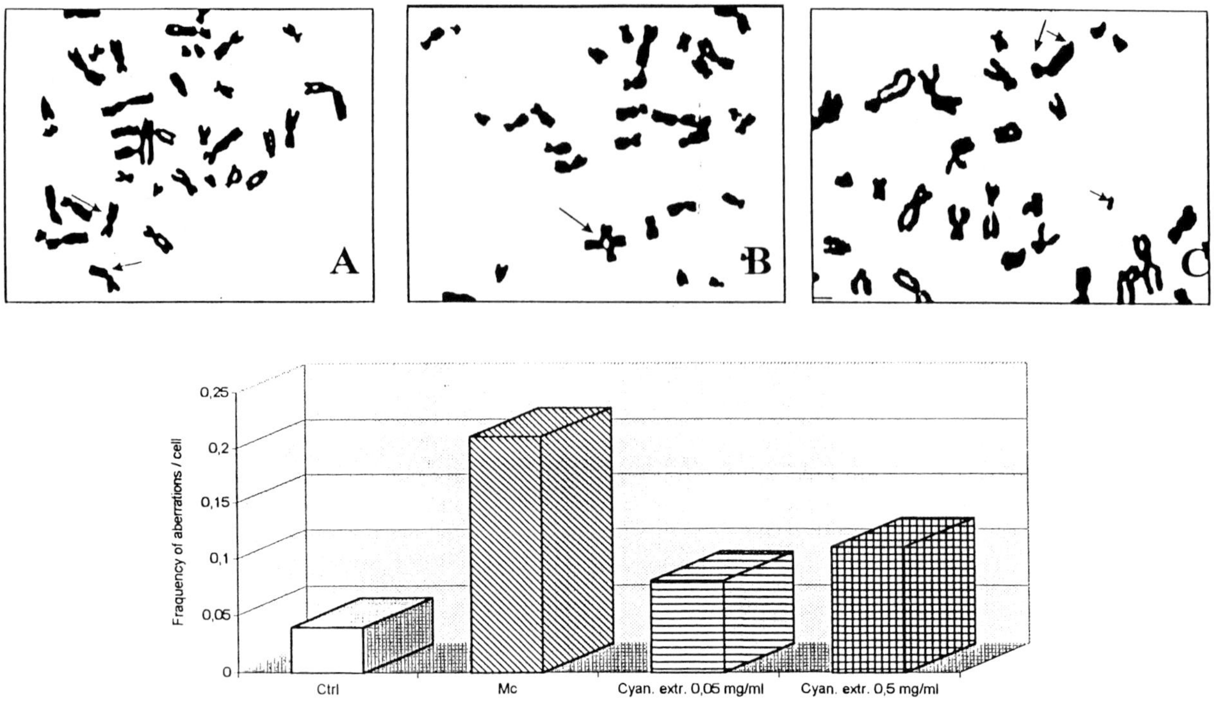

Figure 1. Chromosomal aberration induced by extract from a cyanobacterial bloom from Sulejow reservoir on *in vitro* human lymphocytes: (a) chromatic breaks, (b) chromatid exchanges, (c) dicentric chromosomes and acentric fragment, (d) frequency of aberrations in human lymphocytes., Ctrl – control, Mc – mechloretamina 10-7 M – model mutagen (Kontek et al., 1997).

What does amplification of chances mean in the case of conservation of freshwater resources? The key is to use ecosystem properties, created by evolutionary processes, as the management tool. Crucial properties are the integration of the resistance/robustness and resilience of ecosystems against various forms of human impact (Zalewski et al., 1997).

The dynamics of biota in freshwater ecosystem are established as a consequence of evolutionary adaptations to hydrological processes. Thus, the efficient restoration of eutrophic reservoirs needs to be based on understanding and integration of hydrological and ecological processes.

Proposals for the holistic restoration of reservoirs and their catchments

Recent evidence has accumulated that the most adverse symptoms of eutrophication of lakes and reservoirs are the appearance of toxic and carcinogenic cyanobacterial blooms, which are more potent than strychnine. For instance, concentration of microcistine-LR exceeding 900 μg g^{-1} phytoplankton dry mass (Tarcsynska et al., submitted), which has been often recorded in the Sulejow Reservoir, is considered dangerous for human health (Codd et al., 1994). Malformation of human chromosomes caused by toxins produced by cyanobacteria, even in small concentrations, are similar to those found when using highly toxic pesticides (Figure 1).

Even non-toxic algal blooms are dangerous for drinking water quality, as their occurrence requires increased doses of chlorine during water treatment. This in turn may lead to the increase of concentrations of carcinogenic products in the potable water. Both cyanobacterial toxins and halogens in potable water define the restoration limit, and should be related to measurable values for nutrient retention time, morphometry and thermal conditions.

The first and most fundamental step towards efficient restoration should be the reduction or elimination of phosphorus from point sources of pollution in the catchment of the reservoir. Phosphorus loads from different sources to Sulejow Reservoir catchment are shown in Figure 2.

Phosphorus removal may be achieved by construction, upgrading of existing sewage treatment plants and building of new sewers which transfer sewage from the reservoir catchment to a treatment plant be-

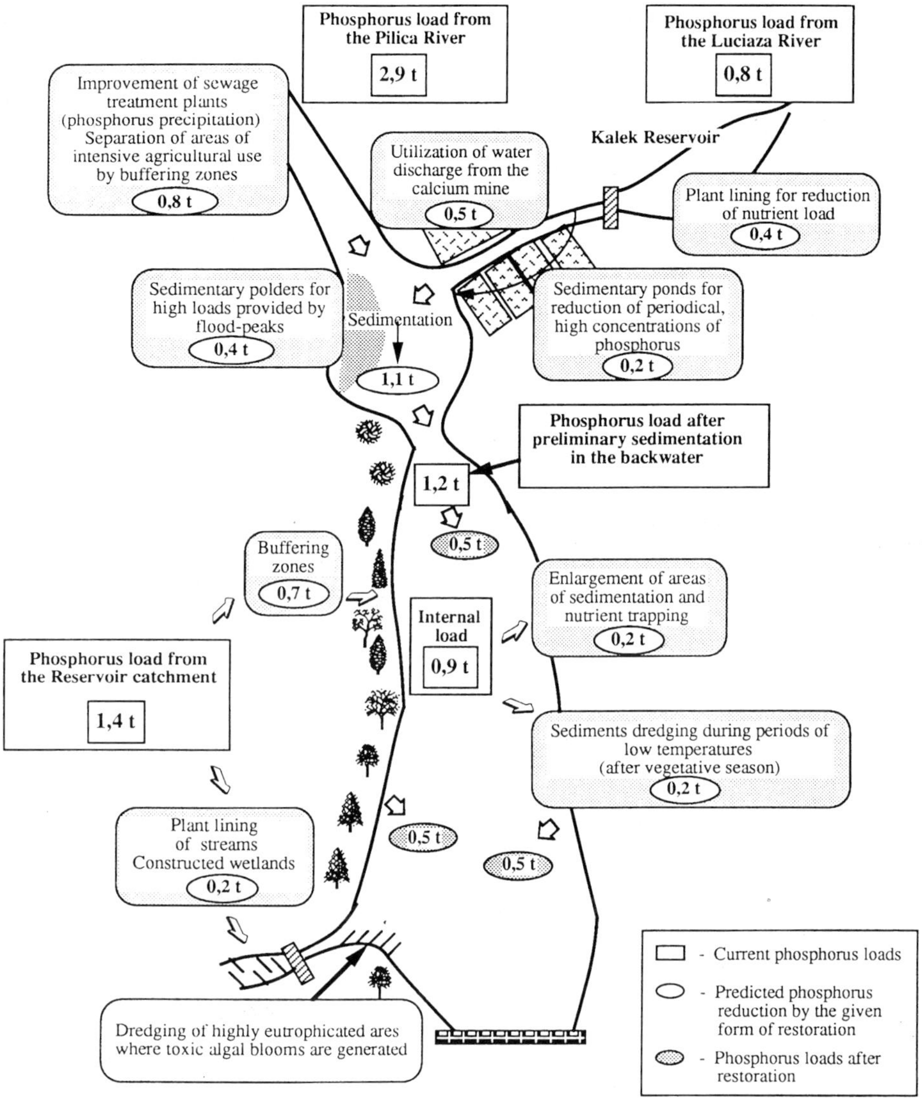

Figure 2. Quantitative evaluation of the efficiency of various methods for reduction in phosphorus input to Sulejow reservoir on the example of phosphorus load in July – the most critical period due to the intensive recreational use and appearance of toxic blooms.

low the reservoir. In parallel, dredging of areas of mud deposition is necessary, in late autumn when the temperature is low. The main reason for the dredging has been toxic cyanobacterial blooms. In Sulejow Reservoir the *Microcistis* bloom was generated in the highly eutrophicated stream delta near the dam, and transported up into the middle reservoir by wind (Figure 3). Such areas could be dredged or alternatively inactivated by constructed wetlands.

The second stage should be control of biogeochemical processes in the catchment and biotic interactions in the aquatic medium. Among the most important, is the reduction of non-point source pollution by land/water ecotone management and, where necessary, restoration. It has been demonstrated that along the river or reservoir edge, a 50 m buffer zone of coniferous trees reduces 48% P and 95% N (Ryszkowski et al., 1996), but similar efficiency can be achieved by mixed broad leaf forest 16 m wide. In small tributaries, constructed wetlands which possess high nitrogen and phosphorus trapping capacity (Mitsch, 1996) should be developed. However the increase of retention and self-purification capacity by creation of ecotones of intermediate complexity which retain phosphorus ef-

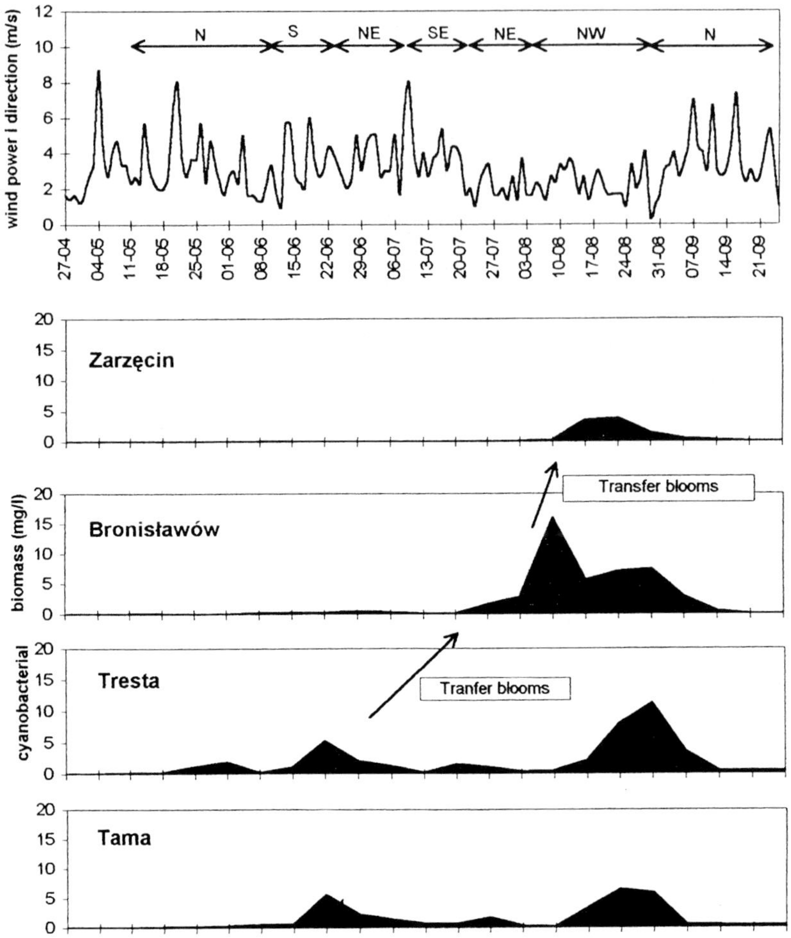

Figure 3. The generation of a toxic cyanobacterial bloom in a eutrophic bay and its wind-induced transfer to upper sites (Izydorczyk & Tarczynska, in press).

ficiently in a diversified stream channel (Figure 4) and maximize the energy flow through the trophic pyramid also has been important (Zalewski et al., 1994; Zalewski et al., in press).

The third phase of restoration should be the control of hydrological processes to achieve optimal timing of nutrient supply and/or release to and from the reservoir. It is well known that the intensity of algal blooms in summer in some reservoirs depends on the pattern of the water release from the reservoir at the end of the previous winter (Straskraba, 1994).

The fourth phase should be the use of hydrological processes to regulate important biotic interactions. A crucial one is control of water level at the spawning period of planktivorous fish to regulate the trophic cascade toward reduction of algal blooms. Regulation of water level during spawning period is an efficient way for reducing an excessive density of young fish, and so for maintaining a high density of filtering zooplankton. In Sulejow, reducing the juvenile fish density at the beginning of the summer to a level less than 5 m^{-2}, stabilised the zooplankton biomass around 15 mg l^{-1}. This was sufficient to reduce biomass of algae by 80% and avoid toxic algal blooms (Zalewski et al., 1990).

The final phase, which may be achieved by controlling the hydrological conditions, should be the use of biological processes for increased resistance and resilience of reservoir ecosystems against excessive nutrient loads, e.g. creation of a vegetated zone as a nutrient trap, which is possible if water level is not too variable. An example of a large functioning constructed wetland is the Kis Balaton, a shallow reservoir covered by reeds at the delta of the River Zala, which reduces up to 80% of the nutrient load into Balaton

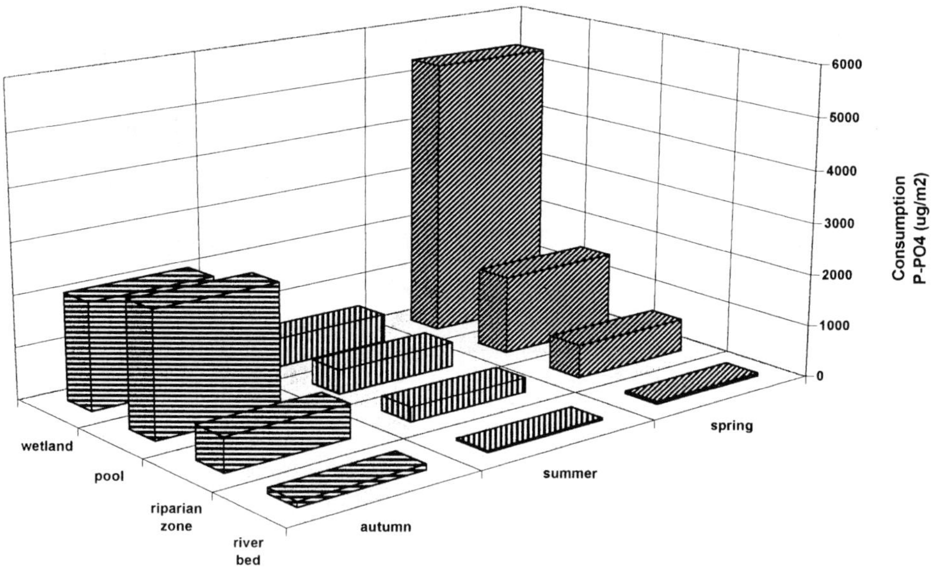

Figure 4. Phosphate consumption by sediments of different ecotone types (Zalewski et al., 1998).

Lake (Pomogyi, 1993). Such hydro-constructions incorporate natural elements of the landscape to function as traps for excessive nutrient loads transported, especially by floods. If they are combined in the river system with ecotone buffer zones, which retain nutrients, reduce erosion and transport of eroded material downstream, such measures may improve water and aquatic environment quality at very low cost.

Evidence is accumulating that, in the aquatic environment, management and conservation of biotic structure can be an important factor increasing tolerance to human impact, thus potentially reducing the costs of water quality management. For instance, in lakes with a well-developed macrophyte zone, dense algal blooms appear at 2–3 times higher concentrations of phosphorus, than in lakes or reservoirs where littoral macrophytes cannot develop due to water level fluctuations (Scheffer et al., 1993). One reason may be that aquatic plants reduce the tool of recirculating nutrients and additionally create refuge areas for zooplankton thus stabilising the effect of the trophic cascade.

Internal loading often occurs in lakes as a consequence of resuspension of bottom sediments by wind, algal recruitment (Hansson, 1996) bioturbation of bottom sediments by benthic feeders (bream, carp) and recirculation of phosphorus in epilimnion due to zooplankton excretion (Ejsmont-Karabin et al., 1993) and enzymatic activities (Chrost, 1994). Resuspension can be minimised by sediment removal or conversion to shallow wetlands. Bioturbation can be controled by selective fish community exploitation: at Sulejow Reservoir, anglers are very efficient at the selective capture of bream using small spherical feeding baskets (5 cm ø) at the end of a fishing rod line to which one to several hooks on short lines are attached. Use of nets for selective fish removal is not appropriate because pikeperch is the fish most vulnerable to such exploitation yet one of the most valuable element controlling planktivores for management of the trophic cascade.

Integration of restoration measures

The example of two processes, one of which controls nutrient recirculation and the other the intensity of algal bloom by strong ‘top down’ pressure are illustrated in Figure 5 (Tarczynska et al., submited). At the most representative station in the middle of the reservoir, the three stages of equilibrium between nutrients, phyoplankton and zooplankton can be distinguished. The first was around 31 May, when zooplankton biomass at 60 mg l^{-1} was sufficent to maintain phytoplankton biomass below 2 mg l^{-1}, but the grazing process caused the release of a large pool of phosphorus (80 μg l^{-1}). The second was when most of the phosphorus had accumulated in zooplankton, to 146 mg l^{-1}, which maintained phytoplankton below 5 mg l^{-1}. At this stage the great density of zooplankton generated two potential effects:

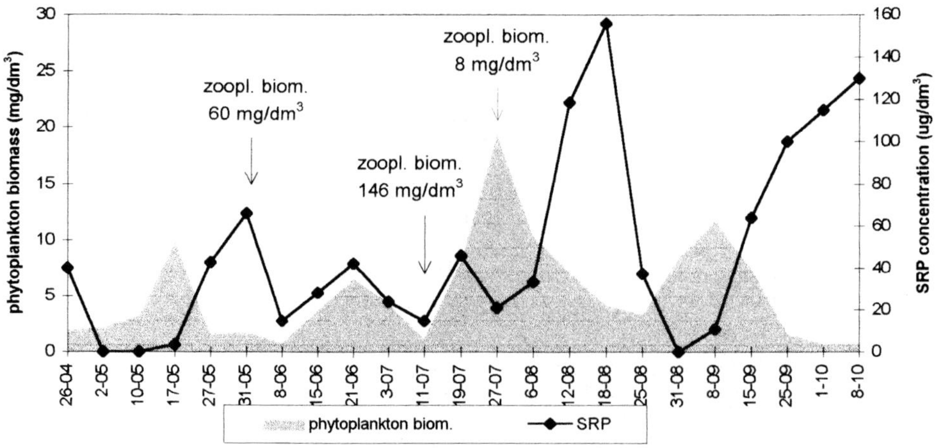

Figure 5. Comparison of soluble reactive phosphorus with phytoplankton and zooplankton biomass in a representative station of Sulejow reservoir in 1996 (Tarczynska et al., submitted).

1. A breakdown in the overcrowded *Daphnia* population due to low food supply; and
2. A 'threshold shift' among the food preferences of fish: in periods of high food density even benthivorous feeders such as bream (*Abramis brama*) eat zooplankton).

In the above case, one or both of these consequences stimulate the transfer into the third stage of equilibrium, in which drastic reduction of zooplankton to 8 mg l^{-1} occurs (27 July). In parallel, the dynamic development of an algal bloom to 20 mg l^{-1} and decline of phosphorus in the photic zone to a level of a few micrograms occurred. Such instability of the cascading effect has been characteristic of the system where, in Spring, the phosphorus level is well above 100 μg l^{-1} (Zalewski, 1995). In shallow lakes, when phosphorus reaches this Spring level, the cascade effect usually stabilises phytoplankton at low concentrations (Jeppesen et al., 1990). For this reason, the reduction of nutrient supply from the catchment of the reservoir has been fundamental.

Figure 6 illustrates the value of integrating so many phases of restoration. In the past the general assumption was that to improve water quality it was sufficient to invest money only in sewage purification plants. However, Statzner & Sperling (1993) demonstrated that in many situations, e.g. for small settlements, it is less expensive and more efficient to integrate moderate technology treatment plants with biological subsystems (e.g. nutrient trapping constructed wetland) than upgrade treatment plants by expensive technology.

This example shows that the use of ecological and hydrological processes as a tool for upgrading the quality of freshwater resources, can be an efficient way to reduce costs and increase the efficiency of protection and restoration activities, thus improving the sustainability of the systems.

The strategies proposed here for restoration of water quality in Sulejow Reservoir are now being implemented, since low water quality not only affects the costs of treatment, but also reduces the popularity of the area for recreation. This in turn reduces the income of local people and subsequent taxes which might themselves be available for reinvestment in environmental improvement. The stimulation of economic with recreational use, balanced against the ecological carrying capacity (increased by restorative measures) represents a substantial step towards holistic management of a reservoir catchment.

Acknowledgements

This research was supported by the Committee of Scientific Research of Poland Grant 6 PO4F 01014, grant 6 PO4C 09410, Voivodeship Fund of Environmental Protection and Water Management in Lodz, Department of Civil Ingineering of the City of Lodz. The authors wish to thank Dr David Harper for valuable discussions on the manuscript.

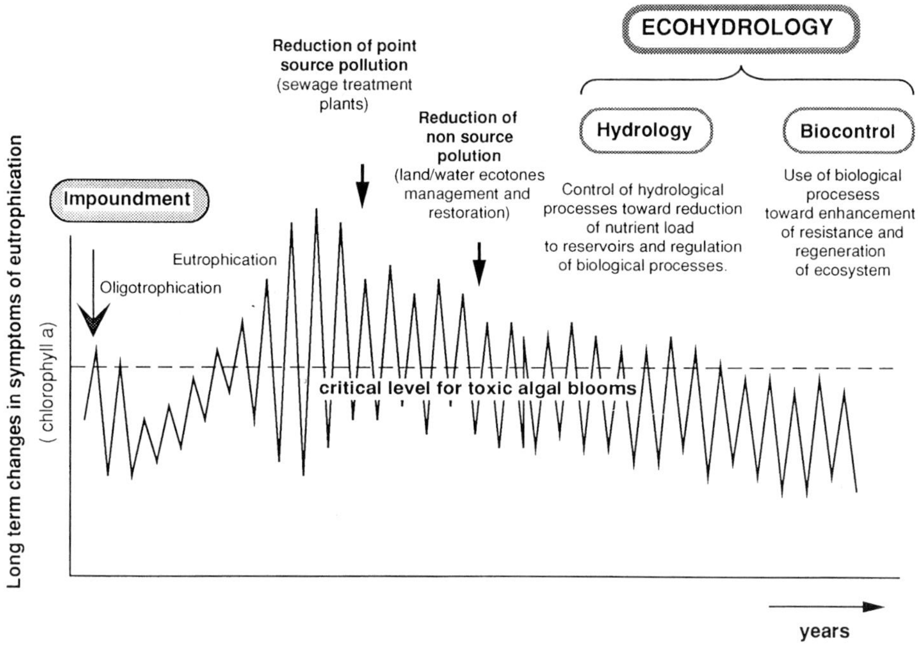

Figure 6. Hypothetical scenario of the application of ecohydrology to the restoration of Sulejow reservoir (Zalewski et al., 1997).

References

Chrost, R. J., 1994. Microbial enzymatic degradation and utilization of organic matter. In J. Overbeck & R. J. Chrost (eds), Microbial Ecology of Lake Pluásee. Springer-Verlag, New York: 118–174.

Cood, G. A., T. M. Jeffries, C. W. Keevil & E. Potter, (Eds), 1994. Detection Methods for Cyanobacterial Toxins. The Royal Society of Chemistry, Special Publication 149, Bookcraft, Bath, 191 pp.

Cowx, I. (ed.), 1994. Rehabilitation of Freshwater Fisheries. Fishing News Books. Blackwell Scientific Publ. Ltd., Oxford: 152–160.

Ejsmont-Karabin, J., J. Krolikowska & T. Weglenska, 1993. Patterns of spatial distribution of phosphorus regeneration by zooplankton in a river-lake transitory zone. In A. Hillbricht-Ilkowska & E. Pieczynska (eds), Nutrient Dynamics and Retention in Land Water Ecotones of Lowland, Temperate Lakes and Rivers. Hydrobiologia 251: 275–284.

Gunderson L. H., C. S. Holling & S. Light, 1995. Barriers and Bridges to Reneval of Ecosystems and Institutions. Columbia University Press, New York.

Hansson, L.-A., 1996. Algal recruitment from lake sediments in relation to grazing, sinking and dominance patterns in the phytoplankton community. Limnol. Oceanogr. 41(6): 1312–1323.

Harper, D., 1992. Eutrophication of Freshwaters. Principles, problems and restoration. Chapman & Hall. London, New York, Tokyo, Melbourne, Madras, 327 pp.

Izydorczyk, K. & M. Tarczynska, Identification of areas of toxic algal bloom generation in Sulejow Reservoir. Hydrobiologia (submitted).

Jeppesen, E., J. P. Jensen, M. Sondergaard, E. Mortensen, O. Sortkjaer & K. Olrik, 1990. Fish manipulation as a lake restoration tool in shallow, eutrophic temperate lakes 2: threshold levels, long-term stability and conclusions. Hydrobiologia 200/201: 219–227.

Kolodziejski, J., 1995. Uwarunkowania ekorozwoju w okresie transformacji systemowej Polski. In S. Kozlowski (ed.), Prognoza Ostrzegawcza Zmian Srodowiskowych Warunkow Zycia W Polsce Na Poczatku XXI Wieku. Zeszyty Nauk. PAN. Komitet Naukowy przy Prezydium PAN 'Czlowiek i Srodowisko' 10: 9–34 (in Polish).

Kontek, R., Osiecka, R., Blaszczyk, A., Tarczynska, M., Zalewski, M. & P. Luchniak, 1997. Toxins produced by freshwater cyanobacteria as potential environment mutagens. Proceedings of Spring symposium on Plant Cytogenetics, 19–22 May 1997, Cieszyn, Poland.

Mitsch, W.J., 1996. Ecological engineering: a new paradigm for engineers and ecologists. In P. C. Schulze (ed.), Engineering with Ecological Constrains. National Academy Press, Washington, DC: 111–128.

Pomogyi, P., 1993. Nutrient retention by the Kis-Balaton Water Protection System. In A. Hillbricht-Ilkowska & E. Pieczynska (eds.), Nutrient Dynamics and Retention in Land Water Ecotones of Lowland, Temperate Lakes and Rivers. Hydrobiologia 251: 309–320.

Ryszkowski L., I. Zyczynska-Baloniak & B. Szpakowska, 1996. The influence of bio-geochemical barriers on the expansion of non-point pollution. In M. Kraska & R. Blazejewski (eds), Constructed Wetlands for Wastewater Treatment. II Scientific-Technical Conference, Poznan, 2–3 September 1996: 147–156.

Scheffer, M., S. H. Hosper, M-L. Meijer, B. Moss & E. Jeppesen, 1993. Alternative equilibria in shallow lakes. Trends in Ecology & Evolution 8: 275–279.

Straskraba, M., 1994. Ecotechnical models for reservoir water quality management. Ecological Modelling 74: 1–38.

Statzner, B. & F. Sperling, 1993. Potential contribution of system-specific knowledge (SSK) to stream management decisions: ecological and economic aspects. Freshwat. Biol. 29: 313–342.

Tarczynska, M., Izydorczyk K., Trojanowska A., Beattie K., Codd G. A., Romanowska-Duda Z. & M. Zalewski, (submitted). The dominant role of biotic interactions in dynamics of toxic cyanobacterial bloom in shallow reservoir.

Zalewski, M., B. Berwinska-Zaras, P. Frankiewicz & S. Kalinowski, 1990. The potential for biomanipulation using fry communities in a lowland reservoir: concordance between water quality and optimal recruitment. Hydrobiologia 200/201: 549–556.

Zalewski, M., 1995. Freshwater habitat management and restoration in the face of global changes. In N. B. Armantrout (ed.), Condition of the Worlds Aquatic Habitats. Proceedings of the World Fisheries Congress, Theme 1. Oxford & IBH Publishing Co. Pvt. Ltd., New Delhi: 170–194.

Zalewski, M., W. Puchalski, M. Nowak, P. Frankiewicz & B. Bis, 1994. Riparian ecotones and fish communities in rivers – intermediate complexity hypothesis. In I. Cowx (ed.), Rehabilitation of Freshwater Fisheries. Blackwell Scientific Publications, Oxford.

Zalewski M., G. S. Janauer & G. Jolankai (eds), 1997. Ecohydrology – A new Paradigm for the Sustainable Use of Aquatic Resources. International Hydrological Programme, UNESCO, Paris: 58 pp.

Zalewski M., Bis B., Lapinska M., Frankiewicz P. & W. Puchalski, 1998. The importance of the riparian ecotone and river hydraulics for sustainable basin-scale restoration scenarios. Aquatic Conservation: Marine and Freshwater Ecosystems. 8: 287–307.

Section Three

New and Refined Ecological Tools for Lake Management

Hydrobiologia **395/396:** 117–121, 1999.
D.M. Harper, B. Brierley, A.J.D. Ferguson & G. Phillips (eds), The Ecological Bases for Lake and Reservoir Management.

Do reservoirs need ecological management?

D.R.H. Price
Anglian Water, Services plc, Endurance House, Vision Park, Histon, Cambs CB4 47Y, U.K.

Key words: rainfall, water supply reservoirs, eutrophication, water treatment

Abstract

Eastern England, the driest part of England, has the fastest population growth and water demand in the UK. Existing supplies taken from groundwater have to be supported by surface water, in pumped storage reservoirs supplied from highly eutrophic rivers. These reservoirs are managed primarily by technical means to produce high quality drinking water but there is recognition that intensive recreational use has to be a compatible part of their management. In the future progress towards a more sustainable, ecologically-based management is possible.

Introduction

Eastern England is the driest part of the U.K. with average annual rainfall of only 610 mm. It also has a faster population growth rate than any other region. These factors have created a need for major water resource development schemes over the last 40 years, including the construction of a number of large water supply reservoirs, which today have virtually all the characteristics of natural lakes. Indeed, there are no truly natural lakes of any consequence in the region, as even the Norfolk Broads which form a very important aquatic environment, are man-made and result from peat digging in medieval times.

Against this background, the water supply reservoirs of the region assume a special importance in both the provision of adequate water supplies and a lake environment with the many benefits which that brings. The majority of the reservoirs are owned and operated by Anglian Water, although Essex and Suffolk Water also have reservoirs in the eastern part of the region.

Within England and Wales, roughly one third of water supplies comes from upland catchments, one third from aquifers and one third from lowland river systems. However, 52% of the water which Anglian Water supplies to its customers, comes from lowland river systems. There are no significant uplands in eastern England, and because of this flat topography, the vast majority of the river water is pumped into large storage reservoirs before being abstracted again for treatment and supply. In this way, some two million people receive surface water derived supplies from Anglian Water.

The water supply reservoirs

The current policies for managing these man-made lakes have their origins largely in the approach adopted during the planning and design of Grafham Water near Huntingdon in the early 1960s. This pumped storage reservoir fed from the river Great Ouse is some 635 hectares in area and has a volume of 59 million cubic metres at top water level.

Grafham Water was designed to provide very full public access for a wide variety of recreational uses and also to provide important areas for nature conservation. This was in marked contrast to many earlier water supply reservoirs where public access was severely restricted or totally prohibited. However, more extensive – and hence more costly – water treatment was required in these circumstances, but in practice this also proved necessary to deal with the eutrophic waters which are characteristic of eastern England.

When Rutland Water in Leicestershire, the region's largest reservoir (surface area 1255 ha and volume 124×10^6 m^3) and Alton Water in Suffolk were constructed some years later, similar approaches were adopted, as they subsequently have at the other reservoirs managed by Anglian Water.

The principal recreational uses which the reservoirs support are:

1. General recreation (walking, cycling, taking picnics etc.).
2. Angling.
3. Sailing and windsurfing.
4. Bird watching.
5. Education (e.g. local history and culture, environmental studies).

Conservation of flora and fauna has become an increasingly important feature of such reservoirs and they are particularly noted for their waterfowl. At Rutland Water the nature reserve extends to 210 ha, at Grafham Water to 163 ha and at Pitsford Reservoir to 194 ha. All three are Sites of Special Scientific Interest and Rutland Water is also a Ramsar site.

These man-made lakes today play a very important part in the life of the region combining water supply use with many others. Yet ironically 45 years ago the sites of reservoirs such as Pitsford, Grafham and Rutland were farmland, and when proposals were made for their construction, there was considerable opposition raised. One could speculate on the degree of opposition there would be today to any proposal to drain the reservoirs and return the land to agriculture.

Co-existence

When lakes are being called upon to meet a variety of different uses, the question must inevitably be asked as to whether they can satisfy such a range of uses, or whether multiple use creates major conflicts leaving all dissatisfied.

Certainly there is considerable potential for conflict. Nature conservation interests call for constancy of water level and minimum disturbance. Yet the primary use of the reservoir for water supply, of necessity, leads to fluctuating water levels. The reservoirs are filled mainly in the winter to avoid the need to abstract heavily from rivers during the summer, and thus cause unacceptable environmental harm. Recreational users wish to circumnavigate the reservoirs on foot or bicycle and this in turn has the potential to cause serious disturbance in the nature reserve areas. And, of course, sailors and anglers both use boats in rather different ways with the conflicts which this can bring.

A degree of zoning and of compromise provides much of the answer, with prohibition on the majority of reservoirs of some activities (such as power boating and water skiing) which cannot be accommodated without conflict.

The success of the policies adopted can be seen in the results achieved. During the intensely hot dry spell, which was experienced in the U.K. in 1995, all water companies were stretched to maintain supplies to their customers. The large reservoirs operated by Anglian Water played a crucial role and despite being the driest region in the country, supplies were maintained without imposing any bans on water use by customers.

The recreational use of Anglian Water's reservoirs continues to grow with over one million visitors a year. Angling is a particularly important use which continues to attract large numbers because of the excellent trout fishing that several of the reservoirs provide. In 1995/96 the total trout catch by anglers for the reservoirs at Grafham, Pitsford, Ravensthorpe and Rutland was 85,191 fish. Habitat creation has been developed actively in the nature reserves, and in 1994 Rutland Water was the runner up in the British Trust for Ornithology/Zeneca Challenge for the number of bird species recorded during the year at a large site with a figure of 183 species.

A relatively recent issue has been the increasing number of cormorants present and the damage that they create to the fish stocks. The anglers wish to see cormorant numbers controlled by shooting, whereas the conservationists are opposed to such measures. In the event the problem has been mitigated to some degree by the following measures:

1. Deferring spring stocking until mid-March to allow the winter migrant cormorant population to disperse.
2. Increasing the average size of trout stocked into the reservoir so making them less vulnerable to predation.
3. Distributing the introduced trout more widely over the reservoirs.
4. Removing coarse fish from Grafham Water and thus reducing the availability of food for cormorants.

Active and sympathetic management of the reservoirs is essential if serious conflict is to be avoided and a sensible compromise achieved. It would be wrong to suggest there are no conflicts or at least potential ones. They do arise and clearly, ways must be sought to overcome them. The lakes remain, first and foremost, water supply reservoirs, but conflicts are sometimes more apparent between the secondary uses rather than between primary and secondary uses.

Water supply use

The reservoirs were constructed to provide a reliable source for public water supply – one which is buffered against short-term variations in both flow and quality in the river from which the water originates. They are filled principally by pumping as they only have very limited natural catchments.

A consequence is that reservoirs are generally drawn down more than most natural lakes. This clearly has important implications for the ecology of shallow areas but is an inescapable consequence of their water supply role. However, river ecology is protected by having filled the reservoirs at times of higher river flows and thus having removed the need to abstract at times of low river flow.

A key issue for water supply is the quality of stored water in the reservoir. There are marked seasonal patterns of water quality in the rivers from which the water is abstracted. For example, sediment loads and nitrate concentrations tend to be greatest in winter, whereas phosphorus and sulphate concentrations are normally at their highest in summer. Thus, the majority of water entering the reservoirs will tend to be relatively high in nitrate and in suspended matter.

However, extended storage for a period of several months brings very real benefits in terms of water quality. Apart from the obvious effect of smoothing out short-term quality variations seen in river waters, the natural physical and biological processes taking place in the reservoir also generate significant quality changes.

The effect upon nitrate concentrations is particularly important as the biological processes in the reservoirs allow the limit of 50 mg l^{-1} (or 11.3 mg l^{-1} as N) set in the European Drinking Water Directive to be met without further nitrate removal. The magnitude of the reduction in nitrate concentration in Grafham Water can be seen in Figure 1. The concentrations of ammonia and coliform bacteria also decline markedly (Figures 2 and 3). These changes are important in facilitating the subsequent water treatment, and traditionally the storage of water in a reservoir for an extended period (in some instances for over 12 months) has been regarded as a valuable first stage of water treatment.

However, not all changes which take place in reservoir quality are beneficial to the water supply use. Some water quality characteristics can vary very considerably through the seasons as a result of primary production and water treatment processes must therefore be adequate to cope with this wide range of variation. Today, treatment plants taking water from these lowland reservoirs are extremely sophisticated and complex. They must be capable of removing algae and zooplankton effectively and also very low concentrations of organic substances which could otherwise pose a threat to health or impart unpleasant taste and odour to the water.

The water treatment and supply difficulties which can arise from heavy algal growths include the following:

1. Filter blockage.
2. Carry-over into supply of iron salts used in water treatment.
3. Unpleasant tastes and/or odours.
4. Production of trihalomethanes (THMs) during water treatment and distribution.
5. Potential presence of algal toxins.
6. Presence of mucopolysaccharides causing problems for some industrial water users.
7. De-oxygenation of the hypolimnion with the potential for the release into the water of various constituents in the sediments.

The problems are addressed in a number of ways. The use of microstrainers to remove zooplankton and the use of double layer (anthracite/sand) or triple layer (anthracite/sand/garnet) filters has improved filtration processes. Anglian Water introduced granular activated carbon treatment at all surface water treatment plants in the 1980s to control taste and odour problems. In the early 1990s, these were uprated and ozonation was added, primarily to remove very low concentrations of pesticides to meet the very stringent European limit of $0.1\mu g\ l^{-1}$. This additional treatment also has the important benefit of removing a wide range of trace organic compounds.

Despite these important (and costly) improvements to treatment it has been necessary to look to other ways to deal with some of the concerns. Some opportunity normally exists to change, to a limited degree, the depth at which water is abstracted from a reservoir or even use an alternative draw-off point elsewhere in the reservoir. However, beyond this, one must look to reservoir management to address the problem.

Anglian Water has the facility to de-stratify all of its reservoirs and these facilities have been improved considerably in recent years. This is done by the injection of air on the bed of the reservoir to induce an upward water flow from the hypolimnion. Today, these facilities are run throughout the summer period to prevent stratification developing.

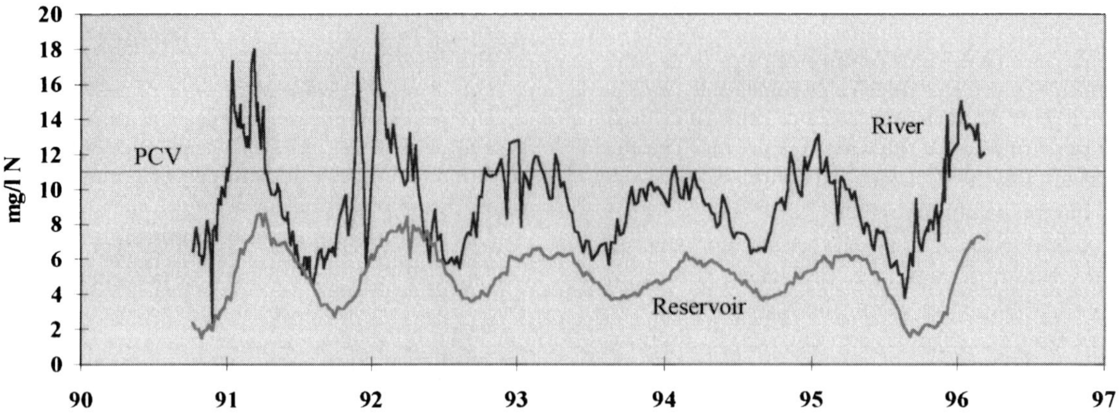

Figure 1. Nitrate concentration compared between Grafham Water and the River Ouse, its supply river.

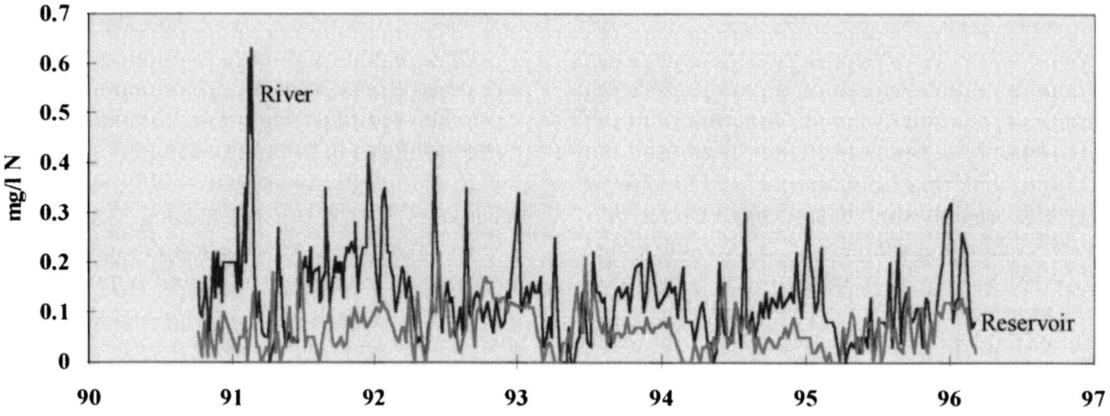

Figure 2. Ammonia concentration in Grafham Water and the River Ouse.

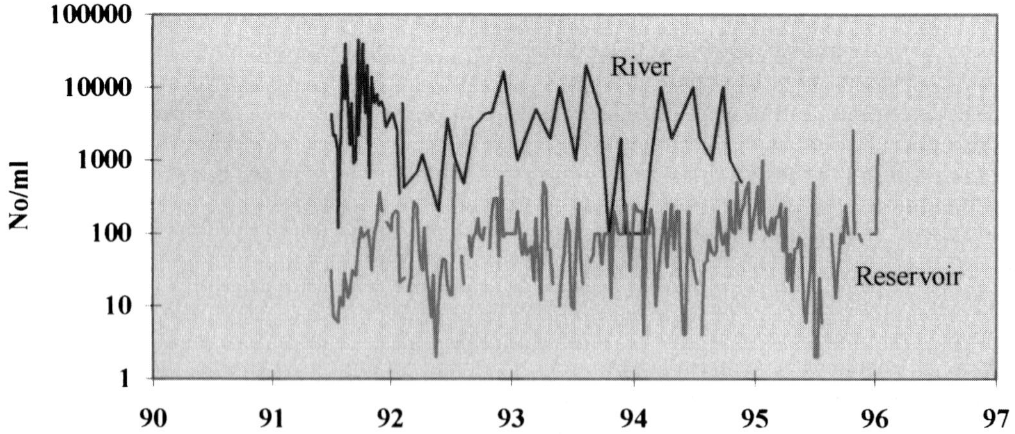

Figure 3. Coliforms concentration in Grafham Water and the River Ouse.

The other major measure is to control orthophosphate concentrations in the reservoirs. Throughout the past decade, this has been done by injecting ferric sulphate solution into the water pumped into the reservoirs from the river, or in some special cases, by the introduction of ferric sulphate solution from a boat.

Iron salts were first used by Anglian Water for reservoir phosphorus control in 1981 at Foxcote Reservoir near Buckingham. The technique was subsequently extended to some other waters, and following massive blue-green algal blooms in 1989, it was introduced at the larger reservoirs including Rutland Water. In 1989, there was serious concern for both the water supply and recreational use of the reservoirs and following confirmation that the heavy cyanobacteria scums at the shoreline were toxic, all Anglian Water's reservoirs were temporarily closed to recreational use. The view was taken that phosphate concentrations in the reservoirs should be reduced as rapidly as possible. Thus, the quantities of iron needed initially to reduce orthophosphate concentrations were relatively high but they have subsequently fallen dramatically.

The management target adopted was to reach and maintain orthophosphate concentrations at or below 10 μg P l^{-1}. In this the strategy has been largely successful (Brierley & Harper, 1999; Randall et al., 1999). The exception was Grafham Water, where prior to the control programme, orthophosphate concentrations were of the order of 1000 μgP l^{-1}. It proved impracticable to achieve the target without elevating sulphate levels to the point where the European drinking water limit of 250 mg l^{-1} was at risk of being exceeded. Background sulphate concentrations in the river waters are fairly high and can exceed 200 mg l^{-1}, and further sulphate is added as part of the water treatment processes.

The speed of ecological response to phosphorus control has varied from reservoir to reservoir but following an initial lag phase, shifts have been seen both in algal species composition and chlorophyll *a* concentrations (Daldorph, 1999). Treatability of the water has been substantially improved, and recreational uses have not again been curtailed by cyanobacterial scums.

Experiences gained in the management of these reservoirs have also demonstrated the importance of careful planning at the design stage to minimise subsequent operational problems. For example, reservoir depth and the location of water draw-off points in relation to the prevailing wind direction both have major influences upon the magnitude of the problems caused by algal growth.

The Future

Current methods of phosphorus control are expected to continue at several reservoirs, but efforts are being made to find alternative means of phosphorus control, and indeed alternative ways of reducing algal growth in reservoirs. Experimental phosphorus removal has already been undertaken at a number of large sewage treatment works discharging to the upstream river catchments, and in accordance with the EC Urban Waste Water Treatment Directive, phosphorus removal has been undertaken routinely at these works from 1998. Anglian Water first installed phosphorus removal at sewage treatment works in the Norfolk Broads area in 1977 and experience gained from this work has proved to be valuable in planning and design elsewhere.

If all sewage-derived phosphorus were removed from river systems (which in any event is not practicable with existing technology) then the orthophosphate concentrations in the water pumped into the reservoirs would still be some 50–100 times the current target of 10 μg l^{-1}. Therefore some treatment of the pumped water would still be necessary to achieve such a target – albeit with a lower iron usage.

The phosphorus control programme is one which has ben followed in consultation with both the National Rivers Authority (now succeeded by the Environment Agency) and English Nature. Both organisations have issued formal consent for the work to be undertaken. Looking to the future the three organisations have formed a working group to develop in partnership a strategy for managing water quality into the 21st century, which will move closer to the shared ideal of ecologically managed, sustainable water supply reservoirs.

References

Brierley, B. & D. Harper, 1999. Ecological principles for management techniques in deeper reservoirs. Hydrobiologia 395/396 (Dev. Hydrobiol. 136): 335–353.

Daldorph, P. W. G., 1999. A reservoir in management-induced transition between ecological states. Hydrobiologia 395/396 (Dev. Hydrobiol. 136): 325–333.

Randall, S., D. Harper & B. Brierley, 1999. Ecological and ecophysiological impacts of ferric dosing in reservoirs. Hydrobiologia 395/396 (Dev. Hydrobiol. 136): 355–364.

Hydrobiologia **395/396:** 123–131, 1999.
D.M. Harper, B. Brierley, A.J.D. Ferguson & G. Phillips (eds), The Ecological Bases for Lake and Reservoir Management.

Modelling phytoplankton dynamics and its application to lake management

C. S. Reynolds
Institute of Freshwater Ecology, The Ferry House, Ambleside, Cumbria LA22 0LP, U.K.

Key words: phytoplankton, dynamics, modelling

Abstract

Conceptual models assist the understanding of complex, multivariate problems. Management models should offer reliable predictions of the outcome of alternative approaches to problems which assist the manager to decide the optimal course of action. It is rare for one model to fulfil both purposes. Excessive phytoplankton production in lakes, reservoirs and rivers presents, at best, a potentially conspicuous detraction from water quality or, at worst, a lethal cocktail which must be excluded from recreational waters and potable supplies. Thus, the difficulties may relate as much to the type of organisms present and to the problems they may cause as they do to the biomass that may be achieved. Examples will be cited of some particular problems of lake and reservoir management that have been confronted in recent years. The range of model solutions available to plankton biologists is reviewed. The philosophies of these are unsympathetic to specific management problems; the models are shown to be unhelpful and potentially misleading in the context of the questions usually asked. Even when quite general questions are submitted to generalised models, imprecision can lead to erroneous judgements. Approaches to making much more process-based models and expert systems are advocated. The ability to identify and quantify the principal regulatory processes in operation, including the effects of light, turbidity and physical mixing and those relating to the trophic structure, is highlighted. Dynamic simulations, based upon the population ecology of several selected species simultaneously can give reasonable fits to observable phenomena. Applying altered model components to simulate viable options can be tested for their likely comparative impacts. A yet more recent approach to modelling lake metabolism is introduced for its potential as a guide to determining management impacts and priorities at particular sites. The objective of the paper is to encourage the development of site-specific functional models which are oriented to both conceptual and management issues. We have to overcome the widespread but naive "my lake is phosphorus-limited" syndrome if we are to learn how to better manage our standing waters.

Introduction

There can have been few periods in the past when public interest in the United Kingdom over the management of the nation's water resources was more acute. Popular concern over health-related aspects of recreational water and the quality of product delivered to the tap is expressed frequently. In the past, quality of water was a matter for the engineers – to design and operate purification plants to the capacity of the water to be supplied and against the magnitiude of the defilement of the raw resource that nature might throw at them. Meanwhile, the academics continued to bolster their applications for grant-aid with the justification that their research 'would benefit the water industry', even if nobody was quite sure how or, if it were true, how the technology could be passed on. Then water became a matter for the accountants, who identified 'efficiency savings' if they could reduce the scale of treatment problems, and for the public relations representatives whose job is to assure the wider public that the toxic cyanobacterial scums that kill their dogs cannot possibly affect the water drawn from the kitchen tap, even if both come from the same reservoir. Over the last 5 years or so, there has been an explosion of interest in the causes of poor quality of raw source waters in lakes and reservoirs and how these might be overcome or avoided in the first place. At the very time that ecology was set to figure more prominently in management, the corresponding efficiency drive within the academic community had, by then, depleted its knowledge base and eroded its core capability so far that qualified advice has been in short supply and even that is perceived to be expensive.

It is against this background that the urgent need for managers and academics to evolve ecologically based strategies for the management of lakes and reservoirs should be approached. They need to understand each other's problems and they have to talk in the same language (This is so self-evident to need saying!). The widespread availability and power of the personal computer provide the vital bridge between those who think they know what happens and those who wish to monitor and regulate it appropriately: 'can we make a model of that?'. A well-developed artificial intelligence is surely the best assistance to the manager on a day-to-day basis, provided that it is robustly based and the extent of its reliability is clearly known. Not only is it uncertain whether these provisions are always met, there is also a further difficulty in applying models if the author's objectives differ significantly from those of the user or if the variance inherent in biological systems is not accommodated.

This paper begins by comparing the probable perceptions about models of their authors and their users, with special reference to phytoplankton growth in source waters. The nature of the specificity of the problems encountered by managers is briefly illustrated by some summarised case histories, while the generalised nature of familiar models are shown to provide little guidance to managers about how the problems might be overcome. The paper goes on to advocate models which are dynamically based and driven by components whose worth is strongly site-sensitive. Some recent models which have adopted this approach are reviewed and are suggested to hold great promise for managers in the near future.

What do we want from models?

The ability to make a mathematical model which simulates a natural system or process is a measure of the modeller's understanding of that system. Nowadays, however, models are popularly equated with computer outputs. Certainly, the power of the modern PC allows models to be quickly assembled, run and rejected. The last is important, for the power of a modern PC also exceeds, often considerably, the wisdom of the operator, to the extent that the simulation of any but the simplest responses of selected aquatic biota to a single stimulus can scarcely represent the almost infinite and stochastic variability of the fluid, three-dimensional environment. How this issue is resolved is greatly influenced by what is wanted from the model. Among reservoir managers, for instance, there is a strong desire for site-specific models which will process relevant information about the condition of the system and a few key inputs to yield a probabilistic prediction of events. The black box of internal processes is acknowledged to be complicated and to be within the province of the physiological ecologist. In order to be able to generate the appropriate algorithms, however, the ecologist may follow one of two further approaches: either to probe exhaustively the relevant internal workings, component by component, and then build simulations of component functions; or to predict the broad function of the box by analysis of the outputs of a statistically significant number of such boxes in other systems. These limitations are quite familiar to modellers, as the three approaches correspond to Levins' (1966) classification of model-building strategies. The first is a Type-I model, that simulates so far as it can the precise inputs and responses, but loses general applicability. The second has the Type-II attribute of limited inputs to isolated processes which do not yield real-world solutions. Providing its generalism through breadth rather than precision, the third is a Type-III approach.

While there is no fundamental reason why approaches should not be combined, case studies tend not to have done so: there is still a hiatus between the requirements of the manager and the instinctive guardedness of the scientist's predictions. The latter's desire to support concepts and hypotheses is culturally different from the manager's desire for precision and reliability. In the next two sections, an overview of the types of problem with which managers and regulators have to deal is contrasted with the range of main-stream, empiricised conceptual models on offer. To protect commercial confidentialities, however, real names have largely been omitted.

Problems with phytoplankton

Planktonic autotrophs are essential to the function of pelagic (open water) ecosystems. Their production sustains the base of the pelagic food webs, usually culminating in fish; and, in contributing a net oxygen flux into the system, the photosynthetic gas exchanges normally aerate the surface waters. It is also well-understood that the production is often seasonally regulated, mostly through the interaction of temperature and the aggregate underwater photoperiod (as modelled by Reynolds, 1989). Its biomass accumulates accordingly, up to limits broadly related to the supply of inorganic nutrients and to the extent of consumption by planktonic animals.

Most (but by no means all) of the problems associated with phytoplankton arise from the biomass it is or may be able to achieve. As hinted above, the planktonic biomass that forms in storage reservoirs has to be removed in treatment for potability, at a cost representing a substantial (25–35%) fraction of the price of the delivered product (OFWAT, 1994). Elsewhere, public reaction to green-coloured amenity waters ranges from disquiet to hostility, fuelling demands for corrective action. The decomposition of accreting algal products in the deep waters of lakes not only leads to anoxia and low redox dissolution of metal–phosphate complexes but simultaneously reduces habitat tenability to cold-water demersal salmonid fish (Baroudy & Elliott, 1994), as has been infamously demonstrated in the case of the Windermere charr (see Elliott & Reynolds, 1996). Where progressive nutrient enrichment appears to be the cause of the consequentially additional plant production (together the qualification for the term 'eutrophication' (Rohlich, 1969; Vollenweider, 1968)), managers need to know whether proposed corrective action is likely to be effective. Other frequent questions relate to waters already enriched and known to be supportive of potentially high phytoplankton crops: 'what feasible alternative methods for retarding plankton production would be most cost-effective?; 'were I to modify an existing impoundment or create a new one, what problems can I anticipate and could I design out?'. Models have also been sought in respect of proposed coastal barrages and for which the statutory impact assessments must include: 'what kind of problems will phytoplankton cause us, and how might we mitigate them?'.

As often as not, the questions relate to likely species composition, for it is increasingly acknowledged that all planktonic algae do not equal all other planktonic algae and certainly not so in the extent of the nuisance they may create. Odour-, flavour-, slime- and foam-production are traits which vary specifically. In water-treatment processes assembled around flocculation and upward-flow filtration, instances of penetration by planktonic cyanobacteria are a familiar headache; large populations in the raw water may well lead to very serious filtration difficulties and plant shut-down, if their intake into the works cannot be reduced. Not surprisingly, the industry seeks scientific guidance when it asks 'how can we avoid having these species?'. Since the summer of 1989, when the capacity of certain species of cyanobacteria to form surface scums and to produce acutely toxic microcystins and anatoxins was at last taken seriously in the UK (NRA, 1990), the question has become more pressing. Both suppliers and regulators are anxious to have models that will identify waters that are or will become susceptible to phases of cyanobacterial abundance, that will predict the onset of bloom conditions and that can evaluate alternative control strategies.

Current model availability

If the expectations of models are so clear, is it not then a simple matter for science to satisfy them? Well, no. So far as scientists are concerned, they might be dubbed 'the wrong kind of questions'. The tried and tested models currently used by plankton scientists are generalised, based on the mathematical fit (or conformity to an empirical behaviour) of collected data to a long-range behaviour. They come up either with Type-III probabilistic solutions with predictable margins of variability (which those used to dealing with precision judge to be 'chaos') or with Type-II outcomes that are insensitive to other attributes. Both are fundamentally inappropriate to the Type-I problems rehearsed above. Even where the problem raised is a general one: 'at what kinds of site will we be liable to problems with plankton algae?' – the solutions are reluctant and wrapped in so many caveats that they cannot be proved wrong under any circumstance. Indeed, a weakness in the conceptual basis of some of these models is often exposed.

It was not within the scope of my brief to review the structure and predictive reliability of all the models available. Besides, such a thankless and rapidly outdatable exercise has recently been undertaken by Jørgensen (1995) and it would be difficult to surpass the authority of either his overview or his critiques. It is interesting that he should comment that models which seek generality in application and precision in prediction are so complex that they cannot be supplied with all the data to drive them and that anything approaching an attestable simulation of a particular site requires complete re-evaluation of the interaction of the variables at the next site. I take this to mean that our systems models are too inflexible to accommodate the fact that no two sites function in the same way. So far as this essay is concerned, I take a few, central model tenets and question the validity of their application to the main issues of management issues. Later, I shall seek to show how they might yet contribute through new approaches to common management problems.

The principal general models relating to the ecology and dynamics of phytoplankton are directed towards describing (note, not necessarily predicting) how much plankton biomasss lakes support, how quickly they can furnish it and what kinds of algae are present. There is no great mismatch there with the manager's questions: why should it then be inappropriate to apply the one to the other? Let us look at each of these approaches, beginning with the question, how much will be produced?

Capacity models

Plants of the plankton need light and carbon dioxide to photosynthesise every bit as much as do plants on land. Moreover, photosynthetic rate *(P)* is relatively easy to measure by gas-exchange and, between 1950 and 1965, a great many data relating to the photosynthetic productivity of phytoplankton were amassed. It was very clear that the productivity of the biomass *(P/B)* related first to the availability of underwater downwelling light energy (*I*). There has been great interest in the solution of the *P/I* relationship and thus to the calculation of the maximum light-supportable biomass, $B(I)_{max}$. Solutions of Talling (1957) and Vollenweider (1965) are just as valid 30 years on but experience has shown that, although pelagic plant biomass scarcely exceeds the predicted capacity, it is also rarely attained – other factors intervene. Even when the physical conditions allow, the production reaches a ceiling imposed by the supply of carbon or the availability of nutrients. By coincidence, concerns about the impacts of eutrophication were then running high: during the next decade and a half (1965–1980), attention switched quickly to the biomass yields of available nutrients. The work of Vollenweider (1968; 1976), especially, established the statistically robust linkage between the average phytoplankton biomass supported in lakes, $B(P)_{mean}$ and the morphometry- and flow-weighted phosphorus loads, $\Delta(P)$ (Vollenweider & Kerekes, 1980). Because few other elements in lakes exercise this critical control or because shortages of other elements are inexclusive in their biological limitation (e.g., nitrogen, silicon), phosphorus has tended to be regarded as the critical factor in the level of biomass supported. Among lakes where all the free phosphate is consumed, the maximum biomass, $B(P)_{max}$, has been shown to be a predictable function of the soluble fraction of the element supplied (Reynolds, 1992).

By itself, neither approach has much relevance to the manager. Either will confirm what he knows already – the capacity to produce too much phytoplankton is there – but can give him no more than the imprecise idea that if he could reduce substantially the light dose or the phosphorus load he should have less biomass on average, but he still does not know when or how often it will be troublesome. Neither will the regression tell him by how much the loading should be reduced. His position is analogous to the student of social trends who is trying to forecast the number of houses that will be built in a year from the number of bricks that are supplied to the builders. One may assume that builders do not stockpile reserves of bricks indefinitely and that, realistically, such a rough, Type-III correlation exists but it makes no allowance for land availability, labour, weather – all of which could influence the outcome and, at times, be more critical to production than the number of bricks. How much must the brick supply be reduced before the number of houses built is a continuous function of the brick supply?

Rate models

The question of how quickly plankton biomass can be recruited to and lost from the open water has been addressed in a variety of modelling approaches, each based on aspects of the investigations into the population dynamics of phytoplankton, which became fashionable in the mid- to late-seventies. Topics explored experimentally and resolved by mathematical equations included: the relationship between *in-situ* growth rates and photosynthesis (see Forsberg, 1984; Peterson, 1978); that between the rates of nutrient uptake by plankton and the concentrations available, founded essentially on the insights of Dugdale (1967) and of Droop (1973); and the rates at which plankton is lost to settlement and to grazers (Reynolds et al., 1982). The difficulty here is that the controlling conditions are not faithfully reproduced by those who wish to apply the models or for how long the relationships might apply. It can be easily demonstrated that the capacities for photosynthetic carbon fixation and nutrient uptake far exceed the rates of cell replication, unless substantially reduced to the point where it can be shown that the cell replication rate depends exclusively upon the flux of one or other of the resources (Reynolds, 1994). Without the satisfaction of this condition, none of the widely embraced concepts about the growth of populations being either light- or nutrient-limited actually hold. Something else, usually the time that is required to assemble the raw materials at that temperature, determines the rate of growth. In this way Type-II models of single processes do not assist the manager until he knows that the control is operative. To return to the house-building analogy, we are trying to gauge the rate of completions from the number of bricks actually on site. The Dugdale/Droop formulation recognises that deliveries of bricks determine that the number of bricks on site fluctuates with time but the prudent builder ensures a frequency of supply that maintains a sufficient stock to meet his anticipated needs. Burmaster's (1979) theory that the completion of houses and the supply of bricks achieve equivalence is also honoured. However, the rate of completion is not determined by the number of bricks on site (as suggested by the Droop model), until the supply to the site is slower than they are being incorporated into new houses. Neither is their erection halted until the stockpile is exhausted.

Composition models

There is, as yet, no tradition underpinning the philosophy of species selection: Shapiro (1990) was still able to argue the relative merits of a dozen factors claimed to be crucial to the dominance of plankton assemblages by just one group of cyanobacterial genera. The most prominent assumption is the Darwinian one of continuous competition among species and the selection of the best adapted. Within this context, Tilman's (e.g. 1982) resource-based competition model is the most influential and is founded on data from experiments with planktonic algae (Tilman & Kilham, 1976). In essence, this showed that the supply of nutrients that would impose the control of growth rate was not the same for all species and, thus, it was possible to nominate concentrations at which one species was nutrient-limited and another not. He further showed that this applied to algal growth against other nutrients and he contrived experiments with two algae at different concentrations of two resources. There were combinations at which one would grow faster than the other but others where the relative growth rates were reversed. Most interesting of all, he could generate combinations wherein the species co-existed because one was limited by the first resource and the other was simultaneously limited by the second. Hence, adjustment of the ratio between the two resources allows one to escape its constraint and, thus, to select for its dominance.

One popular deduction is that the ratio between nutrient resources (N:P is a great favourite) determines the outcome of 'competition' between different types of phytoplankton, (for instance, cyanobacteria versus other, eukaryotic genera), though rarely with any regard for the Tilman–Kilham condition that both elements are simultaneously limiting or with any recognition of the rates of replication achieved by either. The spuriousness of models embracing resource ratios as determinants of competitive outcome can be shown up by returning, yet again to the house-building analogy. What we are now saying is that there are two kinds of house, both using bricks and cement but one design uses a greater percentage of bricks than the other, this latter one using relatively more cement than the first (this is an analogy to 'seventies' thinking!). If one is short of cement, one should build the first kind; if bricks are the problem, go for the cement jobs. If both are in short or erratic supply, one can probably manage to build a few of each. But if neither material is limiting the rate of house building, one retains the ability to build both. Moreover, if brick ones can be completed more quickly than cement ones, the probability is that more brick ones than cement ones will have been built, unless cement ones happen to sell faster. The point is that none of these outcomes is ratio driven. Absolute quantities or other factors are always responsible for the differences in completion rates.

The lesson is important – even where the between-species differentiation is more sharply defined than their respective capabilities for uptake of nitrogen and phosphorus (for instance, losses to selective grazing) – species ascendancy and dominance must be modelled on empirical inputs. Many who apply Tilman's resource ratio model forget that it holds only for limiting supplies – growth rate is dependent upon the nutrients concerned. We still need to know how fast one can build houses.

Descriptive models

The fourth kind of model is a complex development of the curve-fitting exercise of Type-I models, essentially based on multiple regression analysis and the identification of the sources of variation. To formulate these, a first-class mathematical modeller is often required. The approach has been used to great and elegant effect: I have great admiration for the mathematical description of the seasonal fluctuations in the concentration of phytoplankton in the River Thames (Whitehead & Hornberger, 1984). Such models are research tools, inso far as they assist investigators to determine the key variables driving a highly variable system. Principal components analysis does this too, while canonical co-variance analysis assists the researcher to compare a number of sites. Without the experience and intuition to interpret the apparent relative influence of the quantified vectors, we are in danger of misunderstanding how they might be applied to management or just how much of what is found out even describes other systems. The model still lacks the generality that the manager requires but the site-to-site adaptation of the methodology would overcome its inflexibility.

Some recent modelling approaches

Lest this account be taken as too critical about models or too despondent about what they can contribute to management, I want to draw attention to some promising results from process-based simulations. Common to these modelling approaches is their initiation with the organismic capability. If it is known what the organism requires to sustain a given performance and it is known quantitatively what the contribution is of each of the various factors in sustaining that performance, then it is possible both to determine which factor is critical and to decide by how much it should be altered to better manage the output. Resorting once more to the problem confronting the social analyst – 'how many houses can be built in the year?' – the basic unit is the time it takes to build one house under the optimal conditions we can contrive. The number of persons engaged in the building is important but there is a point when they restrict each other's activities or when one task awaits the completion of a precursor task, or when they all have to wait for the paint to dry. Let us say that twenty men in fine weather could finish a house in 2 weeks. Then a labour force of one million could build 25000 houses per week. A smaller or larger workforce could build fewer or more, but let us stay with the 25000. Can land be released and prepared at the rate of 1000 ha per week? Do we have the capacity to make enough bricks, quarry enough sand, supply enough of every component to be able to match the production? If not, which is the limiting factor?

In order to determine how sensitive lakes are to cyanobacterial bloom formation and to which standard form of control they might readily respond, the Environment Agency is using a model called PACGAP (Prediction and Assessment of Cyanobacterial Growth and Production). PACGAP has by analogy exactly the above logic. On the basis of its morphometry, thermal structuring, flushing time and nutrient loads, it calculates the supportive capacity of each factor, $B(I)_{\max}$ $B(P)_{\max}$ in turn and finds the lowest. It also requires to know something about the carbon relations and the present levels of algal chlorophyll produced, from which it characterises the habitat as being either often, seasonally or rarely likely to meet the requirements of bloom-forming cyanobacteria.

So far as I am aware, it is tantalisingly accurate in stating what is provable – what the lake supports now. Its predictive powers have not, as yet, been generally verifiable. The routine can be used to experiment with management approaches: running the programme with altered nutrient loads, faster flushing rates, deeper or more frequent mixings allows the operator to judge which strategies for avoiding or eradicating bloom populations might be more successful.

The operating instructions with PACGAP insist that it is a guide and not suitable as the basis for strategic decision making. Yet much the same philosophy lies in the coding of the day-by-day simulations of phytoplankton growth constructed by the PROTECH-family of models. The original PROTEC (Phytoplankton Responses to Environmental Controls) was developed jointly by the Freshwater Biological Association and the former Welsh Water Authority to demonstrate the predictions of the scale and composition of phytoplankton in the proposed Cardiff Bay Barrage. The management questions posed to the modellers were clear enough – 'will we have algal growth?' 'how much?' 'when?' 'will they form blooms?' The programme, like the long-hand predictions, considered how fast different species would grow in situ, for given light and nutrient inputs at given temperatures. The model could recalculate the biomass for each of eight species as many as 15 times per hour given the day of the year, the temperature and the rate of riverine input. It was set for the 'worst-case' scenario, everything grew at its fastest sustainable rate, all nutrients from grazed species were instantaneously regenerated, and being eaten or washed into the Bristol Channel were the only mortalities. The high flushing rate in fact kept the algal population down to the few micrograms of chlorophyll per liter that occur in the Taff and Ely rivers but in dry weather, the nutrient richness would favour the very rapid growth of small green algae, which would increase biomass sharply to the light capacity until a growing zooplankton, responding to the food resource, caught up sufficiently to clear algae faster than they could be recruited. Prolonged low flows were predicted to lead to *Oscillatoria* dominance at populations of over 100 μg chlorophyll l^{-1} which flushing but not grazers could remove. However, not even the prediction that *Microcystis* was unlikely ever to grow in the lake, even were it the only species, would cheer up the sponsors: 'how do we know we can believe it?'. In fact, the model's authenticity was handsomely displayed when tested on a dataset for the periodically well-flushed, shallow coastal lagoon, Slapton Ley: it was substantially correct in the scale of biomass supported, its periodicity and in its summer dominance of *Peridinium* and *Anabaena* (FBA, 1989).

PROTEC P was a direct and rather more sophisticated development of the flushed lagoon. It concentrated on better descriptions of in-lake properties, and included more variables, which extended its suitability to lakes and reservoirs whose retention times are in the order of months rather than days. The structure of this model and its operation were described by Hilton *et al.* (1992).

Two lessons about real lakes with longer retention times were reinforced: settling becomes a dominant process and internal nutrient recycling assumes major metabolic importance. The next variant to be developed became known as PROTECH2 to acknowledge that 'C' now meant 'CHange' and '2' refers to a second, vertical dimension. For this is a multi-layered model, inspired by the physical model DYRESM (Imberger & Patterson, 1981), which not only allows integration and separation of water mixing or stratifying according to the interaction of wind-forcing and buoyancy but also lets algae sink, float or swim according to their specific abilities. The model is used exclusively by the EA to predict the onset of bloom conditions in lakes and reservoirs. Finally, some preliminary work has been undertaken which simulates the accumulation and diagenesis of sedimentary material and under appropriate shear conditions will resuspend and redisperse the sedimented material and the nutrients leached therefrom. It is possible that these two additional features will bring still greater authenticity to these simulation models and greater confidence in their sensitivity to alterations in basin management.

Other bottom-up process models of phytoplankton have been in use for some time. That of Jiménez Montealegre et al. (1995) simulates the growth of one cyanobacterium in a shallow lake system. Steel's dynamic simulations have been evolved carefully and in their latest form (Steel, 1995; Steel & Duncan, 1999) tackle complex issues like light-adaptation of populations. Like the *Oscillatoria* model and many other previous modelling attempts, Steel begins further still towards the bottom with algal photosynthesis and nutrient uptake. Selection is incorporated by reference to typical properties of phylogenetic representatives. The model is easy to use and apply and produces its predictions as plots developing before the operator's eyes. It also makes it easy to vary inputs and re-run with altered scenarios.

Lastly, I draw attention to a first water-quality management model for Windermere. This is not directly concerned with what will grow so much as when the growth of certain organisms (or the consequences thereof) will become, or cease to be, a nuisance in the lake. The villain here is sewage-derived phosphorus and, over 30 years, it has led to a steady eutrophication of the lake's South Basin. From a careful analysis of 50 years' data on the lake, the phosphorus supply was computed. The fate of this load has been reconstructed from the database. It is then assumed that all the residual, assimilable phosphorus generates phytoplankton carbon in strict stoichiometric ratio, this carbon being then allocated to sediment and outflow. That which sediments consumes hypolimnetic oxygen in the stoichiometric ratio. Thus, a given phosphorus load will generate a prediction of the degree of hypolimnetic oxygen tension, the effect on fish survival and breeding success, and the potential to support bloom-forming cyanobacteria. This model is fully authenticated by the historical data which confirm where in the eutrophication of the lake, each of these symptoms developed. Happily, as the restoration proceeds, the symptoms are disappearing pretty well as the model provides (unpublished reports to North West Water Ltd and the NRA North West Region).

Conclusion

This brief tour of the types of models of phytoplankton dynamics that have been developed by ecologists finds that they have not proved to be astonishingly helpful to managers with specific problems. This owes most to the philosophy of generalisation of the models and the natural reluctance of scientists to sacrifice accuracy of prediction in responding to a problem of precision. However, the power of computers opens the way to more elaborate process models which build up scenarios in the same way that natural phytoplankton populations develop in the waters we seek to manage: from modest inocula and within the constraints set by the capacities of the environmental factors. Similar models will be useful in helping all of us to improve our perception of the dynamic structure of lake ecosystems on a site-by-site basis. This is essential to proper management of lakes and reservoirs. We simply have to overcome the naive and erroneous 'my-lake-is-phosphorus-limited' syndrome which has so distracted our progress in phytoplankton ecology.

Acknowledgements

I wish to express my appreciation to the sponsors who, among them and at various times, have supported the development of models referred to in this paper: Cardiff Bay Development Corporation, Bristol Water Company, Essex and Suffolk Water, Thames Water, North West Water and, especially, the National Rivers Authority. The Natural Environment Research Council and the U.K. Department of the Environment have supported the acquisition of the original knowledge-base. I am grateful to many colleagues and friends who have supported this work over many years, especially Tony Irish, Glen George and John Hilton. The views I have expressed here are my own. I accept full responsibility for these and I do not assume that they are shared by colleagues, either at The Freshwater Biological Association or at The Institute of Freshwater Ecology.

References

Baroudy, E. & J. M. Elliott, 1994. Tolerance of parr of Arctic charr, *Salvelinus alpinus*, to reduced dissolved oxygen concentration. J Fish Biol. 44: 736–738.

Burmaster, D., 1979. The continuous culture of phytoplankton: mathematical equivalence among three steady-state models. Amer. Nat. 113: 123–134.

Droop, M. R., 1973. Some thoughts on nutrient limitation in algae. Journal of Phycology 9: 264–272.

Dugdale, R. C., 1967. Nutrient limitation in the sea: dynamics, identification and significance. Limnol. and Oceanogr. 12: 685–695.

Elliott, J. M. & C. S. Reynolds, 1996. Lake enrichment and Windermere charr. Biol. Sci. Rev. 8: 17–20.

FBA (Freshwater Biological Association), 1989. The FBA-Welsh Water simulation model of phytoplankton dynamics in a flushed system – authentication in a coastal lagoon (Cyclostyled). Freshwater Biological Association, Ambleside.

Forsberg, B. R., 1985. The fate of planktonic primary production. Limnol. Oceanogra. 30: 807–819

Hilton, J., A. E. Irish & C. S. Reynolds, 1992. Active reservoir management: a model solution. In: Sutcliffe, D. W. & J. G. Jones (eds), Eutrophication, Research and Application to Water Supply. Freshwater Biological Association, Ambleside: 185–196.

Imberger, J. & J C. Patterson, 1981. A dynamic simulation model – DYRESM:5. In Fisher, H. B. (ed.), Transport models for inland and coastal waters. Academic Press, New York: 310–361.

Jiménez Montealegre, R., J. Verreth, K. Steenbergen, J. Moed & M. Machiels, 1995. A dynamic simulation model for the blooming of *Oscillatoria agardhii* in a monomictic lake. Ecolo. Model. 78: 17–24.

Jørgensen, S. E., 1995. State-of-the-art management models for lakes and reservoirs. Lakes and Reservoirs: Res. Man. 1: 79–87.

Levins, R., 1966. The strategy of model building in population ecology. Am. Sci. 54: 421–431.

NRA (The National Rivers Authority of England and Wales), 1990. Toxic blue-green algae. Water quality Report No. 2. HMSO, London.

OFWAT (The Office of Water Services in England and Wales), 1994. Report on the cost of water delivered and sewage collected. OFWAT, Birmingham.

Peterson, B. J., 1978. Radio-carbon uptake: its relation to net particulate carbon production. Limnol. Oceanogra. 23: 179–184.

Reynolds, C. S., 1989. Physical determinants of phytoplankton succession. In: Sommer, U. (ed.), Plankton Ecology. Brock-Springer, Madison: 9–55.

Reynolds, C. S., 1992. Eutrophication and the management of planktonic algae. What Vollenweider couldn't tell us. In: Sutcliffe, D. W. & J. G. Jones (eds), Eutrophication, Research and Application to Water Supply. Freshwater Biological Association, Ambleside: 4–29.

Reynolds, C. S., 1994. The role of fluid motion in the dynamics of phytoplankton in lakes and rivers. In Giller, P. S., A. G. Hildrew & D. G. Raffaelli (eds), Aquatic Ecology, Scale, Pattern and Process. Blackwell Scientific Publications, Oxford: 141–187.

Reynolds, C. S., J. M. Thompson, A. J. D. Ferguson & S. W. Wiseman, 1982. Loss processes in the population dynamics of phytoplankton maintained in closed systems. J. Plank. Res. 4: 561–600.

Rohlich, G. A., 1969. Eutrophication: causes, consequences, correctives. U.S. National Academy of Sciences, Washington.

Shapiro, J., 1990. Current beliefs regarding dominance by blue-greens: the case for the importance of CO_2 and pH. Verh. Int. Ver. Theor. angew. Limnol. 24: 38–54.

Steel, J. A., 1995. Modelling adaptive phytoplankton in a variable environment. Ecol. Model. 78: 117–127.

Steel, J. A. & A. Duncan, 1999. Modelling the ecological aspects of bankside reservoirs and implications for management. Hydrobiologia 395/396 (Dev. Hydrobiol. 136): 133–147.

Tailing, J. F., 1957. The phytoplankton population as a compound photosynthetic system. New Phytol. 56: 133–149.

Tilman, D., 1982. Resource Competition and Community Structure. Princeton University Press, Princeton.

Tilman, D. & S. S. Kilham, 1976. Phosphate and silicate uptake and growth kinetics of the diatoms *Asterionella formosa* and *Cyclotella meneghiniana* in batch and semicontinuous culture. J. Phycol. 12: 375–383.

Vollenweider, R. A., 1965. Calculation models of photosynthesis-depth curves and some implications regarding day rate estimates in primary production. Mem. Ist. ital. Idrobiol. 18 (Suppl.): 425–457.

Vollenweider, R. A., 1968. Scientific Fundamentals of the Eutrophication of Lakes and Flowing Waters, with Particular Reference to Nitrogen and Phosphorus As Factors In Eutrophication. OECD, Paris.

Vollenweider, R. A., 1976. Advances in defining critical loading levels for phosphorus in lake eutrophication. Mem. Ist. ital. Idrobiol. 33: 53–83.

Vollenweider, R. A. & J. Kerekes, 1980. The loading concept as basis for controlling eutrophication philosophy and preliminary results of the OECD programme on eutrophication. Prog. Wat. Technol. 12: 5–38.

Whitehead, P. G. & G. M. Hornberger, 1984. Modelling algal behaviour in the River Thames. Wat. Res. 18: 945–953.

Hydrobiologia **395/396:** 133–147, 1999.
D.M. Harper, B. Brierley, A.J.D. Ferguson & G. Phillips (eds), The Ecological Bases for Lake and Reservoir Management.

Modelling the ecological aspects of bankside reservoirs and implications for management

J.A. Steel[1] & A. Duncan[2]
[1]*5 Fort View, Ardbrack, Kinsale, Co Cork, Ireland*
[2]*Royal Holloway College, University of London, Egham, Surrey TW20 0EX, U.K.*

Key words: reservoir, quality management, modelling, algae, zooplankton, fish

Abstract

Bankside storage reservoirs are used as a major water supply resource in the lower Thames Valley, England. They form the link between the River Thames and the water treatment works of the Greater London area. The reservoirs act as both a water reserve in times of low river flows, and a quality 'buffer' between the river and the treatment works. The load on the water treatment works (particulate material, physico-chemical characteristics) primarily reflects the water qualities of the reservoirs. Management of such reservoirs thus seeks to reduce the adverse impacts which would otherwise arise from direct river use, and to ensure as far as possible that the ecological processes within the reservoirs do not introduce new challenges to the water treatment. Reservoir management clearly needs a good understanding of those ecological processes and their interactions, and, hopefully, a means to exploit that understanding in hindcasting to explain past events, in forecasting near- or far-future events, and to help in exploring operational options to ameliorate any foreseable difficulties. The reservoirs consist of a variety of configurations, physical dimensions and operational circumstances. They have, importantly, basically simple morphologies, known hydraulic regimes and physico-chemical qualities. Nonetheless, they appear to behave essentially as small (1–50 Mm^3), eutrophic lakes; and various aspects of their ecology has been studied for the past 65 years. Their attributes and operational involvement make them ideal candidates for ecological modelling, which has been applied to them in varying extents for the past 30 years. The major conclusion which may be drawn from these studies is that even in such relatively simple water bodies, current (and probably future) models can only encompass their broad ecological characteristics. Detailed operational needs have to be met by a variety of modelling approaches, mainly predicated on the basis of only being able to know a lot about a little or a little about a lot. The operational needs for modelling fall into the following broad types: (a) understanding: why did those events occur, or where is our ignorance greatest? (b) short-term forecasts: how will the current situation develop in the short-term (weeks)? (c) what-if considerations: what would happen if some management facility were employed or used differently? (d) optimisation: what are the optimal volume– quality supply arrangements? (e) long-term prediction: what is the longer-term (years) outlook under foreseeable scenarios? (f) projective evaluation: how would potential, as yet non-existant reservoirs behave under prescribed circumstances? Examples of how these needs have been met are outlined, with examples ranging from simple models of the diatom ecology of the reservoirs to much broader trophic–dynamic descriptions which can allow expression of fish–zooplankton–phytoplankton interactions. This is crucial for present and future management of cyanobacterial phases. It is clear that considerable management insight and control can result from modelling assistance, but only if the appropriate questions are asked. Whilst simple short-term modelling is less demanding, any attempt to model the full complexity of the ecology of even these relatively simple water-bodies is probably doomed to founder on complexity–understanding difficulties, unless these are resolved to much more constrained system aspects. This is particularly so for the qualitative biology. The best that may presently be foreseen is for development of the newer multi-biological type models, with reasonably realistic and dynamic physical and chemical environment sub-models, being able to manifest the general characteristics of the ecosystem in question. Despite such difficulties, new reservoir management insights and approaches will inevitably be founded on critical modelling of those ecosystems.

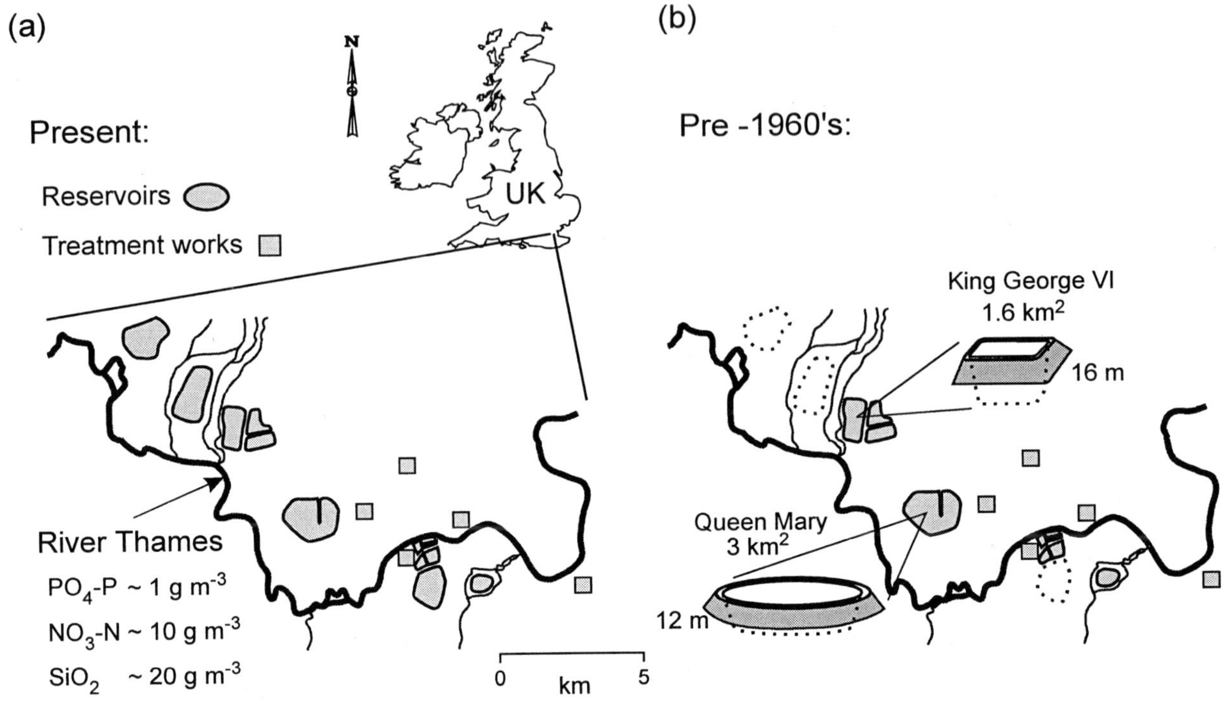

Figure 1. Reservoirs and treatment works in the lower Thames Valley, South-East England. (a) The system at present, and order of magnitude nutrient concentrations in the River Thames. (b) The system in the period before 1960, and the different configurations of Queen Mary and King George VI reservoirs.

Introduction

The reservoirs being considered lie in the south-east of Britain, near London and are distributed along the banks of the River Thames (Figure 1a).

As they are close geographic neighbours, they experience the same climatic conditions. They also all have the same source water, which is the eutrophic River Thames, with nutrient concentrations of around 1 mg PO_4-P l^{-1}, 10 mg NO_3-N l^{-1} and 20 mg SiO_2 l^{-1}. The main reservoirs were built during the past 65 years and, between 1963 and 1974, the total capacity doubled from about 100 to 200 Mm^3. The reservoirs are designed to provide bank-side storage, and they all possess similar features of simple morphologies, steep sides and relatively uniform depths, and are operated with known quantities of throughput water when in supply. The surface areas of the main reservoirs range from 1.5–3.0 km^2; they have volumes of 10–40 Mm^3 and depths of 10–25 m. In general, when in supply, their retention time is within the range 10–100 days.

These reservoirs cannot properly be considered in isolation as single water bodies since they form part of a water supply chain linking a series of different (eco)systems: river–reservoir–treatment works–water supply (Steel, 1972; 1975). It is important to realise that the fundamental ecological processes of these reservoirs are not different from those in lakes. The reservoirs may, of course, differently manifest the effects of those processes because their unique morphological/operational characteristics can influence relative process magnitudes. However, also most importantly, these are bodies of water which are manageable and measurable, which allows considerable opportunity to identify and quantify those fundamental processes.

Earlier studies during the International Biological Programme 1966–72 showed that loadings of organic material upon the treatment works was wholly dependent upon the quality of the reservoir water (Steel, 1975; Steel et al., 1972) (Figure 2). In the absence of a reservoir stage in the water chain, the works' loads would be both much greater and more variable. Nevertheless, algal crops could be limited by appropriate reservoir management, despite the presence of concentrations of phosphorus and nitrogen so high in relation to algal growth requirements that early studies showed there was virtually no discernible planktonic nutrient N and P uptake in the reservoirs.

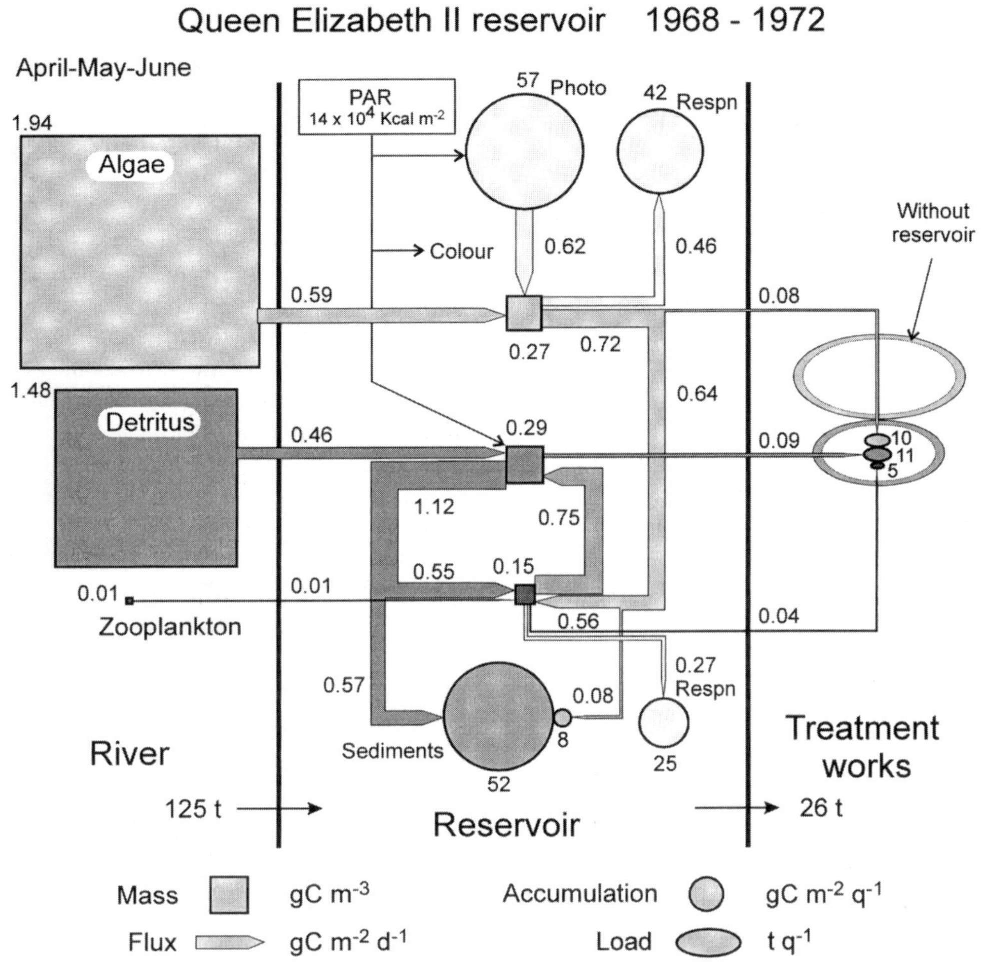

Figure 2. Measured values of various ecosystem quantities, for the productive quarter April–June, 1968–1972.

An ability to control the quantities of plankton plants in the reservoir provides important benefits for treatment technology, investment requirements and ease of operation. The primary overall objective for this research was the absolute need to maintain an adequate supply of potable drinking water. It was clear that both understanding the ecological interactions within the reservoirs and quantifying their magnitudes was, and is, crucial to an informed management. Operational need was therefore the main driving force for the present exercise.

This example of the application of increasing ecological understanding, and its associated modelling approach, is drawn from attempts to answer some major business questions posed during developments that occurred during the past 35 years. In the 1950–60s, consideration of new storage resources (Figure 1b) included questions such as 'how deep should these reservoirs be?' In the lower Thames Valley, the main options were to build a basin of large area but shallow depth (as Queen Mary reservoir) or one that was deep and small in area (as King George VI reservoir). For land already owned, the former would be cheaper, but for maximum resource efficiency the latter choice was desirable. In either event there were great reservations about water qualities.

In Queen Mary reservoir, long-term deep deoxygenation was not experienced, despite occasional, transient thermal stratification, but large crops of diatoms and blue–greens were a regular feature. These were treated by either mass dosing with $CuSO_4$(ca 1 mg $l^{-1} \equiv$ c 30 tonnes of $CuSO_4$ in Queen Mary reservoir), a regime of continuous low level $CuSO_4$ dosing (0.1–0.3 mg l^{-1}) or by closure and switch of supply to another reservoir, usually King George VI reservoir. Annual appearance of a stable thermal stratification, with regular de-oxygenation and formation of H_2S was the dominant characteristic of King George VI Reservoir. These were often also accompanied by large algal populations, particularly diatoms, and cyanobacteria and *Ceratium* in the epilimnion under thermally stratified conditions (Figure 3a). This reservoir was mainly managed by enclosure, with relatively occasional use. Despite having multiple, shoreline draw-off facilities, a stable supply of good quality water for abstraction was rarely possible because of internal wave motions associated with the thermal stratification.

In summary, the considerable expenditure required for the extra, deep storage did not ensure a usable supply, mainly because of thermal stratification. Detailed experimental work by White et al. (1955) suggested that thermal stratification would be controlled by internal mixing with submerged water jets. The first of the new reservoirs with such a mixing facility was therefore designed to be fairly deep (17 m), but still similar enough to the current experience, in case mixing was unsuccessful, or introduced unforeseen quality effects. In the event, jet mixing provided an outstanding control of thermal stratification. Figure 3b shows that, in Queen Elizabeth II reservoir, isothermal conditions could be easily maintained throughout the year, with consequent oxygenation of the complete water column and, possibly, the sediment surface.

This raised the next business question: 'could the remaining new reservoirs be even deeper?' There was little information in the literature as to what happens to the algae in such deep, mixed basins, with unlimited nutrient supplies. Several important subsidiary questions emerged: would they grow large crops throughout their depths? would there be a changeover to less easily treatable types of algae? Would there occur some alteration in the algal sequencing?

Modelling was seen as a major tool in attempting to provide some answers, at least semi-quantitative, to these questions. As there was a firm conviction that an understanding of the reservoirs' biological dynamics was sought, an early decision was for an 'ecologically' structured model, rather than attempting a multi-parameter regressional type model. As previously indicated, the ecological interactions to be considered were of general applicability and not unique to these reservoirs, although the simple basin morphologies and operational attributes does make them easier to study. This generality also suggested that appropriate lake and marine studies could provide a fruitful beginning.

A simple model and developing ecological understanding

The simple model

Talling's (1957a; b) algal model was recognised as an entirely appropriate starting point. It considers the population's photosynthetic carbon gain against its respiratory carbon loss in an homogeneously mixed water column. Initial quantification of the local reservoir values of the variables contained in Talling's model led to a swift development of a simple, mixed system algal model (Steel, 1972). For diatoms in deep, fully mixed reservoirs, this model predicted that:

(1) algal growths begin later;
(2) their growth rates are constrained;
(3) maximum attainable biomasses could be energy-limited rather than nutrient-limited;
(4) potential maximum biomasses would be significantly reduced.

Some of these model effects are illustrated in Figure 4a. Under similar conditions, these effects were largely determined by a combination between mixed depth (Z_{mix}) and light attenuation (ϵ_q), because depth offsets transparency and turbidity offsets depth. In part, what the model predicts is that column maximal algal biomasses (mg chlorophyll-a m^{-2}) would be linearly, inversely related to mixed depth-light attenuation, as plotted in Figure 4b. The field observations in Figure 4b broadly agreed with such a prediction. Occasionally, however, greatly reduced crops were observed, which implied that some other factor or factors needed to be included in the model. Even so, the application of this simple model was already a major step on the path to answering some of the difficult questions previously posed.

Addition of a nutrient term

The most obvious deficiency of the simple model was its supposition of unlimited nutrient availability, although in the nutrient context of these reservoirs, only SiO_2 was remotely likely to cause limitation in diatoms. For completeness, however, nutrient effects were included by simple Michaelis–Menten nutrient characteristics for SiO_2 and phosphorus, with subsequent addition of Droop's (1965) 'quota'-type effects (Steel,

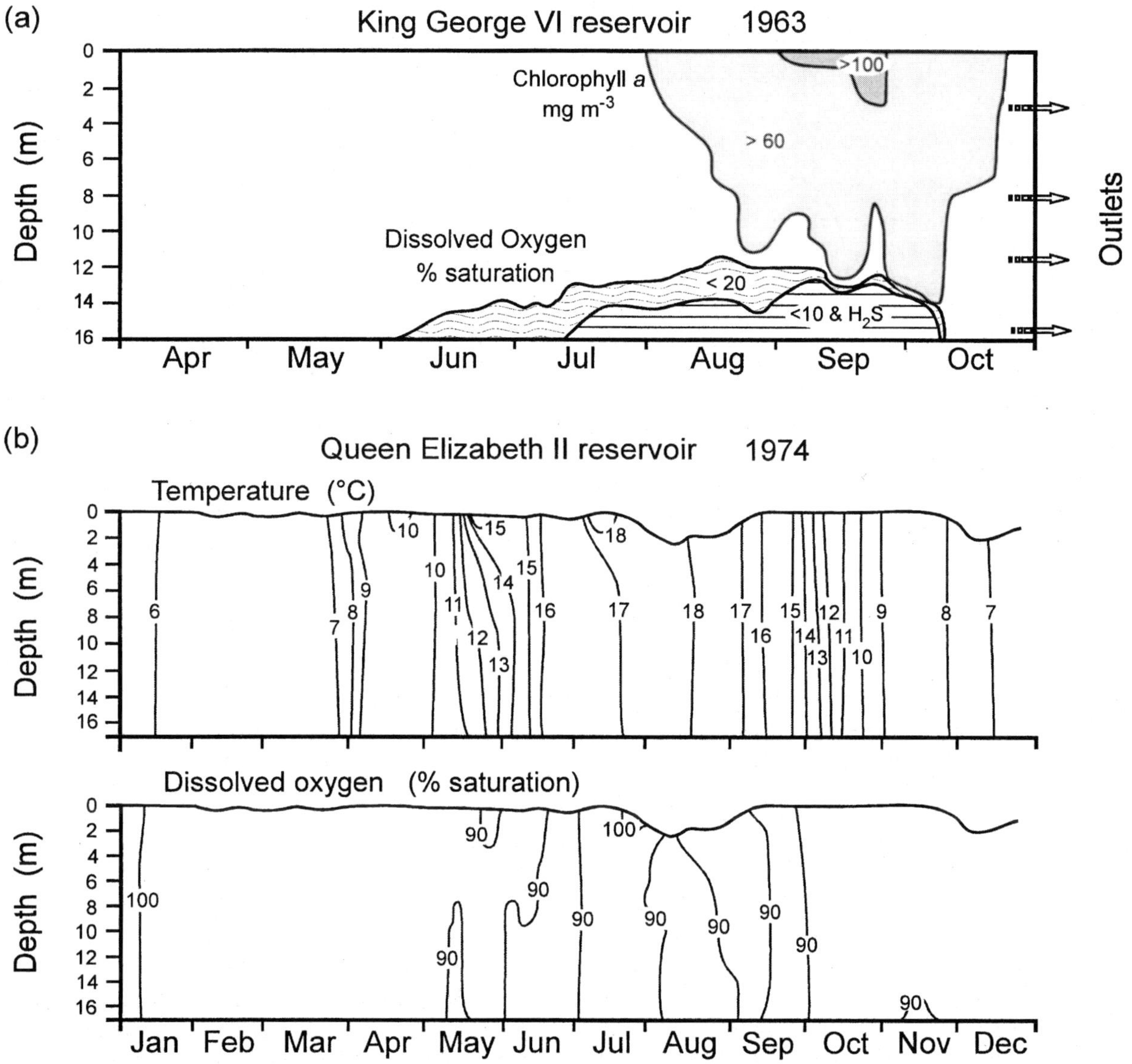

Figure 3. (a) Thermal stratification in King George VI reservoir during the growing season of a representative year, with consequences to algal crops and de-oxygenation. (b) Results of jet mixing in Queen Elizabeth II reservoir in 1974, illustrating the degree of isothermy and oxygenation achieved.

1978). Some indication of the results of this model for typical reservoir waters is given in Figure 5a. It is clear that phosphorus limitation, for example, is extremely unlikely.

In this form, the model indicated that phosphorus-stripping from the existing 1000 mg PO_4-P m^{-3} or so to, perhaps, levels of the order of 10 mg PO_4-P m^{-3} would be necessary to achieve any further, significant reduction in potential algal crops. Even if such a reduction was achievable, the costs involved far out-stripped any potential savings. The business therefore decided not to undertake P-stripping, even of only the reservoir inlet water: another major investment decision. Similarly, SiO_2 was never low enough in the early part of the year to limit diatom crops, and the model indicated that most diatom crops were maximal long before SiO_2 was potentially limiting. Analysis of the experience with $CuSO_4$ treatment showed that it did not reduce either algal growth rates or maximum crops to levels below that achieved by en-

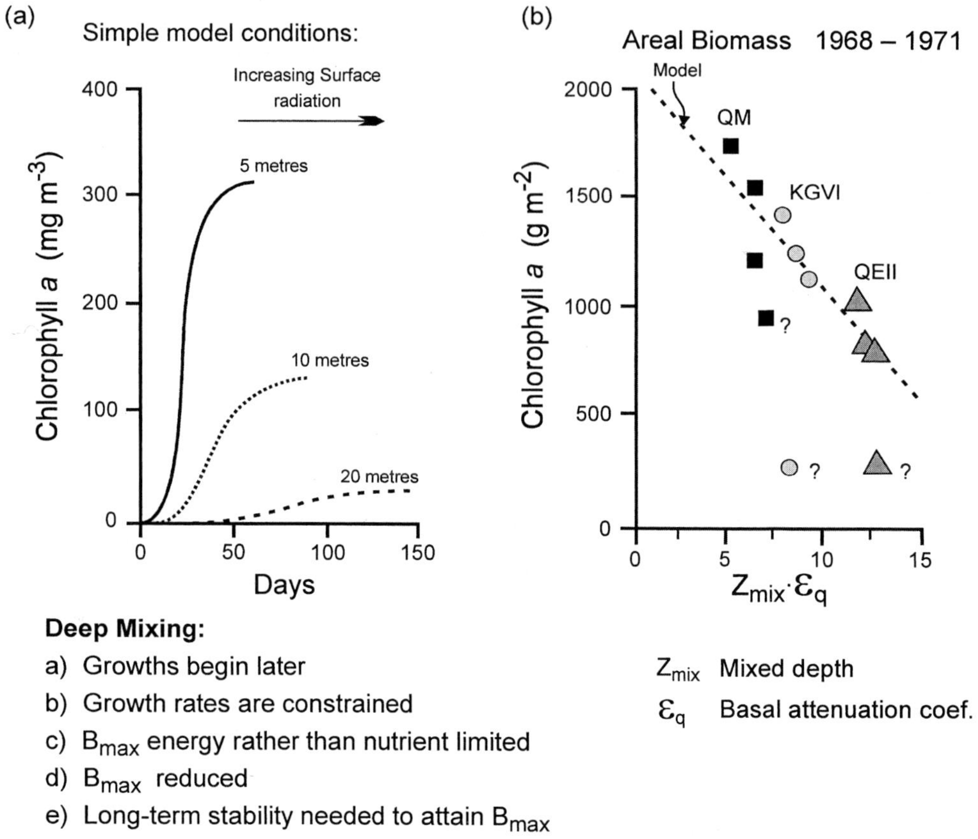

Figure 4. Chlorophyll-*a* concentration outputs from a simple, 'Talling-type' model for typical early season conditions. (a) Algal growth under full-depth mixing conditions in basins of different depth. (b) Inverse linear relationship between maximal algal biomass and the combined mixed depth–light attenuation coefficient, validated by observed data points from three reservoirs. See text for explanation of the exceptional points.

ergy limitation by full-depth mixing. Copper treatment was therefore discontinued as a management tool: yet another major quality decision with significant, beneficial, financial implications. Furthermore, studies were also beginning to show that $CuSO_4$ dosing also had adverse effects on the reservoir cladoceran populations

More detailed model investigations of mixed depth and the water's light attenuation suggested that mixing the potential epilimnetic algal populations through a greater, de-stratified water column offers a way to severely limit the algal crops, and potentially to a far greater extent than due to mere 'dilution' (Figure 5b). So the lesson for the business of water supply is to make reservoirs as deep as it is technically and financially feasible, along with mixing and draw-off facilities at least as efficient as those in the Queen Elizabeth II reservoir.

However, there were still some unresolved questions. Sedimentation studies and modelling suggested that mixing was not reducing sedimentation losses and that SiO_2-stress enhanced sedimentation could potentially remove large diatom crops very rapidly. However, sedimentation was not the reason why, sometimes, virtually no crop appeared, despite SiO_2 being copiously available. As far as could be ascertained, algal parasitism did not appear to be the cause. This raised the question 'could cladoceran grazing be sufficient to suppress the diatom crops; especially if freed from the effects of $CuSO_4$ poisoning?'.

An intermediate model: addition of a grazing equation

On the basis of measurements of zooplankton biomass, feeding and respiratory rates in the reservoirs, particularly in Queen Elizabeth II reservoir (Andrew, 1976; Chalk, 1981; Duncan, 1975), a simple size-

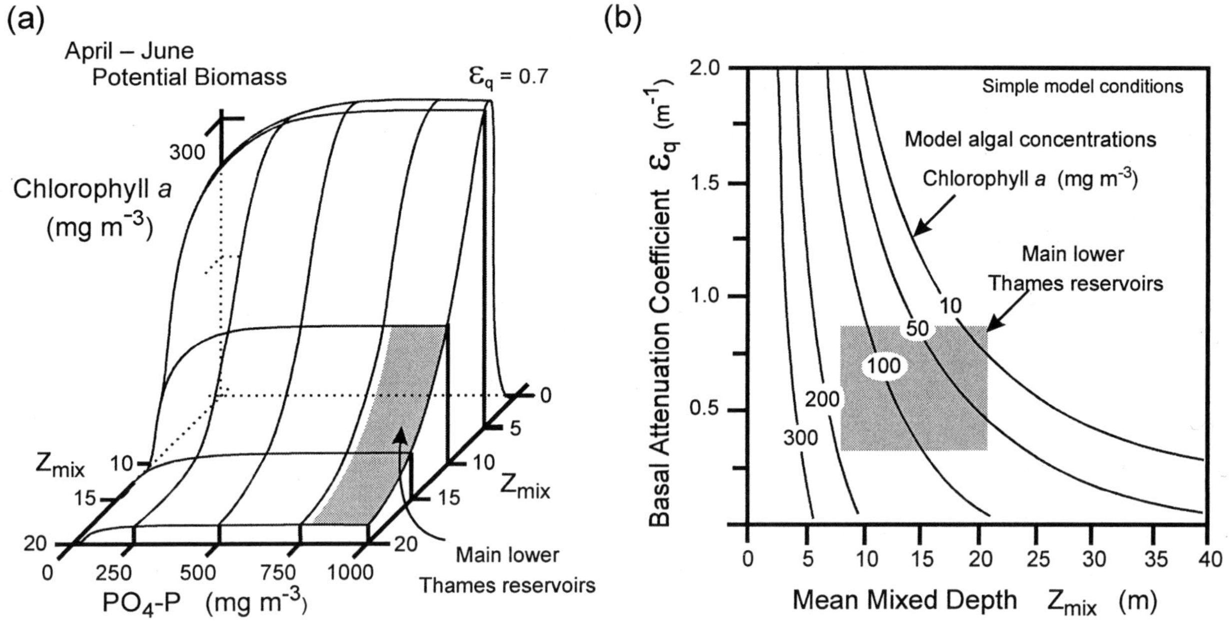

Figure 5. (a) Output from the simple model with a nutrient term for various mixed depths and typical sub-surface light attenuation. Also implied is the degree of P-stripping required to reduce potential algal crops under the conditions of the main lower Thames Valley reservoirs. (b) Output from the simple model of how mixed depth and light attenuation combine to reduce potential epilimnetic algal crops, and the range of conditions in the main lower Thames Valley reservoirs.

related grazing equation was incorporated into the model, based upon a large, 2 mm length daphnid filter feeder, and running simultaneously with the algal growth equation (Steel, 1975). This model, now of intermediate complexity, predicted that the grazeable algae of deep, fully mixed reservoirs would be much more susceptible to a given grazer pressure than if they had been in shallower waters (Figure 6a). This pressure could be continued to quite low algal concentrations due to the daphnids' ability to maintain feeding rates, despite reduction in food concentrations to an incipient limiting level. These grazing effects would be most marked in the deeper mixed reservoirs, where the later and slower algal growths tended to achieve their maximal levels at a time when the daphnid populations could take advantage of them – in late April when the water was warming to 10 °C. If these predicted effects were real, then there should be observable relationships between the algal and zooplankton crops in the deeper, mixed reservoirs.

Figure 6b is a plot of observed algal and zooplankton crops in Queen Elizabeth II reservoir for the period 1968–73, expressed as averages for the quarter April–June which incorporates the population maxima. This plot appears to show grazer-sensitive algal crops, with a changeover from maximal energy-limited algal crops in 1968–69 to markedly reduced crops from 1971 onwards associated with increasing zooplankton biomasses. The 1973 value for zooplankton biomass is the same as that for the 1969 algal crops, which is energetically impossible without some additional zooplankton food source. One source is likely to be the period of intense, small diatom production that occurs in the River Thames during the period April–June. In relatively small surface-area reservoirs, input of riverine water with such algae will add considerably to both the reservoir's productivity and potential daphnid food supply. This is a significant source as the river crop concentration is usually much greater than is possible in the deep, mixed reservoirs – and can be up to an additional 50% in apparent productivity when river crops are large.

Addition of a term for river algal immigration

Incorporation of possible levels of algal immigration into the reservoir during the April–June quarter raised the complexity of the model by another stage. It also introduced a dynamic daphnid-type grazer component which predicts the general form of the empirical algae–zooplankton relation shown in Figure 6b for different possible levels of river phytoplankton

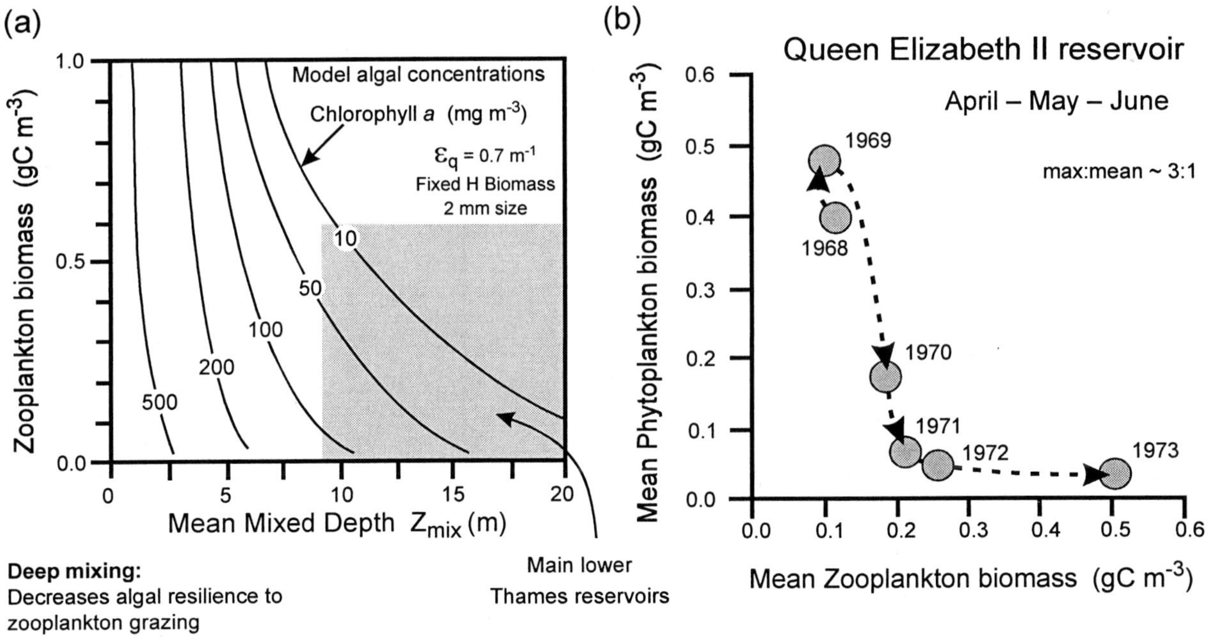

Figure 6. (a) Output from the intermediate model with a simple size-related grazing equation. With deep mixing, algal resilience to zooplankton grazing is reduced. The relevant range of conditions in the lower Thames Valley reservoirs is indicated. (b) The observed relation between phytoplankton and zooplankton biomass means in the jet-mixed Queen Elizabeth II reservoir during April–June 1968–1973.

(Figure 7). Superimposed upon these plots is the observed data set from 1968–1988 which conform more or less to the general prediction. However, careful examination of Figure 7 shows that most data points fall within a characteristic cluster and that the data of 1968, 1969 and 1973 are extraordinary. Large scale, natural fish mortalities occurred in these reservoirs during the later 1960s and it makes it appropriate to ask: 'were fish–zooplankton interactions yet a further factor in these events?' To answer this question, and to improve the predictive power of the rough grazer model needed much more information on zooplankton sizes and associated dynamics, coupled with more critical size-related algal modelling.

Detailed observations on the daphnid communities of Queen Elizabeth II reservoir in 1972 showed very high absolute population sizes as biomass, as well as an extraordinary proportion (>50%) of large animals in the population (Duncan, 1975) (Figure 8a). Present-day observations from another deep, mixed reservoir confirmed the same picture of large crops of large-sized daphnids (Santos, 1989) (Figure 8b). This raises a major question: 'how can these large-bodied grazer populations exist in the reservoir and be supported? To start to address these latter two questions needs considerably more realistic climatic-environmental and biological-ecological detail in the model.

A complex model

At present, the model incorporates these environmental effects by use of sub-system 'mimics', together with the facility to impose particular conditions unique to the reservoirs, such as mixing. The advantage of having the sub-systems is that changing a property, such as depth, also causes appropriate, concomitant changes to the thermal and mixing conditions to occur. In its current form, the model outputs can illustrate the general behaviour of various, environmentally adaptive algal groups (Steel, 1995), and the size-related, structured zooplankton populations (Figure 9). It can also be used to explore in more realistic detail many of the ecological interactions previously more simply considered – a thinking tool for the manager or ecologist.

An area of particular concern is: 'are the zooplankton and/or mixing seriously reducing potential cyanobacterial populations?'. These appear to be much less prolific now, even in those reservoirs which

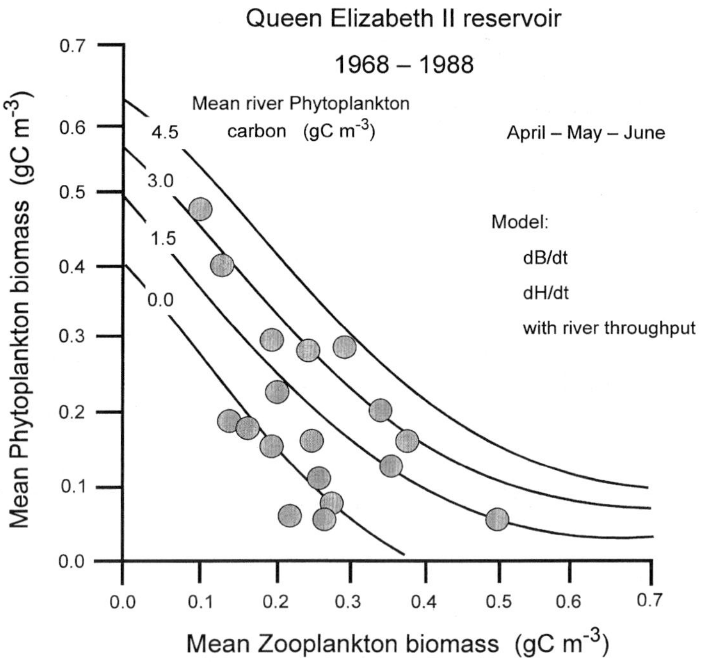

Figure 7. Output from the intermediate model with grazing and an algal immigration term. The observed mean phytoplankton and zooplankton biomasses for the April–June period from 1968–1988 fall between the predicted limits for realistic levels of riverine algal immigration. Apart from three exceptional points (see text), most data points form a cluster with narrow ranges.

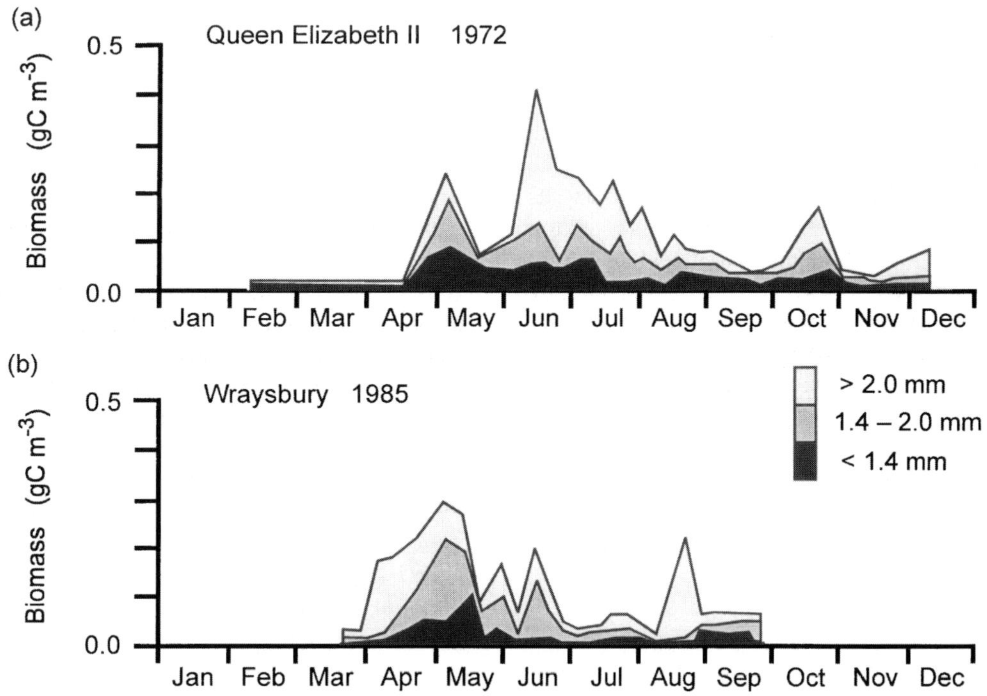

Figure 8. Observed data. The size structure of populations of *Daphnia* species in two jet-mixed reservoirs: (a) Queen Elizabeth II reservoir, 1972; (b) Wraysbury reservoir, 1985.

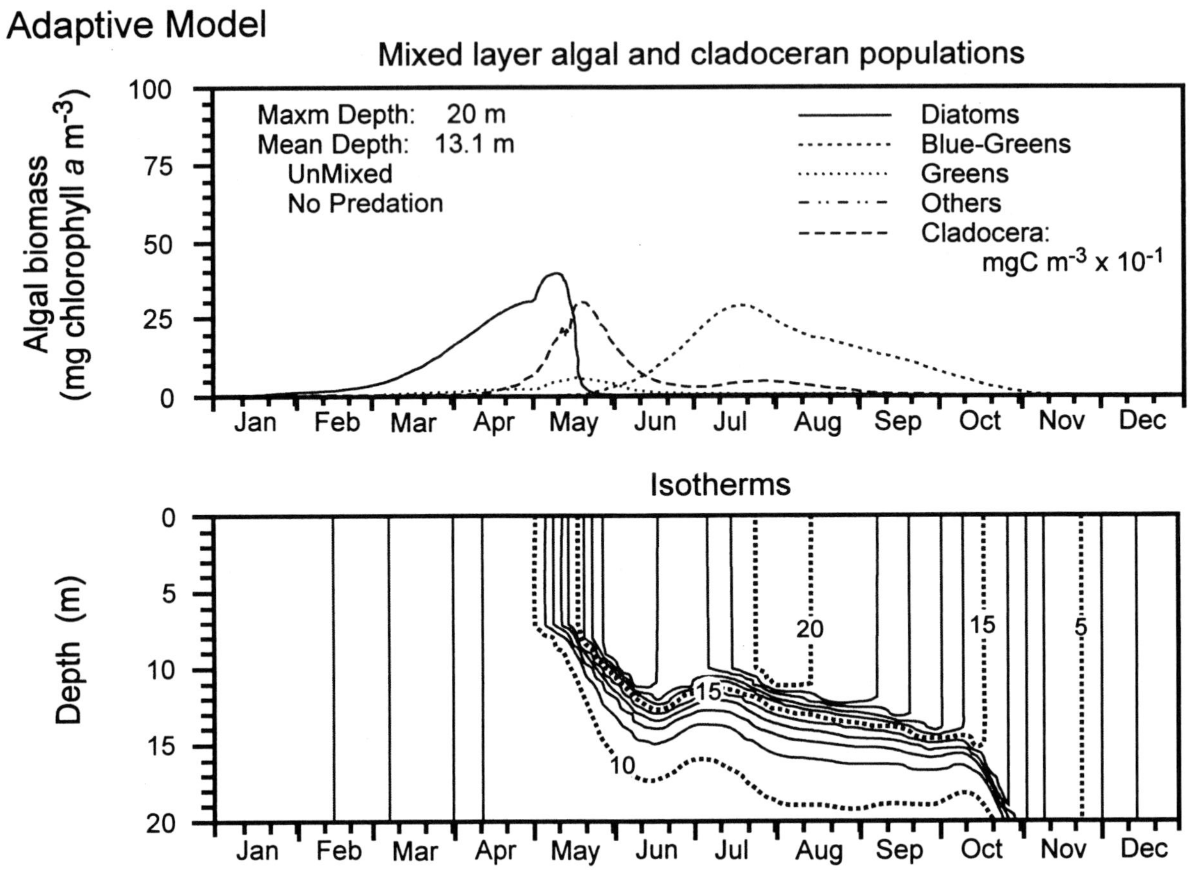

Figure 9. A sample of the output from the complex, environmentally and biologically adaptive model.

were notorious for them 30–40 years ago. What the model suggests in response to this question is that, as long as there is a reasonable, large-sized zooplankton population present, and adequate mixing when the cyanobacterial initiates appear, then sufficient grazing can markedly reduce any later, non-grazeable sized cyanobacteria. However, all this is as yet untested, and will be experimentally difficult directly to support or refute!

Future developments

Fish–zooplankton interaction

Studies during 1993 (Figure 10a) provided a good characterisation of three of the reservoirs' fish populations and associated zooplankton species composition and size structure. (Kubecka & Duncan, 1994; Renton et al., 1995; Seda & Duncan, 1994). The extraordinarily low fish biomass in Wraysbury reservoir is associated with the highest proportion of large-sized daphnids in the zooplankton, in contrast to the much larger fish biomass of Queen Mary reservoir which is associated with a much smaller proportion of large daphnids in the zooplankton, with an intermediate position in Queen Elizabeth II reservoir. There were quality differences also. In Wraysbury, the fish biomass was virtually all perch and ruffe with no cyprinids, whereas in Queen Mary reservoir it was largely bream and roach (cyprinids) plus some perch. The differences in zooplankton size structure applied to species as well as individuals. The large-sized *Daphnia magna* and *D. pulicaria* co-existed all year with the smaller *D. galeata* in Wraysbury reservoir (as in Queen Elizabeth II reservoir, but with different proportions), whereas only *D. galeata* existed in the open water zooplankton of Queen Mary reservoir. It is probable that the total zooplankton biomass data of Figure 10a are not significantly different.

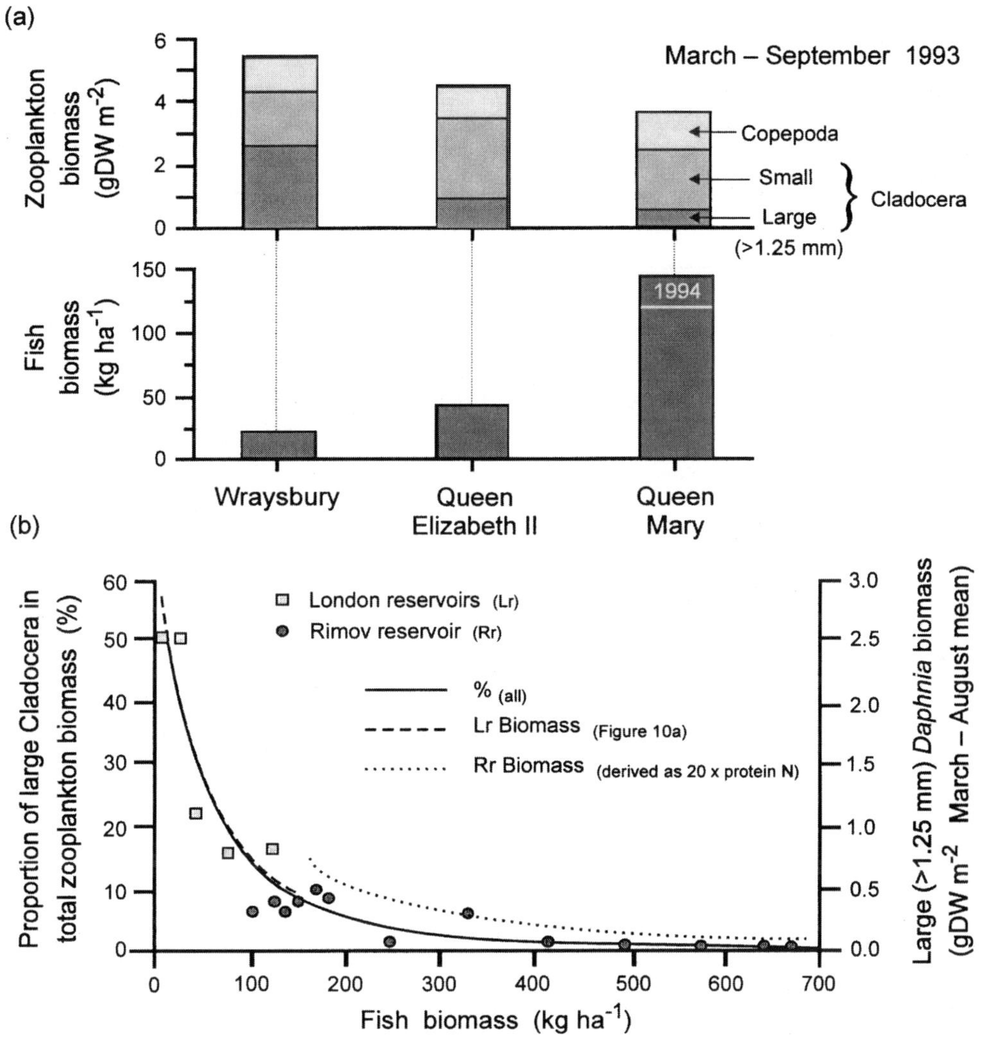

Figure 10. Observed data on the fish-zooplankton interaction. (a) The fish and size-structured zooplankton biomasses determined simultaneously in three well-mixed reservoirs during 1993. (b) The observed relation between the size-structured zooplankton biomass and fish biomass in three lower Thames Valley reservoirs and the Rimov reservoir, Czech Republic.

The lower fish biomasses of the present reservoirs fit well with a similar set of data available for the Czech Rimov reservoir (Kubecka, 1989; Seda et al., 1989) (Figure 10b) in which the fish biomass was deliberately reduced over a period of 14 years from its highest levels (650 kg ha^{-1}) down to 100–50 kg ha^{-1}, below which it proved impossible to go. Figure 10b also contains a rough estimate of large cladoceran biomass derived from nitrogen measurements given in Kubecka and Duncan (1994). The combined data suggest that there might be a crucial fish biomass of the order of 100 kg ha^{-1} above which the large cladocera are under such severe predation pressure they can no longer constitute a substantial proportion of the zooplankton biomass. A secondary consequence seems to be a significant diminution of large (>1.25mm) cladoceran biomass. It is interesting to note in these reservoirs of similar trophy, the comparable form and disposition of the proportion and large cladoceran biomasses, which is suggestive of relative constancy (of the order of 5 g dw m^{-2}) in the total zooplankton biomass, despite the large range of fish biomasses involved.

A combination of these measurements with simultaneous measures of algal crops in the three reservoirs and the graphical model in Seda & Duncan (1994) al-

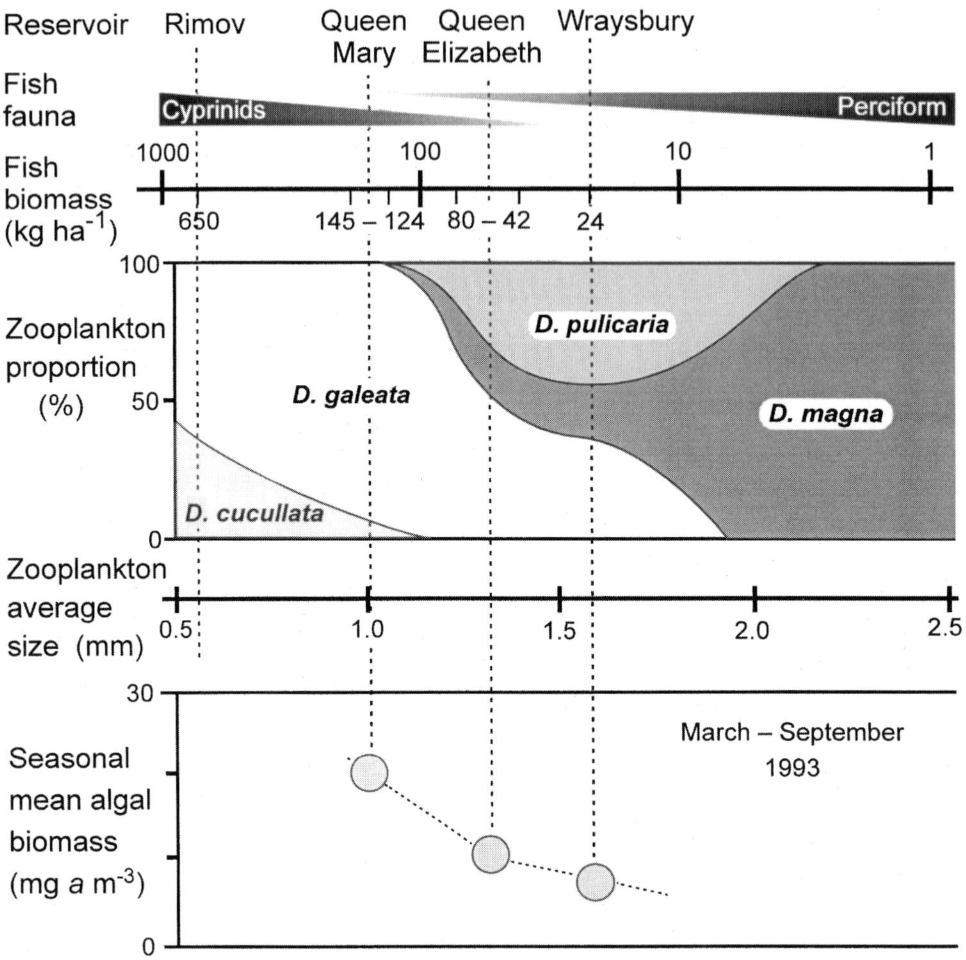

Figure 11. Quantification of top-down effects in the trophic cascade hypothesis in three deep, well-mixed reservoirs. Data from the Czech Rimov reservoir is included.

lows a first quantification of the top-down effects of the trophic cascade hypothesis (Carpenter & Kitchell, 1993) in these deep, well-mixed basins of similar trophic status with natural fish populations, and is shown in Figure 11. The seasonal maximum:mean algal crops are of the order of 3:1. Importantly, the daphnid speciation implies within it a particular adult size, characteristic of the species. The average size covers all individuals, not just adults.

An important question, relevant to the business, now is: 'are these ecological conditions stable without further intervention'? The Wraysbury fish appear unlikely to develop a significant cyprinid fauna because the reservoir does not have suitable spawning sites (Duncan & Kubecka, 1995; Kubecka & Duncan, 1994). The same limitation applies to Queen Elizabeth II reservoir with its similar concrete margins, but spawning sites are presently enhanced by the net-sides of empty fish cages. In Queen Mary reservoir, however, bream and roach are already breeding and growing, so this reservoir may well be able to support a much larger cyprinid population. Earlier observations (1975) show that the zooplankton of Queen Mary reservoir containing a 40:60 proportion of *Daphnia cucullata* (an even smaller daphnid species) to *Daphnia galeata* (from Figure 11) was indicative of perhaps 500 kg ha^{-1} fish biomass and was associated with a seasonal mean chlorophyll-*a* of 40 mg chlorophyll-*a* m^{-3} (equivalent to 135 mg chlorophyll-*a* m^{-3} maximum). Allied to a deepening of Queen Mary from 12 to 15 m, this suggests potential future seasonal algal crops of 30-35 mg chlorophyll-*a* m^{-3} (100–120 mg chlorophyll-*a* m^{-3} max.). As these are likely to be cyanobacteria populations, meas-

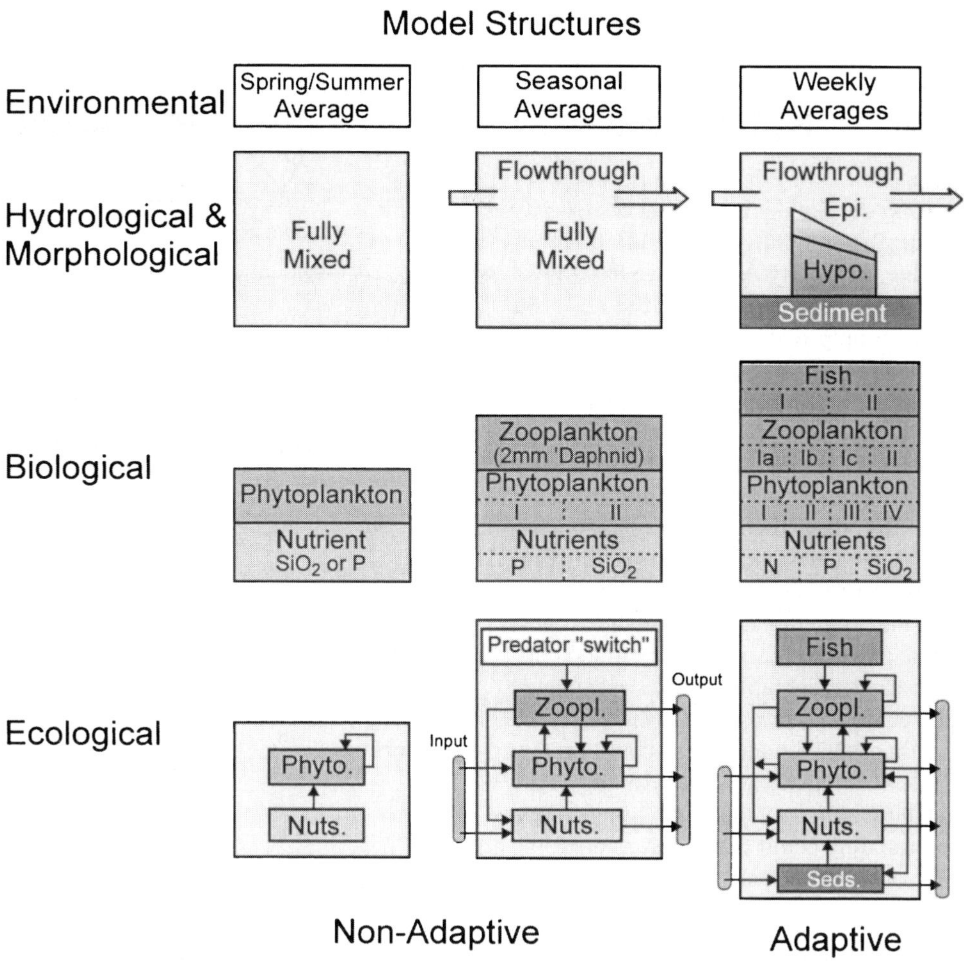

Figure 12. Diagram illustrating the historical evolution of the simple algal model and its environment towards greater complexity as ecological understanding developed in parallel.

ures would also be needed to prevent troublesome downwind aggregations being drawn into the reservoir outlets.

The empirical relationship between the seasonal mean proportion of large cladocerans in the total zooplankton (or biomass of *Daphnia* larger than 1.25 mm) and fish biomass illustrated in Figure 10b provides a start to the possibility of modelling a fish predation term. For Queen Mary reservoir, whose fish biomass in 1994 was near to the threshold value of 150 kg ha^{-1} and whose water quality was worsening, the management implications are reduction of the cyprinid fish by biomanipulative measures (Renton et al., 1995). The present development of the complex model is incorporating not only some of the direct, quantitative information of Figure 11, but also as much of the qualitative effects as current information allows. Again, at this level of development, gaining information of the necessary type and quality becomes increasingly difficult.

Conclusion

The developments of the reservoir models have reflected the management questions which have arisen throughout the past 30 years or so, the measurements which it has been possible to make and the growing understanding of the ecosystems in question. Figure 12 gives a diagrammatic summary of the model structure and nature, the main constituents and some of their interactions, over that time. Even the earliest, most simple models gave answers of considerable worth to the water supply business, and showed where ecological ignorance was particularly acute. The models currently being developed are, unavoidably, substantially more complex than their predecessors. Whilst the models may be of significant help in further understanding and, perhaps, quantification of these ecosystems, their increasing complexity can cause considerable difficulty to their full interpretation and resulting credibility. Experience has shown that there is no one, 'best' model, and the reservoir management has simply adopted the least complex model which will serve the immediate needs.

The development of the ecological knowledge of the lower Thames reservoirs and its application has been an iterative process. Figure 13 illustrates some of the complementary aspects of modelling reservoir quality dynamics and the growth in ecological knowledge. In this particular instance, the initial impetus came from management questions. In the beginning, ignorance about the problems addressed in those questions was great, but relatively straight forward experimentation and measurement rapidly gave considerable insight into the ecological factors involved. This led to simple modelling which nonetheless could greatly enhance the understanding of the system by providing quantified, testable prediction about the system. That enhanced understanding could then be translated into an answer to the original question posed.

Inevitably, once the simple questions have been addressed, the more complex issues become even more pressing, and the main cycle is traversed again. Figure 13 tries also to illustrate some of the limitations involved. For example, more complex business questions may be considered as the problem of ignorance is reduced ('what is it, in ecosystem terms, that is implied by this particular question': 'in the present instance, what do we need to know to answer the question about whether the current reservoir algal conditions are long-term stable'?). However, such ignorance will never be entirely eradicated. Also, as the more complex questions are addressed, there can be exponential increase in the difficulty of the associated, necessary measurement and modelling. This increased complexity does not however necessarily translate into similarly enhanced understanding of the system in question. There is probably a limit to the system insight that can be attained. Overly complex, or ill-conceived models may even so cloud the whole issue that credibility in the capacity of the models to provide insights into the system is diminished – to the detriment of the overall enterprise!

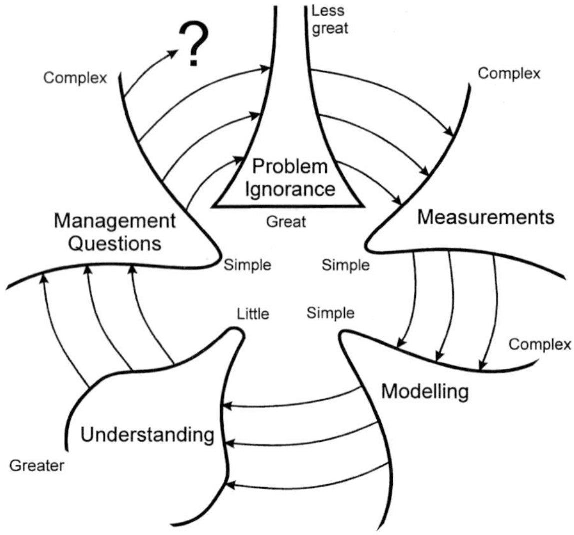

Figure 13. Diagrammatic representation of the main inter-related elements and pathways involved in the development of a predictive model, ecological understanding and management application. Degrees of difficulty and magnitudes are also indicated.

References

Andrew, T. E., 1976. The production and respiration ecology of reservoir populations of zooplankton, with special reference to daphnids. PhD. Thesis, Royal Holloway College, Univ. London: 296 pp.

Carpenter, S. R. & J. R. Kitchell, 1993. The Trophic Cascade in Lakes. Cambridge Univ.Press: Cambridge, 385 pp.

Chalk, E. A., 1981. Cladoceran filter feeding in a Thames Valley reservoir. PhD Thesis, Central London Polytechnic & Thames Water Authority: 266 pp.

Duncan, A., 1975. The importance of zooplankton in the ecology of reservoirs. In: The Effects of Storage on Water Quality. Water Research Centre, Medmenham U.K.: 247–272.

Duncan, A. & J. Kubecka, 1995. Land/water ecotone effects in reservoirs on the fish fauna. Hydrobiologia 303 : 11–30

Kubecka, J., 1989. Development of the ichthyofauna of the Rimov Reservoir and its management. Arch. Hydrobiol. Beih. Ergebn. Limnol. 33: 611–613.

Kubecka, J. & A. Duncan, 1994. Low fish predation pressure in the London reservoirs: I. Fish species composition, density and biomass. Int. Rev. ges. Hydrobiol. 79: 143–155.

Renton, P. J., A. Duncan, J. Kubecka & J. Seda, 1995. The management implications of low fish stocks in the London reservoirs. J. Water SRT-Aqua 44: 72–79.

Santos, L. C., dos, 1989. The effects of food limitation on the population dynamics, production and biological interactions of three *Daphnia* species co-existing in a London reservoir. PhD Thesis, Royal Holloway and Bedford New College, Univ. London: 236 pp.

Seda, J., J. Kubecka & Z. Brandl, 1989. Zooplankton structure and fish population development in the Rimov Reservoir, Czechoslovakia. Arch. Hydrobiol. Beih. Ergebn. Limnol. 33: 605–609.

Seda, J. & A. Duncan, 1994. Low fish predation pressure in London reservoirs. II. Consequences to zooplankton community structure. Hydrobiologia, 291: 179–191.

Steel, J. A., 1972. The application of fundamental limnological research in water supply system design and management. Symp. Zool. Soc. Lond. 29: 41–67.

Steel, J.A., 1975. The management of Thames Valley reservoirs. In: Effects of Storage on Water Quality. Water Research Centre, Medmenham: U.K.: 371–419.

Steel, J. A., 1978. Reservoir algal productivity. In: A. James (ed.), The Use of Mathematical Models In Water Pollution Control. University of Newcastle-upon-Tyne, Newcastle: 107–135.

Steel, J. A., 1995. Modelling adaptive phytoplankton in a variable environment. Ecol. Model. 78: 117–127.

Steel, J. A., A. Duncan & T. E. Andrew, 1972. The daily carbon gains and losses in the seston of Queen Mary Reservoir, England. during early and mid 1968. In: Kajak J. & A. Hillbricht-Ilkowska (eds), IBP-UNESCO Symp. Productivity Problems of Freshwaters. Kazimierz Dolny, Poland: 515–527.

Talling, J. F., 1957a. Photosynthetic characteristics of some freshwater plankton diatoms in relation to underwater radiation. New Phytol. 56: 29–50.

Talling, J. F., 1957b. The phytoplankton population as a compound photosynthetic system. New Phytol. 56: 133–149.

White, C. M., S. L. Harris & P. Cooley, 1955 (unpublished). The hydraulic aspect of stagnation in reservoirs.

Hydrobiologia **395/396:** 149–159, 1999.
D.M. Harper, B. Brierley, A.J.D. Ferguson & G. Phillips (eds), The Ecological Bases for Lake and Reservoir Management.

The importance of palaeolimnology to lake restoration

Richard W. Battarbee
Environmental Change Research Centre, University College London, 26 Bedford Way, London, U.K.

Key words: sediment, diatoms, phosphorus, acidification, eutrophication

Abstract

Palaeolimnology has developed rapidly over the last two decades to deal with problems of eutrophication, and acidification. This paper reveiew the techniques for coring, dating and interpreting sediments. The applications of palaeolimnology in interpreting the past through 'transfer functions' calculated from biological indices are reviewed. Rates of change, the causes of change, and the restoration of lakes to some predefined target are reviewed and the direction of future developments considered.

Introduction

Background

Palaeolimnology as a science has developed rapidly over the last two decades. It is no longer an esoteric subject concerned with long-term studies of lake evolution, but an increasingly robust methodology for understanding contemporary lake dynamics. This shift in focus has been stimulated by three key developments, which are:

(1) accurate methods for dating recent sediments, especially ^{210}Pb and ^{137}Cs (e.g. Appleby et al., 1986);
(2) coring techniques for the precise sampling of surface and near-surface sediments (e.g. Renberg, 1981);
(3) numerical techniques, especially transfer functions, for reconstructing past lake-water chemistry from the microfossil record (e.g. Birks et al., 1990).

A further key stimulus has been the need to generate information about lake history, first with respect to problems of surface water acidification in the 1980s (e.g. Battarbee et al., 1990), and most recently with respect to nutrient loading and eutrophication (e.g. Bennion et al., 1996a,b).

This paper is not concerned with methods, that have been fully described and reviewed elsewhere (e.g. Battarbee, 1991; Berglund, 1986), nor with the details of the environmental issues that palaeolimnological science has been addressing. Instead it focusses upon the generic questions that are asked about environmental change and lake management that palaeolimnology is most suited to answer (cf. Smol et al., 1990).

Characteristics of lake sediments

The characteristics of lake sediments vary according to the lake and its catchment. The sediments most frequently used by palaeolimnologists who are interested in lake history are those that accumulate in deep water. The sediments are a mixture of material, both organic and inorganic, derived chiefly from the lake and the lake catchment, but material in trace quantities, is also derived from the atmosphere (Figure 1). Current dating techniques enable sediment cores to be dated with considerable accuracy over the last 100 years, and enable sediment accumulation rates to be calculated. Whilst accumulation rates differ within and between lake basins, they are usually sufficiently rapid (> 1 mm per year) to allow high temporal resolution analyses of changes in water quality to be carried out.

Lake sediments can be analysed using an increasingly diverse range of physical, chemical and biological methods. Although the most complete understanding of lake and lake catchment history requires a comprehensive analysis of all attributes, most attention has focussed on the biological record and the extent to which changes in the biological (fossil) record can be used to reconstruct changes in water quality, es-

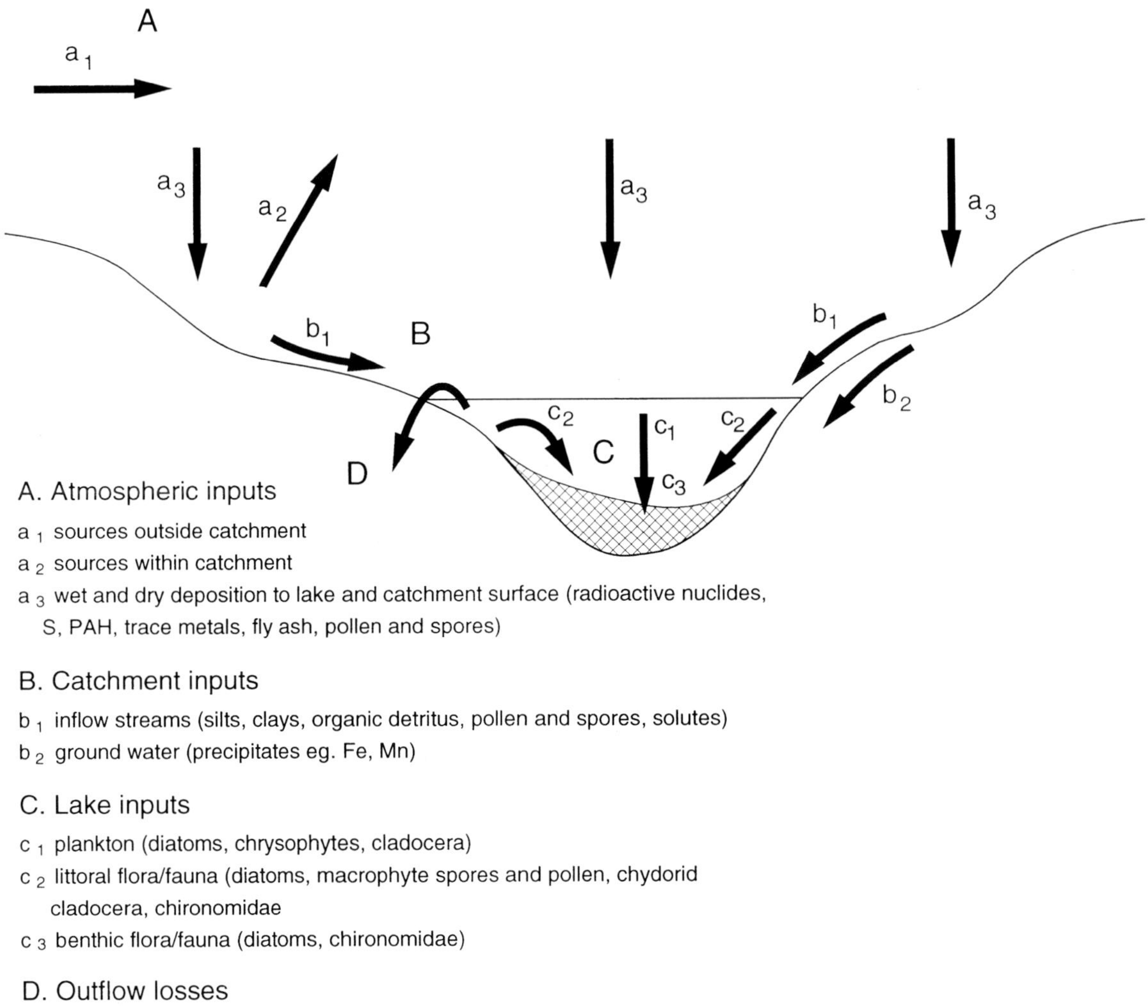

Figure 1. Schematic cross-section of a lake and its catchment showing sources and pathways of material found in lake sediments (from Battarbee et al., 1988).

pecially with respect to acidity and nutrients. Methods are continually being improved, but diatom analysis in particular is now sufficiently well-established to allow quantitative reconstruction of pH and total phosphorus for almost all lake types (Battarbee, 1997; Bennion et al., 1996; Birks et al., 1990).

Problems and pitfalls

Whilst palaeolimnologists have developed a robust set of procedures that can be used for lake management, successful application of those procedures depends on continuing validation of the models (e.g. Bennion et al., 1996) and on an understanding of the strengths and weaknesses both of the approach itself and of the various methods that are used (Anderson & Battarbee, 1994). Most natural lakes have excellent sediment records, but in some cases the sediments may be disturbed and the fossil record may suffer from poor preservation. Consequently it is advisable to assess the quality of the sediment record at study sites before embarking on detailed sediment core analysis. Furthermore, whilst many applications only require analysis of a single core it is often very useful to obtain a range of cores from different locations in the lake basin so that the best core can be chosen for detailed analysis. This would normally be the core with the least disturbance and the most rapid sediment accumulation rate (cf. Allott et al., 1992).

Applications

General considerations

Some lakes have annually laminated sediments that provide the highest quality records with respect to the accuracy of the chronological framework and the temporal resolution of the historical record itself (e.g. Simola, 1990). Most frequently sediments are not laminated and palaeolimnological techniques in general are most suited to questions of change on decadal and longer time-scales. Applied palaeolimnology, therefore, is usually concerned with attempts to establish baseline values for lake ecosystem status, quantifying and evaluating directional changes in lake status, and evaluating why change has occurred. Such knowledge is a pre-requisite for intelligent management that sets targets for restoration, uses dynamic models to assess different management scenarios and monitors the outcome. Figure 2 illustrates these ideas.

For almost all lakes, palaeolimnology is the only means of obtaining data on past and present trends. Alternative historical records needed to establish past baselines are usually lacking or inadequate, whereas most monitoring systems, designed to identify longer-term environmental change, are subject to the variations imposed by episodic events and seasonality that often obscure the more important underlying trends. Data from Loch Dee (Figure 3) illustrate both points. The gradual acidification of Loch Dee by approximately 0.5 pH over the last 100 years is clearly recorded by shifts in the diatom assemblages in the lake sediment (Figure 3c). However, the onset of acidification pre-dates the beginning of the monitoring programme on the lake, and the rate of long-term pH change at this site is too low to be revealed by monitoring that is influenced more by seasonal processes (Figure 3b) and episodic events (Figure 3a), both of which have an amplitude of variation (> 1 pH unit) greater than the mean shift in pH over the last 100 years.

Defining baseline values of pH and total phosphorus

The development of robust diatom-based transfer functions for both pH (e.g. Birks et al., 1990) and total phosphorus (e.g. Bennion et al., 1996) in recent

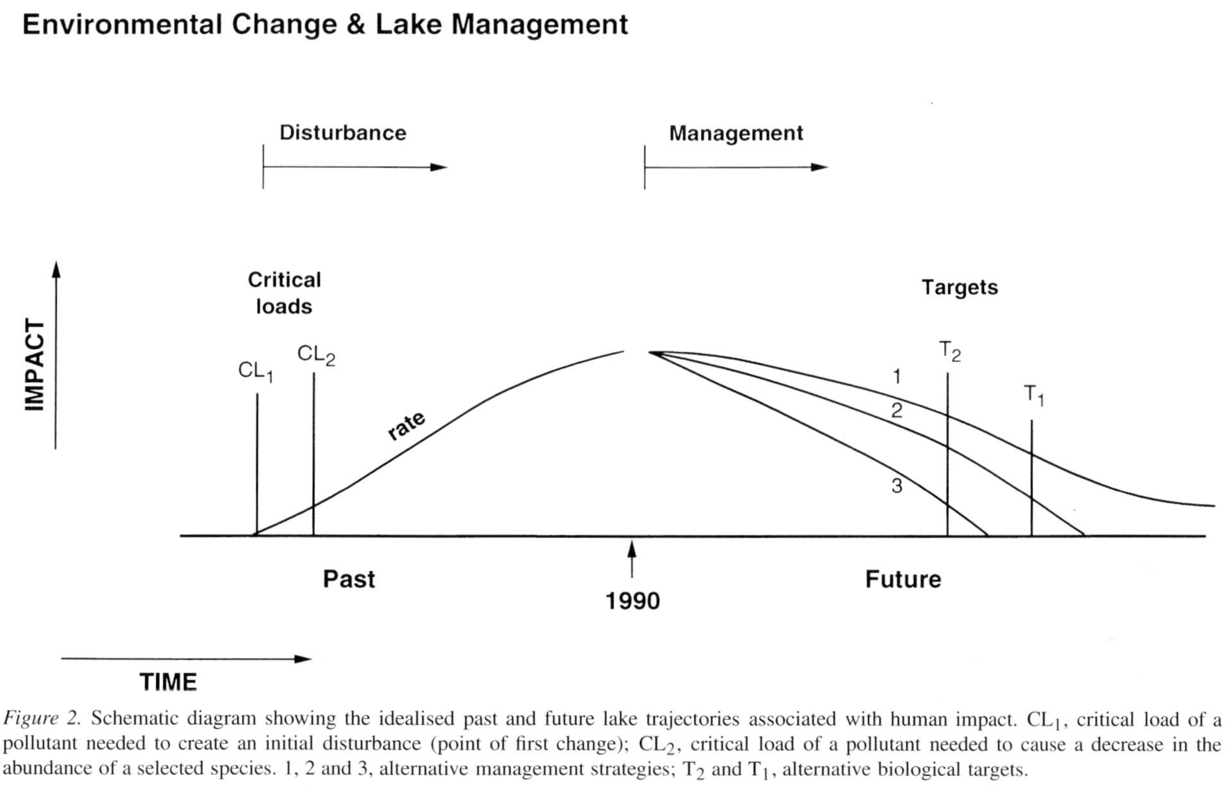

Figure 2. Schematic diagram showing the idealised past and future lake trajectories associated with human impact. CL_1, critical load of a pollutant needed to create an initial disturbance (point of first change); CL_2, critical load of a pollutant needed to cause a decrease in the abundance of a selected species. 1, 2 and 3, alternative management strategies; T_2 and T_1, alternative biological targets.

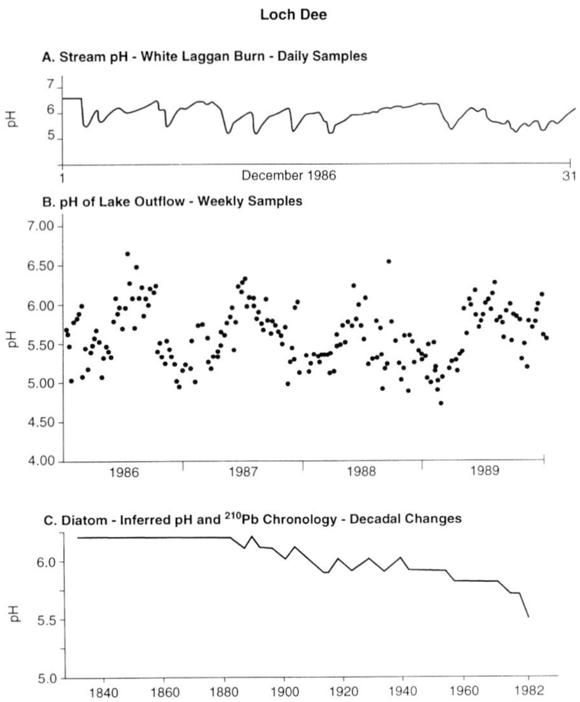

Figure 3. pH variability of Loch Dee on different time-scales: (1) hourly and daily changes in response to rainfall events, (2) annual changes in response to seasonal hydrological changes and photosynthetic activity; (3) decadal changes in response to acid deposition (modified from Anderson & Battarbee, 1994).

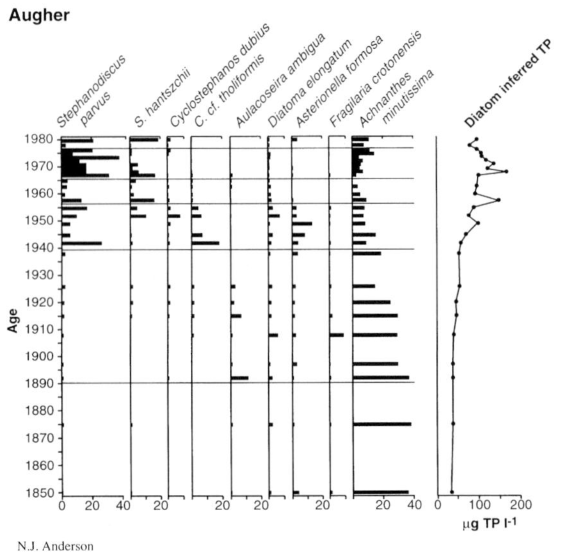

Figure 4. Diatom diagram (planktonic taxa only) and diatom-inferred total phosphorus reconstruction from a Lough Augher sediment core (from Anderson, pers. comm.).

years allows the recent pH and total phosphorus history of lakes to be reconstructed. If the cores have been dated it is possible to define a baseline value for pH or total phosphorus for any past time-period, and such a baseline value can then be used either to assess the degree of change between the baseline and the present day, or used to select a target for the future (see below). Figure 4 provides an illustration from Lough Augher in Northern Ireland (Anderson et al., 1990). This lake has been polluted by effluent from a creamery. The analysis shows that total phosphorus levels rose to between 100 and 150 mg l^{-1} by the 1960s, against a probable natural baseline at the beginning of the twentieth century of approximately 30 mg l^{-1}. Recent removal of the effluent has allowed some recovery to take place.

One question concerns the date of an appropriate baseline. For most lakes this is relatively simple. As significant human disturbance of lakes did not occur until the nineteenth century, 1850 is probably the most appropriate baseline for both acidification and eutrophication. For acidification the first clear signs of the impact of acid deposition mainly post-date 1850 (Battarbee, 1984), and for cultural eutrophication the first signs normally post-date 1900 and are associated with the development of early water supply and sewerage systems (Battarbee, 1978; 1986). Some exceptions to this latter generalisation are small deep lakes in the lowlands where there has been a long history of agriculture in the lake catchment, and eutrophication may have been a progressive and complex process taking place over hundreds to thousands of years (e.g. Fritz, 1989). In these cases the choice of a baseline date is entirely a subjective one.

Providing early warning signals

One of the most remarkable aspects of the research on 'acid rain' and its effects in the 1980s was the discovery that surface waters in the UK had been acidifying for over 100 years without this being noticed. Whilst this can be attributed partly to the lack of environmental awareness and the pre-occupation with other issues (such as urban smogs and sewage effluents) in the first half of this century, it highlights the need for, and importance of, early warning systems for environmental change. Lake sediments can be used in such a way, to identify trends in potentially toxic contaminants, such as trace metals or trace organic compounds (e.g. Boyle, 1997; Grimalt, 1997) and to identify changes in the microfossil record indicative of early signs of biological change. For example, recent work

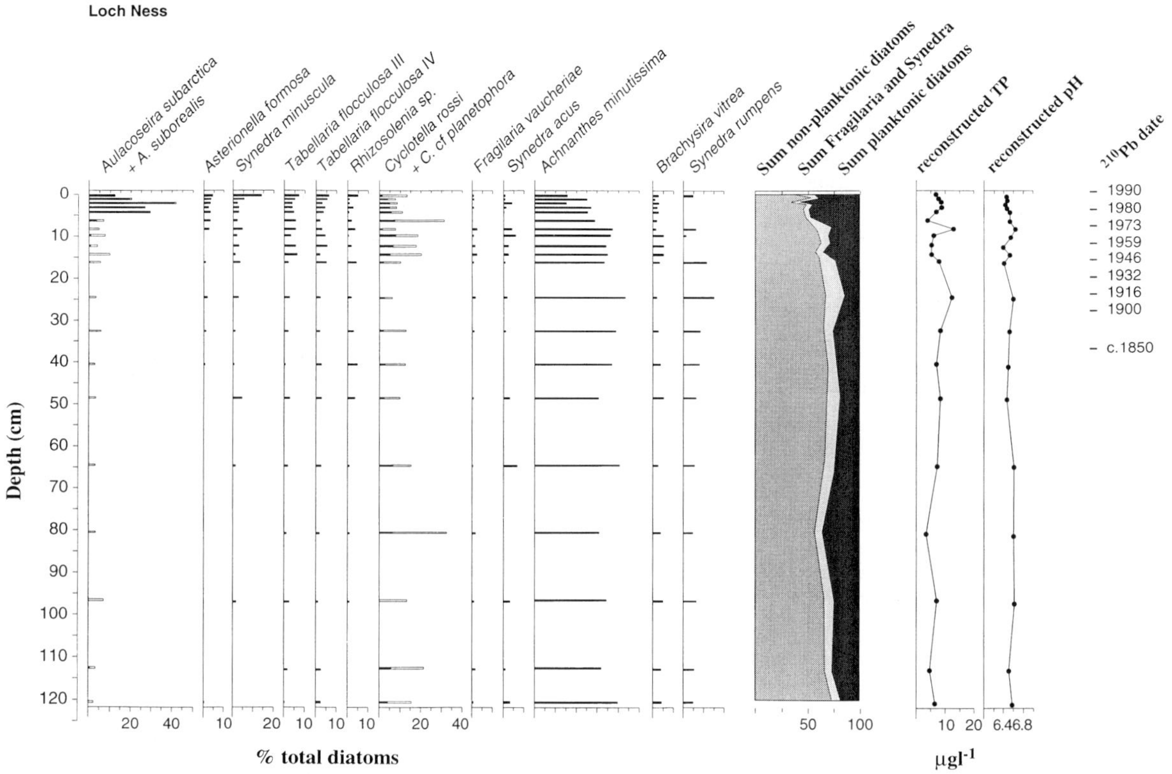

Figure 5. Diatom diagram (selected taxa) and diatom-inferred total phosphorus and pH from an Loch ness sediment core (from Jones et al., 1997).

on Loch Ness sediments (Jones et al., 1997) indicates that there have been some highly significant changes in the composition of the planktonic diatom flora over the last few decades probably reflecting an increase in nutrient loading on the lake (Figure 5). However, the lake is still ultra-oligotrophic and, as yet, there has been no observed increases in total phosphorus concentration. The change in the diatom record cannot be fully explained, but the data provide an early warning, and indicate that an ecologically important threshold, probably related to human disturbance of the lake ecosystem, has now been passed.

Assessing the naturalness of lakes

Ratcliffe (1971) proposed 'naturalness' as one of the main criteria for the selection of nature reserves. This is a difficult criterion to apply to many ecological systems because of the lack of knowledge of the past. For lakes, however, the degree of naturalness can be assessed from an examination of the sediment record, supplemented where possible from a study of the lake's catchment history from documentary and other sources. The entire physical, chemical and biological information contained in the sediment record could be used for this purpose, although in practice, much can be gained from the analysis of key indicators, such as diatoms. Information is needed not only for changes in water chemistry associated with the impact of acidifying or enriching pollutants, but also for the degree to which there have been changes in habitat as a result of water level changes, catchment erosion, etc., and whether there have been species introductions and extinctions. These largely independent effects can be visualised using a 'naturalness matrix' (Figure 6) that represents the degree to which a lake has changed compared to its natural baseline (Battarbee, 1997). For nature conservation or lake management purposes, where restoration to a baseline state might be the objective, the target ecosystem is one with minimum pollution and minimum habitat disturbance, represented by the 'target box' (Figure 6a).

Deriving a state-changed classification

Moss et al. (1996; 1997) have argued cogently that

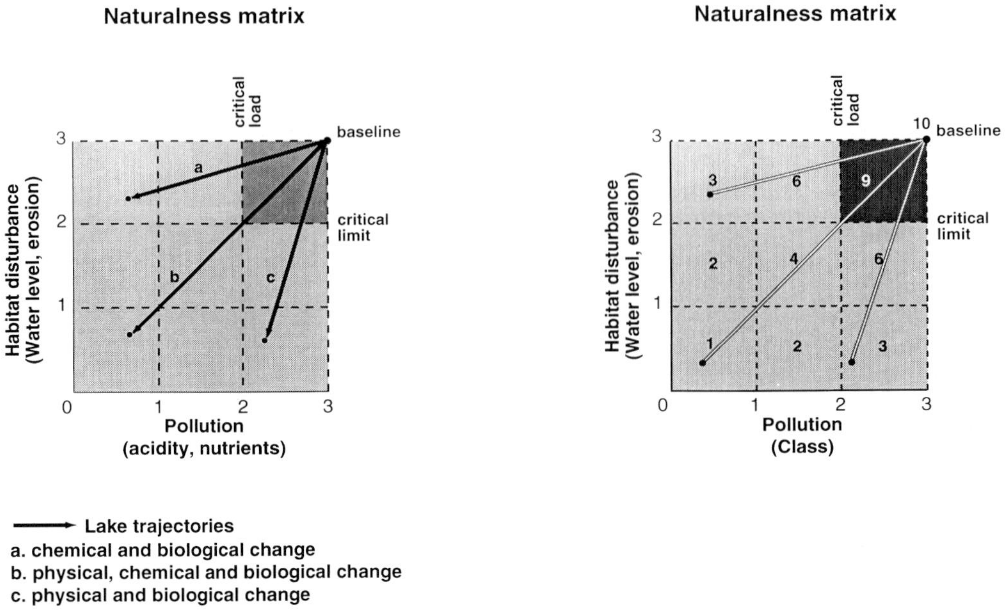

Figure 6. Naturalness matrix showing: (1) hypothetical change over baseline for three lakes (a, b, & c); (2) a 'state-changed' classification system related to change over baseline, woth 1=most changed, 9=least changed and 10=pristine (from Battarbee, 1997).

the classification of lakes for management purposes should be based on the extent to which a lake has changed compared with a baseline state. They call this a 'state-changed' scheme. The approaches contrasts with the more usual approach to classification that relies on assessing the contemporary differences between lakes (a 'spatial state' scheme). Use of a 'state-changed' scheme requires a knowledge of lake history and the definition of a baseline state in the past. This may be the natural baseline, as defined above or may be a date in the past set by environmental managers and that accepts some degree of pollution or disturbance. For example, Moss et al. (1996) have argued that 1930 is an appropriate baseline date. Although some change may have occurred before 1930, this pre-dates the major increase in energy consumption, intensification of agricultural land-use, and introduction of synthetic detergents that have had major impacts on water-bodies after the Second World War.

The main difficulty of a state-changed classification is the acquisition of accurate historical information. However, as long as a lake in question has a good sediment record then palaeolimnological techniques are the most effective means of providing the information (Battarbee, 1997). The naturalness matrix can then be modified to give a classification system (Figure 6b), based not only on the degree of pollution (acidification, eutrophication), but also on the degree of habitat disturbance (e.g. water level change). In this 10-point scheme (Figure 6b), class 10 is pristine and the most degraded standing water is class 1, and relates to strongly polluted, highly disturbed systems, such as a Pennine reservoir that may be both strongly acidified and suffer from excessive water level change.

Defining current trends and rates of change

Simple comparison between the present state and some previous state (either the natural baseline or a subjectively chosen one) is useful for classification purposes, but of little value in assessing which direction, and how rapidly a lake is moving at the present. Again, unless a lake is being continuously monitored, palaeolimnology is the most effective technique for this purpose. Depending on the rate of sediment accumulation, use of lake sediments in this context requires careful sampling of the uppermost sediment and core sub-sampling at 2–5-mm intervals. Where bioturbation is a problem trends may be masked or muted, but there are several studies where analysis of this kind has generated important information, for example about the extent of recovery following management intervention (Anderson & Rippey, 1994; Battarbee et al., 1988; Dixit et al., 1989; Flower et al., 1996). In Figure 7, high resolution analysis of the uppermost sediment of Swan Lake, a soft-water lake close to the

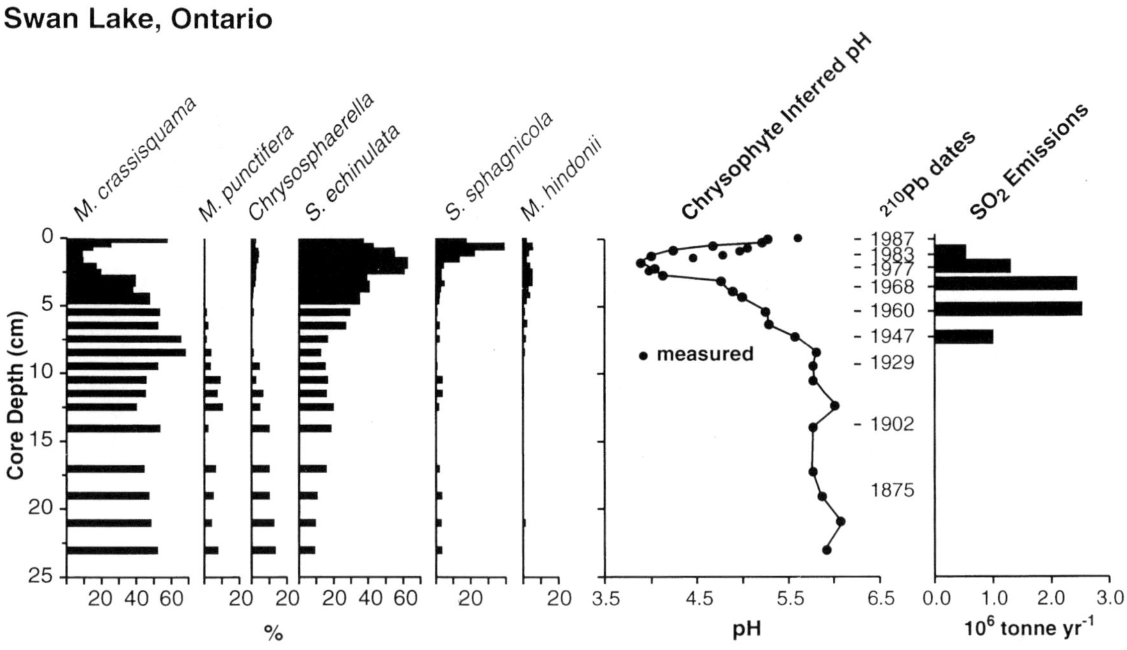

Dixit *et al* **(1989)**

Figure 7. Chrysophyte scale diagram for Swan Lake, Ontario, showing response to SO_2 emissions from the Sudbury nickel smelter (from Dixit el al., 1989).

Sudbury nickel smelter in Ontario shows the acidifying impact of the smelter and striking recovery after a reduction in S deposition.

Identifying the causes of change

One of the most powerful uses of palaeolimnology is its role in identifying probable causes of change, either by reconstructing the history of individual sites or by comparison between the histories of sites. This role has been put to most telling effect in the lake acidification debate of the 1980s when several competing hypotheses were put forward to explain the current low pH values and poor fish populations of many upland lakes in Britain and Scandinavia (e.g. Battarbee et al. 1985). The main hypotheses were that the current status was due to the following causes (Battarbee, 1990):

(1) natural acidification processes;
(2) land-use changes, such as catchment afforestation;
(3) acid deposition.

As all these relate to changes that have, or may have, occurred on long (decadal and greater) timescales, palaeolimnological methods are needed to discriminate between them. A very simple example concerns forestry in the U.K. Although catchment afforestation can exacerbate surface water acidification (Harriman & Morrison, 1985) it is clear that this is not the principal cause of the acidification of Loch Dee. Figure 8 shows that acidification began at this site at the end of the nineteenth century, well before the catchment was afforested in the 1960s (Flower et al., 1987). This proposition can be equally disproved by showing that recent acidification has occurred at non-afforested sites (Flower & Battarbee, 1983).

Defining restoration targets

If a polluted and/or disturbed lake, or any other, ecosystem is to be restored, a management target is needed. Whilst it is not always desirable, or sometimes even possible, to restore a lake to its previous undisturbed condition a knowledge of that condition is essential for two principal reasons:

(1) it defines one of the options for management; and
(2) it dictates the limits of what can be achieved.

Until now most attention in applied palaeolimnological research has been devoted to hydrochemical reconstruction and the definition of pH and total phosphorus baselines (above). In both the case of acidifica-

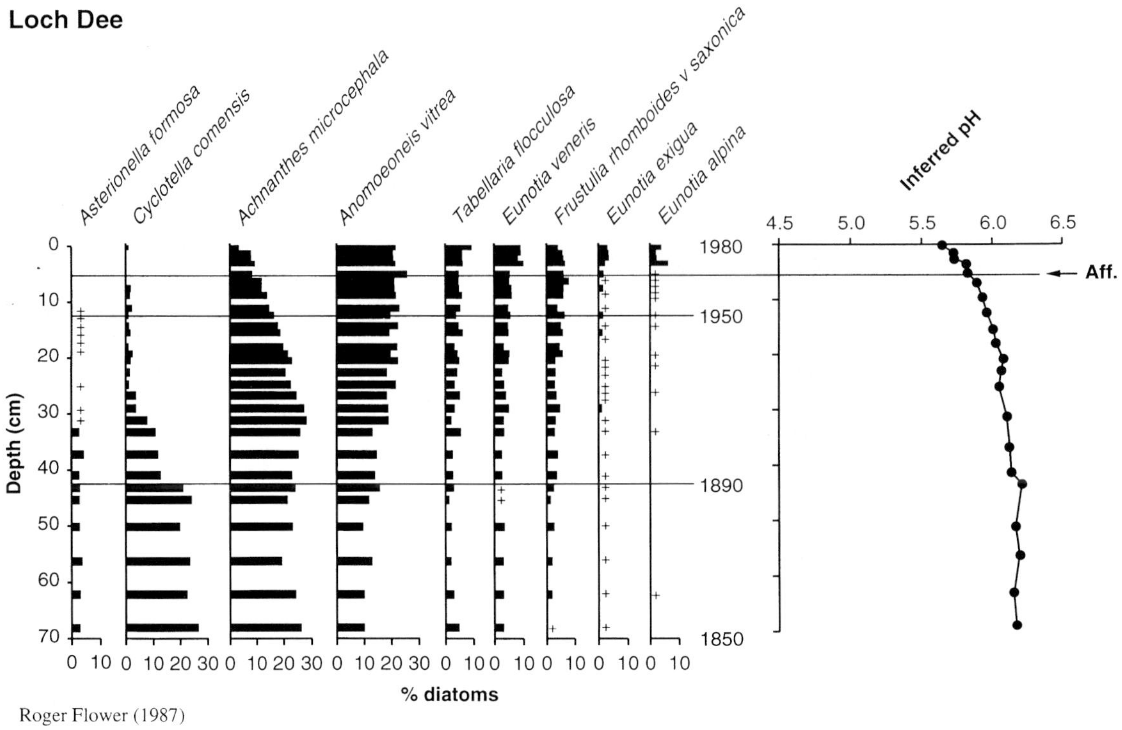

Figure 8. Diatom diagram for Loch Dee, Galloway, showing changes before and after catchment afforestation (aff.) (modified from Flower et al., 1987).

tion and of eutrophication, these baselines can be used as recovery targets, and indicate the maximum achievable recovery. For example, they ensure that total phosphorus targets are not set too low for enriched lakes that are naturally eutrophic, or that pH targets are not set too high for acidified waters that are naturally acidic. This latter example is most apposite in the case of liming. In Sweden non-acidified, naturally acidic lakes have been needlessly limed, causing pollution of a different nature (Renberg, personal communication), and in many other instances acidified lakes have been over-limed creating pH levels higher than the natural baseline. Figure 9 shows how the excessive liming of Loch Fleet in Scotland caused an increase in lake-water pH to 6.5, approximately 1 pH unit higher than the natural baseline, and stimulated the development of a planktonic diatom flora that had not previously occurred in the lake (Cameron, 1995).

Although not yet fully developed as a management method, palaeolimnology also has the potential to define biological as well as chemical targets for recovery, using the full range of microfossil and macrofossil remains in lake sediments. These include, for example, pollen and macrofossil evidence of past aquatic plant assemblages, diatoms and other algae, including the chlorophyll and carotenoid pigment record; zooplankton and chydorid cladoceran remains; chironomid remains and, in hard water lakes, ostracod and molluscan records. The richness of this record potentially allows many of the structural and functional characteristics of pre-pollution lake ecosystems to be reconstructed and thereby inform restoration management decisions. It can in addition be used to identify lakes that are currently relatively unpolluted that can act as modern analogues for the polluted lakes. In these cases where a close analogue can be identified (e.g. Flower et al., 1997) contemporary limnological surveys can be carried out to identify additional key characteristics that the currently polluted lake may have possessed. Such surveys may be especially important for fish and benthic macroinvertebrate populations as these are the most important structural components of lake ecosystems that are most poorly recorded by lake sediments.

Validating predictive dynamic and other models

The ultimate goal of much applied environmental research is the generation of robust predictive models

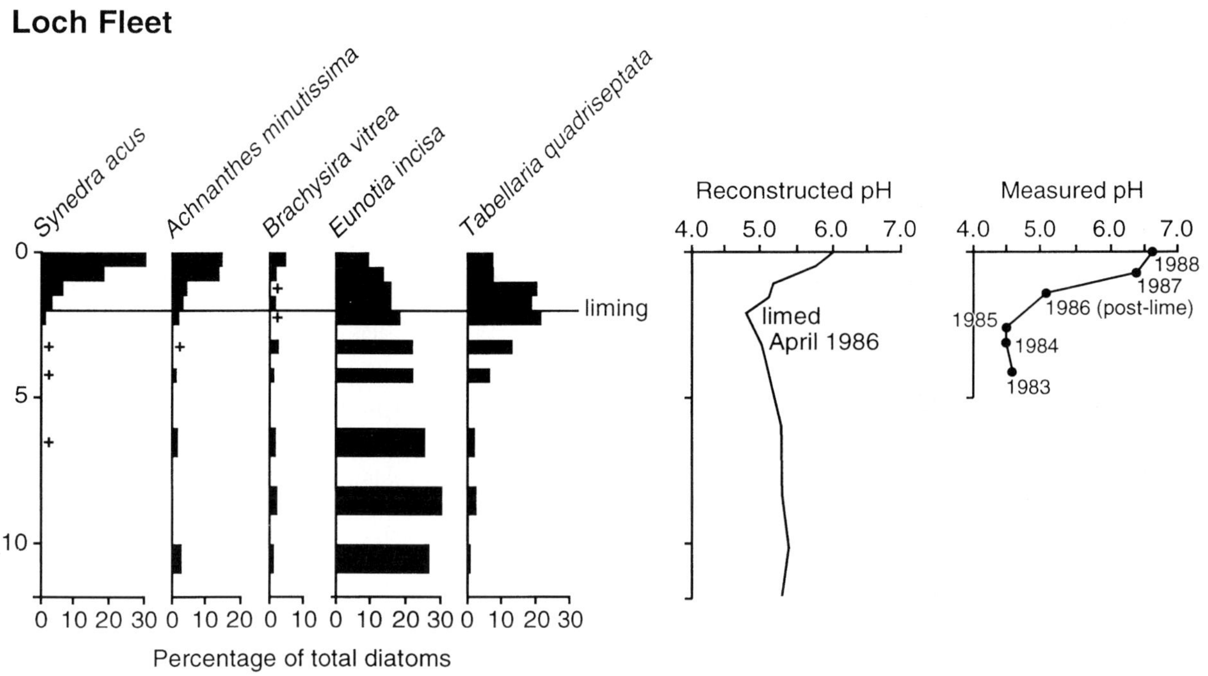

Figure 9. Diatom diagram for Loch Fleet, Galloway, showing changes associated with catchment liming in 1986 (from Cameron, 1995).

that can be used for management purposes, especially scenario testing. Such models, if they are to be relied on, need rigorous validation whereby model output is compared with experimental or field observations. The validation of models that are designed to simulate change over long time-scales (e.g. decades or greater) requires equally long time-scale observational data that can, in the case of lakes, be provided by sediment records. Appropriate dynamic nutrient models are not yet available, but this approach has been used frequently for dynamic acidification models such as MAGIC (Model of Acidification of Groundwaters in Catchments), which simulates the physical and chemical processes that occur in response to sulphur loading on catchments (Cosby et al., 1985).

The model can be set up for any individual site given a knowledge of rainfall and runoff quantity and quality, sulphur deposition history, and soil characteristics (especially bulk density, depth, cation exchange capacity, sulphate adsorption rates and base saturation, temperature and carbon dioxide levels). Once set up the model is run to reconstruct past acidification trends and to predict future pH and water chemistry according to specified future sulphur deposition scenarios.

The output of MAGIC can be validated by comparing model-based pH reconstructions with diatom-based reconstructions (Jenkins et al., 1990, 1997; Sullivan et al., 1996). Figure 10 shows model output compared with diatom-inferred past pH values for two sites in Galloway, S.W. Scotland. The relatively close agreement between these two independent methods gives some confidence in the use of the model to predict future changes following planned reductions in sulphur emissions in Europe.

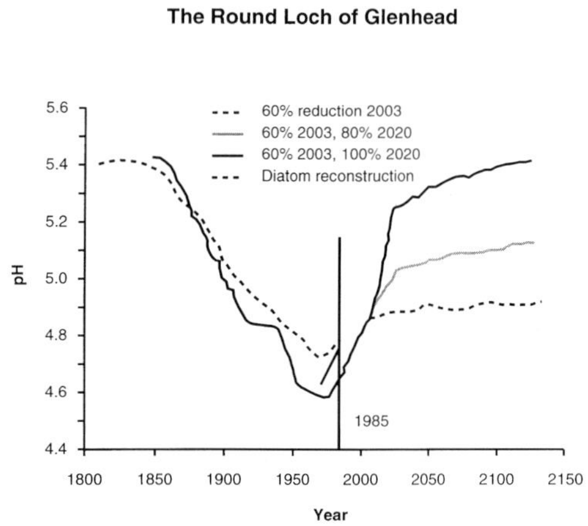

Figure 10. Comparison of diatom-inferred pH reconstructions and MAGIC model pH hindcasts and forecasts for the Round Loch of Glenhead (from Battarbee, 1993).

Conclusions

Whilst there can be difficulties at some sites, for example, if there are problems of sediment disturbance, palaeolimnology can now be used expertly as a tool in lake management. It can provide essential information that is otherwise unavailable or of limited quality, relating to questions associated with the definition of pre-pollution baselines and natural variability, and the evaluation of the magnitude, direction and causes of change. It can also help to inform management strategies, especially by helping to define chemical and biological targets for lake restoration and by constraining and validating other models.

Whilst the full analysis of sediment records needed to tackle all issues requires a significant commitment of resources, the technique can be used in simple, inexpensive ways or with increasing levels of sophistication appropriate to the questions that need to be answered. For example, high quality information relating to questions of baseline definition, early warning, naturalness and lake classification (above) can be acquired from the diatom analysis of a small number of samples from a single dated core. Such an analysis would at the same time allow an assessment of the suitability of the site for further, more detailed and sophisticated palaeolimnological work to be carried out.

Techniques continue to be refined and developed. Diatom training sets are becoming increasingly large and comprehensive, providing more reliable estimates of species optima and distribution, and, with the development of new expert systems the basic taxonomic, statistical and computational methods needed to generate diatom-based pH and total phosphorus reconstructions will become more accessible. However, as biodiversity, biological conservation and lake restoration issues also assume greater importance more attention needs to be directed towards the analysis and use of the complete fossil record.

Acknowledgements

I should like to thank members of the ECRC for many stimulating discussions on these ideas, and I am grateful to John Anderson for permission to reproduce Figure 4. Catherine Dalton and Cath Pyke helped expertly in the preparation of this manuscript.

References

Allott, T. E. H., R. Harriman & R. W. Battarbee, 1992. Reversibility of acidification at Round Loch of Glenhead, Galloway, Scotland. Envir. Pollut. 77: 219–225.

Anderson, N. J. & R. W. Battarbee, 1994. Aquatic commmunity persistence and variability: a palaeolimnological perspective. In Giller, P. S. A. G. Hildrew, & D. Rafaelli (eds), Aquatic Ecology: Scale, Pattern and Processes. Blackwell Scientific Publications, Oxford.

Anderson, N. J & B. Rippey, 1994. Monitoring lake recovery from point-source eutrophication: the use of diatom-inferred epilimnetic total phosphorus and sediment chemistry. Freshwat. Biol. 32: 625–39.

Anderson, N. J., B. Rippey, & A. C. Stevenson, 1990. Change to a diatom assemblage in a eutrophic lake following point-source nutrient re-direction: a palaeolimnological approach. Freshwat. Biol. 23: 205–17.

Appleby, P. G., P. J. Nolan, D.W. Gifford, M. J. Godfrey, F. Oldfield, R. W. Battarbee & N. J. Anderson, 1987. ^{210}Pb dating by low background gamma counting. Hydrobiologia 143: 21–27

Battarbee, R. W., 1978. Observations on the recent history of Lough Neagh and its drainage basin. Phil. Trans. r. Soc. London B 281: 303–345.

Battarbee, R. W., 1984. Diatom analysis and the acidification of lakes. Phil. Trans. r. Soc. London B 305: 451–477.

Battarbee, R.W., 1986. Diatom analysis. In B. E. Berglund (ed.), Handbook of Holocene Palaeoecology and Palaeohydrology. John Wiley & Sons, Chichester: 527–570

Battarbee, R.W., 1990. The causes of lake acidification, with special reference to the role of acid deposition. Phil. Trans. r. Soc. London B 281: 303–345.

Battarbee, R.W., 1991. Recent palaeolimnology and diatom-derived environmental reconstruction. In L. C. Shane, & E. J. Cushing (eds), Quarternary Landscapes, University of Minnesota Press, Minneapoliis: 129–174

Battarbee, R. W., 1993. Surface water acidification: evidence, effects and recovery. In N. Roberts (ed.), Global Environmental Change: Geographical Perspectives. Blackwell Scientific Publications, Oxford: 213–241

Battarbee, R. W., 1997. Freshwater quality, naturalness and palaeolimnology. In P. J. Boon & D. L. Howell (eds), Freshwater Quality: Defining the Indefinable?, The Stationery Office, Edinburgh: 155–71.

Battarbee, R. W. & T. E. H. Allott, 1993. Lake acidification: evidence, effects, management and reversibility. Mem. Ist. Ital. Idrobiol. 52: 319–40.

Battarbee, R. W., R. J. Flower, A. C. Stevenson, & B. Rippey, 1985. Lake acidification in Galloway: a palaeoecological test of competing hypotheses. Nature 314: 350–352.

Battarbee, R. W., N. J. Anderson, P. G. Appleby, R. J. Flower, S. C. Fritz, E. Y. Haworth, S. Higgitt, V. J. Jones, A. Kreiser, M. A. R. Munro, J. Natkanski, F. Oldfield, S.T. Patrick, N. G. Richardson, B. Rippey & A. C. Stevenson, 1988. Lake Acidification in the United Kingdom 1800–1986: Evidence from the Analysis of lake Sediments. London, Ensis.

Battarbee, R. W., J. Mason, I. Renberg, & J. F. Talling (eds), 1990. Palaeolimnology and Lake Acidification. London, The Royal Society.

Bennion, H., 1994. A diatom-derived phosphorus transfer function for shallow, eutrophic ponds in south-east England. Hydrobiologia 275/276: 391–410.

Bennion, H., S. Wunsam & R. Schmidt, 1995. The validation of diatom-phosphorus transfer functions: an example from Mondsee, Austria. Freshwat. Biol. 34: 271–283.

Bennion, H., S. Juggins & N. J. Anderson, 1996. Predicting epilimnetic phosphorus concentrations using an inproved diatom-based transfer function, and its application to lake eutrophication management. Envir. Sci. Technol. 30: 2004–2007.

Berdie, L., J. O. Grimalt, P. Fernandez, R. Vilanova, D. Pastor, R. Psenner, R. Hoffer, B. O. Rosseland, P. G. Appleby & R.W. Battarbee, submitted. Fractionation and cold trapping of airborne organochlorine compounds in high altitude lakes. Nature.

Berglund, B. E. (ed.), 1986. Handbook of Holocene Palaeoecology and Palaeohydrology. John Wiley & Sons, Chichester.

Birks, H. J. B., M. Line, S. Juggins, A. C. Stevenson, & C. J. F. ter Braak, 1990. Diatoms and pH reconstruction. Phil. Trans. r. Soc. London B 327: 263–278.

Boyle, J. F., A. W. Mackay, N. L. Rose, R. J. Flower, P. G. Appleby & Granina (in press) Sediment heavy metal record in Lake Baikal: natural and anthropogenic sources. J. Paleolimnol.

Cameron, N. G., 1995. The representation of diatom communities by fossil assemblages in a small acid lake. J. Paleolimnol. 14: 185–223.

Cosby, B. J., R. F. Wright, G. M. Hornberger & J. N. Galloway, 1985. Modelling the effects of acidic deposition: assessment of a lumped-parameter model of soil and streamwater chemistry. Wat. Resources Res. 21: 51–63.

Dixit, S. S., A. S. Dixt, & J. P. Smol, 1989. Lake acidification recovery can be monitored using chrysophycean microfossils. Can. J. Fish. aquat. Sci. 46: 1309–12.

Flower, R. J. & R. W. Battarbee, 1983. Diatom evidence for recent acidification of two Scottish lochs. Nature 20: 130–33.

Flower, R. J., R. W. Battarbee & P. G. Appleby, 1987. The recent palaeolimnology of acid lakes in Galloway, south-west Scotland: diatom analysis, pH trends and the role of afforestation. J. Ecol. 75: 797–824.

Flower, R. J., S. Juggins & R. W. Battarbee, 1997. Matching diatom assemblages in lake sediment cores and modern surface sediment samples: the implications for conservation and restoration with special reference to acidificed system. Hydrobiologia 344: 27–40.

Fritz, S. C., 1989. Lake development and limnological response to prehistoric and historic land use in Diss, Norfolk, UK. J. Ecol. 77: 182–202.

Harriman, R. & B. R. S. Morrison, 1982. The ecology of streams draining forested and non-forested catchments in an area of central Scotland subject to acid precipitation. Hydrobiologia 88: 251–263.

Jenkins, A., P. G. Whitehead, B. J. Cosby & H. J. B. Birks, 1990. Modelling long-term acidification: a comparison with diatom reconstructions and the implications for reversibility. Phil. Trans. r. Soc. London B 327: 209–214.

Jenkins, A., M. Renshaw, R. Helliwell, C. Sefton, R. Ferrier & P. Swingewood, 1997. Modelling surface water acidification in the UK: an application of the MAGIC model to the Acid Waters Monitoring Network. IH Report No 131, Institute of Hydrology, U.K.

Jones, V. J., R. W. Battarbee, N. L. Rose, C. Curtis, P. G. Appleby, R. Harriman, & A. J. Shine, 1997. Evidence for the pollution of Loch Ness from the analysis of its recent sediments. Sci. Tot. Envir. 203: 37–49.

Juggins, S., R. J. Flower, & R. W. Battarbee, 1996. Palaeolimnological evidence for recent chemical and biological changes in UK Acid Waters Monitoring Network Sites. Freshwat. Biol. 36: 203–219.

Moss, B., P. Johnes & G. Phillips, 1996. The monitoring of ecological quality and the classification of standing waters in temperate regions: a review and a proposal based on a worked scheme for Britsh waters. Biol. Rev. 71: 301–319.

Moss, B., P. Johnes & G. Phillips, 1997. New approaches to monitoring and classifying standing waters. In P. J. Boon & D. L. Howell (eds), Freshwater Quality: Defining the Indefinable? The Stationery Office, Edinburgh: 118–133.

Radcliffe, D. A., 1971. Criteria for the selection of nature reserves. Adv. Sci. 27: 294–296.

Renberg, I., 1981. Improved methods for sampling, photographing and varve-counting of varved lake sediments. Boreas 10: 255–258.

Simola, H., I. Hanski & M. Luikkonen, 1990. Stratigraphy, species richness and seasonal dynamics of plankton diatoms during 418 years in Lake Lovojarvi, South Finland. Ann. Bot. Fenn. 27: 241–276.

Smol, P., 1990. Palaeolimnology: recent advances and furture challenges. Mem. Ist. Idrobiol. 47: 253–276.

Sullivan, T. J., B. J. Cosby, C. T. Driscoll, D. F. Charles, & H. Hemond, 1996. Influence of organic acids on model projections of lake acidification. Wat. Air Soil Pollut. 91: 271–282.

Hydrobiologia **395/396:** 161–172, 1999.
D.M. Harper, B. Brierley, A.J.D. Ferguson & G. Phillips (eds), The Ecological Bases for Lake and Reservoir Management.

Biological control of cyanobacteria: principles and possibilities

D.C. Sigee, R. Glenn, M.J. Andrews, E.G. Bellinger, R.D. Butler, H.A.S. Epton & R.D. Hendry
School of Biological Sciences, University of Manchester, 1.800 Stopford Building, Oxford Road, Manchester, U.K.

Key words: actinomycetes, biological control, bluegreen algae, cyanobacteria, protozoa, Streptomyces

Abstract

A range of naturally occurring organisms are available for the biological control of cyanobacteria: including viruses, bacteria, fungi, actinomycetes and protozoa. Development of these organisms as biological control agents involves isolation from environmental samples, characterisation of anti-cyanobacterial activity, microcosm and large-scale field experiments and final development of a biological control lake management strategy. Two groups of antagonist are considered in detail – actinomycetes (e.g. *Streptomyces exfoliatus*, mode of action by production of a lytic agent) and protozoa (*Nuclearia delicatula* and *Nassula tumida*, mode of action by predation). The effectiveness of biological control agents in the lake environment depends on a range of biological and physico-chemical factors. Various strategies can be implemented to optimise their activity.

Introduction

Blooms of freshwater cyanobacteria, particularly in the genera *Microcystis* and *Anabaena*, have caused increasing problems in recent years. These have frequently resulted in a deterioration of water quality, with adverse effects on lake ecology, livestock, human water supply and recreational amenity. The most direct form of control involves the application of algicides, but this is expensive and potentially damaging to the environment. Even where chemical treatments have no immediate damaging effects on lakes there is the risk of accumulation of harmful concentrations in bottom sediments (Mason, 1996). An alternative approach for the direct elimination of nuisance algae involves the application of biological control agents. In this context, they may be defined as microbial organisms that have the ability to destroy, or limit the growth of, target algae. This paper will consider the investigation and potential use of such agents in the control of cyanobacteria in the lake environment.

Range and mode of action of potential biocontrol agents

Activity of potential biocontrol agent ranges from highly specific parasitism to non-specific release of extracellular lytic products. Potential biocontrol agents include viruses, bacteria, fungi, actinomycetes and protozoa.

Viral agents

The first isolation and partial purification of a virus of cyanobacteria (cyanophage) was by Safferman & Morris (1963). The cyanophage was called LPP-1 because it lysed species of the genera *Lyngbya, Plectonema* and *Phormidium*. Subsequent searches for cyanophages on a world-wide scale have made it clear that they are extremely widespread in both freshwater and marine environments (Ohki & Fujita, 1996; Padan et al., 1967). The role of cyanophages in determining the cyclic blooms of cyanobacteria, and their potential use as biocontrol agents, has been suggested ever since these viruses were discovered.

The rapid generation time of cyanophages makes them attractive agents for controlling blooms of cyanobacteria. In the case of LPP-DUN1, for example, the generation time is 10 h and the burst size is about 100 phage particles for each infected cell making up the *Plectonema boryanum* filament (Daft et al., 1970).

An important consideration in the potential use of cyanophages as biocontrol agents is the rapid appearance of resistant host mutants (Padan & Shilo, 1973). These may have changes in the algal cell envelope, preventing phage adsorption (Padan et al., 1967). Barnet et al. (1981) isolated two types of *Plectonema* resistant to wild-type LPP-DUN1, however resistant host strains were susceptible to attack by mutant cyanophage strains and it was suggested that mutant phages may present a method for the control of cyanobacteria. The high degree of host specificity, occurrence of resistant host mutants and effect of environmental factors all contribute to the complexity and unpredictability of cyanobacterial/phage interactions in the field. Difficulties involved with producing large amounts of active inoculum also present problems in the effective use of cyanophages as biocontrol agents in the lake environment.

Bacterial agents

Bacteria which can lyse cyanobacteria have been associated with sudden declines in cyanobacterial biomass (Fallon & Brock, 1979), although no conclusions were reached as to whether the bacteria acted as the primary causative agent (pathogens) or whether they were acting as saprophytes, decomposing dead algal material resulting from other primary processes. Daft et al. (1975) found a direct statistical correlation (5% level) between the chlorophyll-*a* concentration of eight bodies of water and the abundance of lytic bacteria. Cyanobacteriolytic bacteria appear to act in three major ways: production of extra-cellular products; contact lysis and entrapment lysis.

The culture filtrate from a *Bacillus* sp., isolated by Reim et al. (1974), was found to lyse seven genera of cyanobacteria (including *Anabaena* and *Microcystis*). The heat stability and small molecular size of the diffusible inhibitory factor suggested that it was an antibiotic substance. In a later study, complex volatile products released by *Bacillus* sp. were shown to be particularly inhibitory to filamentous cyanobacteria (Wright & Thompson, 1985). Wright et al. (1991) subsequently implicated isoamyl alcohol (3-methyl-1-butanol) as one cyanobacteriolytic volatile product. Recently, two species of *Flexibacter* (*F. flexilis* and *F. sancti*) were isolated from domestic sewage and found to lyse the cyanobacterium *Oscillatoria williamsii* (Sallal, 1994). Inhibition of photosynthetic electron transport reactions, glycolate dehydrogenase and nitrogenase activity were observed. An extracellular metabolite in the cell filtrate caused growth inhibition (Sallal, 1994). Slab gel electrophoresis was used to identify the compound as a lysozyme.

Daft et al. (1985) questioned the significance of generally released extracellular lytic compounds in nature due to rapid dilution and subsequent loss of activity. More specific bacterial/cyanobacterial associations, where the activity of lytic compounds is more localised, may be more efficient. Two particular types of association, bacterial attachment (contact lysis) and entrapment of cyanobacteria (entrapment lysis), have been studied in detail.

Daft & Stewart (1971) isolated four strains of Myxobacteriales (designated CP 1-4) which caused lysis of over 40 strains (from all orders) of cyanobacteria, causing disruption within 20 min of contact. Although no detectable extracellular products were produced by the CP isolates, it was suggested that enzymes on the bacterial surface might be effective in causing lysis. Lysis of bacterial cell walls has been demonstrated using an enzyme extracted from *Myxobacter* (Ensign & Wolfe, 1965).

The effectiveness of contact lysis depends on a number of factors, including population levels of the algae and myxobacteria, and nutrient status of the water medium. When introduced into laboratory cultures or field samples of cyanobacteria, at least 10^6 cells ml^{-1} myxobacteria were required to cause significant algal cell lysis (Daft & Stewart, 1971; Daft et al., 1975; Shilo, 1970). Fraleigh & Burnham (1988) postulated that above 10^6 cells ml^{-1} predation will occur, with little dependence on inorganic nutrient concentrations or host density. Below this population level the growth of the predator population may be related to the inorganic nutrient status of the water. Fallon & Brock (1979) suggested that lytic bacterial populations of $<10^3$ cells ml^{-1} were insufficient to cause significant cell lysis in natural host populations, and that low densities of lytic bacteria encountered in the environment may be due to insufficient inorganic fertility (Fraleigh & Burnham, 1988).

Burnham et al. (1981) described cyanobacterial lysis by a member of the *Myxococcus* group (*Myxococcus xanthus*, designated strain PC02), in which the cyanobacterium *Phormidium luridum* var. *olivacea* was entrapped and encapsulated within colonies of the bacterium. The colonial spherules entrapped the cyanobacterial prey and then lysed the cells by the release of a 'lysozyme'-like enzyme.

Fraleigh & Burnham (1988) tested the effectiveness of myxococcal predation in natural cyanobacterial populations over a range of natural densities and nutrient concentrations against two sensitive cyanobacteria (*Nostoc muscorum* and *Phormidium lur-*

idum). The results suggested a threshold myxococcal density for significant cell lysis by *M. fulvus* and that concentrations of inorganic nutrients in eutrophic lakes might be insufficient to support growth of myxococcal populations to this threshold density. Burnham *et al.* (1984) suggested that when initial predator densities were less than the threshold density, the myxococci may derive their necessary organic nutrients from secretions produced by cyanobacterial hosts or by preying on the cyanobacteria without significantly reducing the number of cells in the host population.

Daft et al. (1985) considered myxococci to be potentially the best bacterial biological control agents of bloom-forming phytoplankton on the basis of seven attributes which they considered would make an organism a good predator. These were:

(1) adaptability to variations in physical conditions;
(2) ability to search or trap;
(3) capacity and ability to multiply;
(4) prey consumption;
(5) ability to survive low prey densities;
(6) wide host range;
(7) ability to respond to changes in the host.

Fungal pathogens of cyanobacteria

Parasitism of a cyanobacterium (*Oscillatoria agardhii* var. *isothrix*) by the chytridiaceous fungus *Rhizophidium planktonicum* was first demonstrated by Canter & Lund (1951). More recently Daft *et al.* (1985) considered the chytridiaceous fungi to be of limited use in the biological control of bloom-forming organisms because of the apparent obligate nature of these parasites and difficulties in their large scale culture.

Safferman & Morris (1962) tested 142 cultures of non-chytrid fungi and found that 4.2% formed products which demonstrated specific antagonistic effects on either green algae or cyanobacteria. The study of non-chytrid fungi antagonistic to cyanobacteria was continued by Redhead & Wright (1978). Sixty-two out of 70 pure cultures able to lyse cyanobacteria were fungi representing the genera *Acremonium*, *Emericellopsis,* and *Verticillium*. All isolates lysed *Anabaena flos-aquae* and, in most cases, several other filamentous and unicellular cyanobacteria.

Lysis of cyanobacteria by *Acremonium* and *Emericellopsis* sp. was associated with the formation of diffusible heat-stable extra-cellular factors. Redhead & Wright (1980) extracted and partially purified the β-lactam antibiotic cephalosporin C from liquid cultures of the fungi *Emericellopsis salmosynnemata* and *Acremonium kiliense*. These authors suggested that the isolates probably produced small quantities of antibiotic, which affect cyanobacterial cells only when they are in close proximity to the fungus (antibiotics may concentrate to antagonistic levels in the mucous sheath surrounding many cyanobacteria), as occurs in the aggregates produced in agitated liquid culture.

Antagonism of cyanobacteria by actinomycetes

Actinomycete antagonists of cyanobacteria have been isolated from both freshwater and soil environments. Investigations on reservoirs have shown *Streptomyces* to be the most common genus of actinomycete found in freshwater sites (Silvey, 1973), occurring in the bottom sediments associated with vegetation and along the shoreline frequently associated with mats of cyanobacteria or green algae. Growth in the various types of aquatic environment is also dependent on the natural biological productivity of the water. Studies of South-western United States reservoirs have shown that the greater the primary productivity, as measured by C_{14} uptake and ATP measurements, the more copious the actinomycete development (Silvey & Roach, 1975). Though clearly associated with cyanobacteria (and able to use algae as a nutrient source) the exact place that actinomycetes occupied in the biodynamic cycle could not be completely evaluated. Daft et al. (1985) also mentioned the lack of detailed studies of cyanobacteria-actinomycete interactions.

Soil isolates of actinomycetes may also have potent anti-cyanobacterial properties. Safferman & Morris (1962) assayed the culture filtrate of soil-isolated actinomycetes for anti-microbial properties and found that 213 out of 403 strains tested showed inhibitory effects; 90% of these agents displayed specific effects against the cyanobacteria. Various actinomycete isolates could effectively eliminate growth of cyanobacteria (including species of *Anacystis*, *Fremyella*, *Lyngbya*, *Nostoc*, *Phormidium* and *Plectonema*). Whyte et al. (1985) found that *Streptomyces achromogenes* was the main species in soil that lysed *Anabaena cylindrica* and *Tolypothrix tenuis*. Al-Tai (1982) described an actinomycete (AN6) isolated from Iraqi soil which had a very wide host range. The extracellular products of AN6 were able to lyse cyanobacteria, fungi, bacteria and green algae. Resistant cyanobacterial cells were not found.

In a general survey, the lytic properties of various isolated strains of *Streptomyces* have been compared with those of other actinomycetes and myxobacteria (Salton, 1955). Relatively little applied work has been carried out on the potential use of actinomycetes as biocontrol agents. Work by Martin (1976), in attempts to develop cheap and environmentally safe algal control procedures, resulted in the isolation of an actinomycete strain which was specifically lytic against cyanobacteria but not against several bacterial test species. It was found that cell-free filtrates from the actinomcyete had lytic activity against three *Anabaena* strains. A few studies have been carried out on the effectiveness of actinomycetes against bloom-forming cyanobacteria (Bershova et al., 1968; Rubenchik et al., 1965). Actinomycetes may also be indirectly involved as biocontrol agents in the inhibition of cyanobacteria by barley straw, where they are part of a wide range of micro-organisms that produce anti-cyanobacterial substances (Newman & Barrett, 1993)

Protozoan predators of cyanobacteria

Within aquatic ecosystems protozoa have an important role in the reduction of phytoplankton populations by grazing (Canter et al., 1990; Pace & Orcutt, 1981). Cyanobacteria in particular provide a suitable food source for several genera of protozoa, including the ciliate *Nassula* (Canter et al., 1990), the flagellate *Ochromonas* (Cole & Wynne, 1974) and the amoebae *Acanthamoeba* (Wright et al., 1981), *Mayorella* (Laybourne-Parry et al., 1987) and *Nuclearia* (Yamomoto, 1981).

Predation of cyanobacteria by protozoa has been observed in samples taken directly from the natural environment (Canter et al., 1990; Cook et al., 1974; Laybourn-Parry et al., 1987), in laboratory experiments (Dryden & Wright, 1987; Yamomoto, 1981) and in biocontrol field experiments (Brabrand et al., 1983). Such work has contributed to our understanding of cyanobacteria–protozoan interactions, providing information on the range of protozoa capable of grazing cyanobacteria (Dryden & Wright, 1987) and their potential role in biocontrol (Brabrand et al., 1983). The effectiveness of protozoa as biocontrol agents will depend upon a number of factors – including protozoan growth and grazing rates, predation specificity, cyanobacterial growth rates and predation rates by higher organisms such as copepods on the protozoa (Brabrand et al., 1983).

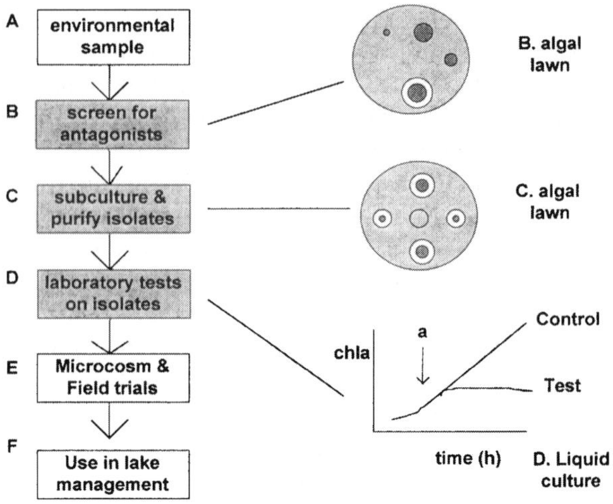

Figure 1. General scheme for the testing and development of biocontrol agents to be used against lake cyanobacteria. The central part of the scheme (work carried out in the laboratory) is illustrated to the right of the flow diagram.

Case studies on two types of potential biocontrol agent: actinomycetes and protozoa.

A general scheme for the testing and development of biocontrol agents to be used against lake cyanobacteria is outlined in Figure 1. This scheme is broadly similar to the development of biocontrol agents against other target organisms (e.g. plant pathogenic bacteria; Sigee, 1993), and follows a well-defined sequence. This involves: isolation of naturally occurring antagonistic strains from environmental samples (Figure 1a,b); subculture, purification and identification of particular isolates (Figure 1c); characterisation of anti-cyanobacterial activity in the laboratory (Figure 1d); small-scale (microcosm) and large-scale field trials (Figure 1e); and finally development of a biocontrol strategy and use of the biocontrol agents in lake management. The final stage (Figure 1f) would almost certainly also involve commercial production of the biocontrol agents, with assessment of cost-effectiveness and bulk production.

The following section considers examples of two particular groups of antagonist, with quite different modes of action: actinomycetes and protozoa.

Actinomycetes

The ability of selected isolates of actinomycete antagonists to control populations of cyanobacteria has been examined in both laboratory (algal lawn and liquid culture experiments) and field (microcosm) experiments.

Laboratory studies provide a range of useful information on the activity of potential biocontrol agents, including the range of algae that are lysed, changes in cyanobacterial and actinomycete populations during the interaction sequence, level of antagonist inoculum required for complete lysis and mode of action.

Actinomycetes were isolated by applying dilute samples of soil or freshwater to agar cultures of cyanobacteria (algal lawns), then streaking material from areas showing lysis onto yeast–dextrose medium (10 g yeast extract; 10 g D-lucose; 1000 ml distilled water). Actinomycetes were purified by picking single colonies and repeatedly sub-culturing. Pure strains were then re-tested against agar cultures of the same cyanobacterial species. One specific isolate, *Streptomyces exfoliatus* strain RG12, proved particularly effective in laboratory experiments and was used as the major test organism.

Algal lawns provide a rapid and convenient method to determine the range of algal species that are lysed by particular actinomycete isolates, and have been widely used by previous workers (Al-Tai, 1982; Burnham et al., 1981; 1984; Daft et al., 1975; Martin, 1976; Stewart & Daft, 1977; Yamamoto & Suzuki, 1990).

The algal lawn assay can be used in various ways. Details of one commonly used approach are shown diagrammatically in Figure 2a, with an example illustrated in Figure 2b. Replicate cores (0.5 cm diameter) were cut from 3-week-old *S. exfoliatus* agar cultures and inverted onto lawns of cyanobacteria, so that antagonist and cyanobacterium were in direct contact. Cores cut from un-inoculated plates were used as control. Preparations were incubated at 20°C, continuous 500 lux light and examined after 3 days, any yellowing or clearing of the treatment compared to control being recorded as positive. The speed of development and diameter of the lysis zone also gives a measure of antagonistic effectiveness under the prevailing experimental conditions.

Nine out of 13 cyanobacteria tested were lysed within 3 days by *S. exfoliatus*, including members of the major bloom forming genera *Anabaena*, *Microcystis* and *Oscillatoria*. Filamentous and unicellular cyanobacteria were equally susceptible. Colonies of *S. exfoliatus* were observed growing over the cyanobacterial lawn. Clear lysis zones were seen beyond the leading edge of actinomycete growth, indicating an extracellular lytic process (Figure 2b).

Antagonist isolates were tested against a range of cyanobacteria in liquid culture, under different physical conditions and algal/antagonist ratios. Results from a typical liquid culture experiment are shown in Figure 3. *S. exfoliatus* was cultured for 3 days in inorganic salts–starch medium then 5ml (10^6 CFU ml^{-1}) added to replicate flasks containing 100-ml cultures of *Anabaena cylindrica*. Incubation was at 20°C, continuous light (1000 lux) with continuous (100 r.p.m.) stirring. As controls, 5 ml sterile actinomycete growth medium were added to replicate *A. cylindrica* cultures.

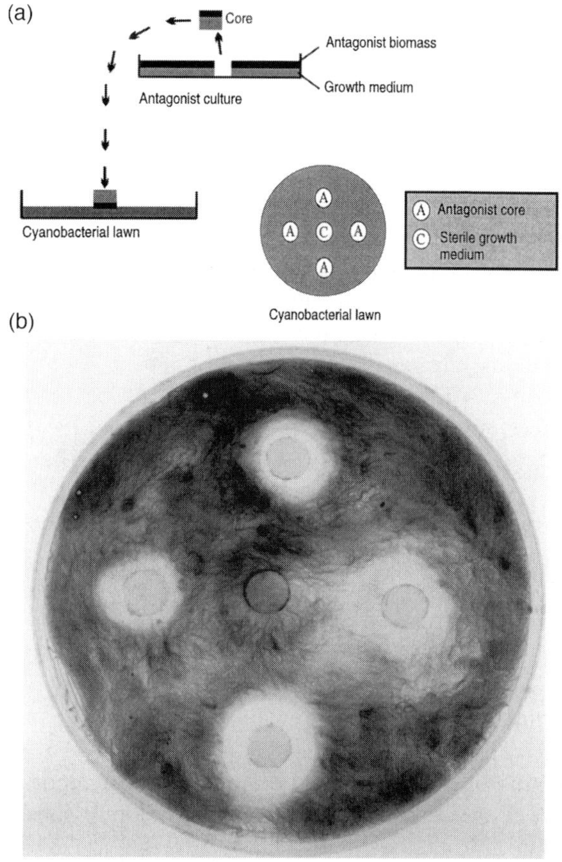

Figure 2. Algal lawn assay for testing actinomycete isolates. (a). Diagram of technique, using antagonist cores. The cyanobacterial lawn comprises a uniform layer of algae overlying the agar and was normally used within 24 h of production. Cores from actinomycete cultures were inverted onto the algal lawn and their activity assessed in terms of lytic activity. (b) Appearance of lawn assay, 5 days after inversion of cores from S. exfoliatus culture onto a lawn of *Lyngbya* sp. Note lysis zones extending from the edge of actinomycete growth. The central control core (no actinomycete) has no lysis zone.

Colonies of *S. exfoliatus* were observed as discrete floccules in the algal culture, frequently directly associated with algal filaments. Over a period of 7 days the presence of actinomycete antagonist sharply reduced the biomass of *A. cylindrica*, as measured by

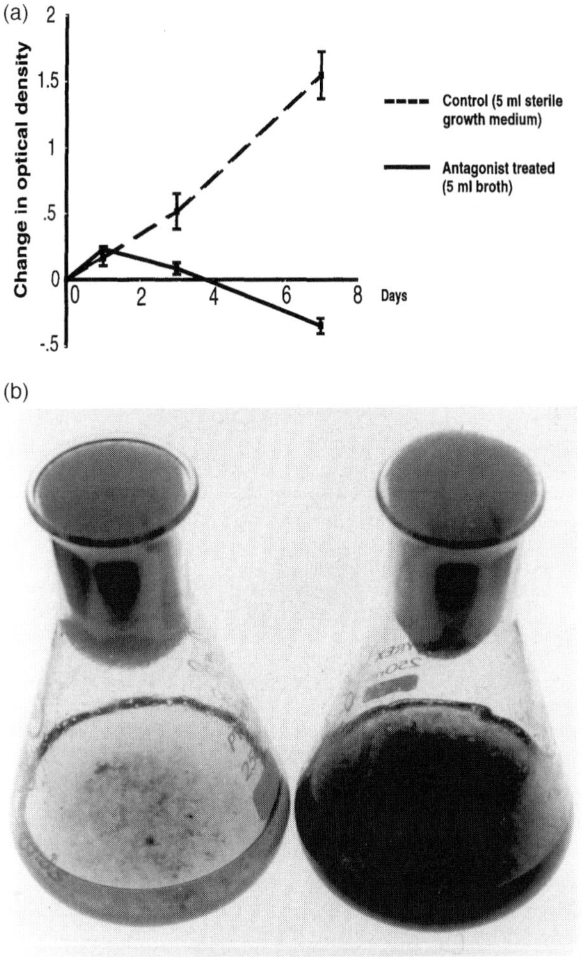

Figure 3. Liquid culture assay for testing actinomycete isolates. (a) Effect of *Streptomyces exfoliatus* strain RG12 on the growth (change in optical density) of *Anabaena cylindrica* in 100 ml of medium. (b) Appearance of antagonist treated (left flask) and control (right flask) 7-day cultures of *Anabaena cylindrica* described in (a). Lysis of the cyanobacterium in the presence of antagonist results in almost complete clearing of the algal culture.

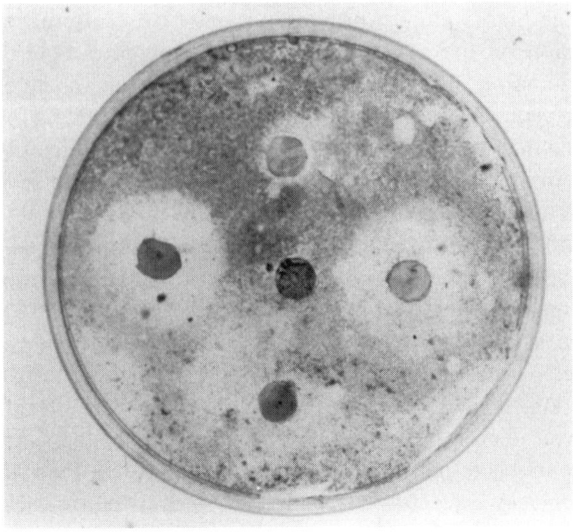

Figure 4. Testing of actinomycete isolates on environmental lawns. Testing of four actinomycete isolates on an algal lawn obtained from an environmental (bloom) sample. The lawn (composed largely of *Microcystis aeruginosa*), shows large lysis zones around two antagonist cores (right/left), but the top and bottom cores have little activity (see also Figures 2a, b).

optical density (Figure 3a) and chlorophyll-*a* concentration, leading to almost complete clearing of the cyanobacterial culture (Figure 3b). *S. exfoliatus* initially resulted in filament fragmentation, followed by cell lysis. Heterocysts were apparently unaffected by antagonist treatment and accumulated as single cells in lysed cultures. In parallel experiments, lysis of *Anabaena* cultures was accomplished by the addition of as little as 0.05 ml *S. exfoliatus* inoculum (10^6 CFU ml^{-1}) in 100 ml of algal culture.

The possible use of antagonists for environmental control of cyanobacteria can be tested progressively using a sequence of experimental approaches: laboratory testing of pure cultures of environmental isolates, laboratory testing of freshly-collected environmental (phytoplankton) samples (Figure 4), small (microcosm) and large (mesocosm) enclosed-volume experiments in lakes and finally testing in the open water against natural blooms. In the early stages of antagonist testing there is probably most value in controlled, replicated laboratory or microcosm experiments. Simply adding biological control organisms to open waters may create unforseen problems due to variables not recreated in laboratory experiments.

Environmental testing of *S. exfoliatus* has so far progressed via intermediate laboratory studies to the microcosm stage. Laboratory trials with freshly isolated field samples, using both lawn (Figure 4) and liquid culture assay, have demonstrated clear lysis of lake phytoplankton – including *Microcystis* and *Anabaena* bloom samples.

Microcosms provide a useful intermediate step towards larger-scale environmental experiments. They involve antagonist testing under physical and biological conditions that approximate to the natural state and that the effect of these agents can be quantified within a defined volume of lakewater. The volumes of microcosms used to test *S. exfoliatus* varied from very small (350 ml) to 5 l, with all experiments plus controls carried out in triplicate. Addition of antagonist culture to such microcosms containing dominant cy-

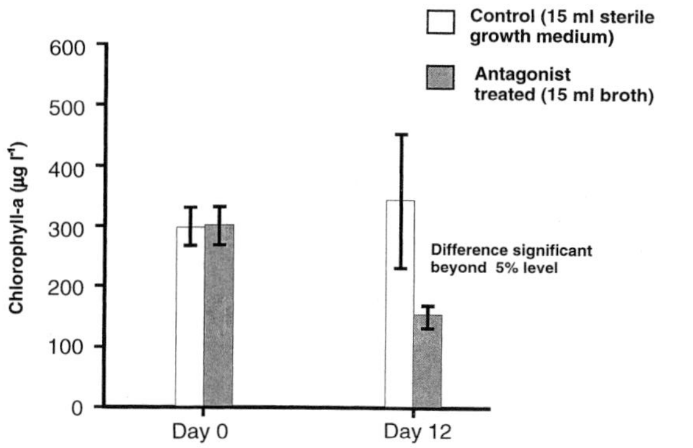

Figure 5. *In situ* (microcosm) lake experiment. *Streptomyces exfoliatus* strain RG12 was added to laboratory-cultured *Microcystis aeruginosa* within dialysis tubing (350 ml). Algal biomass (determined as chlorophyll-*a*) showed a 60% decrease in the presence of antagonist.

anobacteria typically resulted in a significant reduction in algal population, normally decreasing to about 50% of the initial level over a 1–2-week period (Figure 5).

Protozoa

Studies are currently in progress to assess the potential biocontrol activity of culture collection strains of two species of protozoa: the amoeba *Nuclearia delicatula* (CCAP 1552/1, Figure 6a) and the ciliate *Nassula tumida* (CCAP 1650/2, Figure 6b), both of which are known to feed on cyanobacteria. So far, the investigation has mainly involved laboratory studies to determine predation range and the dynamics of protozoan/algal interactions, though some small-scale preliminary field work has been carried out.

The potential use of *N. delicatula* and *N. tumida* as biocontrol agents of different nuisance algae was assessed in liquid culture. Cultures of different algae were inoculated with protozoa and maintained under static conditions at 20°C, with a diurnal regime of 15 h light/8 h dark to simulate the natural environment. Changes in the population of algae were monitored through optical density with a colorimeter and the protozoan population determined by direct counts on a Sedgewick Rafter counting cell.

The results have shown that *N. delicatula* grazes a range of cultured and environmental isolates of *Oscillatoria* sp. and *Anabaena* spp. (but not *Microcystis aeruginosa*), whereas *N. tumida* has only proved active against *Oscillatoria* spp. The feeding preferences of the amoeba and the ciliate appear to be filamentous cyanobacteria, shown in Figure 6a, b, where

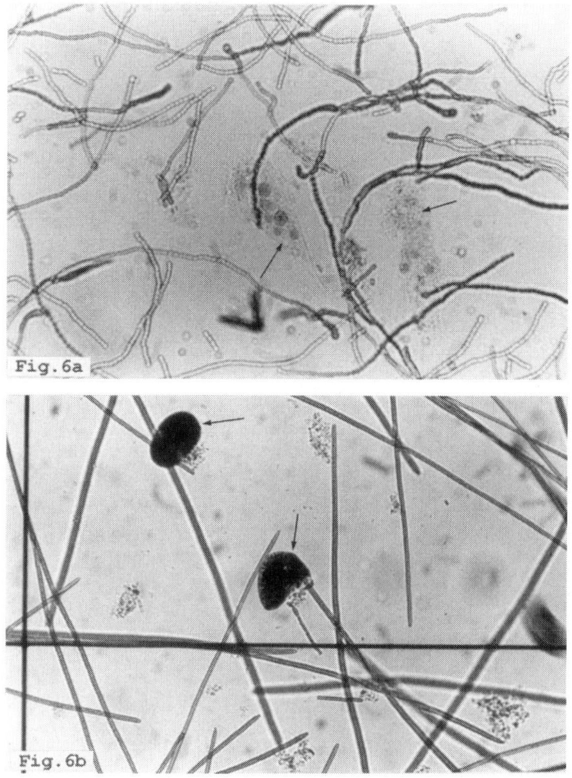

Figure 6. Protozoan antagonists of cyanobacteria. (a) Light micrograph of *Nuclearia delicatula* grazing on *Anabaena flos-aquae*. (×300); (b) Light micrograph of *Nassula tumida grazing* on *Oscillatoria* sp. (×150).

both species of protozoa are attached to the end of a cyanobacterial filament which is being ingested. Lack of ingestion of other cyanobacteria may relate to factors such as the non-filamentous habit, cell size, insufficient nutrient source and production of algal toxins.

The grazing efficiencies of the protozoa were investigated in laboratory cultures by varying the absolute and relative populations of algae and protozoa under different environmental conditions. Results obtained from one of these experiments, involving *N. delicatula* and an environmental isolate of *Anabaena* sp., are shown in Figure 7. The experiment was carried out in a growth cabinet under similar conditions to those noted earlier, with three replicate flasks for each combination of algal and protozoan populations. Daily measurements of optical density, chlorophyll '*a*' and protozoan counts were made over the 10-day experimental period. Initial population levels of *N. delicatula* (100, 500 and 1000 cells ml^{-1}) were tested against three starter levels of *Anabaena* sp. (40, 204 and

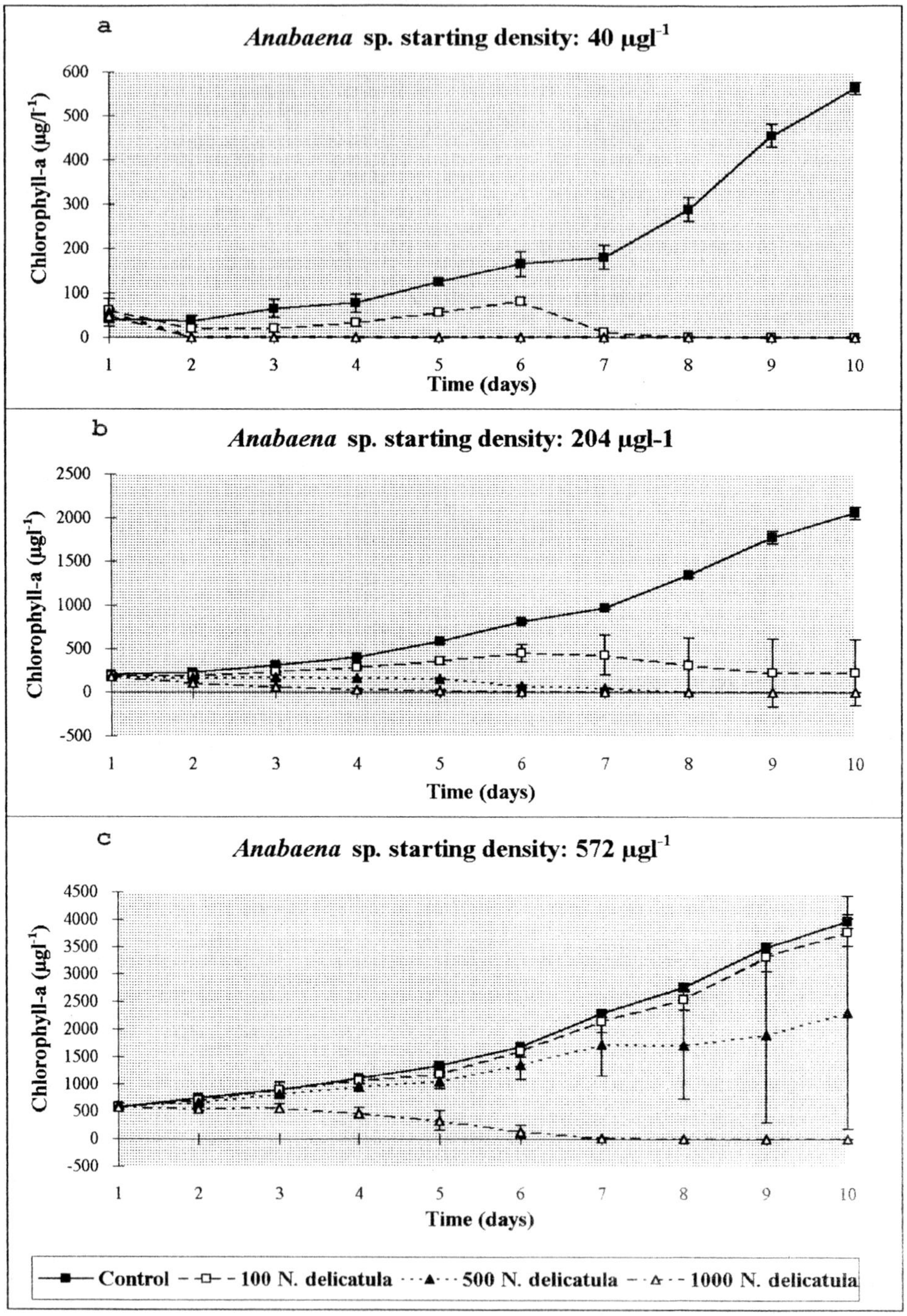

Figure 7. Predator–prey experiments, using laboratory cultures. The ability of different inoculum levels of *Nuclearia delicatula* to control the growth of *Anabaena* sp. in laboratory culture is shown for three starter levels of algal population, with respective chlorophyll '*a*' concentrations of 40 μg ml^{-1} (a), 204 μg ml^{-1} (b) and 572 μg ml^{-1} (c). In each case, protozoa were added to give overall initial population levels of 100, 500 and 1000 cells ml^{-1}. Algal population levels are given as the mean chlorophyll-'*a*' concentration ($\pm$ SD) of three replicate flasks.

572 μg chlorophyll '*a*' l^{-1}, Figure 7a, c). The highest chlorophyll '*a*' level used was far in excess of that normally occurring in an algal bloom. The results show that the ability of *N. delicatula* to reduce cyanobacterial population levels under laboratory conditions depends on the initial populations of both organisms. As expected, *N. delicatula* was typically most effective at high protozoan and low algal population levels, with rapid reduction of all three algal biomass levels at 10 000 protozoa ml^{-1}. Experiments such as this give some idea of the protozoan inoculum required to control different cyanobacterial populations in a lake environment.

As with the actinomycete studies, microcosm testing allows different species or strains of antagonist and algae to be brought together at defined initial population levels and under controlled nutrient conditions, but interacting under ambient environmental conditions of temperature and light regime. Preliminary experiments have been carried at a local freshwater site using submerged dialysis tubes with a capacity of 350 ml. These were suspended from a wire mesh floating platform, and were separately inoculated with the cyanobacteria *Oscillatoria* sp. (chlorophyll '*a*' 645 μg l^{-1}) or *Anabaena flos-aquae* (chlorophyll '*a*' 678 μg l^{-1}) plus either sterile protozoan medium (control) or protozoan culture (treatment). The results showed a clear reduction in the populations of *Oscillatoria* sp. and *Anabaena flos-aquae* following addition of *N. delicatula*. As with the laboratory-based experiments, the initial algal levels were substantially higher than those normally encountered under bloom conditions. In each experiment, the protozoan population showed a marked increase, which was most pronounced during the initial phase of rapid algal decline.

Role of biological control in the lake environment

Biological control of cyanobacteria represents a potential short-term measure to reduce the population, or prevent the build-up, of nuisance algal populations. This is in contrast to long term control strategies such as nutrient limitation (bottom-up control) and biomanipulation (top-down control), which involve more fundamental alterations to the lake ecosystem. The biocontrol agents considered in this paper have either been recently isolated from environmental samples (actinomycetes), or were originally derived from the freshwater environment (protozoa). These native organisms have not undergone any deliberate genetic manipulation or gene enhancement. The addition of such agents to the lake can thus be viewed as an alteration to the existing balance of organisms already present within the ecosystem. This alteration is invariably temporary, with reversion of the ecosystem to its original state within a relatively short period of time.

Biological control of cyanobacteria, like biological control of other nuisance organisms (plant pathogenic bacteria and fungi, insect pests, weeds) has a number of advantages over the other major type of immediate control, chemical control. Biological control can be highly specific to the target organism, with no destruction of other organisms and with no direct chemical pollution that might affect humans. Potential disadvantages include limited destruction of the target organism, limited survival of the agent or its removal by other organisms and problems with large-scale production, storage and application of the biocontrol agent.

Laboratory studies are an important aspect of the general scheme for development of biocontrol agents outlined in Figure 1, providing useful information on aspects such as mode of action, effect of physicochemical factors, progress of control activity and required antagonist inoculum levels. In the case of *Streptomyces exfoliatus*, for example, laboratory studies showed that activity involved a lytic diffusible agent, production of which was genetically stable—persisting over repeated sub-culturing. This is in contrast to *Streptomyces* isolates obtained by Redhead & Wright (1978) which lost cyanobacteriolytic activity after four transfers. Although formation of the lytic agent in *S. exfoliatus* occurs independantly of the presence of cyanobacteria, the ability to destroy these organisms is probably an advantage to the antagonist since actinomycete growth is promoted.

Although laboratory studies have an important part to play in biocontrol work, results obtained should be viewed with caution if they are to be interpreted in the lake context. Laboratory data cannot simply be extrapolated to the freshwater environment. This is shown with both of the case studies presented in this paper, where the anti-cyanobacterial effectiveness of the antagonists in the microcosm environment was much less than in laboratory cultures. In the case of *Streptomyces exfoliatus*, addition of 0.05 ml *Streptomyces* culture ($\sim 10^6$ CFU ml^{-1}) to 100 ml of laboratory *Anabaena cylindrica* culture (50 μg l^{-1}chlorophyll-'*a*') achieved complete inhibition within 12 days.

Similar (2000-fold) dilutions of antagonist were not effective in the lake, where inoculum levels of

250 ml in 5 l lakewater were required for significant reduction in the cyanobacterial population. Differences between the effectiveness of cyanobacterial antagonists in laboratory cultures and the lake environment may relate to a number of general factors. The important ones are:

(1) the effects of the lake environment (physico-chemical and biotic) on antagonist populations, growth rate and antagonistic activity;
(2) effects of the lake environment on cyanobacterial response to antagonists;
(3) Differences in the distribution of cyanobacteria and antagonists within the lake ecosystem.

Separation in the populations of these two groups of organisms (e.g. loss of antagonists from surface waters due to sinking, replacement of surface algae from the hypolimnion) will limit the biocontrol activity of antagonists.

The most important test for any potential biocontrol agent must be its effectiveness in the environment, where the above factors are encountered under the prevailing lake conditions. In spite of these limitations, both of the antagonists (actinomycetes and protozoa) detailed in this paper did have a significant biocontrol effect in lake microcosms at the appropriate inoculum level. Other workers have also carried out direct environmental testing, particularly with viral and bacterial antagonists. Using 5000-gallon tanks Jackson & Sladecek (1970) examined interaction of cyanophage LPP-1 with *Plectonema boryanum*. They concluded that cyanophages had the potential to prevent growth of their hosts under natural conditions. Daft et al. (1975) tested 15 fresh algal blooms from various waters against *Myxobacter* CP-1, in each case obtaining clear lysis. Two environmental experiments subsequently confirmed the effectiveness of this antagonist. In the first, 3 l of *Microcystis aeruginosa* were taken from the top 0.5 m of the reservoir and added to each of four polythene trays which were floated semi-submerged on the water. A log phase culture of CP-1 was added to two of the trays to give an initial concentration of 10^6 ml^{-1}. Within 96 h chlorophyll-'*a*' in the treatments had declined to negligible amounts. In the second experiment, 200 l of reservoir was portioned off using polythene sheeting. One area was sprayed with CP1 to 10^6 ml^{-1}. Within 60 h there was an almost complete loss of chlorophyll from the treated area.

Various field studies have also been carried out on the use of barley straw to control cyanobacteria, the activity of which may involve microbial action. Although the use of straw has been proposed as an inexpensive and environmentally acceptable form of control (Ridge & Barrett, 1992; Ridge et al., 1999) practical results have been erratic, with field studies proving successful in some cases (Everall & Lees, 1996) but not others (Kelly & Smith, 1996).

Practical applications

In cases where a particular biocontrol agent is shown to be consistently effective in environment testing, the next phase would be full-scale field trials (including complete environmental analysis), commercial cost assessment and evaluation of how the biocontrol agent could be used in lake management. In assessing the practical use of such antagonists for control of cyanobacteria in the lake, aspects such as frequency and timing of application, mode of application (formulation and method of dispersal) and implications for water processing are all important.

The volume of inoculum required per unit area of lake surface is also relevant. In the case of *Streptomyces exfoliatus*, minimum inoculum levels obtained in laboratory and microcosm experiments (see above) would translate to required levels of 5×10^3 and 5×10^5 l of antagonist inoculum per hectare (treatment of top metre of water column, total volume approximately 10 000 m^3 ha^{-1}). These amounts of biocontrol agent required do not sound encouraging, either in terms of cost or ease of production and application, and would suggest that simple addition of antagonist culture to control established growths of cyanobacteria in a large body of lakewater is not feasible. The potential use of such biocontrol agents in lake management should not be dismissed, however, since smaller volumes may be possible in some circumstances. The following may be appropriate:

1. Control of developing (pre-bloom) cyanobacterial populations. Application of such agents may be most effective during build-up of blooms, rather than under full bloom conditions. This would require very careful monitoring of the increase of cyanobacterial populations within the epilimnion, including the movement of these organisms to surface waters from the hypolimnion.
2. Application to small water bodies, including municipal ponds and small amenity lakes, where the overall amount of inoculum required would obviously be less than larger lakes and reservoirs.
3. Improved types of biocontrol agent. The effectiveness of particular biocontrol agents may be

improved in various ways, including the use of genetically-engineered organisms (e.g. with higher growth rates, improved synthesis of anti-cyanobacterial compounds).

4. Improved methods of application. One problem noted in the microcosm experiments with *S. exfoliatus* was that antagonist floccules tended to sink in the water column. Maintenance of the antagonist in suspension, by attachment to a floating substrate such a straw, would ensure that the agent stays in the surface algal zone.
5. Mixture of agents. Combining different types of biocontrol agent might offer greater flexibility in a continuously changing lake environment. Daft et al. (1985) have pointed out that a such mixture, offering different modes of attack, may offer the best chance of preventing problem cyanobacterial blooms in eutrophic lakes.
6. Use with other techniques. Experience with other biocontrol situations has shown that such agents are best used in conjunction with other approaches, and not in isolation. The use of biological control agents in lake and reservoir management should be regarded as just one of a number of available control measures that form part of an integrated management policy.

References

Al-Tai, A. M. S., 1982. Characteristics and microbial interactions of certain actinomycetes from Iraq. Ph.D. Thesis, University of Dundee.

Barnet, Y. M., M. J. Daft, & W. D. Stewart, 1981. Cyanobacteria–cyanophage interactions in continuous culture. J. appl. Bact. 51: 541–552.

Bershova, O. I., Z. P. Kopteva & E. V. Tantsyurenko, 1968. The interrelations between the blue-green algae - the causative agents of 'water bloom' – and bacteria. In A.V Topachevsky (ed.), Tsvetnie Vody: 159–171.

Brabrand, A., B. A. Faafeng, T. Kallquist, & J. P. Nilssen, 1983. Biological control of undesirable cyanobacteria in culturally eutrophic lakes. Oecologia 60: 1–5.

Burnham, J. C., S. A. Collart & M. J. Daft, 1984. Myxococcal predation of the cyanobacterium *Phormidium luridum* in aqueous environments. Arch. Microbiol. 137: 220–225.

Burnham, J. C., S. A. Collart, & B. A. Highison, 1981. Entrapment and lysis of the cyanobacterium *Phormidium luridum* by aqueous colonies of *Myxococcus xanthus* PCO2. Archives of Microbiology 129: 285–294.

Canter, H. M. & J. W. Lund, 1951. Fungal parasites of the phytoplankton. II. (Studies on British Chytrids XII). Ann. Bot. XV 58: 129–159.

Canter, H. M., S. I. Heaney & J. W. Lund, 1990. The ecological significance of grazing on planktonic populations of cyanobacteria by the ciliate *Nassula*. New Phytol. 114: 247–263.

Cole, G. T. & M. J. Wynne, 1974. Endocytosis of *Microcystis aeruginsoa* by *Ochromonas* danica. Journal of Phycology 10: 397–410.

Cook, W. L., D. G. Ahearn, Reinhardt, D. J. & R. J.Reiber, 1974. Blooms of an algophorous amoeba associated with *Anabaena* in a freshwater lake. Wat. Air Soil Pollut. 3: 71–80.

Daft, M. J., J. Begg, & W. D. Stewart, 1970. A virus of blue-green algae from fresh-water habitats in Scotland. New Phytol. 69: 1029–1038.

Daft, M. J., J. C. Burnham, & Y. Yamamoto, 1985. Algal blooms: consequences and potential cures. J. appl. Bact. 175S–186S.

Daft, M. J., S. B. McCord & W. D. Stewart, 1975. Ecological studies on algal lysing bacteria in fresh waters. Freshwat. Biol. 5: 577–596.

Daft, M. J. & W. D. Stewart, 1971. Bacterial pathogens of freshwater blue-green algae. New Phytol. 70: 819–829.

Dryden, R. C. & J. C. Wright, 1987. Predation of cyanobacteria by protozoa. Can. J. Microbiol. 33: 471–482.

Ensign, J. C. & R. S. Wolfe, 1965. Lysis of bacterial cell walls by an enzyme isolated from a Myxobacter. J. Bact. 90: 395–402.

Everall, N. C. & D. R. Lees, 1996. The use of barley straw to control general and blue-green algal growth in a Derbyshire reservoir. Wat. Res. 30: 269–276.

Fallon, R. D. & T. D. Brock, 1979. Lytic organisms and photooxidative effects: Influence on blue-green algae (cyanobacteria) in lake Mendota, Wisconsin. Appl. environ. Microbiol. 38: 499–505.

Fraleigh, P. C. & J. C. Burnham, 1988. Myxococcal predation on cyanobacterial populations: nutrient effects. Limnol. & Oceanogr. 33: 476–483.

Jackson, D. F. & V. Sladeck, 1970. Algal viruses – eutrophication and control potential. Yale Sci. Mag. 44: 16–22.

Kelly, L. A. & S. Smith, 1996. The nutrient budget of a small eutrophic loch and the effectiveness of straw bales in controlling algal blooms. Freshwat. Biol. 36: 411–418.

Laybourn-Parry, J., K. Jones, & J. P. Holdich, 1987. Grazing by *Mayorella* sp. (protozoa: Sarcodina) on cyanobacteria. Funct. Ecol. 1: 99–104.

Martin, E. L., 1976. Biological control of blue-green algae. National Technical Information Service Report PB-288 311. Nebraska.

Mason, C. F., 1996. Biology of Freshwater Pollution, 3rd edn. Longman, Essex.

Newman, J. R. & P. R. F., Barrett, 1993. Control of *Microcystis aeruginosa* by decomposing Barley straw. J. of aquatic Plant Manage. 31: 203–206

Ohki, K. & Y. Fujita, 1996. Occurrence of a temperate cyanophage lysogenising the marine cyanophyte *Phormidium persicum*. J. Phycol. 32: 365–370.

Pace, M. L. & J. D. Orcutt, 1981. The relative importance of protozoans, rotifers and crustaceans in a freshwater zooplankton community. Limnol. Oceanogr. 26: 822–830.

Padan, E. & M. Shilo, 1973. Cyanophages-viruses attacking blue-green algae. Bacteriol. Rev. 37: 343–370.

Padan, E., M. Shilo, & N. Kislev, 1967. Isolation of 'Cyanophages' from freshwater ponds and their interaction with *Plectonema boryanum*. Virology 32: 234–246.

Redhead, K. & S. J. Wright, 1978. Isolation and properties of fungi that lyse blue-green algae. Appl. Environ. Microbiol. 35: 962–969.

Redhead, K. & S. J. Wright, 1980. Lysis of the cyanobacterium *Anabaena flos-aquae* by antibiotic producing fungi. J. Microbiol. 119: 95–101.

Reim, R. L., M. S. Shane, & R. E. Cannon, 1974. The characterisation of a *Bacillus* capable of blue-green bactericidal activity. Can. J. Microbiol. 20: 981–986.

Ridge, I. & P. R. Barrett, 1992. Algal control with barley straw. Aspects appl. Biol. 29: 457–462.

Ridge, I., J. Walters & M. Street, 1999. Algal growth by terrestrial leaf litter: a realistic tool? Hydrobiologia 395/396 (Dev. Hydrobiol. 136): 173–180.

Rubenchik, L. I., O. I. Bershova, & Z. P. Knizhnik, 1965. On the interrelationship of *Anabaena* with bacteria and actinomycetes. In: Ecologia i physiologia sinezelenych vodorosleiy. Nauka Moscow: 223–226.

Safferman, R. S. & M. Morris, 1962. Evaluation of natural products for algicidal properties. Appl. Microbiol. 10: 289–292.

Safferman, R. S. & Morris, M. 1963. The antagonistic effects of actinomycetes on algae found in waste stabilisation ponds. Bacteriol. Proc. 14: A56.

Sallal, A. K., 1994. Lysis of cyanobacteria with *Flexibacter* spp isolated from domestic sewage. Microbios 77: 57–67.

Salton, M. R. J., 1955. Isolation of *Streptomyces* spp. capable of decomposing preparations of cell walls from various microorganisms and a comparison of their lytic activities with those of certain actinomycetes and myxobacteria. J. Gen. Microbiol. 12: 25–30.

Shilo, M., 1970. Lysis of blue-green algae by *Myxobacter*. J. Bacteriol. 104: 453–461.

Sigee, D. C., 1993. Bacterial Plant Pathology: Cell and Molecular Aspects. Cambridge University Press, Cambridge.

Silvey, J. K., 1973. Determination of nuisance organisms. Proceedings AWWA Water Quality Technology Conference, December 1973.

Silvey, J. K. & A. W. Roach, 1975. The taste and odour producing aquatic actinomycetes. CRC Crit. Rev. environ. Control March 1975: 233–273.

Stewart, W. D. & M. J. Daft, 1977. Microbial pathogens of cyanophycean blooms. Adv. aquatic Microbiology 1: 177–219.

Whyte, L. G., A. Maule, & D. R. Cullimore, 1985. Method for isolating cyanobacteria-lysing *Streptomycetes* from soil. J. appl. Bact. 58: 195–197.

Wright, S. J. & R J. Thompson, 1985. *Bacillus* volatiles antagonise cyanobacteria. FEMS Microbiol. Lett. 30: 263–267.

Wright, S. J., K. Redhead & H. Maudsley, 1981. *Acanthamoeba castellanii*, a predator of cyanobacteria. J. Gen. Microbiol. 125: 293–300

Wright, S. J., C. J. Linton, R. A. Edwards, & E. Drury, 1991. Isoamyl alcohol (3-methyl-1-butanol), a volatile anti-cyanobacterial and phytotoxic product of some *Bacillus* spp. Lett. appl. Microbiol. 13: 130–132.

Yamamoto, Y., 1981. Observations on the occurrence of microbial agents which cause lysis of blue-green algae in Lake Kasumigaura. Jpn. J. Limnol. 42: 20–27.

Yamamoto, Y. & K. Suzuki, 1990. Distribution and algal-lysing activity of fruiting myxobacteria in lake Suwa. J. Phycol. 26: 457–462.

Hydrobiologia **395/396:** 173–180, 1999.
D.M. Harper, B. Brierley, A.J.D. Ferguson & G. Phillips (eds), The Ecological Bases for Lake and Reservoir Management.

Algal growth control by terrestrial leaf litter: a realistic tool?

Irene Ridge[1], John Walters[1] & Mike Street[2]
[1]*The Open University, Milton Keynes, U.K.*
[2]*Milton Keynes Parks Trust, Milton Keynes, U.K.*

Key words: algal control; barley straw; leaf litter; FPOM

Abstract

When barley straw and deciduous leaf litter decompose aerobically in water, inhibitors are released that suppress the growth of nuisance algae. Barley straw has been widely used for algal control in small, shallow lakes and we review the advantages and disadvantages of the method. It is particularly effective at promoting the switch from algal to macrophyte domination. Despite its cheapness and apparent safety in the short term, however, the use of barley straw requires considerable management effort and the long-term ecological safety of such un-natural litter inputs is unkown. We therefore recommend it to lake managers primarily as a short-term measure. Deciduous leaf litter from a range of woody species can suppress the growth of *Chlorella* and *Microcystis* very effectively in laboratory bioassays and, in field trials with medium-sized ponds, the addition of leaf litter produced significant inhibition of the filamentous alga *Cladophora glomerata*. We followed the development of algal inhibitory activity over 2.5 years with freshly fallen oak leaves placed in a large tank of aerated water and using *Chlorella* as the test species. Two periods of inhibitor release were identified: 4–90 days (early phase) when soluble, relatively stable inhibitors were present in tank liquor, probably generated from oxidized tannins; and 120–900+ days (late phase) when inhibitors were relatively unstable in solution and were associated primarily with fine particulate organic matter (FPOM). Late phase inhibitors may, as suggested for barley straw, be generated during the oxidative breakdown of lignin. The prolonged and powerful anti-algal properties of these natual litter inputs offer possibilities for low-effort, sustainable management of lakes and catchments so as to reduce the problem excessive algal growth.

Introduction

Reversing the effects of cultural eutrophication, in particular the frequent occurrence of dense algal blooms, has become one of the primary aims of lake and reservoir management. The long-term goal remains that of reducing nutrient inputs but, even if this is achieved, continued management will often be necessary to maintain lakes in the 'desired' state. Management of natural litter inputs, either in inflow streams or directly into lakes, is one option that can help to minimize the likelihood of excessive algal growth.

Whilst nutrient levels are being reduced, however, or in situations where the chances of doing so are extremely remote, emergency measures may be required to deal with algal problems. Phosphate precipitation by the addition of ferric iron is one method currently used in reservoirs (Daldorph, 1999; Brierley & Harper, 1999). Another method which has been widely used in small lakes, drainage ditches and ponds is the addition of barley straw (Welch et al., 1990). The advantages and disadvantages of this method and its suitability for larger lakes are discussed here.

The use of both barley straw and natural inputs of leaf litter to reduce algal growth depends on the fact that they generate inhibitors during aerobic decomposition in water (Gibson et al., 1990; Ridge et al., 1995). The precise chemical nature of these inhibitors is still unknown but there is good circumstantial evidence that they are or derive from oxidised poylphenolics which originate mainly from lignin (Pillinger et al., 1994, 1995; Ridge & Pillinger, 1996). Oxidising conditions are essential for inhibitor activity (Pillinger et al., 1994) and inhibitors are also relatively unstable, so that particulate litter must be present constantly to sustain activity. These characteristics influence strongly the practical use of litter to suppress algal growth.

Barley straw: an emergency method

Barley straw was the first type of plant litter shown to inhibit algal growth (Welch et al., 1990) and it remains the only type in widespread use. Typical sites where straw is used are relatively shallow lowland lakes and an estimated 70% of trout lake managers in England have used the technique (Salmon and Trout Association & J. Newman, pers. comm.). Newman (1994) provides details of practical application, the single most important requirement being maintenance of straw in a submerged but aerobic state. Packing straw loosely within plastic netting (using Christmas tree wrapping machines, for example) together with buoyancy devices has proved an effective way of meeting this requirement, the straw 'sausages' being towed into position, anchored and allowed to float just below the water surface.

From laboratory bioassays and monitored field trials, the range of algae whose growth is inhibited by barley straw appears to be wide, encompassing both filamentous types and phytoplankton (Gibson et al., 1990) and including cyanobacteria (Newman & Barrett, 1993). Based on the finding that diatoms may occur in tanks of rotting, active barley straw, it was suggested earlier that these algae are particularly resistant to straw inhibitors (Ridge et al., 1995). However, two common planktonic species, *Asterionella formosa* and *Tabellaria flocculosa* were strongly inhibited in laboratory bioassays (D. Martin, unpublished), so that insensitivity is not a universal feature of the Bacillariophyta. Different sensitivity of algal species to barley straw inhibitors could certainly influence relative abundance and seasonal succession. It remains possible, moreover, that some potential nuisance algae are relatively insensitive and might increase in abundance if barley straw were used continuously for several years.

The last point is clearly a possible disadvantage of using barley straw. A particular concern for water managers is that barley straw might selectively encourage more toxic strains of bloom-forming cyanobacteria. To date there is no evidence that this has happened but more work on the relative sensitivity of different algal strains and species is clearly desirable. Other disadvantages are of a practical nature. Straw sausages can be unsightly and may interfere with activities such as fishing and water sports. There is also the problem of their disposal once the straw has ceased to be anti-algal (which usually occurs after a period of 6–12 months). Allowing straw residues and plastic netting to accumulate on the lake bed is clearly undesirable and straw bales partially buried in anaerobic sediments have been reported to release products toxic to koi carp (Wildgoose, 1994). However, removing heavy, water-logged straw sausages from water and disposing of them on land requires considerable effort.

Against these actual or potential disadvantages, the main advantages of the barley straw method are that it is relatively cheap, effective and, so far as is known, safe. Effectiveness has been demonstrated for the shallow lowland water bodies described earlier, most of which have a pH close to neutral. Too few trials have been carried out in acidic water bodies to generalise, however. And similarly there have been few trials in larger lakes and reservoirs, only two of which have been carefully monitored. For one of these trials Barrett et al. (1996) applied barley straw to a reservoir in the Grampian Region of Scotland (area 25 000 m^2, normal capacity 250 000 m^3) which had a long history of diatom and cyanobacterial blooms. Following a March (1993) application of 38 g m^{-3}, algal counts remained low throughout summer with no early-autumn *Anabaena* bloom. Further applications of 6.5 and 7.6 g m^{-3}barley straw were made in December 1993 and June 1994. During 1994, the spring diatom bloom was much reduced (maximum cell count 8700 cell cm^{-3}compared with maxima of 67 500, 38 000 and 29 000 in the three previous years) and no cyanobacterial bloom occurred. This is the only report of an apparent effect of barley straw on diatoms in the field. For the second trial, Everall & Lees (1996) applied barley straw (50 g m^{-3}) in April 1994 to a Derbyshire reservoir of 55 000 m^3. They found a significant reduction in total chlorophyll-*a* levels and in cyanobacterial dominance during summer compared with the previous three summers in this reservoir and with an untreated, adjacent reservoir. Neither study proves conclusively that barley straw treatment works infallibly in larger water bodies, as the authors of these two reports pointed out. However they indicate that under the climatic conditions of 1994 barley straw contributed to reduced algal growth in two medium-sized reservoirs in the UK.

Regarding safety, the best that can be said is that no adverse effects of using barley straw have been reported so far. Barrett et al. (1996) used standard techniques for potable supplies to monitor the Grampian reservoir where straw was applied and reported no increase in taint-producing substances such as geosmin. Everall & Lees (1996), however, found low concentrations of a wide range of organic compounds, including simple phenolics and long-chain fatty acids, within 0.1 m of barley straw in the Derbyshire reservoir, although most of these compounds were not present 150 m downstream of straw. These authors found no adverse effects of barley straw on invertebrate populations with 18 species recorded as colonising straw and a bloom of planktonic rotifers from mid April to late May. No adverse effects have been published for trout fisheries where straw has been used and, in a preliminary study using replicated artificial ponds, we have found no effects of barley straw on the development of common newt (*Triturus vulgaris*) or common frog (*Rana temporaria*), other than a small decrease in the mean size of frogs at metamorphosis (Ridge et al., unpublished). It must be emphasised, however, that all these studies have been short-term. Adverse effects on wildlife from long-term use of barley straw cannot be ruled out and this is one reason for advocating the technique primarily as a short-term, emergency measure rather than a long-term management strategy.

In shallow water bodies where there has been a switch from macrophyte to algal domination with increasing eutrophication, the use of barley straw, even without reductions in nutrient levels, can facilitate macrophyte restoration within a period of four to five years (Newman et al., 1999). If macrophyte domination is a management aim, such short-term application of barley straw is particularly recommended. For larger and deeper reservoirs and lakes, the technique is worth serious consideration as an interim measure.

Natural litter inputs and sustainable management

Previously we have described the anti-algal effects of three kinds of tree litter that may be natural inputs to lakes: brown-rotted wood, deciduous leaf litter and coniferous leaf litter (Pillinger et al., 1995; Ridge et al., 1995). Of these only deciduous litter is worth considering as a management aid since brown-rotted wood is a scarce and valuable ecological resource and conifer litter has a strongly acidifying effect (although it is still active in algal bioassays in buffered neutral media). Like barley straw, all these types of litter require oxidising conditions for activity.

Laboratory bioassays

Table 1 lists the deciduous species whose litter was tested in algal bioassays. Freshly fallen leaf litter was dried in air, finely ground and autoclaved with algal medium (modified Jaworksi's, pH 8.2) before inoculation with the test alga, *Chlorella vulgaris* Beijerinck (CCAP 211/12). Cells were counted after 3 days and results are expressed as percentage of controls incubated without litter. The results show that litter from the majority of species released substances inhibitory to *Chlorella* after autoclaving, ash being a notable exception. Similar results were obtained using the cyanobacterium *Microcystis aeruginosa* (CCAP 1450/6) as a test species. Tannins are known to be algicidal (Hussein-Ayoub & Yankov, 1985) and, since ash leaves are thought to be tannin-poor, whereas the most inhibitory species such as oak are tannin-rich, oxidised tannins leached from litter were considered a likely source of algal inhibitors. Subsequent work generally supported this view but showed also that far more significant for algal suppression are inhibitors generated over a long period during aerobic decomposition and derived, most probably, from lignin.

Simulated field conditions

Results from laboratory bioassays cannot be extrapolated with any degree of confidence to the field but, before carrying out field trials, we set up a long-term experiment to investigate the nature of algal inhibition by deciduous litter in conditions that approximated those of a shallow, well-aerated pool. Freshly fallen, dry oak litter, *Quercus robur* (5.5 mg cm^{-3}) was placed in a 218 l tank containing aged, continuously aerated tap-water in an unheated greenhouse. Samples of both leaves and liquor in the tank were removed at intervals and tested in bioassays using *Chlorella*. Leaves were assayed after drying in air, grinding and autoclaving in algal medium (1 mg cm^{-3}). After 15 days, unfiltered liquor, which contained particulate material (<0.5 mm), and liquor filtered through glass-fibre paper (GF/C, pore size 1.4 μm) were assayed separately. Results for a 2.5-year period are shown in Figure 1.

For the first 90 days, both leaf material and filtered and unfiltered liquor from the tank were strongly inhibitory to *Chlorella*, apart from the first 4 days when liquor promoted algal growth. During this period, shown as 'Early Phase' in Figure 1, algal inhibitors leached from the litter were relatively stable, i.e.

Table 1. Inhibition of *Chlorella* by freshly fallen leaf litter. Material was air-dried, finely ground and autoclaved in algal medium at 1 mg cm^{-3}. Results are for a single assay with five replicates except for $*$ where two or more assays were performed. NS, no significant inhibition

% of control growth (cell counts)			
<15	16–30	31–45	NS
Horse Chestnut (*Aesculus hippocastanum*)	Aspen (*Populus tremula*)	Goat willow (*Salix caprea*)	Ash (*Fraxinus excelsior*)*
Field Maple (*Acer campestre*)	Beech (*Fagus sylvatica*)	Scots pine (*Pinus sylvestris*)	
Oak (*Quercus robur*)*	Crab apple (*Malus sylvestris*)	Sycamore (without leaf blackspot) (*Acer pseudoplatanus*)	
Sycamore (*Acer pseudoplatanus*)* with leaf blackspot	Rowan (*Sorbus aucuparia*)		
	White poplar (*Populus alba*)		

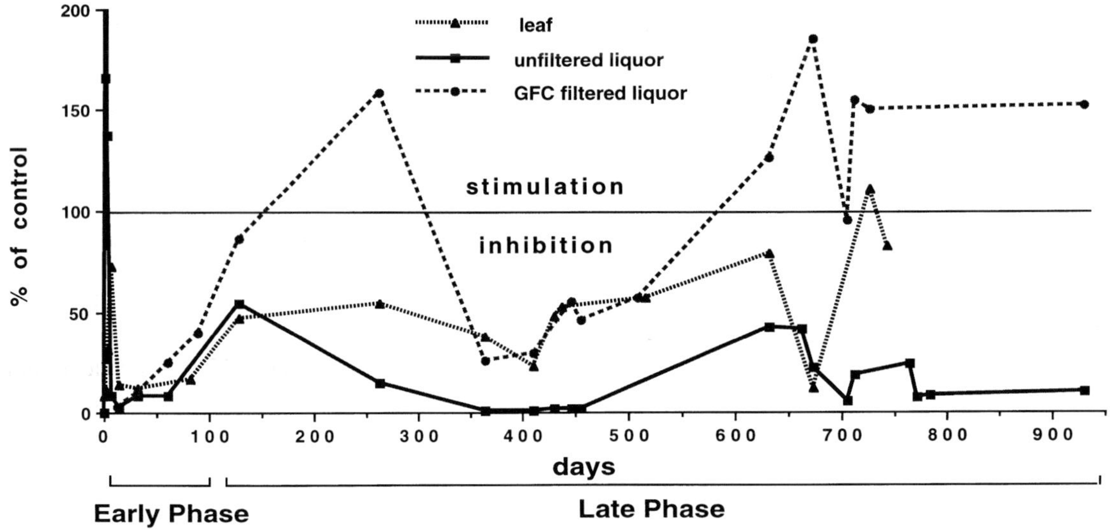

Figure 1. Effects on *Chlorella* growth of oak litter (*Quercus robur*) decomposing in a tank of aerated water. Samples were tested in 3-day bioassays (cell counts) and results are expresssed as percent of controls. For bioassays, leaves were air-dried, ground and autoclaved in algal medium (1 mg cm^{-3}); algal nutrients were added directly to the tank liquors.

filtered liquor left for up to 10 days at 20°C still inhibited growth of *Chlorella* during a 3-day bioassay (data not shown). We have argued elsewhere (Pillinger et al., 1994) that oxidised tannins are the primary source of algal inhibitors from fresh oak litter. Neither tannin-poor ash litter nor oak litter that was collected after prolonged natural leaching in the field show Early Phase inhibition by tank liquor.

After 80–90 days there was a progressive loss of anti-algal activity by both both leaf material and tank liquor (filtered and unfiltered) During June and August (days 129–263) the filtered liquor did not inibit algal growth whereas the unfiltered liquor, which contained increasing amounts of fine particulate matter (FPOM), became increasingly inhibitory. Between days 364 and 456 the unfiltered liquor inhibited *Chlorella* growth in bioassays by 98–99% and remained inhibitory, although to a lesser extent, for the rest of the experiment. During this Late Phase (Figure 1), the filtered liquor showed some anti-algal activity, roughly parallel to, but much less than, that of the unfiltered liquor up to day 509. This suggests that algal inhibitors were associated with the finely particulate matter (FPOM) in unfiltered liquor and were relatively unstable in solution, i.e. lost activity during a 3-day bioassay. Such behaviour parallels that of barley straw, where algal inhibition in bioassays also requires the physical presence of straw and filtered liquor from tanks of rotting straw is commonly inactive. We tentatively ascribe algal inhibition by tank liquor during this Late Phase to oxidized lignin released during microbial decomposition of the oak litter. Leaves (dried and autoclaved) show fairly constant inhibitor release (50–60% inhibition in bioassays) from days 128–516, followed by a gradual loss of activity.

The link between FPOM produced during aerobic decomposition of oak litter and algal inhibition was confirmed after isolating FPOM by GF/C filtration: it inhibited *Chlorella* growth when added to algal medium or to filtered tank liquor containing algal nutrients (Figure 2). There was no correlation between the particle size of FPOM in the range 4–150 μm and inhibitory activity: for three fractions (<2.8 mm >125 μm, <125 μm >63 μm, <63 μm >1.2 μm) activity depended only on the dry weight of FPOM present. Unless autoclaved, whole or large fragments of decaying oak leaves showed little or no activity in bioassays but became active after freezing rapidly in liquid nitrogen and grinding to create artificial FPOM. Mechanical fragmentation during leaf decomposition in water thus plays an important role in the generation of algal inhibitors. Furthermore, autoclaving greatly reduced FPOM activity so that the activity of dried, autoclaved leaves probably reflects a balance between inhibitor release and destruction. FPOM from oak litter has been shown in bioassays to inhibit the growth of several other species of algae in addition to *Chlorella*, which include *Microcystis aeruginosa*, *Cladophora glomerata* and *Asterionella formosa*.

Field trial

To assess the effects of adding leaf litter on algal growth in the field, we used three ponds excavated from clay in 1992 and situated in amenity grassland adjacent to Howe Park Wood, Milton Keynes, UK. Pond 1 was used as a control; mixed deciduous leaf litter (mostly oak, beech and sycamore) was added to pond 2 and pond 3 was treated with barley straw. Table 2 gives details of the ponds and treatments which, because of the shallow depth, were made by spreading material freely on the surface. Shallow channels connected the three ponds and water flowed from pond 1 to 2 to 3 when levels were high. All three supported moderate growths of the filamentous green alga *Cladophora glomerata*. Algal growth was measured in 1993 using suspended tiles and in 1994 within clear plastic cylinders covered at each end with fine nylon mesh and containing a uniform inoculum of laboratory-grown *Cladophora*. The tiles or cylinders, 20 per pond, were attached to an anchored wooden frame, with polystyrene floats attached, so that they lay 5 cm below the water surface. Results for two sampling occasions are shown in Table 3: barley straw strongly suppressed growth of *Cladophora* and there was substantial inhibition in pond 2 to which leaf litter had been added. We cannot rule out the possibility that

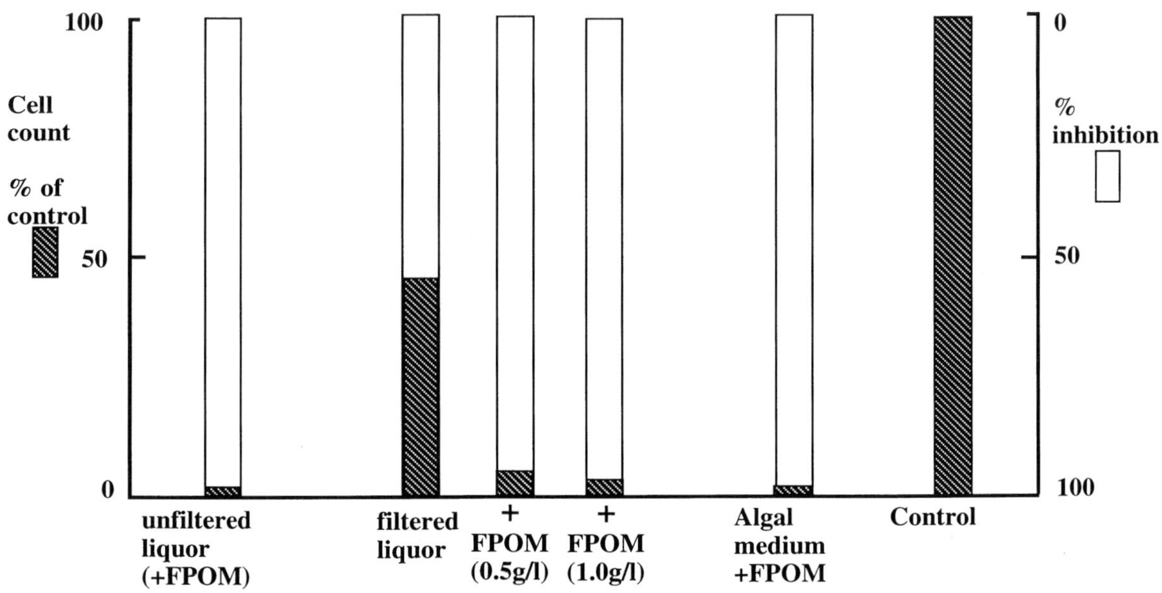

Figure 2. Effects on *Chlorella* growth of filtered and unfiltered liquor and of fine particulate organic matter (FPOM, 1.2 μm to 1.4 mm). Samples were removed from the tank containing oak litter on day 456.

Table 2. Ponds adjacent to Howe Park Wood, Milton Keynes: schedule of treatments

Pond	Av. vol. (m^3)	Max. depth m	Additions kg dry wt. (dose g m^3) 6/93	10/93	12/93	6/94
1 Control	230	1.0	–	–	–	–
2 + leaf litter	260	1.2	29.7 (114)	17.7 (68)	37.5 (144)	3.7 (14)
3 + barley straw	400	1.0	20 (50)	–	20 (50)	10 (25)

Table 3. Effects on *Cladophora* growth of deciduous leaf litter and barley straw addition to ponds at Howe Park, Milton Keynes. Algae grew on suspended tiles (1993) or were enclosed in transparent chambers (1994) and results are expressed as percent of growth in the control pond ± SD

	% of growth in control pond	
	Dry wt.	Chlorophyll '*a*'
+ leaf litter		
9/6 – 3/8/93		34.5 ± 6.2
22/6 – 7/7/94	47.2 ± 11.0	65.4 ± 7.7
22/6 – 3/8/94	27.5 ± 11.5	57.2 ± 9.9
+ barley straw		
9/6 – 3/8/93		16.6 ± 6.9
22.6 – 7.7.94	7.9 ± 1.3	6.7 ± 2.8
22/6 – 3/8/94	6.8 ± 3.4	16.8 ± 10.6

decreased light penetration caused by brown staining after litter addition contributed to algal suppression. However, this is unlikely since the growth of submerged macrophytes was exceptionally vigorous in pond 2 at depths well below that of the algal chambers and we find no inhibition from this cause in laboratory bioassays. The presence of leaf litter appears, therefore, to reduce *Cladophora* growth in field conditions and probably does so through the release of inhibitors as in laboratory bioassays and in the tank experiments.

Leaf litter and management

The Early Phase suppression of algae by fresh leaf litter is likely to be of limited significance for lake managers since leaves fall in autumn when algal growth is slowing down. In contrast, the powerful and prolonged inhibition in Late Phase, mediated by fine particulate leaf material, is potentially useful. The advantages of leaf litter over barley straw are that it is a natural input, so there are no problems from the introduction of foreign material; and it is available wherever catchments contain deciduous trees. The problems are how to facilitate retention of leaf litter in conditions where it can decompose aerobically and from which FPOM can be released or washed into lakes during periods of active algal growth.

Constructing or encouraging natural debris dams where litter is trapped on inflow streams, releasing FPOM that washes into lakes, is one possibility although an obvious problem is that stream flow and FPOM inputs are likely to be minimal when algal growth is maximal. Simply planting deciduous trees upwind of lakes and along inflow streams is another. Neither of these measures will 'solve' the problem of excessive algal growth but they are worth considering as part of wider plans for lake improvement and as an element in long-term, sustainable management programmes.

Artificial addition of deciduous leaf litter either to traps on inflow streams or direct to shallow lakes can also be envisaged in the same way that barley straw is now used. This, however, would be a high-effort, emergency form of management rather than a low-effort sustainable type. Furthermore, since no lake-scale field trials have been carried out, no guidance can be given as to the best ways of adding leaf litter artificially other than reiterating that aerobic conditions are necessary for leaf decompostion and the action of algal inhibitors.

Summary and conclusions

Litter inputs to streams and rivers have long been regarded as valuable sources of food and shelter. Their significance to lakes and, in particular, their chemical effects on algal populations have received much less attention. Based on experiments with large tanks, we have shown that oak litter in well-aerated water generates inhibitors of algal growth in two stages. Leaching leads to the generation of relatively stable inhibitors within a few days derived, we believe, from oxidised tannins. Interestingly, Bärlocher (1992) found that water-soluble extracts from various types of fresh deciduous litter inhibited the radial growth of four aquatic hyphomycetes on agar medium. The duration of this early phase inhibition depends on the amounts of soluble tannin present and the rate of leaching but probably rarely exceeds 3-4 months. As leaves decay and fragment, algal inhibitors associated with fine particulate organic matter (FPOM) are generated. In the case of oak litter, this late phase begins after 3-4 months and, provided that conditions remain aerobic, lasts for 2-3 years. The inhibitors are unstable in solution and may derive, as with barley straw, from oxidised lignin or its breakdown products.

Based on these findings, together with evidence that algal inhibitors can be extracted by autoclaving litter from a wide range of species and that mixed-species litter inhibited *Cladophora* growth in a large pond, we have suggested that natural inputs of deciduous leaf litter could influence the amount of algal growth in slow-flowing or static water bodies. For small, shallow lakes, direct inputs may be significant; for larger lakes and reservoirs, any effects are likely to arise via inputs to inflow streams. Other than the pond trial described here, there is only anecdotal evidence to support this view; for example, small lakes, ponds and drainage ditches that have received regular litter inputs over a period of years have been observed to support less algal growth than similar, nearby water bodies (with similar nutrient levels in the case of ponds and drainage ditches). Moderate litter inputs that do not give rise to anoxia in the water column by no means suppress algal growth entirely: some algae are always present; species vary in their sensitivity to FPOM; and benthic diatoms were common in tanks of rotting oak leaves, suggesting that certain species are highly resistant to the inhibitors. Submerged macrophytes grew vigorously in Pond 2 and in small, artificial ponds to which leaf litter had been added. Thus litter inputs are perhaps best described as a buffer system, reducing the likelihood of excessive growth to produce blooms by one or a few species of algae and increasing the likelihood of maintaining macrophyte rather than algal domination. More work is clearly desirable to determine the extent and nature of algal inhibition from deciduous litter inputs. Nevertheless, the available evidence is sufficiently strong to recommend the incorporation into long-term management plans of measures that will increase natural inputs.

Artificial addition of leaf litter, especially to small shallow lakes, is an option worth considering as a short-term measure to reduce algal growth. Barley straw, however, is already widely used in this way and is a more reliable option in that practical details of how to apply it are available. Moreover, it appears to work as well in medium-sized reservoirs as in small lakes and ponds, to have no adverse effects on wildlife or water quality in the short term, and to facilitate the switch from algal to macrophyte domination. The use of barley straw is thus recommended to lake managers. It is best regarded as a short-term rather than long-term strategy because, in addition to the high level of management effort required, barley straw is not a natural litter input and the effects of continuous use over many years are not known. In the longer term, a better understanding of the effects on lakes of natural litter inputs and catchment management to increase these inputs and facilitate aerobic decomposition offer greater promise.

Acknowledgements

We thank Jonathan Newman and Pip Barrett for many helpful discussions and Judith Pillinger for invaluable advice and support. Financial support from the Open University Research Committee and from English Nature for the installation of artificial ponds is gratefully acknowledged.

References

Bärlocher, F., 1992. Effects of drying and freezing autumn leaves on leaching and colonization by aquatic hyphomycetes. Freshwater Biol. 28: 1–7.

Barrett, P. R. F., J. C. Curnow & J. W. Littlejohn, 1996. The control of diaom and cyanobacterial blooms in reservoirs using barley straw. Hydrobiologia 340: 307–311.

Brierley, B. & D. Harper, 1999. Ecological principles for management techniques in deeper reservoirs. Hydrobiologia 395/396 (Dev. Hydrobiol. 136): 335–353.

Daldorph, P. W. G., 1999. A reservoir in management-induced transition between ecological states. Hydrobiologia 395/396 (Dev. Hydrobiol. 136): 325–333.

Everall, N. C. & D. R. Lees, 1996. The use of barley-straw to control general and blue-green algal growth in a Derbyshire reservoir. Wat. Res. 30: 269–276.

Gibson, M. T., I. M. Welch, P. R. F. Barrett & I. Ridge, 1990. Barley straw as an inhibitor of algal growth II: laboratory studies. J. Appl. Phycol. 2: 241–48.

Hussein-Ayoub, S. M. & L. K. Yankov, 1985. Algicidal properties of tannins. Fitoterapia 56: 227–229.

Newman, J. R., 1994. Control of algae with straw. Information sheet 3, Centre for Aquatic Plant Management, Reading, RG4 0TH, UK.

Newman, J. R. & P. R. F. Barrett, 1993. Control of *Microcystis aeruginosa* by decomposing barley straw. J. Aquatic Plant Manag. 31: 203–206.

Newman, J. R., P. R. F. Barrett & G. Cave, 1999. The use of barley straw to control algal growth in drainage ditches: an ecological survey. In Harpley, J. (ed.), Nature Conservation and the Management of Drainage System Habitat, in press.

Pillinger, J. M., J. A. Cooper & I. Ridge, 1994. Role of phenolic compounds in the antialgal activity of barley straw. J. Chem. Ecol. 20: 1557–1569.

Pillinger, J. M., I. Gilmour & I. Ridge, 1995. Comparison of the antialgal acitivity of brown-rotted and white-rotted wood and *in situ* analysis of lignin. J. Chem. Ecol. 21: 1113–1125.

Ridge, I. & P. R. F. Barrett, 1992. Algal control with barley straw. Aspects Appl. Biol. 29: 457-62.

Ridge, I. & I. M. Pillinger, 1996. Towards understanding the nature of algal inhibitors from barley straw. Hydrobiologia 340: 301–305.

Ridge, I., J. M. Pillinger & J. Walters, 1995. Alleviating the problems of excessive algal growth. In: Harper, D. M. & A. J. D. Ferguson (eds), The Ecological Basis for River Management, John Wiley, Chichester: 211–218.

Welch, I. M., P. R. F. Barrett, M. T. Gibson & I. Ridge, 1990. Barley straw as an inhibitor of algal growth I: studies in the Chesterfield Canal. J. Appl. Phycol. 2: 231–239.

Wildgoose, W., 1994. Barley straw and its effects on ornamental fish and algae. Koi Health Q. 3: 19–23.

Hydrobiologia **395/396:** 181–190, 1999.
D.M. Harper, B. Brierley, A.J.D. Ferguson & G. Phillips (eds), The Ecological Bases for Lake and Reservoir Management.

Aquatic macrophytes as tools for lake management

Arnulf Melzer
Technical University of Munich, Arcisstrasse 21, D-80290 München, Germany

Key words: macrophytes, macrophyte index, bioindication, water quality, lakes

Abstract

Aquatic macrophytes can serve as useful indicators of water pollution along the littoral of lakes. In Bavaria, the submerged vegetation of about 100 lakes has been investigated by SCUBA diving over the past decade to evaluate the state of nutrient pollution. All lakes are marl lakes located in the northern calcareous alps and the prealpine region. The lakes differ in size, morphology, water residence time, nutrient loading, trophic status, recreational activities, and other characteristics. In all cases the entire shoreline of the lakes has been investigated. Among the investigated lakes are the three biggest Bavarian lakes, i.e. Lake Chiemsee, Lake Starnberg and Lake Ammersee. Mapping of the submerged vegetation occurred in four different depth zones, and variable shoreline sections. The length of each section was determined by the uniformity of the vegetation; as it changed, a new section was designated. Within each section and zone species were recorded and abundance of all observed macrophytes was estimated semi-quantitatively on a five-point scale. Nine different groups of macrophytes were recognised, including, in total, 45 different species of macrophytes. On the basis of this catalogue of indicator species, in combination with the abundance of the species, a 'macrophyte index' was devised, which ranges from 1 (unpolluted) to 5 (heavily polluted). Six groups of values of the macrophyte index, each represented by a different colour or grey-scale (in this publication), are presented to allow a clear illustration of the results. Important information for the successful restoration of lakes in Upper Bavaria has been obtained from the distribution patterns of the submerged vegetation. Many unknown waste water inflows or diffuse sources could be detected due to abrupt changes in the macrophyte index. Furthermore, the success of waste water removal by 'ring canalisation', resulting in a re-oligotrophication of many Bavarian lakes can be followed by changes in the macrophyte index.

Introduction

Aquatic macrophytes in the littoral zones of lakes have two fundamental properties which make them attractive as limnological indicators. In the first instance, they react slowly and progressively to changes in nutrient conditions, in contrast to bacteria and microalgae, i.e. over several years. Macrophytes therefore function as integrators of environmental conditions to which they are subjected and thus can be used as long-term indicators with high spatial resolution.

Secondly, the littoral zone may experience patterns of nutrient (and pollutant) concentrations (Dave, 1992; Drake & Heaney, 1987), caused by natural or artifical inflows as well as by diffuse, non-point sources. The latter are difficult to localize and quantify but even chemical analysis of point sources is often neglected in limnological routine work. Since most important nutrient loading occurs in the land–water ecotone, programmes for the restoration of lakes are dependent on reliable data which give insight into quality and quantity of nutrient loading. Early investigations in Scandinavia have shown that aquatic macrophytes reflect the nutrient status of their immediate habitat by their presence/absence and abundance and thus can be effectively used as biological indicators (Suominen, 1968; Uotila, 1971).

In Central European lakes, the disappearance of a number of species as a consequence of rapid eutrophication has been recorded over the past three decades (Lachavanne & Wattenhofer, 1975; Lang, 1968; Melzer, 1981). Concern over the widespread effects of species loss, in comparison with historical records, and realisation of the need to understand precise impacts of nutrient pollution in the littoral zone in order to direct restoration, led to the development of a method for macrophyte community evaluation

Methods

Mapping of macrophytes

Mapping of the submerged vegetation of lakes can be done with the help of mechanical equipment, however, investigation by SCUBA diving is more precise. The entire shoreline of about 100 calcareous lakes in Upper Bavaria has been mapped by divers to obtain comprehensive and reliable data on the distribution of submerged macrophytes. To improve exactness, mapping occurred in four different depth zones: 0–1 m, 1–2 m, 2–4 m and >4 m. By this procedure depth preferences of single species and changes of maximal depth distribution of macrophytes along the shoreline could both be evaluated.

Mapping was carried out in shoreline sections of variable breadth. The boundaries of each section were determined by the uniformity of the vegetation; as this changed, a new section was designated. The breadth of a mapping section ranged between 2 km at maximum and 20–30 m in the case of small point sources. Within a mapping section and depth zone all macophyte species (spermatophyta, bryophyta and charophyta) were recorded and abundance estimated on a five-point scale following Tüxen & Preising (1942), where 1= very rare, 2= infrequent, 3= common, 4= frequent and 5= abundant. However, the scale is not linear and so the abundance value was cubed ($y = x^3$) to develop a 'macrophyte index' (see below).

The occurrence of macrophyte species reflects the degree of nutrient enrichment within a shoreline section. A catalogue of nine 'indicator groups' of macrophyte species (Table 1) exhibiting different sensitivity towards nutrient enrichment was developed from the lake survey. In total 45 species of submerged macrophytes and free floating plants were included. Species allocated to 'indicator group' 1 are restricted to oligotrophic lakes, while those in 'indicator group' 5 are restricted to eutrophic, nutrient-rich lakes or sections of lakes. The remaining seven groups represent transitions between these two extremes.

Macrophyte index

The 'macrophyte index' (MI) was calculated separately for each surveyed section, using the 45 indicator species. It is based on the values of 'indicator group' (I) of the species (a–z) and the 'quantity' (Q) of these species. MI is calculated by the following formula:

$$\mathrm{MI} = \frac{\sum_{a-z} I_{a-z} \cdot Q_{a-z}}{\sum_{a-z} Q_{a-z}},$$

where MI = macrophyte index; I_{a-z} = indicator group of the species a-z; Q_{a-z} = quantity of species a-z.

Six classes of the macrophyte index were designated, each representing different degrees of nutrient pollution. These six classes are assigned different colours (shown as grey scale, Table 2) to allow a clear illustration of the results. The application of the macrophyte index in a series of case studies demonstrates its value in directing cost-effective restoration.

Results

Osterseen lakes

The Osterseen lakes are situated south of Lake Starnberg, into which they drain, are 20 individual lakes which offer the ideal site for a 'natural experiment' and illustrate the kind of survey data which was enabled the macrophyte index to be constructed. The lakes are connected in series with each other by natural channels. Water flows through them from south to north. Anthropogenic influence is concentrated around the southern-most lakes, at the beginning of the lake chain. Natural self-purification as well as dilution occurs along the chain and the nutrient levels gradually decline. Mean total phosphorus concentration of the southern-most Lake Waschsee is 80 μg l^{-1} P, this of the following lake 65 μg l^{-1} P, that of the third 40 μg l^{-1} P. Phosphorus concentrations drop below 15 μg l^{-1} P in the northern lakes. Lake Lustsee is a lake in the north of the area without through flow. It possesses many nutrient-poor groundwater inlets and as a consequence total phosphorus concentrations are below 10 μg l^{-1} P. This results in a high Secchi-disk transparency (annual mean >10 m). The northern lakes are characterized by a flora typical for nutrient poor lakes. Here *Chara hispida, C. aspera, C. tomentosa, C. intermedia, C. contraria, Potamogeton coloratus, Utricularia ochroleuca, U. minor* and *U. australis* prevail. In Lake Lustsee *Chara* meadows cover the surface of the sediment down to 17 m, the deepest point of the lake.

Exclusively restricted to the southern lakes, as well as to Lake Fischkaltersee in the south-east are

Table 1. Classification of 45 submerged and free-floating macrophyte species into nine indicator groups

Group 1.0	Group 1.5	Group 2.0	Group 2.5	Group 3.0	Group 3.5
Chara hispida	*Chara aspera*	*Chara delicatula*	*Chara contraria*	*Chara vulgaris*	*Myriophyllum verticillatum*
Chara polyacantha	*Chara intermedia*	*Chara tomentosa*	*Chara fragilis*	*Myriophyllum spicatum*	*Potamogeton berchtoldii*
Chara strigosa	*Utricularia minor*	*Potamogeton alpinus*	*Nitella opaca*	*Potamogeton filiformis*	*Potamogeton lucens*
Potamogeton coloratus			*Nitellopsis obtusa*	*Potamogeton perfoliatus*	*Potamogeton praelongus*
Utricularia ochroleuca			*Potamogeton gramineus*		*Potamogeton pusillus*
			Potamogeton natans		
			Potamogeton × *zizii*		

Group 4.0	Group 4.5	Group 5.0
Fontinalis antipyretica	*Callitriche cophocarpa*	*Ceratophyllum demersum*
Hippuris vulgaris	*Elodea canadensis*	*Zannichellia palustris*
Lagarosiphon major	*Elodea nuttallii*	*Potamogeton mucronatus*
Potamogeton pectinatus	*Potamogeton crispus*	*Sagittaria sagittifolia*
	Potamogeton obtusifolius	*Lemna minor*
	Ranunculus circinatus	*Spirodela polyrhiza*
	Ranunculus trichophyllus	*Potamogeton nodosus*

Table 2. Relationship between index class, degree of nutrient status and assigned colour

Index class	Degree of nutrient enrichment	
1.00–1.99	Slight	Dark blue
2.00–2.49	Low	Pale blue
2.50–2.99	Moderate	Green
3.00–3.49	Immense	Yellow
3.50–3.99	Heavy	Orange
4.00–5.00	Massive	Red

macrophyte species designated to 'indicator groups' 4.5 and 5.0, respectively (*Elodea canadensis, Potamogeton crispus, Ceratophyllum demersum, Zannichellia palustris* and *Lemna minor*). Floristic transitions between these two extremes are to be found in the middle group of lakes. Mapping of macrophytes was carried out in 1974, 1981 and 1994, respectively. In Figure 1 only the results of the most recent investigation are shown. With the exception of the southern-most lakes changes in macrophyte index in between the years were negligible. Lakes Waschsee, Schiffhüttensee and Sengsee in the south and Lake Fischkaltersee in the south-east were exposed to a substantial nutrient enrichment until 1983. As a consequence, values of the macrophyte index in 1974 and 1981 were higher.

Until 1983 waste water from the village of Iffeldorf, situated in the south of the Osterseen lakes area between Lake Waschsee and Lake Fischkaltersee, was cleaned insufficiently by underground-seepage. From that time on it has been collected by a canal and pumped to a treatment plant ouside the catchment area of the lakes. In addition, the intensity of agriculture was reduced in the surroundings of the lakes which were declared a natural conservation area after 1981. Results of the macrophyte mapping provided decisive information for these restorative measures. The remaining nutrient gradient within the chain of lakes is clearly reflected by macrophyte index. Although nutrient enrichment of the southern lakes was reduced, additional efforts have to be undertaken to obtain total improvement.

Lake Starnberg

Lake Starnberg (5636 ha, 2998 $\times$ 10^6 m^3, length of shoreline 49.2 km, max. depth 128 m, average depth 73 m) is located 25 km south of Munich. In this lake submerged macrophytes clearly indicated that meso-eutrophic conditions prevailed by the end of the 1970s (Figure 2, left). In 1979 the construction of a 'ring canal' for sewage, surrounding almost the entire lake, and reducing waste water pollution by 95%, was completed. In that year mapping of the macrophytes recorded the lake's status. Species of indicator group 3.5 (*Potamogeton pusillus*), 4.0 (*Potamogeton pectinatus*), 4.5 (*Ranunculus circinatus*) and 5.0 (*Zannichellia palustris*) occurred frequently. Maximum abundance of these species was demonstrated for parts of the west bank between the villages Unterzeismering and Niederpöcking. The macrophyte index for these sections indicated 'immense', and 'heavy' nutrient enrichment. Species indicating moderate or low nutient pollution only occurred infrequently. These species exhibited maximum growth in the southern part of the lake. Along parts of the east bank no 'macrophyte index' could be evaluated due to exteme morphology resulting in sparse growth of plants.

Figure 2 (right) reflects the changed situation after 1979. The situation improved markedly up to 1989, especially along the formerly enriched section between Unterzeismering and Niederpöcking. In 1989 shorter shoreline sections could be used than in 1979, which led to a higher resolution. The cases where 'heavy' and 'massive' nutrient enrichment still occurred could be traced to point sources. The Rötlbach and Seeseitenbach brooks were polluted by fish ponds. On the bank of one small section near Unterzeismering is a camp-site with direct waste water inflow into the lake. Within the villages of Bernried, Tutzing, Feldafing and Niederpöcking there are overflows of the ring canal which discharge into the lake in cases of overload caused by heavy rainfall. The water authorities which financed the macrophyte mapping successfully carried out further measures to remove these remaining sources of pollution.

Lake Chiemsee

Lake Chiemsee is the largest lake in Bavaria (7990 ha, 2048 $\times$ 10^6 m^3, length of shoreline 63.9 km, max. depth 73.4 m, average depth 25.6 m). Recovery of the lake by the completion of a ring canal started in 1986 did not begin until in 1991. The macrophyte vegetation had been mapped 1 year earlier. Figure 3 shows the diverse pattern of nutrient enrichment revealed by the macrophyte index. Massive enrichment occurred in the south-east bay where the river Tiroler Ache flows into the lake. 90 tonnes of phosphorus were imported annually by this river at the time of macrophyte

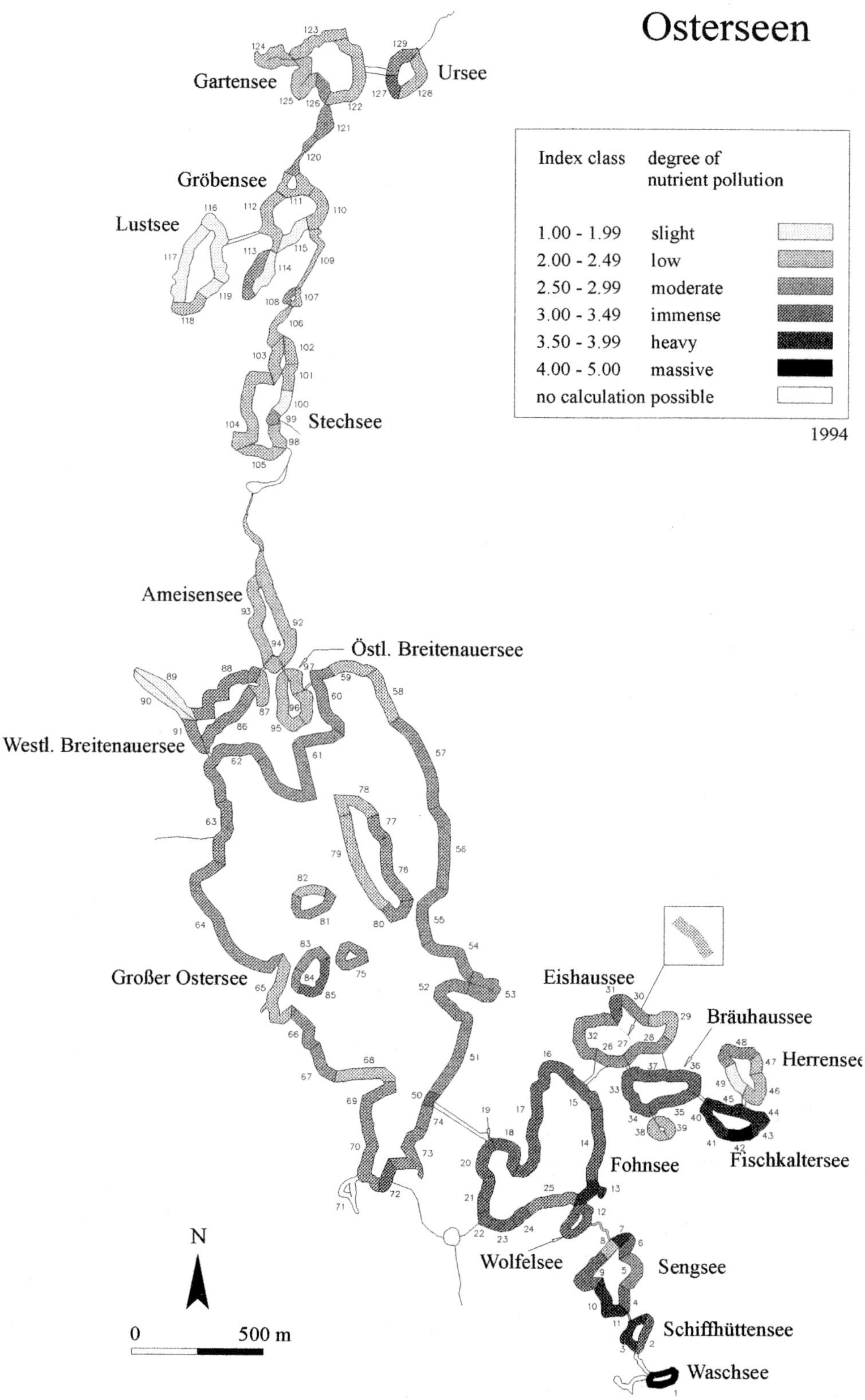

Figure 1. Macrophyte index of the Osterseen Lake area.

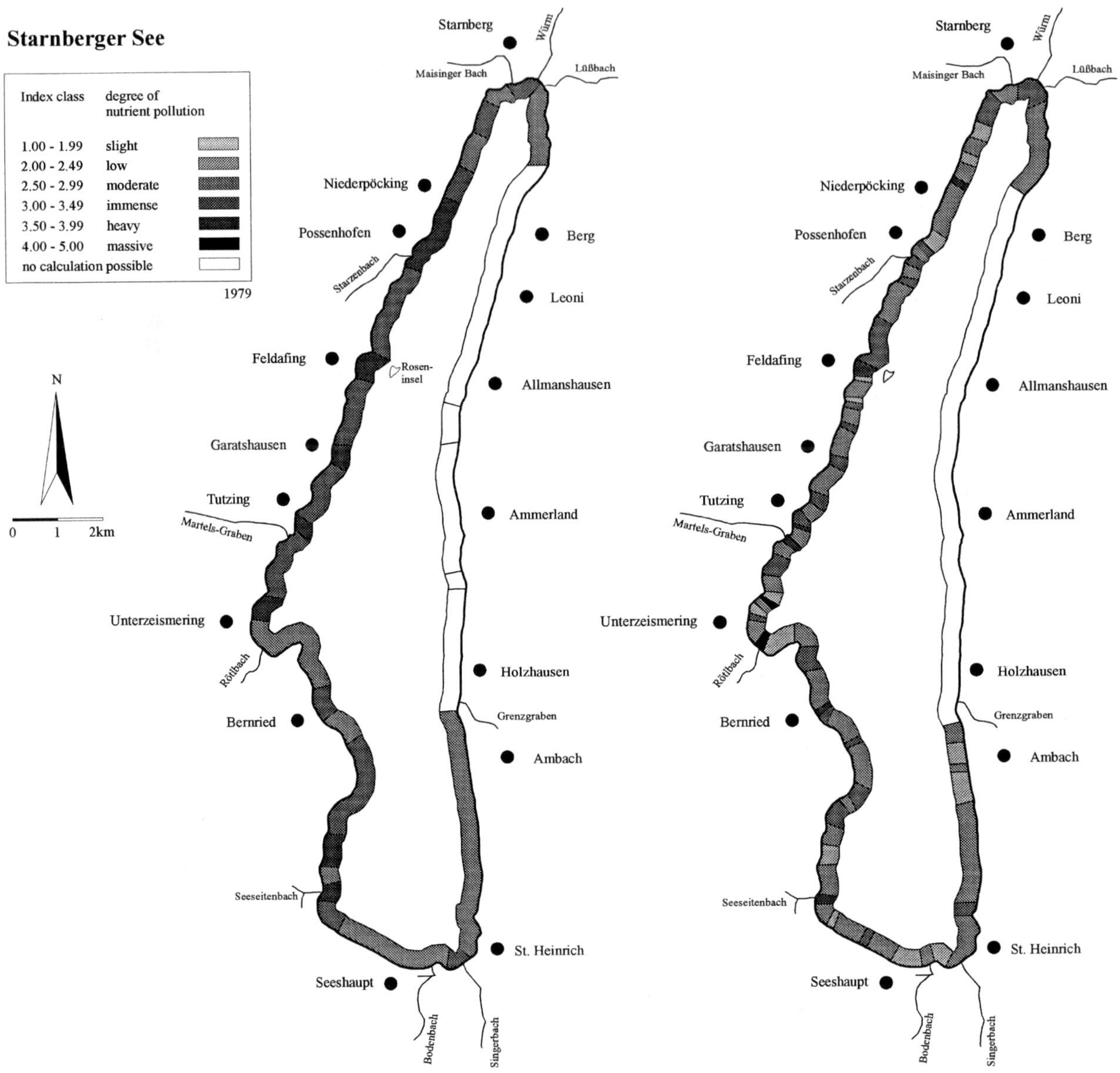

Figure 2. Macrophyte index of Lake Starnberg, 1979 (left) in comparison to 1989 (right).

mapping. The eastern parts of the lake were negatively influenced by this nutrient import. Massive enrichment was also caused by the inflow of all other rivers and brooks, except Krebsbach brook which is mainly fed by groundwater. The river Alz in the north is the outlet of the lake.

A low and moderate nutrient enrichment was registered for Herreninsel island, which is relatively sparsely populated. However, in three sections nutrient pollution was higher due to inflow of drainage ditches. In contrast Fraueninsel island is densely populated, although to reduce nutrient enrichment a sewage treatment plant with chemical phosphorus elimination had been constructed in 1976 with its effluent led into the hypolimnion. However, the effluent pipe was broken at 2 m depth and the (unknown) leakage was detected by macrophytes indicating ‘heavy’ and ‘massive’ enrichment. These species grew in a circle around this point source in the north-west section of the island.

In the area of the village Prien approximately 30 unknown pipes could be deteced due to changes in the composition of macrophytes. These pipes carried

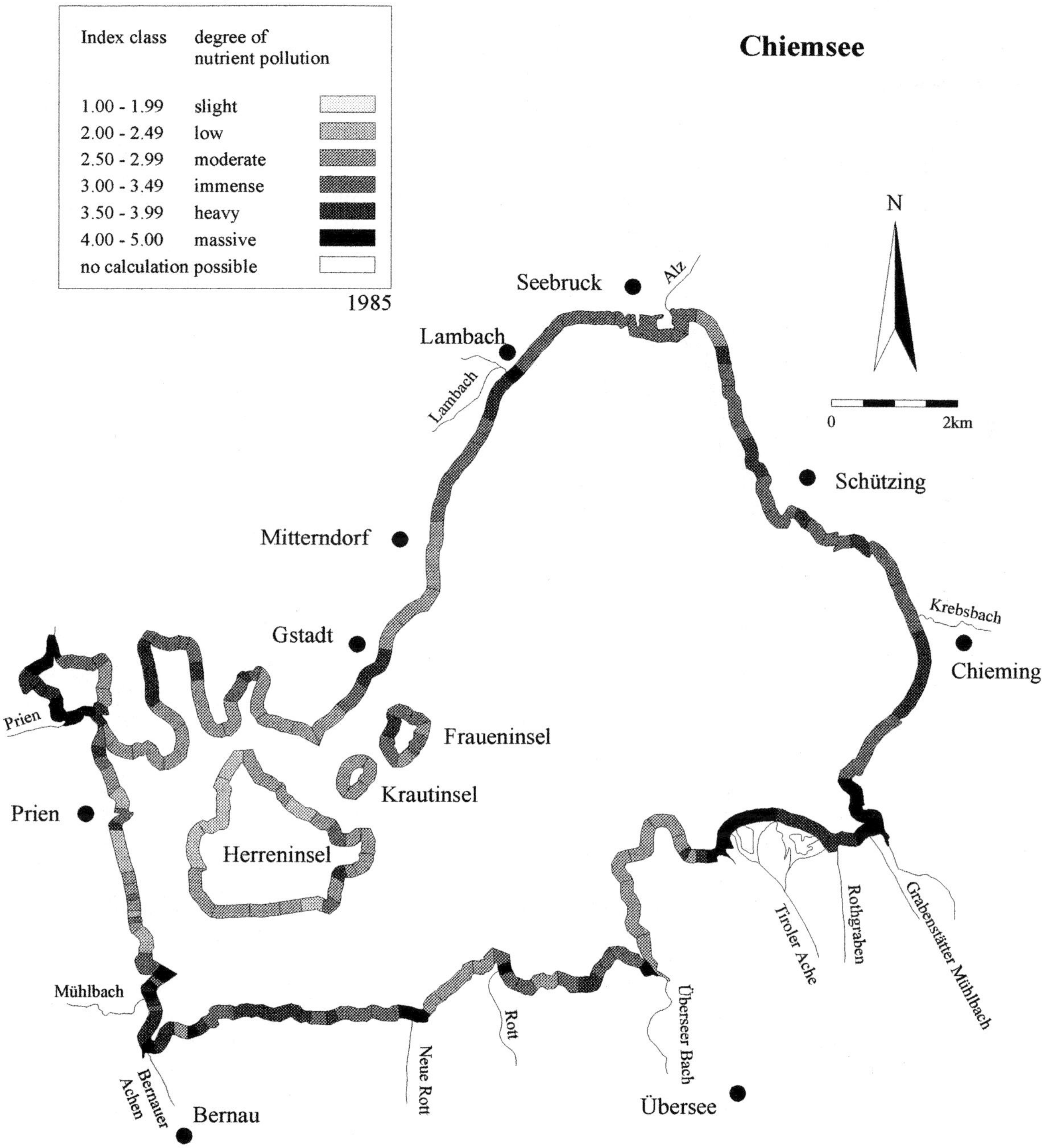

Figure 3. Macrophyte index of Lake Chiemsee.

either mechanically purified waste water, surface run-off or drainage water into the lake. In the area of the village Gstadt as well as in other sections nutrient-rich groundwater inlets caused by seepage of waste water were recognized by the growth of *Zannichellia palustris*. The success of the restoration of Lake Chiemsee following discovery and rectification of these varied sources will be revealed by re-survey in 1998.

Discussion

Natural lakes in Upper Bavaria are glacial in origin. They are all located in the alpine foothill plateaux and the northern calcareous Alps and therefore are hard-water lakes. In such lakes a characteristic flora exisits, differing from that occuring in soft water lakes. Although changes in the composition of the submerged macrophyte flora due to eutrophication have been reported in both types of lakes (Lang, 1968; Roelofs, 1983), the different vegetational characteristics restrict application of macrophyte index to calcareous lakes.

The macrophyte index can be applied to lakes of the southern calcareous Alps (Melzer, unpubl. data), although investigations of calcareous lakes in north Germany (Trapp, 1995) and north-east Germany (Krausch, 1964; Melzer, unpubl. data), respectively, have shown that some species exhibit a different behavior towards nutrient pollution. One example is *Ceratophyllum demersum*. In Bavarian lakes this species (indicator group 5) is restricted to highly polluted lakes or sections of lakes where it forms typical free-floating mats (Succow & Kopp, 1985). However, it grows well in oligo- and mesotrophic lakes of north and north east Germany, where this rootless species is loosely attached to the substrate and descends to greater depths depending on lake trophy (Trapp, 1995). *Ceratophyllum demersum* may even occur together with *Chara* species, for example in Lake Stechlin (Krausch, 1964), a combination which has never been observed in Bavarian lakes.

Comparison of observations between the two areas indicate that differences in nutrient tolerance also exist for *Myriophyllum spicatum*, *Potamogeton pectinatus* and some *Chara* species. It is not clear whether this contrasting behaviour results from differences in climate, in biotic interactions or in the physico-chemical properties of the lakes. From recent investigations it is known that aquatic macrophytes may genetically vary within a habitat, for example *Vallisneria americana* along a depth profile in Detroit River (Lokker et al., 1994). Different genotypes within a lake have also been shown for *Phramites australis* (Neuhaus et al., 1993). *Phragmites australis* is mainly propagated vegetatively (Sculthorpe, 1967) as are many submerged macrophytes and this property would support the formation of genetically uniform clones. Whether the observed differences in ecological behaviour can be traced back to genetically varying ecotypes is not yet known.

Thus, as a consequence of the regionally differing behaviour of some species, an overall application of the macrophyte index is not possible without a screening of the composition of characteristic macrophyte communities and a re-evaluation of the 'indicator groups'.

The question arises whether macrophyte distribution really expresses anthropogenic nutrient enrichment, or whether the occurrence of eutrophic species is a result of a natural aging and nutrient enrichment of lakes. Many observations affirm the former question. In the case of Lake Starnberg there are two lines of supporting evidence.

Firstly, a survey of the macrophyte flora by mechanical means was carried out at the turn of the century (Brand, 1896). At that time, water transparency was high with a minimum of 5 m during summer (Ammann, 1912). Without doubt this was due to low nutrient concentrations. Charophyte species of 'indicator groups' 1.0; 1.5 and 2.0, respectively (*Chara hispida*, *C. aspera*, *C. tomentosa* and *Nitella syncarpa*) formed extensive meadows reaching down to a depth of about 20 m. These species were extinguished completely by 1978. On the other hand, *Potamogeton mucronatus* and *Zannichellia palustris*, the only two species indicating 'massive' nutrient enrichment ('indicator group' 5) in Lake Starnberg in 1978 were absent in 1896 (Melzer, 1981).

Secondly, from 1978 onwards anthropogenetic nutrient enrichment dropped continuously due to waste water removal. Total phosphorus concentrations in the lake were reduced from 25 to 30 μg l^{-1} P in 1978 to half that value by 1991 (Henschel & Melzer, 1992). In 1989 a second survey of the macrophyte flora showed a marked increase in abundance of charophytes, and decline of eutrophic species. *Chara hispida, C. tomentosa* and *Nitella syncarpa* have not yet reappeared in the lake, however. These species have been replaced by *Chara contraria, Chara globularis* and *Nitellopsis obtusa*. Future investigations must show if the former species, requiring a lower nutrient status than the latter (Table 1) will return to the lake in the course of its continuing re-oligotrophication.

Lake Chiemsee and the Ostersee lakes also invalidate the hypotheses that macrophyte distribution solely expresses the natural aging and corresponding nutrient accumulation of a lake. In lake Chiemsee point sources cause striking changes in macrophyte composition and in the Osterseen lakes which are all of the similar age and morphology an anthropogenic gradient in nutrient enrichment developed within the chain of lakes and as a consequence different vegetational elements established. Those typical for eutrophic lakes have diminished since enrichment was reduced. Such changes in macrophyte flora are not exceptional, and are mirrored in other hard-water lakes. Examples are lake Constance (Schmieder, pers. comm.), lake Lugano (Lachavanne et al., 1992) and lake Botshol in the Netherlands (Simons et al., 1994). In all cases nutrient reduction of the lakes was followed by a recovery of charophytes.

Despite the current knowledge of recovery from several countries, the mechanisms of enrichment-controlled changes in macrophyte vegetation are not fully understood. These changes may depend on direct or indirect nutrient effects in which a uniform reaction among species is unlikely. Different strategies in nutrient uptake and assimilation lead to diverse reactions. In contrast to terrestrial plants, aquatic macrophytes are able to assimilate nutrients via both roots and shoots. For example, acquisition of phosphorus by roots is between 50 and approximately 100% (Bristow & Whitcombe, 1971; Carnigan, 1982; Carnigan & Kalff, 1980; Twilley et al., 1977), in which the proportion may vary seasonally (Smith & Adams, 1986). However, some genera (e.g. *Ceratophyllum, Utricularia*) are rootless and uptake is restricted to the shoot. Nitrogen assimilation is not only split between root and shoot uptake but moreover depends on the form of inorganic nitrogen in sediment and water. In general, ammonia prevails in sediment while nitrate does in water. In the latter, however, ammonia increases with eutrophication. In *Myriophyllum spicatum* a suppression of nitrate uptake has been observed with ammonia concentrations in the ambient water down to 30 $\mu g\ l^{-1}$ (Nichols & Keeney, 1976), although, this reaction may not be general. Different species of Lemnaceae deviate in their capacities of nitrate uptake and nitrate reductase activities when exposed to increasing concentrations of ammonia (Melzer & Müller, 1983). A high diversity in the capacity of nitrate uptake and nitrate reductase activities of different coexisting macrophyte species in a river, exposed to identical ambient nutient conditions has also been shown (Melzer & Kaiser, 1986). These findings support the assumption of specific nutritive strategies of submerged plants and they may explain differences in the reaction of plants following changes in nutrient enrichment.

Nutrients in sediment as well as water are increased by eutrophication and, as a consequence, sediment fractions of total phosphorus and nitrogen, together with diatom records, are often used to evaluate the history of lake trophy (Christie & Smol, 1993; Reavie et al., 1995). However, little is known about the time course of nutrient enrichment or depletion in littoral sediments following eutrophication or restoration of lakes. It may be assumed that, after successful control of pollution, a decrease of nutrients in littoral sediments occurs, albeit with some delay. The relatively slow reaction in qualitative and quantitative aspects of macrophyte composition may be traced back to this separation in nutrient status of water and sediment.

Other important environmental factors in lakes also have uneven effects upon macrophytes, e.g. light (Dale, 1986; Vant et al., 1986); epiphytes (Blindow, 1987; Sand-Jensen & Sondergaard, 1981); intra- and interspecific competition (Agami & Waisel, 1985; Moen & Cohen, 1989), and herbivory (Bailey, 1988; Lodge, 1991). Despite the diverse reactions of single species to such environmental conditions, the similarity in the reaction of macrophyte species to nutrients, at least in one lake district, probably indicates their overall limiting effect on growth. This forms the basis for the macrophyte index and its successful application to lake management.

References

Agami, M. & Y. Waisel, 1985. Inter-relationships between *Najas marina* L. and three other species of aquatic macrophytes. Hydrobiologia 126: 169–173.

Ammann, H., 1912. Physikalische und biologische Beobachtungen an oberbayerischen Seen. Thesis, Königl. Techn. Hochsch. München.

Bailey, R.C., 1988. Correlations between species richness and exposure: freshwater molluscs and macrophytes. Hydrobiologia 162: 183–191.

Blindow, J., 1987. The composition and density of epiphyton on several species of submerged macrophytes – The neutral substrate hypothesis tested. Aquat. Bot. 29: 157–168.

Brand, F., 1896. Über die Vegetationsverhältnisse des Würmsees und seine Grundalgen. Bot. Centralbl. 65: 1–13.

Bristow, J.M. & M. Whitcombe, 1971. The role of roots in the nutrition of aquatic vascular plants. Am. J. Bot. 58: 8–13.

Carnigan, R., 1982. An empirical model to estimate the relative importance of roots in phosphorus uptake by aquatic macrophytes. Can. J. Fish. Aquat, Sci. 39: 243–247.

Carnigan, R. & J. Kalff, 1980. Phosphorus sources for aquatic weeds: water or sediments? Science 207: 987–989.

Christie, C.E. & J.P. Smol, 1993. Diatom assemblages as indicators of lake trophic status in southeastern Ontario lakes. J. Phycol. 29: 575–586.

Dale, H.M., 1986. Temperature and light: The determination factors in maximum depth distribution of aquatic macrophytes in Ontario, Canada. Hydrobiologia 133: 73–78.

Dave, G. 1992. Sediment toxicity and heavy metals in eleven lime reference lakes of Sweden. Wat. Air Soil Pollut. 63: 187–200.

Drake, J.C. & S.I. Heaney, 1987. Occurence of phosphorus and its potential remobilization in the littoral sediments of a productive English lake. Freshwat. Biol 17: 513–523.

Henschel, T. & A. Melzer, 1992. Die limnologische Entwicklung des Starnberger Sees im Fortgang der Abwasserfernhaltung unter besonderer Berücksichtigung der Makrophytenvegetation, Informationsber. Bayer. Landesamt f. Wasserwirtsch. München 3/92: 1–117.

Krausch, H.-D., 1964. Die Pflanzengesellschaften des Stechlinsee-Gebietes. I. Die Gesellschaften des offenen Wassers. Limnologica 2: 145–203.

Lachavanne, J.-B. & R. Wattenhofer, 1975. Contribution à l'étude des macrophytes du Léman. Commission internat. pour la protection des eaux du Léman et du Rhône contre la pollution, Geneva: 1–147.

Lachavanne, J.-B., J. Perfetta & R. Juge, 1992. Influence of water eutrophication on the macrophyte vegetation of Lake Lugano. Aquat. Sci. 54: 351–363.

Lang, G., 1968. Vegetationsveränderungen am Bodenseeufer in den letzten hundert Jahren. Schrift. Ver. Gesch. Bodensees 86: 295–319.

Lodge, D.M. 1991. Herbivory on freshwater macrophytes. Aquat. Bot. 41: 195–224.

Lokker, C., D. Susko, L. Lovett-Doust & J. Lovett-Doust, 1994. Population genetic structure of Vallisneria americana, a dioecious clonal macrophyte. Am. J. Bot. 81: 1004–1012.

Melzer, A., 1981. Veränderungen der Makrophytenvegetation des Starnberger Sees und ihre indikatorische Bedeutung. Limnologica 13: 449–458.

Melzer, A. & R. Kaiser, 1986. Seasonal variations in nitrate content, total nitrogen, and nitrate reductase activities of macrophytes from a chalk stream in Upper Bavaria. Oecologia 68: 606–611.

Melzer, A. & M. Müller, 1983. In vivo-nitrate reductase activities of different species of Lemnaceae and comparison with endogenous nitrate depletion. In Proceedings of the International Symposium on Aquatic Macrophytes Sept. 1983 Nijmegen: 139–144.

Moen, R. A. & Y. Cohen, 1989. Growth and competition between *Potamogeton pectinatus* L. and *Myriophyllum exalbescens* Fern. in experimental ecosystems. Aquat. Bot. 33: 257–270.

Neuhaus, D., H. Kühl, J.-G. Kohl, P. Dörfel & T. Börner, 1993. Investigation on the genetic diversity of *Phragmites* stands using genomic fingerprinting. Aquat Bot. 45: 357–364.

Nichols, D.S. & D.R. Keeney, 1976. Nitrogen nutrition of *Myriophyllum spicatum*: uptake and translocation of ^{15}N by shoots and roots. Freshwat. Biol. 6: 145–154.

Reavie, E.D., R.I. Hall & J.P. Smol, 1995. An expanded weighted-averaging model for interferring past total phosphorus concentrations from diatom assemblages in eutrophic British Columbia (Canada) lakes. J. Paleolimnol. 14: 49–67.

Roelofs, J.G.M., 1983. Impact on acidification and eutrophication on macrophyte communities in soft water lakes in the Netherlands, I. Field observations. Aquat. Bot. 17: 139–155.

Sand-Jensen K. & M. Sondergaard, 1981. Phytoplankton and epiphyte development and their effect on submerged macrophytes in lakes of different nutrient status. Int. Rev. ges. Hydrobiol. 6: 529–552.

Sculthorpe, C. D., 1967. The Biology of Aquatic Vascular Plants. Edward Arnold, London.

Simons, J., M. Ohm, R. Daalder, P. Boers & W. Ripl, 1994. Restoration of Botshol (The Netherlands) by reduction of external nutrient load: recovery of a characean community, dominated by *Chara connivens*. Hydrobiologia 275/276: 243–253.

Smith, C. S. & M. S. Adams, 1986. Phosphorus transfer from sediments by *Myriophyllum spicatum*. Limnol. & Oceanogr. 31: 1312–1321.

Succow, M. & D. Kopp, 1985. Seen als Naturraumtypen: Petermanns Geogr. Mitt. 3: 161–170.

Suominen, J. 1968. Changes in the aquatic macroflora of the polluted Lake Rautavesi, SW-Finland. Ann. Bot. Fenn. 5: 65–81.

Trapp, S., 1995. Wasserpflanzen Bremer Seen und ihr Verhältnis zur Gewässergüte. Abh. Naturw. Verein Bremen 3: 165–177.

Tüxen, R. & E. Preising, 1942. Grundbegriffe und Methoden zum Studium der Wasser- und Sumpfpflanzengesellschaften. Dtsch. Wasserwirtsch. 37: 10–17 & 57–69.

Twilley, R. B., M. M. Brinson & G. J. Davis, 1977. Phosphorus absorption, translocation, and secretion in *Nuphar luteum*. Limnol. Oceonogr. 22: 1022–1032.

Uotila, P., 1971. Distribution and ecological features of hydrophytes in the polluted Lake Vanajavesi, S-Finland. Ann. Bot. Fenn. 8: 257–295.

Vant, W. N., R. J. Davies-Colley, J. S. Clayton, & B. T. Coffey, 1986. Macrophyte depth limits in North Island (New Zealand) lakes of differing clarity. Hydrobiologia 137: 55–60.

Hydrobiologia **395/396:** 191–198, 1999.
D.M. Harper, B. Brierley, A.J.D. Ferguson & G. Phillips (eds), The Ecological Bases for Lake and Reservoir Management.

The central role of fish in lake restoration and management

Eddy H. R. R. Lammens
Institute for Inland Water Management and Waste Water Treatment, P.O. Box 17, 8200 AA Lelystad, The Netherlands

Key words: fish stock management; gill-net fishery; planktivory; benthivory; piscivory; fish community development; trophic levels; *Daphnia* spp.; bream; carp; pikeperch; roach; inedible prey; individual based model; sediment resuspension

Abstract

The central role of fish in lake restoration and management has a practical purpose: fish are much easier to manipulate than nutrients, phytoplankton and zooplankton, and therefore they are a relatively easy (additional) instrument in restoration and management. The management of the fish stock may be a measure of water quality, of fish stock composition or a measure of both and may vary from very drastic removal of planktivorous and benthivorous fish to a more gradual change in the population by continual predator management and less drastic reduction of inedible prey. For lake restoration, drastic removal is the most efficient in order to obtain clear water and vegetation and a subsequent fish community adapted to this. Continual management will result in a more gradual change and may be more acceptable to the interest of both fishermen and water quality managers.

Introduction

There is an essential difference between the role of fish in lake management and the role of fish in ecosystems. In restoration and management it has the meaning of practicality: fish are much easier to handle than nutrients or phytoplankton and therefore they are a relatively easy instrument in restoration and management. The role of fish in ecosystems is not more than being part of the ecosystem, clearly influencing it, but there is no reason that the role should be central. Therefore it should be clear that the role of fish in management relates to practicality and the effects of changing the fish stock on the ecosystem.

The goal of restoration and management should be well defined, clarifying the (future) use of the lake. Water quality objectives may conflict with the wishes of fishery management or the wishes of nature conservation demanding specific food for visiting birds. So the management of fish stocks may serve many purposes depending on the wish of the water manager. This paper focusses on three aspects of fish management:

1. Fishery management as a measure of water quality.
2. Fishery management as a measure of fish stock composition.
3. Fishery management as a measure of both.

Firstly the effect of fish on the structure of the ecosystem is considered in general; discriminating between planktivorous, benthivorous and piscivorous fish. Then the development of a fish community in an eutrophic open water system and its effect of the successive stages on algal development is described. A case study where fishery and water quality management were combined concludes this paper.

The effect of fish on the structure of the ecosystem

There have been many experimental studies showing the effect (removal or addition) of fish on the structure of the ecosystem (Kerfoot & Sih, 1987; Carpenter, 1988; Gulati et al., 1990; Jeppesen et al., 1997). Characteristic is the cascading effect through all trophic levels and the more-or-less drastic change of the ecosystem, particularly at the appearence or disappearence of vegetation. Many of these studies proved to be useful for management purposes and were the basis for a restoration handbook in the Netherlands (Hosper & Meijer, 1992) and later in the U.K. (Moss et al., 1996). In this section, the effect of removal and addition of fishes will be described and some case studies used to illustrate the seperate effects of planktivory, benthivory and piscivory.

Total removal

The most drastic effects are found by (almost) complete removal of the fish stock. The most famous example is lake Zwemlust. In this lake the total fish stock was removed by pumping out the water and removing 100% of the fish. The effects of the experiment on all trophic levels are very well described (Van Donk et al., 1990b). The main result was a drastic switch from turbid to clear water caused by a high grazing pressure of *Daphnia* spp. and the subsequent development of vegetation and slight decrease in nutrients (Meijer et al., 1994b; Van Donk & Gulati, 1995). Afterwards the lake was stocked with pike (*Esox lucius*) and rudd (*Scardinius erythrophtalmus*). Because pike could not control the rudd population, the fish population increased and the system started to oscillate between clear and turbid. Many more birds visited the lake and influenced the succession of the vegetation (Van Donk & Gulati, 1995).

In many other lakes, large scale removals have been applied, not total, but usually 50–80% of the total fish stock. The effect in the first year is usually a marked positive effect on transparency and development of the *Daphnia* spp. populations. If no vegetation develops, the system easily reverts to the former situation, but if vegetation develops, the system becomes more stable and nutrients slightly decrease in concentration. One of the buffering mechanisms of the vegetation is providing refuges for daphnids. During the day, daphnids hide among the vegetation as protection against predation from visual feeders and come out during the night to feed in the open water (Timms & Moss, 1984; Schriver et al., 1995; Jeppesen et al., 1996). However, oscillaton between clear and turbid is difficult to prevent (Van Donk et al., 1990b; Jeppesen et al., 1990, 1995; Persson et al., 1993; Scheffer et al., 1993; Meijer et al., 1994a, b).

Stocking with planktivorous, benthivorous or piscivorous fish

Usually the fish stock is composed of many species and size-classes with specific selectivities regarding their food organisms. Specific effects of planktivory, benthivory or piscivory can not usually be discriminated and so can only be determined in experimental conditions.

Planktivorous fish

The general effect of planktivorous fish on zooplankton is not that all zooplankton species are reduced, but mainly the larger specimens and species. Within the *Daphnia* spp. group *D. magna* and *D. pulex* are the first to disappear (Van Donk et al., 1990). At very high predation pressure *D. galeata* also disappears and usually *D. cucullata* dominates together with other small cladorans such as *Bosmina longirostris* and *Chydorus sphaericus*. The smaller species have a lower grazing efficiency than the large daphnids and have therefore a lower grazing impact on phytoplankton (Brooks & Dodson, 1965). The effect of planktivores on zooplankton, particularly daphnids, has been demonstrated in many fields and experimental studies, of which the studies of Hrbaceck et al. (1961) and Brooks & Dodson (1965) are most famous. There is no doubt that evidence for the effect of planktivorous fish, particularly the size and species composition of zooplankton is overwhelming (Andersson, et al., 1978; Stenson et al., 1978; Nilsson & Pejler, 1973; Anderson, 1984; Lammens et al., 1985; Jeppesen et al., 1990; Meijer et al. 1990a; Berg et al., 1994).

A good example of an experimental study on the effect of planktivorous fish is shown by the study of Meijer et al. (1990a). Ten (0.1 ha) ponds were divided into halves and each half stocked with planktivorous fish (young-of-the-year roach and bream) with the other half remaining fishless. During the spring peak of the daphnid population, the biomass of the fish was still small and no difference was found in size and species composition of zooplankton. Only in summer when predation became high because of increasing biomass of the planktivorous fish, *Daphnia* spp. populations were suppressed, but the other zooplankton species increased. In these latter halves the algal population increased and transparency decreased in comparison to the fishless halves.

Benthivorous fish

Benthivorous fish feed on bottom-dwelling organisms. The consequence of this feeding is that nutrients are transported from the benthic to the pelagic phase and sometimes (in case of bream (*Abramis brama*) and carp (*Cyprinus carpio*)) there may also be a considerable resuspension of sediments because of their feeding behaviour (Lammens & Hoogenboezem, 1991). Other effects may be that they prevent macrophyte seedlings developing (Ten Winkel & Meulemans, 1984). Benthivorous fish are usually larger than planktivorous fish and species such as bream and carp are not vulnerable to predation by piscivorous fish (Lammens, 1989). Several studies of bream and carp (Lamarra, 1975; Anderson, 1978; Meijer et al., 1990a, b; Breukelaar et al., 1994) have shown a combined effect of high concentrations of suspended solids and high concentrations of nutrients. The number of field studies demonstrating these effects is much less than those on planktivorous fish.

The effect of benthivorous fish may be difficult to determine in the field, because of the interacting effects of wind and the presence of planktivorous fish. An elegant experimental study was carried out by Breukelaar et al. (1994). In a series of fish ponds, varying amounts of benthivorous bream and carp were stocked. Highly significant correlations between fish biomass and concentration of resuspended sediment were found, but also with chlorophyll '*a*' and nutrient concentrations (Breukelaar et al., 1994). The feeding behaviour of bream and carp clearly influenced turbidity by the direct action of feeding and thus resuspension of sediments, but also due to the release of nutrients from bottom sediments and from digested benthos displacing nutrients from the benthic to the pelagic phase.

Piscivorous fish

Piscivorous fish feed on fishes which are usually 10–40% of their body length. In Europe pike, pikeperch *(Stizostedion lucioperca)* and perch (*Perca fluviatilis)* are the most abundant predators (Willemsen, 1977; Popova & Sytina, 1977). Pikeperch is most successful in open and turbid water because it hunts by search, whereas for pike the opposite is true – it lives in overgrown, clear water and hunts by ambush. Perch is intermediate as it prefers open and clear water and it usually hunts by search in groups (Dziekonska, 1954; Savino & Stein, 1989; Chapman et al., 1989; Eklov & Persson, 1996).

Stocking with piscivorous fish may have pronounced effects, providing both the prey population and the habitat are suitable for the predator. The introduction of pikeperch in a Norwegian lake forced the roach to move from the open water and effects on chlorophyll and transparency were demonstrated (Brabrand & Faafeng, 1993). Another example of predator stocking was shown in lake Bautzen where short-term effects of annual stocking with 0+ pike and pikeperch were apparent as increase in clarity because of a reduction of the planktivorous fish and increase in density of *Daphnia galeata*. Longer-term effects, however, were apparent as an increase in *Microcystis* blooms and increase in toxicity (Benndorf et al., 1984; 1988). Similar short-term effects of 0+ pike stocking were found in some Danish lakes (Jeppesen et al., 1996) . However, there has been no study to date which demonstrates an effect of pike, perch or pikeperch which is similar to the complete removal of fish in eutrophic lakes. Piscivore stocking has only been successful when done permanently; it has not proved possible to prevent subsequent successful recruitment of cyprinids (Benndorf, 1990).

Fish stock development after complete removal

One of the arguments against biomanipulation is that the ecosystem effects are not stable and that the system will return to its original situation, that is, full of fish and full of algae. If, in the first years after removal, no vegetation develops, the original community will indeed return quickly (Van Donk et al., 1990b; Meijer et al., 1994a). Depending on the success of the original removal this period may vary from one to ten years (Van Donk & Gulati, 1995). If vegetation does develop, then the conditions for fish, zooplankton and benthos will also differ and then a different fish community will develop (Meijer et al., 1996).

A good example is the well-studied development of the fish community in the eutrophic lake Volkerak (5000 ha, average depth 5 m) which changed from a marine into freshwater community. It is illustrative because it shows the effect of (unintended) stocking a lake with fish (larvae) and the effect that has had on the total ecosystem. The development of the fish community is characteristic of situations when nutrient levels are high (>0.1 mg P l^{-1}), but there is no fish management at all. The managers were afraid that the lake would quickly turn into an algal-dominated system (Ligtvoet, 1993).

In its first year of transition from marine to freshwater, the lake was flushed from the adjacent eutrophic rivers (Hollands Diep/ Haringvliet). The chloride concentration decreased to 400–500 mg l^{-1} and the lake was suitable for all freshwater fish. The first species in-

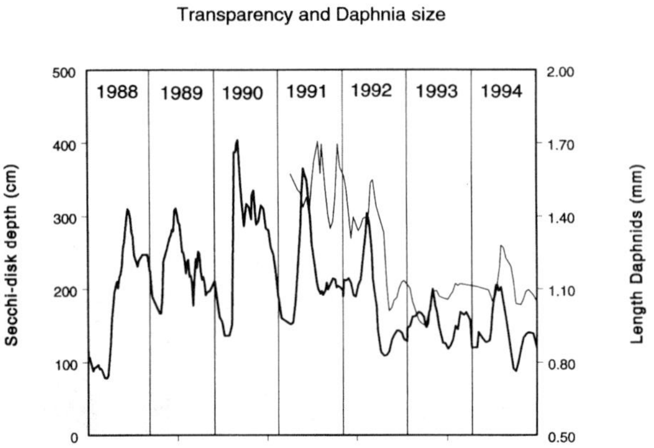

Figure 1. Transparency of Lake Volkerak and mean length of daphnids (thin line).

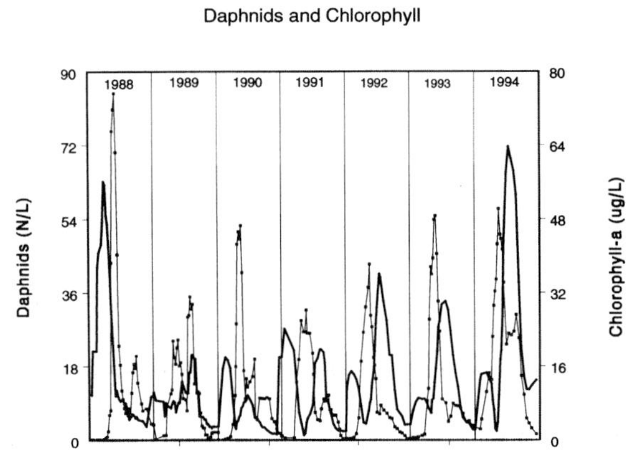

Figure 2. Chlorophyll 'a' in Lake Volkerak and density of daphnids (thin line).

vading the system were perch and pikeperch, entering the lake as larvae (adult fish did not enter). Because the larvae of these species are pelagic in their early stages, they are easily distributed through the open water, whereas cyprinids (roach and bream) are born in littoral regions and are not passivily transported to the open water. Therefore percids dominated over cyprinids in the first years. The fear of the lake manager that the lake would change quickly to an algal-dominated system did not come true for the first four years. The lake had a transparency of 2–4 m (Figure 1). Fish biomass was low and dominated by perch, growing extremely rapidly on a diet of *Neomysis integer* and *D. pulex* (Houthuizen et al., 1993). The zooplankton was dominated by *D. pulex* and the algal biomass was $<20\ \mu g\ l^{-1}$, although nutrient levels were not limiting (Figure 2). This community is typical for mesotrophic ($<50\ \mu g$ P l^{-1}) conditions (Persson et al., 1991).

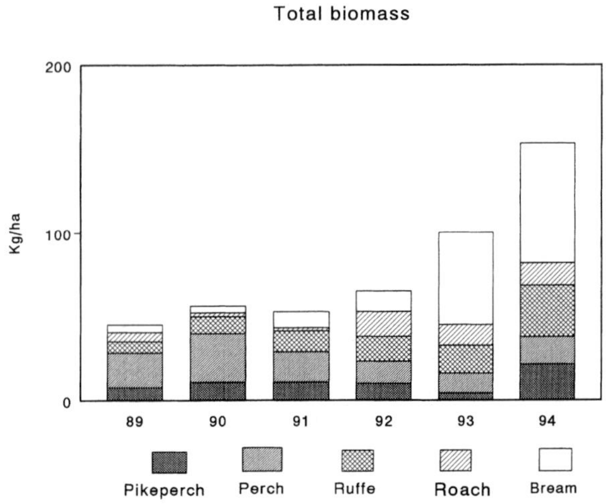

Figure 3. Development of fish biomass in Lake Volkerak.

The recruitment of cyprinids was limited as long as the recruitment was dependent on small numbers of incoming larvae, but as soon as the roach population became adult (after three years) a potentially high recruitment was possible. By the fifth year the recruitment of roach was very strong and could no longer be prevented by perch and pikeperch. In that year *D. pulex* disappeared and the transparency decreased from 3 to 1 m (Figure 1). During this summer *Microcystis aeruginosa* become dominant. *D. pulex* was replaced by *D. galeata* and did not return. This implied that the average size of the daphnids decreased from 1.5 to 1 mm, although the density did not change (Figures 1 and 2). The roach population remained high and the bream and pikeperch population increased as well (Figure 3). Although the recruitment of bream and pikeperch was usually much lower than that of roach and perch, their survival was much higher and after seven years the bream and pikeperch population had become dominant within the fish community. This stage of community development is characteristic for eutrophic lakes and is quite stable in the absence of management (Lammens, 1989; Persson et al., 1991)

The development of the fish community attracted large numbers of piscivorous birds, mainly grebes and cormorants (Figure 4). In the last few years their numbers have increased to several thousands and the total fish removal was considerable, almost similar to the consumption by piscivorous fish (Figure 5). Because the annual recruitment, growth and mortality of all species was known (Ligtvoet et al., 1994), the dynamics of the total fish community could be similated and the largest part of the mortality could be explained as a

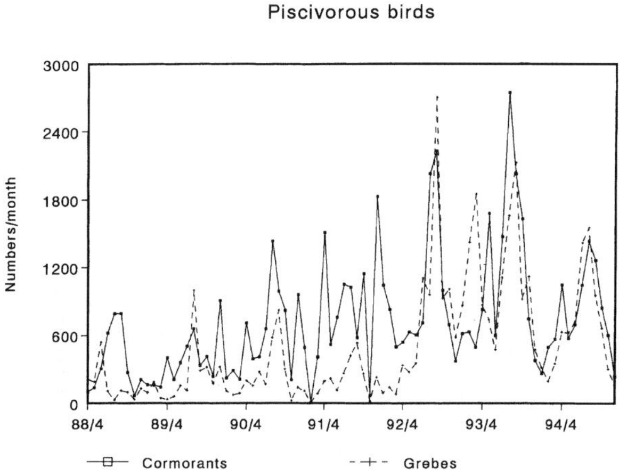

Figure 4. Piscivorous birds in Lake Volkerak.

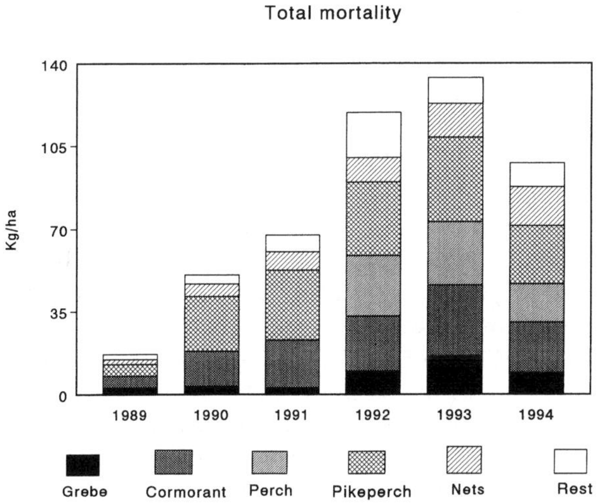

Figure 5. Total removal of cyprinid fish, Lake Volkerak.

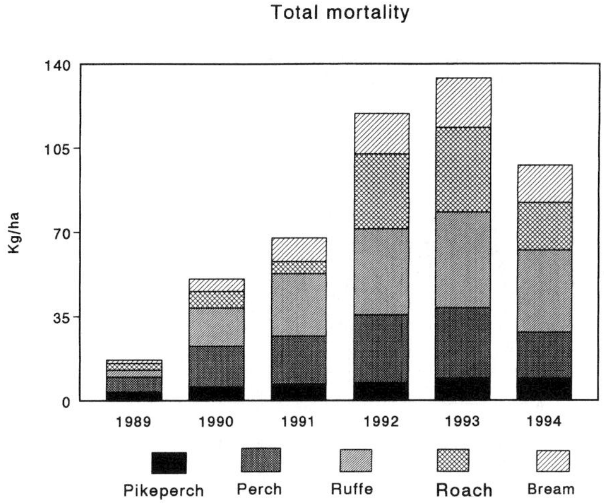

Figure 6. Total mortality by species, Lake Volkerak.

result of predation by fish and birds. For the simulation I used an Indivudual Based Model (PISCATOR, Lammens et al., 1995) and compared the total mortality in the lake with mortality caused by the sum of the main predators, that is predation by piscivorous fish, birds, fishery and other causes. It was surprising to see that the yearly annual mortality equalled the total biomass (Figures 3 and 6) and that, with the knowledge of the annual recruitment and selectivity of piscivorous fish and birds, the development of the fish community could fairly accurately be predicted.

It is clear that in eutrophic conditions the development of the fish population cannot be stopped by piscivorous fish, particularly when long-lived species such as bream (and carp) are present. After a few years' growth they become only slightly vulnerable to predation and because of their long lifespan they can build up a considerable biomass. Only a few successful recruitments of roach and bream are sufficient to escape the predation pressure. With our current knowledge, it cannot be predicted how successful the recruitment of a particular species will be and therefore it cannot be anticipated by predator stocking. Therefore it will be impossible to prevent a fish community development, typical for specific conditions. However, it is possible to change the fish community in a more desirable composition by management, as explained in the next section.

Management of predator and invulnerable prey

Instead of doing nothing or severely reducing the fish stock, management can be performed with the help of a commercial fishery if lakes are too large for complete fish removal. In the Frisian lakes (10 000 ha), an experimental fishery was set up for 5 years (1990–1995) to change the predator (pikeperch) population in such a way that the population would be dominated by relatively small individuals (Lammens & Klein Breteler, 1995). The hypothesis was that removing the relatively large specimens would increase the survival chances of the smaller ones similar to that found in pike (Grimm, 1981a, b) and as was more or less apparent from earlier fishery periods (Lammens et al., 1990). A pikeperch population consisting of relatively small individuals would exert a higher predation pressure on the small planktivorous fish and would therefore be beneficial for water quality.

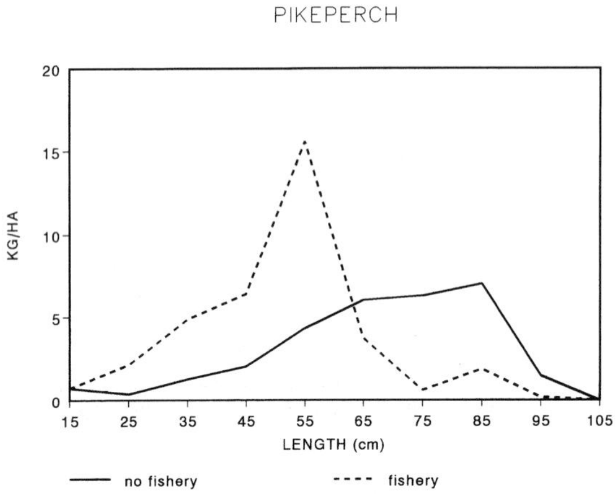

Figure 7. Pikeperch biomass per length class under conditions of fishing and no fishing.

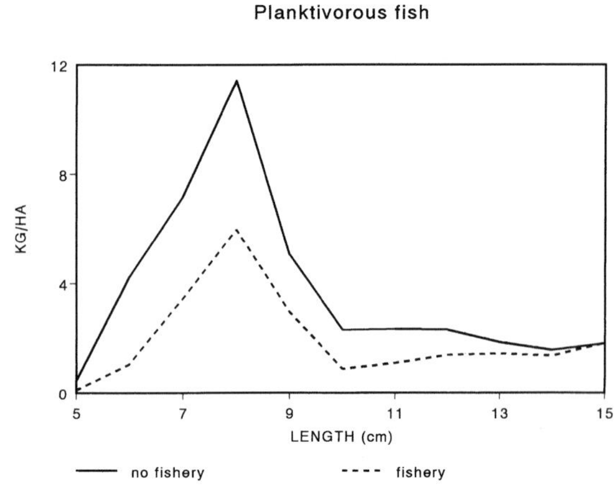

Figure 8. Planktivorous fish biomass per length class under conditions of fishing and no fishing.

An intensive gill-net fishery was started in 1989 to remove pikeperch >60 cm. The profit from the yield of pikeperch (which have a high market value) was used to pay a seine net fishery to remove bream >20 cm (which have a low market value) . Removing bream would benefit water quality (Breukelaar et al., 1994) and create better feeding conditions for eel (Lammens et al., 1985) and thus benefit the commercial fishermen. In order to evaluate the experimental fishery, a registration system was set up to to document all the catches and a monitoring program to estimate the biomass and size, and species composition of the fish was continued for 5 years. This monitoring program had already started 5 years previously and so a good reference situation was present.

The pikeperch population changed from a population dominated by large specimens into one dominated by small specimens. The biomass before 1990 had been dominated by the size-class 60–90 cm and after 1990 by the size-class 40–60 cm (Figure 7). The total biomass increased slightly even though 5–8 kg ha^{-1} was harvested each year. Therefore the original goal of changing the population structure was successful. It may be questionable whether a period of 5 years is sufficient to cover the large variation in recruitment and thus to predict this effect for a longer period: the change, however, was sufficient to establish an indirect effect on the biomass of small planktivorous fish. A greater than two-fold increase of 40–60 cm pikeperch was probably mainly responsible for a two-fold decrease of planktivorous fish <15 cm (Figure 8). So both the direct and indirect effects were clear regarding this pikeperch fishery.

The direct effect of the bream fishery on the total population was only visible in early summer, shortly

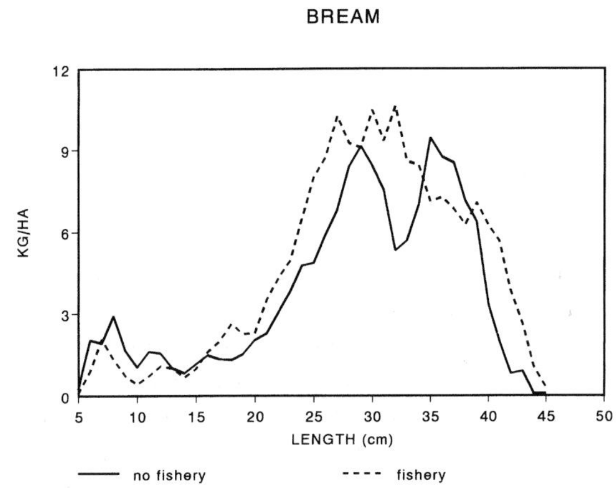

Figure 9. Bream biomass per length class under conditions of fishing and no fishing.

after the winter fishery stopped. This reduction (30–50 kg ha^{-1}, 30–35% of the total benthivorous part of the poplation) was compensated during the summer by indivual growth, not by an increase in numbers. The population gradually increased in biomass over the course of the summer after the recovery of spawning at the end of June (Lammens et al., 1985). Feeding activity and release of nutrients must have been 30–35% lower in spring and early summer than in preceeding years and have affected the nutrient concentration (Breukelaar et al., 1994). So an indirect effect of the fishery was an decrease in release of nutrients. Because of the improved growth and increase in maximal size of fish, the reduction in numbers was compensated for, and the total biomass at the end of the year hardly changed (Figure 9). Another indirect effect was a decrease in bream <15 cm because of increased predation comparable to the general decrease of planktivorous fish.

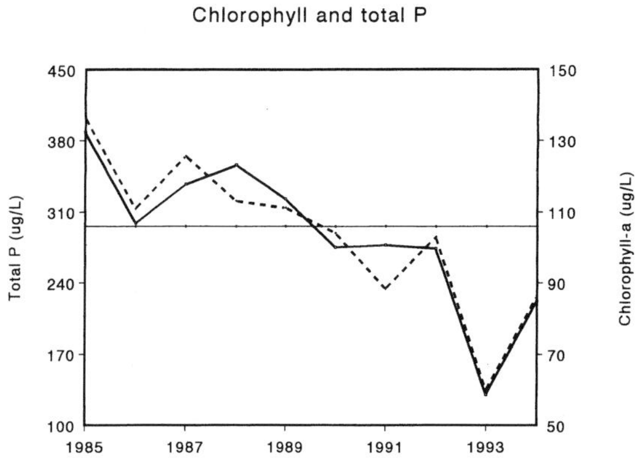

Figure 10. Annual change in chlorophyll 'a' and total phosphorus in the Frisian lakes.

By the total reduction of 250 kg bream ha^{-1} over 5 years a total amount of 0.1 g P m^{-2} was removed and, although this corresponded to the extra reduction of phosphorus in these 5 years (Figure 10), it is not necessarily the direct cause of this reduction. It seems most likely that the decrease in phosphorus and chlorophyll concentration was caused by the decreased activity of the (smaller) bream population in spring and early summer in combination with an increased grazing pressure of daphnids.

So, although it is difficult to discriminate all the separate effects and determine how much of the water quality improvement is caused by management of predator and inedible prey separately or is caused by the interaction of measures, there are strong indications that fish management contributed substantially and that it is worthy of consideration in large lakes.

Conclusions

The central role of fish in lake restoration and management may largely have a practical purpose: fish are more easy to manipulate than nutrients, phytoplankton and zooplankton, and are therefore an easy (additional) instrument for management. Management measures may vary from very drastic removal of planktivorous and benthivorous fish to a more gradual change in the population by continual predator management and less drastic reduction of inedible prey. For lake restoration, drastic removal is the most efficient in order to obtain clear water and vegetation and a subsequent fish community adapted to this. At high nutrient concentrations the old situation may easily return. Continual management is less drastic and will result in a more gradual change: this may be more acceptable to the interests of both fisherman and water quality managers.

References

Andersson, G., H. Berggren, G. Cronberg & C. Gelin, 1978. Effects of planktivorous and benthivorous fish on organisms and water chemistry in eutrophic lakes. Hydrobiologia 59: 9–15.

Andersson G., 1984. The influence of fish on eutrophic lake ecosystems. In 1st. Nordic Symposium on Limnology Apr. 1984. Norwegian Soc. Limnol., Fagerfjell, Norway.

Benndorf, J., H. Kneschke, K. Kossatz & E. Penz, 1984. Manipulation of the pelagic food web by stocking with predacious fishes. Int. Rev. ges. Hydrobiol. 69: 407–428.

Benndorf, J., H. Schultz, A. Benndorf, R. Unger, E. Penz, H. Kneschke, K. Kossatz, R. Dumke & U. Hornig, 1988. Food-web manipulation by enhancement of piscivorous fish stocks: long-term effects in the hypertrophic Bautzen Reservoir. Limnologica 19: 97–110.

Benndorf, J., 1990. Conditions for effective biomanipulation; conclusions derived from whole-lake experiments in Europe. Hydrobiologia 200/201: 187–203.

Berg, S., E. Jeppesen, M. Sondergaard & E. Mortensen, 1994. Environmental effects of introducing whitefish, *Coregonus lavaretus* (L), in Lake Ring. Hydrobiologia 276: 71–79.

Jensen, J. P., E. Brabrand & B. Faafeng, 1993. Habitat shift in roach (*Rutilus rutilus*) induced by pikeperch (*Stizostedion lucioperca*) introduction – predation risk versus pelagic behaviour. Oecologia 95: 38–46.

Breukelaar, A. W., E. H. R. R., Lammens & J. G. Breteler, 1994. Effects of benthivorous bream (*Abramis brama*) and carp (*Cyprinus carpio*) on sediment resuspension and concentrations of nutrients and chlorophyll-*a*. Freshwat. Biol. 32: 113–121.

Brooks, J. L. & S. I. Dodson, 1965. Predation, body size and composition of plankton. Science. 150: 28–35.

Carpenter, S. R. (eds), 1988. Complex Interactions in Lake Communities. Springer-Verlag, New York: 283 pp.

Cazemier, W. G., 1983. The growth of bream *Abramis brama* in relation to habitat and population density. Hydrobiol. Bull. 16: 269–278.

Chapman, L. J., W. C., Mackay & C. W., Wilkinson, 1989. Feeding flexibility in northern pike (*Esox lucius*): fish versus invertebrate prey. Can. J. Fish. Aquat. Sci. 46: 666–669.

Dam, E. M. van & C. Breukers. 1995. Natuurontwikkelingen Volkerak-Zoommeer 1993. RIZA Nota 95.022.

Dziekonska, J., 1954. The feeding characteristics of mature pike, perch and pikeperch in lakes. Polskie Archievum Hydrobiologii 2: 165–183.

Eklov, P. & L. Persson. 1996. The response of prey to the risk of predation: Proximate cues for refuging juvenile fish. An. Behav. 51: 105–115.

Gulati, R. D., E. H. R. R. Lammens, M.-L. Meijer & E. van Donk (eds), 1990. Biomanipulation – Tool for Water Management. Developments in Hydrobiology 61. Kluwer Academic Publishers, Dordrecht, 628 pp. Reprinted from Hydrobiologia 200/201.

Grimm, M. P., 1981. Intraspecific predation as a principal factor controlling the biomass of northern pike (*Esox lucius* L). Fish. mgmt 12: 77–80.

Grimm, M. P., 1981. The composition of northern pike (*Esox lucius* L.) populations in four shallow waters in the Netherlands, with special reference to factors influencing 0+ pike biomass. Fish. mgmt 12: 61–76.

Hosper, S. H. & M.-L., Meijer., 1992. Handleiding Aktief Biologisch Beheer.

Hosper, S. H. & M.-L., Meijer, 1993. Biomanipulation, will it work for your lake? A simple test for the assessment of chances for clear water, following drastic fish–stock reduction in shallow, eutrophic lakes. Ecol. Eng.: 63–72.

Houthuizen, R. P., J. J. Backx & A. D. Buijse. 1993. Exceptionally rapid growth and early maturation of perch in a freshwater lake recently converted from an estuary. J. Fish Biol. 43: 320–324.

Hrbacek, J., M. Dvorakova, V. Korinek & L. Prochazkova, 1961. Demonstration of the effect of the fish stock on the species composition of zooplankton and the intensity of metabolism of the whole plankton association. Verh. Int. Ver. Theor. Angew. Limnol. 14: 192–195.

Jeppesen, E., K. Olrik & P. Kristensen, 1994. Impact of nutrients and physical factors on the shift from cyanobacterial to chlorophyte dominance in shallow Danish lakes. Can. J. Fish. aquat. Sci. 51: 1692–1699.

Jeppesen, E., M. Søndergaard, O. Sortkjaer, E. Mortensen & P. Kristensen, 1990. Interactions between phytoplankton zooplankton and fish in a shallow hypertrophic lake a study of phytoplankton collapses in lake Sobygard Denmark. Hydrobiologia 191: 149–164.

Jeppesen, E., J. P. Jensen, M. Søndergaard, T. L. Lauridsen, L. J. Pedersen & L. Jensen, 1997. Top–down control in freshwater lakes: the role of nutrient state, submerged macrophytes and water depth. Hydrobiologia 342/343 (Dev. Hydrobiol. 119): 151–164.

Kerfoot, W. C. & A. Sih, 1987. Predation: Direct and Indirect Impacts on Aquatic Communities. University Press of New England, Hanover, 386 pp.

Lamarra Jr., V. A., 1975. Digestive activities of carp as a major contributor to the nutrient loading of lakes. Verh. int. Verein. Limnol. 19: 2461–2468.

Lammens, E. H., H. W. de Nie, J. Vijverberg & W. L. T. van Densen, 1985. Resource partitioning and niche shifts of bream (*Abramis brama*) and eel (*Anguilla anguilla*) mediated by predation of smelt (*Osmerus eperlanus*) on *Daphnia hyalina*. Can. J. Fish. aquat. Sci. 42: 1342–1351.

Lammens, E. H., 1989. Causes and consequences of the success of bream in Dutch eutrophic lakes. Hydrobiol. Bull. 23: 11–18.

Lammens, E. H., W. L. T. van Densen & R. Knijn. 1990. The fish community structure in Tjeukemeer in relation to fishery and habitat utilization. J. Fish Biol.36: 933–945.

Lammens, E. H. & W. Hoogenboezem, 1991. Diets and feeding behaviour. In I. Winfield & Nelson (eds), The Biology of Cyprinid Fishes. Chapman and Hall, London: 353–376.

Lammens, E. H., M. Scheffer & E. van Nes, 1996. PISCATOR, a model for the interaction between fish stock and fishery in Ijsselmeer and Markermeer. Version 0.0. Werkdocument 96, 123 pp.

Ligtvoet, W., 1993. Visstandsontwikkeling Volkerak-Zoommeer 1987–1997: scenario voor het eutrof ieringsproces?: 108–115.

Ligtvoet, W., et al. 1990, 1991, 1992, 1993, 1994, 1995. Omvang en samenstelling van de visstand in november 1989, 1990, 1991, 1992, 1993, 1994. Rapport Boz. 81.1, 81.2, 81.3, 81.7, 81.17, 81.20. Witteveen and Bos Raadgevende ingenieurs b.v., Deventer.

Meijer, M.-L., E. H. Lammens, A. J. P. Raat, M. P. Grimm & S. H. Hosper, 1990. Impact of cyprinids on zooplankton and algae in ten drainable ponds; preliminary results. Hydrobiologia 191: 275–284.

Meijer, M.-L., M. W. De Haan, A. W. Breukelaar & H. Buiteveld, 1990. Is reduction of the benthivorous fish an important cause of high transparency following biomanipulation in shallow lakes? Hydrobiologia 200–201: 303–316.

Meijer, M.-L., E. H. Van Nes, E. H. Lammens, & R. D. Gulati, 1994a. The consequences of a drastic fish stock reduction in the large and shallow lake Wolderwijd, the Netherlands – Can we understand what happened? Hydrobiologia 276: 31–42.

Meijer, M.-L., E. Jeppesen, E. van donk, B. Moss, M. Scheffer, E. Lammens, E. Vannes, J. A. Vanberkum, G. J. Dejong, B. A. Faafeng & Jensen, 1994b. Long-term responses to fish–stock reduction in small shallow lakes – interpretation of 5-year results of biomanipulation cases in the Netherlands and Denmark. Hydrobiologia 276: 457–466.

Meijer, M.-L, E. H. Lammens, J. P. G. Klein-Breteler & M. P. Grimm, 1996. Development of fish communities after biomanipulation. Neth. Jo. of aquat. Ecol. 29: 91–103.

Moss, B., J. Madgwick & G. Philips, 1996. A guide to the restoration of nutrient enriched shallow lakes. Broads Authority / Environment Agency.

Nilsson, N.-A. & B. Pejler, 1973. On the relation between fish fauna and zooplankton composition in North Swedish lakes. Rep. Inst. Freshw. Res. Drottningholm 53: 51–78.

Nilsson N. A., 1980. The role of size-based predation in competition and interactive segregation in fish. In S. D. Gerking (ed.), Ecology of Freshwater Fish Production. Blackwell Sci. Publ., Oxford: 303–325..

Persson, L., S. Diehl, L. Johansson & G. Andersson, 1991. Shifts in fish communities along the productivity gradient of temperate lakes: patterns and the importance of size-structured interactions. J. Fish Biol. 38: 281–294.

Persson, L., L. Johansson, G. Andersson, S. Diehl & S. F. Hamrin, 1993. Density dependent interactions in lake ecosystems: whole lake perturbation experiments. Oikos 66: 193–208.

Popova, O. A. & L. A. Sytina, 1977. Food and feeding relations of Eurasian perch (*Perca fluviatilis*) and pikeperch (Stizstedion lucioperca) in various waters of the USSR. J. Fish. Res. Bd Can. 34: 1559–1570.

Savino, J. F. & R. A., Stein, 1989. Behavior of fish predators and their prey: habitat choice between open waters and dense vegetation. Envn. Biol. Fishes 24: 287–293.

Scheffer, M., S. H. Hosper, M.-L. Meijer & B. Moss, 1993. Alternative equilibria in shallow lakes. Trends in Ecology & Evolution: 275–279.

Schriver, P., J. Bogestrand, E. Jeppesen & M. Sondergaard, 1995. Impact of submerged macrophytes on fish-zooplankton-phytoplankton interactions: large-scale enclosure experiments in a shallow eutrophic lake. Freshwat. Biol. 33: 255–270.

Stenson, J. A. E., T. Bohlin, L. Henrikson, B. I. Nilsson, H. G. Nyman, H. G. Oscarson & P. Larsson, 1978. Effects of fish removal from a small lake. Verh. Int. Ver. Limnol. 20: 794–801.

Ten Winkel, E. H. & J. T. Meulemans, 1984. Effects of fish upon sumerged vegetation.. Hydrobiol. Bull. 18: 157–158.

Timms, R. M. & B. Moss, 1984. Prevention of growth of potentially dense phytoplankton populations by zooplankton grazing in the presence of zoo planktivorous fish in a shallow wetland ecosystem. Limnol. Oceanogr. 29: 472-486.

Van Donk, E., M. P. Grimm, R. D. Gulati, P. G. M. Heuts, W. A. de Kloet & E. van Liere, 1990. First attempt to apply whole-lake food-web manipulation on a large scale in the Netherlands. Hydrobiologia 200/201 (Dev. Hydrobiol. 61): 291–303.

Van Donk, E., M. P. Grimm, R. D. Gulati & J. P. G. Klein Breteler, 1990. Whole-lake food-web manipulation as a means to study community interactions in a small ecosystem. Hydrobiologia 200/201 (Dev. Hydrobiol. 61): 275–290.

Van Donk, E. & R. D. Gulati, 1995. Transition of a lake to turbid state six years after biomanipulation: mechanisms and pathways. Wat. Sci. Technol. 32: 197–206.

Willemsen, J., 1977. Population dynamics of percids in Lake Ijssel and some smaller lakes in the Netherlands. J. Fish. Res. Bd Can. 34: 1710–1719.

Hydrobiologia **395/396:** 199–210, 1999.
D.M. Harper, B. Brierley, A.J.D. Ferguson & G. Phillips (eds), The Ecological Bases for Lake and Reservoir Management.

The practical importance of the interactions between fish, zooplankton and macrophytes in shallow lake restoration

Martin R. Perrow[1], Adrian J.D. Jowitt[1], Julia H. Stansfield[2] & Geoff L. Phillips[2]
[1]*ECON, Biological Sciences, University of East Anglia, Norwich, NR4 7TJ, U.K.*
[2]*Environment Agency, Eastern Area, Anglian Region, Cobham Road, Ipswich, U.K.*

Key words: macrophytes, refuges, *Rutilus rutilus*, *Perca fluviatilis*, *Daphnia*

Abstract

Zooplankton, using macrophytes as refuges from predation by zooplanktivorous fish, are believed to be important in maintaining the macrophyte-dominated state in shallow lakes. Their grazing upon phytoplankton is also believed to be instrumental in preserving water transparency in the establishment phase of macrophytes which follows an attempt to restore a shallow lake from the effects of eutrophication. This paper interprets the results of fish, zooplankton and macrophyte interactions from an intensive 3-year study in the Norfolk Broads of eastern England. In the presence of even a low density of 0+ fish (0.2 m^{-2}), *Daphnia* spp. which typically dominate the cladoceran community in spring, and through grazing are responsible for producing clear water, are typically reduced to very low levels. This decline may become protracted, but not eliminated, by the presence of macrophytes, implying a refuge effect. The efficacy of any refuge effect appears to increase with an increasing proportion of the water column occupied by macrophytes (PVI). As macrophytes develop, the grazing role is taken over by *Ceriodaphnia* spp. and *Simocephalus* spp., and may be at a sufficient rate to maintain clear water at least within macrophyte stands and possibly in intervening open water areas through diel migration of the zooplankton. A macrophyte PVI of 30–40% may provide an adequate refuge for these species through the mechanism of predation-free space, although this depends on fish density and community structure. At high densities (1 m^{-2}) of a suite of fish species, including efficient foragers in open water (roach, *Rutilus rutilus*) and within macrophytes (perch, *Perca fluviatilis*), any refuge effect is nullified. In stable macrophyte-dominated lakes, the shift in fish community structure towards a higher proportion of piscivorous compared to zooplanktivorous fish may have a role in promoting the refuge effect through changing the distribution of zooplanktivorous fish. Predation upon zooplankton may also be reduced through the provision of alternative prey in the form of macrophyte-associated macroinvertebrates for an alternative fish stock dominated by perch, rudd (*Scardinius erythrophthalmus*) and tench (*Tinca tinca*).

Introduction

Shallow lakes, particularly as a component of a larger wetland system, are of high ecological and socioloeconomic importance (e.g. Finlayson, 1992; National Research Institute, 1992; Broads Authority 1994). The value of such lakes has typically been compromised by nutrient enrichment leading to the replacement of clear water, dominated by submerged macrophytes supporting diverse communities of invertebrates, fish and birds, with turbid, algal-dominated water and associated depauperate animal communities (de Nie, 1987). The resultant loss of biodiversity and decline in water quality has led to many attempts to restore macrophyte-dominated conditions (Meijer et al., 1994). Restoration typically focuses on the reduction of nutrient loading, both through external and internal control (Phillips et al., 1999; Sas, 1989), although experience shows that this has rarely been successful as a single measure (Jeppesen et al., 1990; Phillips et al., 1999; Sas 1989). It has been demonstrated that buffering mechanisms, particularly control of grazing zooplankton by zooplanktivorous fish, tend to preserve algal domination even where the nutrient concentration is suitably low for stable macrophyte populations to exist (Irvine et al., 1989; Scheffer et al., 1993). It is now widely accepted that biomanipulation, the top-down manipulation of fish communities (i.e.

enhancement of piscivores or reduction of zooplanktivores and/or benthivores) is required as an extreme perturbation to push one stable equilibrium to another (Perrow et al., 1997).

Following the initial creation of clear water, an extensive and dense macrophyte community may quickly become established (Ozimek et al., 1990; Meijer et al., 1994). However, this situation may not remain stable in the long term (Lauridsen et al., 1994; Perrow et al., 1994; van Donk & Gulati, 1995), leading to an increase in algal populations once more. Alternatively, macrophytes may be slow to develop, for a variety of possible reasons (Perrow et al., 1997). Where macrophytes are present, knowledge of the processes responsible for the maintenance of clear water and low algal populations is critical if the ultimate goal of sustainable restoration is ever to be achieved, particularly in a cost-effective manner (Perrow et al., 1997). Further, such knowledge may help in the maintenance of clear water in the development phase when macrophytes are slow to colonise.

Although several mechanisms have been suggested for the maintenance of macrophyte communities in restored lakes (Perrow et al., 1997), research has thus far tended to focus on the provision of macrophytes as refuges for zooplankton (Phillips et al., 1996; Schriver et al., 1995), as this mechanism may be applicable both when macrophytes have become rapidly established or where they suffer some lag in development. The migration of grazing Cladocera from macrophyte refuges to open water has the potential to limit phytoplankton growth outside of macrophyte beds, thus providing a potential mechanism for the stabilisation of macrophytes (Timms & Moss, 1984). However, it is unclear whether other mechanisms such as nitrogen-limitation (van Donk et al., 1993) or allelopathy (Wium-Anderson et al.,1982) could influence the transparency of intervening open water areas. Moreover, experience shows that the successful recruitment of zooplanktivorous fish in the developing phase is often the cause of macrophyte decline and the regression of favourable conditions (Meijer et al., 1994). In contrast, macrophyte-dominated lakes frequently contain a high biomass of fish (Carpenter & Lodge, 1986; de Nie, 1987) with apparently little impact on the maintenance of clear water. An equilibrium of fish/zooplankton must exist and, as argued by Shapiro (1990), refuges for zooplankton from predation are likely to be a critical factor in the maintenance of this equilibrium.

In this paper, the hypothesis that macrophytes act as refuges for zooplankton is explored, and that this refuge effect is an important mechanism contributing to the maintenance of clear water and ultimately to the stability of the macrophyte-dominated state. The hypothesis is tested on data from the Norfolk Broads, a series (around 50) of small (1–124 ha) shallow (0.5–2 m) lakes in eastern England (see Phillips et al. (1999) for further details) and discussed in the light of studies elsewhere. The key questions of the zooplankton, macrophyte and fish interaction are: Do grazing Cladocera need a refuge from fish predation? Do macrophytes act as refuges for Cladocera against fish predation? Under what conditions does any refuge effect occur? What is the mechanism behind any refuge effect? Are refuging Cladocera able to control phytoplankton populations?

Methods

Five broads (Belaugh, Cockshoot, Cromes, Hoveton Little/Pound End and Upton), varying in trophic status and macrophyte cover and diversity, formed the basis of investigations conducted from 1993 to 1995 under the Broads Authority/Environment Agency EU LIFE project (Table 1). A stepwise programme of monitoring, each step building on the experiences of the last, was adopted. By the final year, fish, macrophytes and zooplankton were sampled, in a fully integrated manner, monthly from April to September.

Zooplanktivorous fish and small piscivorous pike (*Esox lucius*) were sampled by point-abundance electrofishing (PAS) (Copp & Peňáz, 1988; Perrow et al., 1996a). During fish sampling at between 100-160 points, the macrophytes cover (percent volume infested, PVI, Canfield et al., 1984) at each sampling point was assessed visually, with water depth and height of macrophytes measured by use of a graduated anode. Zooplankton samples were taken at a subset of fish sampling points in a stratified random manner. This strategy ensured that samples were taken across the range of macrophyte densities on each occasion in the different lakes. A sample (>5) of each principal zooplanktivorous fish species of any age <10 cm, was taken from each broad on each occasion for dietary analysis. The distribution of piscivorous pike was determined by radio tracking (see Perrow et al., 1996b), with locational fixes on every hour. PVI or vegetational composition of the littoral margin where applicable was recorded at each location.

Table 1. Biological and management characteristics of the study sites.

	Hoveton Little Broad	Pound End	Belaugh Broad	Cockshoot Broad	Cromes Broad	Upton Broad
Size (ha)	14.3	5.5	1.9	5.5	2.0	5.6
Mean depth (m)	1.7	1.1	0.9	1.0	1.0	1.2
Riverine connection	River Bure	River Bure via Hoveton Little	River Bure	None	None	None
Management history	Dredged 1990	Dredged 1990 Biomanipulated 1990–	Dredged 1987	Isolated and Dredged 1981 Biomanipulated 1989–	Dredged 1988 Isolated 1992	None
Macrophytes						
% occurrence	0	6	64	85	80	91
mean % PVI	–	37	52	38	66	23

Further details of methodologies throughout the study period are given in Phillips et al. (1996), Stansfield et al. (1997) and Perrow et al. (in press).

Results and discussion

Do grazing Cladocera need a refuge?

In most of the broads studied *Daphnia* spp. had an early summer peak of abundance followed by a rapid decline (Perrow et al., in press; Phillips et al., 1996; Stansfield et al., 1997). For the refuge effect to be meaningful, it must be established that this decline was a result of predation and not other factors, such as food limitation. There are a number of pieces of evidence which suggest this is the case. Firstly, such a decline corresponds closely to the timing of recruitment and recruitment strength of 0+ fish (e.g. Cryer et al., 1986; Townsend & Perrow, 1989; Perrow et al., 1994). Even a low (0.2 m^{-2}) density of fish is enough to cause large cladoceran populations to collapse (Figure 1). Indeed, *Daphnia* spp. may only persist through the summer months where fish recruitment is virtually absent (Perrow et al., 1994). The importance of 0+ recruitment appears to be a particular characteristic of the Broads, as in shallow lakes in The Netherlands, spring peaks of *Daphnia* spp. do not always occur due to predation from a high overwintering fish biomass (Lammens et al., 1992). Controlling the production of 0+ is thus the basis of biomanipulation in the broads (Perrow et al., 1997).

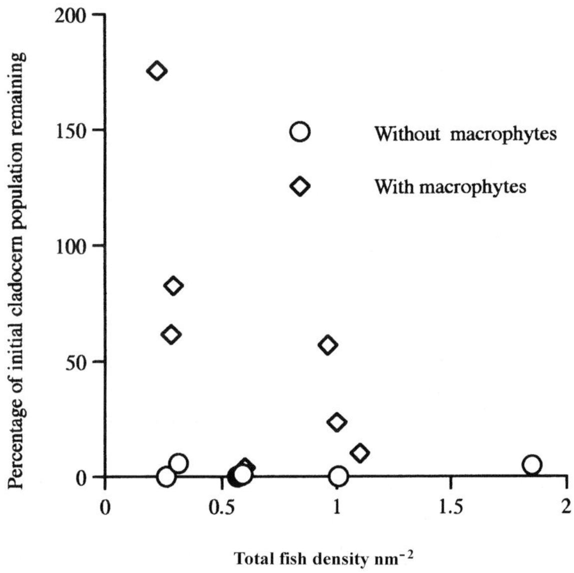

Figure 1. Proportional (%) change in grazing cladoceran density from spring (April/May) to late summer (August/September) in relation to the density (n m^{-2}) of all zooplanktivorous fish for a selection of broads with and without macrophytes. Data from the following sites were used: with macrophytes, Belaugh 1995, Cromes 1994, Cockshoot 1994 & 1995, Pound End 1994 & 1995, Upton 1995; without macrophytes, Alderfen 1994, Barton 1979, Hoveton Little Broad 1994, Ranworth 1991 & 1994, South Walsham 1990, Wroxham 1981 & 1991.

Secondly, the relatively high concentrations of nutrients in the Broads (Moss, 1983), means that the abundance of edible algae (expressed as chlorophyll '*a*') is never apparently limiting, perhaps apart from in

Upton Broad (Phillips et al., 1996). In lakes with macrophytes, *Daphnia* spp. is always quickly replaced by high densities of other species, which is not consistent with food limitation (Stansfield et al., 1997). It therefore seems reasonable to conclude that predation by 0+ fish is the principal cause of the decline in populations of *Daphnia* spp.

Do macrophytes act as refuges for Cladocera against fish predation?

Stansfield et al. (1997) found that the decline in *Daphnia* spp., even where 0+ fish recruited successfully (to 0.8 m^{-2}), tended to be delayed in the presence of macrophytes (Figure 1). The density of *Daphnia* spp. was positively associated with macrophyte cover in such circumstances (Phillips et al., 1996). This contrasts with the general impression of the pelagic nature of the group, with several authors suggesting that they typically avoid macrophytes (Dorgelo & Heykoop, 1985; Pennak, 1973). The recent experimental studies of Lauridsen & Lodge (1996) with *Daphnia magna* in the laboratory, provides an explanation for this apparent paradox. Here, *D. magna* preferred to stay in open water until exposed to the risk of predation from zooplanktivorous fish, whereupon it migrated to macrophytes. A trade-off between predation risk and repellency clearly exists. However the repellancy of macrophytes is still apparent as *Daphnia* spp. tend to aggregate at the edges of a macrophyte stand or in patches between macrophytes within the stand (Jeppesen, pers. comm.). Dense swarming of *Daphnia* spp. amongst littoral emergent macrophytes (Cryer & Townsend, 1988) suggests these forms may even be preferred to submerged macrophytes (Figure 2). It is thus feasible that, prior to recruitment of 0+ fish, *Daphnia* would be found in open water, but as they enter the diet of these fish and risk of predation increases, they are forced to take refuge in macrophytes. However, it is unlikely that *Daphnia* are able to feed efficiently in structured environments (such as amongst high densities of filamentous algae; Gliwicz, 1990) and, if confined to a small habitat, may deplete the available phytoplankton.

Diel horizontal migration (DHM) of *Daphnia* from stands of macrophytes to open water under the cover of darkness, initially suggested by Timms & Moss (1984) has been confirmed by several other studies (Davies, 1985; de Meester et al., 1993; Lauridsen & Buenk, 1996) and appears to be generally attributable to fish (Lehtovaara & Sarvala, 1984; Walls et al., 1990). With this strategy, it is puzzling that *Daphnia* virtually decline to extinction in the presence of a high density of zooplanktivorous fish (Schriver et al., 1995; Stansfield et al., 1997). The answer appears to lie in the fact that many common fish species, particularly 0+ cyprinids, are capable of feeding in darkness (Townsend & Risebrow 1982), which suggests the risk of predation, although reduced, is still present.

A zooplankton community shift from *Daphnia* to forms such as *Simocephalus vetulus*, *Ceriodaphnia* spp. and *Sida* spp. is typical in the presence of macrophytes (Figure 3). Of these species, *Simocephalus* spp., appears to be dependent to macrophytes (Lauridsen & Buenk, 1996). *Ceriodaphnia* spp. on the other hand, appear less dependent on macrophytes, and may be present in open water following the disappearance of the larger *Daphnia* spp. This suggests some competitive suppression by the latter (Vanni, 1986), even with diel horizontal migration (Lauridsen & Buenk, 1996). Both the differential habitat use by *Ceriodaphnia* spp. according to fish density, and the positive association of *Ceriodaphnia* and *Simocephalus* with increasing macrophyte cover, are consistent with the refuge effect (Stansfield et al., 1997). The ultimate consequence of the presence of macrophytes is the maintenance of large populations (typically >50 l^{-1}) of grazing Cladocera throughout the summer months (Stansfield et al., 1997).

Under what conditions does the refuge effect occur?

As fish density increases, the use of macrophytes as refuges by all groups of Cladocera (both pelagic and macrophyte-associated) increases. However, Schriver et al. (1995), working in experimental enclosures within Lake Stigsholm in Denmark, observed that the refuge effect may be nullified by a high fish density (2 m^{-2}), mainly of sticklebacks (*Gasterosteus aculeatus*). Recent evidence from the Broads supports this view, cladocerans declining to very low levels as the fish density increased to around 1 m^{-2} (Figure 1). The difference may be due to differences in fish community structure, the highly efficient zooplanktivore roach (*Rutilus rutilus*) being dominant in the Broads.

This may also account for differences in the macrophyte cover required to operate as an effective refuge, where fish density is below a critical threshold. Schriver et al. (1995) demonstrated that a PVI of only 15-20% was sufficient to allow cladocerans to persist, whereas in the Broads, >40% seemed to be required. A high density of macrophytes also appeared to pre-

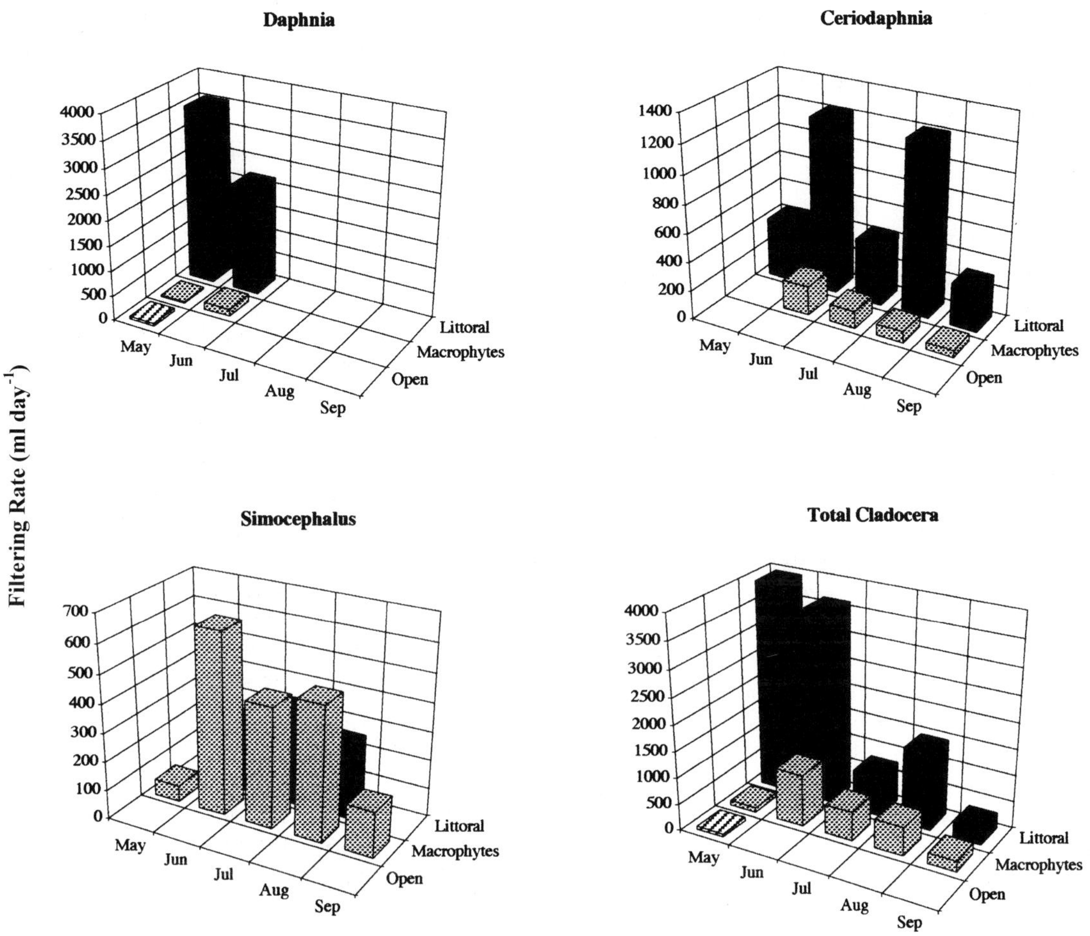

Figure 2. Filtering rates (ml day^{-1}) of different Cladocera in different habitats over the summer period (May–September 1994) in Cromes Broad.

clude the need for extensive coverage. For example in July 1995 at Belaugh, with a mean density of macrophytes of 42% where they occurred, a refuge effect was operating when the occurrence of macrophytes was patchy (30% of points fell upon bare sediment) (Perrow et al., in press).

Lauridsen et al. (1996) suggested that small dense macrophyte beds may be better refuges than larger beds of the same density for pelagic species such as *Daphnia.* This is because *Daphnia* tend to aggregate at the edge of the bed when seeking refuge and the density of potential competitors in the form of macrophyte-associated species, such as *Simocephalus* spp., increases with increasing bed size.

What is the mechanism behind the refuge effect?

Although some authors have suggested that small fish may associate with macrophytes as a result of enhanced foraging opportunity caused by the concentration of cladocerans in this habitat (Carpenter et al., 1987; Hall & Werner, 1977; Venugopal & Winfield, 1993). Engel (1988) suggested fish had difficulty in penetrating dense (>300g DW m^{-2}) macrophytes. More recent evidence suggests fish avoid dense macrophytes as a response to predation risk (Persson & Eklov, 1995; Savino & Stein, 1982; Werner et al., 1983). Stansfield et al. (1997) pointed out that even though fish may be present in macrophytes, if they are not feeding they do not present a threat to zooplankton. The absence of all fish or feeding fish results in a reduced predation pressure in dense macrophytes (Phillips et al., 1996), suggesting that the formation of predation (if not predator-free space) is the mechanism responsible for the refuge effect (Stansfield et al., 1997).

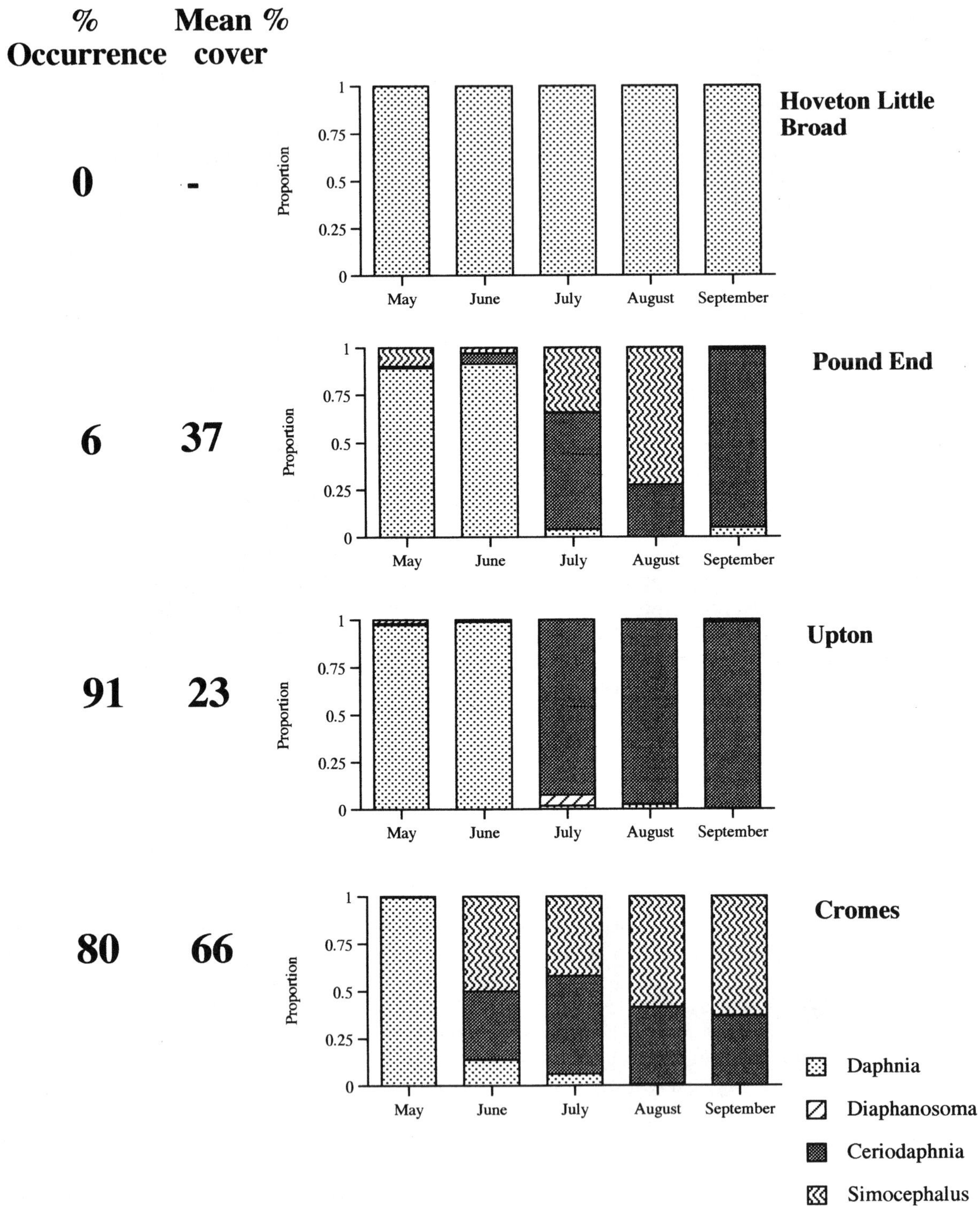

Figure 3. Changes in the relative proportion of different large-bodied grazing cladocerans over the summer season (May–September 1994) for four broads, varying with macrophyte coverage. The percent of macrophytes and mean percent cover where they occurred is shown.

It is clear that the composition of the fish community has great bearing on the consumption of zooplankton in different habitats. For example, from species in the mixed community of young roach (0.41 m^{-2}), rudd (*Scardinius erythrophthalmus*) (0.09 m^{-2}), bream (*Abramis brama*) (0.01 n m^{-2}), tench (*Tinca tinca*) (0.15 m^{-2}), perch (*Perca fluviatilis*) (0.15 m^{-2}) and ruffe (*Gymnocephalus cernuus*) (0.01 m^{-2}) in Cromes Broad, only perch and roach contributed significantly to the predation pressure on the zooplankton (Figure 4). Moreover, predation by roach occurred predominately in habitats with <50% PVI, whereas perch predation amongst submerged macrophytes was highest in the densest category (>75% PVI) (Figure 4). This reflects the foraging abilities of the species concerned. Whereas 0+ perch may show high feeding efficiency in structured environments, the efficiency of 0+ roach declines dramatically in such circumstances (Winfield, 1986). Such is the magnitude of this effect that submerged vegetation may serve as a competitive refuge for perch (Persson, 1993).

There is good evidence in the Broads, that as macrophyte coverage develops or persists, fish communities shift rapidly in favour of species such as perch, rudd and tench (Perrow, 1990; Perrow & Jowitt, in press; Perrow et al., 1994). As yet, however, there are no examples of recruitment of these latter species together exceeding 0.7 m^{-2}, even where their potential competitor, roach, was absent, which is

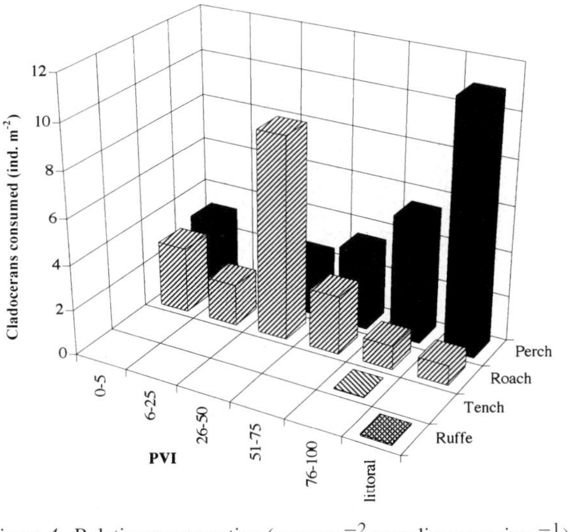

Figure 4. Relative consumption (mean m^{-2} sampling occasion $^{-1}$) of grazing cladocerans by different fish species in different habitats in Cromes Broad between June and August 1995.

likely to allow zooplankton populations to persist (Figure 1). It is possible that the predation pressure from fish communities in macrophyte-dominated systems is not of sufficient magnitude to prevent refuges being effective.

Several factors may operate to reduce the predation pressure of macrophyte-associated fish assemblages still further. First, perch are known to undergo ontogenetic shifts in response to the availability of zooplankton prey, which is often mediated by roach (Persson, 1986). Thus, where zooplankton populations are low, perch tend to take a higher proportion of macroinvertebrates (Persson, 1986) (Figure 5). Where macrophytes dominate, even if cladoceran numbers remain high after the decline in *Daphnia*, through the development of *Ceriodaphnia* or *Simocephalus* populations, it is unlikely that perch are efficient predators of these forms, on account of the small size of *Ceriodaphnia* and the low mobility of *Simocephalus*. Perch may then prefer to feed on non-grazing zooplankton such as copepods (Furnass, 1979), or switch to macroinvertebrates, which are typically at far greater density within macrophytes than in open water (Krull, 1970; de Nie, 1987). Therefore, the mechanism of macrophytes offering alternative prey also appears to be important in the maintenance of any refuge effect.

Often the proportion of piscivores within the fish community increases in macrophyte-dominated lakes (Meijer et al., 1994). In the Broads, this results from enhanced recruitment of pike (*Esox lucius*) populations (Perrow & Jowitt, in press; Phillips et al., 1996). Conceivably, the risk of predation from piscivores may influence the distribution and/or behaviour of zooplanktivorous fish which, in turn, may influence the distribution of zooplankton, perhaps reinforcing the refuge effect of macrophytes.

The experiments of Jacobsen et al. (1997) provide some support for this hypothesis. In an unreplicated experiment within enclosures, predation risk from large piscivorous perch appeared to confine 0+ zooplanktivorous perch and roach to macrophytes (see also Savino & Stein, 1982) and allow *Daphnia* to utilise the open water. Without piscivores, zooplanktivorous perch frequented the open water zone and zooplankton were forced to take refuge within macrophytes. Moreover, in experimental ponds, roach responded to the attack capabilities of the two predators used in a predictable manner, i.e. in the presence of pike, an ambush predator striking outwards from within stands of macrophytes (Bean & Winfield, 1995) roach took refuge in open water, and in the pres-

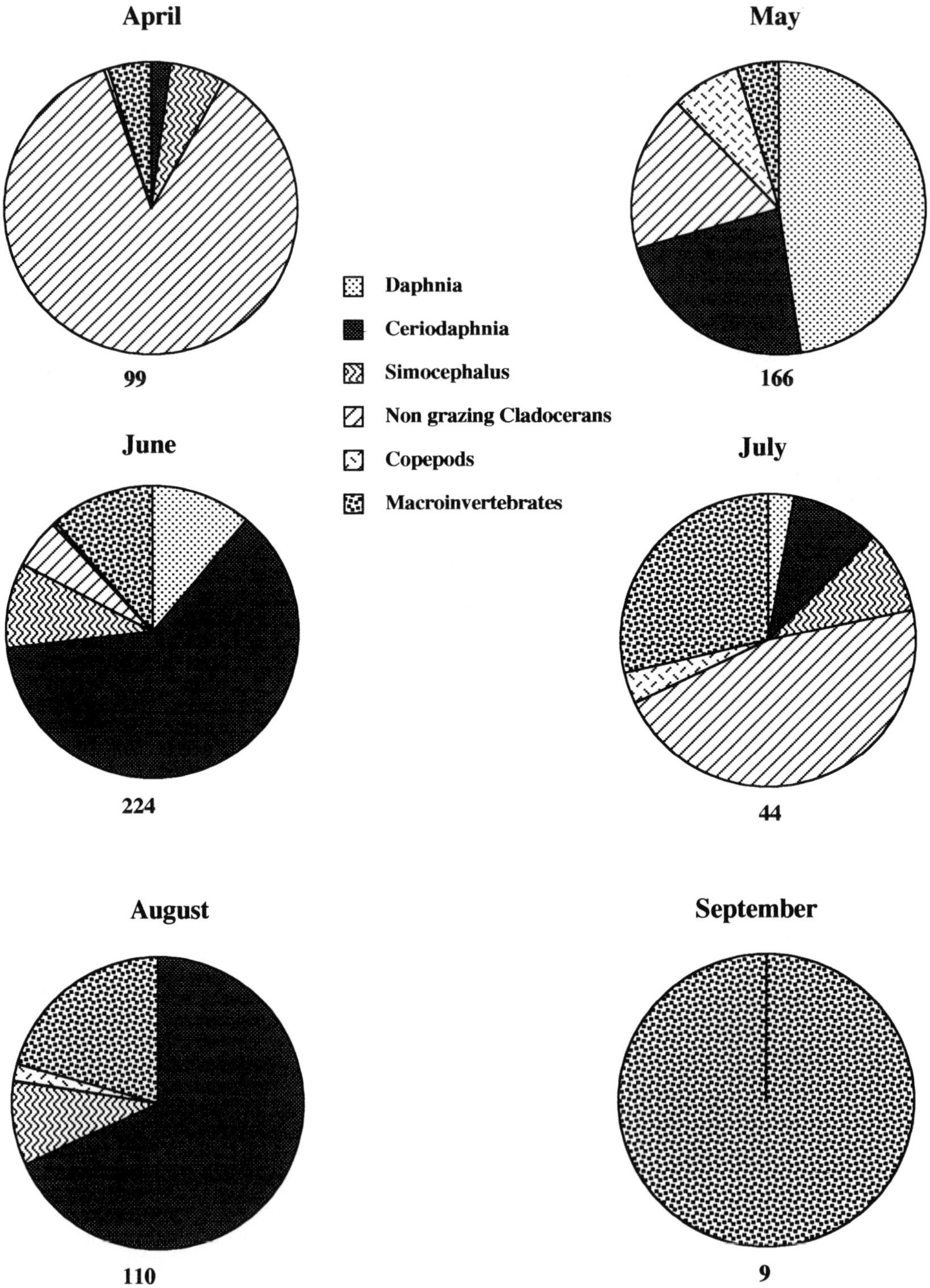

Figure 5. Seasonal change in dietary composition of 0+/1+perch (<110 mm) in Cromes Broad in 1995. The mean number of items per stomach ($n = 4$ to 11) is shown.

ence of perch, an open water pursuit predator (Eklöv, 1992) roach took refuge in macrophytes. However, in both cases no corresponding effect on the distribution of *Daphnia* spp. could be detected. Dietary analysis showed that the number of *Daphnia* spp. consumed by roach was very small in the experiments and it may be that the predation pressure on zooplankton was insufficient to instigate any response.

Whole-lake investigations on the distribution of zooplankton in relation to zooplanktivorous and piscivorous fish in four macrophyte-dominated broads provided limited evidence to suggest that the effects of predation risk may cascade from piscivores to zooplankton (Perrow et al., in press). At Cromes Broad, the distribution of pike was significantly positively related to the distribution of zooplankton over the whole season. In contrast, the opposite effect was observed in Upton Broad (Table 2). No effects were observed at the other two sites. Cromes Broad had a much higher zooplanktivorous fish density (0.82 m^{-2}) than Upton Broad (0.22 m^{-2}), as well as a higher density of pike (0.05 compared to 0.006 m^{-2}) and it may be that the relative densities of predator and prey are critical in determining the direction and strength of any relationship. Further experimental work is needed to clarify the causes and effects of this interaction.

Are refuging Cladocera able to control phytoplankton populations?

The value of any refuge effect hinges upon the cladoceran grazers present controlling phytoplankton populations and promoting clear water. Stansfield et al. (1997) illustrated that in broads with reasonable numbers of cladoceran zooplankton (>30 l^{-1}), chlorophyll *a* concentrations remained low (<25 μg l^{-1}). But whether this was attributable to grazing, particularly by species other than *Daphnia*, is not clear. Similar relationships between length and filtering capacity (Mourelatos & Lacroix, 1990) in species such as *Ceriodaphnia,* suggest that if they are at high enough density they may be able to graze out phytoplankton crops in the manner observed by Turner & Mittelbach (1992).

Calculation, or preferably direct measurement, of zooplankton community filtration rates should be accompanied by measurements of phytoplankton productivity to determine the exact impact of the zooplankton community filtration rate. In the absence of such measures, the studies of Thompson et al. (1982) and Garnier & Mourelatos (1991) suggest the threshold for clear water is in the region of a filtration rate of 300–400 ml l^{-1}day^{-1}. In the study of Stans-

Table 2. Parameters of statistically significant subset models from stepwise multiple regression to explain the distribution of large-bodied grazing cladocerans over the summer period (May–September 1995) at four broads. Significance values where $^{*}P < 0.05$, $^{**}P < 0.01$, $^{***}P < 0.001$ are shown.

Site	Variable	Model r^2	Coefficient	Std. coeff.	*P*
Belaugh	PVI	0.135	0.127	0.432	**
	pike >30 cm		0.338	0.244	NS
Cockshoot	–				*
Cromes	PVI	0.455	0.116	0.399	***
	time		−0.170	−0.290	*
	pike >30 cm		0.483	0.333	*
	0+ density				***
Upton	PVI	0.593	0.075	0.294	**
	time		0.106	0.241	*
	pike >30 cm		−0.512	−0.491	***
	0+ density		−1.955	−0.157	NS

Variables entered in the regression were time, PVI category, distribution of pike >30 cm (from radiotelemetry) and densities of pike <30 cm and 0+ and >0+ zooplanktivorous fish (from PAS by electrofishing). Fish density data was log +1 transformed, whereas data for large pike were arcsine transformed.

field et al. (1997) this rate was generally exceeded until August in Cromes Broad, firstly by *Daphnia* spp. in early season then by *Simocephalus* spp. and to a lesser extent *Ceriodaphnia* spp. populations (Figure 2).

It seems plausible that the filtration rate of refuge-using grazing Cladocera is enough to control phytoplankton crops (Jeppesen et al., 1998) perhaps until mid-summer at least. If the entire lake is to remain clear the distribution of grazing pressure must encompass open water areas as well as within stands of macrophytes. For this to occur, either phytoplankton must move into the plant bed or Cladocera must move out into the open water. Phytoplankton could be carried into the macrophyte bed by mass flow of water, through wind or wave action, though the resistance to water movement in the bed is likely to be high, thus limiting the feasibility of this explanation. The movement of zooplankton through DHM is the most tenable option (Jeppesen et al., 1998). However, it is clear that only *Daphnia* spp. and perhaps *Ceriodaphnia* spp. are capable of undertaking such movements.

The decline in grazing rate below a threshold necessary to maintain clear water, and the elimination of virtually all species of zooplankton at high fish densities ($>1 m^{-2}$) even in macrophyte-dominated sites (Figure 1) after mid-summer, do not necessarily imply that phytoplankton populations will increase. Provided the lake is heavily vegetated at this stage, other mechanisms such as limitation of nitrogen or phosphorus or both for algal growth may help maintain clear water (van Donk et al., 1993).

Concluding remarks

The establishment of a sparse cover of patchily-distributed macrophytes following an attempt to restore a lake is analogous to that the development of macrophytes in early season within a macrophyte-dominated lake. Assuming nutrients are not limiting, in both cases, grazing by *Daphnia* spp. will probably be essential to maintain clear water. As the season develops, *Daphnia* spp. populations will decline, often to extinction, even in the presence of a low fish stock ($0.2\ m^{-2}$). This decline may be retarded, but not eliminated, by the presence of macrophytes, the likelihood of a refuge effect increasing with increasing macrophyte cover. At this stage, *Daphnia* may undergo DHM, but as many fish species are capable of feeding in darkness, this strategy appears to be unsustainable. Within macrophytes, *Ceriodaphnia* spp. and *Simocephalus* spp. and perhaps other macrophyte-associated species take over the principal grazing role, which may be at a sufficient rate to maintain clear water. *Ceriodaphnia* spp. but not *Simocephalus* spp. may also undertake diel movements and provide some grazing pressure in open water areas. Fish community structure appears to have an important role determining whether the refuge effect is maintained. Open water feeders such as roach provide little threat in dense vegetation which probably accounts for the preference of grazers for dense macrophytes and the observations that a macrophyte PVI of 30–40% may provide an adequate refuge. Species such as perch on the other hand, being more efficient foragers within structured environments, are capable of depleting zooplankton even within dense macrophyte stands. This may be buffered to some extent however, by perch switching onto macrophyte-associated macroinvertebrate prey as they grow. In mixed predator communities, should fish density exceed a critical threshold (perhaps $1\ m^{-2}$), the refuge effect appears to collapse and all large zooplankton species are likely to be virtually eliminated. At this stage, or in fact as soon as the potential to maintain a grazing rate above that of phytoplankton productivity is exceeded, other mechanisms such as nutrient limitation may become important in maintaining clear water in well vegetated lakes.

Acknowledgements

This study was undertaken as part of the Broads Authority/Environment Agency, formerly National Rivers Authority (NRA) EC LIFE programme project (LIFE 92-3/UK/031 and NRA Project 475) on the Restoration of the Norfolk Broads: Development of Biomanipulation Techniques and Control of Phosphorus Release from Sediments (Project Managers Jane Madgwick and Dr Geoff Phillips). Further financial support was provided by the Soap and Detergents Industry Association. We thank the numerous landowners for access to sites and the staff of the Environment Agency and ECON for assistance with field sampling and laboratory analysis. The experience gained while one of us (MRP) worked with Erik Jeppesen and the lake group of NERI in Denmark also proved invaluable in structuring the work undertaken in this study.

References

Bean, C. W. & I.J. Winfield, 1995. Habitat use and activity patterns of roach (*Rutilus rutilus* (L.)), rudd (*Scardinius erythrophthalmus* (L.)), perch (*Perca fluviatilis* L.) and pike (*Esox lucius* L.) in the laboratory: the role of predation threat and structural complexity. Ecol. Freshwat. Fish 4: 37–46.

Broads Authority, 1994. The Broads Plan – No Easy Answers. Broads Authority Consultation Document, Broads Authority, Norwich, U.K.

Canfield, D. E. Jr., J. V. Shireman, D. E. Colle, W. T. Haller, C. E. Watkins & M. J. Maceina, 1984. Prediction of chlorophyll *a* concentrations in Florida lakes: importance of aquatic macrophytes. Canadian Journal of Fisheries & Aquatic Sciences 41: 497–501.

Carpenter, S. R., J. F. Kitchell, J. R. Hodgson, P. A. Cochran, J. J. Elser, M. M. Elser, D. M. Lodge, D. Kretchmer, X. He, & C. N. von Ende, 1987. Regulation of lake primary productivity by food web structure. Ecology 68: 1863–1876.

Carpenter, S. R. & D. M. Lodge, 1986. Effects of submersed macrophytes on ecosystem processes. Aquat. Bot. 26: 341–370.

Copp, G. H. & M. Peňáz, 1988. Ecology of fish spawning and nursery zones in the flood plain, using a new sampling approach. Hydrobiologia 169: 209–224.

Cryer, M. & C. R. Townsend, 1988. Spatial distribution of zooplankton in a shallow eutrophic lake, with a discussion of its relation to fish predation. J. Plankt. Res. 10: 487–501.

Cryer, M., G. Peirson & C. R. Townsend, 1986. Reciprocal interactions between roach, *Rutilus rutilus*, and zooplankton in a small lake: Prey dynamics and fish growth and recruitment. Limnol. Oceanog. 31: 1022–1038.

Davies, J., 1985. Evidence for a diurnal horizontal migration in *Daphnia hyalina lacustris* Sars. Hydrobiologia 120: 103–105.

de Meester L., S. Maas, K. Dierckens & H. J. Dumont, 1993. Habitat selection of patchiness in *Scapholeberis*: horizontal distribution and migration of *S. mucronata* in a small pond. J. Plankt. Res. 15: 1129–1139.

de Nie, A. W., 1987. The Decrease in Aquatic Vegetation in Europe and its Consequences for Fish Populations. EIFAC Occasional Paper 19. FAO, Rome: 88 pp.

Dorgelo, J. & M. Heykoop, 1985. Avoidance of macrophytes by *Daphnia longispina*. Verha. Int. Ver. Limnol. 22: 3369–3372.

Eklov, P., 1992. Group foraging versus solitary foraging efficiency in piscivorous predators: the perch *Perca fluviatilis*, and pike *Esox lucius*, patterns. Anim. Behav. 44: 313–326.

Engel, S., 1988. The role and interactions of submerged macrophytes in a shallow Wisconsin lake. J. Freshwat. Ecol. 4: 229–241.

Finlayson, C. M. (ed.), 1992. Integrated Management and Conservation of Wetlands in Agricultural and Forested Landscapes. Proceedings of the International Waterfowl and Wetlands Reserarch Bureau workshop, Trebon, Czechoslovakia, 25–31 March 1992. IWRB Special Publication 22: 104 pp.

Furnass, T. I., 1979. Laboratory experiments on prey selection by perch fry (*Perca fluviatilis*). Freshwat. Biol. 9: 33–43.

Garnier, J. & S. Mourelatos, 1991. Contribution of grazing in phytoplankton overall losses in a shallow French lake. Freshwat. Biol. 25: 515–523.

Gliwicz, Z. M., 1990. Why do cladocerans fail to control algal blooms? Hydrobiologia 200/201 (Dev. Hydrobiol. 61): 83–97.

Hall, D. J. & E. E. Werner, 1977. Seasonal distribution and abundance of fishes in the littoral zone of a Michigan lake. Trans. Am. Fish. Soc. 106: 545–555.

Irvine, K., B. Moss, & H. Balls, 1989. The loss of submerged plants with eutrophication II. Relationships between fish and zooplankton in a set of experimental ponds, and conclusions. Freshwat. Biol. 22: 89–107.

Jacobsen, L., M. R. Perrow, F. Landkildehus, M. Hjørne, T. L. Lauridsen, & S. Berg, 1997. Interactions between piscivores, zooplanktivores and zooplankton in submerged macrophytes: preliminary observations from enclosure and pond experiments. Hydrobiologia 342/343 (Dev. Hydrobiol. 119): 197–205.

Jeppesen, E., J. P. Jensen, P. Kristensen, M. Søndergaard, E. Mortensen, O. Sortkjær & K. Olrik, 1990. Fish manipulation as a lake restoration tool in shallow, eutrophic, temperate lakes 2: threshold levels, long-term stability and conclusions. Hydrobiologia 200/201 (Dev. Hydrobiol. 61): 219–227.

Jeppesen, E., T. L. Lauridsen, T. Kairesalo & M. R. Perrow, 1998. Impact of submerged macrophytes on fish-zooplankton interactions in lakes. In Jeppesen, E., Søndergaard, Mo., Søndergaard, Ma. & Christoffersen, K. (eds), The Structuring Role of Submerged Macrophytes in Lakes. Ecological Studies Series, Springer-Verlag, New York: 171–189.

Krull, J. H., 1970. Aquatic plant-macroinvertebrate associations and waterfowl. J. Wildlife Manage. 34: 707–718.

Lammens, E. H., Boesewinkel, P. J. de Bruyn, H. Hoogveld & E. Van Donk, 1992. P-load, phytoplankton, zooplankton and fish stock in Loosdrecht Lake and Tjeukemeer: confounding effect of predation and food availability. Hydrobiologia 233: 87–95.

Lauridsen, T. L. & I. Buenk, 1996. Diel changes in the horizontal distribution of zooplankton in the littoral zone of two shallow lowland lakes Arch. Hydrobiol. 137: 161–176.

Lauridsen, T. L. & D. M. Lodge, 1996. Avoidance by *Daphnia magna* of fish and macrophytes: chemical cues and predator-mediated use of macrophyte habitat. Limnol. Oceanogr. 22: 805–810.

Lauridsen, T. L., E. Jeppesen, & M. Søndergaard, 1994. Colonization and succession of submerged macrophytes in shallow Lake Væng during the first five years following fish manipulation. Hydrobiologia 275/276 (Dev. Hydrobiol. 94): 233–242.

Lauridsen, T. L., L. J. Pedersen, E. Jeppesen & M. Søndergaard, 1996. The importance of macrophyte bed size for composition and horizontal migration of cladocerans in a shallow lake. J. Plankt. Res. 18: 2283–2294.

Lehtovaara, A. & J. Sarvala, 1984. Seasonal dynamics of total biomass and species composition of zooplankton in the littoral of an oligotrophic lake. Verh. Int. Ver. Limnol. 22: 885–891.

Moss, B. 1983. The Norfolk Broadland: experiments in the restoration of a complex wetland. Biol. Rev. 58: 521–561.

Mourelatos, S. & G. Lacroix, 1990. In situ filtering rates of Cladocera: effect of body length, temperature, and food concentration. Limnol. Oceanogr. 35: 1101–1111.

Meijer, M.-L., E. Jeppesen, E. van Donk, B. Moss, M. Scheffer, E. H. Lammens, E. van Nes, J. A. van Berkum, G. L. de Jong, B. A. Faafeng, & J. P. Jensen, 1994. Long-term responses to fish-stock reduction in small shallow lakes: interpretation of five-year results of four biomanipulation cases in The Netherlands and Denmark. Hydrobiologia 275/276 (Dev. Hydrobiol. 94): 457–466.

National Research Council, 1992. Restoration of Aquatic Ecosystems – Science, Technology and Public Policy. National Academy Press, Washington D.C, U.S.A.: 552 pp.

Ozimek, T., R. D. Gulati & E. van Donk, 1990. Can macrophytes be useful in biomanipulation of lakes? The Lake Zwemlust example. Hydrobiologia 200/201 (Dev. Hydrobiol. 61): 399–407.

Pennak, R. W., 1973. Some evidence for aquatic macrophytes as repellents for a limnetic species of *Daphnia*. Int. Rev. ges. Hydrobiol. 58: 569–576.

Perrow, M. R., 1990. Biomanipulation in Broadland. In K. T. O'Grady, Butterworth, A. J. B., Spillett, P. B. & Domaniewski, J. C. J. (eds), Fisheries in the Year 2000. Proceedings of the 21st anniversary conference of the Institute of Fisheries Management: 335–337.

Perrow, M. R., B. Moss, & J. H. Stansfield, 1994. Trophic interactions in a shallow lake following a reduction in nutrient loading: a long term study. Hydrobiologia 275/276 (Dev. Hydrobiol. 94): 43–52.

Perrow, M. R., A. J. D. Jowitt, & L. Zambrano González, 1996a. Sampling fish communities in shallow lowland lakes: point-sample electrofishing versus electrofishing within stop-nets. Fisheries Manage. Ecol. 3: 303–313.

Perrow, M. R., A. J. D. Jowitt, & S. R. Johnson, 1996b. Factors affecting the habitat selection of tench (*Tinca tinca* (L.)) in a shallow eutrophic lake. J. Fish Biol. 48: 859–870.

Perrow, M. R., M.-L. Meijer, P. Dawidowicz & H. Coops, 1997. Biomanipulation in shallow lakes: state of the art. Hydrobiologia 342/343 (Dev. Hydrobiol. 119): 355–65.

Perrow, M. R. & A. J. D. Jowitt, in press. The influence of macrophytes on the structure and function of fish communities. Broads Authority/Environment Agency, EC LIFE project 92–3/UK/031.

Perrow, M. R., J. H. Stansfield, A. J. D. Jowitt & L. D. Tench, in press. Macrophytes as a refuge for grazing zooplankton from fish predation. Broads Authority/Environment Agency, EC LIFE project 92–3/UK/031.

Persson, L., 1986. Effects of reduced interspecific competition on resource utilization in perch (*Perca fluviatilis*). Ecology 67: 355–364.

Persson, L. 1993. Predator-mediated competition in prey refuges: the importance of habitat dependent prey resources. Oikos 68: 12–22.

Persson, L. & P. Eklov, 1995. Prey refuges affecting interactions between piscivorous perch and juvenile perch and roach. Ecology 76: 70–81.

Phillips, G. L., M. R. Perrow & J. H. Stansfield, 1996. Manipulating the fish-zooplankton interaction in shallow lakes: a tool for restoration. In S. P. R. Greenstreet & Tasker, M. L. (eds), Aquatic Predators and their Prey. Blackwell Scientific Publications Ltd., Oxford, England: 174–183.

Phillips, G., A. Bramwell, J. Pitt, J. Stansfield & M. Perrow, 1999. Practical application of 25 years' research into the management of shallow lakes. Hydrobiologia 395/396 (Dev. Hydrobiol. 136): 61–76.

Sas, H., 1989. Lake restoration by reduction of nutrient loading: expectations, experiences, extrapolations. Academia Verlag: 497 pp.

Savino, J. & R. A. Stein, 1982. Predator-prey interaction between largemouth bass and bluegills as influenced by simulated, submersed vegetation. Trans. am. Fish. Soc. 111: 255–266.

Scheffer, M., S. H. Hosper, M.-L. Meijer, B. Moss & E. Jeppesen, 1993. Alternative equilibria in shallow lakes. Trends Ecol. Evol. 8: 275–279.

Schriver, P., J. Bøgestrand, E. Jeppesen, & M. Søndergaard, 1995. Impact of submerged macrophytes on fish-zooplankton–phytoplankton interactions: large-scale enclosure experiments in a shallow lake. Freshwat. Biol. 33: 255–270.

Shapiro, J., 1990. Biomanipulation: the next phase-making it stable. Hydrobiologia 200/201 (Dev. Hydrobiol. 61): 13–27.

Stansfield, J. H., M. R. Perrow, L. D. Tench, A. J. Jowitt, & A. A. Taylor, 1997. Submerged macrophytes as refuges for grazing Cladocera against fish predation: observations on seasonal changes in relation to macrophyte cover and predation pressure. Hydrobiologia 342/343 (Dev. Hydrobiol. 119): 229–240.

Thompson, J. M., A. J. Ferguson, & C. S. Reynolds, 1982. Natural filtration rates of zooplankton in a closed system: the derivation of a community grazing index. J. Plankt. Res. 4: 545–560.

Timms, R. M. & B. Moss, 1984. Prevention of growth of potentially dense phytoplankton populations by zooplankton grazing, in the presence of zooplanktivorous fish, in a shallow wetland system. Limnol. Oceanogr. 29: 472–486.

Townsend, C. R. & M. R. Perrow, 1989. Eutrophication may produce population cycles in roach, *Rutilus rutilus*, by two contrasting mechanisms. J. Fish Biol. 34: 161–164.

Townsend, C. R. & A. J. Risebrow, 1982. The influence of light level on the functional response of a zooplanktivorous fish. Oecologia 53: 293–295.

Turner, A. M. & G. G. Mittelbach, 1992. Effects of grazer community composition and fish on algal dynamics. Can. J. Fish. aquat. Sci. 49: 1908–1915.

van Donk, E. & R. D. Gulati, 1995. Transition of a lake to turbid state six years after biomanipulation: mechanisms and pathways. Wat. Sci. Technol. 32: 197–206.

van Donk, E., R. D. Gulati, A. Iedema, & J. T. Meulemans, 1993. Macrophyte-related shifts in the nitrogen and phosphorus contents of the different trophic levels in a biomanipulated shallow lake. Hydrobiologia 251: 19–26.

Vanni, M. J., 1986. Competition in zooplankton communities: suppression of small species by *Daphnia pulex*. Limnol. Oceanogr. 31: 1039–1056.

Venugopal, M. N. & I. J. Winfield, 1993. The distribution of juvenile fishes in a hypereutrophic pond: Can macrophytes potentially offer a refuge for zooplankton? J. Freshwat. Ecol. 8: 389–396.

Walls, M., M. Rajasilta, J. Sarvala, & J. Salo, 1990. Diel changes in horizontal microdistribution of littoral Cladocera. Limnologica 20: 253–258.

Werner, E. E., J. F. Gilliam, D. J. Hall, & G. G. Mittelbach, 1983. An experimental test of the effects of predation risk on habitat use in fish. Ecology 64: 1540–1548.

Winfield, I. J., 1986. The influence of simulated aquatic macrophytes on the zooplankton consumption rate of juvenile roach, *Rutilus rutilus*, rudd, *Scardinius erythrophthalmus*, and perch, *Perca fluviatilis*. J. Fish Biol. 29: 37–48.

Wium-Anderson, S., U. Anthoni, C. Christophersen, & G. Houen, 1982. Allelopathic effects on phytoplankton by substances isolated from aquatic macrophytes (Charales). Oikos 39: 187–190.

Hydrobiologia **395/396:** 211–226, 1999.
D.M. Harper, B. Brierley, A.J.D. Ferguson & G. Phillips (eds), The Ecological Bases for Lake and Reservoir Management.

Use of remote sensing and geographical information systems in developing lake management strategies

Serwan M. J. Baban
GRRU Group, Geography, School of Natural and Environmental Sciences, Coventry University, Coventry, CV1 5FB, U.K.

Key words: lake management, remote sensing, GIS

Abstract

This work examines the holistic approach to lake management in terms of relevant in-lake and catchment area parameters. The basic concepts of remote sensing and GIS technologies and the possible use of both in lake studies are explored. The role of remote sensing as a potential data source and of GIS as an analytical tool in developing a management system based on this approach have been investigated through their application to management issues in the Norfolk Broads, England.

Introduction

Hutchinson (1957), some 40 years ago, indicated clearly that lakes are dynamic in nature and experience gradual changes over time. Since then, the slow gradual changes, in many lakes around the world, has been replaced by a more rapid change and swift degradation, mainly due to the additions of plant nutrients, organic matter or sediment. These additions usually produce increased algal and rooted plant biomass, increased turbidity and, typically, decreased lake volumes.

In the U.S.A., about 30% of all lakes and reservoirs assessed during 1988 were either eutrophic or hypereutrophic, whilst 23% were mesotrophic and 14% oligotrophic. The trophic status of 30% of the lakes was unknown (USEPA, 1990). Accelerated eutrophication is also widespread throughout Europe (Gulati et al., 1990), as well as other continents where scientific surveys have reported (e.g. South Africa, Australasia). Minimising the damage requires a clear understanding of the relationships between the loadings and the lake's response.

Elements in lake management strategies

The elements of lake management are well known and it is now widely realised that both control of catchment sources of nutrient (usually phosphorus) as well as in-lake management are necessary (Cooke et al., 1993; Phillips et al., 1999).

Lake management is a relatively new science. Thus, there is still much uncertainty in estimating the cost-effectiveness of some techniques. Remote sensing and Geographical Information Systems (GIS) can be used to provide a rapid or a large-scale understanding of lake change and in developing lake management strategies. The combination of remote sensing as an information source and GIS as an analytical tool enables managers to capture and update all the relevant information (e.g. water quality parameters), plan, simulate/implement, compare, visualise and evaluate the outcome resulting from simulating various management scenarios.

Basic concepts in remote sensing

The main objective of remote sensing, in terms of lake studies, is to record images of lakes and their catchment area by using electromagnetic radiation energy sensors mounted on satellites or aircraft (Figure 1).

The wavelengths that are of most interest to lake studies are the visible (0.4–0.7 μm), where in pure water the majority is transmitted, the near infra-red (0.7–2.0 μm), where the majority is absorbed and thermal infra-red (3–50 μm) (Curran, 1985) (Table 1).

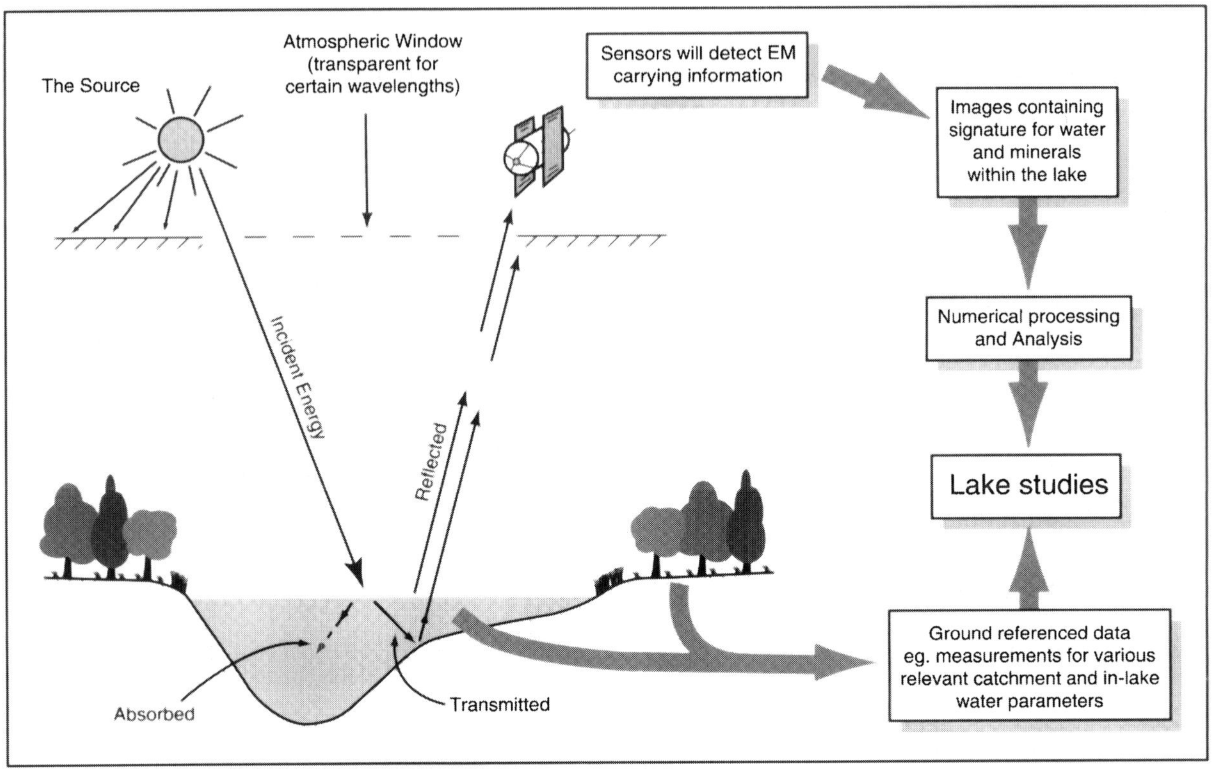

Figure 1. Schematic representation of the collection and use of remotely sensed data.

Table 1. The wavebands recorded by the Thematic Mapper sensor (after Curran, 1985)

Band number	Band name	Band width	Characteristics
TM1	Blue/green	0.45–0.52	Good water penetration, strong vegetation absorbance
TM2	Green	0.52–0.6	Strong vegetation reflectance
TM3	Red	0.63–0.69	Very strong vegetation absorbance
TM4	Near infrared	0.76–0.90	High land/water contrasts, very strong vegetation reflectance
TM5	Near middle infrared	1.55–1.75	Very moisture sensitive
TM6	Thermal infrared	10.4–12.5	Very sensitive to soil moisture and vegetation
TM7	Middle infrared	2.08–2.35	Good geological discrimination

The factors that affect the spatial variability in reflectance of a lake are usually determined by its physical configuration, water quality and climate. The three most important factors are: water depth, suspended materials (i.e. non-organic sediments, tannin and chlorophyll) and the surface roughness (Lillesand & Kiefer, 1994).

The interaction between electromagnetic energy, the water and the materials contained within it, results in various processes which depend on the wavelength and the atomic structure of water and its contents. The energy balance relationship is expressed thus:

$$E_I(\lambda) = E_R(\lambda) + E_A(\lambda) + E_T(\lambda),$$

where the incident energy (E_I) is a function of the reflected (E_R), absorbed (E_A) and transmitted (E_T). In this equation all the energy components are a function of wavelength, which has two important implications for lake studies. Firstly, the proportions of energy reflected, absorbed and transmitted will vary for different lakes, depending on their environmental and ecological conditions. Secondly, for a particular characteristic of a lake (i.e. water depth), the proportion of reflected, absorbed, and transmitted energy will vary at different wavelengths. Thus two features (i.e. shallow and deep water) may be indistinguishable in one spectral band and be very different in another spectral band. Thus each lake and its catchment may provide characteristic spectral signatures for analysis.

The use of remote sensing in lake management is based on the fact that the consequences of eutrophication and an increase in productivity will be associated with a change in the optical properties of the water mass. Increases in chlorophyll '*a*' are associated with a decrease in the relative amount of energy in the blue wavelength (0.45–0.52 μm) and increases in the green wavelength (0.52–0.60 μm) (Clarke et al., 1970). Increases in suspended solids are associated with an increase in the reflected energy, and the peak of reflectance will move towards the longer wavelengths (Lillesand & Kiefer, 1994). These changes are measurable by remote sensing techniques (Baban, 1996).

Most Earth observation from space is carried out by satellite-mounted instruments which have a number of sensors operating at different wavelengths. The popular instruments such as the Landsat Thematic Mapper (TM), System pour l'Observation de la Terre (SPOT), and the Advanced Very High Resolution Radiometer (AVHRR), have suites of sensors operating in most wavelengths relevant to lake studies (the visible and infrared regions). These should provide the staple diet of limnologists, hydrologists and ecologists, since their spectral coverage corresponds to regions of high energy output from the Sun and Earth, and suitable atmospheric windows, where the electromagnetic energy receives least distortions. More recently, Synthetic Aperture Radar (SAR) imaging devices have come to the fore, because even the visible and infrared atmospheric windows are shuttered when cloud is present. The microwave wavelengths of radar beamed down from satellite antennae readily penetrate cloud, to bounce off the lake surface and return to the spacecraft. These echoes are used to construct images containing information about morphology and composition.

The following concepts are important in selecting satellite imagery (Barrett & Curtis, 1992):

(1) Spatial resolution: this is the smallest feature size that can be studied in an image. This translates into a measure of the smallest ground objects that can be distinguished as a separate entity, usually as a single pixel on the imagery.

(2) Spectral resolution: the sensor capability to record spectral information at a particular range of wavelengths. Some sensors are broad-band and capable of recording energy across a range of wavelengths. Others are finely adjusted perhaps to the discrete reflectance frequency of a particular pollutant in a lake.

(3) Radiometric resolution. The amount of electromagnetic energy falling on a satellite sensor varies more or less continuously but the actual range of energies that can be recorded, and the precision of the measurement, depends on the operating range and radiometric resolution of the instrument.

(4) Temporal resolution. Satellites are placed in regular orbits so that they can regularly revisit a given location on the Earth. The time between visits depends on the angular field of view of the sensor, the latitude and longitude of the area of interest, and the characteristics of the spacecraft orbit.

Sensor design is largely concerned with reaching a compromise between these various kinds of resolution, since an increase in performance in one respect generally means a deterioration in another. The balance is met by optimising the sensor for the purposes for which it is expected to be used. The limnologist will need to consider the relative importance of spectral, radiometric, spatial and temporal requirements for a given objective, as well as the budget and economic viability, before deciding on which imagery to obtain.

Sensors produce a digital image containing the signatures of the water and its contents in the lake as well as the relevant attributes in the catchment area. This raw image needs to be adjusted and numerically processed through number of stages (e.g. correcting the image for any distortion and degradation, increasing the apparent distinction between the features to optimise visual interpretation, and to maximise the contrast between features of interest) producing an optimised image for use by a limnologist. Finally, the image needs to be calibrated with a number of measured parameters on the ground, the 'ground truth' samples.

Remote sensing and lake management

Remotely sensed data can never replace conventional hydrological observation networks in lakes, but they can provide essential information for developing lake management strategies. This is mainly due to:

(1) the provision of information that resembles reality as closely as possible (high resolution area-based synoptic data collected on a regular basis instead of conventional point measurements);

(2) the ability to deliver composite measurements (radiation/ reflectance), which integrates several relevant ecological characteristics of lakes;

(3) the ability to provide information concerning land categorisation and classification as well as some geomorphic elements required for lake management;

(4) the digital format of remotely sensed data makes it easy to retrieve and analyse large amounts of information at low cost in a short period of time.

Remote sensing has successfully extracted and provided relevant information on catchment characteristics, e.g. mapping the extent and the change in agricultural land over a particular period (Pattie, 1993) as well as providing the necessary in-lake water quality data for management including; chlorophyll '*a*', total phosphorous, Secchi disk depth, suspended solids, salinity and temperature (Baban, 1993a, 1994; Smith & Addington, 1978).

Water circulation patterns can be detected through a study of the distribution of suspended sediments, which can act as natural tracers to indicate flow directions. Recent examples have used both reflective and thermal bands to map the spatial and temporal distribution of sediments and to deduce water circulation patterns in lakes from these maps (Baban, 1993a). Sediment distribution patterns deduced from Landsat imagery were also found to be positively correlated with sedimentation patterns in lakes by Smith et al. (1983).

Various mathematical algorithms have been used to estimate water depth in lakes using information from the 0.45–0.70 μm region. Such algorithms have been developed, tested and used to produce bathymetric charts for lakes (Baban, 1993b; Hathout, 1985). Therefore, remote sensing can also provide information about the shape of the lake basin. Shape has a major effect on lake productivity, types of organisms, water chemistry and upon the choices available to manage and restore it (Cooke et al., 1993). The thermal patterns in many lakes have been studied using remotely sensed information from the thermal infrared band (Baban, 1993c). Information relating temperature patterns to lake depth are essential.

Incorporation of Geographic Information Systems

In terms of lake studies, GIS can be defined as a powerful set of tools for capturing, storing, checking, manipulating, combining, analysing and, displaying data which are spatially referenced to lakes and to their catchment areas (Burrough, 1986; DeMers, 1997). Although different GIS software may vary in capabilities, most contain the following components (Martin, 1991):

(1) Data collection and input. Operations concerned with receiving data into the system, including manual digitizing, scanning, keyboard entry of attribute information, and online retrieval from other database systems.

(2) Data storage and retrieval mechanisms which organize the spatial data into a form that permits it to be quickly retrieved by the user for subsequent analysis, as well as allowing for rapid and accurate updates and corrections to be made to the spatial database.

(3) Data manipulation and analysis represents a group of techniques which converts data through user-defined aggregation rules, or produces estimates of parameters and constraints for various space-time optimization or simulation models.

(4) Data reporting, which cover a range of operations concerned with the display of all or part of the original database, as well as manipulated data, and the output from spatial models in tabular or map form.

Data acquisition represents an important part of a water resource based GIS project. It is labour-intensive and can account for up to 80% of the total cost of a GIS. Various data acquisition options for hydrologic oriented GIS are presented by Van Blargan & Ragan (1991) including, manual digitizing, commercial digitizing, satellite products, and automated scanning. Data storage in GIS is established by using either raster or vector data structures. A raster structure describes the geographical space in terms of a two-dimensional grid of rows and columns, whilst a vector system describes the space in terms of points, lines and areas. A raster-based GIS has advantages over a vector-based GIS in lake studies because virtually all types of relevant data, including remotely sensed data and scanned maps, are represented in raster form. However which format to use depends on two aspects of the specific application; first, the nature of the phenomenon the data represents and secondly, the processes that need to be performed on the data in order to realise all the necessary objectives.

Applications of GIS to lake management can be classified into two main areas, inventory studies and modelling.

Inventory studies

Inventory is the collection of base-line data and, its subsequent storage in a GIS for monitoring and management purposes. For example, prior to conducting an analysis of data, the process requires a number of lengthy and time consuming preparations, including collecting data on, lake topography, water quality, sources of pollution, climatic, etc. Followed by compilation, storage, retrieval and manipulation. GIS has evolved as a highly sophisticated database management system to put together and store the voluminous spatial data typically required in hydrologic modelling (Bhaskar et al., 1992; Vieux et al., 1988).

Modelling

The ability of a GIS to characterize and model the spatial variations in hydrological processes makes it an effective aid for managing the use of land within a drainage basin (Stuart & Stocks, 1993). Modelling capabilities of GIS can also be used to simulate various lake processes and to predict the outcome of development actions, natural hazards or environmental change within the contributing basin. Sharma & Anjaneyulu (1993), for example, have successfully integrated the use of GIS geo-referenced overlays with satellite imagery to study the degradation in water quality in Vembanad lake, India. The lake is endangered with two problems. First, it is becoming brackish due to an ingress of salt water from the Lakshadweep sea during high tide whenever the Tannir Mukkom barrage is opened for navigation. Secondly, it is becoming silted due to sediments contributed by streams from the south and east. The study used GIS to model the quality of water front movements during January, March and May.

GIS offers an effective spatial data handling tool that can enhance various hydrological lake models through interfacing. For example, the lake eutrophication model, PROTEC (Reynolds, 1999) models the growth rates of a number of algal types according to the nutrient levels and light climate within the lake. Input requirements to this model are extensive, requiring daily flow rates and nutrient concentrations of each major inflow to the lake. Wright (1994) has investigated the possibility of using a GIS system to estimate this set of inputs, with the aim of reducing field sampling efforts required to run the model.

The use of a model interactively within a GIS has a number of advantages such as; a short running time and an almost immediate turnaround of results. Smith et al. (1992) used GIS to calibrate and verify a continuous simulation model as well as an event-driven model. They found it beneficial in enhancing the modelling efforts due to increased accuracy, and minimization of human errors and time. Penning-Rowsell (1996) created and tested a spatially and temporally distributed deterministic model using a GIS. This model deals with phosphorus sources and loads from both non-point sources in the surrounding catchment and from sewage treatment works to streams flowing into lake Bassenthwaite and predicted the daily total phosphorous loads entering this lake.

A holistic approach to lake management

The large number of lakes suffering from eutrophication, often clustered in lake districts, makes conventional management difficult and yet favours the development of a holistic approach using remote sensing and GIS. Landsat imagery has been available for over 20 years in a relaible format, and is widely used. Smith & Blackwell (1980) developed an information system by integrating digital landsat data, digital terrain data, conventional maps and ground truth data for Lake Tahoe and its catchment. The system allowed them to cross correlate the remotely sensed information and topographic data with a variety of environmental data relating to such parameters as surface runoff, drainage basin area, and terrain configuration. Parameters were evaluated and compared for each drainage basin defined by the regional planning agency.

The holistic approach will allow managers to observe and model, for instance, the effects of a particular ecological scenario on a large number of small lakes or the full spatial extent of large lakes concurrently and over appropriate periods of time. This capability will facilitate experiment with various management scenarios through the GIS simulation facilities. It will also provide data for areas which have no ground measurements based on the interpolation of area-based information from a number of sampled lakes/locations, subject to their size and the spatial resolution of the imagery.

The Norfok Broads

The Norfolk Broads, a series of about 40 shallow lakes in eastern England (Phillips et al., 1999) (Figure 2) are taken as an example of these procedures.

The majority of the Broads are connected to a river by narrow dikes, while some are isolated. This feature, and the low flow rate of rivers, have been largely responsible for the rapid sedimentation in the Broads. As a result water depth ranges from 21 cm at Surlingham to 366 cm at Fritton Decoy (Mason & Bryant, 1975).

The water in the Broads is well mixed with little possibility of thermocline formation (Moss & Timms, 1982). Most of the Broads are no longer clear water and are without submerged plants, but frequently have a growth of unwanted and high concentrations of algae. The Broads in addition to low river flows suffers from depleted groundwater (water resources), the threat of increased saltwater incursion and flooding. The Broads Authority (1997) sets out an integrated management plan based on the central issues of water resources, water quality and flood protection. Clearly, the plan recognises that the quantity and quality of water in the Broads are influenced by water management in the Broads as well as landuse practices within their catchment areas. In recent years a number of the broads damaged by eutrophication have been restored using techniques for nutrient control, ecological engineering using fish manipulation, plant and fish re-establishment (Moss et al., 1996; Phillips et al., 1999).

The Landsat Thematic Mapper (TM) is a scanner carried on Landsat 4 & 5 satellites. It records 256 radiance levels in seven wavebands, operating in the range from blue to thermal infra-red (Table 1). The spatial resolution of all bands is 30 m apart from the thermal band which is 120 m. A Landsat TM imagery was used to estimate chlorophyll *a*, total phosphorus, suspended solids, salinity and temperature (Tables 2–4; Figures 3–6) in the Norfolk Broads using algorithms designed for water quality output from TM data (Baban, 1993, 1994, 1996). It was not possible to locate a control set of water quality data corresponding to the remotely sensed data as the National Rivers Authority did not then have a comprehensive sampling programme. Equivalent field data available for five Broads (Table 5) show general correspondence, although as data were not concurrently collected statistical comparison cannot be made.

Inputting the estimated data into the Carlson Trophic State Index (TSI) (Table 6) allows the Broads to be trophically ranked (Table 7, Figure 7).

The available Landsat TM images for the Norfolk Broads region (path 210, row 23) are listed in Table 8. As the availability of these images is hindered by cloud cover, calculating the amount and seasonal variation in susnshine hours is essential. The amount and seasonal variation in sunshine hours (a direct indication of cloud cover), taken from the Weather Log, Royal Meteorological Society, together with hours of daylight at latitude 50° N (Pearce & Smith, 1984), were used to calculate percentage possible sunshine (PS) for each month (Table 9). The outcome indicates that the lowest percentage of sunshine is during January (19.7%), while the highest is in May (44.6%). The Broads during the months from April to September when biological water quality information is most critical have a reasonable chance of obtaining regular and useful coverage.

A management information system could thus be built up based upon the concepts described above within a GIS framework (Figure 8) (Ragan & Fellows, 1980) and maintained up-to-date through an appropriate combination of satellite images and concurrent 'ground truthing'.

The spatial, thematic and temporal characteristics of information provided by satellite imagery can provide the best spatial and temporal coverage of lakes and their catchment area. The management, visualisation, modelling and simulation capabilities of the GIS make a powerful and unique medium to take full advantages of the richness of digital remotely sensed information in terms of developing, testing and simulating and examining the consequences of implementing of any individual management strategy on lakes and their catchment areas.

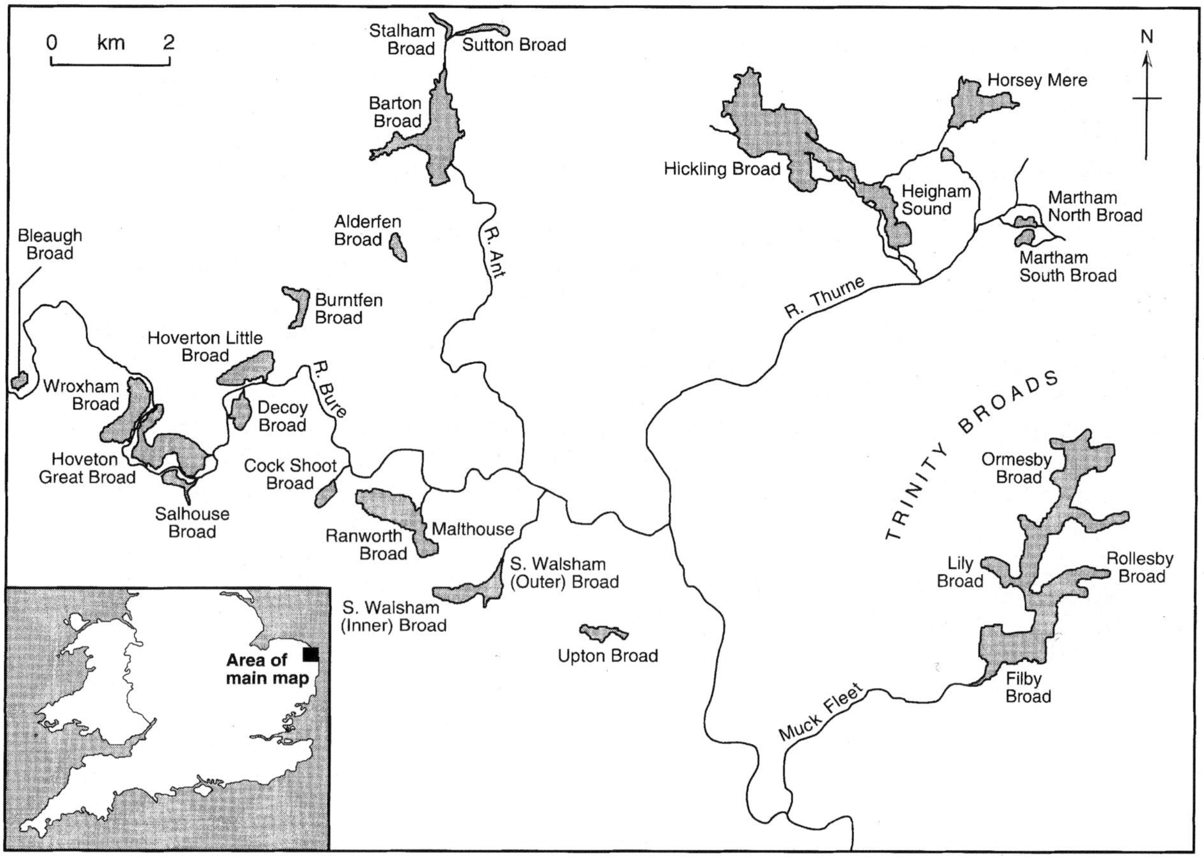

Figure 2. The Norfolk Broads.

Table 2. TM bands 1–4 and their combinations correlated with ground referenced data represented by the correlation coefficient (r) and the coefficient of determination (r^2). All these relationships are significant at the 99% confidence level apart from the ones which are marked by a (*) where they are only significant at the 95% level or less (after Baban, 1993a).

	Chlorophyll-*a* (μg/l)		Total phosphorus (μg/l)		Secchi depth (μg/l)		Suspended solids (μg/l)		Salinity (μg/l)	
Variables	r	r^2	r	r^2	r	r^2	r	R^2	r	r^2
1) TM3	0.85	0.72	0.75	0.56	0.76	0.58	0.77	0.59	0.75	0.56
2) TM2–TM3	−0.24	– *	−0.17	– *	0.44	0.19 *	−0.33	– *	0.07	– *
3) TM2/TM3	−0.52	– *	−0.17	– *	0.44	0.19 *	−0.55	– *	−0.10	– *
4) TM1/TM2	−0.73	0.53 *	−0.69	0.48 *	0.48	0.23 *	−0.33	– *	−0.38	– *
5) TM2	0.85	0.72	0.78	0.61	−0.58	0.34 *	0.67	0.45 *	0.51	– *
6) TM3/TM1	0.85	0.72	0.77	0.59	−0.8	0.64	0.64	0.41 *	0.76	0.58
7) (TM2+TM3)/2	0.85	0.72	0.77	0.59	−0.73	0.53 *	0.74	0.55	0.65	0.42 *
8) $(TM2)^2$	0.85	0.72	0.73	0.53 *	0.67	0.45 *	0.52	0.27 *		
9) $(TM3)^2$	0.85	0.72	0.73	0.53 *	−0.81	0.66	0.64	0.41 *	0.82	0.67
10) TM1	0.76	0.58	0.68	0.46 *	−0.83	0.69	0.88	0.77	0.66	0.43 *
11) (TM1+TM2)/2	0.83	0.69	0.75	0.56	0.82	0.67	0.84	0.71	0.62	0.38 *
12) (TM1+TM3)/2	0.81	0.66	0.72	0.52 *	−0.83	0.69	0.85	0.72	0.71	0.50 *
13) TM4	0.56	– *	0.43	– *	0.01	– *	0.63	0.40 *	0.51	0.26 *

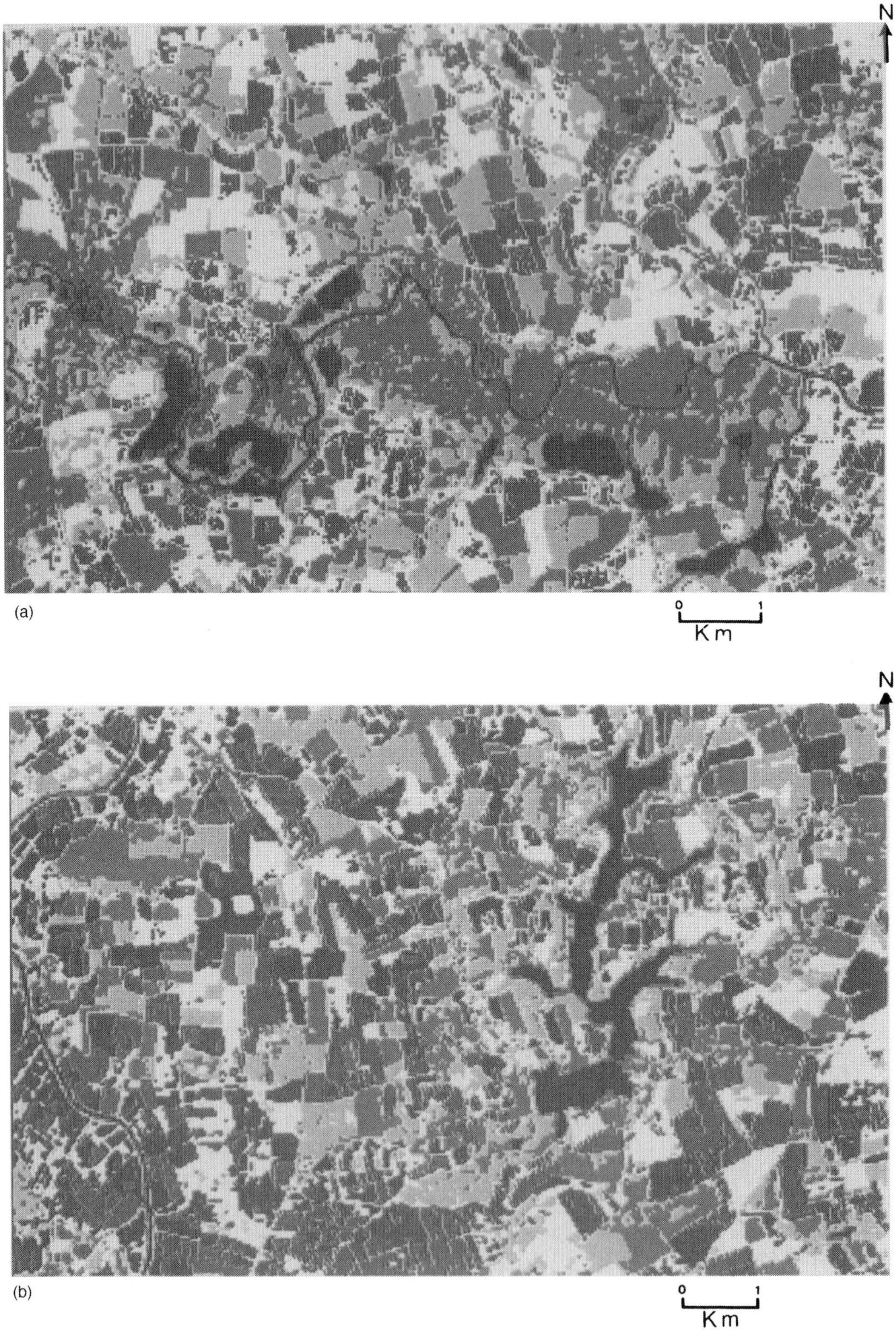

Figure 3. Satellite images of (a) the River Bure Broads and (b) the Trinity Broads.

Table 3. Predicted water quality parameters for the River Bure Broads and its surrounding Broads for June 1979 and June 1980 (after Baban, 1993a).

Broad	Pred. chlor-*a* (μg/l)	Pred. total P. (μg/l)	Pred. Secchi depth (m)	Pred. S. solids (μg/l)	Pred. salinty (μg/l)	Pred. temp. (°C)
Malthouse	117.4	271.3	0.4	38.7	94.0	17.55
Ranworth	129.4	296.1	0.6	25.5	89.7	17.90
Cockshoot	47.1	230.1	0.7	11.1	64.6	18.07
Decoy	40.3	179.9	0.6	17.7	64.6	17.20
Salhouse	40.2	190.7	0.6	19.2	68.9	17.65
Hoveton Gt.	37.5	166.1	0.6	23.2	72.2	16.85
Burntfen	69.4	212.5	0.6	26.3	75.5	16.53
Belaugh	61.5	173.5	0.5	31.0	79.3	15.1
Hoveton Lt.	96.3	215.1	0.5	31.6	87.4	17.2
S. Walsham (outer)	69.5	199.6	0.5	29.3	78.7	17.2
S. Walsham (inner)	51.1	165.5	0.5	32.1	75.5	17.55

Table 4. Predicted water quality parameters for the Trinity Broads, River Thurne and River Ant Broads for June 1979 and June 1980 (after Baban, 1993a).

Broad	Pred. chlor-*a* (μg/l)		Pred. total P (μg/l)		Pred. S. solid (μg/l)		Pred. Salinity (μg/l)		Pred. Secchi depth (m)		Pred. temp. (C°)	
	Mean	S.D.	Mean	S.D.	Mean	S.D.	Mean	S.D.	Mean	S.D.	Mean	S.D.
Ormesby	56.7	8.3	101.1	21.4	47.6	4.4	89.3	0.7	0.4	0.1	16.27	0.2
Rollesby	58.2	9.2	122.7	19.9	42.6	2.8	86.4	3.3	0.4	0.1	16.38	0.2
Filby	110.4	9.5	182.1	20.6	51.2	1.3	103.1	2.7	0.3	0.1	16.62	0.4
Lily	53.7	–	168.4	–	32.5	–	76.5	–	0.5	–	16.32	0.2
Hickling	222.8	29.6	375.4	17.1	45.7	10.9	124.8	15.1	0.4	–	16.32	0.2
Horsey Mere	210.4	13.1	470.8	18.7	56.7	5.5	136.5	6.1	0.2	0.1	16.15	–
Black Feet	83	–	249.2	–	21.6	–	74.4	–	0.6	–	14.75	–
Heigham Sound	135.2	–	261.7	–	38	–	99.5	–	0.5	0.1	16.15	–
Martham	37	–	13.7	–	34.1	–	75.5	–	0.5	–	16.15	–
Sutton	169.2	–	462.4	–	40.6	–	102.4	–	0.4	–	16.85	–
Barton	77.2	17.5	178.6	30	35.7	1.1	95.5	9.8	0.4	0.1	16.62	0.2
Crome's	87.1	–	255.4	–	23.6	–	75.8	–	0.6	–	16.67	–

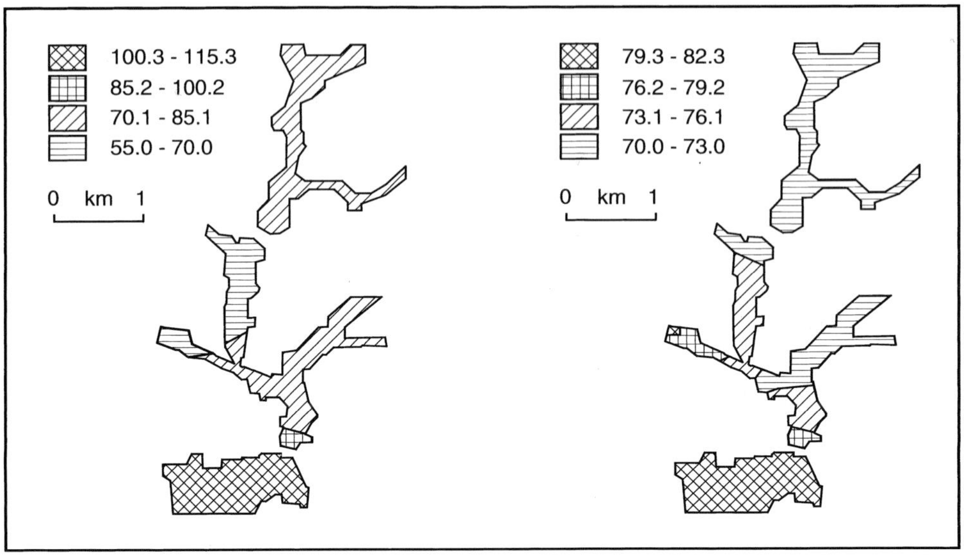

Figure 4. The distribution of estimated chlorophyll-*a* and total phosphorus (both in $\mu g\ l^{-1}$) in the Trinity Broads from June 1979 and 1980 images (after Baban, 1993a).

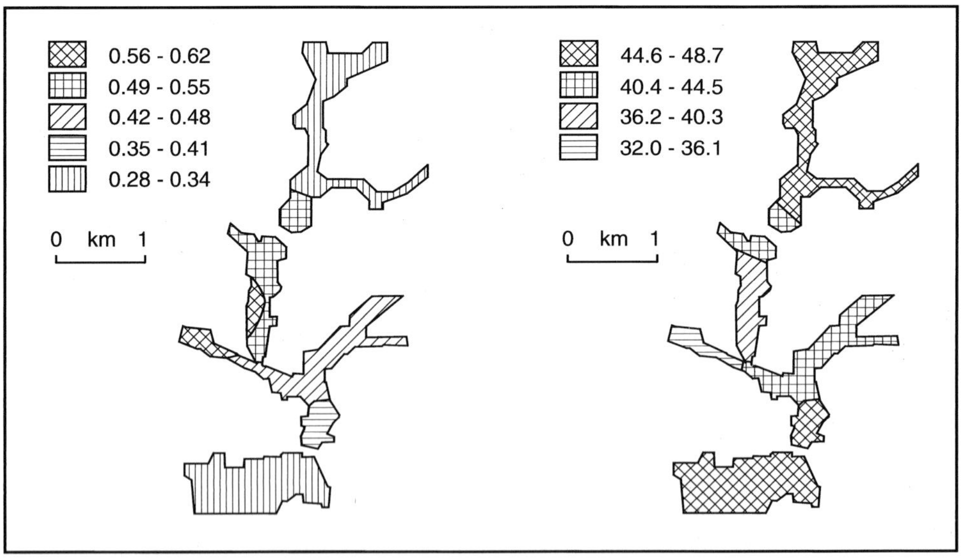

Figure 5. The distribution of estimated Secchi depth in m and suspended solids in mg l^{-1} in the Trinity Broads from June 1979 and June 1980 images (after Baban, 1993a).

Table 5. Ground truth data (mean value for June 1979 and June 1980) for water quality parameters in the Broads.

Broad	Concentration (μg/l)				Secchi depth (m)	Temp. (°C)
	Chlor-*a*	Total P.	S. solid	salinity		
Malthouse	165.5	390.5	34.45	116	0.365	17.33
Ranworth	122.5	281.5	24.95	89.5	–	–
Cockshoot	34.5	191	12	79.5	–	–
River Ant	190	332	61	114.5	0.26	–
Decoy	31.5	202	10.15	62.5		–
Salhouse	84.5	159.5	22.75	64.5	0.74	17.5
Hoveton Gt.	87	203.5	25.8	62.5	0.535	17.5
Burntfen	38.5	152	10.9	87	0.65	15.9
Belaugh	31.6	300.5	31.8	59.5	0.545	15.1
Hoveton Lt.	94.7	342	27.9	72	0.6	17.17
S. Walsham (outer)			36.55		0.415	17.5
S. Walsham (inner)			41.15		0.44	17.5

Table 6. Carlson trophic state index (TSI) values for the Trinity Broads, River Ant and River Thurne Broads, using different waterquality parameters from Table 4 (after Baban, 1996).

Broads	TSI		
	Chlor.-*a*	Total P.	Secchi depth
Ormesby	70.2	70.7	73.6
Rollesby	70.4	73.5	74.3
Filby	76.7	79.2	78.4
Lily	69.6	78.1	69.7
Hickling	83.6	89.7	74.3
Horsey Mere	83.0	92.9	80.6
Black Feet	73.9	83.8	66.9
Heigham Sound	78.7	84.5	70.3
Martham	66.0	74.6	70.9
Sutton	80.9	92.7	72.5
Barton	73.2	79.0	71.5
Crome's	74.4	84.1	67.4

Table 7. Trophic categorisation of the Norfolk Broads based on the effect of nutrient loads (after Baban, 1996).

Group A:

There is nutrient loading but it is not the major influence for algal growth due to opposed influence of the replacement flushing rate (water residence time) and sediment coefficient is the loading of solids carried by water from the surrounding land into a water body and depends on two things: the size of the catchment in relation to the area of the lake and the rate of removal of the substance per unit area of catchment). The outcome wll be similar calculated TSI values for both Secchi disk depth and chlorophyll-*a* and much higher index values for total P. Typical examples for this group include Barton Broad (Table 6).

Group B:

Nutrient loading is the major influence for algal growth; this will result in variable TSI values for all parameters. The magnitude of the differences between these values is mostly proportional to the phosphorus activity. The following Broads represent good examples of this group: Hickling and Heigham Sound (Table 6).

Group C:

These is no loaded nutrient (or may be very little), the index value for all the parameters are similar. Typical examples include: Ormesby, Rollesby and Filby (Table 6).

Table 8. Availability of Landsat TM for the Norfolk Broads, 1984–1994 .

Date	Date
14.05.1984	25.04.1991
21.10.1984	07.07.1991
02.06.1985	21.07.1991
14.12.1986	30.07.1991
14.10.1987	05.08.1991
19.02.1988	31.08.1991
06.08.1988	22.01.1992
21.02.1988	29.01.1992
21.02.1989	14.02.1992
21.05.1989	20.05.1992
28.05.1989	22.01.1992
24.07.1989	21.04.1993
15.07.1989	07.05.1993
12.03.1990	07.05.1993
06.04.1990	23.05.1993
16.06.1990	08.06.1993
02.07.1990	10.07.1993
03.08.1990	27.08.1993
20.09.1990	19.02.1994

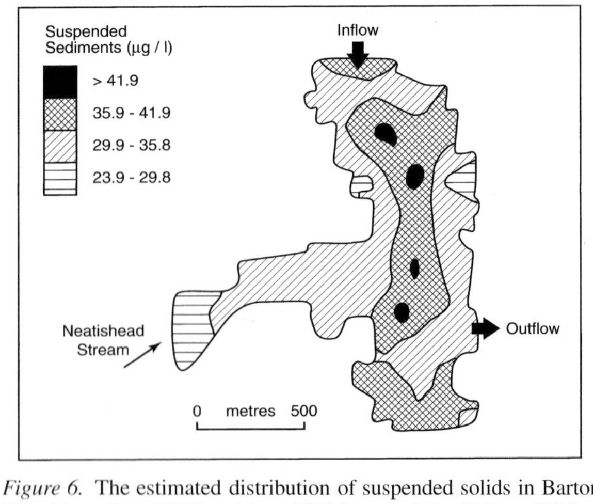

Figure 6. The estimated distribution of suspended solids in Barton Broad from June 1979 and June 1980 images (after Baban, 1994).

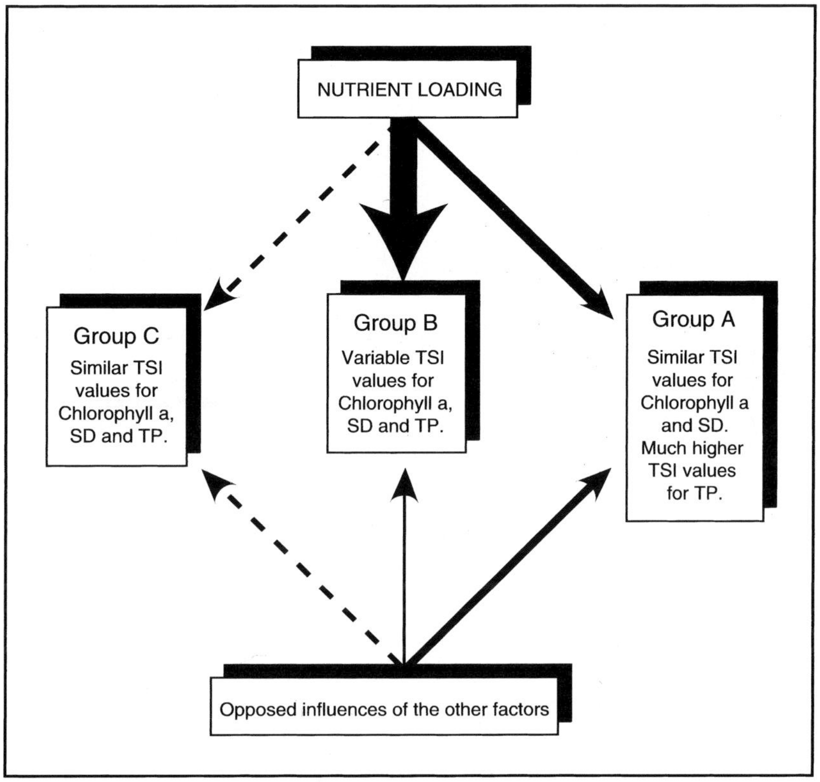

Figure 7. The alternative effects of nutrient loading on trophic state in the Broads (after Baban, 1996).

Table 9. Average sunshine, the length of day and percent possible sunshine in the Norfolk Broads area (calculated from data provided in the Weather Log, Royal Meteorological Society data and Pearce & Smith, 1984).

Month	Av. sunshine	Hours of daylight	Percent possible sunshine
January	50.82	8.30	19.75
February	66.67	10.07	22.83
March	119.01	11.48	33.44
April	170.37	13.44	42.45
May	210.48	15.22	44.61
June	216.05	16.21	44.42
July	202.20	15.38	42.41
August	182.24	14.33	41.02
September	150.45	12.42	40.38
October	114.03	10.47	35.13
November	60.21	9.06	22.15
December	50.00	8.05	20.04

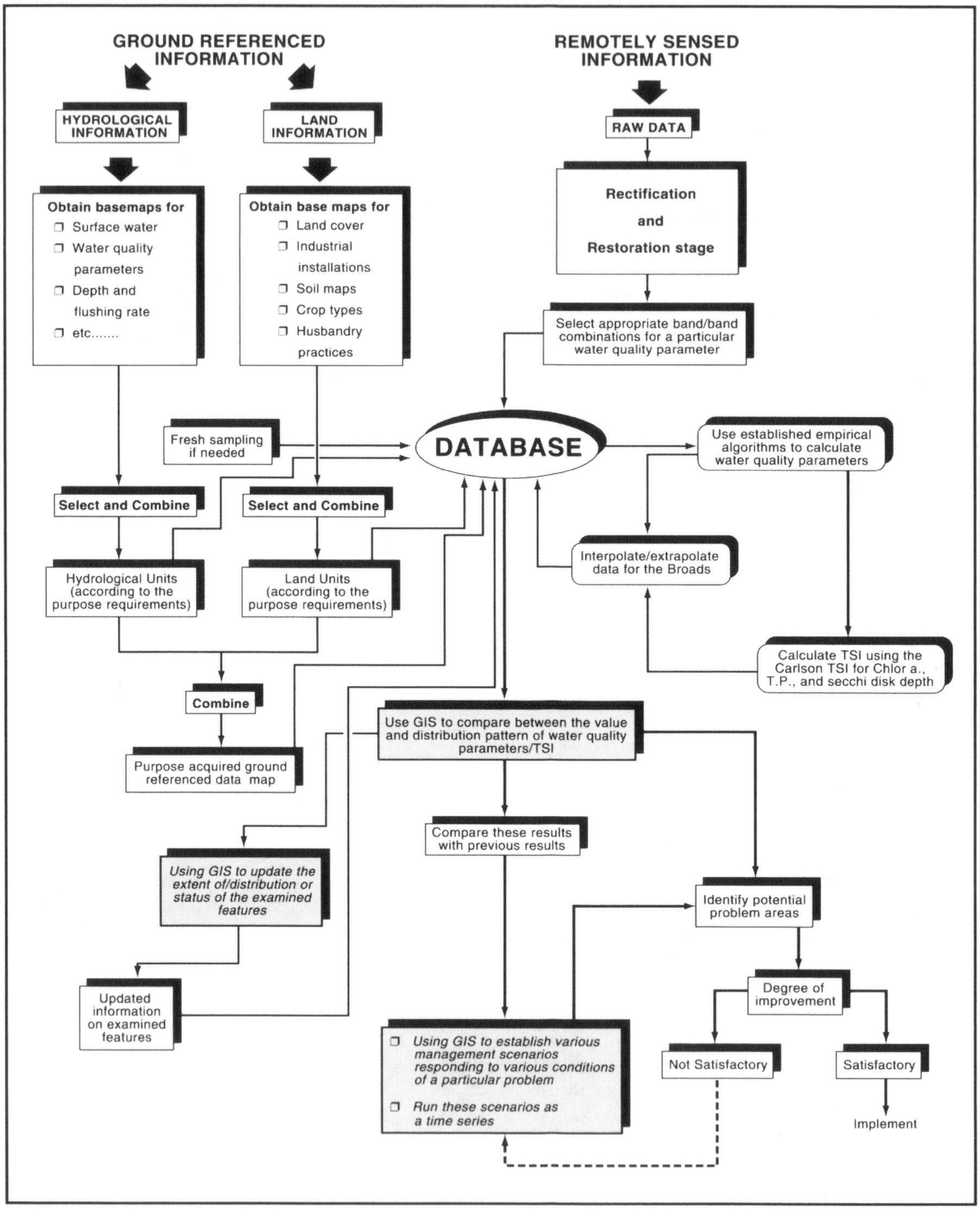

Figure 8. A proposed management system for the Norfolk Broads.

Acknowledgments

The author wishes to thank Professor Ian Foster for commenting on the paper and the Environment Agency for providing the actual water quality data.

References

Baban, S. M. J., 1993a. Landsat imagery and the detection of water quality parameters in Norfolk Broads, U.K. Int. J. Remote Sensing 14: 1247–1267.

Baban, S. M. J., 1993b. The evaluation of different algorithms for bathymetric charting of lakes using Landsat imagery. Int. J. Remote Sensing 14: 2263–2273.

Baban, S. M. J., 1993c. Detecting and evaluating the influence of water depth, volume and attitude on the variations in the surface temperature of lakes using Landsat Imagery. Int. J. Remote Sensing 15: 2747–2758.

Baban, S. M. J., 1994. Mapping turbidity, surface temperature and water circulation patterns with the aid of satellite imagery. J. Ins. Wat. Environ. Manage. 8: 197–204.

Baban, S. M. J., 1996. An application of Landsat imagery to the trophic classification for lakes. Hydrol. Sci. J. 41: 939–957.

Baban, S. M. J., 1998. Adopting a holistic view to improve student learning of Geographic Information Systems. Geography 83: 257–265.

Barrett, E. C. & L. F. Curtis, 1992. Introduction to Environmental Remote Sensing, 3rd edn. Chapman & Hall, London: 426 pp.

Bhaskar, N. R., W. James & R K. Devulapalli, 1992. Hydrologic parameter estimation using Geographic Information Systems. J. Wat. Resources Plan. Manage. ASCE 118: 492–513.

Broads Authority, 1997. Broads Plan. Broads Authority, Norwich: 176 pp.

Burrough, P. A., 1986. Principles of Geographical Information Systems for Land Resources Assessment. Oxford University Press, Oxford: 194 pp.

Carlson, R. E.., 1977. A trophic state index for lakes. Limnol. Oceanog. 22: 361–369.

Clarke, G. L., G. C. Ewing & C. J. Lorenzen, 1970. Computer derived coastal waters classification via spectral signatures. Proc. Int. Symp. Remote Sensing Environ. 9: 1213–1239.

Cooke, G. D., E. B. Welch, A. P. Peterson & P. R. Newroth, 1993. Restoration and Management of Lakes and Reservoirs., 2nd edn. Lewis Publications, USA: 548 pp.

Curran, P. L. 1985. Principles of Remote Sensing. Longman, London.

De Mers, M. N., 1997. Fundamentals of Geographic Information Systems. Wiley, New York, U.S.A. 486 pp.

Fee, E J. 1979. A relationship between lake morphometry and primary productivity and its use in interpreting whole-lake eutrophication experiments. Limnol. Oceanogr. 24: 401–416.

Gulati, R. D., E. H. R. R. Lammens, M.-L. Meijer & E. van Donk (eds), 1990. Biomanipulation – Tool for Water Management. Developments in Hydrobiology 61. Kluwer Academic Publishers, Dordrecht, 628 pp. Reprinted from Hydrobiologia 200/201.

Groves, J. R., R. M. Ragan & R. B. Clapp, 1983. Development and testing of a remote sensing based hydrological model. In Hydrological Applications of Remote Sensing and Remote Data Transmission, Proceedings of the Hamburg Symposium, IAHS Publications 145: 601–612

Hathout, S., 1985. The use of enhanced Landsat imagery for mapping lake depth. J. Environ. Manag. 20: 253–261.

Hutchinson, E., 1957. Treatise on Limnology, Volume 1. Wiley, New York.

Lillesand, T M., & R W. Kiefer, 1994. Remote Sensing and Image Interpretation, 3rd edn. John Wiley & Sons, New York, U.S.A.: 750 pp.

Martin, D., 1991. Geographic Information Systems and their Socioeconomic Applications. Routledge, London: 182 pp.

Mason, C. F. & R J. Bryant, 1979. Changes in the ecology of the Norfolk Broads. Freshwat. Biol. 5: 257–270.

Moss, B. & M. Timms, 1982. Studies on Hoveton Great Broad and the limnology of the associated waterway. Final report to the Nature Conservancy Council, HF 3/03/133, Peterborough, U.K.

Moss, B., F. J. Madgwick & G. L. Phillips, 1996. A Guide to the Restoration of Nutrient-enriched Shallow Lakes. Broads Authority, Norwich, U.K, 180 pp.

Pattie, N., 1993. Farmwatch UK. GIS Europe, November 22–24.

Pearce, E. D. & C. G. Smith, 1984. The World Water Guide. Hutchinson & Co. Ltd., London: 480 pp.

Penning-Rowsell, K. R., 1996. Creation and Testing of a Model to Predict Phosphorous Loads in Streams Flowing into Bassenthwaite Lake. Unpublished MSc Thesis, Dept. of Geography, University of Edinburgh, U.K.

Ragan, R. M. & J. D. Fellows, 1980. Computer-aided watershed analysis using remote sensing-based regional information systems. In V. V. Salomonson & P. D. Bhavsar (eds), The Contribution of Space Observations to Water Resources Management. COSPAR, Advances in Space Exploration 9. Pergammon Press, U.K.: 280 pp.

Sharma, S. K. & D. Anjaneyulu, 1993. Application of remote sensing and GIS in water resource management. Int. J. Remote Sensing 14 17: 3209–3220.

Stuart N. & C. Stocks, 1993. Hydrological modelling within GIS: an integrated approach. Proceedings of the Vienna Conference in Hydrology GIS 93: Applications of GIS in Hydrology and Water Resources, IAHS: 319–320.

Smith, A. Y. & J. D. Addington, 1978. Water quality monitoring of lake Mead: a practical look at the problems encountered in the applications of remotely sensed data to analysis of temporal change, 5th Canadian Symposium on Remote Sensing, Victoria B.C., Canada.

Smith, A. Y. & R. J. Blackwell, 1980. Development of an information data base for watershed monitoring. Photogrammetric Eng. Remote Sensing 46: 154–165.

Smith, S E., K H. Mancy, K. H. Latif & E. A. Fosnight, 1983. Assessment and monitoring of sedimentation in Aswan High Dam using Landsat data. In Hydrological Applications of Remote Sensing and Remote Data Transmission, Proceedings of the Hamburg Symposium, IAHS Publ. 145: 499–508.

Smith, R H., S N. Sahoo & L W. Moore, 1992. A GIS-based synthetic watershed sediment routing model. In Proceedings of the Water Resources Planning and Management Conference, ASCE, Baltimore, 2–6 August: 200–207.

U.S. Environmental Protection Agency, 1990. National Water Quality Inventory. 1988 Report to Congress, EPA 440/4-90-003.

Van Blargan, J. E. & R. M. Ragan, 1991. Automated estimation of hydrologic parameters. In Proceedings of the ASCE National Conference Civil Engineering Applications of Remote Sensing & Geographic Information Systems Washington, DC, 14–16 May: 278–285.

Vieux, D. E., V. F. Bralts & L. J. Segerlind, 1988. Finite element analysis of hydrologic response areas using Geographic Information Systems. In Proceedings of the ASCE International Symposium Chicago, IL: 437–447.

Wetzel, R. G., 1983. Limnology, 2nd edn. W.B. Saunders, Philadelphia, U.S.A.

Wright, J., 1994. An Investigation into The Use of GIS for the Estimation of Variables Required for the Prediction of Algal Growth Rates and Chlorophyll Levels in Freshwater Lakes. Unpublished M.Sc. Thesis, Department of Geography, University of Edinburgh, U.K.

Hydrobiologia **395/396:** 227–237, 1999.
D.M. Harper, B. Brierley, A.J.D. Ferguson & G. Phillips (eds), The Ecological Bases for Lake and Reservoir Management.

Modelling catchment-scale nutrient transport to watercourses in the U.K.

Barry Shepherd[1], David Harper[1] & Andrew Millington[2]
[1]*Department of Biology,* [2]*Department of Geography, University of Leicester, Leicester LE1 7RH*

Key words: catchment modelling, hydrology, eutrophication, field survey, remotely sensed data, SWAT, reliability estimation, mollifier

Abstract

Water resource managers require catchment-scale nutrient transport models to help in the investigation and control of diffuse pollution. Many different models exist for water and pollutant movement, causing problems of selection because each model has often been constructed for specific purposes, and the choice of an appropriate one is difficult: wrongly applied models can lead to unreliable results and misleading information. The selection of a model for estimating phosphorus loss from a lowland English catchment is described. 14 models, whose underlying principles had been published, were evaluated. 'SWAT' was found to be the most suitable. It has several favourable qualities, but also several shortfalls which may impair its predictive ability. Their identification enables the model code to be re-written. Reliability estimation is needed to ensure that the ranges of model prediction are known.

Introduction

Eutrophication is a continuous problem for water resource managers (ICWE, 1992; Harper & Pacini, 1995; Tim, 1996b). The most important nutrients in eutrophication are phosphorus, nitrogen and at times silica (Heathwaite et al., 1990; Harper, 1992). Phosphorus is more often limiting in freshwater systems, whereas nitrogen is more important in marine systems (Thomann & Mueller, 1987). In the past, nutrients from point source effluents have been the main contributors (Harper, 1992). However, even though tertiary treatment has become more widespread and detergents contain fewer poly-phosphates (McGarrigle, 1993), the limnological problem has not diminished. Attention has now switched to diffuse nutrient sources from intensive agriculture because there is increasing evidence of phosphorus runoff, as soil phosphorus capacity is exceeded (De Willigen et al., 1990; Heathwaite et al., 1990; Burt & Heathwaite, 1993; Morse et al., 1993). The pollutants are of a diffuse, chronic nature, and so the overall impact on a river or lake can only be evaluated by examination of the whole catchment over a long period of time. The catchment system includes spatial and temporal variables, and is difficult if not impossible to describe without models (Styczen & Nielsen, 1989).

Models assist the collation and interpretation of information, upon which decisions can be made for the management of water resources (Shaw & Falco, 1990; He et al., 1993; Tim & Jolly, 1994; Refsgaard & Abbott, 1996; Tim, 1996a). Their applications are diverse (e.g. pesticide movement, hydrograph production), and used by many disciplines. They provide inexpensive, repetitive and non-invasive questioning about hypothetical scenarios and can support education and research into physical processes (Oliver et al., 1990; Shaw & Falco, 1990; Refsgaard et al., 1996; Tim, 1996b). Prediction of the movement of runoff through a catchment has been the most fundamental requirement for water resource management, and has been developed more than other aspects of water management. The growing concern about agricultural nutrients entering water bodies and causing deterioration in water quality, has however resulted in the need for nutrient transport models. Nutrient transport within a watershed depends almost entirely upon hydrology. Some nutrient models have been described as an integral part of a larger modelling system (Nearing et al., 1986; Young, et al., 1989; Thorsen et al., 1996). Hydrological modelling has been more successful than nutrient modelling, and although improvements are needed in both, the latter still remains to be adequately validated.

Model characteristics

There are several nutrient transport models available (Donigian & Rao, 1990), but it is not clear whether they are applicable to any catchment and for any scenario. Development of a new model is a complex task but often utilises specific databases, or applies to a specific catchment or operation. Consequently, models may be extremely limited in their use and have their own assumptions. Most models have therefore been developed but not tested (validated) as fully as they should be.

Generally two categories of model are recognised, conceptual and physically-based. Conceptual models are based on experimental observations and do not try to mimic the processes of nutrient transport. Physically-based models describe the physical, chemical and biological processes perceived to be the important ones affecting hydrology and nutrient transport. However, it is accepted that the distinction between the two is vague and that the modelled processes are only concepts of the system.

The distinction between applicability of the model and reliability of the predictions is important. Almost any complex model can have many of its parameter values adjusted to obtain a fit between its predicted and observed values (calibration) (Klemes, 1986; Beven, 1996; Fisher et al., 1997). Such adjustments may produce values outside of the known ranges occurring in the watershed. This may provide reliable results within a range of external variables, but may give a false impression of how applicable the perceived processes are to the watershed. Physically-based models are expected to help in understanding the watershed processes and predict the effects of land-use and climate change. This necessarily involves parameter values outside those involved with the model's calibration. If the processes are not truly applicable, yet are used due to the apparently reliable results, predictions could be very unreliable. Conversely, careless calibration may mask how well the algorithms reflect the system processes. Calibration may still be needed but must be carried out with caution to avoid the above problems. Calibration of conceptual models is required and is acceptable for management purposes provided the limitations are observed. Conceptual model output is 'fitted' to observed values through calibration and therefore relies heavily on the relationship of the adjusted parameter values and the observed values. For this reason they strictly must not be used when watershed variables fall outside those within the calibration data set. For example the U.S. export coefficient model of Beaulac & Reckhow (1982) applied by Johnes (1996) to the U.K., is a modelling system that requires calibration due to the abstracted way in which the data are derived. Export coefficients are used to account for the many sources of phosphorus within a watershed. These are then serially adjusted until the predictive output matches the measured values at several points within the river or lake catchment. This system requires calibration because the parameter values are generally not measured in the watershed and it is highly unlikely that pre-calibrated output would compare well with measured values. This system appears to work well even though the export coefficients may not relate to those measured in the U.K. Export coefficient modelling is strongly dependent on vast amounts of input data as parameters and instream data for calibration. Output is limited to lumped output and cannot be considered reliable for changes in environmental variables such as an increase in rainfall.

Reliability estimation

Confidence intervals are seldom reported for predictions from large-scale catchment modelling as they are for simple regression analysis for example. Reliability estimation of the predictive output is essential to help in identifying how well the model performs (Beven, 1989, 1993; Beven & Binley, 1992) and how well the algorithms apply to the system. There are several methods available for assessing predictive reliability and model applicability. Two that have been applied to erosion and hydrological models are the Generalised Likelihood Uncertainty Estimation (GLUE) (Beven & Binley, 1992) and the mollifier-method (Keyzer & Sonneveld, 1997). Both estimate reliability of predictive output, but the mollifier method is also reported to assess error in the input. In order to assess the predictive output it needs to be compared with measured data. For the hydrological component, flow would have to be measured at the outlet of the catchments. The fractions of phosphorus considered by the model must be measured also. Alternatively, total phosphorus can be measured in the river, at the outlet, as a sum of the fractions predicted by the model. Instream data need to be intensive enough to ensure that all rainfall events are sampled, and underestimation of flow and phosphorus loading avoided (Burn, 1991). Sub-optimal sampling will result in comparison of model output with inaccurate data. Flow must be measured constantly and

Table 1. Potential variables measurable with RS data and the sensors used

Parameter	Platform/Sensor	Method (& key reference)
Crop/Land Cover type	AVHRR, ATSR2, CASI	Multispectral classification of multidate data.
LAI/Ground cover likely	AVHRR, ATSR2,	Calibration of Vegetation Indices, most
	CASI	SAVI (Elvidge & Chen, 1994)
Soil moisture	ERS2 SAR	Predicted from MIMICS (Ulaby et al., 1990)
Soil organic matter	CASI	Curran et al. (1990)
Soil texture[a]	ERS2 SAR	Tansey et al. (1997)
Eroded/non-eroded soil proportions	AVHRR,ATSR2, CASI	Steglik (1982) and Mather (1987).

[a]Only possible when fields are bare.

samples taken for total phosphorus analysis at suitable intervals depending on size & dynamics of river.

All catchment-scale modelling is input-driven, so the quality (accuracy) and quantity (resolution) of that input affect the model's performance (Klemes, 1986; van Genuchten et al., 1990; Beven, 1993; Tim, 1996a). Collection of data by point measurements during field survey is limited by resources and does not provide information about the whole area modelled. In addition, interpolative techniques between point measurements can be unreliable. To overcome some of these problems, some models suggest that parameter values can be estimated. Unless very experienced users are involved, who know their study catchment well, this practise is clearly unsatisfactory. Remotely sensed (RS) data offer some answers to this predicament in large-scale modelling (Schultz, 1988; de Jong, 1994; Beven, 1996; De Troch et al., 1996; Tim, 1996a) for the reasons outlined below. Table 1 lists some parameters that have been successfully extracted from RS data.

This paper describes the preliminary stages of model selection for the estimation of phosphorus runoff from an intensive agricultural region of lowland eastern England, using remotely-sensed data.

Incorporation of remotely-sensed data into models

The advantages of remote-sensing over field survey methods of data collection may be summarised as follows (De Jong, 1994):

1. They cover large areas.
2. They collect data at short, regular intervals.
3. They have sufficient spatial and temporal resolution.
4. They provide data on the required attributes accurately.
5. They provide data in a form suitable for further processing.
6. They are less expensive than field collection of the same intensity.

To assess the accuracy of models using remotely-sensed data, output from a model parameterised with field survey data, must be compared with the same output when part-parameterised with RS data. Reliability estimation as outlined above will help in this comparison.

Selecting a model

The choice of model is very important, and is largely influenced by the initial reasons for using the model. Boutwell et al. (1985) described a flowchart to help in deciding if a model is required and another to assess whether empirical or physically-based modelling is needed (Figures 1 and 2).

Once it has been established that a model is required and whether it should be empirical or physically-based, the expected goals of the modelling must be identified. The requirements used here are to understand the magnitude and spatial pattern of phosphorus loss from English lowland agricultural catchments draining to water supply reservoirs, so that future changes in nutrient runoff under different management scenarios can be modelled. In order to understand the processes within the catchment, as well

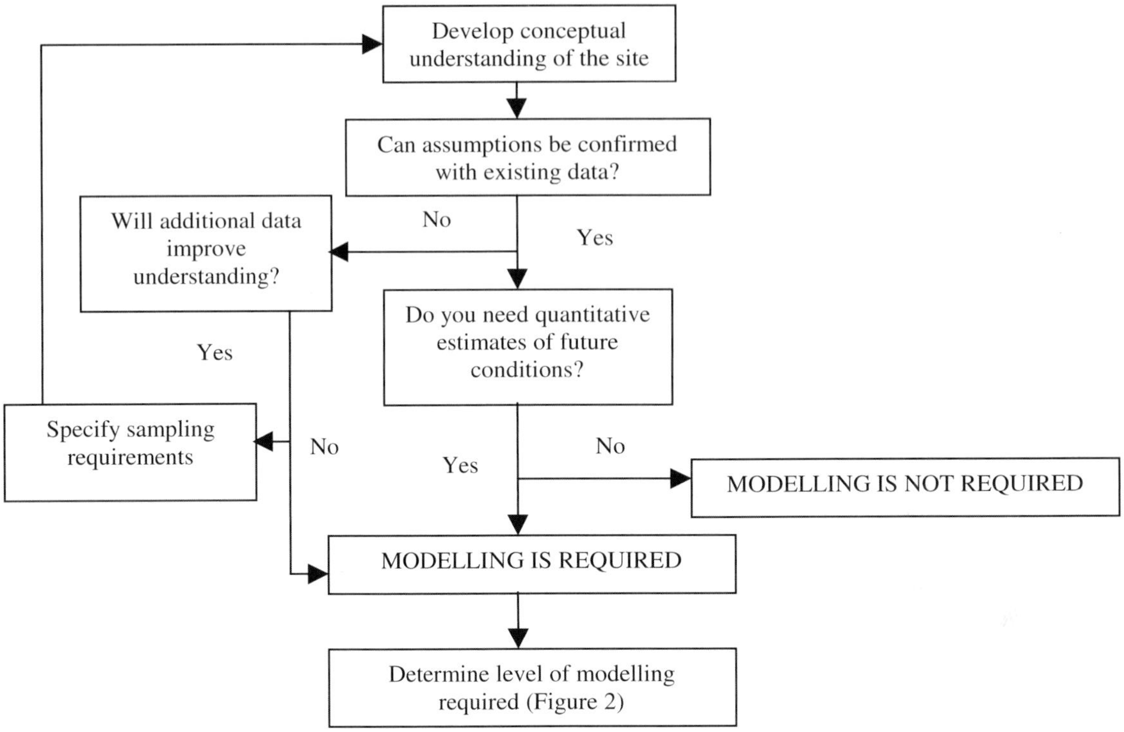

Figure 1. Decision flowchart to determine if modelling is required (from Boutwell et al., 1985).

as the nutrient output from it, the model is planned to run on two catchment scales with a nested approach. The larger catchment, the river Welland, covers an area of 700 km^2 and the smaller catchment, the Stonton Brook, covers an area of 75 km^2. Both catchments are in Leicestershire and the Stonton Brook is a subcatchment of the Welland. A successful model will thus require the following general characteristics:

1. To be applicable to catchments in lowland England, with areas greater than 50 km^2.
2. To be able to identify management practices and/or geographical regions that are responsible for the greater part of nutrient transport to surface waters.
3. To help in identification of land-water processes that affect eutrophication.
4. To simulate hypothetical scenarios (i.e. climate change) in risk assessment exercises.

Table 2 shows 13 models that have been developed and used in runoff modelling, together with their outline characteristics. Clearly there are differences in structure, scale of operation and processes covered.

Each model can be evaluated in greater detail against the specific requirement of these lowland catchments. Seven criteria are needed for this evaluation:

1. Scale of application.
2. Associated sub-models.
3. Characteristic of catchment to be studied.
4. Advantages and disadvantages of physical or conceptual?
5. Advantages and disadvantages of lumped versus distributed?
6. Type of output.
7. Model complexity and data requirements.

1. Scale of application

Processes that affect flow and nutrient transport at one scale do not necessarily apply at another scale. Application of a small-scale model of the plot or field scale across a whole catchment, would also involve prohibitive quantities of input and processor time. Scale is therefore an important issue and is probably more relevant when applying models below their intended scale than above. The scale required is dependent upon the catchments to be modelled. A model has to be applied across a range of different sized catchments to assess its spatial scale performance. Temporal scale is generally fixed within the model as short or long term. Some models only simulate for short periods, e.g. single storm events, whilst others can be set to

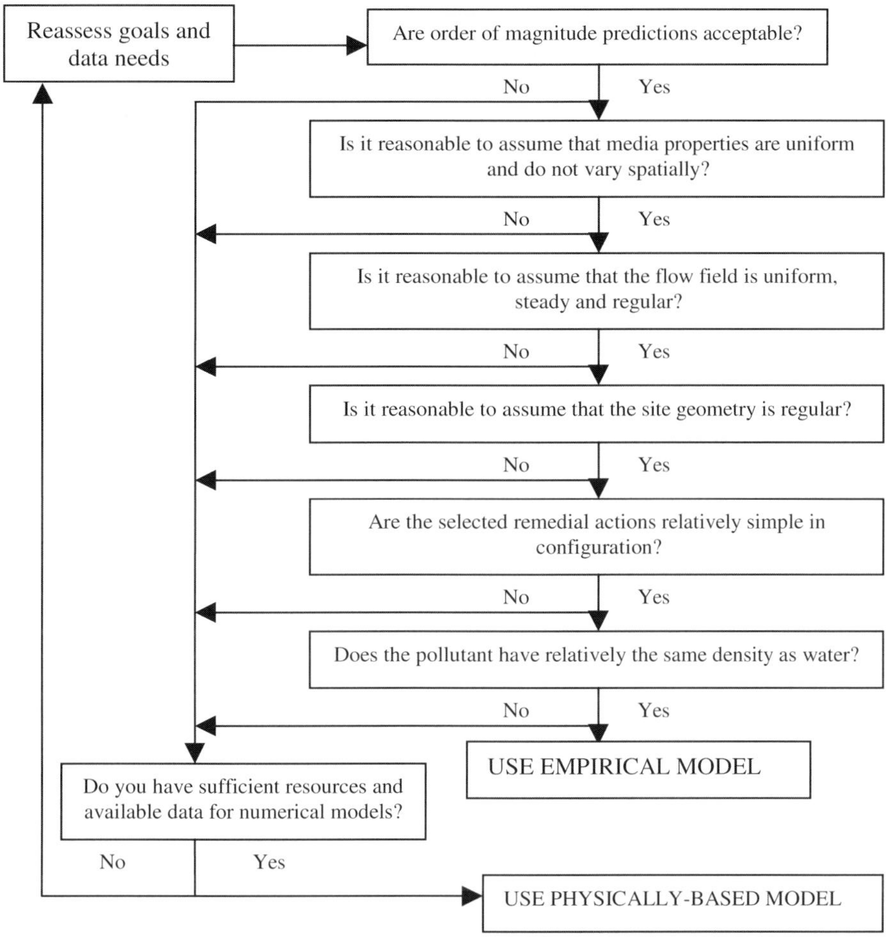

Figure 2. Decision flowchart to determine the type of model required (from Boutwell et al., 1985). This flowchart refers specifically to pollutant transport modelling, but is relevant to nutrient transport modelling also.

run for periods in excess of one month to hundreds of years. Within these time intervals, the resolution (time-step) at which the models run vary from hourly to daily. Some conceptual models, including the export coefficient model, are based upon a gross input to the system and provide lumped output covering a given period of time. These models do not work on a time series basis. If seasonal detail and river nutrient flux is required, longer simulation runs are needed with at least daily time steps.

Based on the above, there are several models that can be eliminated. CREAMS, GLEAMS and SWAT (U.K.) operate at the field scale and POPPIE is based on national databases to consider the regional scale. SWRRB is suitable for the Stonton Brook but not for the River Welland. At present, AGNPS will simulate for single storm events and can therefore be excluded, although a continuous version has been promised.

2. *Associated sub-models*

Most of the phosphorus entering water bodies is transported through overland flow during storm events often adsorbed onto particulate matter. Proportionately less phosphorus will be leached out through the soil, but a proportion of that will travel via artificial sub-surface drains where they have been employed. Sub-surface drains are employed extensively throughout the U.K., including the River Welland and Stonton Brook. Therefore a routine for phosphorus transport should accompany a routine for soil erosion and include overland flow, sub-surface flow and artificial sub-surface drainage routines. Nutrient management practices should also be flexible and include applications of chemical and organic fertilizers.

Of those models that are left ANSWERS, IHDM, TOPMODEL and WEPP do not include nutrient trans-

Table 2. Models available and their characteristics

Model and Acronym	Physical	Empirical	Lumped	Distributed	Continuous	Event-based	Nutrient	Scale
Agricultural Non-Point Source (AGNPS), Young et al., 1989.	X			X		X	X	Field-catchment
Areal Non-point Source Watershed Environmental Response Simulation (ANSWERS), Beasley et al., 1980.		X		X		X		Small watershed
Chemicals, Runoff and Erosion from Agricultural Systems (CREAMS), Knisel, 1980.		X	X		X		X	Field
Export Coefficient approach, Beaulac & Reckhow, 1982.		X	X				X	WATERSHED
Groundwater Loading Effects of Agricultural Management Systems (GLEAMS), Leonard et al., 1987.		X	X		X			Field
Hydrological Simulation Program-Fortran (HSPF), Donigian et al., 1985.	X			X	X		X	Small to large watershed
Institute of Hydrology Distributed Model (IHDM), Beven et al., 1987.	X			X		X		Hillslope catchments
MINDER, Johnson et al., 1992.		X		X	X		X	catchment
Prediction of Pesticide Pollution In the Environment (POPPIE)		X		X	X			Regional-national
MIKE SHE, Refsgaard & Storm, 1995.	X			X	X	X	X	catchment
SWAT (U.K.), Brown & Hollis, 1996.		X		X	X		X	Field
Soil and Water Analysis Tool (SWAT)(U.S.A.) Arnold et al., 1996.	X			X	X		X	Catchment watersheds
Simulator for Water Resources in Rural Basins (SWRRB), Williams et al., 1985.	X			X	X		X	Small catchment simple watersheds
TOPMODEL, Beven et al., 1995.	X			X	X	X		Hillslope-catchment
Water Erosion Prediction Project (WEPP), Lane & Nearing, 1989.	X			X	X			catchment

port sub-model. SHE was originally a hydrological model but has now diversified into a number of different modelling systems, of which MIKE SHE has a nutrient sub-model (Refsgaard & Storm, 1995).

3. Characteristic of catchment to be studied

The catchment characteristics have a large influence on the choice of model. All internal and external variables have to be taken into account and the closer the match upon that which the model was originally based the better. For example, lowland soils and undulating terrain in the U.K. have very different mechanics than highland terrain or some soils not found in the U.K. Few models have been developed in the U.K., but if the model reflects general mechanisms of soil dynamics it should apply to similar soils outside of the developmental area. In addition to the geological, topographical and climatological variables, land management considerations must be taken into account e.g. artificial sub-surface drainage above. Many of the models in Table 2 were developed in the U.S., where artificial sub-soil drainage systems are uncommon and are not therefore included in the models. Lag in hydrological flow is likely to be overestimated and phosphorus is likely to be underestimated, should this not be corrected. Even though SHE and MINDER were developed in the U.K., artificial sub-surface drainage elements appear to be absent. The

U.S. models HSPF and SWAT also do not have components for artificial sub-surface drainage and is one factor that will have to be accepted, but considered carefully during application.

4. Lumped or distributed?

The catchment is a geographical entity and all data relating to modelling systems of the type described in this paper are spatial in nature. Valuable information is lost if the distributed nature of the collected data is lumped together as many simple models do. Lumping the input data and output offers few insights into spatial heterogeneity of factors making up 'nutrient hotspots'. If distributed input and output are used, an area within the study catchment can be isolated from the catchment as a whole and simulation runs executed. This can examine the effects of scale on the model. To simplify the mathematics and limit the data, the catchment is generally partitioned into discrete units (discretization) based on natural boundaries such as those of sub-catchment hydrological boundaries or artificial units, i.e. a grid. Within these units, data and output is lumped and routed to adjoining units. However, Fisher et al. (1997) indicated that AGNPS and CREAMS do not exhibit spatial distribution of their outputs although it is claimed that they do. This is worrying because many of the USDA models (i.e. WEPP and SWAT) are based on similar algorithms. It is possible however, that this may be an artefact of spatial randomisation of variables, rather than a reflection of their spatial insensitivity. All the remaining models are distributed in their algorithm processing. Distributed and quantitative output from models are well documented in Abbott & Refsgaard (1996), which do offer some spatial information (see Beven, for a rationalisation of distributed models).

5. Physical or conceptual?

The flow chart in Figure 2 will indicate whether a physical or conceptual model is required. In order to use models, to help in understanding the processes affecting phosphorus transport, it is necessary for the model to attempt to mimic the processes thought to occur in the watershed. As outlined above, conceptual models do not offer this provision. Physically-based models can give insight into catchment processes, providing the algorithms reflect these processes. Future land-use change and climate change will include variables outside those currently measured. Output from conceptual models that include climate change must be regarded with caution. Physically-based models are the only alternative (Beven, 1992). MINDER and the export coefficient system is based on empirical relationships of phosphorus input to measured outputs and is therefore excluded.

6. Type of output

The form of output required can be very important. Several authors note how important stochasticity can be (Plate & Duckstein, 1990; Vauchad et al., 1990), and is more often associated with large spatial and temporal scale models (Rose et al., 1990). However, deterministic models can be assessed stochastically by performing Monte Carlo analysis (Quinn, pers. comm.) whereby input values are varied sequentially within a given range and output is calculated for each change. The result is a clustering of predictions. The most probable results will be found in the densest area of the scattergram. Achievement of this is entirely dependent on few parameters or highly accurate measurements. Therefore this is not so much a requirement of the model as of the data.

7. Model complexity and data requirements

The simpler a model is the better it is for data collection and management. Complex models with many parameters are more prone to equifinality than models with fewer parameters. WEPP offers methods of estimating parameters should data be unavailable (WEPP manual). Such techniques may compromise reliability when compared to simpler models where all parameters can be measured. Scale of input will also impact on the performance of the output. The more data there is available for input the better, hence the reason for utilising RS data. Some field survey data will always be required and the intensity of sampling will largely be based on the available resources. When the scale issue is considered and either the model or sampling needs to be investigated across scales, the intensity of field survey data is an important factor to consider. In the case of assessing the model's applicability across scales, the sampling intensity needs to be comparable between the scales, whereas, if input scale is to be investigated, the intensity of sampling will differ between scales in order to identify the minimum data requirements. Scale of the data input is resource-dependent but will decrease in intensity as the number of parameters requiring measurement increases. SHE has a very high number of parameters and is generally used via consultancy rather than by individuals

(Kilsby, pers. comm.). Additionally, MIKE SHE has a high commercial cost. Unfortunately HSPF and SWAT have high numbers of parameters, and will require more effort to apply than simpler systems. Geographic Information Systems (GIS) allow large quantities of spatial data to be analysed very efficiently (He et al., 1993; Tim & Jolly, 1994; Adamus & Bergman, 1995; Deckers & Te Stroet, 1996; Tim, 1996a). The use of GIS with the model can therefore overcome some of the problems mentioned above. SWAT has been linked to GIS and also has a good graphical user interface (GUI). HSPF has not been linked to a GIS and requires a detailed knowledge of Fortran. This will be acceptable for research but would not be popular for application.

Final model selection

SWAT therefore appears to be the most suitable model.

A full description of SWAT is contained in Arnold et al. (1996). SWAT is a logical progression from, and amalgamation of, other hydrology and diffuse pollutant models such as CREAMS, GLEAMS and ROTO developed by the United States Department of Agriculture, Agricultural Research Service (USDA-ARS). The main reasons for development were to model large catchments for several aspects of land management, such as soil erosion and agricultural pollution. Many thousands of square kilometres can be modelled, and to limit processing time, the catchment is divided into sub-basins. Each sub-basin is then individually discretised in a number of ways depending on extent of heterogeneity. Parameter values are stored in database files which are opened as needed and then closed to aid in efficient use of computer time. It has a working Graphical User Interface (GUI) and is available for PC, Solaris and IRIX operating systems. To further aid efficient data processing, there is a link programme between SWAT and the GRASS GIS. SWAT is a fairly complex system and requires evaluation of many parameters. Notwithstanding the technical and theoretical problems associated with this, there are also the resource requirements for collection of data. Many of the parameters used in SWAT are available from existing databases relevant to the U.S. but not the U.K. Digitised soil maps are available, and there are some data available from Ministry of Agriculture Fisheries and Food (MAFF) from its Agricultural Census. However, most of the data will have to be pre-processed for inclusion in SWAT. RS data can be processed in a form suitable for use in GRASS which should then be available to SWAT. It is inevitable that some data will have to be estimated, but the detrimental effects of this can be reduced by identification through sensitivity analysis of the parameters that are most influential (sensitive) in the model. In addition to a hydrological model, SWAT combines a series of sub-models to simulate processes of interest within agricultural catchments. These are phosphorus and nitrogen transport, soil erosion, pesticide transport, crop growth and lake water quality. Phosphorus is the only sub-routine of interest here but the others will have advantages for water quality managers.

In the U.S., SWAT has been applied to two watersheds for validation at Riesel, Texas of 17.7 km^2, and the Lower Colorado River of 8927 km^2 (Arnold et al., 1990). The former was validated against measured water and sediment yields and the latter was validated against water yields only. Bingner (1996) also used SWAT to simulate water yield on the Goodwin Creek watershed (21.31 km^2). The latter study discretised the watershed into sub-basins that were assumed to be homogeneous. Each sub-basin outlet was served by a streamflow-measuring station. Results for hydrological response were good within each sub-basin for ten years' data using rainfall and flow. This indicates that SWAT is spatially sensitive even when using the coarsest discretization method (sub-basin). Results show that the data was only reliable with this format of the model when comparing the lumped output, rather than individual events. It may become more responsive to individual rainfall events if grid discretization was used together with more measured data from the sub-basins rather than estimates. This amount of information cannot be considered adequate to assess a model's validity across a range of catchment types, nor the accuracy of its sub-routines. No information has been found regarding its predictive performance of phosphorus transfer. However, the above studies indicate adequate flow predictions on small and large-scale catchments.

SWAT operates on a daily timestep and is designed to simulate for as many as a hundred years. It will therefore provide detailed flow and flux charts over almost any period of simulation, but the study by Bingner (1996) may indicate that it is more suited to longer runs. In order to make an assessment of SWAT across scales in the U.K., two catchments have been selected on which to apply the model. These catchments, detailed above, fall into the category small to medium which appears to suit SWAT. By using nested catchments the effects of scale and the model's ability to process distributed data will be assessed.

Phosphorus transport in SWAT is based initially upon its partitioning into soluble and particulate phases. Water flow is divided into overland flow and through-flow using the Soil Conservation Service (SCS) curve number. Output from these routines are then used in conjunction to calculate particulate phosphorus via erosion using the Modified Universal Soil Loss Equation (MUSLE), and soluble phosphorus through surface runoff. These are the only routes for phosphorus transport included in SWAT. Phosphorus is also transported through the soil to water bodies and via soil drains, but is ignored. This is likely to underestimate phosphorus, especially if applied in areas with sub-surface drains. To overcome this problem the model code must be modified, or as a temporary measure the output adjusted to account for known phosphorus levels entering water bodies through the soil and via sub-surface drains.

SWAT was developed for a broad application across much of the U.S. and is based on general principles such as the water balance equation shown below (from Arnold et al., 1996):

$$SW_t = SW = \sum_{i=1}^{t} (R_i - Q_i - ET_i - P_i - QR_i,$$

where SW is soil water content minus the 15-bar water content, t is time in days, R, Q, ET, P and QR are daily amounts of precipitation, runoff, evapotranspiration, percolation and return flow. This physical basis should allow SWAT to be based on U.K. catchments. There is provision in SWAT to estimate many of the parameter values, based on U.S. measurements, but this could lead to erroneous results if used in the U.K. A reasonable appraisal can be made of SWAT in the U.K. provided as many of the variables as possible are measured rather than estimated.

SWAT claims to be physically-based even though some of the principles are conceptual such as the SCS curve number. In the main however, the routines are based on current understanding of the catchment processes. The number of parameters is high and this can obscure clarity in understanding the system, and reduces the ability to isolate error. To counter these problems sensitivity analysis will ensure that the most influential parameters are identified.

Output from SWAT is distributed and simulations for individual areas can be requested rather than just at the catchment outlet. In addition, it uses a continuous time scale based on a daily time step. This will allow flux graphs to be generated for the period of simulation.

Although the SCS curve is U.S. based, it has been applied in the U.K. (Morse et al., 1994). The USDA-ARS has plans to include the Green and Ampt principles as an alternative percolation model to the SCS curve (Arnold, pers. comm.). Such an alternative would help for the U.K. scenario. For long term simulation runs, climate variables can be generated as input to SWAT, based on geographical co-ordinates. It is not documented that this system works outside of the U.S., but there is an extensive weather data archive in the U.K. from which future climate input can be generated.

Conclusion

The basic characteristics of catchment-scale nutrient transport models and the reasons for their use have been described. There are many models available and many are difficult to assess. The choice of model for a given scenario is an important one, and can affect the performance of it in predictive output. However, the choice is not a simple process. A project in which catchment-scale nutrient transport modelling is needed is used as an example to outline the selection of a suitable model. From a table of 14 models, three appear suitable in application in a research sense: HSPF, SHE and SWAT. There are two advantages that SWAT has over HSPF; a GIS link and a user-friendly Graphical User Interface. SHE would be the obvious selection due to its European base, but its complexity and cost are too great for this project. SWAT was selected as the preferred option and a more detailed appraisal made. SWAT contains all the physical characteristics of the ideal model such as scale and distributed nature. It has a fully functioning GUI and GIS link, that are advantages in this application. Where it falls short of the project requirements, however, is in the physical processes of hydrology and phosphorus transport. There is no routine for sub-soil drainage which is prevalent in the U.K. Additionally, it is assumed that phosphorus is transported only via overland runoff. Whilst it is true that most of the phosphorus is transported in this fashion, it is not the only route. However, the programming code for this model is available and could be altered to account for these two factors. This shows that all of the considerations discussed are required to ensure correct application of models. Assumptions of the models are not always as obvious as this and they are not normally documented in scientific literature. Finally, without reliability estimation, it is impossible to know how good the model's performance actually is. Although reliability estimation is being used in the stated project to assess the efficacy of RS data, it could equally be used on its own to assess the normal working performance of a model.

Acknowledgements

The project is funded as a CASE Ph.D. grant from NERC and the EA to Barry Shepherd, Ref. GT22/96/EO/3. The authors are grateful to Peter Fisher, Steven Anthony, Paul Quinn & Fraser Gemmell for their kind guidance on modelling and related subjects, and to Bill Brierley, Nick Haycock & Peter Buckland for their help in this project.

References

Abbot, M. B. & J. C. Refsgaard, 1996. Distributed Hydrological Modelling. Kluwer Academic Publishers, Dortrecht.

Adamus, C. L. & M. J. Bergman, 1995. Estimating nonpoint source pollution loads with a GIS screening model. Water Resources Bulletin 31(4): 647–655.

Arnold, J. G., R. Srinivasan & K. W. King, 1996. SWAT: Soil and Water Assessment Tool. USDA, ARS, Temple, Texas.

Beasley, D. B., L. F. Huggins, & E. J. Monke, 1980. ANSWERS: a model for watershed planning. Transactions of the ASAE 23(4): 938–944.

Beaulac, M. N. & K. M. Reckhow, 1982. An examination of land-use-nutrient export relationships. Water Resources Bulletin 18: 1013–1024.

Beven, K. J., 1989. Changing ideas in hydrology – the case of physically-based models. Journal of Hydrology 105: 157–172.

Beven, K. J., 1993 Prophecy, reality and uncertainty in distributed hydrological modelling. Advances in Water Resources 16: 41–51.

Beven, K. J., 1996. A discussion of distributed hydrological modelling. In M. B. Abbott & J. C. Refsgaard (eds). Distributed Hydrological Modelling. Kluwer Academic Publishers, Dortrecht.

Beven, K. J. & A. M. Binley, 1992. The future of distributed models: model calibration and predictive uncertainty. Hydrological Processes 6: 279–298.

Beven, K. J., A. Calver & E. Morris, 1987. The Institute of Hydrology Distributed Model. Institute of Hydrology Report 98, Wallingford, U.K.

Beven, K. J., R. Lamb, P. Quinn, R. Romanowicz & J. Freer, 1995. TOPMODEL. In V. P. Singh (ed.), Computer Models of Watershed Hydrology. Water Resources Publications, Colorado, U.S.

Bingner, R. L., 1996. Runoff Simulated from Goodwin Creek Watershed Using SWAT. Transactions of the ASAE 39(1): 85–90.

Boutwell, S. H., S. M. Brown, B. R. Roberts & D. F. Atwood, 1985. Modelling remedial actions at uncontrolled hazardous waste sites. Anderson-Nichols and Co., Inc., Palo Alto, CA., EPA/540/2-85/001. Prepared for Environmental Protection Agency, Hazardous Waste Engineering Research Laboratory, Cincinnati, OH.

Burn, D. H., 1991. Water quality sampling for nutrient loading estimation. In L. C. Wrobel & C. A. Brebbia (eds), Water Pollution: Modelling, Measuring and Prediction. Proceedings of the First International Conference on Water Pollution. Elsevier Applied Science, London.

Burt, T. P. & A. L. Heathwaite, 1993. A scale-dependent approach to the study of nutrient export from basins. Institute of Hydrology Report 120: 82–92.

Curran, P. J., G. M. Foody, K. Y. Kondratyev, V. V. Kozoderov & P. P. Fedchenko, 1990. Remote Sensing of Soils and Vegetation in the USSR. Taylor and Francis, London.

Deckers, F. & C. B. M. Te Stroet, 1996. Use of GIS and database with distributed modelling. In M. B. Abbott & J. C. Refsgaard (eds), Distributed Hydrological Modelling. Kluwer Academic Publishers, Dortrecht.

de Jong, S. M., 1994. Applications of Reflective Remote Sensing for Land Degradation Studies in a Mediterranean Environment. Netherlands Geographical Studies 177.

de Troch, F. P., P. A. Troch, Z. Su & D. S. Lin, 1996. Application of remote sensing for hydrological modelling. In M. B. Abbott & J. C. Refsgaard (eds), Distributed Hydrological Modelling. Kluwer Academic Publishers, Dortrecht.

de Willigen, P., L. Bergström & R. G. Gerritse, 1990. Leaching models of the unsaturated zone: their potential use for management and planning. In D. G. Decoursey (ed.), Proceedings Of An International Symposium On Water Quality Modelling Of Agricultural Non-Point Source Pollution. Logan, UT ARS-81 USDA-ARS, Washington DC, U.S.

Donigian, A. S. & P. S. C. Rao, 1990. Selection, application and validation of environmental models. In D. G. Decoursey (ed.), Proceedings International Symposium on Water Quality Modelling of Agricultural Non-Point Sources. Logan, UT ARS-81 USDA-ARS, Washington D.C.

Elvidge, C. D. & Z. Chen, 1995. Comparison of broad-band and narrow band red and near-infrared vegetation indices. Remote Sensing of the Environment 54: 38–48.

Elser, J. J., E. R. Marzolf, & C. R. Goldman, 1990. Phosphorus and nitrogen limitation of phytoplankton growth in the freshwaters of north America: a review and critique of experimental enrichments. Can. J. Fish. aquat. Sci. 47: 1468–1477.

Fisher, P., R. J. Abrahart, & W. Herbinger, 1997. The sensitivity of two distributed non-point source pollution models to the spatial arrangement of the landscape. Hydrological Processes 11: 241–252.

Franks, S. W., K. J. Beven, P. F. Quinn, & I. R. Wright, 1995. On the sensitivity of soil-vegetation atmosphere transfer SVAT schemes: equifinality and the problem of robust calibration. Agriculture, Forestry and Meteorology.

Harper, D. M., 1992. Eutrophication of Freshwaters: Principles Problems and Restoration. Chapman & Hall, London.

Harper, D. M. & N. Pacini, 1995. Eutrophication Control Methodologies. National Rivers Authority Peterborough, U.K.

He, C., J. F. Riggs & Y-T. Kang, 1993. Integration of geographic information systems and a computer model to evaluate impacts of agricultural runoff on water quality. Water Resources Bulletin 29(8): 891–900.

Heathwaite, A. L., T. P. Burt & S. T. Trudgill, 1990. The effect of land use on nitrogen, phosphorus and suspended sediment delivery to streams in a small catchment in south-west England. In J. B. Thornes (ed.), Vegetation and Erosion. John Wiley and Sons Ltd.

ICWE, 1992. The Dublin Statement and report of the conference. International Conference on Water and the Environment: Development issues for the 21st century. 26–31 January 1992, Dublin, Ireland.

Johnes, P. J., 1996. Evaluation and management of the impact of land use change on the nitrogen and phosphorus load delivered to surface waters: the export coefficient modelling approach. Journal of Hydrology 183: 323–349.

Keyzer, M. A. & B. G. J. S. Sonneveld, 1997. Using the Mollifier-Method to Characterise data sets and models: the case of the Universal Soil Loss Equation. Centre for World Food Studies.

Kleme, V., 1986. Dilettatism in Hydrology: Transition or Destiny? Water Resources Research 22: 177S–188S.

Knisel, W. G. (ed.), 1980. CREAMS: A Field Scale Model for Chemicals, Runoff, and Erosion from Agricultural Management Systems. Conservation Research Report, 20, USDA-ARS, Washington DC, U.S.

Lane, L. J. & M. A. Nearing, 1989. USDA–Water Erosion Prediction Project: Hillslope profile model documentation. NSERL Report No.2, USDA-ARS, National Soil Erosion Research Laboratory, West Lafayette, Indiana 47907.

Leonard, R. A., W. G. Knisel, & D. A. Still, 1987. GLEAMS: Groundwater loading effects of agricultural management systems. Transactions of the ASAE 305: 1403–1418.

Mather, P. M., 1987. Computer Processing of Remotely Sensed Images. John Wiley & Sons, Chichester, U.K.

McGarrigle, M. L., 1993. Aspects of river eutrophication in Ireland. Ann. Limnol. 29(3-4). 355–364.

Morse, G., A. Eatherall & A. Jenkins, 1994. Managing agricultural pollution using a linked Geographical Information System and non-point source pollution model. Journal of the Institute of Water and Environmental Management 8: 277–286.

Morse, G. K., J. N. Lester & Parry, R. 1993. The Economic and Environmental Impact of Phosphorus Removal from Wastewater in the European Community. Selper Publications.

Nearing, M. A., G. R. Foster, L. J. Lane & S. C. Finkner, 1990. A process-based soil erosion model for USDA – Water Erosion Prediction Project. Transactions of the ASAE 32(5): 1587–1593.

Oliver, G., J. Burt, & R. Solomon, 1990. The use of surface runoff models for water quality decisions-a user's perspective. In D. G. Decoursey (ed.), Proceedings of an International Symposium on Water Quality Modelling of Agricultural Non-Point Source Pollution. Logan, UT ARS-81 USDA-ARS, Washington DC, U.S.

Plate, E. J. & L. Duckstein, 1990. Stochastic aspects of water quality modelling for nonpoint sources. In D. G. Decoursey (ed.), Proceedings of an International Symposium on Water Quality Modelling of Agricultural Non-Point source Pollution. Logan, UT ARS-81 USDA-ARS, Washington DC, U.S.

Quinn, P. F. & K. J. Beven, 1993. Spatial and temporal predictions of soil moisture dynamics, runoff, variable source areas and evapotranspiration for Plynlimon, mid-Wales. Hydrological Processes 7: 425–448.

Refsgaard, J. C. & M. B. Abbott, 1996. The role of distributed hydrological modelling in water resources management. In M. B. Abbott, & J. C. Refsgaard, (eds), Distributed Hydrological Modelling. Kluwer Academic Publishers, Dortrecht.

Refsgaard, J. C. & B. Storm, 1995. MIKE SHE. In V. P. Singh, (ed.), Computer Models of Watershed Hydrology. Water Resources Publications, Colorado, U.S.

Refsgaard, J. C., B. Storm & M. B. Abbott, 1996. Comment on "A discussion of distributed hydrological modelling" by K. Beven. In M. B. Abbott & J. C. Refsgaard (eds), Distributed Hydrological Modelling. Kluwer Academic Publishers, Dortrecht.

Rose, C. W., W. T. Dickinson, H. Ghadiri & S. E. Jøregnsen, 1990. Agricultural nonpoint-source runoff and sediment yield water quality NPSWQ models: modeller's perspective. In D. G. Decoursey (ed.), Proceedings of an International Symposium on Water Quality Modelling of Agricultural Non-Point Source Pollution. Logan, UT ARS-81 USDA-ARS, Washington DC, U.S.

Schultz, G. A., 1988. Remote sensing in hydrology. Journal of Hydrology 100: 239–265.

Shaw, R. R. & J. W. Falco, 1990. A management perspective on water quality models for agricultural nonpoint sources of pollution. In D. G. Decoursey (ed.). Proceedings of an International Symposium on Water Quality Modelling of Agricultural Non-Point Source Pollution. Logan, UT ARS-81 USDA-ARS, Washington DC, U.S.

Steglik, O., 1982. A technique for retrieving soil erosion based on remote sensing data. Studies of the Earth from Space 3: 92–95.

Styczen, M. & S. A. Nielsen, 1989. A view of soil erosion theory, process-research and model building: possible interactions and future developments. Quaderni Di Scienza Del Suolo 2: 27–45.

Tansey, K. J., A. C. Millington, K. White & A. Battikhi, 1997. Comparison of modelled backscatter response and ERS-1 SAR data for desert surfaces, the eastern Badia of Jordan. Proceedings Third ERS Symposium on Space at the Service of our Environment, Florence 1997, ESA SP-414, Vol 1: 21–26.

Thomann & Mueller, 1987.

Thorsen, M., J. Feyen & M. Styczen, 1996. Agrochemical modelling. In M. B. Abbott & J. C. Refsgaard (eds), Distributed Hydrological Modelling. Kluwer Academic Publishers, Dortrecht.

Tim, U. S., 1996a Emerging technologies for hydrologic and water quality modelling research. Transactions of the ASAE 39(2): 465–476.

Tim, U. S., 1996b. Coupling vadose zone models with GIS: emerging trends and potential bottlenecks. J. Envir. Qual. 25: 535–544.

Tim, U. S. & R. Jolly, 1994. Evaluating agricultural nonpoint-source pollution using integrated GIS and hydrological/water quality model. J. Envir. Qual. 231: 25–35.

Ulaby, F. T., K. Sarabandi, K. McDonald, M. Whitt & M. C. Dobson, 1990. Michigan microwave canopy scattering model. International Journal of Remote Sensing 117: 1223–1253.

Vachaud, G., M. Vauclin & T. M. Addiscott, 1990. Solute transport in the vadose zone: a review of models. In D. G. Decoursey (ed.), Proceedings of an International Symposium on Water Quality Modelling of Agricultural Non-Point Source Pollution. Logan, UT ARS-81 USDA-ARS, Washington DC, U.S.

van Genuchten, M. T., S. M. Gorelick & W. W-G. Yeh, 1990. Application of Parameter estimation techniques to solute transport studies. In D. G. Decoursey (ed.), Proceedings of an International Symposium on Water Quality Modelling of Agricultural Non-Point Source Pollution. Logan, UT ARS-81 USDA-ARS, Washington DC, U.S.

Williams, J. R., A. D. Nicks & J. G. Arnold, 1985, Simulator for Water Resources in Rural Basins. ASCE Journal of Hydraulic Engineering 11: 970–986.

Young, R. A., C. A. Onstad, D. D. Bosch & W. P. Anderson, 1989. AGNPS: A nonpoint-source pollution model for evaluating agricultural Watersheds. Journal of Soil and Water Conservation.

Section Four

Management Strategies for Different Lake Types

Hydrobiologia **395/396:** 241–251, 1999.
D.M. Harper, B. Brierley, A.J.D. Ferguson & G. Phillips (eds), The Ecological Bases for Lake and Reservoir Management.

Saline lakes: integrating ecology into their management future

F. A. Comín[1], M. Cabrera[2] & X. Rodó[1]
[1]*Department of Ecology, University of Barcelona, Diagonal 645, 08028 Barcelona, Spain*
[2]*Departamento de Agricultura y Medio Ambiente, Diputación General de Aragón, Paseo Ma Agustín 65, 50075 Zaragoza, Spain*

Key words: saline lake, fluctuations, management, legislative

Abstract

Defining the key temporal scales of variability in ecological processes is fundamental to saline lake management because the physical, chemical and biological characteristics of saline lakes undergo fluctuations at different time scales, compared to those of freshwater lakes. A long-term study of Lake Gallocanta (25 km^2 area inland saline lake in NE Spain) provides the ecological basis for its management. The coupling of the administrative framework with its ecological processes is examined. Limnological and climatological parameters were measured using standard methods during 1977–1995. Complete desiccation of the lake occurred in 1983–86 and in 1994–95. Water level fluctuations were strongly correlated with meteorological events, particularly rainfall for seasonal and decadal time scales. The relationship between salinity and water volume follows the typical model described for playa lakes. However, the intensity and frequency of the salinity and water volume fluctuations in Lake Gallocanta also change over time. Most species show pulsing population dynamics at the larger time scales (annual, decadal). However, many of these show a continuous population development at shorter time scales. Examples are provided for several phytoplankton and zooplankton species. Fluctuations in water level also influence phyto and zooplankton community structure. The combination of the number of species, S, and the number of individuals, N, as $K=\log S/\log N$ indicates that the community structure also fluctuates at different time scales. Most of the present laws established for the management of saline lakes are centered on the conservation of waterfowl. Although some generic laws attempt to preserve the overall natural ecosystem, only those dealing with the protection of wild flora and fauna, and threatened species, are implemented. However, ecological knowledge now suggests that the major objective for the conservation of saline lakes should be to preserve the fluctuation of the hydrological balance, avoiding groundwater extraction in the catchment area and surface water. Generic laws regulate water use in Spain and in many other countries. These laws are not useful for saline lake management because they do not offer solutions to the frequent conflicts arising between the water demand for intensive agriculture and nature conservation, which are common in the semi-arid zones where saline lakes are located. Integrated management of the land and water resources in the catchment area is required to preserve saline lake characteistics. Recent legislation aims to promote such management. A model of conservation and development of the overall ecosystem is presented.

Introduction

Saline lake ecosystems are suffering an accelerating process of degradation all around the world (Williams, 1992). The lack of management plans for saline lakes and their catchment area contrasts with the considerable efforts put into preserving and managing freshwater lakes. Furthermore, experience in managing freshwater lakes is much more extensive (De Bernardi et al., 1996), perhaps because human settlements have developed with more frequency and intensity in the vicinity of freshwater lakes than of saline ones. This difference bears no relation to the quantity of water in either type of lake which is quite similar (Vallentyne, 1972), nor to their respective value (economic, social, recreative, aesthetic) which is very high for saline lakes (Williams, 1981a). There is, then, no significant reason for the existing difference between management and legislative regulations affecting freshwater lakes and those affecting saline lakes. However, for the last 20 years, knowledge about saline lakes has increased in concert with their use, environmental impact and management problems (Williams, 1981b; Hammer, 1983; Melack, 1988; Comín & Northcote, 1990; Hurlbert, 1993).

Many physical, chemical and biological characteristics, apart from the salt content, distinguish saline from freshwater lakes and constitute an essential part of their natural resources (Comín et al., 1992). These differential characteristics should be taken into account when drawing up their management and legislative plans. For example, most saline lakes are very shallow and their waters do not stratify vertically. In deep saline lakes, stratification of the water occurs through differences in salt content between the upper and lower layers. Moreover, the physical processes controlling the dynamics of the water column are not stimulated by the exchange of heat or other type of energy between the lake and atmosphere, but are due to wind-induced currents in shallow saline lakes and to changes in the density of the water mass because of changes in salt content in deep saline lakes. Currently, major water inflows take place in most saline lakes *via* groundwater discharges. This is the reason why environmental impact on the groundwater of the catchment area is greater in saline lakes than in freshwater lakes, where major inflows are, in general, surface runoff.

However, the most striking and common characteristic of saline lakes is the high number of different time-scale fluctuations followed by all their characteristics, due to the fact that saline lakes are located in regions subjected to stronger climatic fluctuations than the zones where freshwater lakes are located. This is the most important feature in understanding the functional ecology of saline lakes (Comín et al., 1991). It is also a key feature of all biological systems and in certainly freshwater lakes, but with considerable fewer temporal scales or, at least, scales which appear to have less resolution than in saline lakes. However, these fluctuations have not yet been incorporated into the framework for the conservation and management of lakes; nor can they be easily incorporated into management plans for lakes. Yet it is essential to do so, as the success of any lake management will only be achieved if both spatial and temporal lake heterogeneity is preserved.

This paper presents the ecological fluctuations occurring in Lake Gallocanta (NE Spain), the largest saline lake in Spain, and its present management and legislative framework as a model for other saline lakes in S. Europe. Because of the wide range of ecological fluctuations occurring in Lake Gallocanta, this model can be useful to develop future strategies of management and legislative planning, which are also discussed here.

General background

Lake Gallocanta is the hydrological terminus of the surface waters from a closed catchment (area 520 km^2) in the Iberian Range (40° 50′ N, 2° 11′ W) (Figure 1). Most of the catchment is used for agricultural purposes, a small part of it is naturally forested. There are no industrial activities in the area. The human population (about 2000 inhabitants) is distributed in 12 villages, the largest one with 500 inhabitants. The

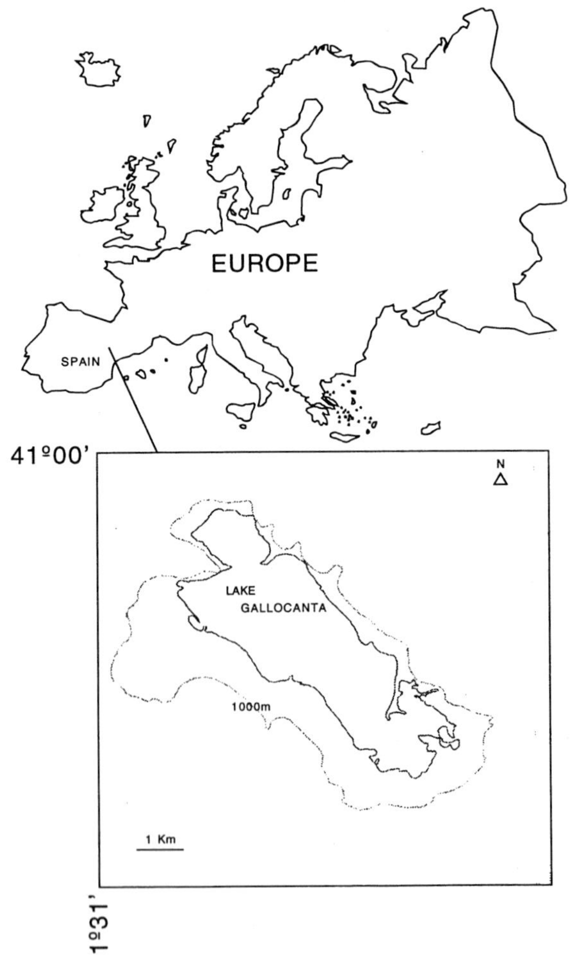

Figure 1. Geographical setting of Lake Gallocanta in Europe.

socio-economic status of the zone is typical of mountain areas in inland Spain. A high percentage of the population migrated during the sixties and seventies, and nowadays most people are elderly (50% over 60 years old).

The lake is the largest saline lake in Spain and, very probably, in Europe (see Williams, 1981a). It was recognized by the Ramsar Convention as a valuable wetland because of the high number of bird species and individuals staying in the zone at some periods of the year. However, no management plans relating to the natural resources or the development of the socio-economical structure of the site have been prepared or implemented. Specific public investments for some activities (agriculture) have been made in some parts of the catchment. General laws regulating uses of land affect this zone too. However, these do not include impacts on, or relationship with, the natural environment.

Limnological data were collected using the following methods. Water level was recorded every five days between 1974 and 1994 by visual observations of graduated stick by rangers of the regional Nature Consevancy Department. Water samples were collected from the same location in the central basin of the lake at different intervals: monthly during 1981–1990 and weekly during 1990–94. Total dissolved solids were determined from filtered (Whatman GF/C filters) samples by drying 110 °C and weighing the remaining solids. Water subsamples were stored frozen for analysis. Analyses were performed using standard methods (Grasshoff, 1975). Phytoplankton samples were preserved with Lugol's and counted (at 840× magnification) using an inverted microscope (Utermohl, 1958). Zooplankton samples were collected with a Neeskens bottle, filtered through a 55 μm mesh, preserved in formalin (4%) and counted at 140× magnification with an inverted microscope.

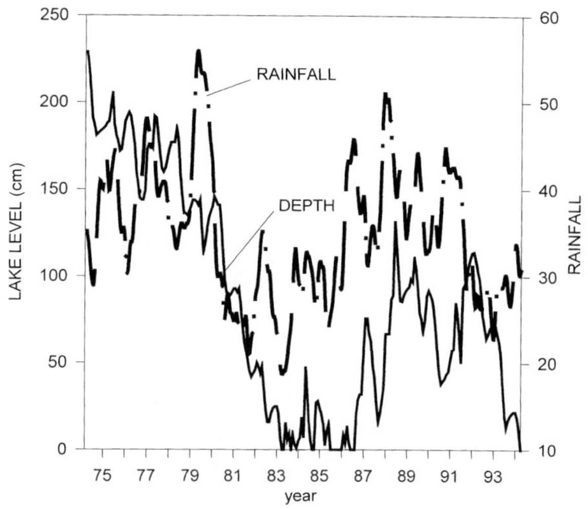

Figure 2. Temporal changes of water level and rainfall in Lake Gallocanta.

Hydro-ecological features

Fluctuations of the water level occur on different time scales (Figure 2). A typical annual fluctuation was observed every year from 1975 to 1984 and from 1989 to 1994, characterised by an increasing water level during the rainy periods of the year (most frequently autumn and spring) and a decreasing water level during the dry periods of the year (always in summer, but also in winter in some years). This pattern was not clearly observed during extremely dry years (1984–86), nor during some years of irregular distribution of the annual rainfall (1987–89, 1991–92).

Superimposed to the annual fluctuation, there have been rising trends for some years and falling trends in others. Although there are no data before 1975, many old people living in the villages surrounding Lake Gallocanta remember that the lake dried up completely (no water at all above the sediment of the Lake basin for at least four months) six times during the present century. Recently, evidence supporting this information was obtained by paleolimnological studies of the lake sediment (Rodó et al., submitted). This means that water level also fluctuates at an interannual scale (with an approximate frequency of 14 years).

Water level also fluctuates in Lake Gallocanta on other time scales. Longer time-scale fluctuations than those mentioned above occurred in the past as a consequence of climatic changes (Comín et al., 1991; Gracia, 1995). Fluctuations at time scales of less than one year occur with a frequency of several months. There may also be fluctuations at time scales of less than one month, even one week. However, the frequency of data recording (every 5 days) and the precision of the records (0.5 cm) does not permit interpretation at this scale.

The lake level fluctuations are strongly related to rainfall (Figure 2). A response lag of water level with respect to rainfall of between 0.3–2.4 months was calculated using data corresponding to the period 1975–1983 following Langbein's approach (Langbein, 1961) to the hydrology of closed lakes (Comín et al.,

1991). A time lag of 0 to 1 month has been estimated using data for the period 1975–1995 after time series analysis (Rodó & Comín, unpublished data).

The water level fluctuations correspond to those of the water table in the Quaternary aquifer which is made up of detritic materials, connected to the lake and occupies half of the catchment area, mostly that surrounding the lake. Its average depth is 110 m, and the discharges from this aquifer occur through the hydrographic network of temporary streams and several springs round the lake's shores. A carbonated aquifer, made up of Jurassic (50 m thickness) and Cretaceous materials (450 m thickness), separated by Mesozoic sand deposits (30–180 m thickness) also contributes to the lake. It flows through springs and surface run-off in the upper parts (1050 m a.s.l.) of the catchment and through its discharge into the Quaternary aquifer on the western side of the lake. The Mesozoic materials occupy 40% of the Gallocanta catchment area, which is an important zone for groundwater recharge. The two aquifers are underlain by Triassic (Keuper) clay and gypsum. Several smaller aquifers exist but are quantitatively unimportant. The whole hydrogeological system maintains more than 100 small (less than 5 ha), mostly temporary, water bodies distributed around the Gallocanta catchment. Direct runoff through aboveground streams is important during rainy periods. The porosity of the soil in the catchment area is very high and the surface run-off then reaches the lake through 5 temporary streams.

It has been estimated through a simulation programme, that groundwater abstraction for agricultural irrigation and water supply to the villages, can decrease the lake water level about 15 cm (Serrano, 1993). Surface water flow control in the catchment area also takes place by means of an artificial dam on one of the streams, during alternate years. The amount of water stored in this way could increase the level of the northernmost basin of Lake Gallocanta (Lagunazo de Gallocanta) up to 20 cm.

Although the hydrogeological studies performed have been inconclusive, several theoretical budgets have been proposed for the groundwater system of Lake Gallocanta. All of them attribute major importance to the recharge occuring in the Mesozoic area of the watershed, and some importance to the inputs occurring through the Quaternary materials. The major outputs are attributed to evapotranspiration from the lake and its shores. Water extraction for agricultural and urban purposes was calculated to be 1–2 Hm^3 ann^{-1} during 1987–89 and 0.37–0.77 Hm^3 ann^{-1} during 1978–83, which accounted for up to 20% of the total outputs from the lake's catchment area (Serrano, 1993). Differences between dry and rainy years are about 45% for the water inputs and 20% for the outputs.

Salinity in Lake Gallocanta also fluctuates at different time scales (Figure 3a), which means that salinity changes with different intensity and frequency, within the recorded range 11–350 g TDS l^{-1}. On the interannual time scale, these changes were, in general, inverse to those of water level, i..e. salinity increases as water level decreases, and vice versa, during the period 1975–1995. The same is true for annual periods of time. However, the salinity vs. water level (or water volume) changes follow a typical closed lake model (Figure 3b) (Langbein, 1961) during the drying up and refilling periods.

The other chemical characteristics of the water in Lake Gallocanta do not follow as clear a relationship with the water level as does salinity. An increase in micronutrient concentrations (nitrogen and phosphorus) was observed during periods of increasing water level, and vice versa, at shorter time scales (from weeks to a year) (Comín et al., 1983; Comín & Rodó, 1992). However, the ranges of variation of some chemical characteristics in the open waters of Lake Gallocanta (pH 7–10.5, alkalinity 2–8 meq l^{-1}, dissolved inorganic nitrogen 4–50 μM, soluble reactive phosphorus 0–10 μM) indicate that there are not great changes of these variables, nor are the waters affected by the impact of intensive industrial, agricultural or urban activities. However, the waters in the Quaternary aquifer do contain nitrate (connected with the presence of potassium), at concentrations higher than 50 mg l^{-1} over most of its area, which indicates a non-point source pollution by fertilizers, and nitrites above the concentration permitted by law in the vicinity of two villages located in the South plains of the Lake, which indicate point source pollution by wastewater.

The biological patterns differ between species. Temporal changes in abundance reflect the physical and chemical characteristics of the water and are also affected by interspecific relationships. Figure 4 shows a range of different temporal scales of variability for a group of the most abundant planktonic species. During 1990–94, the population of *Arctodiaptomus salinus* fluctuated at a longer temporal scale (interannual) than *Brachionus plicatilis*, reflecting their respective adaptations to the chemical changes associated with the water level fluctuations. They both also fluctuated over

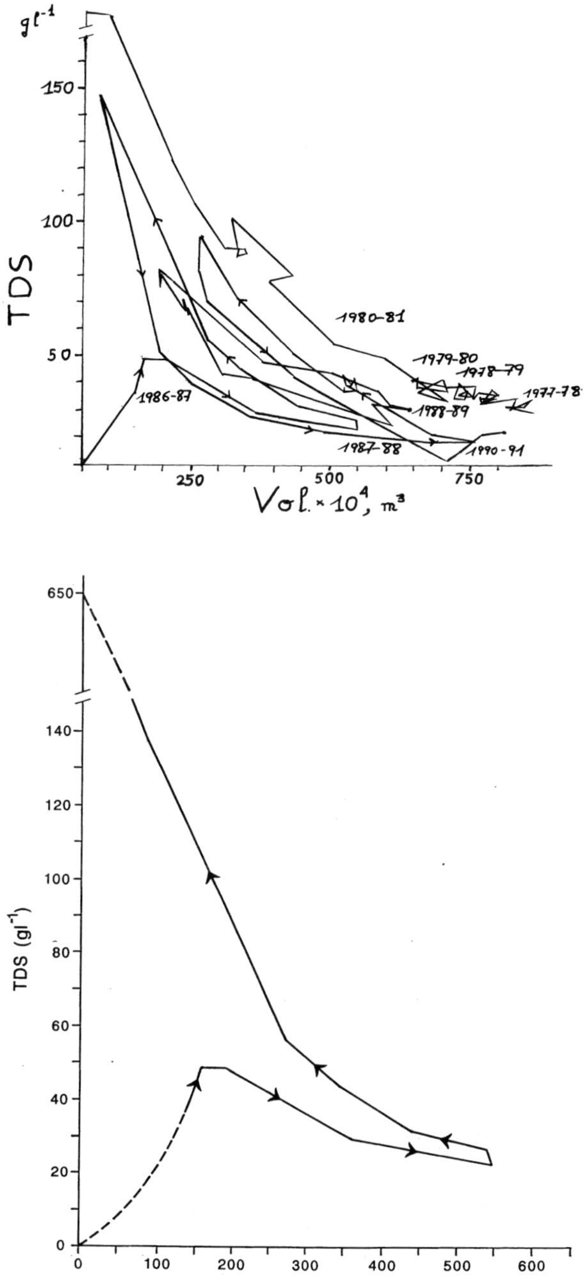

Figure 3. Total dissolved solids vs. water volume fluctuations in Lake Gallocanta during 1977–1991, above. Langbein's model for salinity vs. water volume fluctuations in closed lakes, below.

annual and shorter time scales. The range of temporal scales is completed by *Branchinella spinosa*, which only develops a population in the water column over several years after a total dryness. Similar examples have been observed for phytoplanktonic species. Mac-

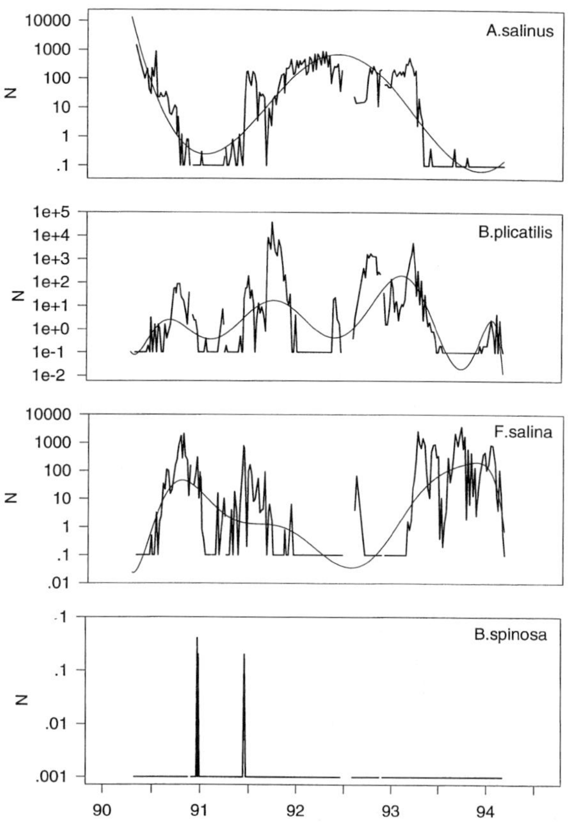

Figure 4. Temporal changes of the abundances of the several populations of zooplankton (*Arctodiaptomus salinus, Brachionus plicatilis, Fabrea salina, Branchinella spinosa*) in Lake Gallocanta during 1990–1994. Cyclic changes of these abundances are indicated in the figures by a simple 10-term moving average.

rophytes (*Ruppia drepanensis* and *Lamprothamnion papulosum*) develop seasonally during periods of relatively deep water column (>20 cm) and low salinity concentrations (<45 g l^{-1}).

Other biological populations fluctuate in Lake Gallocanta on different time scales. It is known as a migrating station for the largest European populations of common crane (*Grus grus*) however, numbers are not strongly related to water level fluctuations because cranesonly use flooded areas as night stopping-places. On the other hand, the total number of waterfowl individuals shows a strong correlation with the water level and with other characteristics (Comín et al., 1992), as they depend mostly both for nesting and feeding on the other aquatic species. Huge numbers of waterfowl (200 000 waterfowl, mostly overwintering birds) were observed in Lake Gallocanta during the seventies, whereas only between a few hundred and a few thousand were observed during the nineties, and none at all during completely dry years (1983–86, 1994–95).

Thus all the characteristics of the Lake display temporal fluctuations with different time scales strongly related to climatic fluctuations; and the overall ensemble of ecological features is only shown if the complete range of fluctuations is permitted to occur naturally. It is likely (and desirable) that fluctuations in the lake's characteristics occur in the future and this may be conceptually modelled by a loop (Figure 5, compared with Figure 3). The shorter the trajectory of the loop, the higher the constancy of the community structure and lake stability. During years of high increase or decrease in the water level, a long loop is shown, which indicates a large change in the lake's characteristics, including the biological community.

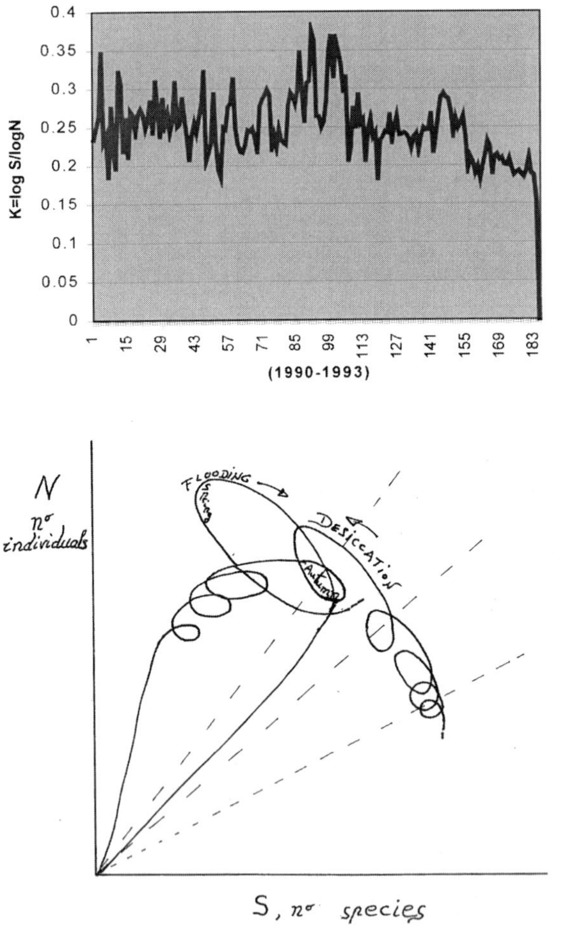

Figure 5. Temporal changes of the diversity (K=log$S$/log$N$) of phytoplankton during 1990–94, above, and a schematic diagram of the K changes during a longer period of time (every loop corresponds to an annual period), below.

It is easy to recognize the different dynamics shown during drying and refilling periods. This approach emphasises that the lake's overall richness is shown where the different time-scale of biological fluctuations occur at the rhythm imposed by the climatic fluctuations.

Present legislative and management framework

A large legislative arsenal exists at international, national and regional levels to regulate the conservation of this type of environment (Figure 6). However, most of these regulations deal with particular aspects of the territory and with species protection and do not consider the overall ecosystem and the physical and biogeochemical processes linking its different components.

The basic Spanish law for water management, which regulates all activities related to the use of water, is the Law of Water of 1995 (LA/95). In addition, the Regulation of the Hydraulic Public Domain, regulates the uses of water and defines the zones which are state property because of their importance. These laws also regulate groundwater extraction and establish conditions for the disposal of wastewaters. However, in practice there are several non-regulated activities or undefined areas with very negative effects on the management of saline lakes. For example, groundwater extraction is not, in reality, regulated because abstraction of up to 7000 m^3 per annum per rural property is permitted without any administrative authorization. Because of this, there are now more than 200 wells used for agricultural irrigation in the Gallocanta catchment area where the amount of water extracted, and its impact on the aquifer and lake is out of the control of the water and environment authorities. Moreover, regulations established an upper limit to the shores of Lake Gallocanta based on the average depth of the water for a limited number of years, without taking into account the fluctuating water level as a dynamic, essential, part of the lake's ecology. If adopted, these regulations would eliminate all the natural resources and ecological functions associated with much of the lake's shores. It would be better to define the upper limit of the hydraulic public domain in saline lakes as the highest elevation reached by the lake waters during the present climatic-ecological- conditions. Incorporating this definition would mean establishing a clear conservation objective for this type of habitat.

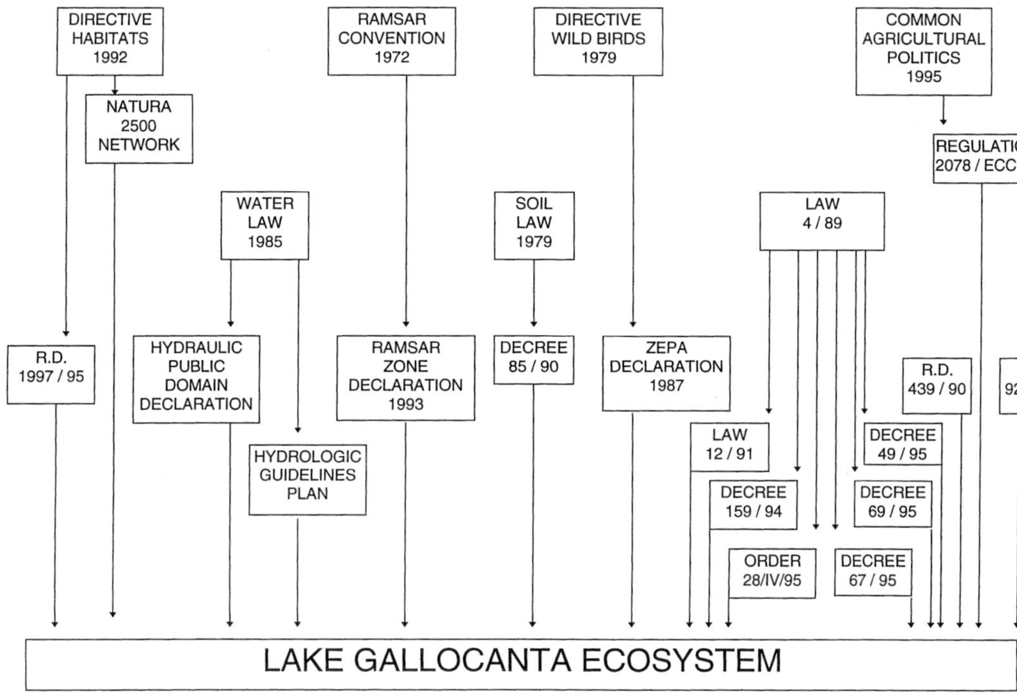

Figure 6. Diagram showing the present legislative framework affecting Lake Gallocanta ecosystem. The laws are distributed in three levels: the upper one contains international laws, the intermediate contains national laws and the lower one regional laws.

Regional Decree 85/1990, based on the Law of Land (Ley del Suelo), protects zones of high environmental value against the effects of urban developments by defining boundaries fot these zones. This decree also establishes that the authorities in charge of nature conservation must report offficially to the Provincial Urban Commissions, which take the final decision on all urban development projects, about the effects of these projects. In fact, these reports should be binding when they impose restrictions on urban development.

Birds are one of the most important natural resources of saline lakes: Lake Gallocanta was included in the Ramsar Convention in 1993 because of the high richness and abundance of bird species living at or visiting the lake. Furthermore it was declared an special zone for the Protection of Birds (ZEPA) based on EC Directive 79/409/ECC, after which the Spanish Government declared the zone a Wild Fauna Refuge (Decree 69/1995). However, these regulations focused only on the protection of animals and against hunting. They did not include the definitions of protection and conservation declared by IUCN-WWF (1980) and EC Directive 92/43/ECC, which are concerned with the conservation of natural habitats, and of wild flora and fauna. A new law passed in 1989, Law 4/89 on the Conservation of Natural Areas and Wild Flora and Fauna, was a significant advance in the protection of species and their habitats and in planning natural resources. This law was in line with the principles of the Spanish Constitution. A National and Regional Registry of Threatened Species was opened, which gave those species which required special protection a separate category. In the case of Gallocanta, *Puccinellia pungens* and *Botaurus stellaris* are endangered species by law. The common crane was given a special interest category. However, these regulations, in practice, mostly concerned the preservation of species rather than habitats. There was no consistent conservation of the ecosystem.

This was the objective of another initiative, called Rea Natura 2000, stemming from the same EC Directive 92/43/CEE and Spanish Royal Decree 1997/1995, which tries to develop the missing points of Law 4/89 for the protection and preservation of special areas. The aim is to declare, as Special Zones of Conservation, areas of ecological value and interesting habitats. Inexplicably, inland saline lakes and their associated habitats, which are one of the most representative and valuable ecosystems in Europe, are not included in the list of this Royal Decree. However, Lake Gallocanta will be able to take advantage of the Royal Decree 1997/95, because it was declared a special protected zone for birds (ZEPA).

During the 1980s and 1990s, a long conflict arose between farmers and conservationists because of the feeding habits of the common crane population, which ate cereal seeds on the arable land surrounding Lake Gallocanta and other zones. This arose because of the slow reaction of the regional administration, which compensated the farmers for the loss of productivity caused by cranes, but with lack of consistent assessment criteria for a few years. Management practice was also very inefficient because the costs of calculating the loss of productivity (calculated by insurance company valuers) were much higher than the loss of productivity itself. This situation was improved recently by an agreement between the administration and farmers to pay compensation for the year of lost productivity on a satisfactory basis for the farmers. Because of this and other experiences, different regional and national regulations have been established during the last decade in order to promote nature conservancy by farmers by avoiding extensive uses of natural areas and intensive uses of agricultural land which may have negative effects on natural areas. The National Law 4/89, and the regional Law 12/1992 regulating hunting activities in Aragón, included rules to compensate for damage to farming activities by wild or protected fauna. More recently three laws (Decree 159/1994 and Order 28 April 1995, both approved by the Regional Administration, and the Royal Decree 928/1995) have been passed to favour sustainable activities (e.g. cultivation of plants consuming low amounts of water, no groundwater extraction, over planting of seeds, preservation of wetland pastures) compatible with the preservation of natural areas of interest. Many of these measures are co-funded by the European Commission, which in this way makes practical its clear interest in the preservation of natural habitats, by the Spanish State and the Regional Government. 40 applications (pieces of land) were made under one of these measures during its first year in force and 400 during the second year, which is an indicator both of the possibility of integrating nature conservancy and socio-economic development, and of the potential interest of local people in participating actively in the management of land and water uses, which include natural processes in development plans aimed at preserving the overall ecological system.

A new law, the regional Decree 67/1995 to begin to establish a Plan of Regulation of the Natural Resources (PORN) in the ZEPA area of Lake Gallocanta, has been recently added to the legislative framework for the Lake Gallocanta ecosystem. This decrees that all activities requiring administrative authorization should also have a positive report from the authority in charge of the preservation of the natural environment. In practice, this law is the consequence of the social and political situation after the conflicts between some farmers and conservation groups. This fact shows the considerable social and political pressure on the administration and the lack of knowledge of the lake's ecosystem, including the basis for a sustainable long-term socio-economic development, and also the lack of clear policies for nature conservation.

The present legislative framework (represented in Figure 6) is in practice very inefficient as no clear statment about the conservation of the lake's ecosystem has been achieved. It is quite chaotic with different Governments (local, regional, national) involved on the management of different parts of the territory. Figure 7 (above) shows schematically this situation and the lack of consideration of ecological information in the management.

Future management framework

The history of the management steps followed in Lake Gallocanta, as in many other saline lake ecosystems, is characterized by two outstanding features – the lack of a planned objective and the imposition of initiatives by outside agencies. The objectives of the different management activities performed during the last 23 years were to protect some physical components of the ecosystem (e.g. birds, some particular plant species, promotion of a type of agricultural activity) and were based on such a general legislative framework that it was very difficult to apply them efficiently. Nevertheless, specific laws designed to provide economic stimulation of local activities, mostly farming, were very succesful. However, this is not an efficient way of pemanently managing an ecosystem because people's activities are not integrated in a general management strategy, but are just compensated not to do one thing for a certain time, which they can start again at any time.

However, opinion is growing, both among the catchment's inhabitants and external groups, in favour of integrating the conservation of the lake ecosystem into the socio-economic development of the catchment area, and also of the preservation of the natural re-

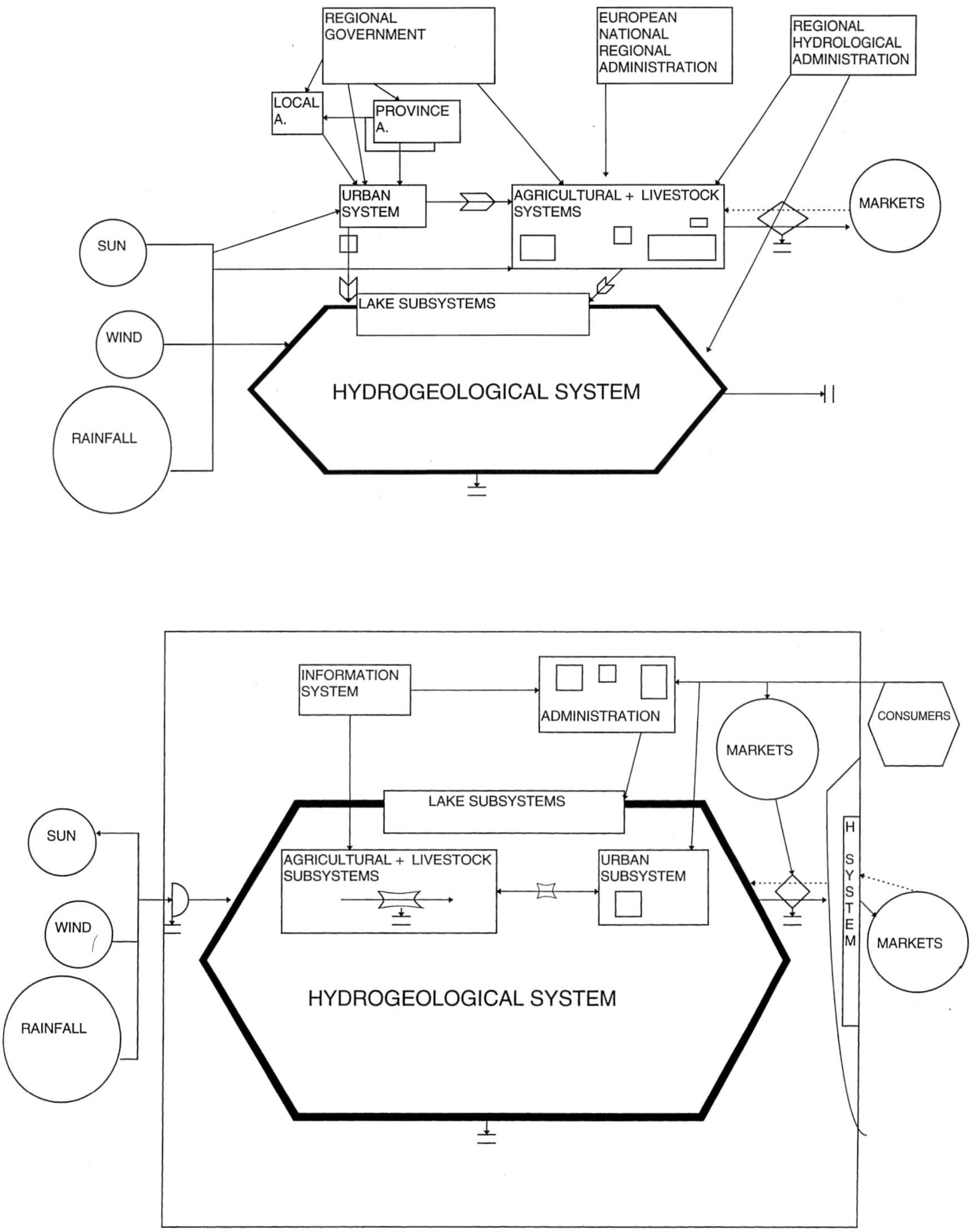

Figure 7. Flow diagram showing the present relationships between different compartments of Lake Gallocanta ecosystem and the management structure, above. Flow diagram of Lake Gallocanta ecosystem showing an idealized view which incorporates the ecological knowledge in its management structure, below.

sources by the inhabitants themselves. This is based on two facts: – no significant collective improvement in the relative (compared to other culturally close territories) standard of living has taken place in the catchment area for the last 23 years and a number of initiatives carried out by some people living in the catchment area and its vicinity have revealed the efficiency of adopting new practices as alternatives to the traditional socio-economic activities developed in the zone.

Undoubtedly, the conservation of the Lake Gallocanta ecosystem, and similar areas in S. Europe is a major objective to be achieved as soon as possible, and one which has been assumed by different general documents from the European Commision (Europe Environment, 1996) Participation of local people is vital in defining the objectives and, also, to make them aware of the importance of the natural resources of the area and the role they must play. However, the arsenal of legislation on water and natural areas is not based on the ecological knowledge of this type of ecosystem which could contribute to its preservation and management. In the case of Lake Gallocanta, it is clear that a protected natural area should be established, including the entire catchment area with zoning, providing different levels of protection. Two basic requirements must be accomplished to improve the management of the Lake Gallocanta ecosystem (and other saline lake ecosystems):

1. Planning the objectives should include local participation. Because the lake ecosystem's functioning is mostly dependent on groundwater flows, this planning should include inhabitants of the whole hydraulic catchment area. Their participation in the hydrogeological system should also be stimulated. The cultural heritage of local people is of inestimable value because it has evolved under and adapted to the same fluctuations as the lake and because they are also part of the lake ecosystem. Figure 7 shows schematically the ideas of integrating the overall ecosystem and incorporating the ecological knowledge in the management structure of this type of environments.

2. Inclusion of the lake ecosystem's fluctuations in the future management and legislative framework should be uncontroversial, since the overall values of the ecosystem are only of value if the fluctuations occur in relation with the climatic fluctuations and are not negatively affected by human disturbance. The lake and the ground waters should be permitted to fluctuate in line with natural climatic fluctuations and activities negatively affecting these fluctuations should be restricted. For example, an increased groundwater extraction of 3.5 and 4.5 Hm^3 ann^{-1} for agricultural irrigation would cause a lake water fall of 4, 12 and 15 cm, respectively, (Serrano, 1993), which is enough to produce a complete change in the structure of the biological community and the ecological functioning of the lake, during the periods of time when the lake water level is lower than 1 m. Moreover, the present groundwater abstraction for intensive irrigation in the catchment area can take the overall lake ecosystem to more frequent and persistent desiccation periods with a clear loss of natural resources (Figure 8). Alternative activities for the socio-economic development of the zone, which integrate the conservation of the lake's ecosystem and its natural resources, should be promoted and stimulated both from within the catchment and from outside.

Acknowledgments

This paper was made possible by field and administrative assistance of the Diputación General de Aragón-Servicio del Medio Natural (the paper does not put forward the ideas of DGA, but those of the authors). Funds were received from CICYT AMB95-0935-E and EC-ENV4-CT95-0147.

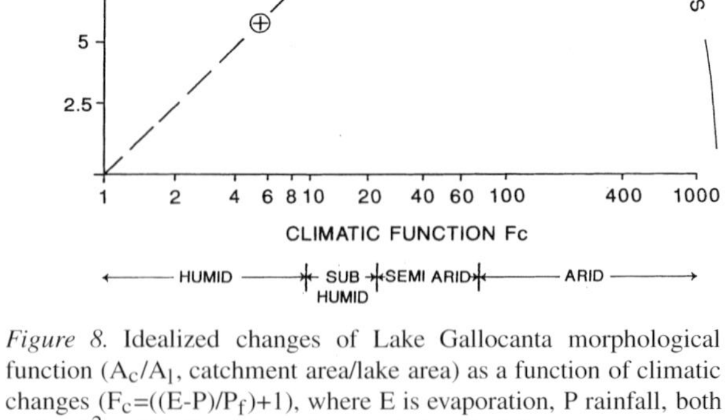

Figure 8. Idealized changes of Lake Gallocanta morphological function (A_c/A_l, catchment area/lake area) as a function of climatic changes (F_c=((E-P)/P_f)+1), where E is evaporation, P rainfall, both as $L\,m^{-2}$ and P_f the water volume entering the lake per square meter of catchment area) and man made water abstraction.

References

De Bernardi, R., A. Calderoni & R. Mosello, 1996. Environmental problems in Italian lakes, and lakes Maggiore and Orta as succesful examples of correct management leading to restoration. Verh. int. Ver. Limnol. 26: 123–138

Comín, F. A. & T. G. Northcote (eds), 1990. Saline Lakes. Developments in Hydrobiology 59: 308 pp.

Comín, F. A., R. Julia & P. Comín, 1991. Fluctuations, the key aspect for the ecological interpretation of saline lake ecosystems. Oecol. Aquatica 10: 127–136.

Comín, F. A. & X. Rodo, 1992. Seguimiento y Control de las Comunidades Biológicas de la Laguna de Gallocanta. DGA-Servicio del Medio Natural, Unpublished Report 101 pp.

Comín, F. A., X. Rodó & P. Comín, 1992. Lake Gallocanta (Aragón, NE Spain). a paradigm of fluctuations at different scales of time. Limnetica 8: 79–96.

Comín, F. A., M. Alonso., P. López & M. Comelles, 1983. Limnology of Gallocanta Lake, Aragón, Northeastern Spain. Hydrobiologia, 105: 207–222.

De Bernardi, R. (in press). Lake management and restoraton experiences. Verh. int. Ver. Limnol. 26.

Europe Environment, 1996. European Commission: communication on water policy. Europe Environment, 472, 23 pp.

Gracia, F. J., 1995. Shoreline forms and deposits in Gallocanta Lake (NE Spain). Geomorphology 11: 323–335.

Grasshoff, K. 1975. Methods of Seawater Analysis. Verlag Chimie, 426 pp.

Hammer, U. T. (ed.), 1983. Saline Lakes. Developments in Hydrobiology 16. Martinus Nijhoff / Dr W. Junk Publishers, The Hague, 274 pp. Reprinted from Hydrobiologia 105.

Hurlbert, S. H. (ed.), 1993. Saline Lakes. Developments in Hydrobiology 87. Kluwer Academic Publishers, Dordrecht, 335 pp. Reprinted from Hydrobiologia 197.

Langbein, W. B., 1961. Salinity and hydrology of closed lakes. U.S.G.S. Prof. Paper 412, 46 pp.

Melack, J. M. (ed.), 1988. Saline Lakes. Development in Hydrobiology 44. Kluwer Academic Publishers, Dordrecht, 325 pp. Reprinted from Hydrobiologia 158.

Serrano, J., 1993. Inventario y Control Piezométrico del Sistema Acuífero de la Laguna de Gallocanta. DGA.-Consejería de Agricultura, Ganadería y Montes (Exp. Z-20768), 133 pp.

Rodó, X., S. Giralt, F. Burjachs, F. A. Comín, R. G. Tenorio & R. Julia (submitted). High-resolution saline lake sediments as enhanced tools for relating proxy paleolake records to recent climatic data series.

Utermöhl, H., 1958. Zur Vervollkomnung der quantitativen Phytoplankton-Methodik. Mitt. int. Ver. theor. angew. Limnol. 9: 1–38.

Vallentyne, J. R., 1972. Freshwater supplies and pollution: Effects of the demophoric explosion on water and man. In N. Polunin (ed.), The Environmental Future. Macmillan, London.

Williams, W. D., 1981a. Inland salt lakes: an introduction. Hydrobiologia 81: 1–14.

Williams, W. D. (ed.), 1981. Salt Lakes. Developments in Hydrobiology 5. Martinus Nijhoff / Dr W. Junk Publishers, The Hague, 458 pp. Reprinted from Hydrobiologia 81/82.

Williams, W. D., 1992. The worldwide occurrence and limnological significance of falling water levels in large, permanent saline lakes. Verh. int. Ver. Limnol. 25: 2132–2138.

Hydrobiologia **395/396:** 253–263, 1999.
D.M. Harper, B. Brierley, A.J.D. Ferguson & G. Phillips (eds), The Ecological Bases for Lake and Reservoir Management.

Understanding deep oligotrophic subalpine lakes for efficient management

Nico Salmaso[1], Fabio Decet[2] & Paolo Cordella[1]
[1]*Dipartimento di Biologia, Università di Padova, via U. Bassi 58/B, 35131 Padova, Italy*
[2]*Presidio Multizonale di Prevenzione - ULSS 1, via S. Andrea 1, 32100 Belluno, Italy*

Key words: Lake Garda, deep lakes, mixing depth, oligomixis, chemistry, phytoplankton

Abstract

Lake Garda – along with other large lakes located on the southern edge of the Alps – belongs to a distinct typology of lakes characterised by high depths and large volumes. Owing to its morphometric features (max. depth 350 m), the water column of L. Garda undergoes complete mixing only in harsh winters. During the period 1991–1995 this was only once, in 1991; in other years the mixing depth lay between 140 to 180 m. There is a correlation between lake volume at overturn with algal abundances and water transparencies. The extent of the overturn appears to control the interannual variations of the algal crop around a baseline trophic status determined by the whole total phosphorus content in the water column. Full understanding of the limnological features of the lake are thus important for the most suitable management practices. Future plans should try to reduce the discharge of sewage into deep waters to a minimum; it is stressed that the only reliable solution for preservation of water quality in deep lakes of this typology is reduction of nutrient loads. As far as L. Garda is concerned, this appears fairly pressing, considering the increase of total phosphorus concentrations in the water column observed since the 1970s.

Introduction

Lake Garda is the largest (49×10^9 m^3) and most extensive (368 km^2) of the Italian lakes. Along with lakes Orta, Maggiore, Lugano, Como, Iseo and Idro, it is part of the group of deep lakes (Insubrian lakes) located south of the Alps in the most densely populated and highly productive area of Italy. With a total volume of over 124×10^9 m^3 these lakes constitute one of the largest freshwater supplies in Europe. Their waters are used for agricultural and industrial purposes, fishery and as supplies of drinking water. Moreover, they represent an important resource for recreation and tourism due to the attractive landscape and high water quality.

The deep southern subalpine lakes, once oligotrophic, have in recent decades experienced progressive deterioration, indicated by an increase in nutrient concentrations reflected in palaeolimnological studies (Guilizzoni et al., 1983; Ambrosetti et al., 1992). However, detailed historical and continuous limnological records are available only for lakes Maggiore and Lugano, as the other lakes have been studied less systematically. In Lake Garda, detailed limnological research, with monthly sampling, started only in the late eighties, 20 years after the last study (IRSA, 1974), (e.g. Chiaudani & Premazzi, 1990; Salmaso & Cordella, 1994; Provincia Autonoma di Trento, 1994). During this interval, however, annual chemical analyses were carried out in the periods of greatest overturn (spring; e.g. Bonomi et al., 1979; Ambrosetti et al., 1983; 1992).

Construction of water treatment plants, reduced phosphorus in detergents and the laws passed against water pollution in the seventies and eighties led to an improvement in the water quality of lakes Maggiore and Lugano (Ambrosetti et al., 1992); in Lake Garda it is more difficult to define the trophic evolution due to the lack of detailed and continuous historical information. However, the continuous increase of spring phosphorus concentrations in the water column (Ambrosetti et al., 1992), recent episodes of cyanobacterial blooms (Salmaso et al., 1994) and the high phytoplankton density and biovolume recorded in recent years (Salmaso & Cordella, 1994) seem to indicate a worsening of the lake's trophic condition. Interpretation of annual variations of algal production in Lake Garda, however, is not so simple, due to the lake's oligomictic character, and thus the different degree of the water circulation. The importance of this characteristic in explaining the interannual quantitative phytoplankton variations observed in the deep subalpine lakes, should not be underestimated considering the high nutrient concentrations (e.g. Mosello et al., 1991; Barbieri & Mosello, 1992; Garibaldi et al., 1995) that characterise their deep waters.

This paper reports an account of the limnological research carried out from 1991 to 1995 in Lake Garda and assesses the role of the extent of spring water circulation in explaining interannual variation in the algal community. It then examines, in the light of the lake's limnological features, the problems related to management policies.

The lake and catchment

Lake Garda has two basins, separated by an underwater ridge connecting the Sirmione peninsula with Punta S. Vigilio (Figure 1). The northwestern (NW) basin is the larger and deeper of the two; in its northern part the shores are steep and the bottom extends over 20 km to depths between 300 and 350 m. The shallower southeastern (SE) basin, with a maximum depth of 81 m, represents only a small portion of the lake's overall volume (less than 7%).

Ambrosetti et al. (1992), pointed out that, because of its geographical position, L. Garda, along with other Italian deep subalpine lakes, should be a warm monomictic basin; however, owing to their great depth, Insubrian lakes are in fact oligomictic. In the deeper NW basin of L. Garda, the water column undergoes a complete overturn only in very cold and windy winters, whereas in the shallower basin overturn occurs annually. A detailed analysis of the causal factors determining overturn extent has been carried out in L. Maggiore by Ambrosetti et al. (1979); wind, solar radiation, superficial water temperature and air temperature were shown to be the fundamental meteorological parameters in determining the mixing processes.

Lake Garda has a long theoretical water renewal time (27 years for the whole basin), due to its low catchment (lake included)/lake surface ratio (around 6) and to low annual rainfall (790–1150 mm, Giorgi

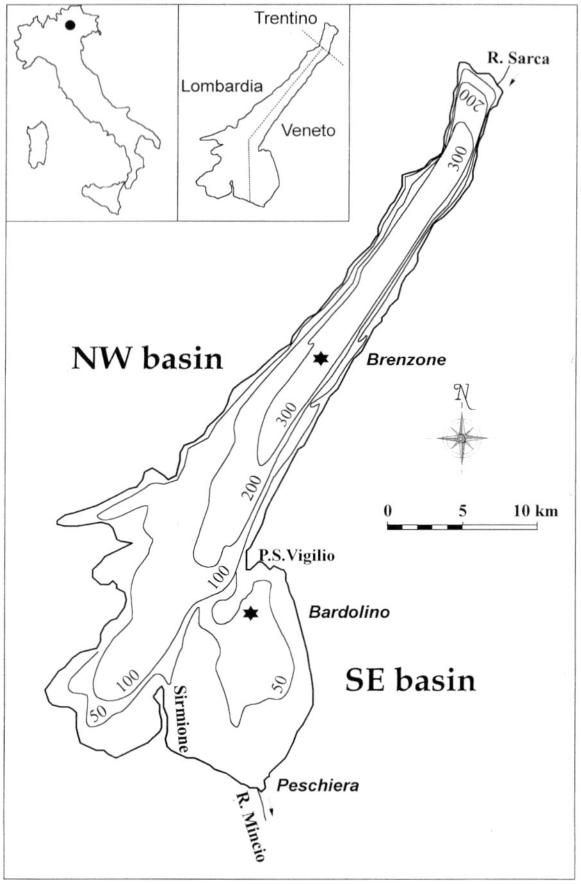

Figure 1. Bathymetric map of Lake Garda and location of the sampling stations (from Barbanti, 1974, modified).

et al., 1974). The drainage basin covers 1922 km^2 (excluding the lake), most of which is formed by sedimentary rocks (limestones, dolomites, marly and dolomitic limestones, morainic and fluvial deposits). Igneous (granites, granodiorites, quartz diorites, quartz porphyries) and metamorphic rocks (quartz phyllites, mica schists and paragneisses) are abundant only in the northwestern part of the catchment.

The main inflow is River Sarca (at the northern edge of the lake), the outflow is River Mincio (Figure 1); other tributaries are of less importance and mainly flow towards the western and northern shores.

Methods (1991–95)

The data considered in this study refer to samples collected between spring 1991 and December 1995 in two pelagic stations situated in the deepest zone of the NW (Brenzone) and SE (Bardolino) basins (Figure 1). During these five years, other stations were sampled, but the data do not cover the whole study period.

In 1991 to mid-1992, samples were carried out every two weeks between July and September, and at intervals of 14–28 days in other months; afterwards they were carried out monthly.

Detailed chemical analyses are available only for the last year of the series (1995) when water samples were collected in upper layers (strata 0–2 and 19–21 m) in both stations and at depths of 100 m and 200 m in Brenzone and 40 m in Bardolino. In other years the information is more scanty. Oxygen data (Winkler method) were determined at Brenzone at 200 m for the whole study period and at 100 m and 350 m (maximum depth) from 1993 onwards. At Bardolino, oxygen data for the lake bed have been available since the start of the research. Chemical analyses were carried out following the methods described in APHA et al. (1989), as summarised in Decet et al. (1995).

Samples for phytoplankton analyses (see Salmaso, 1996, for details) were collected at the surface layer (0.3 m, 0–2 m) from 1991 to 1995 in Bardolino and in the periods 1991–May 1992 and June 1994–1995 in Brenzone; in both stations integrated samples (0–20 m) were taken from June 1992 to May 1994. Additional samples are available, in 1995, for the stratum 19–21 m.

Transparency was estimated, with the aid of a bathyscope, using a Secchi disk. Temperature, pH, conductivity and oxygen measurements were carried out on the water column with an underwater multiparameter probe. A sequence of depth profiles of pH, conductivity and oxygen was utilised for determining the time and extent of mixing in L. Garda. Depth of mixing was taken to be the maximum point at which the upper layer of uniform values for conductivity, O_2 and pH met a lower layer of rapidly changing values (Goldman & Jassby, 1990). The reported values of mixing depths represent only approximate estimates owing to their dependence on the sampling frequency and on non-perfect synchrony, in some years, in depth and time changes of all the selected variables.

Results and discussion

Thermal structure, oxygen values and overturn

On the surface the lowest temperatures were reached in January and March (8–10 °C), the highest in July and August (22–27 °C). From late spring to autumn,

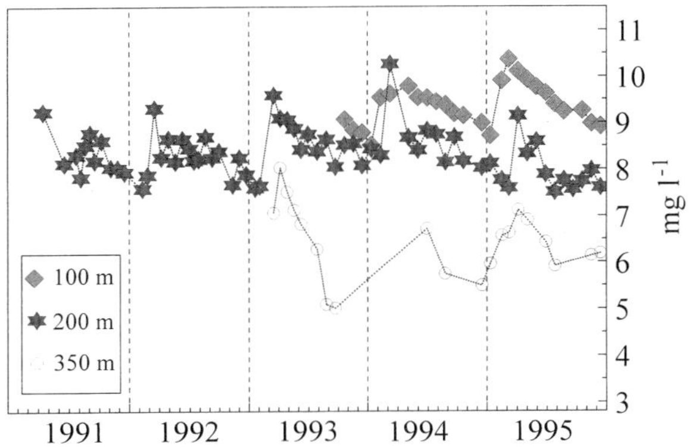

Figure 2. Oxygen concentrations at Brenzone; 350 m values refer to samples taken one metre above the bottom at maximum depth.

there was marked stratification, with a maximum deepening of the thermocline to 35–40 m. In the deep waters of the NW and SE basins, the temperatures ranged from 7.6–8.6 °C (100–350 m) and 7.0–9.5 °C (70 m).

At the surface, oxygen concentrations in the warmest months (9–10 mg O_2 l^{-1}) were greater than the theoretical values, yet generally within 120% of saturation. In the coldest months oxygen saturations were around 90%, with values nearly 10.5–11.5 mg O_2 l^{-1}at the time of the maximum overturn which occurred in March or early April. At the surface, the oxygen dynamics of the two basins were similar.

Temporal change in oxygen concentrations showed a clear separation of the deepest waters from the superficial ones (Figure 2). There has been a clear oxygen gradient since 1993, with bottom values between 5–8 mg O_2 l^{-1} (40–70% of saturation). The inversion of the oxygen gradient in spring 1994 (as evidenced by the values at 100 and 200 m) must be explained by mixing mechanisms, different from convective motions which are generally responsible for the cooling phase in lakes and bring about multilayered structures in the water column (see Ambrosetti et al., 1983; 1992). At the time of maximum overturn, oxygen values at 100 m (85–90%) approached the equilibrium with respect to the surface values.

From the middle of April 1991 to 1992, oxygen values at 200 m ranged from 7.6 to 9.3 mg O_2 l^{-1} (65–80% saturation), never approaching the surface values. However, the data collected at the beginning of March 1991 with the multiparameter probe, clearly demonstrated a complete process of reoxygenation and homogenisation of the water column in this month. The occurrence of complete water overturn in

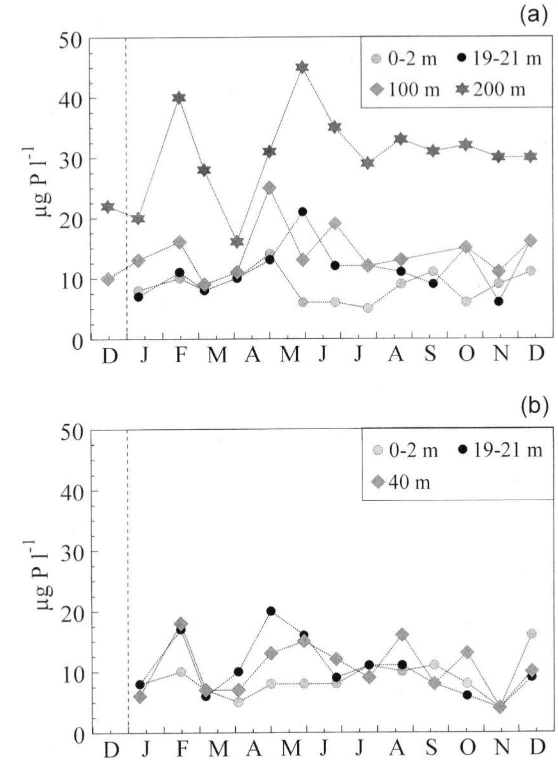

Figure 3. Total phosphorus concentrations at (a) Brenzone and (b) Bardolino.

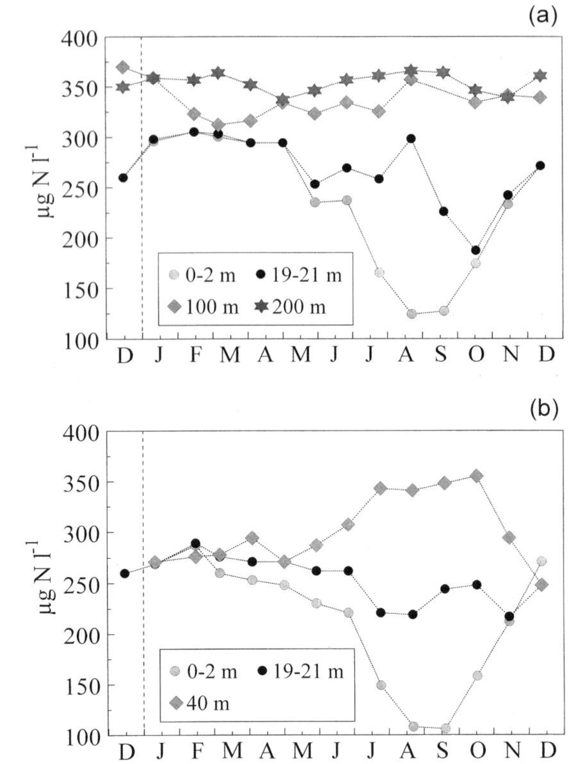

Figure 4. Nitrate (as N) concentrations at (a) Brenzone and (b) Bardolino.

the deepest basin in 1991 is also confirmed by observations made by other authors (Provincia Autonoma di Trento, 1991; Ambrosetti et al., 1992). In the later years, 1992–1995, the depth profiles of O_2, conductivity and pH show that the layer involved in complete overturn ranges between 140 and 180 m.

The shallower SE basin underwent a complete overturn every year from February to March, with O_2 concentrations being homogeneous over the whole water column (10–11 mg $O_2 l^{-1}$; 85–90% saturation); at the bottom, the lowest O_2 values (3.0–4.5 mg O_2 l^{-1}; 25–40% of saturation) were recorded towards the end of the thermal stratification, from September to December (Salmaso & Cordella, 1994).

Algal nutrients

The lake's oligomictic character has profound consequences, not only for oxygen distribution, but also on the concentration of algal nutrients on the water column. In 1995, four years after the last complete overturn, phosphorus, nitrogen and silica compounds showed a remarkable increase in the deepest waters (100 m and/or 200 m) of the NW basin (Brenzone).

In the epilimnion (0–2 m and 19–21 m strata) the total phosphorus concentration was always less than 21 μg P l^{-1} (Figure 3a,b) and the annual average concentration (10 μg P l^{-1}) was the same at both stations. A temporal trend for this nutrient was hardly observable; only replenishment from deeper strata soon after the period of maximum overturn (March–April) is rather evident at 20 m; this increase is less evident at the surface, probably due to algal utilisation in the absence of light limitation and to the sinking of algal particles. Total phosphorus concentrations at 100 m (NW) and 40 m (SE) were comparable to those found in the upper strata; on the other hand, concentrations at 200 m were considerably higher (15–45 μg P l^{-1}). Minor differences over the entire water column were present only during the limited period of maximum overturn. Soluble phosphate concentrations in the epilimnion (0–2 m, 19–21 m) of the two basins and at 40 m of the SE basin were generally lower than 5 μg P l^{-1} but were between 4–13 μg P l^{-1} and 12–30 μg P l^{-1} at 100 m and 200 m respectively in the NW basin.

Steep concentration gradients also characterised nitrate nitrogen (NO_3-N) and silica (SiO_2-Si), with

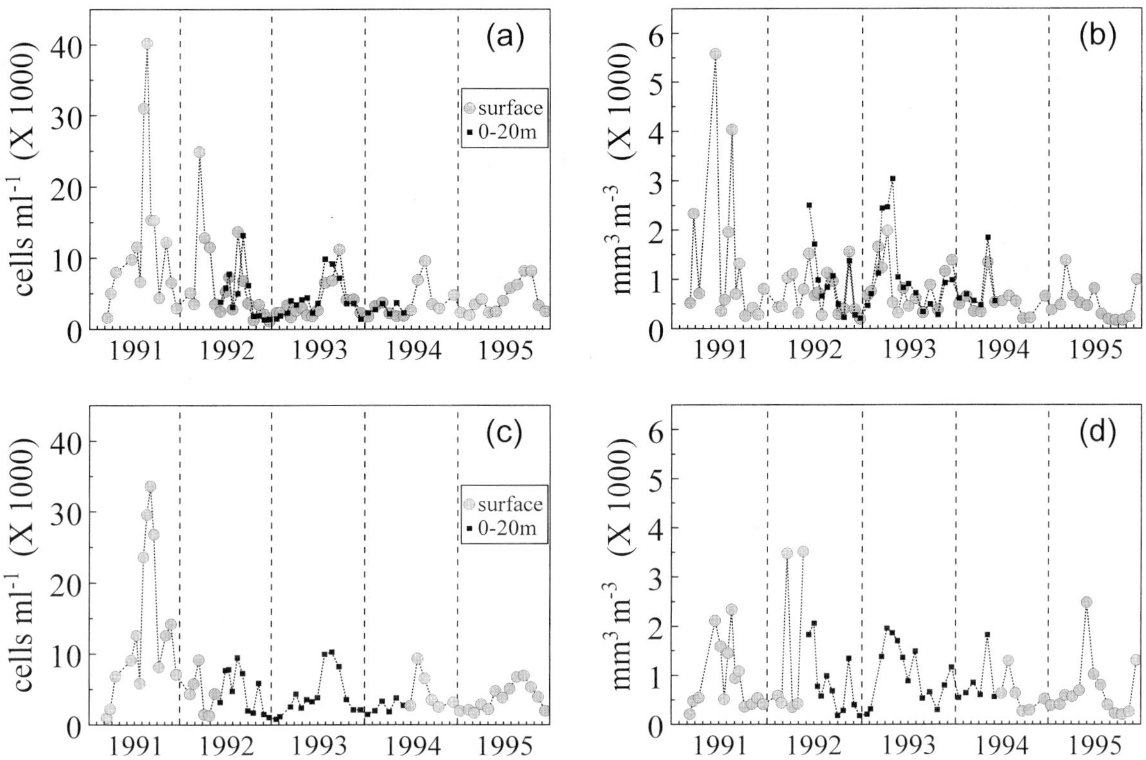

Figure 5. Phytoplankton density and biovolume at Bardolino (a, b) and Brenzone (c, d) respectively.

some differences and a more regular temporal pattern in comparison with phosphorus compounds.

At the surface (0–2 m) and at 19–21 m, nitrate concentrations were similar in the two basins (Figure 4a,b); moreover they showed similar temporal trends, indistinguishable in the 0–2 m layers (r=0.98, P<0.01). A strong reduction of nitrate in the warmest months was particularly evident (with values ranging from 100 to 175 μg N l^{-1} in July–October) in the first two metres of both stations due to the increase and activity of the algal community (see next section). Nitrate concentrations at 100 m (310–370 μg N l^{-1}) were comparable or a little lower than those found at 200 m (335–370 μg N l^{-1}); temporal trends at these depths were slight. Minor differences in the water column occurred in spring during the maximum overturn, with complete homogenisation of the waters in the first 40 m in the SE basin (Figure 4b) but with significant differences persisting in the deepest basin (Figure 4a).

Si-SiO_2 concentrations in the surface layers (0–2 m, 19–21 m) ranged between 0.05–0.60 mg Si l^{-1} with similar values and temporal trends in the two basins. The lower values in these layers occurred in the warmest months. At 100 m and 200 m of the NW basin, silica concentrations ranged between 0.50–0.90 mg Si l^{-1} and 0.80–1.35 mg Si l^{-1}, respectively.

The algal community

The moderate trophic level, suggested by the low epilimnetic phosphorus concentrations is confirmed by moderate phytoplankton development. However, since 1991, the two variables expressing algal abundance – density and biovolume – have shown a progressive reduction (Figure 5) in surface samples (only those for the southern stations (Bardolino) are available for the entire study period). The first year of observation was characterised by high values of the two variables, with summer peaks of over 30 000 cells ml^{-1} and 3000 mm^3 m^{-3} and annual mean values of 12 200–13 800 cells ml^{-1} and 1400–950 mm^3 m^{-3} respectively for Bardolino and Brenzone. Successive years showed a decrease, rather regular for density values, of the amplitude of the seasonal cycles of algal abundances, up to annual mean values of 4250–3850 cells ml^{-1} and 500–700 mm^3 m^{-3} respectively, for the SE and NW stations in 1995. At Bardolino, the integrated 0–20

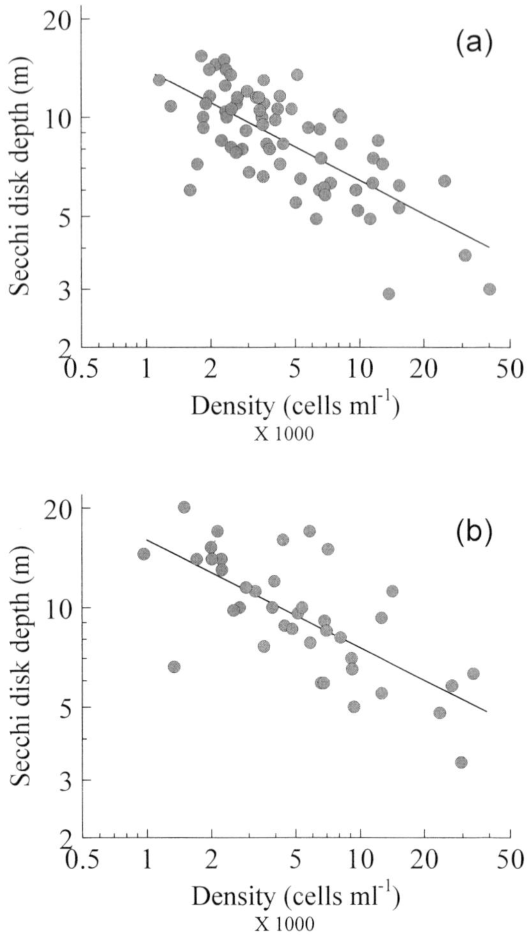

Figure 6. Relationship between Secchi disc depth and surface phytoplankton density at Bardolino (a) and Brenzone (b); after logarithmic transformation, both regression are highly significant ($P<0.01$).

m algal abundances appeared to be in line with the surface ones, particularly as far as density values are concerned.

The reduction in algal particles in the lake, mirrors the trend in Secchi disk depths (Figure 6; after logarithmic transformation both relationships, with r^2 equal to 0.50 and 0.48, are highly significant). Minimum summer transparencies between 1991 and 1994–1995 doubled from 3 m to 5–6 m, while the corresponding annual mean values rose from 7.0 m to 9.5–10 m in Bardolino and from 8.7 to 11.5 m in Brenzone.

The trends observed in density and biovolume values of algae are almost exclusively due to seasonal and interannual variations of the dominant groups (Figure 7). The most abundant classes were Chlorophyceae, Cyanobacteria and Bacillariophyceae (the others classes had fewer than 1300 cells ml^{-1}). All three classes considered contributed to the high summer density peaks observed in 1991 (Figures 7a,b,c) with a major development of Chroococcales (*Aphanothece*, *Coelosphaerium*) and Oscillatoriales (*Planktothrix*, *Planktolyngbya*) among the cyanobacteria, Chlorococcales and Tetrasporales among the chlorophytes and small *Cyclotella* spp. among the diatoms. In the following years these three groups underwent a significant reduction (rather gradual in the case of chlorophytes, Figures 7a) in their summer density values. Summer 1991 was also characterised by two extensive cyanobacteria blooms of *Anabaena lemmermanni* P. Richt. in the upper few centimetres of the water column; a similar bloom of the same species was observed in summer 1992 (Salmaso et al., 1994).

The dominant classes by biovolume were Chlorophyceae, Bacillariophyceae and Dinophyceae (Figures 7d,e,f); (other classes never reached 300 mm^3 m^{-3} (Cyanobacteria and Chrysophyceae) or 800 mm^3 m^{-3} (Cryptophyceae)). The highest (>5000 mm^3 m^{-3}) biovolume peak observed in 1991 in the SE basin (Figure 5b) was due to the development of a dinoflagellate (*Ceratium hirundinella* (O.F. Müll) Dujardin, Figure 7f); in the same year other peaks were due to spring development of colonial diatoms (*Tabellaria*, *Aulacoseira*, *Fragilaria*), to small centric summer diatoms, to filamentous chlorophytes (*Mougeotia*) and to cryptophytes (*Cryptomonas* sp.).

Every year, the coldest months had weak development, both in density and biovolume, of filamentous cyanobacteria and chlorophytes (Oscillatoriales, *Mougeotia* sp.), colonial diatoms and cryptophytes. Despite the decrease in algal crop recorded, seasonal phytoplankton community followed a similar pattern with comparable assemblages and clear trends in the community change rate every year (Salmaso et al., 1995; Salmaso, 1996).

The link between algal development and internal physical factors

The reduction in algal quantities, since 1991, is difficult to explain with classical models relating annual algal crops to nutrient loads. The great lakes are slow to react to reductions in nutrients (for example, the case histories of lakes Lugano and Maggiore: Manca et al., 1992; Barbieri & Mosello, 1992); moreover, models including renewal time (e.g. OECD, 1982) appear imprecise owing to the lake's oligomictic character.

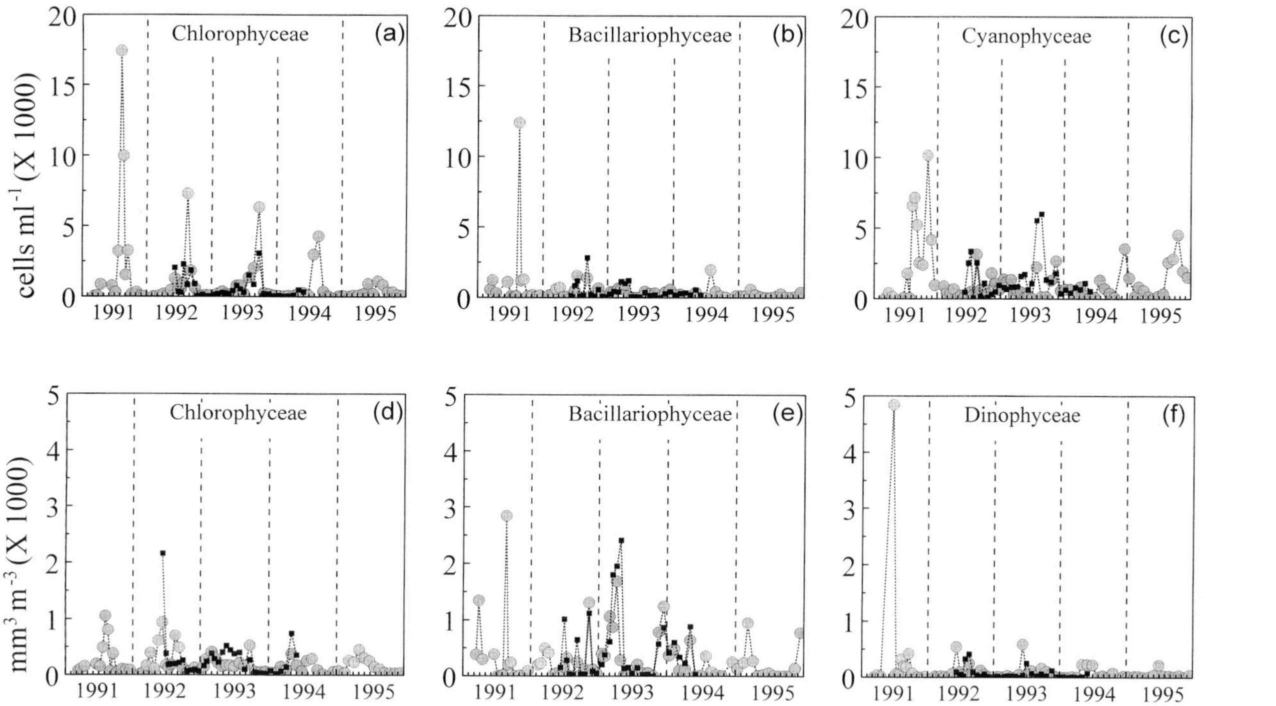

Figure 7. Density (a, b, c) and biovolume (d, e, f) of the most abundant phytoplankton classes at Bardolino. Circles and squares indicate surface and integrated 0–20 m samples respectively. With minor differences, the trends found in Brenzone are comparable to those reported for Bardolino.

Owing to the distinct gradients of nutrients at depth, the extent of the mixed water column could have a strong influence on nutrient replenishment to euphotic layers. The importance of mixing events in fertilising surface waters and in modifying primary productivity, was recognised more than 50 years ago in the oceans and since then also in deep lakes (Viner, 1985; Goldman & Jassby, 1990). However, despite its acknowledged importance, there are few quantitative examples (e.g. Goldman & Jassby, 1990) demonstrating the potential role of spring circulation as a determinant of annual primary production in deep lakes.

In Figure 8, mean algal abundances and minimum Secchi disk values are plotted along with the lake volumes involved, every year, at spring overturn. Starting from the mixing depths, the mixed lake volumes are calculated in Figure 8a, by means of a polynomial interpolation of fourth degree, on the basis of discrete hypsographic data reported in Barbanti (1974). In Figures 8b and 8c, the values refer to average densities and biovolumes relative to the period affected by the overturn, between the end of the circulation and the beginning of the next one (i.e. from March to December). The reported data refer to surface values; in the case of Brenzone (cf. Figures 5c and 5d), the averages in 1992 and 1994 were computed also on the basis of the 0–20 m samples whereas 1993 averages include only 0–20 m samples. The data collected at Bardolino (for which both the surface and 0–20 m series are available, cf. Figure 5a,b), suggest complete comparability of the two series in estimating averages in the case of density values (with differences less than 15%), but an overestimation of biovolume values in the case of 0–20 m data (with differences up to 30% in 1993).

The figures show a striking influence of the extent of water circulation on the three variables. Mixed lake volumes have a significant direct association with average density (r=0.93, P<0.01) and biovolume (r=0.70, P<0.05) values regarding the two sampling stations; moreover, due to the strong influence of algal particles on water transparency (Figure 6), they also have an inverse association with the minimum Secchi disk values (r=0.71, P<0.05). The lower association of biovolume averages compared with the density ones, is due to the lower values recorded in 1991 in Brenzone and, probably, to an overestima-

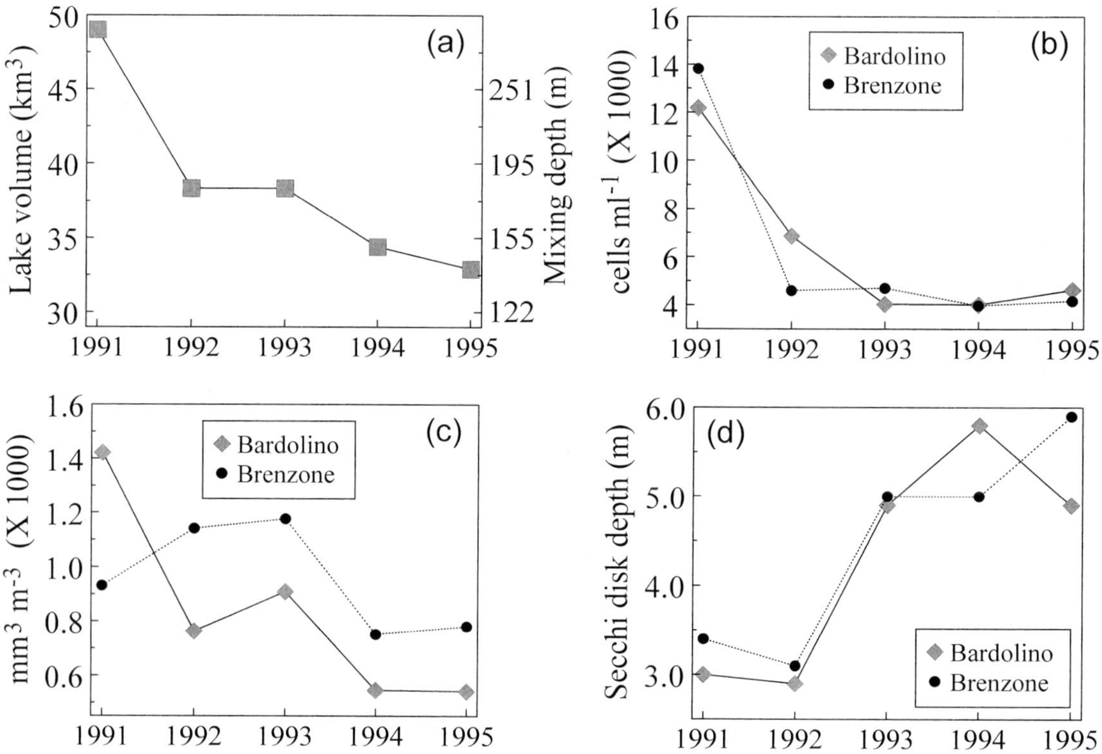

Figure 8. Spring mixed lake volumes (a) compared with the annual trends of average phytoplankton density (b) and biovolume (c) calculated, for Bardolino and Brenzone, for the March–December period and with the annual minimum Secchi disk values (d) always recorded in summer months.

tion of biovolume in the years from 1992 to 1994 due to inclusion of 0–20 m data when calculating the averages.

However, owing to the influence of other factors (primarily the interannual variability of meteorological conditions during the spring–autumn algal growth), the verification and understanding of the influence of the intensity of nutrient recycling on the biological development in the lake will require further continuous observations.

The contribution of the deep waters to nutrient enrichment of surface ones can be estimated by calculating the nutrient concentration in the mixed layers, obtainable by more or less extended overturns of the water column. On the basis of the first three 1996 samplings (carried out in January, February and March, with analyses available at 0–20–100–150–200–300 m), and in the case of an homogeneous circulating layer reaching 175 m or 350 m, total phosphorus concentrations would reach about 15 and 18 μg P l^{-1}, respectively. These data give only an approximation of the real concentrations that could be found in the upper layers because of possible incomplete homogeneity and due to assimilation of nutrients by algal cells; however, the smaller numbers of chemical measurements taken in the year of complete overturn (1991) seem to indicate a high total phosphorus content in surface waters in comparison to subsequent years (Provincia Autonoma di Trento, 1991; Salmaso & Cordella, 1994).

Anthropomorphic and climatic factors as determinants of the lake's trophic state

At the present time, the lake appears to be in an oligo-mesotrophic condition on the basis of the usual trophic parameters (OECD, 1982). This situation is determined by the low phosphorus concentrations in the euphotic zone. However, bearing in mind the oligomictic character and high nutrient concentrations at the deeper layers, the lake's actual trophic state is not so easily definable.

There has been a continuous, significant increase in the overall content of total phosphorus in the lake judging from the data collected since the seventies at the times of the lake's maximum overturn (Ambrosetti et al., 1992; Mosello & de Bernardi, 1996). Average spring concentrations of total phosphorus for the entire water column rose progressively from about 10 μg P l^{-1} at the end of 1970s to about 17 μg P l^{-1} in 1995.

These findings highlight a worsening of the L. Garda's overall trophic condition; however, the lake's total phosphorus content is only potentially utilised by algal cells, the replenishment of the upper layers being mediated also by the mixing events. Water volumes involved by the overturn appear to effect interannual variations of the algal crop around a baseline trophic status determined by the whole total phosphorus content in the water column. Anthropic pressures affect the long term trophic fate of the lake, whereas year to year climatic conditions influence annual fluctuations.

The role of the mixing depth in explaining interannual quantitative variations of the algal community may be important also for other deep Insubrian lakes, considering the high nutrient concentrations that characterise their deep waters (e.g. Como: Mosello et al., 1991; Lugano: Barbieri & Mosello, 1992; Iseo: Garibaldi et al., 1995; Idro: Garibaldi et al., 1996).

Management problems and the role of limnological research

One of the most urgent problems for the lake arises from economic and tourist development in the catchment, which is not matched by efficient control of waste waters. Recent investigations have attributed a considerable part of total loads to urban discharges (Chiaudani & Premazzi, 1990). Between the early 1970s and 1980s the population increased from nearly 160 000 (Calamari & Vighi, 1974) to *circa* 180 000 (Chiaudani & Premazzi, 1990). Since the 1960s, the number of days spent by tourists in the lake district has undegone a progressive and nearly linear annual increase of about 300 000 d ann^{-1}, reaching at present a figure between 12 and 13 million d ann^{-1} (cf. Pasini, 1992), mostly spent along the shore region in the warm season.

The upper catchment of the lake, in the Trentino region, has its own sewage treatment plants, whereas a ring trunk sewer runs along the Lombard and Venetian shores of the lake carrying waste waters to the treatment plant at Peschiera, downstream of the lake. However, recent investigations (Ministero dell'Ambiente et al., 1993) have pointed out various operational problems in the system, the most severe being, probably, its inadequacy in terms of size due to underestimation of tourist and urban expansion and discontinuity of the sewers; moreover, the system was originally intended only for sewage, but many boroughs also discharge storm sewage into it. In order to protect bathing areas from bacterial pollution, excess sewage is discharged into the lake, at various depths. These discharges seem to have decreased biological pollution along the shores (e.g. Bonù et al., 1994) but to the detriment of water quality in the hypolimnion. Along the Veneto coast, it has been estimated that nearly 50% of the sewage collected by the trunk sewer is discharged into the lake (Ministero dell'Ambiente et al., 1993), but the exact load seems unknown.

Of all the known techniques for treatment of eutrophication (Ryding & Rast, 1989; Harper, 1992; Cook et al., 1993) only control of the external phosphorus load constitutes a suitable solution for conservation of the deep Insubrian lakes; the large volumes and depth of these great waterbodies make in-lake eutrophication control methods quite ineffective and non-feasible. Encouraging results obtained in Lake Maggiore in the control of nutrient loads (Calderoni et al., 1993), provide an example of sensible management. At present, surface water quality seems to be entrusted to the lake's oligomictic character and thus to the isolation of nutrient-rich waters at depth, out of the euphotic zone. If the increase of phosphorus compounds in the entire water column continues, the effect of complete homogenisation of the lake on algal crops could become more evident.

Future management of this lake should be oriented towards its preservation and this is possible only by taking into account its ecological framework, as defined by continuing limnological research.

Conclusions

Lake Garda – along with other Insubrian lakes – belongs to a distinctive type characterised by great depths and large volumes. Owing to their morphometric features, the water column of these lakes undergoes complete homogenisation only in harsh winters.

Correct interpretation of the trophic evolution of Lake Garda must take both anthropic and climatic factors into account. Currently the deep waters of the lake function as a deposit for decomposing organic matter, thus deferring complete replenishment of nutrients in the surface layers until years of complete overturn. Full understanding of these limnological features appears important to guide the most suitable management practices. Future plans should try to reduce to a minimum the discharge of surplus sewage into deeper waters. The only reliable solution for preservation of the lake's water quality is reduction of nutrient loads; this is rather urgent in the light of the increase of total phosphorus concentrations observed since the seventies.

Managers must be urged to work, not only for the short-term perspective, but also for long-term lake quality conservation. If the deep waters are allowed to worsen, evidence of reduced quality at the surface might come too late; Lake Garda has considerable inertia, due to its large volume and long renewal time, so that reversing the downward trend could become a very difficult task with serious consequences for the economy of the entire region.

Acknowledgements

We wish to express our thanks to all the staff of the Presidio Multizonale di Prevenzione - ULSS 20 of Verona for critical discussions and their support in the field. We are grateful to Dr David Harper of the University of Leicester for helpful comments and the revision of the English text and to an anonymous referee for suggestions. This work was partially funded by the Veneto Region.

References

Ambrosetti, W., L. Barbanti & A. Rolla, 1979. Mescolamento parziale o totale nel Lago Maggiore nell'ultimo trentennio. Mem. Ist. ital. Idrobiol. 37: 197–208.

Ambrosetti, W., L. Barbanti, R. Mosello & A. Pugnetti, 1992. Limnological studies on the deep southern Alpine lakes Maggiore, Lugano, Como, Iseo and Garda. Mem. Ist. ital. Idrobiol. 50: 117–146.

Ambrosetti, W., L. Barbanti, R. Mosello, A. Rolla & D. Ruggiu, 1983. Mescolamento, caratteristiche chimiche, fitoplancton e situazione trofica nei laghi profondi sudalpini. C.N.R.-P.F. "Promozione della Qualità dell'Ambiente", AQ/2/20: 151 pp.

APHA, AWWA & WPCF, 1989. Standard methods for the examination of water and wastewater. 17th edn. American Public Health Association, Washington: 1587 pp.

Barbanti, L., 1974. Morfologia della conca lacustre. In M. Gerletti (ed.), Indagini Sul Lago Di Garda. IRSA quaderni 18, Roma: 101–115.

Barbieri, A. & R. Mosello, 1992. Chemistry and trophic evolution of Lake Lugano in relation to nutrient budget. Aquat. Sci. 54: 219–237.

Bonomi, G., A. Calderoni & R. Mosello, 1979. Some remarks on the recent evolution of the deep Italian subalpine lakes. Symp. Biol. Hung. 19: 87–111.

Bonù, A., G. De Luca, S. Stampi & F. Zanetti, 1994. La qualità delle acque del Lago di Garda nel tratto sud-ovest. Inquinamento 8: 62–69.

Calamari, D. & M. Vighi, 1974. Indagine sul carico inquinante del bacino imbrifero del Lago di Garda. In M. Gerletti (ed.), Indagini Sul Lago Di Garda. IRSA quaderni 18 Roma: 73–97.

Calderoni, A., R. de Bernardi & D. Ruggiu, 1993. The changing trophic state of Lago Maggiore. Strategies for lake ecosystems beyond 2000. 5th International conference on the conservation and management of lakes Stresa: 58–61.

Chiaudani, G. & G. Premazzi, 1990. Il Lago di Garda. Evoluzione trofica e condizioni ambientali attuali. Commissione delle Comunità Europee, EUR 12925, Lussemburgo: 196 pp.

Cook, G. D., E. B. Welch, S. A. Peterson & P. R. Newroth, 1993. Restoration and Management of Lakes and Reservoirs. Lewis Publishers 548 pp.

Decet, F., R. Burigo & N. Salmaso, 1995. Materiali per l'idrochimica di alcuni laghi del Parco Naturale di Paneveggio-Pale di S. Martino ed in particolare del L. Calaita (Dolomiti, Alpi Orientali). Boll. Chim. Igien. 46: 139–163.

Garibaldi, L., M. C. Brizzio, A. Varallo & R. Mosello, 1996. Water chemistry of the meromictic Lake Idro (Northern Italy). Mem. Ist. ital. Idrobiol. 54: 77–95.

Garibaldi, L., M. C. Brizzio, V. Mezzanotte, A. Varallo & R. Mosello, 1995. The continuing evolution of Lake Iseo (N. Italy): the appearance of anoxia. Mem. Ist. ital. Idrobiol. 53: 191–212.

Gerletti, M. (ed.), 1974. Indagini sul Lago di Garda. Indagini sul Lago di Garda. IRSA Quaderni 18 Roma: 540 pp.

Giorgi, M., M. Colacino, F. M. Vivona, M. Tonini & G. Zambon, 1974. Climatologia e meteorologia. In M. Gerletti (ed.), Indagini Sul Lago Di Garda. IRSA quaderni 18 Roma: 33–72.

Goldman, C. R. & A. Jassby, 1990. Spring mixing depth as a determinant of annual primary production in lakes. In M. M. Tilzer & C. Serruya (eds), Large lakes. Ecological Structure and Function. Springer-Verlag: 125–132.

Guilizzoni, P., G. Bonomi, G. Galanti & D. Ruggiu, 1983. Relationship between sedimentary pigments and primary production: evidence from core analysis of twelve Italian lakes. Hydrobiologia 103: 103–106.

Harper, D. M., 1992. Eutrophication of Freshwaters. Chapman & Hall, London, 327 pp.

Manca, M., A. Calderoni & R. Mosello, 1992. Limnological research in Lago Maggiore: studies on hydrochemistry and plankton. Mem. Ist. ital. Idrobiol. 50: 171–200.

Ministero dell'Ambiente, Regione del Veneto, Comunità del Garda, 1993. Progetto finalizzato alla formulazione di un piano di gestione e controllo volto al contenimento di eventuali fenomeni di eutrofizzazione delle acque del lago di Garda. Rapporto finale (coordination: G. Chiaudani), 326 pp.

Mosello, R. & R. de Bernardi, 1996. Ricerche del C.N.R. Istituto Italiano di Idrobiologia sulla evoluzione della qualità delle acque dei Laghi Profondi Sudalpini. Convegno "La regolazionedei Grandi Laghi Sudalpini", Gardone Riviera 2–3 Maggio 1996. Regione Lombardia Assessorato all'Agricoltura, Enti Regolatori dei Grandi Laghi Alpini: 196–209.

Mosello, R., D. Ruggiu, A. Pugnetti & M. Moretti, 1991. Observed trends in the trophic conditions and possible recovery of the deep subalpine Lake Como (N. Italy). Mem. Ist. ital. Idrobiol. 49: 79–97.

OECD, 1982. Eutrophication of waters. monitoring, assessment and control. OECD, Paris, 154 pp.

Pasini, P. G., 1992. Lago di Garda. Trent'anni di turismo. Civiltà Gardesana, Studi ed Esperienze, 6: 96 pp.

Provincia Autonoma di Trento, 1991. Lago di Garda. Relazione annuale sui rilevamenti idrobiologici. Anno 1991. Ed. Stazione Sperimentale Agraria Forestale di S. Michele all'Adige, 265 pp.
Provincia Autonoma di Trento, 1994. Quaderno del Garda. 1994. Ed. Stazione Sperimentale Agraria Forestale di S. Michele all'Adige, 181 pp.
Ryding, S.-O. & W. Rast (eds), 1989. The control of eutrophication of lakes and reservoirs. UNESCO & The Parthenon Publishing Group, Paris, 314 pp.
Salmaso, N., 1996. Seasonal variation in the composition and rate of change of the phytoplankton community in a deep subalpine lake (Lake Garda, Northern Italy). An application of nonmetric multidimensional scaling and cluster analysis. Hydrobiologia 337: 49–68.
Salmaso, N. & P. Cordella, 1994. Indagini limnologiche sul Lago di Garda. In P. Cordella & N. Salmaso (eds), Indagini Limnologiche Sui Principali Laghi Della Regione Del Veneto (1987-1992). Regione del Veneto Venezia: 79–124.
Salmaso, N., F. Cavolo & P. Cordella, 1994. Fioriture di *Anabaena* e *Microcystis* nel Lago di Garda. Eventi rilevati e caratterizzazione dei periodi di sviluppo. Acqua Aria 1: 17–28.
Salmaso, N., P. Cordella & M. Manfrin, 1995. Seasonal changing rates of the phytoplankton community in Lake Garda. Giorn. Bot. Ital., 129: 153.
Viner, A. B., 1985. Thermal stability and phytoplankton distribution. Hydrobiologia 125: 47–69.

Hydrobiologia **395/396:** 265–280, 1999.
D.M. Harper, B. Brierley, A.J.D. Ferguson & G. Phillips (eds), The Ecological Bases for Lake and Reservoir Management.

Protecting the oligotrophic lakes of the English Lake District

I. Zinger-Gize, A. Hartland, K. J. Saxby-Rouen & L. Beattie
The Environment Agency, North West Region, P.O. Box 12, Knutsford Road, Warrington WA4 1HG, U.K.

Key words: Elterwater, Lake District, eutrophication, continuous monitoring, nutrients, phytoplankton

Abstract

Continuous monitoring of physico-chemical parameters enable the determination of water quality in lakes of the English Lake District, where nutrient enrichment needs to be managed. In the case of Elterwater, severe deterioration of water quality has resulted in the prolonged deoxygenation of the whole water column of the inner basin. More specifically, the phytoplankton differed between the three basins, with progressive increases in biomass and species indicative of eutrophy, between the oligotrophic outer to the hypertrophic inner basins. Remedial action is proposed for this lake, with continual monitoring of recovery.

Introduction

The protection of still waters in England and Wales is supported by powers and duties invested in the Environment Agency by the Environment Act 1995 (see Moss, this volume). These include taking precautions against pollution, and the chemical and biological monitoring of waters to assess both the degree of pollution and any changes in water quality.

In addition to these powers, the Urban Waste Water Treatment (91/271/EEC) and Nitrate (91/676/EEC), EC Directives have provided a basis for the monitoring of lakes, according to specific criteria. The first Directive requires that proper treatment of major sewage discharges (including nutrient removal) be implemented for population equivalents greater than 10 000 people. For population equivalents of less than 2000 people, as in the case of Elterwater, appropriate treatment of sewage being discharged into an inland water will be in place by the year 2006. The Nitrates Directive is concerned with waters which might be eutrophic, due to nitrates from agricultural sources, or could be at risk of changing trophic status due to agricultural practices unless steps are taken.

Whilst classification systems exist for rivers and estuaries, no systematic monitoring of still waters by the Environment Agency is currently in place for England and Wales (Carvalho & Moss, 1995). A national R&D project into classification of still waters (areas greater than 2 hectares) is being undertaken, concentrating on catchment land use as a whole and monitoring only the inlets and outlets of lakes, excluding sampling within the lakes themselves (see Johnes, this volume).

The need for in-lake surveys still exists, however, particularly in the case of those with complex morphometry, to understand the dynamics of eutrophication and recovery. This paper focusses upon work at Elterwater initiated by the National Rivers Authority and continued by the Environment Agency.

Protecting still waters in the Lake District

The Cumbrian Lakes are an important national resource attracting millions of tourists each year (Figure 1). Along with their aesthetic value, they are of national scientific importance. Bassenthwaite Lake and Derwent Water are the only water bodies in Britain to house the rare vendace (*Coregonus abula*). Brotherswater, Haweswater and Ullswater are amongst the only seven U.K. water bodies where the rare schelly (*Coregonus lavaretus*) is found; both species are protected under the Wildlife and Countryside Act (1981). Esthwaite is a Site of Special Scientific Interest (SSSI – sites designated as of national importance by the U.K. Government's conservation agency, English Nature) and 'Ramsar' site of international status because it is one of the best examples of a moderately nutrient rich (mesotrophic) lake. Bassenthwaite is a

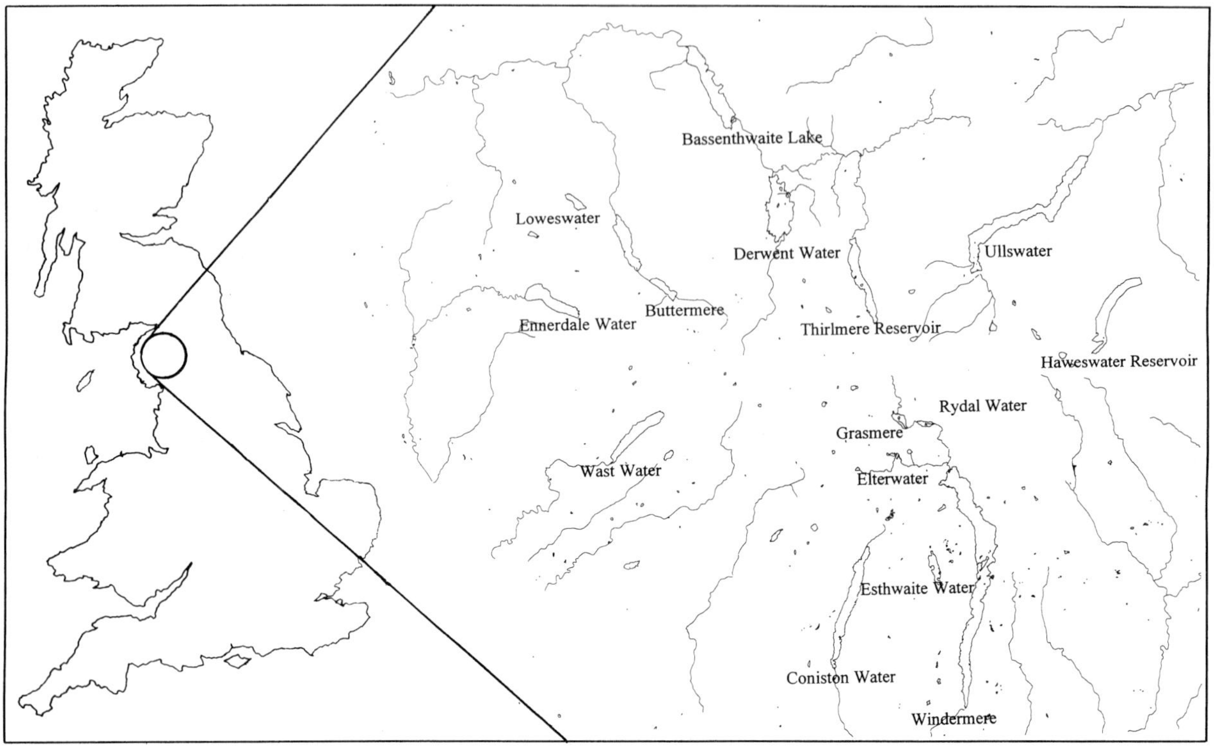

Figure 1. Location of the English Lake District lakes.

SSSI due to its unusual emergent and submerged vegetation. Elterwater is a SSSI because it presents one of the most extensive and least disturbed examples of lake shore wetlands in the region. Wast Water, Buttermere and Brotherswater are all designated as SSSIs because they are examples of extreme oligotrophy.

Many of the Cumbrian lakes, in common with the Cheshire meres (mesotrophic natural lowland glacial lakes) in the North West of England are showing evidence of increasing trophic status, generally due to anthropogenic inputs. Concern for the deterioration of water quality in these still waters led in 1990 to a strategy of the National Rivers Authority (NRA) North West Region which contained the following recommendations:

1. To protect, and where necessary improve, the existing environmental status of all still waters within the region. The principal objectives are to conserve and enhance the natural beauty and amenity value of these habitats and to safeguard the flora and fauna that are dependent on these water bodies.

2. To increase understanding of how these systems function by developing and maintaining an effective monitoring programme for these aquatic environments.
3. To give particular consideration to those waters which contain rare or endangered species, or possess other special habitat characteristics.
4. To have regard to the use and potential use of still waters as a water resource and to consider the effects of water abstraction on the environment.
5. To take into account the value of still waters as a recreational resource.
6. To improve, where appropriate, the visual appearance and general aesthetic quality of still waters.

Independent monitoring had previously occurred, with specific site investigations based on local initiatives and priorities. In 1993, a series of water quality surveys were organised and continuous monitoring established on certain lakes. Those lakes which were identified as being at risk of water quality deterioration included Elterwater, Bassenthwaite, Esthwaite and Grasmere.

A two year project was therefore started in 1994 concentrating on eight key lakes in Cumbria: Bassenthwaite, Derwent Water, Ullswater, Brotherswater, Esthwaite, Coniston, Grasmere and Elterwater.

Between April 1994 and February 1996, fortnightly monitoring of nutrients in inlet tributaries and outlets, flow recordings at gauging stations and by spot samples were collected, with more extensive water quality surveys of the eight lakes in July and September, and analyses for spatial and temporal distribution of phosphate and metals in the lakes sediments and algal taxonomy (Beattie et al., 1996).

In addition to the two year study, investigations into using airborne remote sensing to provide information on the physical, chemical and biological characteristics of lakes were carried out (George & Charlton, 1996).

The following case study compiles some of the findings derived from over three years' chemical and planktonic data collection. It also demonstrates practical management use of the information acquired.

Elterwater lake

Elterwater is one of the smaller lakes of the Lake District, on the eastern side of the Langdale Fells, with minor industrial activity in the catchment (only slate quarries). It consists of three shallow interconnected basins, of overall length 0.9 km (Figure 2, Table 1). The two main inflows to the lake are Great Langdale Beck and Little Langdale Beck, both entering the Outer basin. The Inner basin receives secondary treated sewage effluent from Langdale Sewage Treatment Works (STW). Small tributaries feed in from the surrounding hills. The main outflow is the River Brathay, which ultimately discharges to the northern end of Lake Windermere.

Due to the unusual hydrology of the lake, the Inner basin and to a lesser extent the Middle basin are poorly flushed, receiving only a small amount of surface run off and sporadic floodwater from the Outer basin. The yearly average rainfall is less than 2.2 m. Flows were measured in the interconnected narrows from September 1994 to December 1995. Most flow between basins occurred as distinct 'events', with the observed flow directions either going from the Inner to the Middle basins (occasionally), or from the Middle to the Outer basins. Other surveys have shown that flood water may pass from the Middle basin into the Inner basin, which may increase the Inner basin's volume by over 18% (Stockdale, 1991).

Table 1. Physical charateristics of the three basins of Elterwater

	Inner	Middle	Outer
Surface (m^2)	35 860	74 070	83 260
Mean depth (m)	2.7	2.3	2.5
Maximum depth (m)	7.4	6.0	7.5
Maximum length (m)	250	390	400

North West Water plc's Langdale sewage treatment works was first commissioned to discharge treated sewage effluent into a tributary, feeding the Inner basin of Elterwater in 1973/74. The works treat domestic sewage for a population equivalent of 1752 (Elterwater village and Chapel Stile), but this figure excludes the increase due to tourism in the summer months. Although the works suffers from severe water infiltration, it complies with its discharge consent of of 25 mg/l BOD and 30 mg/l suspended solids. Since the size of the population served by the sewage treatment works is just below 2000, appropriate treatment under the EC UWWTD would not be required until the year 2006.

The average daily hydraulic inflow into the Inner basin was estimated at 790 m^3 d^{-1} with substantial loads of phosphorus (0.87 kg soluble P d^{-1}) and BOD (9.48 kg d^{-1}) (Stockdale, 1991). The discharge accounts for 43–66% of the main daily hydraulic input to the Inner basin, and 70% of ortho-phosphate input from the catchment.

The morphology and hydrology of the lake are such that monitoring the inlets and outlet alone would not be representative of its water quality. An initial survey July 1993 (Anon., 1994) indicated that the water quality differed significantly between the three basins. Monitoring was extended in 1994 and 1995 to examine more closely the poor condition of the Inner basin, and gather more evidence to confirm the impact of the treated sewage effluent.

Methodology

Continuous monitoring instruments were deployed in Elterwater Inner basin from 1994. A multi-parameter probe (Yellow Springs Instrument Company, Ohio, YSI-6000) recorded pH (low ionic strength probe), temperature dissolved oxygen (rapid pulse probe) and specific conductivity at 15 minute interval near the surface (about 1.5 m depth) and near the bottom (about 1 m above sediments). A chlorophyll fluorometer ('Aquatraka', Chelsea instruments Ltd) recorded near-surface algal biomass by fluorimentry. At two weekly

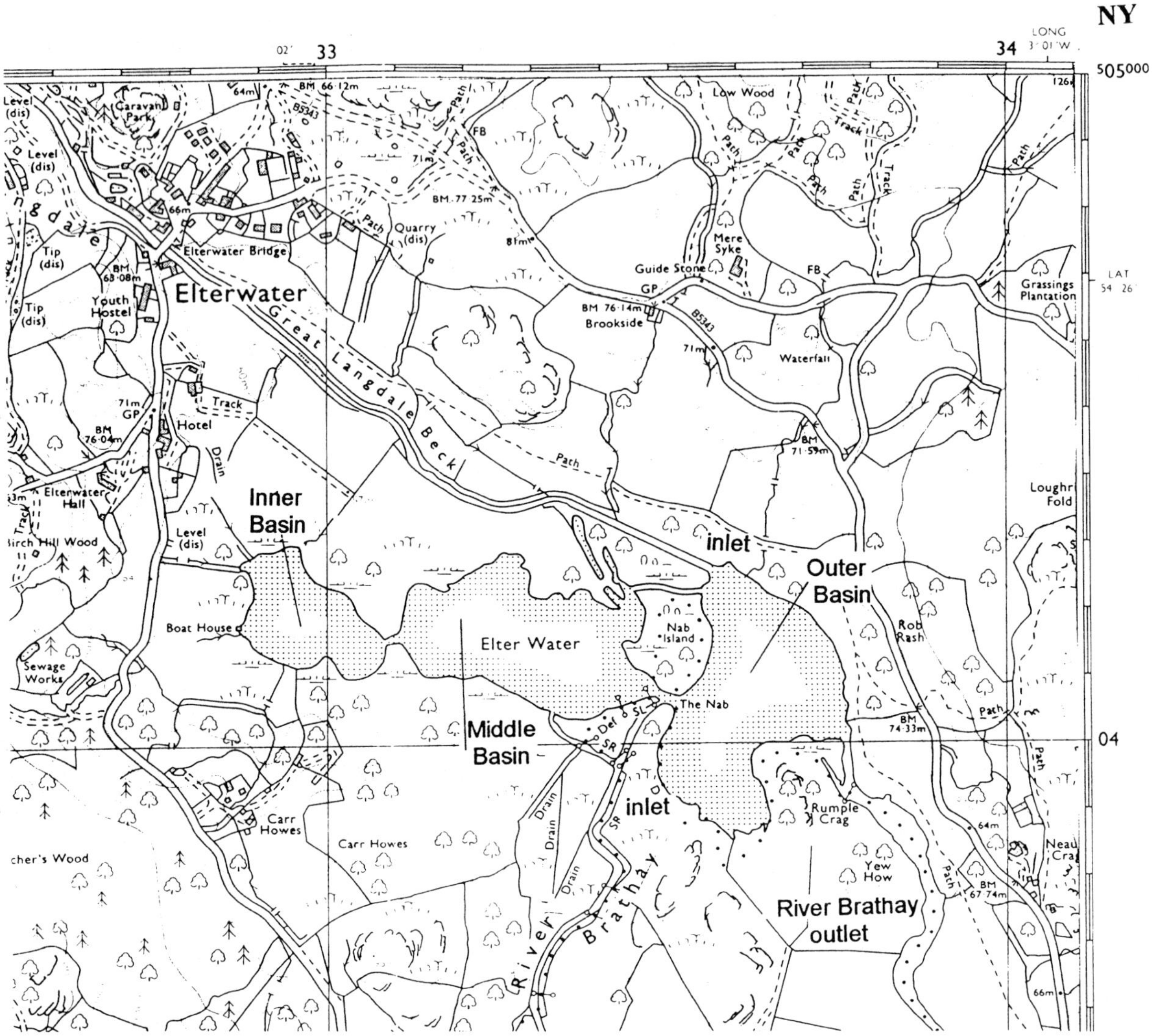

Figure 2. Location of Elterwater.

intervals, Secchi disc extinction was recorded and a 1 litre sample for laboratory chlorophyll *a* determination were taken during instrument calibration and downloading.

Spatial surveys were also conducted in July and September 1994 and July 1995 at fifteen sites selected to represent the whole lake and located precisely by hand-held satellite global positioning system (GPS). Secchi disc extinction was recorded at each location. Water samples from about 20 cm below the surface, and at bottom were collected using a dual line Casella sampler (Casella instruments, Bedfordshire) for the determination of ammonia, nitrate, nitrite, total phosphorus, ortho-phosphate and silicate (Anon 1988; 1993). Surface samples were also analysed for suspended solids, ash solids, chlorophyll *a* and phaeophytin concentrations (Anon., 1980a,b). All samples were stored in cool boxes on site for transport to the laboratory. Measurements of depth, temperature, dissolved oxygen, pH, specific conductivity and light penetration were made *in situ* at 1 m intervals (more frequently at gradients) using a 'Windermere' profiler 2 (FBA, Ambleside, Cumbria), calibrated before each survey.

Algal samples (1 per basin) were collected in a 5 m integrated tube sample (Hall et al., 1992) and transferred to a 1-litre plastic bottle with Lugol's iodine. Prior to counting, each sample was concentrated by a factor of 25 to 45 by serial passive sedimentation in glass measuring cylinders. The concentrate was mixed and counts made using a Lund Chamber (Lund, 1959) at a magnification of 250×. The area of the counting chamber used for each count was dependent on the taxa involved and their concentration. The standard area tallied yielded a limit of sensitivity of 1.0–1.2 algal units ml^{-1}. The algal concentration was expressed in two forms. Firstly, as the number of algal units ml^{-1}, and secondly, as a weighted abundance taking into account the size (volume) as well as number of individual taxa. The algal units used are listed in Table 3. The weighted abundance was not intended to give an accurate estimate of phytoplankton biovolume, but rather to give a broad indication of the major contributing taxa. Each taxon was allocated one of nine categories according to the estimated volume of the algal unit recorded (cell/length/colony). Category 1 related to algal units of volume equivalent to a 2–5 μm diameter sphere, category 9 to algal units equivalent to a 40–50 μm^3 sphere; the minimum, maximum and mid-point volumes being 4, 65 and 34.5 μm^3 respectively. A weighting factor was attached to each category relative to category 1: (mid-point volume of category x)/(mid-point volume of category 1). The weighting factor multiplied by the number of algal units ml^{-1} gave the weighted abundance ml^{-1} (in units of $\times$ 34.5 μm^3 ml^{-1}). Results are presented for the month of July in 1993, 1994 and 1995.

Calibration of the 'Aquatracka', which provided continuous fluorometry data, was by acetone-extracted chlorophyll *a* using standard methods (Lorenzen, 1966; Anon., 1980). However the relationship shows considerable scatter (Figure 5) with a regression coefficient of 37.56%. This may be explained by the fact that samples are collected fortnightly at different times throughout the year, and represent various algal communities. Strong linear relationships have been found between Aquatraka and laboratory results over shorter time scales (Mills & Tett, 1990), or by use of single species cultures.

Results

Both temperature and oxygen showed marked stratification in 1994 and 1995 with deoxygenation in the hypolimnion (below 3 m depth) from May to September (Figure 3). Specific conductivity differences also defined the two water layers.

The inner basin differed from the two other basins (Figure 4) with de-oxygenation below 4 m depth in the Middle basin, while only slight decrease in dissolved oxygen levels occurred in the outer basin. The temperature difference between the surface and the bottom exceeded 10 °C in the Inner basin, and about 7 °C in the Middle basin.

Specific conductivities differed between each basin, with 80 μS^{-1} cm in the Inner, 55 μS^{-1} cm in the Middle and 40 μS^{-1} cm in the outer surface waters. Below the thermocline they increased to 140 μS^{-1} cm in the Inner and about 90 μS^{-1} cm in the Middle basin.

The pH varied between 5.50–6.67 in July 1994; 6.20–7.23 in September 1994 and 5.77–7.87 in July 1995. The pH values always became more acidic with depth down to thehypolimnion, but increased slightly in the deeper water.

Nutrient levels also differed between basins, with greater concentrations usually observed in the inner basin followed by middle and then outer (Table 2). Bottom samples of ammonia had up to eight times higher levels of ammonia than surface and usually slightly higher phosphate. Nitrate concentrations were lower in the Middle basin than either inner and outer basins (100 vs 300 $\mu g\ l^{-1}$). Silicate levels ranged between 519 and 2870 $\mu g\ l^{-1}$ at the surface and 266 and 4670 $\mu g\ l^{-1}$ near the bottom, with no noticeable trend between sites, basins, or surveys.

Transparency increased progressively, from the inner basin (1 m) to the outer basin (up to 5 m). Chlorophyll *a* concentrations mirrored this gradient, with 30–70 $\mu g\ l^{-1}$ in the inner, 10–20 $\mu g\ l^{-1}$ in the middle and <5 $\mu g\ l^{-1}$ in the outer basin. Diurnal variations in chlorophyll *a* concentrations were observed, usually with minimum during the day and maximum at night, (Figure 5), probably due to the migration of motile species (Salonen et al., 1984). The progressive increase in phytoplankton chlorophyll between the basins was supported by a parallel increase in numbers and biovolume.

The phytoplankton in the outer basin was initially dominated by the colonial green alga *Volvox globator* (July 1993), comprising approximately one-third of the phytoplankton biovolume. In the following July, *Volvox* was present in similar numbers although co-dominated with *Rhodomonas* sp. The larger *Cryptomonas,* present in only moderate numbers, and the filamentous *Anabaena solitaria* contributed significantly to the overall phytoplankton biovolume in 1994. One year later, *Volvox* was virtually absent, the

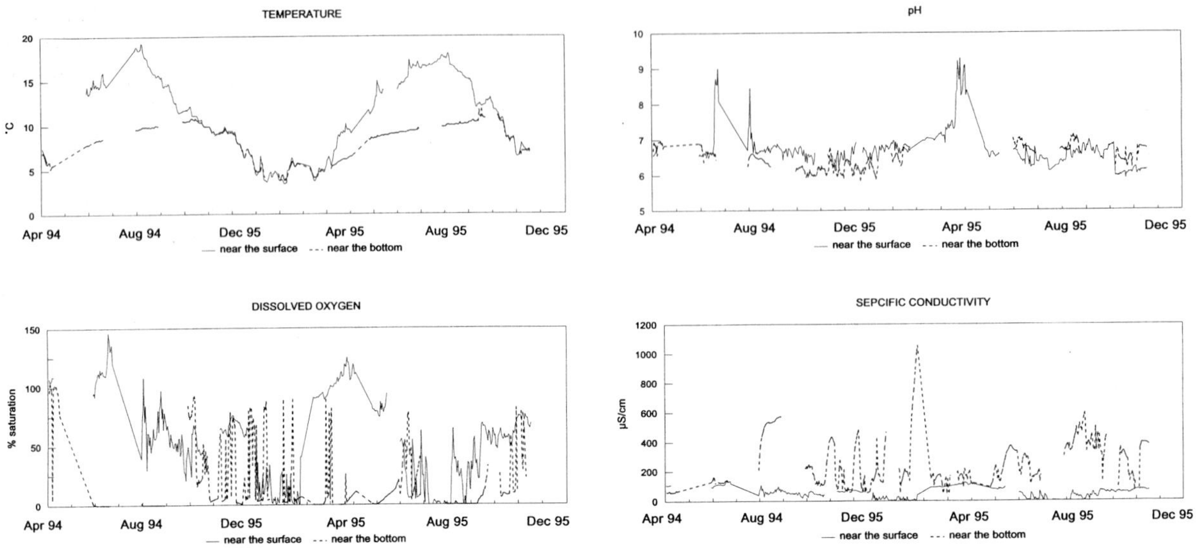

Figure 3. Continuous monitoring data from the inner basin, readings at 1300 hours.

Table 2. Summary of the nutrient state of the three basins of Elterwater

		Susp. Solids mg/l	Solids-ash mg/l	Secchi m	Chlorophyll µg/l	*Phaeophytin* µg/l	Total P mg/l	Nitrate µg/l	Nitrite µg/l	Ammonia µg/l	ortho-PO_4 µg/l	Silicate µg/l
July 1994												
INNER	Surface	4.67	< 10	2.00	34.70	13.54		430.00	23.47	477.67	111.13	1318.00
	Bottom						0.74	34.10	14.83	1929.50	131.90	965.00
MIDDLE	Surface	1.77	< 10	2.35	13.80	2.12		148.67	7.99	51.47	2.37	696.67
	Bottom						0.16	127.67	8.87	209.00	4.66	367.00
OUTER	Surface	1.30	< 10	4.50	1.60	0.27		330.00	3.07	23.10	1.87	1428.00
	Bottom						0.06	386.33	2.68	56.00	4.00	935.33
September 1994												
INNER	Surface	5.67	< 10	2.13	78.75	33.20	288.00	394.67	27.80	3693.33	108.80	2790.00
	Bottom						909.67	408.00	33.00	922.50	162.00	3533.33
MIDDLE	Surface	1.77	< 10	1.90	21.80	12.23	63.00	214.00	10.70	209.00	22.43	2233.33
	Bottom						68.00	304.00	14.79	343.50	22.05	2300.00
OUTER	Surface	1.30	< 10	2.75	2.85	1.75	33.00	334.75	1.41	12.50	1.69	2290.00
	Bottom						33.00	302.25	1.84	24.45	3.63	2250.00
July 1995												
INNER	Surface	3.33	6.67	1.20	30.83	16.67	145.00	301.67	30.33	219.67	81.80	979. 67
	Bottom						394.33		19.33			1510.00
MIDDLE	Surface	2.00	6.67	2.13	9.67	5.93	31.44	94.20	5.10	33.00	6.80	769.33
	Bottom						189.67	68.20	4.00	170.67	6.07	1477.00
OUTER	Surface	2.00	6.67	3.30	2.42	0.68	18.33	279.80	1.32	16.20	0.84	1658.00
	Bottom							320.40	1.52	54.20	2.66	1900.00

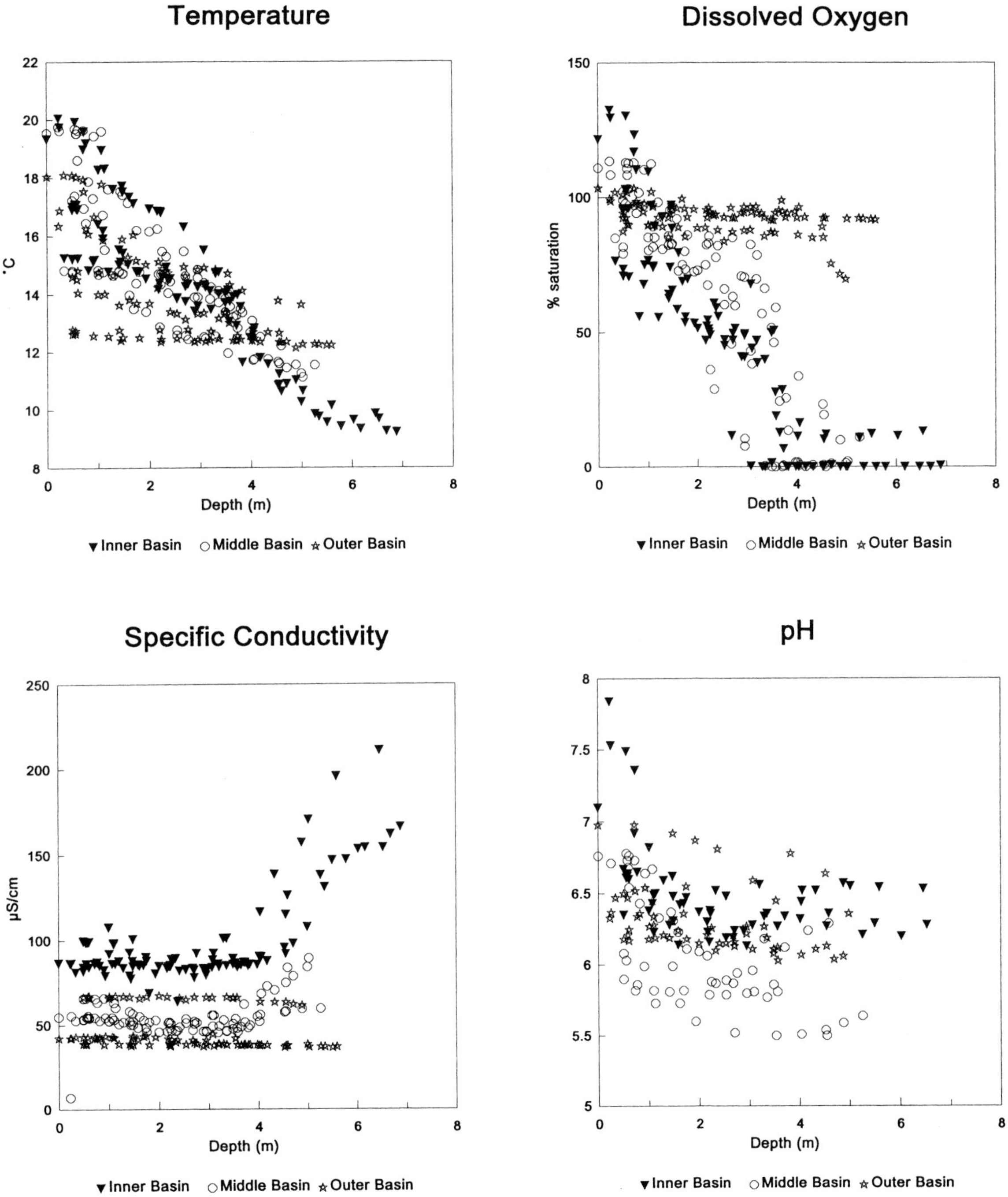

Figure 4. Water column profiles taken in July 1993, 1994, 1995.

Table 3. Phytoplankton species and their abundances

Algal Species	Unit	Type	1995			1994			1993 **		
			INN	MID	OUT	INN	MID	OUT	INN	MID	OUT
Anabaena flos-aquae/circinalis	cell	BG	-	-	-	72	122	X	-	-	-
Anabaena flos-aquae/spiroides	cell	BG	930	3451	133	-	-	-	-	-	-
Anabaena spiroides f. spiroides	cell	BG	79	214	-	-	-	-	-	-	-
Anabaena spiroides f. crassa	cell	BG	158	-	-	-	-	-	-	-	-
Anabaena solitaria	leng	BG	X	0.17	-	6.23	12.4	0.78	0.89	0.98	X
Gleocapsa spp.?	colo	BG	-	-	-			1	-	-	-
Merismopedia spp.	cell	BG	7985	35	24	-	-	-	-	-	-
Microsystis spp.	colo	BG	-	-	-	-	-	-	-	-	X
Oscillatoria spp.	leng	BG	X	-	-	-	-	-	X	X	X
Oscillatoria limnetica?	leng	BG	-	0.14	X	-	-	-	-	-	-
* *Anabaena solitaria*	cell	BG	X	22	-	890	1772	112	-	-	-
* *Anabaena flos-aquaea/circinalis*	fila	BG	-	-	-	4	6	X	X	-	-
* *Anabaena flos-aquaea/spiroides*	fila	BG	9.9	44	4.8	-	-	-	-	-	-
* *Anabaena spiroides f. spiroides*	fila	BG	0.1	3.8	-	-	-	-	-	-	-
* *Anabaena spiroides f. crassa*	fila	BG	0.2	-	-	-	-	-	-	-	-
* *Anabaena solitaria*	fila	BG	X	1	-	21	42	3	3	3.3	X
* *Lyngba/Phormidium*	fila	BG	-	-	-	-	-	-	-	X	X
* *Oscillatoria* spp.	fila	BG	X	-	-	-	-	-	-	-	-
* *Oscillatoria limnetica* ?	fila	BG	-	2	X	-	-	-	-	-	-
Centric diatoms (unidentified)	cell	DI	5.9	15	3.9	2	2	-	6.3	X	-
Pennate diatoms (unidentified)	cell	DI	15	799	32	-	-	-	-	-	-
Navicula/Acnanthes	cell	DI	-	-	-	-	-	-	6.3	X	8.4
Acnanthes spp.	cell	DI	-	-	1	-	-	-	-	-	-
Asterionella formosa (live)	cell	DI	38	40	2.9	177	44	0.7	33.2	9	7.9
Aulacoseira spp.	cell	DI	-	2	-	-	-	-	-	-	-
Cymbella spp.	cell	DI	-	-	-	-	-	1	X	X	2.5
Gomphonema spp.	cell	DI	-	-	-	-	-	-	1.3	-	X
Fragilaria crotonensis	cell	DI	-	-	-	-	-	-	-	-	X
Roicosphenia curvata	cell	DI	-	-	-	-	-	-	-	-	X
Stephanodiscus/ Cyclotella	cell	DI	-	-	-	2	0.6	-	58	1.8	1.4
Synedra spp.	cell	DI	29	5.9	1	-	-	-	4.8	1.8	1.5
Tabellaria fenestrata	cell	DI	1	-	-	-	-	-	-	-	X
Tabellaria flocc. var flocculosa	cell	DI	2	-	2	-	-	-	1.3	14	
Tabellaria flocc. var. asterion	cell	DI	X	-	-	-	-	-	-	-	-
Chlamydomonas spp.	cell	G	10	-	-	-	-	-	-	-	
Chlamydomonas-type spp.	cell	G	-	112	40	-	-	-	-	-	-
Chlamydomonas spp. (v. small)	cell	G	-	-	-	-	-	-	12	-	X
Volvox spp. (individual cells)	cell	G	-	3	2	5018	2386	400	3222	9812	372
Volvox spp.	colo	G	-	-	-	-	-	-	-	-	-
Volvox, daughter colonies	colo	G	-	-	-	4	0.6	-	14	11	X
Volvox (cyst)	cyst	G	-	-	1	0.4	X	0.1	5.2	147	1
Ankistrodesmusd spp.	cell	G	-	6.9	-	-	-	-	-	-	-
Ankyra/Schroederia	cell	G	-	-	-	1	1	-	-	-	
Chlorella-type spp.	cell	G	188	164	109	-	-	-	3.3	1	X
Chodatella spp. (small)	cell	G	-	-	-	-	-	-	-	X	-
Chodatella spp. (large)	cell	G	2	0.5	-	-	-	-	-	-	-
Coelastrum spp.	colo	G	-	1	-	-	-	-	-	-	-
Dictyosphaerium	colo	G	-	2.4	1	1	-	-	1.8	1	-
Elakatothrix spp.	cell	G	X	4.9	2	11	52	1	X	X	X
Kirchneriella spp.	cell	G	12	20	1	-	0.6	-	-	-	-
Monoraphidium spp.	cell	G	127	41	1	X	6	-	3.2	X	4.9

Table 3. contd.

Algal Species	Unit	Type	1995 INN	1995 MID	1995 OUT	1994 INN	1994 MID	1994 OUT	1993 ** INN	1993 ** MID	1993 ** OUT
Oocystis	cell	G	20	12	-	-	-	X	-	-	-
Sphaerocystis spp. (small)	colo	G	2	-	-	0.6	23	1	1.7	5.7	1.3
Sphaerocystis, isolated daughter colonies ?	colo	G	138	-	-	-	-	-	-	-	-
Westella spp. (?)	colo	G	-	16	3.9		-	-	-	-	-
Koliella longiseta	cell	G	1	-	-	-	-	-	-	-	-
Ulothrix spp.	fila	G	-		-	-	-	-	-	X	-
Mougeotia spp.	fila	G	-	-	-	-	-	-	-	-	2.4
Closterium spp.	cell	G-	-	-	-	-	-	-	X	X	X
Cosmarium spp.	cell	G	-	-	-	X	-	-	-	X	5
Desmidium spp.	cell	G	-	-	-	-	-	-	12	-	-
Scenedesmus quadricauda	cell	G	743	71	16	7	10	1	40	5.7	X
Scenedesmus spp.(smooth)	cell	G	282	31	3.9	-	-	-	-	-	-
Spondylosium spp.	cell	G	-	-	-	-	-	-	-	-	-
Staurastrum spp.	cell	G	1	1	X	-	-	-	-	-	X
Cryptomonas spp.	cell	CR	617	389	52	11	4	40	232	96	6
Cryptomonas marsonii	cell	CR	-	-	-	6	15	21	-	-	-
Cryptomonas rostrata	cell	CR	59	3	-	-	-	-	-	-	-
Rhodomonas spp.	cell	CR	11	147	186	4	18	417	18	9	1.4
Dinobryon spp.	cell	CH	-	-	-	0.6	1.3	-	-	-	-
Dinobryon bavaricum	cell	CH	-	1	-	-	-	-	-	-	-
Dinobryon cyclindricum	cell	CH	1	12	2	-	-	-	3.7	50	1.6
Mallomonas spp.	cell	CH	2.4	1	2	-	-	-	9.8	-	-
Syncrypta spp. (?)	cell	CH	-	3	-	2	5	5	21	2.8	-
Ceratium spp.	cell	DN	-	1	-	-	-	-	X	-	-
Gymnodinium spp. ?	cell	DN	-	-	-	-	-	-	-	X	-
Peridinium spp. (small)	cell	DN	22	6.7	1	1	-	1	12	-	-
Phacus spp.	cell	EU	-	-	-	-	-	-	1.7	-	-
Trachelomonas spp.	cell	EU	1	4.9	-	0.6	0.1	-	-	X	-
Xanthophyta – unidentified	cell	XA	-	-	1	0.6	6	1	31	45	3.8
A – unidentified colony	colo	UN	9.9	-	-	-	-	-	-	-	-
B – unidentified filament of 4 cells	fila	UN	8.9	1	-	-	-	-	-	-	-
Flagellates indeterminate	cell	UN	-	-	-	-	-	-	7.5	-	-

* Algal species given in alternative units to allow comparisons with other studies (not included in total counts).
** Figures relate to the mean of 3 or 4 sites per Basin, with 'present' being taken as an abundance of 0.5 units/ml.
leng length (mm)
fila filament
colo colony
X Taxon was observed in casual scans, but not in the formal count
INN Inner Basic
MID Middle Basin
OUT Outer Basin
G Green algae (Chlorophyta)
BG Blue-Green algae (Cyanophyta)
DI Diatoms (Bacillariophyta)
CR Cryptomonads (Cryptophyta)
CH Chrysophyta
DN Dinoflagellates (Dinophyta)
EU Euglenophyta
XA Xanthophyta
UN Unidentified taxa
Tabellaria flocc. var. flocculosa Tabellaria flocculosa var. flocculosa
Tabellaria flocc. var. asterion Tabellaria flocculosa var. asterionelloides

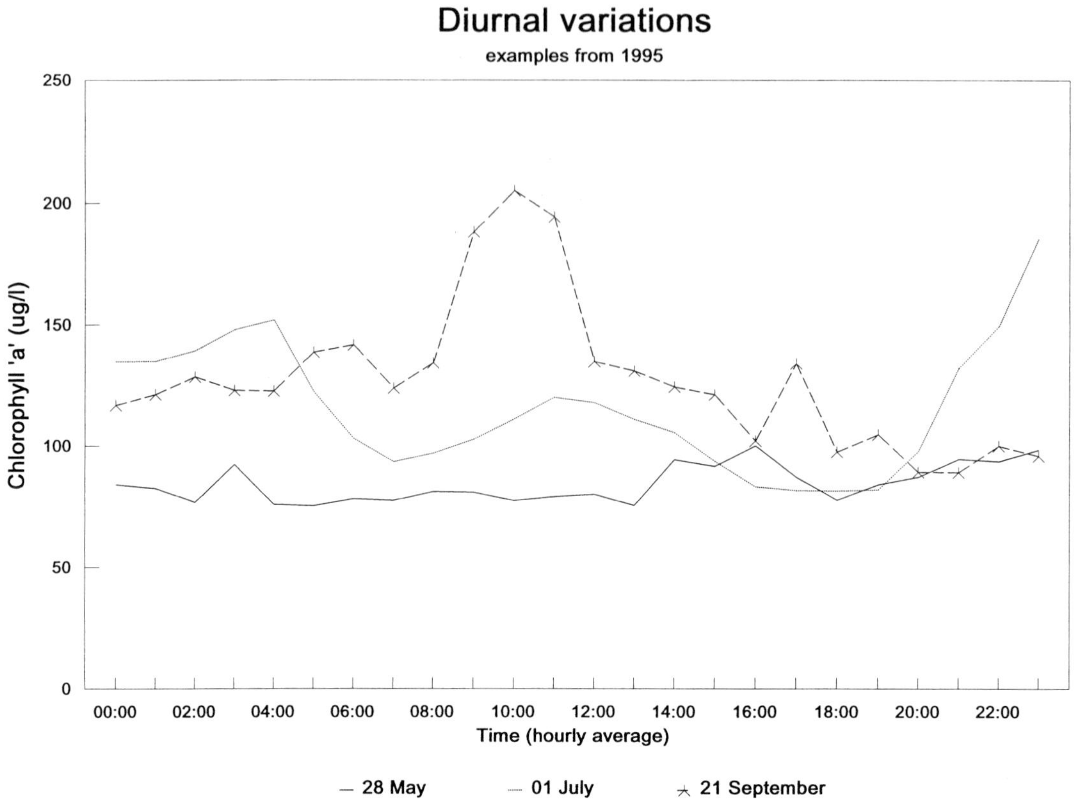

Figure 5. Diurnal variation in chlorophyll *a*, based on 'Aquatraka' and laboratory calibration.

phytoplankton being numerically dominated by other small green algae (*Chlamydomonas* and *Chlorella*) and *Rhodomonas*. (Table 3, Figures 6–8).

The middle basin in 1993 was similarly characterised by *Volvox*. *Cryptomonas*, *Dinobryon cylindricum* and *Bumilleriopsis* (?) were present at more moderate concentration. The only cyanobacterium was *Anabaena solitaria*, present at low concentration. In July 1994, although *Volvox* was again numerically dominant, it contributed only about 10% to the total phytoplankton biovolume compared to 91% in 1993, with *A. solitaria* contributing 65%. The cryptomonad and xanthophyte populations were lower than in 1993, and *Asterionella formosa*, *Elakatothrix*, *Sphaerocystis* and *Scenedesmus* all increased in abundance. In this year also the cyanobacteria *A. flos-aquae (?)* first appeared. In July 1995, with *Volvox* almost absent, *A. flos-aquae* (?) became numerically dominant, with *Cryptomonas* having the largest biovolume. High numbers of pennate diatoms with *Chlamydomonas* and *Chlorella* were present, together with moderate numbers of *Asterionella formosa*, *Rhodomonas*, *Kirchneriella*, *Monoraphidium* and *Scenedesmus*.

The inner basin, with the highest chlorophyll concentration, contained similar taxa and inter-annual changes as the middle basin, but at higher numbers. A further species of *Anabaena, A. spiroides f. crarra* was also present.

Discussion

Earlier studies of Elterwater carried out during 1974–1976 (Lund, 1981), following the installation of the sewage discharge, described the inner basin as having blooms of *Volvox* with often abundant small nannoplanktonic algae-centric diatoms (*Cyclotella*), coccoid chlorophytes and cryptomonads. *Anabaena solitaria* was the sole gas-vacuolate cyanobacterium. Lund described the algal community of the inner basin as "similar to that common in sewage-oxidation ponds", whereas that of the outer basin was clearly of an oligotrophic nature.

Clearly there are similarities between the state of the basins, and the progression between them, as twenty-five years ago. There may possibly have been

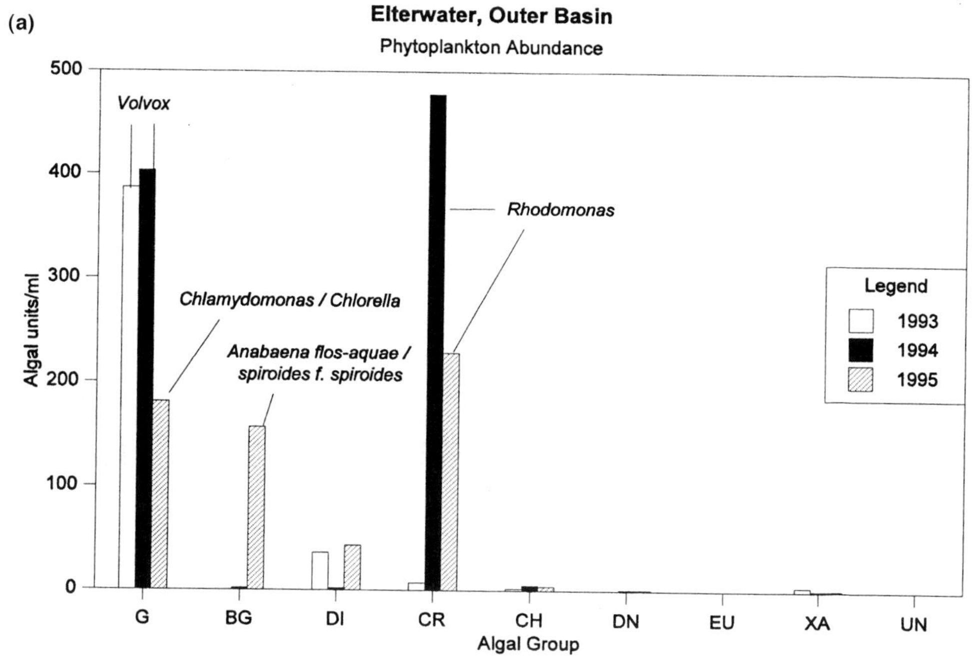

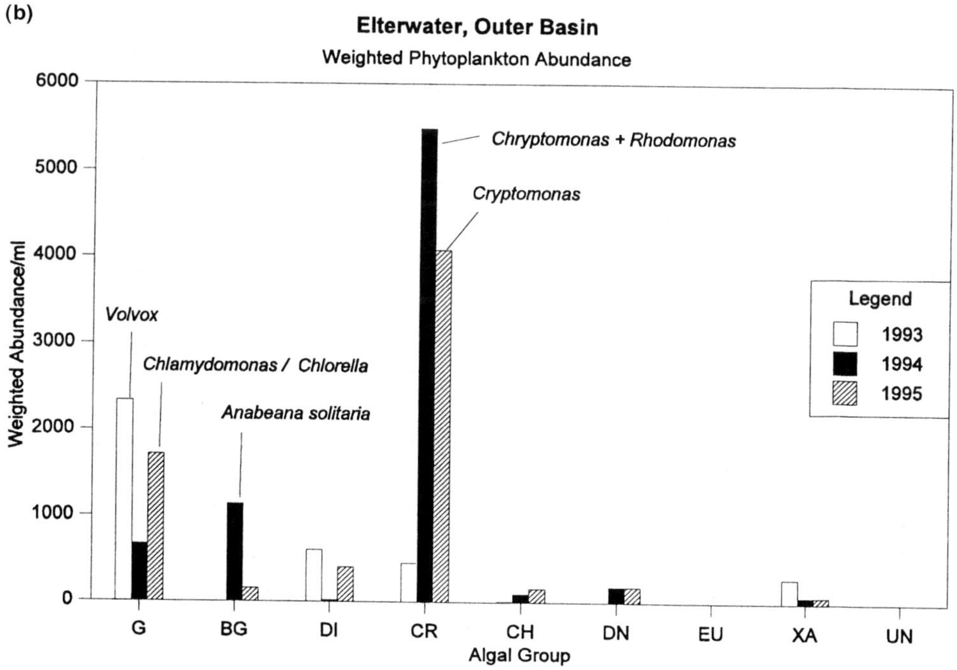

Figure 6. (a) Abundances of phytoplankton, outer basin. (b) Weighted abundance in units of $\times$ 34.5 μm^3 ml^{-1}.

Key:

G	Green algae (Chlorophyta)	BG	Blue-Green algae (Cyanophyta)
DI	Diatoms (Bacillariophyta)	CR	Crytomonads (Cryptophyta)
CH	Chrysophyta	DN	Dinoflagellates (Dinophyta)
EU	Euglenophyta	XZ	Xanthophyta
Un	Unidentified taxa		

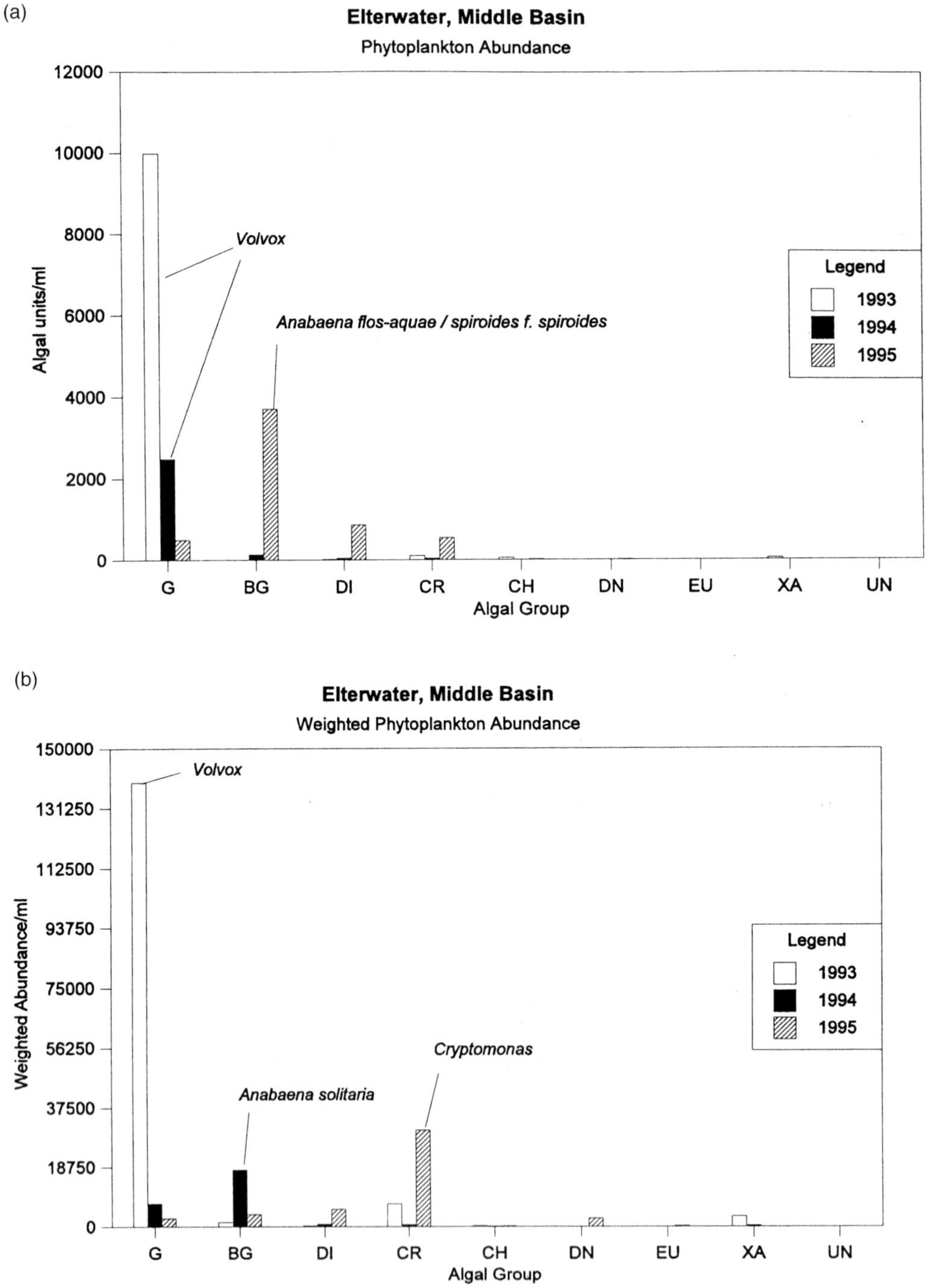

Figure 7. (a) Abundances of phytoplankton, middle basin. (b) Weighted abundance in units of $\times$ 34.5 μm^3 ml^{-1}. See Figure 6 Key.

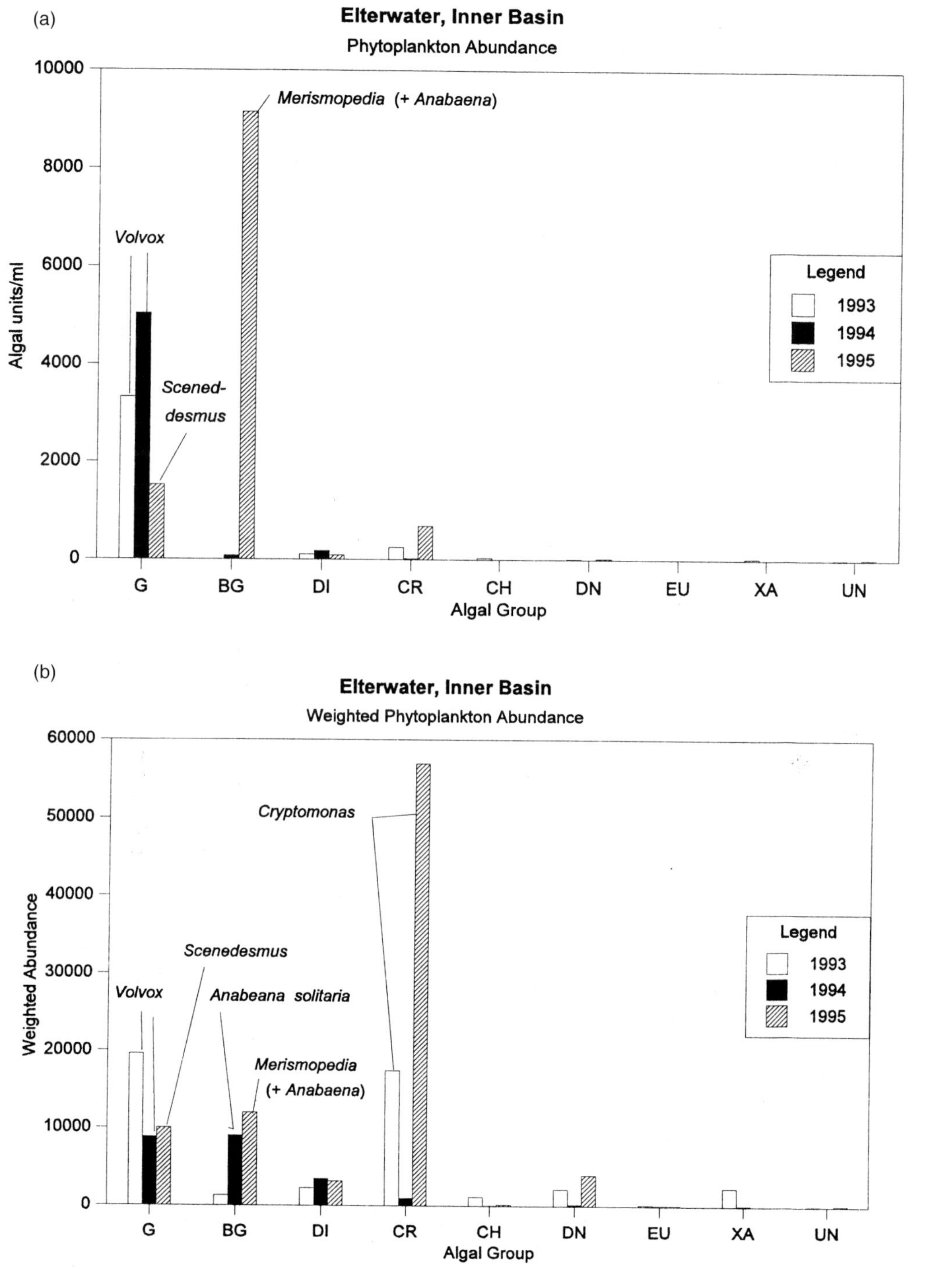

Figure 8. (a) Abundances of phytoplankton, inner basin. (b) Weighted abundance in units of $\times$ 34.5 μm^3 ml^{-1}. See Figure 6 Key.

a slight increase in enrichment in the Inner basin over the time period, if the change in the abundance and composition of the cyanobacteria are significant. However, bloom-forming gas-vacuolate cyanobacteria are often absent from highly enriched water bodies (Whitton, 1973; Lund, 1981), though are commonly found in eutrophic lakes. In the present study, their populations at no time reached the 'warning threshold' for potential bloom formation given in National Rivers Authority management guidance (Anon., 1995) although all but *Merismopedia* are potential toxin-formers (Codd & Bell, 1996). The recent increase in green algae *Scenedesmus*, *Monoraphidium* and *Oocystis*, taxa common in highly enriched waters (Reynolds, 1984), is consistent with such an increase in enrichment.

Elterwater's three basins are still very distinct. Based on the OECD boundaries and UWWTD criteria for trophic categories (Table 4), Elterwater is oligotrophic/ mesotrophic in the outer basin, mesotrophic/ eutrophic in the middle basin, and hypertrophic in the inner basin.

Cranwell et al. (1995) recorded total phosphorus levels between 8007 and 8940 μg g^{-1} in the inner basin sediments, approximately eight times higher than other lakes in the area.

Table 4. OECD trophic state classification of the basins, showing position of Elterwater

Trophic Status	Total P** mean*	Chlorophyll *a* mean*	Chlorophyll *a* max*	Secchi mean*	Secchi min*
Inner Basin					
Ultra-	≤ 4	≤ 1	≤ 2.5	≥ 12	≥ 6
Oligotrophic	≤ 10	≤ 2.5	≤ 8	≥ 6	≤ 3
Mesotrophic	10–30	2.5–8	8–25	6–3	3–1.5
Eutrophic	35–100	8–25	25–75	3–1.5	**1.5–0.7**
Hyper-eutrophic	≥ 100	**≥ 25**	**≥ 75**	**≤ 1.5**	≤ 0.7
Middle Basin					
Ultra-oligotrophic	≤ 4	≤ 1	≤ 2.5	≥ 12	≥ 6
Oligotrophic	≤ 10	≤ 2.5	≤ 8	≥ 6	≤ 3
Mesotrophic	10–30	2.5–8	**8–25**	6–3	**3–1.5**
Eutrophic	35–100	**8–25**	25–75	**3–1.5**	1.5–0.7
Hyper-eutrophic	≥ 100	≥ 25	≥ 75	≤ 1.5	≤ 0.7
Outer Basin					
Ultra-oligotrophic	≤ 4	≤ 1	≤ 2.5	≥ 12	≥ 6
Oligotrophic	≤ 10	**≤ 2.5**	**≤ 8**	≥ 6	≤ 3
Mesotrophic	10–30	2.5–8	8–25	**6–3**	**3–1.5**
Eutrophic	35–100	8–25	25–75	3–1.5	1.5–0.7
Hyper-eutrophic	≥ 100	≥ 25	≥ 75	≤ 1.5	≤ 0.7

*Concentrations in μg/l (except Secchi readins in m)
**Not enough samples taken to comment

During the summer months, when the lake is stratified, a de-oxygenated hypolimnion layer covers over 50% of the basin, but extensive de-oxygenation near the surface also occurs noted at times. Continuing discharge of the present effluent from Langdale Sewage Treatment Works (STW) into the inner basin of Elterwater is therefore not acceptable. If remedial action is not taken, there can only be further deterioration in water quality with increased anoxia and continued sediment enrichment. There is also some evidence from the algal distribution studies that the problems of the inner basin are starting to impact on the middle basin.

Because of the minimal water exchange between the inner and the outer basins, it is unlikely that improved treatment of Langdale STW, i.e. nutrient removal, in itself will prove effective in the long term. Many scenarios have been suggested including: diversion, isolation and nutrient precipitation (Carvalho & Moss, 1995). Diverting Great Langdale Beck into the inner basin combined with widening the channel between the inner and middle basins was also considered, but most of these solutions are too expensive.

There is no doubt that the preferred solution for the long term benefit to Elterwater is to find an alternative discharge location for the Langdale STW's effluent (Beattie et al., 1996). The proposed solution, by agreement with North West Water plc, English Nature and other environmental bodies, is to divert the sewage effluent into the River Brathay downstream of Elterwater, thus eliminating the discharge from the lake itself.

When the scheme is completed in August 1998, monitoring of the water quality will be continued. A programme to monitor improvements to the inner basin and the impact on the River Brathay has been designed. In the inner basin, improvement to the water quality may not be evident for some time, due to the large pool of phosphorus in the sediments which is in itself a source of nutrient to the basin. The natural remediation processes and timescales within the inner basin and the need for intervention methods will be assessed.

Additional options (Boers & van der Molen, 1993) which can be considered include: tertiary treatment (phosphorus removal to 0.2 mg l^{-1}) of the relocated effluent (estimated cost 6500–9000 ECU ha^{-1} ann^{-1}), dredging the sediments (14 000–40 000 ECU ha^{-1} ann^{-1}), and hypolimnetic aeration (400 ECU ha^{-1} ann^{-1}).

Future work in the English Lake District lakes

Following this first management project, continuous monitoring of water quality has been extended to more English Lake District waters. These include Ullswater (north and south basins), Grasmere, Loweswater, Bassenthwaite, Elterwater, Buttermere, Ennerdale and Esthwaite. The monitoring in Ennerdale and Buttermere is designed specifically to investigate concerns at the apparent acidification of their catchments (Fowler et al., 1985).

Apparent increases in nutrient concentrations in Ullswater over the past five years (Beattie et al., 1996) have led to intensive survey work to confirm the upward trend, and assess suitable monitoring and management strategy. Continuous monitoring continued in 1996 and 1997 for the above lakes, with similar results between the years (Anon., 1998). Elterwater remained the most eutrophic (maximum chlorophyll *a* 80 μ l^{-1}, minimum Secchi disc extinction 1.7 m), with Ennerdale and Buttermere the least (maximum chlorophyll *a* 1 μ l^{-1}, minimum Secchi disc extinction 11 m). Complete de-oygenation of the hypolimnion was recorded during the summer in Grasmere, Estwaite, Bassenthwaite and Elterwater.

Monitoring of some of the larger English District lakes may be carried out through remote sensing, building on findings of earlier work (George & Charlton, 1996).

Acknowledgments

The monitoring of all Cumbrian lakes, including Elterwater, has required the help and support from various people inside and outside the Environment Agency. The authors wish to thank all staff involved in the monitoring of Elterwater, the land owners and the Lake District National Park for their support, and English Nature for using their access unto the lake. The authors also acknowledge the Institute of Freshwater Ecology for their time and expertise in assisting the Environment Agency in this area of work. The views expressed are those of the authors and not necessarily the Environment Agency.

References

Anon., 1980a. Suspended, Settleable and Total Dissolved Solids in Waters and Effluents. HMSO, London.

Anon., 1980b. The Determination of Chlorophyll *a* in Aquatic Environments, Methods for the Examination of Waters and Associated Materials. HMSO, London.

Anon., 1988. Discrete and air segmented automatic methods of analysis including robots 2nd edn. HMSO, London.

Anon., 1993. Skalar automated N & P analyser. Skalar, UK.

Anon., 1994. Elterwater Survey – July 1993, Marine and Special Projects, EQ & PC, internal report, MSP-94-003, April. National Rivers Authority, Warrington, UK.

Anon., 1995. Blue-Green Algal Monitoring and Management of Incidents, procedures manual. National Rivers Authority, Warrington.

Anon., 1998. Continuous monitoring of the Lake District, Cumbria: 1997 Results. Environment Agency, North West Region, Warrington.

Beattie, L., I. Gize, S. Berry, R. Webster, A. Fletcher & A. Hartland, 1996. A study of the water and sediment chemistry of key Cumbrian lakes, Environment Agency, North West Region, Warrington.

Boers, P. C. & D. T. van der Molen, 1993. Control of eutrophication in lakes: the state of the art in Europe. European Water Pollution Control 2: 19–25.

Carvalho, L. & B. Moss, 1995. The current status of a sample of english sites of special scientific interest subject to eutrophication. Aquatic Conservation: Marine and Freshwater Ecosystems 5: 191–204.

Codd, G. A. & S. G. Bell, 1996. The Occurrence and Fate of Blue-Green Algal Toxins in Freshwater, National Rivers Authority R&D Report 29, Foundation for Water Research, Medmenham, UK .

Cranwell, P. A., C. S. Reynolds, A. J. Lawlor & J. P. Lishman, 1995. Sediment Analyses of Sensitive Cumbrian Lakes, Institute of Freshwater Ecology Report T04050z1. National Rivers Authority, Warrington.

Fowler, D., J. N. Cape, I. D. Leith & I. S. Paterson, 1985. Acid Deposition in Cumbria. Pollution in Cumbria, Symposium no. 16, Ed. P. Ineson, Institute of Terrestrial Ecology, Grange-Over-Sands, Cumbria, UK.

George, D. G. & F. L. Charlton, 1996. Inland use of Airborne Remote Sensing. Environment Agency R & D Technical Report E5. Foundation for Water Research, Medmenham, UK.

Hall, G. H., P. V. Allen, J. E. Corry, P. A. Cranwell, E. Y. Haworth, D. P. Hewitt & G. H. Jaworski, 1992. The Lakes Tour 1991: a Survey of the Cumbrian Lakes, Institute of Freshwater Ecology report T04051n, National Rivers Authority, Warrington, UK.

Irving, B., 1993. Too Much of A Good Thing: Nutrient Enrichment in the U.K.'s Inland and Coastal Waters. World Wild Fund, Surrey, UK.

Lorenzen, C. J., 1966. Determination of chlorophyll and pheopigment: spectrophotometric equations. Limnol. Oceanogr. 11: 343–346.

Lund, J. W., 1959. A Simple Counting Chamber for Nannoplankton, Limnol. Oceanogr. 4: 57–65.

Lund, J. W., 1981. Investigations on Phytoplankton with Special Reference to Water Usage. Occasional Publication No.13. Freshwater Biology Association, Windermere, UK.

Mills, D. K. & P. B. Tett, 1990. Use of a recording fluorometer for continuous measurement of phytoplankton concentration. Environmental and Pollution Measurement Sensors and Systems, SPIE Proceedings: 1269.

Reynolds, C. S., 1984. The Ecology of Freshwater Phytoplankton. Cambridge University Press, Cambridge, 384 pp.

Rouen, K. J., 1995. Lake Surveys, 1994: Phytoplankton of Coniston Water, Elterwater, Esthwaite Water and Grasmere. Technical memorandum NB 314 01/95. National Rivers Authority, Warrington, UK.

Salonen, K., R. I. Jones & L. Arvola, 1984. Hypolimnetic phosphorus retrieval by diel vertical migrations of lake phytoplankton. Freshwat. Biol. 14: 431–438.

Stockdale, J. M., 1991. A Study of the Hydrology and Water Quality of Elterwater, Cumbria, MSc thesis, Imperial College of Science, Technology and Medicine, University of London.

Vollenweider, R. A., 1968. Scientific fundamentals of the eutrophication of lakes and flowing waters, with particular reference to nitrogen and phosphorus as factors in eutrophication. OECD, Paris.

Whitton, B. A., 1973. Freshwater plankton. In N. G. Carr & B. A. Whitton (eds), The Biology of Blue-Green Algae. Blackwell, Oxford: pp. 353–367.

Hydrobiologia **395/396:** 281–291, 1999.
D.M. Harper, B. Brierley, A.J.D. Ferguson & G. Phillips (eds), The Ecological Bases for Lake and Reservoir Management.

The importance of inter-annual variability for management

Caridad de Hoyos[1] & Francisco A. Comín[2]
[1]*Department of Ecology, University of Salamanca, Salamanca, Spain*
[2]*Department of Ecology, University of Barcelona, Spain*

Key words: oligotrophic lakes, Spain, inter-annual variation, management

Abstract

In regions with high inter-annual climatic variability, lake management strategies should be based on scientific data covering the maximum range of lake dynamics. Lake Sanabria is a monomictic lake located in a gneiss-dominated glacial valley at 1000 m a. m. s. l. in NW Spain. A three-year study was carried out during 1987–89 to obtain the basic ecological knowledge required for a sustainable lake management. Water and phytoplankton samples were collected from several depths of the water column at the deepest zone of the lake. Temperature and Secchi disk visibility, and total and dissolved inorganic nitrogen and phosphorus, chlorophyll *a*, and phytoplankton species composition and abundance (in laboratory) were determined monthly following standard limnological methods. Stratification began in March–April, with the thermocline stabilized at between 6–10 m, and lasted until November–December. The euphotic depth was higher than the mixing depth throughout the stratification period, except in the month previous to the water column mixing. The water residence time, estimated from the volume of the lake and the inflows and outflows, was between 0.48–0.67 years. However, during the months of highest rainfall (September–February) it was up to three times greater than during the same months of low rainy years. Nutrient concentrations in the water were very low throughout the year (52 $\mu g\ l^{-1}$ dissolved inorganic nitrogen, 6.1 $\mu g\ l^{-1}$ soluble reactive phosphorus, and 1.45 mg l^{-1} soluble reactive silica) and dominated by the dissolved organic fractions, which was related to the inputs of organic materials from the densely forested catchment. The colour of the lake, 21–41 mg Pt l^{-1}, is also indicative of the relatively high amount of organic substances dissolved in the waters of the lake. A bloom of cyanobacteria (*Microcystis flos-aquae*, *Aphanothece clathrata*, *Merismopedia tenuisssima*, *Synechococcus* sp.) persisted in most of the water column during August 1988–March 1989, associated with the high residence time of the water. In contrast, the cyanobacteria bloom was limited to the mesolimnion for 3–5 months during 1987 and 1989. A model is proposed for the seasonal variation of the phytoplankton, which differentiates the seasonal trend during years of high water turnover from that of years with long residence time. Lake Sanabria is oligotrophic, based on a range of criteria (Secchi disk visibility, chlorophyll *a* concentration, total phosphorus concentration). However, during years of long water residence time (e.g. 1988), the annual input of total phosphorus is critical to maintain the characteristics of the lake. Short-term strategies for the management of lakes should take into account the inter-annual variability rather than averages as indicators to prevent the impact of increased nutrient inputs derived from changes in land and water use in the catchment area. Long-term strategies should also consider the potential changes in the lake's dynamics, derived from the changes in the rainfall pattern caused by climatic changes.

Introduction

The inter-annual variability of ecological processes in lakes of the temperate zone can be very high, dependent upon changes caused both by climatic fluctuations and anthropogenic activities (Catalán & Fee, 1994). Yet many environmental plans and laws affecting lake management directly or indirectly are established on data averages or a few data, which may not be representative enough of the limnological characteristics upon which lake management is based, even they may contain continuous records covering several years (Kira, 1994). (e.g. Directive 78/659/CEE of the European Community on quality of continental waters (EEC, 1978). Inter-annual variation has been recognized as an intrinsic character of most limnetic ecosystems (Hutchinson, 1957), although it is hardly referred to in textbooks (Goldman & Horne, 1983; Wetzel, 1983).

Regional and local climatic variations are responsible for the differences observed from year to year in mixing depth (Harris & Piccinin, 1980) transparency, nutrient content, chlorophyll concentration, primary production (Dickman, 1969), seasonal succession of plankton and ice breakup timing (Reynolds, 1984; Goldman & Amezaga, 1984; Harris, 1986; Palecki & Barry, 1986).

In many cases, local conditions (e.g. position on an altitudinal or hydrological gradient within the same watershed) may be asynchronous with regional and global climate trends, as demonstrated by Williams et al. (1996) and Webster et al. (1996) for mountain lakes. However, even in lakes located in areas of low climatic variability, the inter-annual variability of some ecological characteristics can be an important issue in their conservation and management.

Climatic variations of relatively low frequency (lower than a decade), such as those related to the influence of the El Niño Southern Oscillation, often correlate well with the observed inter-annual variability of temperate lakes such as Castle Lake (Strub et al., 1985), or the Wisconsin Lakes, (Anderson et al., 1996). Even lakes located in southern Europe are sensitive to the climatic variability originated by ocean-atmosphere interactions in remote areas of the South Pacific (Rodó et al., 1997). Identifying physical, chemical and biological sensors of inter-annual variability will help to evaluate the response of lakes to natural (short climatic) variability as well as long-term global changes caused by human activities.

The analysis of long-term series of data by Sas (1989) included the presentation of long-term inter-annual variations of ecological characteristics of nineteen lakes from Europe. The most general recommendation of this study was to maintain lake monitoring for several years before any conclusions on the effects of nutrient loading reduction on lake restoration could be established unequivocally. The study of inter-annual variability of chemical characteristics is also relevant to recovery of lakes from acidification (Cosby et al., 1985). Reversibility from both eutrophication and acidification may require a long period (more than a decade) and differentiation between a net trend and inter-annual variability will require a long set of annual records.

Species themselves, as well as lake systems, show annual variability, and knowledge of the extent of this is important, particularly for species used as environmental indicators (Smith, 1993). Classical long-term records and experiments (Lund, 1978) highlighted the slight differences in phytoplankton micro-succession taking place in temperate lakes related to annual changes in the availability of nutrients and mixing intensity. Flushing of algae and low productivity was related to heavy annual rainfall by Dickman (1969). Changes in the specific composition of phytoplankton, from cyanobacteria dominance to chlorophytes, cryptophytes and chrysophytes was associated with an increase in conductivity (within the oligohaline range) taking place in prairie lakes over 12 years (Evans & Prepas, 1996).

Lake Sanabria, the largest Spanish lake, is used here to illustrate the importance of inter-annual variability in lake ecology and management. It is located in a geographic area which receives the effects of the North Atlantic Oscillation (Rodó et al., 1997), so it experiences relatively high external inter-annual variability. It has been predicted to be in an area where major impacts of global change on hydrology and water resources will occur during next century (Wetherald, 1991; Hayes, 1991). The data and ideas presented here are mostly focussed on the physical, chemical and phytoplankton characteristics of the lake, which are critical aspects for its management. The paper also considers the hydrological history of the lake, based on a 50-year series of rainfall data, and its future under different climatic scenarios.

Study area

Lake Sanabria is a glacial lake located in the upper part of the Tera River, in the river Duero catchment area, at 1000 m.a.s.l. in NW Spain (42° 07′ N, 06° 43′ W) (Figure 1). The lake's catchment (127 km^2) is composed of gneiss (66.5%), granodiorite (22%), slate (6.5%) and fluvio-glacial deposits (5%). The lake receives water from a mountainous plateau at 1600 m a.s.l. Steep slopes form the transition from this plateau to the lake basin. There is only one natural inflow and one natural outflow.

Annual air temperatures range from below 0 °C between November and May, to 36 °C in summer, with mean monthly temperature 10.4 °C. High inter-annual rainfall variability, most falling between September to February, also occurs. Annual average rainfall for

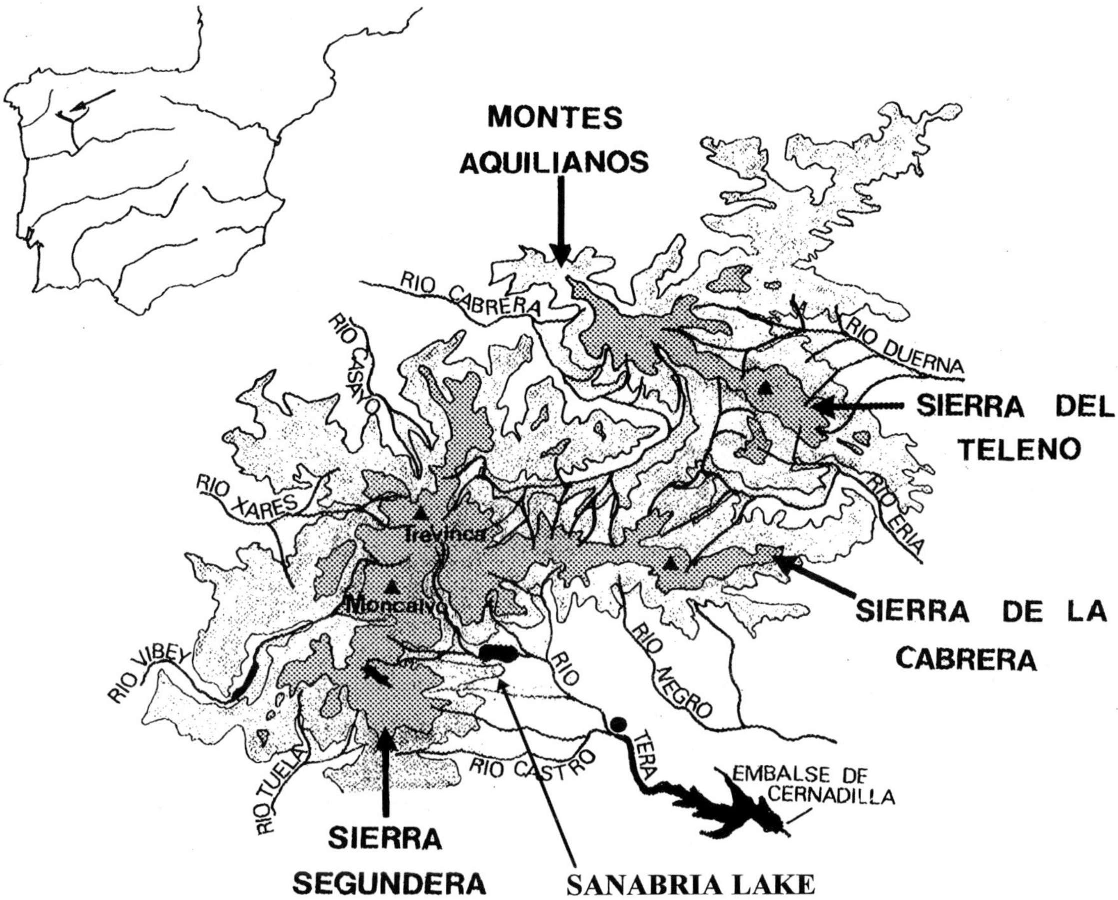

Figure 1. Hydrographic network of Lake Sanabria watershed (NW Spain).

1942–93 was 1422 mm, with range 544–2223 mm and coefficient of variation 42.7% of the average.

Lake Sanabria is almost rectangular with the longest axis in the direction of the water flow. The average depth is 27 m and the maximum depth, in the East basin, is 50 m. The bottom is flat (development of volume 1.6) corresponding to a typical U-shaped glacial valley. The lake area is 3.5 km^2, and its maximum length is twice the maximum width. Other morphometric parameters, are listed in Table 1. The shore of the lake is mostly rocky and dense stands of aquatic macrophytes (*Equisetum fluviatile, Potamogeton natans, Utricularia australis, Myriophyllum alterniflorum, Isoetes velata*) are located only in a small area of the west shore (Aldasoro et al., 1992). Further descriptions are given by Planas (1991), Aldasoro et al. (1992), Vega et al. (1993) and Rieradevall et al. (1993).

Table 1. Morphometric characteristics of Lake Sanabria

Lake area, m^2	3 460 876
Lake volume, m^3	93 297 863
Maximum depth, m	50
Average depth, m	27
Maximum length, m	3178
Maximum width, m	1590
Shoreline development	1.5
Volume development	1.6

The upper part of the catchment is a mountainous plateau of moorland containing a large number of peat bogs and shallow ponds. A few of these ponds were dammed during the 1940s to accumulate water with which to produce energy at an electric power plant located near the lake. Their total capacity is 25.7

Hm^3. The mountain slopes down to the lake shore are densely forested, mainly with oak. Most of the highest catchment is occupied by shrub and pasture. There are three small villages (with a total number of 200 permanent inhabitants) although the number of occasional visitors increases the population 50-fold for a short period in summer. The establishment of a Natural Park in 1978 (initially 5027 ha, including the lake, and later enlarged to 22 370 ha) attracted large numbers of visitors. Forest fires, whether spontaneous or started by shepherds to destroy shrub and wooded areas to obtain pasture, is one of the major environmental problems for the Park. Intensive use of beaches and camping sites in the south and east shores in summer already occurs, as well as some wastewater discharge from the main village to the River Tera.

Study material and methods

Monthly samples of water at 0, 2, 5, 10, 15, 20, 25, 30, 40 and 50 m depth were collected with Van-Dorn bottles from a boat anchored over the deepest part of the lake during 1987, 1988 and 1989. Measurements of temperature, pH, and oxygen were made immediately after collection from the sampling bottle, using electronic apparatus previously calibrated. After this, sub-samples were taken and refrigerated for further chemical and biological analysis. Sub-samples for phytoplankton observations were preserved with a few drops of Lugol's solution.

Alkalinity was determined by titration with weak acid. Dissolved inorganic forms of nitrogen and phosphorus were determined from filtered (Whatman GF/C filters) sub-samples using standard analytical methods (APHA, 1975). Total fractions of nitrogen and phosphorus were determined following the same methods after digestion. Dissolved organic fractions were estimated as the difference between total and dissolved inorganic fractions on filtered sub-samples.

Chlorophyll *a* was determined spectrophotometrically from pigment solutions extracted with acetone (Jeffrey & Humphrey, 1975). Phytoplankton abundance was estimated using the sedimentation method with the inverted microscope (Utermöhl, 1958) and biovolumes calculated by conversion of cell number to volume assuming their similarity to geometric figures.

Table 2. Residence time of the water (years) in lake Sanabria calculated for different periods. Sept.–Feb. means from September of the year indicated in the table to February of the next year. Maximum values corresponded to 1944 and to Sept. 1944–Feb. 1945, minimum values to 1978 and to Sept. 1978–Feb. 1979

1987	1988	1989	1942–92	1942–92 Average	1942–92 Maxi-mum	1942–92 Mini-mum
Annual	0.54	0.67	0.48	0.76	1.82	0.44
Sept.–Feb.	0.30	0.90	0.29	0.62	1.92	0.24

Results

Lake Sanabria had one annual mixing period, which affected the entire water column, lasting from November/December to March/April. The thermocline began in the uppermost levels of the water column in March/April (Figure 2) and stabilised between 6 and 10 m during summer. It deepened to 12–17 m during autumn and broke up every year during late autumn. The maximum water column stability was reached during mid summer. Maximum values of the Brunt-Vaisala frequency (as N^2) were 2.7 10^{-3} s^{-2}, $2.6 \cdot 10^{-3} s^{-2}$ and 3.9 10^{-3} s^{-2} in 1987, 1988 and 1989, respectively. There was a difference of two months between the duration of stratification in 1988 and 1989 (Figure 2).

The depth of the euphotic zone was greater than the depth of the thermocline for most of the stratified period. Only two exceptions were observed during the period of study, just before the beginning of the mixing period in both 1987 and 1988.

The residence time of the water was very similar for the three years (0.54, 0.67 and 0.48 years, respectively). However, it showed clear differences in different periods of the year. Generally, the wettest period of the year was September–February when the residence time was higher during 1988–1989 (0.96 years) than during the other two periods (0.3 years) (Table 2). There was clear difference in the monthly water residence time between the 1988–89 winter and the other two winters. Water turnover was low during 1988–1989 because of the low rainfall, compared to the same periods of 1987–88 and 1989–90 (Figure 3).

Anoxia was not observed at any time during the study but the concentration of oxygen decreased progressively during stratification periods with depth, and minimum values were observed at the end of the stratification period at the deepest point sampled,

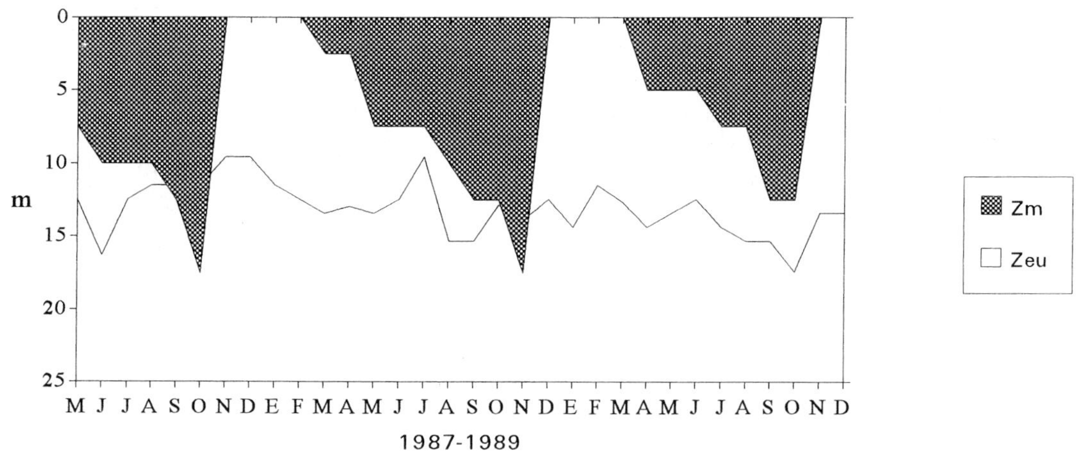

Figure 2. Monthly changes of the mixing (z_m) and euphotic (z_{eu}) depths in Lake Sanabria during 1987–89.

closest to the maximum lake depth. Maximum oxygen concentrations were observed during mixing periods.

The chemical characteristics of Lake Sanabria were typical for soft waters in crystalline areas (Table 3) – slightly acid and with a low mineral content. The alkalinity was very low, 0.045 meq l^{-1} on average. The relative abundance of the major ions $Ca^{2+} > Na^{+} > Mg^{2+} > K^{+}$ and $HCO_3^{-} > Cl^{-} > SO_4^{2}$ indicated that the watershed is influenced by oceanic salts through rainfall.

The concentrations of micronutrients were very low (Table 3) Most of the total combined nitrogen was dissolved organic (71%), the remainder nitrate. Dissolved organic phosphorus was also high (46%).

Table 3. Chemical characteristics of the water in Lake Sanabria during 1987–89. Averages correspond to all the data obtained from all the depths for each variable during the period of study

	Average	Minimum	Maximum
Salinity, mg/l	8.57	7.6	10
Conductivity, μS/cm	14.9	11.8	18
Alkalinty, meq/l	0.045	0.03	0.058
pH	6.23	5.6	8.4
Dissolved oxygen, mg/l	8.22	3	10.5
Nitrate, N-NO_3, μg/l	51.7	0.5	165
Nitrite, N-NO_2, μg/l	0.06	0	2.5
Ammonia, N-NH_4, μ/l	1.27	0	20
Phosphate, P-PO_4, μg/l	6.07	0	34
Silicate, mg/l	1.45	0.9	1.94
Total nitrogen, μg/l	202.7	124	420
Total phosphorus, μg/l	8.7	2.6	46.5

These data indicate that organic substances dissolved in Lake Sanabria play an important role in the lake's ecology. Relatively high values of water colour (21–41 mg Pt l^{-1}) and the fact that red light reaches deeper zones of the water column than other wavelengths during the months of high water inflow confirm the importance of dissolved organic nutrients.

Most of the phytoplankton biovolume, up to 40%, was contributed by Cryptophyta and Chlorophyta (Figure 4). Bacillariophyta were also present at times, 10–30% of the total. Cyanobacteria were alays present at between 1–20% of the total.

Cyanobacteria were the most abundant group of phytoplankters (Figure 4). *Microcystis flos-aquae, Coelosphaerium kutzingianum, Aphanothece clathrata, Aphanocapsa elachista, Merismopedia tenuissima* and *Synechococcus* sp. were the dominant, proliferating during the summer each year, with the 1988 autumn bloom persisting through to winter 1988–89 (Figure 5). The most abundant Chlorophycea were *Sphaerocystis schroeteri, Oocystis submarina, Quadricoccus laevis, Botryococcus neglectus, Radiococcus cf. bavaricus, Crucigenia tetrapedia, Crucigenia quadrata, Monoraphidium komarkovae* and *Chlorella cf. vulgaris*). Desmidacea were not abundant, although a few species (*Staurastrum denticulatum, S. anatinum* and *Staurodesmus jaculiferus*) reached biovolumes higher than 1 mm^3 m^{-3}. Most Dinoflagellata were small species (*Gymnodinium* sp., *Amphidinium elenkinii, Katodinium schilleri*). However, other species (*Peridinium umbonatum, P. cinctum, Ceratium hirundinella*) although present in very low densities, reached an average biovolume higher than 1 mm^3

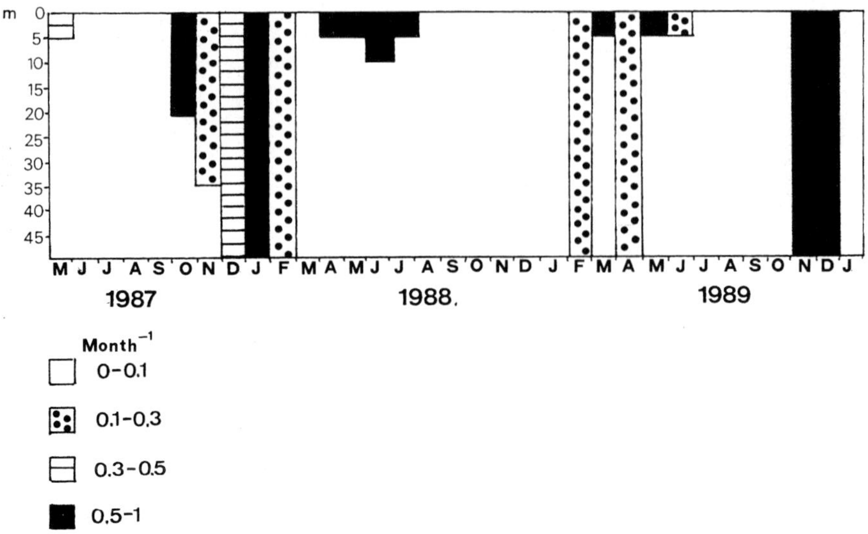

Figure 3. Monthly changes of the water turnover rate (month^{-1}) in Lake Sanabria during the period of study.

m^{-3}. The most abundant Chrysophycea were of the genus *Ochromonas*, *Pseudopedinella* and *Mallomonas*. Cryptophytes were represented by *Rhodomonas minuta, Kathablepharis ovalis* and several species of *Cryptomonas*. The most representative Bacillariophyta were *Aulacoseira distans*, *Cyclotella glomerata* and *Cyclotella stelligera*.

Chlorophyll *a* concentrations were very low. The annual average in the epilimnion was between 1.5 and 1.9 μg l^{-1}. Maximum chlorophyll *a* concentrations occurred during summer (between the thermocline and the lower limit of the euphotic zone) and in autumn (in the entire euphotic zone). Total cell density ranged from 250–37 467 cells ml^{-1}, and biovolume from 50–1000 mm^3 m^{-3}. Relatively high chlorophyll *a* concentrations and biovolumes occurred during winter 1988–89 associated with the relatively low water turnover. The values of the ratio D430/D665 (absorbances at 430 and 665 nm wavelengths) ranged between 1.5 and 50, with and average of 5.2. High values occurred in the hypolimnion and were, most probably, due to pigment degradation. A progressive increase of this index was observed during spring and summer, corresponding to the depletion of epilimnetic nutrients.

The phytoplankton dynamics are represented in Figure 6 as the concentration of chlorophyll *a* in the water and the chlorophyll *a* per phytoplankton cell(both from samples corresponding to the upper 15 m of the water column). During spring, the chlorophyll *a* concentration and the chlorophyll *a* per cell were low (the ratio D430/D665) was high). During this period of the year the phytoplankton was dominated by Cryptophyta or Chlorococcales and Bacillariophyta. During summer, in the upper part of the epilimnion there was a low chlorophyll *a* concentration, low chlorophyll *a* per cell and high D430/D665 ratio. Dinoflagellates and Chrysophyta were abundant, but not cyanobacteria. At the thermocline there was a high chlorophyll *a* concentration, and chlorophyll *a* content per cell (Cyanobacteria, Chlorococcales, diatoms and, in some cases, Cryptophyta, were abundant). Autumn was characterized by an increase of the chlorophyll *a* content per cell, an increase of the chlorophyll *a* concentration in the water and low values of the ratio D430/D665. The algae (including Cyanobacteria) that had accumulated during summer in the thermocline were distributed throughout the water column. At the beginning of winter the ratio D430/D665 was still low. However, the trend during winter was different from one year to another depending on the turnover rate of the water. If the water turnover was high, chlorophyll *a* concentration in the water decreased but the chlorophyll *a* content per cell remained high. Colonial Cyanobacteria disappeared and Volvocales and Cryptophyta increased in abundance. On the other hand, when the water turnover was low (1988–89), the concentration of chlorophyll *a* in the water remained high but the chlorophyll *a* content per cell decreased. Under those favourable conditions of physical stability

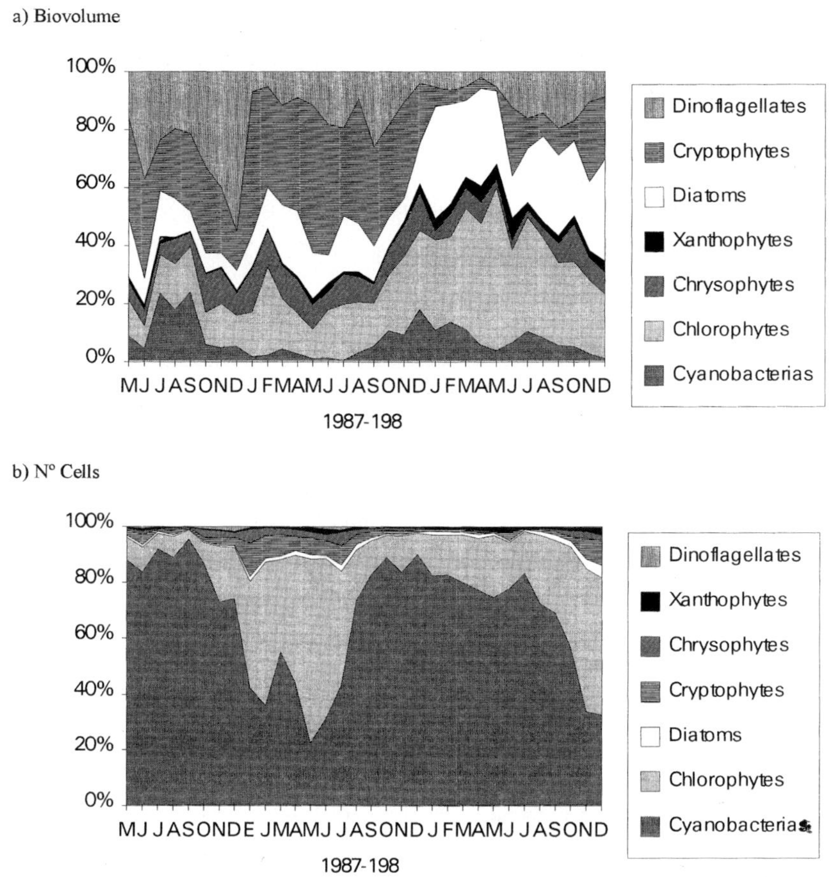

Figure 4. Monthly changes of the relative abundances of (a) biovolume and (b) cell density of different groups of phytoplankton in Lake Sanabria during 1987–1989.

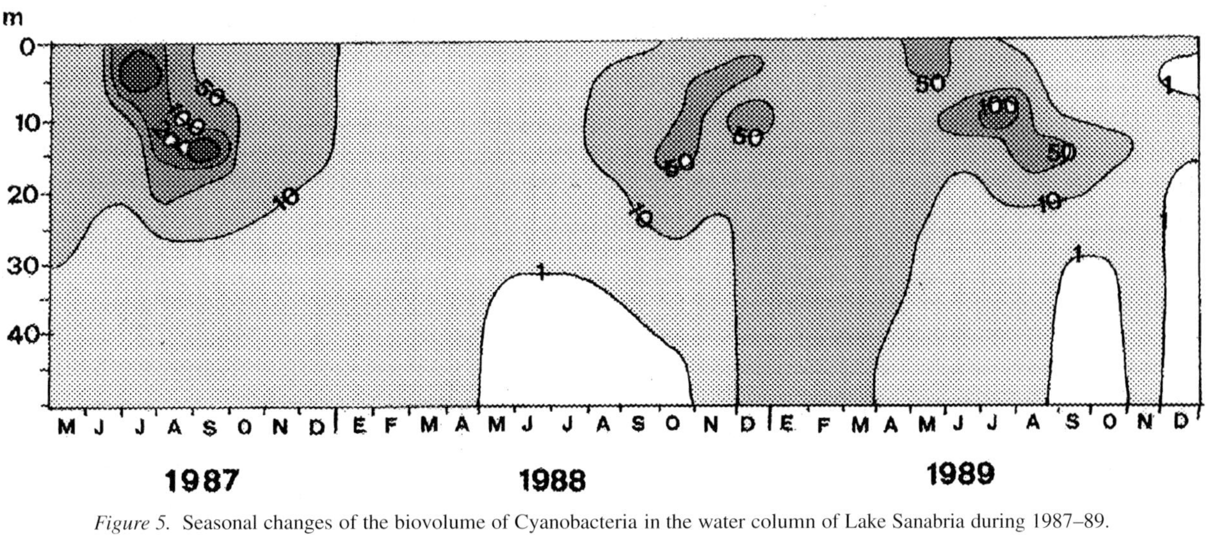

Figure 5. Seasonal changes of the biovolume of Cyanobacteria in the water column of Lake Sanabria during 1987–89.

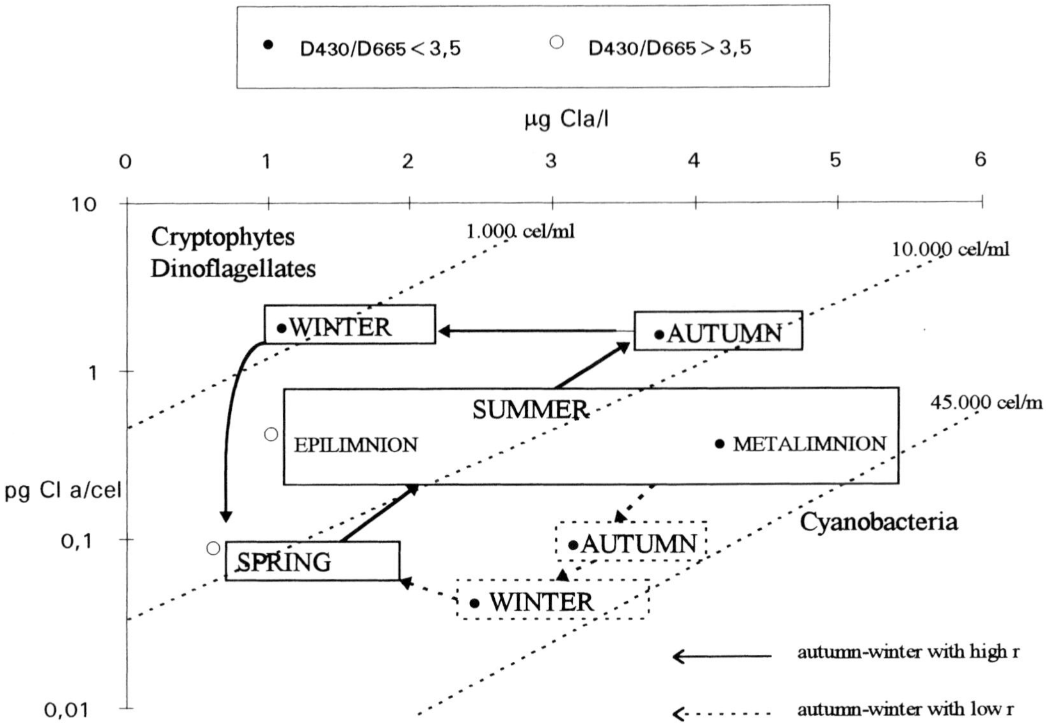

Figure 6. Schematic representation of the seasonal dynamics of phytoplankton in Lake Sanabria represented by the concentration of chlorophyll *a* per water volume (horizontal axis), the concentration of chlorophyll *a* per cell of phytoplankton (vertical axis) and the abundance of cells per ml of water (discontinuous lines). Different values of the pigment ratio D430/D665 are indicated by the type of circle. The temporal trend is indicated by arrows and differentiates periods of relatively high water turnover (r) (solid arrows) from those of relatively low water turnover (discontinuous arrows).

and the quite constant chemical characteristics, Cyanobacteria and Chlorococcales persisted in the water column. Inter-annual variability is thus a major characteristic of the temporal dynamics of phytoplankton due largely to the differences in water turnover rate which, in turn, is caused by climatic changes.

Trophic state indicators (Vollenweider & Kerekes, 1980) place Lake Sanabria as oligotrophic (Table 4), although very close to the mesotrophic conditions based on the phosphorus discharge (Table 5). Acceptable phosphorus discharges (APD) and excessive phosphorus discharges (EPD) into lakes (respectively, the amounts of phosphorus discharged into a lake which does not change the trophic state of the lake, APD, and which will increase the trophic state of the lake, EPD, units mg P m^{-2} yr^{-1}), have been related empirically with the average depth (z) and the water residence time (t) through the following expressions:

$$\text{APD} = 100 + 10z/t$$

$$\text{EPD} = 200 + 20z/t$$

(Vollenweider & Kerekes, 1980).

So, increasing residence time for the water lowers the acceptable and excessive phosphorus discharges; phosphorus discharge into Lake Sanabria through the River Tera from 1st January – 8th December 1989 (water residence time 0.7 years) was estimated as 0.6 g P m^{-2}, between the APD (0.48 g P m^{-2} ann^{-1}) and the EPD (0.97 g P m^{-2} ann^{-1}) (Table 5). The estimated discharge for April–December 1988 (water residence time 1.3 years) coincides with the APD (0.3 g P m^{-2}). Thus phosphorus discharge into Lake Sanabria was just 'acceptable'.

The annual water residence time was estimated for the last fifty years from annual rainfall, assuming no changes in the watershed uses and lake characteristics which is quite reasonable, because this is a remote mountain area with very few landscape changes. Figure 7 shows the annual water turnover rates during 1942–92. The values range from 0.4 years in 1978, the wettest year, to 1.8 years in 1944, the driest year. The APD for these years (Table 5) would have been 0.71 and 0.25 g P m^{-2} ann^{-1} and the EPD 1.4 and 0.5 g P

Table 4. OCDE (1982) standards established to consider a lake as oligotrophic and values of these characteristics in Lake Sanabria

	Mean annual total phosphorus (μg/l)	Mean annual chlorophyll *a* (μg/l)	Maximum chlorophyll *a* (μg/l)	Mean annual Secchi disk (m)
Oligotrophic Lakes	4–10	1–2.5	2.5–8	6–12
Lake Sanabria	8.7	1.89	6	6.9

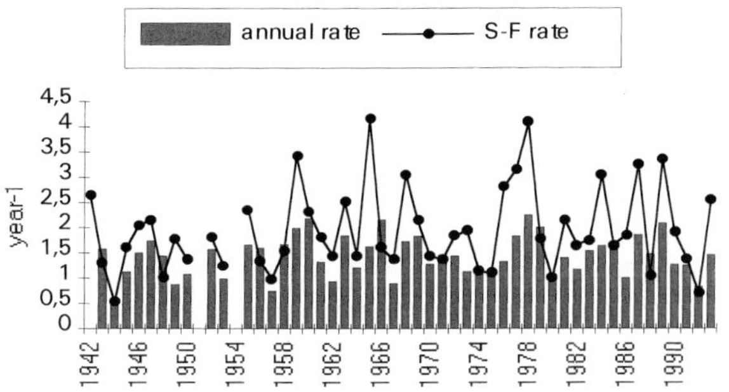

Figure 7. Annual water turnover rates in Lake Sanabria and water turnover rates for every period September–February during the years 1942–1992.

m^{-2} ann^{-1} respectively. The three years (1987–89) of study were relatively wet and the water residence time was below the average for the last 50 years (Table 2). During dry years (e.g., 1944 and others) (Figure 7), the APD and EPD would have been lower than during the years of this study.

Discussion

Water turnover rate is the key variable which drives the other ecological processes of the lake. If the residence time of the water in a lake is lower than 1 year, the accumulation of organic matter will not be enough to increase its trophic conditions (Vollenweider & Kerekes, 1980). In lakes with low hydraulic residence time flushing exerts an important control on both the specific composition and abundance of the phytoplankton (Dickman, 1969; Goldman & Amezaga, 1984; Carmack et al., 1986). Both micronutrients, phosphorus and nitrogen are also strong influences on phytoplankton growth (Planas, 1991). Phosphorus loading to Lake Sanabria could increase through different human activities (e.g. non-controlled development, increase in visitors). Increasing phosphorus loading could enhance the role of nitrogen as a limiting nutrient. If this occurs, the specific composition of the phytoplankton could change and filamentous nitrogen fixing cyanobacteria could bloom, as is currently occurring in a reservoir located 10 km downstream on the River Tera (Negro et al., 1994).

It is important to consider the differences in lake dynamics at different periods of the year, because nutrient supply to phytoplankton in lakes with high and medium water residence times is affected by river supply during the growing season, and may not be directly related to annual nutrient loading (Carmack & Gray, 1982). Carmack et al. (1986) stated that hydrographically variable systems may be considered as short water residence time systems during one period and as long water residence time systems during another.

A plan has been approved by regional authorities that any wastewater from developments will be discharged downstream from Lake Sanabria in order to avoid contaminating the lake. It is hoped however that urbanisation and use of water in the catchment will not take place, so no significant water demand arises that might decrease the turnover of the lake. .Any development should incorporate water use and reuse practices in order not to extract water from the system.

Wood fires, whether natural or deliberate (to provide grazing for cattle), are frequent in the watershed. Those occurring in summer can have a large negative impact on the lake because of the increased runoff occurring in deforested areas (Likens, 1985). The increase of nutrient flux to lakes because of forest fires can be very important (Wright, 1976; Goldman et al., 1990). If fires occur during summer in dry years, which is usual, the negative impact will be higher, because of the accumulation of flushed organic matter in the lake during those dry (high water residence time) years. Prevention and rapid control of fires should be constantly improved within the catchment and incorporated into management plans.

Lake management should also consider other water uses which might affect the lake residence time. Water storage in the reservoirs located in the upper plateau of the catchment to produce electricity in a small electric power plant is a current activity: any proposal for increasing storage here could decrease water turnover in the lake. The driest year should be considered as that limiting the capacity of water to maintain the lake at its present trophic stage and all plans based upon that year.

In the future, it may also be necessary to consider the potential effects of different climatic scenarios on inter-annual variability of the lakes. If rainfall decreases as a consequence of global climatic change, as has been predicted for some parts of Europe (Mitchell, 1986), the residence time of the water will increase and the characteristics of dry years will occur more frequently. The goal of catchment management should be to achieve a sustainable use of the natural resources in the entire watershed.

Acknowledgments

The authors thank J. J. Aldasoro and J. C. Vega for their collaboration during the development of this work and to Junta de Castilla y León, particularly to G. Rodriguez de Lema, for the facilities given to carry out research in the Natural Park of Lake Sanabria.

References

Aldasoro, J. J., C. De Hoyos & J. C. Vega, 1992. El lago de Sanabria. Estudio limnológico. Monografías de la Red de Espacios Naturales de Castilla y León. Junta de Castilla y León: 120 pp.

APHA, 1975. Standard methods for the examination of water and waste water. American Public Health Association, New York, 769 pp.

Anderson, W. L., D. M. Robertson & J. M. Magnuson, 1996. Evidence of recent warming and El Niño-related variations in ice breakup of Wisconsin lakes. Limnol. Oceanogr. 41: 815–821.

Carmack, E. C. & C. B. Gray, 1982. Patterns of circulation and nutrient supply in a medium residence-time reservoir, Kootenay Lake, British Columbia. Can. Water. Resourc. J. 7: 51–70.

Carmack, E. C., R. C. Wiegand, R. J. Daley, C. B. Gray, S. Jasper & C. H. Pharo, 1986. Mechanisms influencing the circulation and distribution of water masses in a medium residence-time lake. Limnol. Oceanogr. 31: 249–265.

Catalán, J. & E. J. Fee, 1994. Interannual variability in limnetic ecosystems: origin, patterns and predictability. In R. Margalef (ed.). Limnology Now, A Paradigm of Planetary Problems. Elsevier, Amsterdam: 81–97.

Comín, F. A. & W. D. Williams, 1994. Parched continents. Our common future? In R. Margalef (ed.). Limnology Now. A Paradigm of Planetary Problems. Elsevier: 473–528.

Cosby, C. J., R. F. Wright, G. M. Hornberger & J. N. Galloway, 1985. Modelling the effect of acid deposition: estimation of long-term water quality responses in a small forested catchment. Water Resources Research 21: 1591–1601.

ECC (European Communities Council), 1978. Directive 78/659/EEC relative to the quality of continental waters which require protection or improvement to be adequate for fish life. ECOD (18 July 1978).

Dickman, M., 1969. Some effects of lake renewal on phytoplankton productivity and species composition. Limnol. Oceanogr. 14: 660–666.

Evans, J. C. & E. Prepas, 1996. Potential effects of climate change on ion chemistry and phytoplankton communities in prairie saline lakes. Limnol. Oceanogr. 4: 1063–1076.

Findenegg, I., 1943. Untersuchungen über die Ökologie und die Produktions-verhältnisse des Planktons in Kartner Seengebeite. Int. Rev. ges. Hydrobiol. Hydrog. 43: 368–429.

Goldman, C. R. & A. Horne, 1983. Limnology. McGraw-Hill Book Company, New York, 464 pp.

Goldman, C. R. & E. Amezaga, 1984. Primary production and precipitation at Castle Lake and Lake Tahoe during twenty four years, 1959–1982. Verh. int. Ver. Limnol. 22: 591–599.

Goldman, C. R., A. D. Jasby & E. de Amezaga, 1990. Forest fires, atmospheric deposition and primary productivity at Lake Tahoe, California-Nevada. Verh. int. Ver. Limnol. 24: 499–503.

Harris, G. P., 1986. Phytoplankton Ecology. Structure, Function and Fluctuations. Chapman & Hall, London, 384 pp.

Harris, G. P. & B. Piccinin, 1980. Physical variability and phytoplankton communities. IV Temporal changes in the phytoplankton community of a physically variable lake. Arch. Hydrobiol. 89: 447–473.

Hayes, J. T., 1991. Global climate change and water resources. In R. L. Wyman (Ed.). Global Climate Change and Life on Earth. Chapman and Hall, New York: 18–42.

Hutchinson, G. E., 1957. A Treatise on Limnology. J. Wiley & Sons, New York, Vol. 1, 1,015 pp.

Jeffrey, S. W. & G. F. Humphrey, 1975. New spectrophotometric equations for determining chlorophylls a, b, c_1 and c_2 in higher plants, algae and natural phytoplankton. Biochem. Physiol. Plant. 167: 191–194.2

Kira, T. (ed.), 1994. Data book of world lake environments. A survey of the state of world lakes. 2. Africa and Europe. ILEC-UNEP, Kusatsu, 852 pp.

Likens, G. E., 1985. Beyond the shoreline. A watershed ecosystem approach. Verh. int. Ver. Limnol. 22: 1–22.

Lund, J. W. G., 1978. Changes in the phytoplankton of an English lake, 1945–1977. Hydrobiological Journal 14: 6–21.

Mitchell, J. F. B., 1986. On modelling the effects of CO_2 on Climate. In A. Ghazi & R. Fantechi (eds). Current Issues in Climate Research. D. Reidel Publishing Company, Dordrecht: 228–239.

Negro, A. I., C. De Hoyos, A. del Rio & R. Le Cohu, 1994. Comparación de las comunidades fitoplanctónicas en dos embalses de reciente creación, Riaño y Valparaiso España. Limnetica 10: 115–121.

OECD, 1982. Eutrophication of waters. Monitoring, assessment and control. Organization for Economic Cooperation and Development, Paris, 154 pp.

Paerl, H.W., 1988. Growth and reproductive strategies of freshwater blue-green algae (cyanobacteria). In C. D. Sandgren (ed.). Growth and Reproductive Strategies of Freshwater Phytoplankton. Cambridge University Press, Cambridge: 261–315.

Palecki, M. A. & R. G. Barry, 1986. Freezeup and breakup of lakes as an index of temperature changes during the transition seasons: A case study fro Finland. J. Clim. Appl. Meteorol. 25: 893–902.

Planas, D., 1991. Factors controlling phytoplankton community structure in an alkaline versus a softwater lake. Oecol. Aquatica 10: 95–112.

Reynolds, C. S., 1984. The ecology of freshwater phytoplankton. Cambridge University Press, Cambridge, 384 pp.

Reynolds, C. S., 1988. Functional morphology and the adaptive strategies of freshwater phytoplankton. In C. D. Sandgren (ed.).

Growth and Reproductive Strategies of Freshwater Phytoplankton. Cambridge University Press, Cambridge: 388–433.

Rieradevall, M. C., I. Ibañez, I. Muñoz, M. Real & N. Prat, 1993. Grupos de macroinvertebrados del bentos del lago de Sanabria y su relación con la granulometría del sedimento. Actas del VI Congreso Español de Limnología Granada: 309–314.

Rodó, X., E. Baert & F. A. Comín, 1997. Variations in seasonal rainfall in Southern Europe during the present century: relationships with the North Atlantic Oscillation and the El Niño Southern Oscillation. Climate Dynamics 13: 275–284.

Sas, H., 1989. Lake restoration by reduction of nutrient loading. Expectations, experiences, extrapolation. Academia Verlag Richarz Publications, 497 pp.

Slobodkin, L., 1960. Ecological energy relationships at the population level. Am. Nat. 94: 213–236.

Smith, A. J., 1993. Lacustrine ostracods as hydrochemical indicators in lakes of the north-central United States. J. Paleolimnol. 8: 121–134.

Sommer, U., 1989. Towards a Darwinian ecology of plankton. In U. Sommer (ed.). Plankton Ecology. Springer-Verlag, Berlin: 1–8.

Strub, P. T., T. M. Powell & C. R. Goldman, 1985. Climatic forcing: effects of El Niño on a small temperate lake. Science, 227: 55–57.

Utermöhl, H., 1958. Zur Vervollkommnung der quantitativen Phytoplankton-Methodik. Mitt. int. Ver. Limnol. 9: 1–38.

Vega, J. C., C. de Hoyos & J. J. Aldasoro, 1992. The Sanabria Lake. The largest natural freshwater lake in Spain. Limnetica 8: 49–57.

Vega, J. C., G. Rodriguéz, M. Rieradevall & N. Prat, 1993. Research and environmental management in the lake Sanabria Natural Park (NW Spain). In Giussani, G. & C. Callieri (eds), Strategies for Lake Ecosystems Beyond 2000. Proceedings of the 5th International Conference on the Conservation and Management of Lakes, Stresa (Italy): 491–494.

Vollenweider, R. A. & J. Kerekes, 1980. The loading concept as a basis for controlling eutrophication philosophy and preliminary results of the OECD programme on eutrophication. Progress in Water Technology 12: 5–38.

Webster, K. E., T. K. Kratz, C. J. Bowser, J. J. Magnuson & W. J. Rose, 1996. The influence of landscape position on lake chemical responses to drought in northern Wisconsin. Limnol. Oceanogr. 41: 977–984.

Wetherald, R. T., 1991. Changes of temperatures and hydrology caused by an increase of atmospheric carbon dioxide as predicted by general circulation models. In Wyman, R. L. (ed.), Global Climate Change and Life on Earth. Chapman and Hall, New York: 1–17.

Wetzel, R. G., 1983. Limnology. Saunders College Publ., Philadelphia, 765 pp.

Williams, M. W., M. Losleben, N. Caine & D. Greenland, 1996. Changes in climate and hydrochemical responses in a high-elevations catchment in the Rocky mountains, U.S.A. Limnol. Oceanogr. 41: 939–946.

Wright, R. F., 1976. The impact of forest fire on the nutrient influxes to small lakes in northeastern Minesota. Ecology 57: 649–663.

Hydrobiologia **395/396:** 293–307, 1999.
D.M. Harper, B. Brierley, A.J.D. Ferguson & G. Phillips (eds), The Ecological Bases for Lake and Reservoir Management.

The management of hypertrophic lochs: case studies in southwest Scotland

Fiona H. Naysmith
Scottish Environmental Protection Agency, Rivers House, Irongray Road, Dumfries DG2 0JE, U.K.

Key words: phosphorus, sediments, eutrophication, Scotland

Abstract

This paper considers the range of management techniques which are currently available for the remediation of eutrophic lakes, with respect to two lochs within southwest Scotland. Preliminary studies of sources of nutrients identified the need for the development of different management prescriptions for each. Such studies will ensure that the management is tailored to meet the specific needs of the site. The two lochs, Carlingwark Loch at Castle Douglas and Castle Loch at Lochmaben, both support algal blooms but have different nutrient sources. Carlingwark Loch appears to be receiving large quantities of nutrients from the surrounding catchment, whereas Castle Loch shows a net loss of nutrients over its surface inflows, indicating release from the sediments and/or input from over-wintering wildfowl. In these cases, if preliminary studies had not been undertaken, the management programmes developed may not have used the most effective control measures.

Introduction

Eutrophication, and the algal blooms which are its main consequence, have been an area of increasing awareness in most countries since the 1960s, with extensive work being carried out to determine the causes (Vollenweider, 1968; Likens, 1972; Schindler, 1977), the results in ecological and recreational terms, and the management principles and tools which can be applied to reduce the effects (Moss, 1983; Phillips, 1984; Phillips & Kerrison, 1991; Phillips & Moss, 1994). A number of lochs within Scotland, subject to algal blooms, have been the focus of much research work, especially Loch Leven in Fife (Bailey-Watts, 1976; Bailey-Watts, 1978; Bailey-Watts & Kirika, 1987; Bailey-Watts et al., 1990). It is widely recognised that eutrophication is usually controlled by the availability of phosphorus within the water column, and as the result of modern water supply, use of detergents and agricultural intensification, available phosphorus has increased markedly. In southwest Scotland, cyanobacterial blooms have received less attention than in the rest of the U.K., although they were identified as affecting three lochs during 1989 (NRA, 1990). An 'algal watch' programme has been undertaken at 12 lochs since 1994 and has highlighted problems at a number of them including two, which are designated as Sites of Special Scientific Interest (SSSI) because of their bird and plant populations (Begg et al., 1996).

Carlingwark Loch at Castle Douglas and Castle Loch at Lochmaben, are high amenity lochs which have both been affected by algal blooms and cyanobacteria in recent years, and at which management and advisory committees have been established to ensure that these lochs remain assets to their local communities.

For lochs with identified algal problems, remediation work requires not only knowledge of the current trophic status of the loch and its catchment, but also understanding of the management strategies which are available and how these can be best applied for maximum benefit.

Management strategies

The first step in developing a loch management programme is to decide what are the overall objectives and what is the optimum total phosphorus concentration for those objectives to be met. Once these decisions have been taken, a number of techniques exist which can be used to try to meet the target total phosphorus concentration. These options can be considered in four separate pathways. While this is not an exhaustive list of potential techniques which could be applied, it covers the main options practically available. Selection of the most appropriate technique will be dependent upon the overall objectives for the loch, the optimal total phosphorus concentration and the provision of available funding and expertise:

1. Prevention of algal blooms.
2. Reduction of total phosphorus inputs.
3. Removal/immobilisation of stored phosphorus.
4. Biomanipulation techniques.

Each technique varies in complexity and cost. Algal blooms may be prevented by the introduction of straw bales into the water column or by the application of an algicide, neither of which is an ideal solution given the uncertainty over the quantity and timing of such applications. Algicides have been widely applied in the past, but environmental concerns regarding the indiscriminate use of chemicals has perhaps resulted in some reluctance for their continued use. The application of straw bales has been popular (see Ridge, this volume), but the mechanism by which they work has yet to be determined. In addition, once decomposition has begun, straw rafts may be effective, but are not always aesthetically pleasing. At one loch in central Scotland, where straw bales have been used in the past, current management practices include the introduction of a 'living system' raft. This technique has been used in the U.S.A. but is a fairly recent development in Scotland and therefore its success has yet to be determined.

Reduction of phosphorus inputs begins with preliminary baseline studies, which may identify previously unknown inlets and point sources where, on further investigation, small changes in current practices could result in marked reductions. In addition, any new developments could be denied Consents to discharge and/or require phosphorus-removal procedures to be implemented. Buffer strips could be introduced and inlets considered for diversion so that they by-pass the loch altogether. The Scottish Environment Protection Agency (SEPA) is currently in the process of producing a strategy for dealing with proposed discharges into inland lochs based on total phosphorus water quality standards (see Fozzard et al., this volume). It is proposed that these take into account the current trophic status and the potential influence of such a discharge.

Removal or immobilisation of stored phosphorus could be achieved in at least two ways. These are:

1. Removing the sediment, similar to the work undertaken in the Norfolk Broads (£15 000 ha^{-1}; see Madgwick, this volume).
2. Adding limestone or ferric chloride to sequester phosphorus.

A loch management programme using biomanipulation techniques would be a last resort, given the cost and level of expertise required to undertake such a project successfully.

Case studies in southwest Scotland

Prior to the development of a management programme which can tackle and attempt to solve the algal problems on any loch, it is necessary to determine the source of the nutrients responsible. In the past, both lochs described here have been subject to direct and indirect inputs of nutrients from sewage and/or industrial effluent, but these have now been halted. In order to determine whether the present algal problem is a result of these historic, or unknown, present day, practices within the catchments, a sampling programme incorporating both water and sediment chemical analysis of phosphorus and nitrogen was devised and undertaken (Begg, 1995).

The main objectives of this study were:

1. To determine the physical and chemical characteristics of the two lochs.
2. To establish the sources of excess nutrient inputs.
3. To enable a programme of action aimed at controlling the algal blooms and restoring habitat quality to be agreed.

Methods

Samples were collected from the main inflowing and outflowing streams (monthly), minor inflowing streams (periodically) and loch waters (spring and autumn, 1994 and spring, 1995). Algal and sediment samples were also collected. Water samples were analysed for both nitrogen and phosphorus in a variety of chemical forms, as well as silicate, chlorophyll *a* and dissolved oxygen. Sediment samples were collected on two occasions in each of the lochs (October, 1994 and February, 1995) for analysis of phosphorus content. In order to determine the availability of the phosphorus within the sediment, a sequential extraction procedure based on that of Hieltjes & Lijklema (1980) was undertaken.

Study sites

Carlingwark Loch forms part of the Threave and Carlingwark SSSI south of the town of Castle Douglas, and covers an area of 42.5 ha with a catchment of 1305 ha (Table 1). It is considered to be a shallow, well-mixed and highly productive loch (Institute of Aquaculture, 1988a, b) and is thought to have been in this state since the turn of the century (West, 1910). The loch receives the majority of its inflowing water from the Gelston Burn to the south east of the loch, with water outflow being through the man-made Carlingwark Lane on the west bank (Figure 1). There are also a number of field and road drains, which run into the loch, either continually or intermittently.

Based on the figures of Bailey-Watts et al. (1992b) (Table 1), mean flow into and out of Carlingwark Loch is 0.32 $m^3\ s^{-1}$. The catchment of Gelston Burn is 1010 ha (77% of the total drainage area of the loch) with an estimated average flow of 0.25 $m^3\ s^{-1}$ (Burns, pers. comm.).

In the past, all of the sewage and waste from industry flowed directly or indirectly into Carlingwark Loch. Human sewage was discharged into Carlingwark Loch until 1939 when the Castle Douglas sewage works became operational. Industrial waste was diverted during the 1960s, when a pipe was laid to the sewage works but it is thought that complete diversion of waste was not achieved until the late 1970s (Smith, 1994). A number of industries are still operational within Castle Douglas, and all waste now passes via treatment into a ditch leading to Carlingwark Lane downstream of the loch.

Present day point sources may include releases from local septic tanks serving lochside houses and the village of Gelston, but the major external source of nutrients to Carlingwark Loch is likely to be run-off from the arable and improved grass within the catchment and urban surface run-off via road drains. Nutrients stored in the sediments of the loch may act as an internal supply of nutrients.

Carlingwark Loch has a catchment area of 1305 ha (Burns, pers. comm.) although determination of this is somewhat complicated by the rolling topography of the surrounding countryside (Bailey-Watts et al., 1992b quoted a figure of 1183 ha). Fifty four percent of this lies between 15 and 183 m above sea level, the main agricultural zone where arable farming and forestry are concentrated (Bailey-Watts et al., 1992a). The remaining 46% lies between 183 and 610 m and is land which is rarely ploughed, and on which afforestation and rough grazing can be extensive (Bailey-Watts et al., 1992a). Land-use within the catchment is predominantly arable (20%) and improved grassland (48%) with 30% rough grazing, 1% afforested and the remaining 1% urban. Export coefficients, as a means of calculating the potential total phosphorus run-off associated with varying land-use types, have been calculated (Dillon & Kirchner, 1975; Harriman, 1978; Bailey-Watts *et al.*, 1987; Harper & Stewart, 1987), suggesting that the total phosphorus run-off to Carlingwark Loch is approximately 262 kg ann^{-1}.

Castle Loch, Lochmaben, covers an area of 78 ha (Table 2) and is an SSSI notified for its over-wintering wildfowl, breeding birds and plant communities which include a number of uncommon species. The loch receives water from two main inflows, the Innerfield Burn (mean discharge 0.10 $m^3\ s^{-1}$) to the east and the Vendace Burn (mean discharge 0.06 $m^3\ s^{-1}$) to the west. The outflow, Valison Burn, drains through a marshy area at the south end of the loch to the River

Table 1. Physical characteristics of Carlingwark Loch

Surface area:	42.5 ha
Mean depth:	2 metres
Maximum depth:	5.2 metres
Volume:	0.87×10^6 m^3
Total drainage area:	1305 ha
Annual rainfall:	1162 mm
Annual evapotranspiration;	400 mm
Annual water input to loch:	10.3×10^6 m^3
Turnover time:	11.8 times per annum
Water retention time:	~1 month

Table 2. Physical characteristics of Castle Loch

Surface area:	78 ha
Mean depth:	2.6 m
Maximum depth:	~5.5 m
Volume:	2.03×10^6 m^3
Total drainage area:	689 ha
Annual rainfall:	1080 mm
Annual evapotranspiration;	450 mm
Annual water input to loch:	5.28×10^6 m^3
Turnover time:	2.6 times per annum
Water retention time:	4.6 months

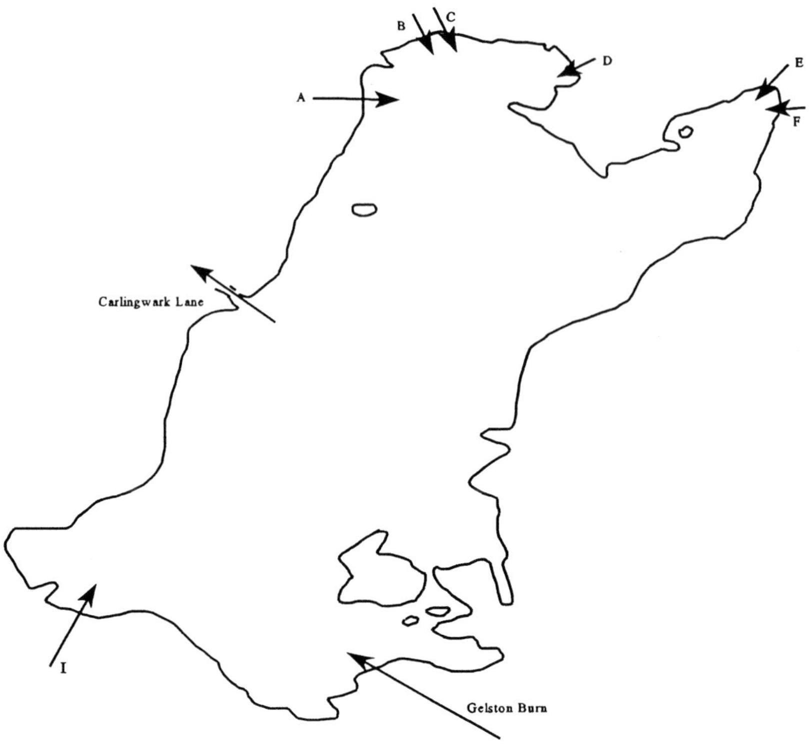

Figure 1. Location of inlets and outflows at Carlingwark Loch.

Annan, 2.6 km away (Figure 2). A number of small irregularly flowing inlets also exist.

Historically, Castle Loch received discharges from Lochmaben sewage works until 1988 or 1989 when the discharge was diverted to the River Annan. At present there is a storm water overflow pipe which discharges screened and settled sewage from the sewage works into Innerfield Burn, at times when the flow is 3–12 times the dry weather flow (DWF). For periods when the flow is >12 times DWF, there is a pipe which discharges screened storm sewage directly into Castle Loch. There is little evidence to suggest that this route has been used in recent times (Olley, pers. comm.).

Castle Loch has a catchment area of 689 ha, 96% of which lies between 15 and 183 m above sea level. Land-use within the catchment is predominantly arable (31%) and improved grassland (58%) with the remainder being coniferous forest (7%), Mill Loch and urban (2%) (Bailey-Watts et al., 1992b). Total phosphorus run-off calculated by export coefficients is 185 kg ann^{-1}. In addition, Castle Loch is known for its over-wintering bird populations, which provide an additional nutrient source to the surface waters of the loch as the result of guano deposits. The number of geese thought to over-winter is approximately 6000 and if each bird is assumed to contribute 0.04 kg ann^{-1}, based on a residence time of 5 months (Bailey-Watts et al., 1992a), an annual deposit of 240 kg of total phosphorus would result.

Results

Figure 3 indicates that for all nutrients, especially nitrogen, the Gelston Burn inflow to Carlingwark Loch had higher concentrations than outflow water in the Carlingwark Lane. In contrast, higher concentrations of chlorophyll *a* occurred in the outflow.

Table 3 summarises the calculated flow, mean total phosphorus concentration and loads for each of the inlets and the Carlingwark Lane outlet. These indicate total phosphorus input was 986 kg $ann^{-1}$from Gelston Burn and 718 kg $ann^{-1}$from the minor inlets, figures well in excess of the 262 kg, calculated using the export coefficients. Phosphorus leaving the loch via Carlingwark Lane was 1110 kg ann^{-1}. This points to a net annual accumulation of phosphorus in Carlingwark Loch of approximately 594 kg or almost 0.7 mg

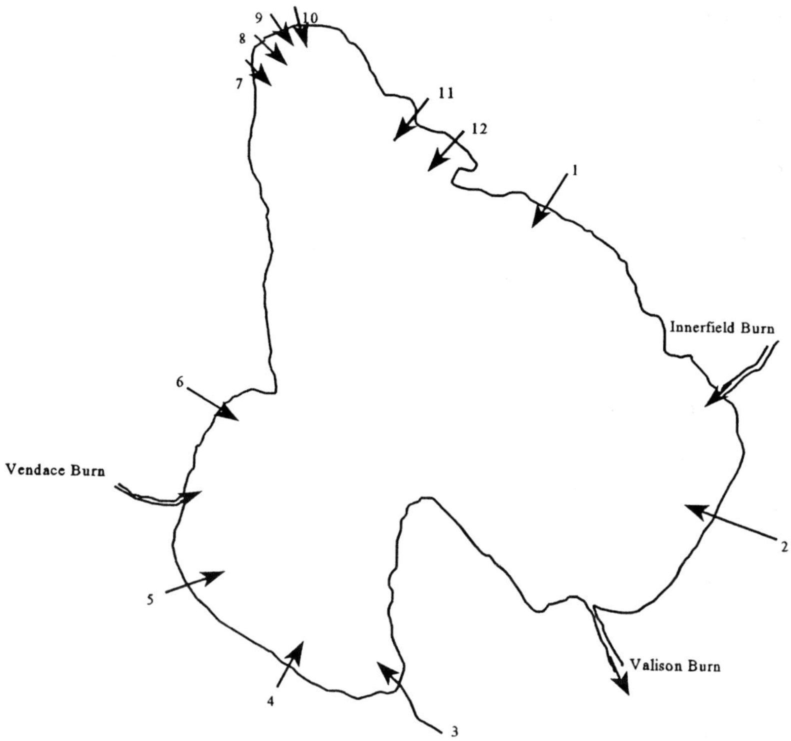

Figure 2. Location of inlets and outflows at Castle Loch.

Table 3. Phosphate loadings in inlets and outlets of Carlingwark Loch

Inlet/Outlet	Mean flow ($m^3\ s^{-1}$)	Mean TP ($mg\ l^{-1}$)	Annual TP loading (kg)
Gelston Burn	0.25	0.125	986
Inlet A	0.003	0.32	30
Inlet B	0.021	0.347	230
Inlet C	0.003	0.74	70
Inlet D	0.024	0.149	113
Inlet E	0.006	1.116	211
Inlet F	0.006	0.31	59
Inlet I	0.009	0.018	5
Carlingwark lane	0.32	0.11	1110

l^{-1}. Phosphorus entering the loch was 72% soluble, but that leaving was 64%.

Carlingwark Loch itself was a fairly well-mixed water body with an even spatial distribution of nutrients both horizontally and vertically (Figure 4). Mean soluble and total phosphorus concentrations were 0.12 and 0.15 mg l^{-1} respectively, giving an in-loch water store of 104 kg soluble and 130 kg total phosphorus. Similar calculations for nitrogen produced mean concentrations of 2.0 and 2.8 mg l^{-1} for nitrate and total nitrogen, respectively, resulting in a potential in-loch water store of 2412 kg total nitrogen of which 1738 kg was nitrate.

Throughout the loch, total phosphorus concentrations were almost identical in January 1994 and February 1995 (mean 84 $\mu g\ l^{-1}$), with concentrations in October 1994 being considerably different (mean 214 $\mu g\ l^{-1}$). Therefore, the in-loch store of total phosphorus in the spring was 73 kg, while in the autumn it rose to 186 kg. The marked decrease (by approximately 150 $\mu g\ l^{-1}$) during the period October–February was matched by an increase in chlorophyll *a*, indicating the uptake of phosphorus by algae in the water column, and resulted in a reduction in the in-loch inorganic store of 113 kg phosphorus.

The sediment from Carlingwark Loch (Figure 5) showed that at all sampling sites, with one exception, phosphate was mainly associated with the Fe/Al oxides and least associated with the calcium fraction (Figure 6). There was little or no difference between the samples collected in October and those collected in February, with both sets of results indicating

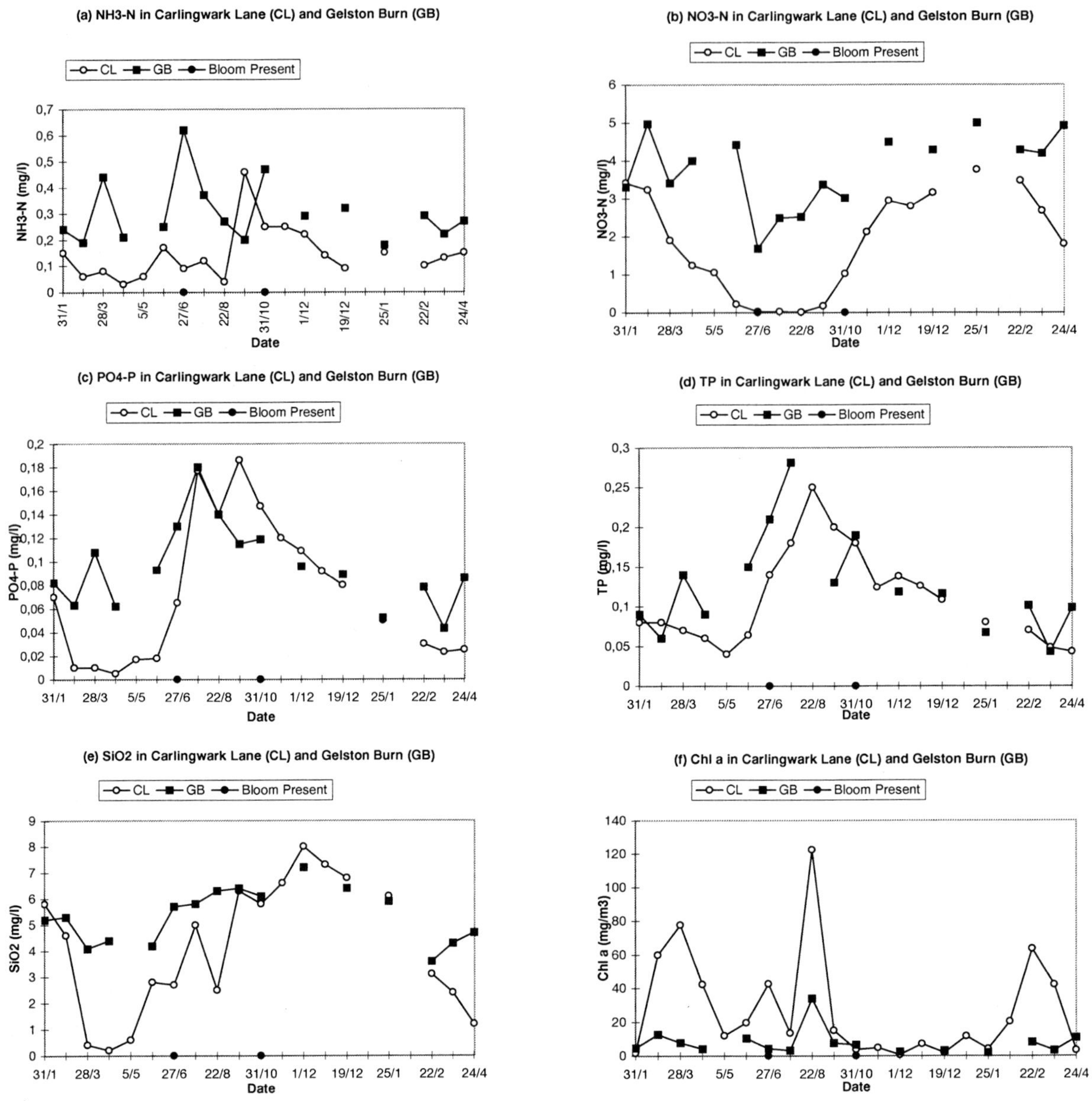

Figure 3. Nutrient concentrations in inflowing and outflowing waters at Carlingwark Loch.

lower total phosphorus concentrations in the samples collected at site 5.

The pool of sedimentary phosphorus was calculated, assuming that the surface was 0–2 cm, with a percentage dry weight of 11.6% and a density of 1.11 g ml^{-1}, to be 1.974 and 0.397 mg g^{-1}dry weight total phosphorus and loosely-bound phosphorus respectively. These figures indicate approximately 2160 kg phosphorus is stored in the sediments with 435 kg loosely-bound and available for uptake (i.e. approx. 20%). However, Sas (1989) stated that the upper 15 cm of sediment can be regarded as being available for release. In that case, the store could be as high as 18 600 kg phosphorus (2300 kg as loosely bound, 12% of total phosphorus). Smith (1994), during diatom studies, estimated that the top 10 cm of sediment in Carlingwark Loch dated from 1987 to present, giving an approximate sedimentation rate of 1.25 cm ann^{-1}. Total phosphorus concentrations down sediment cores were within the range 3–3.8 mg g^{-1}dry weight (av-

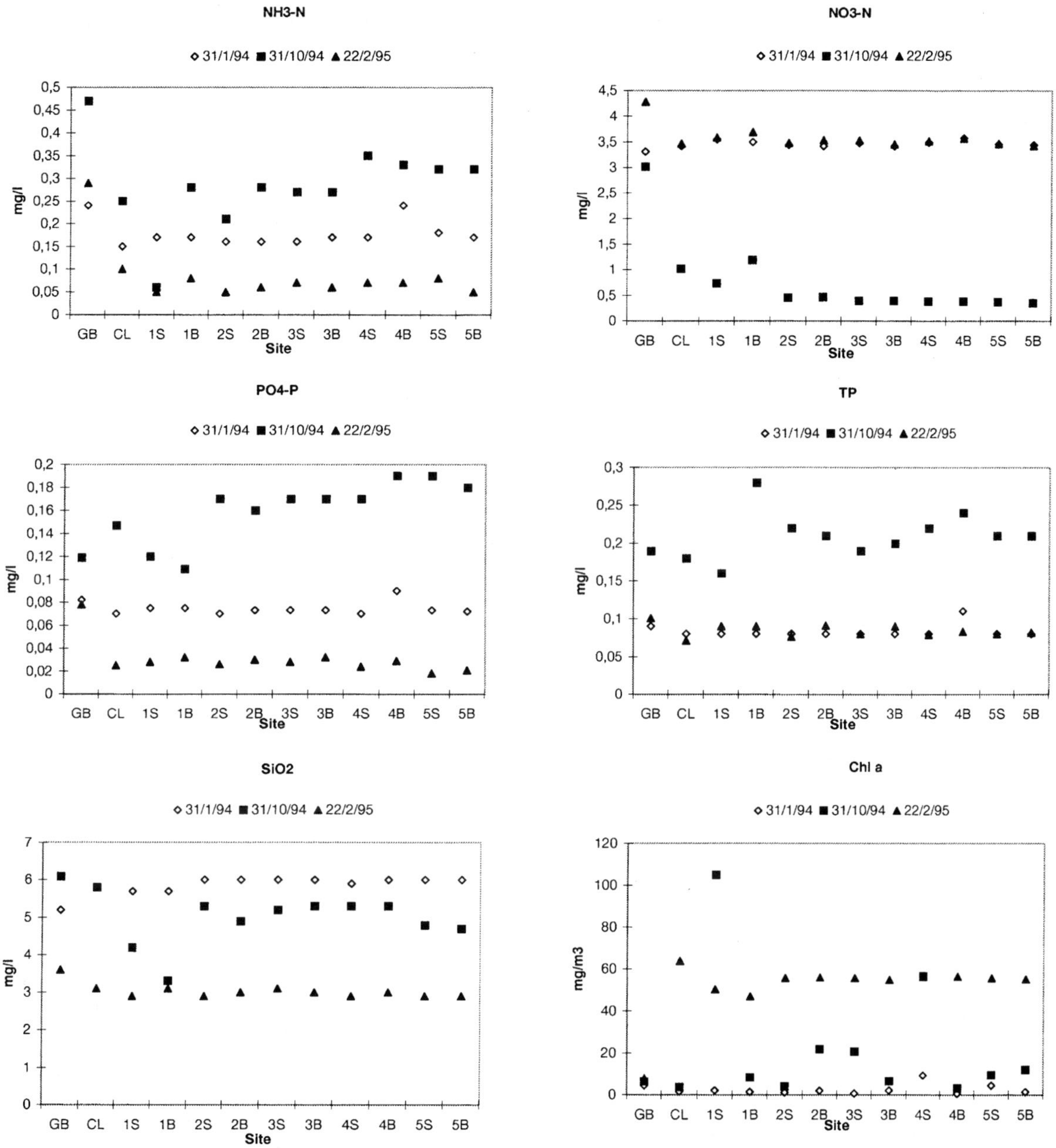

Figure 4. Nutrient concentrations in Carlingwark Loch.

erage 3.4 mg g^{-1} dry weight), values well in excess of those (1 mg P g^{-1}dry weight) thought to result in a release of phosphorus from the sediment (Sas, 1989). Using these figures for sedimentation and mean concentration suggested an annual flux from the water column to the sediments of approximately 1350 to 2330 kg phosphorus has occurred. In reality, it is likely that the sediments experienced bioturbation and that new material reaching the sediment surface was rapidly incorporated into a mixed layer equivalent to the upper 10 cm. Such a high flux does not match the calculated in-loch store of 130 kg (see above). Hence, it is likely that the sediments could at times act as a source of phosphorus to the water column. The figure based on the sediment core data, indicated that five times as much phosphorus could potentially be

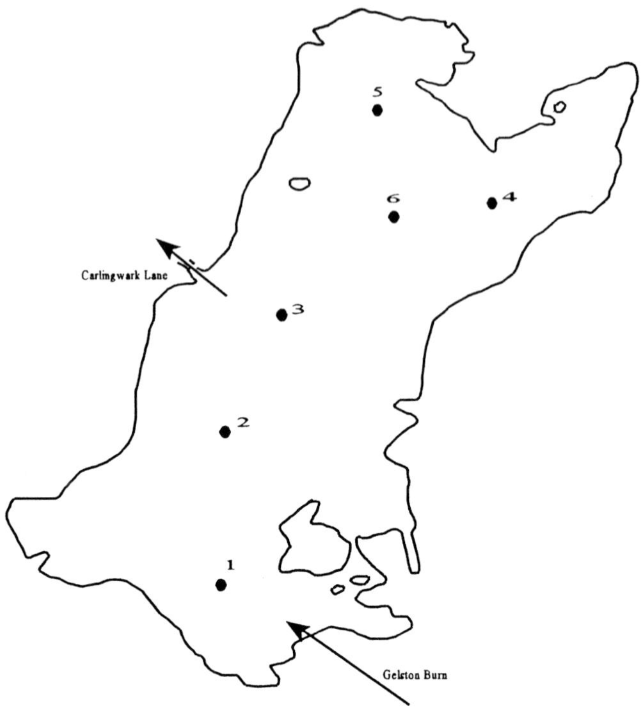

Figure 5. Location of sediment sample sites at Carlingwark Loch.

released into the water column from the sediments of Carlingwark Loch.

Algal counts indicate the presence of cyanobacteria blooms in Carlingwark Loch during four months of the study period. This included a bloom of *Aphanizomenon*, which appeared in August and was still present in October.

Figure 7 and Table 4 summarise the nutrient status of the inflows and outflow for Castle Loch. There is an apparent agreement between the mean flows for the inlets and outlet, and the influence of the minor inlets cannot be calculated in the same way as that carried out for Carlingwark Loch. This does not rule them out as being significant contributors to the overall phosphorus load to Castle Loch, however, as three of them recorded high phosphorus concentrations in July and August. Flow data were not collected at the time of sampling, and therefore it is impossible to determine whether these observed high concentrations introduced significant quantities of phosphorus. The data presented in Table 4 indicate a net outflow of total phosphorus of 306 kg per year. Similar calculations for soluble phosphorus concentrations give annual net losses of 193 kg.

Table 4. Phosphate loadings in inlets and outlets of Castle Loch

Inlet/Outlet	Mean flow (m^3 s^{-1})	Mean TP (mg l^{-1})	Annual TP loading (kg)
Innerfield burn	0.1	0.074	233
Vendace burn	0.06	0.119	225
Valison burn	0.167	0.145	764

There were only slight spatial variations in nutrient concentration of Castle Loch although none with depth (Figure 8). There were distinct temporal differences in nutrient concentrations except for total phosphorus (0.07–0.15 mg l^{-1}). Chlorophyll *a* was at its highest in April when phosphate and silicate were at their lowest. The mean total and soluble phosphorus concentrations were 0.120 and 0.059 mg l^{-1}respectively, giving a potential in-loch store of 118 kg and 244 kg. Similar calculations for nitrogen resulted in mean total nitrogen concentrations of 1.1 mg l^{-1}, of which 0.3 mg l^{-1} was nitrate, and an in-loch store of 2140 kg total nitrogen (560 kg as nitrate).

As at Carlingwark Loch, the sedimentary phosphate (Figure 9) was mainly associated with the

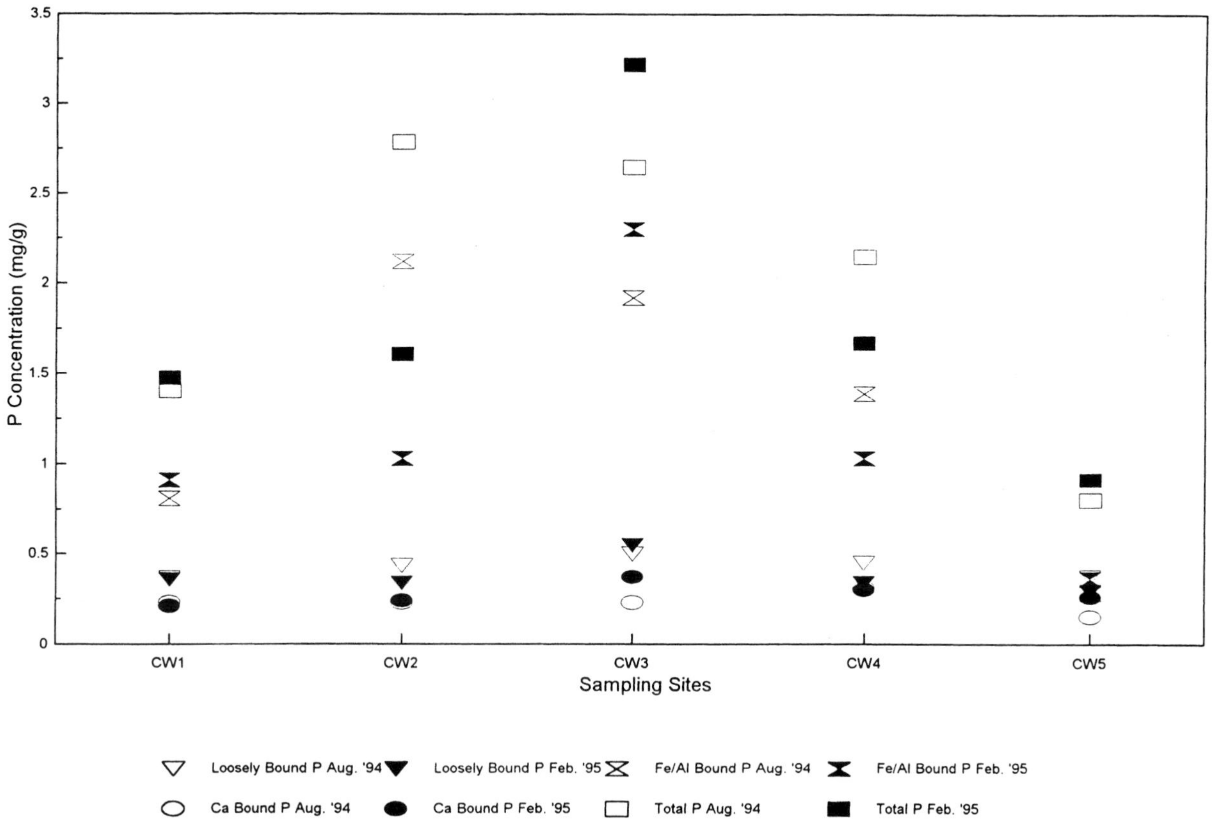

Figure 6. Phosphate associations in Carlingwark Loch sediments.

Fe/Al oxides (74–86%) and the remainder split evenly between the loosely-bound and Ca-bound phosphate fractions (Figure 10). Assuming a percentage dry weight of 13.3%, a density of 1.24 g l^{-1}, and with mean total and loosely-bound phosphorus concentrations of 3.802 and 0.434 mg g^{-1}dry weight sediment respectively, Castle Loch sediments provide a reservoir of 9800 kg total phosphorus and 1120 kg loosely-bound phosphorus. These amounts are considerably higher than those in the water column or in the sediment at Carlingwark Loch. Data from the 10 cm core (mean total phosphorus and soluble phosphorus of 3.912 and 0.408 mg g^{-1}dry weight) results in a pool of sedimentary phosphorus of 50 320 kg of which 5250 kg is loosely bound and readily available.

In Castle Loch, blooms of cyanobacteria occurred in May (*Anabaena* and *Aphanizomenon* sp) and October (*Anabaena*); periods when chlorophyll *a* was relatively high (especially in May when a maximum of 1322 mg $m^{-3}$was observed).

Discussion

Table 5 summarises the findings of this study for total phosphorus concentrations in Carlingwark and Castle Lochs.

Bailey-Watts et al. (1992a, b) had been commissioned by the then Nature Conservancy Council for Scotland (now Scottish Natural Heritage) to undertake desk-based studies to produce limnological profiles for a number of lochs, with special reference to eutrophication and phosphorus. Their study included Carlingwark and Castle Lochs and calculation of the likely annual mean phosphorus concentrations.

For Carlingwark Loch, they found an annual phosphorus loading of 1900 kg which produced a mean total phosphorus concentration of 0.18 mg l^{-1}. The loading was very similar to that calculated here from measured phosphorus concentrations and mean flows, but slightly higher than that calculated from samples collected in 1994 (0.15 mg l^{-1}). Similar calculations for Castle Loch predicted an annual phosphorus loading of 470–600 kg (including a contribution from over-

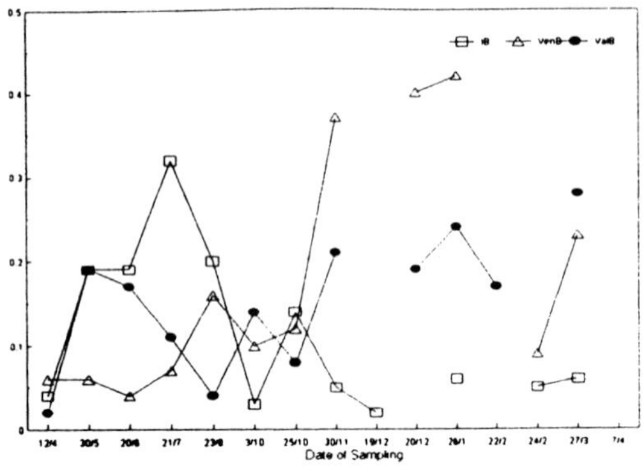

(a) NH_3-N (mg l^{-1}) in Innerfield (IB), Vendace (VenB) and Valison (ValB) Burns

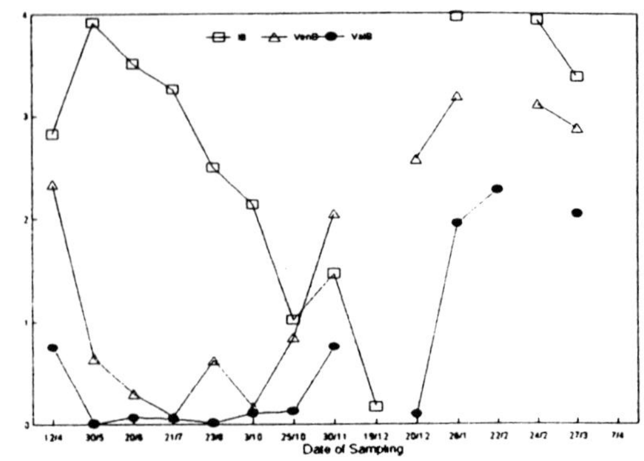

(b) NO_3-N (mg l^{-1}) in Innerfield (IB), Vendace (VenB) and Valison (ValB) Burns

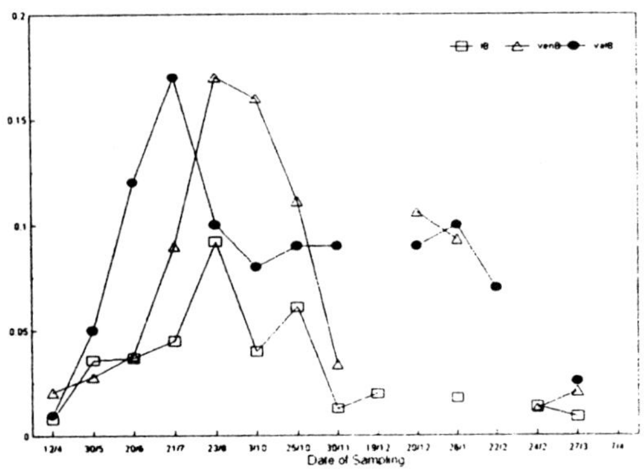

(c) PO_4-P (mg l^{-1}) in Innerfield (IB), Vendace (VenB) and Valison (ValB) Burns

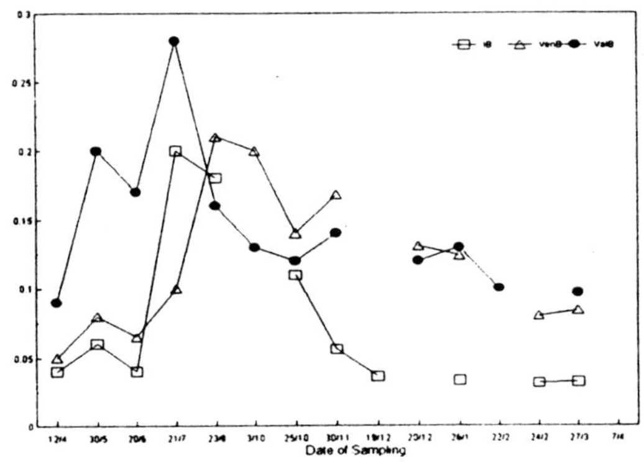

(d) TP (mg l^{-1}) in Innerfield (IB), Vendace (VenB) and Valison (ValB) Burns

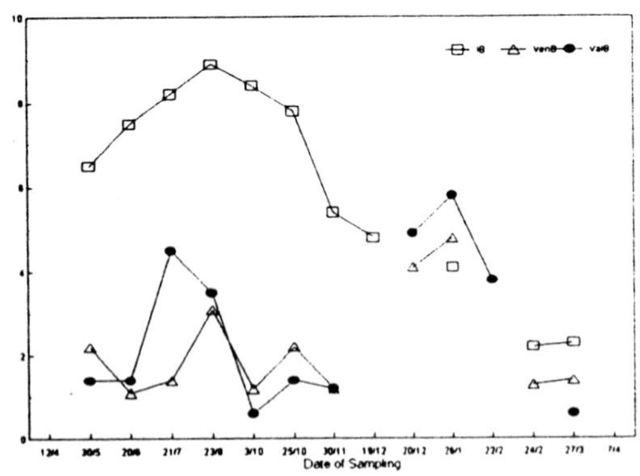

(e) SiO_2 (mg l^{-1}) in Innerfield (IB), Vendace (VenB) and Valison (ValB) Burns

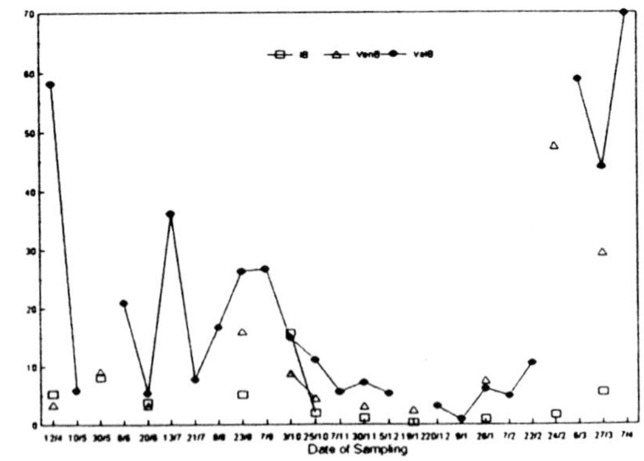

(f) Chl a (mg m^{-3}) in Innerfield (IB), Vendace (VenB) and Valison (ValB) Burns

Figure 7. Nutrient concentrations in inflowing and outflowing waters at Castle Loch.

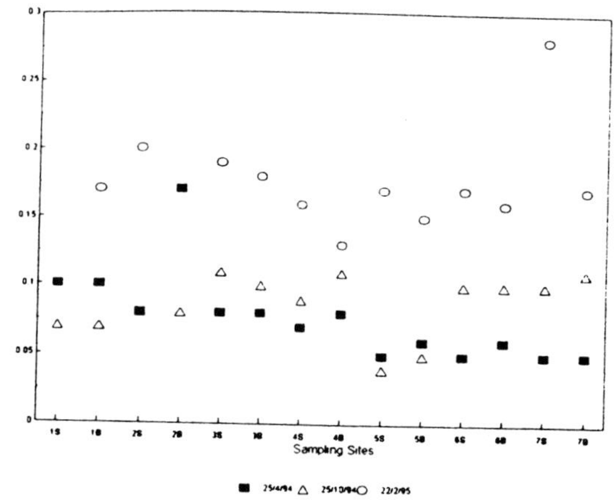

(a) NH_3-N (mg l^{-1})

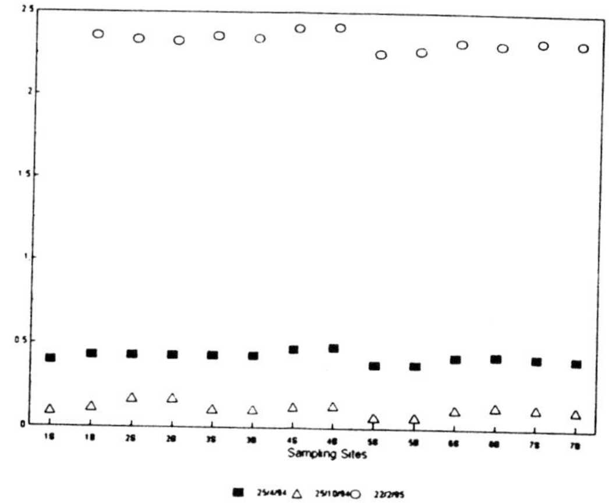

(b) NO_3-N (mg l^{-1})

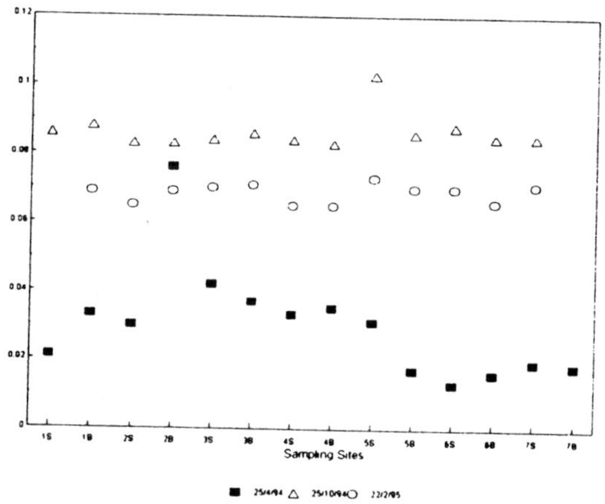

(c) PO_4-P (mg l^{-1})

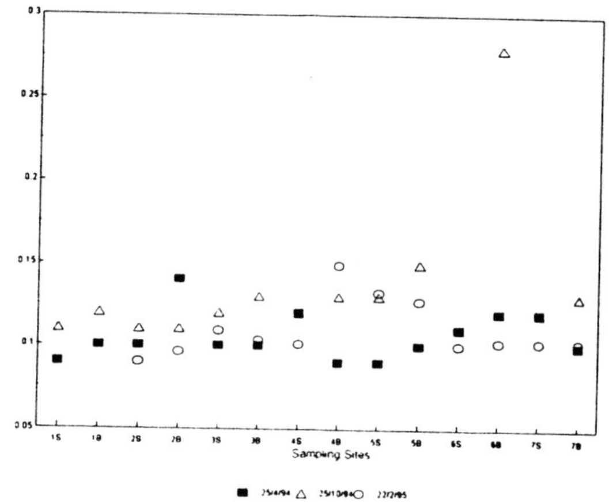

(d) TP (mg l^{-1})

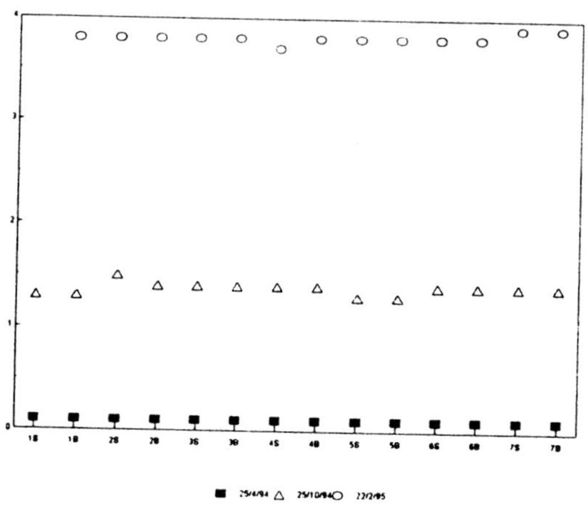

(e) SiO_2 (mg l^{-1})

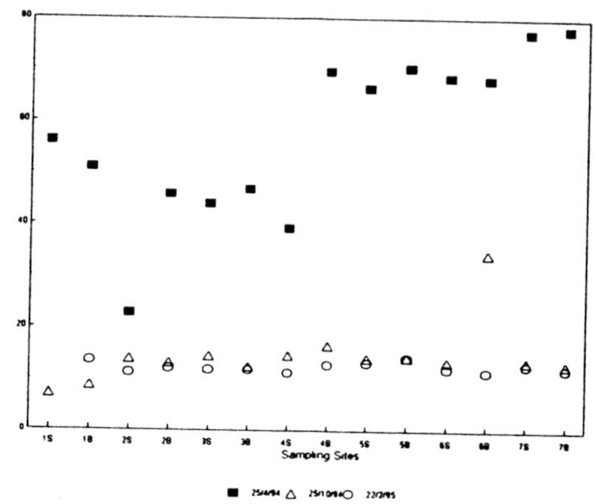

(f) Chl a (mg m^{-3})

Figure 8. Nutrient concentrations in Castle Loch.

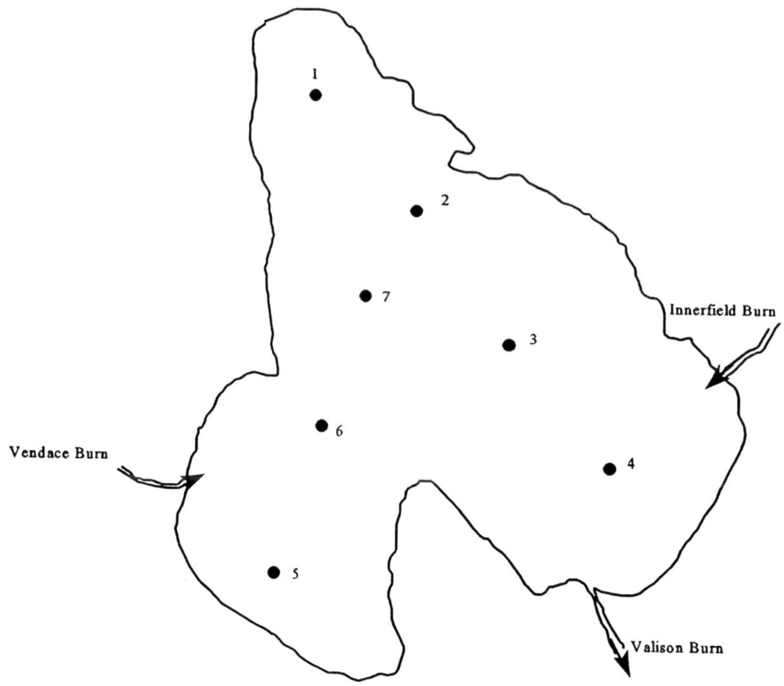

Figure 9. Location of sediment sample sites at Castle Loch.

wintering birds) with annual mean total phosphorus concentrations of 0.064 mg l^{-1}. During the present study, the inputs were 460 kg with a mean total phosphorus concentration of 0.120 mg l^{-1} without adding the contribution from birds. These lower inputs, and higher concentrations, may suggest that release of phosphate from the sediments is occurring.

Problems with algal blooms, as the result of increased nutrient concentrations, have occurred at a number of sites in both England and Scotland. Bailey-Watts et al. (1992a) sampled a wide range of lochs in Scotland during the summer of 1991 for total phosphorus and found concentrations as high as 1.25 mg l^{-1} at Kilconquhar in Fife and eutrophic levels (i.e. >0.05 mg l^{-1}) in Mill Loch (Annandale), Rescobie Loch (Tayside), Balgavies Loch (Tayside) and Loch of Strathbeg (Buchan). These were single summer concentrations (June–August) and are not directly comparable to those for Carlingwark and Castle Lochs, but they do provide a useful comparison. Loch Leven was not included in the above study, but has been extensively studied previously (Bailey-Watts, 1976; Bailey-Watts, 1978; Bailey-Watts & Kirika, 1987; Bailey-Watts et al., 1987; Bailey-Watts et al., 1990). Bailey-Watts et al. (1987) measured in-loch concentrations of 0.029 to 0.134 mg l^{-1} in 1985 with an overall mean annual of 0.063 mg l^{-1} total phosphorus. This is lower than that at Carlingwark and Castle Lochs, but is based on monthly sampling, with the maximum being similar to that observed in Castle Loch.

The most well-known English lakes which are comparable are the Norfolk Broads (Moss, 1983; Phillips & Jackson, 1990; Phillips, 1984). Much work has been undertaken at these sites in an attempt to alleviate the problems (see Phillips et al., 1999). The mean annual total phosphorus in five eutrophic Broads – Martham Broad, Barton Broad, Cockshoot Broad, Ranworth Broad and South Walsham Broad, – are 0.026, 0.143, 0.136, 0.142 and 0.275 mg l^{-1} respectively (Pitt, pers. comm.). Carlingwark and Castle Lochs are comparable to Barton, Cockshoot and Ranworth Broads. Barton Broad is still phytoplankton-dominated even though phosphorus concentrations have been reduced by introducing phosphate-stripping at discharging sewage works, Cockshoot Broad has had surface sediment removed by pumping to remove phosphate-rich sediments and increase depth, while the condition of Ranworth Broad is poor with little or no macrophyte growth (Pitt, pers. comm.). Average phosphorus concentrations in sediments from the Broads ranged from 1.0 mg g^{-1} dry weight in Ranworth and Barton Broads to 2.1 – 2.2 mg

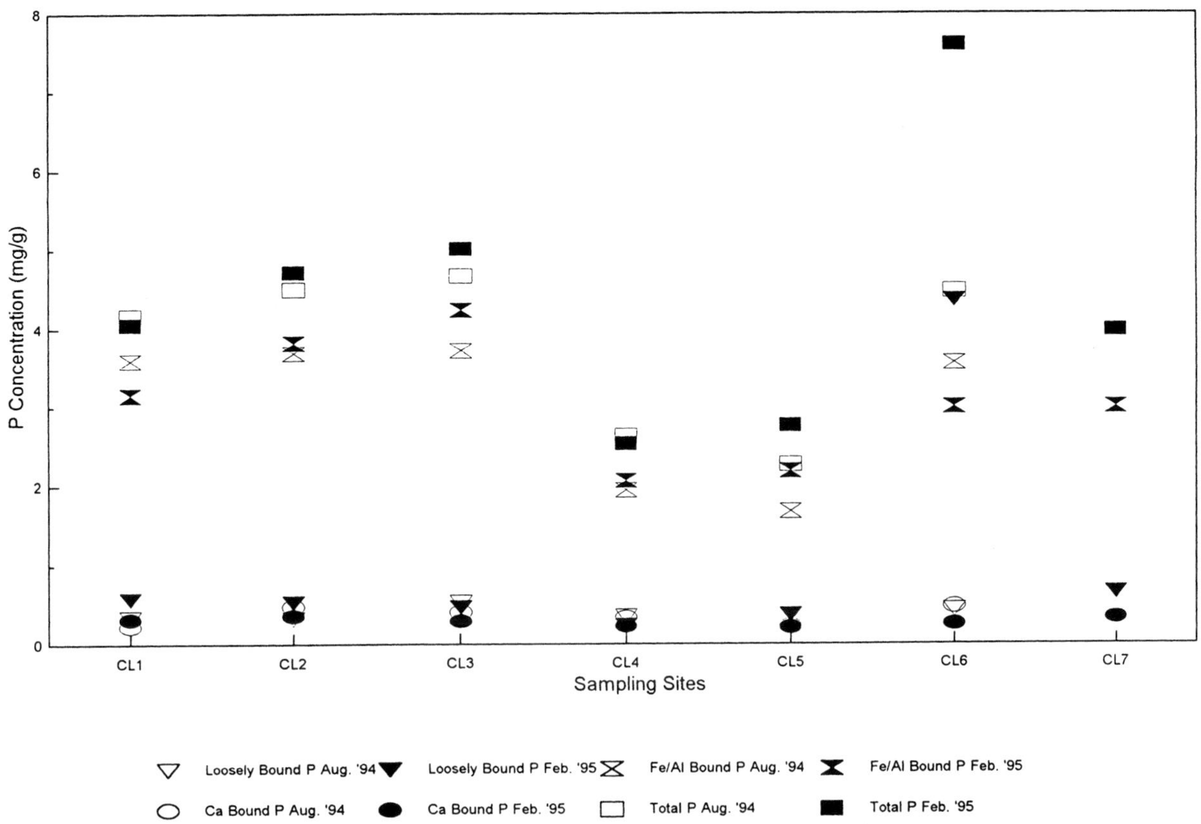

Figure 10. Phosphate associations in Castle Loch sediments.

g^{-1} dry weight in Martham and Cockshoot Broads. The highest concentrations were found in Cockshoot Broad. The mean phosphorus concentrations measured in the sediments (averaged over top 10 cm) of Carlingwark and Castle Lochs were considerably higher than those found in the Norfolk Broads.

At Carlingwark Loch, where it has been identified that inputs are in excess of current outputs, initial stages of the management programme may involve further identification of the sources of phosphorus within the catchments of the Gelston Burn and the minor inlets. Until such studies are undertaken and potential sources identified, it will prove difficult to reduce the phosphorus loading of the inlets. The Carlingwark Loch Advisory Committee includes a representative from SEPA and agreement has been reached that SEPA should play an integral role in studying the phosphorus loading of the inlets in greater detail. Additional options discussed included the establishment of buffer strips and the restrictions of any further developments requesting a discharge to the loch.

Table 5. Summary of total phosphorus budget

Loch	Weight of TP (kg)	Comments
Carlingwark Loch		
Resident	130	73 kg in Spring; 186 kg in Autumn
Inputs	1704	Gelston Burn 986 kg; Minor Inlets 718 kg Theoretical from Export Co-efficients: 262 kg
Outputs	1110	Carlingwark Lane
Sediments	435	Loosely Bound P in top 2 cm
Castle Loch		
Resident	244	217 kg in Spring; 272 kg in Autumn
Inputs	458	Vendace and Innerfield Burn Theoretical from Export Co-efficients: 185 kg from run-off; 240 kg from birds
Outputs	764	Valison Burn
Sediments	1117	Loosely Bound P in top 2 cm

At Castle Loch, inputs from the inlet streams are currently lower than the output levels, and therefore remediation techniques cannot be focused in the same manner as those at Carlingwark Loch. The additional phosphorus in the outflow from the loch could originate either from the sediments and/or the overwintering wildfowl. As the loch is designated an SSSI because of its bird populations, reducing phosphorus inputs by controlling the number of birds using the loch is not a viable option. One of the major incentives in reducing the phosphorus of the loch is SNH's wish to reintroduce vendace to the loch, and the loch's Advisory Committee now have the task of taking the study's findings and determining the best options available to meet this overall objective.

Conclusions

This study was commissioned to determine whether the source of nutrients responsible for the annual algal blooms observed in Carlingwark and Castle Lochs could be identified. A distinction was made between contemporary (run-off from the catchment) and historical (stored in the sediments) sources of the nutrients. The information indicated that Carlingwark Loch was subject to large inputs of phosphorus from both Gelston Burn and minor inlets. The quantity of loosely-bound phosphorus, available in the top 2 cm of sediment, was small in comparison to these inputs.

A different situation occurred at Castle Loch where the chemical data indicated higher concentrations of nutrients leaving the loch than were entering in the two main inlets, while the sediment data indicated a considerable pool of readily-available phosphorus. Large populations of over-wintering birds at Castle Loch may also contribute to the nutrient loading. This would suggest, therefore, that by reducing the inputs into Carlingwark Loch, in-loch phosphorus concentrations should decrease, whereas this would not be the case at Castle Loch.

Acknowledgements

The authors would like to thank Scottish Natural Heritage for the funding for the project. The collection of the samples could not have been achieved without the hard work of the sampling officer and the countryside rangers, while the analysis of the samples could not have been done without the chemistry and biology laboratory staff at the former Solway River Purification Board.

References

Bailey-Watts, A. E., 1976. Planktonic diatoms and silica in Loch Leven, Kinross, Scotland: a one month silica budget. Freshwat. Biol. 6: 203–213.

Bailey-Watts, A. E., 1978. A nine-year study of the phytoplankton of the eutrophic and non-stratifying Loch Leven (Kinross, Scotland). J. Ecol. 6: 741–771

Bailey-Watts, A. E. & A. Kirika, 1987. A re-assessment of the phosphorus inputs to Loch Leven (Kinross, Scotland): rationale and overview of results on instantaneous loadings with special reference to run-off. Transactions of the Royal Society of Edinburgh Earth Sciences 78: 351–367.

Bailey-Watts, A. E., R. Sargent, A. Kirika & M. Smith, 1987. Loch Leven phosphorus loading. Institute of Terrestrial Ecology Project 926 Final Report.

Bailey-Watts, A. E., A. Kirika, L. May & D. H. Jones, 1990. Changes in phytoplankton over various timescales in a shallow, eutrophic loch: the Loch Leven experience with special reference to the influence of flushing rate. Freshwat. Biol. 17: 405–418.

Bailey-Watts, A. E., L. May, A. Kirika & A. A. Lyle, 1992a. Eutrophication Case Studies: Phase II, an assessment based on desk analysis of catchments and summer limnological reconnaissances, Vol. I: Analysis of the whole spectrum of water studied. Institute of Freshwater Ecology, Edinburgh.

Bailey-Watts, A. E., L. May, A. Kirika & A. A. Lyle, 1992b. Eutrophication Case Studies: Phase II, an assessment based on desk analysis of catchments and summer limnological reconnaissances, Vol. II: Limnological profiles of the sites with special reference to eutrophication and phosphorus (P). Institute of Freshwater Ecology, Edinburgh.

Begg, F. H., 1995. Carlingwark and Castle Lochs: A study of their nutrient status. Final Report SNH Contract No. SS/F2A/488. Scottish Natural Heritage, Edinburgh.

Begg, F. H., D. A. Rendall & A. McNeill (eds), (1996). Solway Review: 1954–1996. Solway Offset, Dumfries.

Dillon, P. J. & W. B. Kirchner, 1975. The effects of geology and land-use on the export of phosphorus from watersheds. Wat. Res. 9: 135–148.

Harper, D. M. & W. D. P. Stewart, 1987. The effects of land use upon water chemistry, particularly nutrient enrichment, in shallow lowland lakes: comparative studies of three lochs in Scotland. Hydrobiologia 148: 211–229.

Harriman, R., 1978. Nutrient leaching from fertilised forest watersheds in Scotland. J. appl. Ecol. 15: 933–942.

Hieltjes, A. M. & L. Lijklema, 1980. Fractionation of inorganic phosphates in calcareous sediments. J. Envir. Qual. 9: 405–407.

Institute of Aquaculture, 1988a. Carlingwark Loch, Castle Douglas: An investigation of its current nutrient status and associated weed problems. University of Stirling, Scotland.

Institute of Aquaculture, 1988b. Carlingwark Loch, Castle Douglas: An investigation of its current nutrient status and associated weed problems, Report 2: University of Stirling, Scotland.

Likens, G. E. (ed.), 1972. Nutrients and eutrophication: The limiting nutrient controversy. Special Symposia Vol. 1. American Society of Limnology & Oceanography, Lawrence, Kansas.

Moss, B., 1983. The Norfolk Broadland: Experiments in the restoration of a complex wetland, Biological Reviews 58: 521–561.

National Rivers Authority, 1990. Toxic Blue-Green Algae. Water Quality Series No.2. National Rivers Authority, Peterborough.

Phillips, G. L., 1984. A large-scale field experiment in the control of eutrophication in the Norfolk Broads. J. Inst. Wat. Poll. Cont. 83: 400–408.

Phillips, G. L. & R. Jackson, 1990. The control of eutrophication in very shallow lakes, the Norfolk Broads. Verh. int. Ver. Limnol. 24: 573–575.

Phillips, G. L. & P. Kerrison, 1991. The restoration of the Norfolk Broads: The role of biomanipulation. Mem. Ist. ital. Idrobiol. 48: 75–97

Phillips, G. L. & B. Moss, 1994. Is biomanipulation a useful technique in lake management? R & D Note 276. National Rivers Authority Peterborough: 46 pp.

Phillips, G. L., M. R. Perrow & J. Stansfield, 1997. Manipulating the fish-zooplankton interaction in shallow lakes: A tool for restoration. In Greenstreet, S.P.R. & M.L. Tasker (eds), Aquatic Predators and their Prey. Blackwell Scientific Press, Oxford.

Phillips, G., A. Bramwell, J. Pitt, J. Stansfield & M. Perrow, 1999. Practical application of 25 years' research into the management of shallow lakes. Hydrobiologia 395/396 (Dev. Hydrobiol. 136): 61–76.

Sas, H., 1989. Lake restoration by reduction of nutrient loading. Academia Verlag, Berlin.

Schindler, D., 1977. Evolution of phosphorus limitation in lakes. Science 195: 260–262.

Smith, R. A., 1994. Reconstruction of the productivity history of Carlingwark Loch using fossil diatom analysis. Unpublished undergraduate thesis. University College London.

Vollenweider, R. A., 1968. Scientific fundamentals of the eutrophication of lakes and flowing waters, with particular reference to nitrogen and phosphorus as factors in eutrophication. Tech. Report DA 5/SCI/68.27. OECD Paris.

West, G., 1910. A further contribution to a comparative study of the dominant phanerogamic and higher cryptomic flora of aquatic habit in Scottish lakes. Proceedings of the Royal Society of Edinburgh Session 1909–1910 Vol. XXX Part 11 No.6.

Hydrobiologia **395/396:** 309–323, 1999.
D.M. Harper, B. Brierley, A.J.D. Ferguson & G. Phillips (eds), The Ecological Bases for Lake and Reservoir Management.

Strategies for conservation management of lakes

F. Jane Madgwick
The Broads Authority, 18 Colegate, Norwich, Norfolk, U.K.

Key words: eutrophication, conservation, phosphorus stripping, biomanipulation, Norfolk Broads

Abstract

This paper discusses how the use of nature conservation targets can influence the practice of the management and rehabilitation of shallow lakes. It suggests that managers need to identify realistic targets, based on a sound understanding of the nature and ecological processes operating in the lakes in question. The constraints on effective management towards ecological targets are identified in the context of the rehabilitation of shallow lakes from eutrophication. It is recognised that most commonly, nature conservation is one of several objectives in lake management. Using the Norfolk Broads case study, it identifies a range of ecological and human use influences that need to be assessed in order to form the basis of an integrated, sustainable approach to rehabilitation from the effects of eutrophication. The way in which nature conservation is taken account of through each stage in the planning of broads restoration projects is discussed, and the need for further specific advice for lake managers is indicated. This paper stresses the benefits of large scale experimental management, backed up by research and monitoring as a way of taking forward this applied science. The need for good communication with partner agencies and the local community when undertaking this management work, is also emphasised.

Introduction

The role of lakes as wildlife habitats and the goods and services they provide, such as water supply and recreational opportunities, have tended to be taken for granted. These standing open waters vary in size from tens of thousands of hectares, to ponds just a few metres across. They form an essential component of landscapes in every continent and from the tundra to the equator.

There is a growing international awareness that lakes, as for all types of wetlands, are more valuable in their natural, or slightly modified state than if drained or polluted and increased efforts are being made towards lake restoration and conservation. In their pristine state, most water bodies have nutrient poor water as a result of efficient natural systems for nutrient conservation. Rapid agricultural change and population growth in developed countries, particularly in the past century, has led to widespread eutrophication through the increased and continuous supply of nutrients to river and lake basins in the landscape. Hence eutrophication is now recognised as the most pervasive and extensive cause of lake degradation in Europe (European Commission, 1995). The symptoms and consequences of lake eutrophication have been widely reported elsewhere (e.g. Kling et al., 1995; Moss et al., 1996). In extreme cases, eutrophication results in a dramatic loss of biodiversity linked with the total loss of aquatic macrophytes.

The dominant argument for the conservation of wetlands for nature has historically been their role for supporting birdlife, particularly those on important waterfowl migration routes. The Ramsar Convention (1971) and the EC Wild Birds Directive (1979) aim to reverse the downward trend on an international basis. Through international commitments, measures are required to classify and protect sites and also to incorporate conservation measures in land-use policy. Site selection criteria are well developed now for all types of wetlands and include 'representativeness' and 'rarity' of wetland examples in biogeographical regions, the role a wetland plays in the 'functioning' of river basins or coastal systems, and the value in supporting human communities. Many lakes are part of these designated wetlands and in some countries in western Europe, notably in Denmark, Sweden and the Netherlands, considerable investment in lake restoration and rehabilitation is now in place. This is based on comprehensive national wetland strategies, recognising nature conservation as a major and sometimes primary use.

Sadly in the U.K., as is the case across Europe, lakes seem to have largely fallen through the net of historic nature conservation strategies. This is explained by the dominance of eutrophication as the agent of degradation and the requirement for control through the integrated management of water and land on a catchment basis, delivered by a multiplicity of organisations. Carvalho & Moss (1995) reported that of 90 SSSI lakes in England and Wales, more than three quarters showed 50% change over the last few decades, with only 4.4% remaining in a pristine condition. Many of those which showed the greatest change are considered oligotrophic or mesotrophic, such as those in the uplands of north-west England. The U.K. Biodiversity Action Plan (HMSO, 1995) recognises the historic enrichment of lakes and prioritises action through costed plans for remedial management of 50 mesotrophic lakes by 2010. It is clear that an ambitious programme of action will be needed to make progress with the restoration of degraded lakes (English Nature, 1997).

The European Commission Directives on urban waste water treatment (UWWTD) and nitrogen (Nitrates Directive), have a part to play in combating continued eutrophication, however many sites of high biodiversity significance fall outside the scope of this legislation. The UWWTD only applies to major urban sewage treatment works with population equivalents of over 10 000 and, therefore, will fail to address the nutrient supplies of the most valuable lakes in the U.K., most of which are in rural areas. In the U.K. the Nitrates Directive has only been used to safeguard the quality of drinking water and not, as the Directive allows, to reduce nutrient loading for environmental reasons. Furthermore, the major contribution to eutrophication from nutrients applied to arable farmland is becoming more apparent (Royal Society, 1983; Foy et al., 1995; Johnes, 1996).

There are important policy and strategic planning issues to resolve, and considerable investment to be made if the U.K. commitment to sustainable development and biodiversity are to be fulfilled sufficiently and consistently for the water environment (U.K. Round Table on Sustainable Development, 1997). The U.K. Biodiversity Action Plan endorses the development of a strategy for the control of eutrophication and nutrient standards appropriate for mesotrophic lakes, the latter in association with water quality objectives. Ultimately, it is envisaged that countries within the European Union will define and monitor the ecological of lakes using a standard approach, linked to a common set of environmental objectives (European Commission, 1997). Clearly, the basis on which the ecological status of lakes is assessed in future, will have a profound effect on the ability of relevant agencies to justify regulatory, mitigation or restoration measures.

Attempts have been made to relate physical and chemical parameters to biological communities (e.g. Palmer et al., 1992) but these are incomplete studies. It is apparent, not surprisingly, that a large number of interlinking ecological factors operate, especially in shallow lakes, to make only general guidance possible on which plant and animal communities are likely to be associated with lakes of different environmental conditions. Added to this is the fact that monitoring information is fairly patchy, except in a few 'hotspots' like the Broads and the Lake District, where there has been a long history of interest. Another constraint is the fact that the basic biology and ecological requirements of rare and scarce aquatic species are poorly understood. Some fundamental research is therefore required before appropriate actions can be proposed and justified to secure and enhance particular species populations. All of these constraints make national target setting, using biodiversity standards, an imprecise activity.

Significantly, it has been proposed by the Environment Agency to set lake management targets according to the potential for restoration, based on an assessment of the physical and chemical variables (including nutrient loading) that would be associated with no increase in the current population and a less intensive, pre-war agriculture in the catchment. This 'reference' approach would, for example, draw attention to a small absolute change but extreme proportionate change in the nutrient status of a naturally highly infertile lake. It seems to offer a robust basis for prioritising actions for the conservation and restoration of lakes.

The Norfolk and Suffolk Broads

Over the last 20 years, work has been underway in the Broads wetlands in eastern England to determine ways of restoring the ecological richness and natural beauty of around 40 shallow lakes, (broads) which originated as peat diggings (see Figure 1, and Moss,

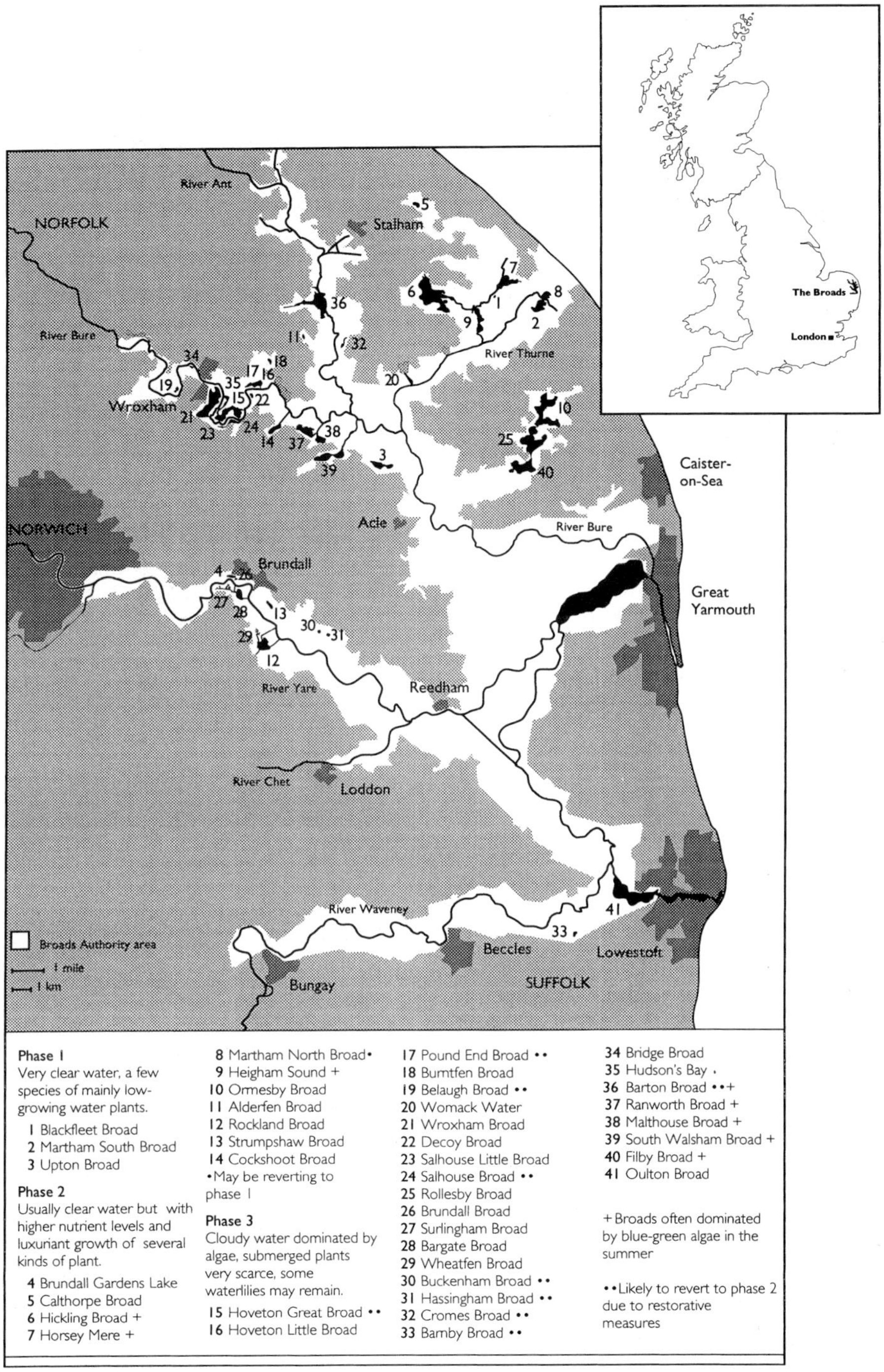

Figure 1. Location of the Broads wetland and individual Broads, grouped according to their ecological condition.

1989; George, 1992; Madgwick & Phillips, 1996 for overview). An initial awareness of the problems was heightened by a report of the then Nature Conservancy (Nature Conservancy, 1965) which drew attention to the catastrophic loss of water plants, decline of the fishery, increased pea-soup algal blooms and erosion of the reedswamp margins of rivers and lakes. These very visible changes in a popular holiday destination led to a strong commitment from a number of agencies to identify and implement practical solutions.

The establishment of a Broads Authority, with responsibilities for coordination of policies and actions for the whole area, enabled an integrated approach to the problem. The aims and working policies are embodied in the Broads Plan (Broads Authority, 1997) and the Environment Agency Catchment Plan for the area (NRA, 1995). The policies and actions relevant to eutrophication control, have been based on the results of a developing programme of research and experimental management. This work was made possible by the cooperation and joint investment of a number of agencies and universities, most recently coordinated by the Broads Authority and Environment Agency and assisted by the E. U. 'Life' Environment Programme. The development of the Broads research investigations have been summarised by Phillips et al. elsewhere in this publication and the recent work is detailed in Madgwick & Phillips (1996). The rest of this paper will consider the strategy for restoring the broads as a case study, showing how practical actions for nature conservation have been proposed, and how they are influenced by environmental constraints and multifunctional uses.

Reasons for broads restoration

Before setting out the influences and constraints on the broads restoration strategy, it is necessary to consider the reasons why a restoration policy has been promoted. Firstly, it is not acceptable to do nothing. Without further action, degradation will continue and further reduce the natural resources and recreational potential. Delay in taking action is only likely to increase the difficulty and cost. Secondly, if restoration actions are successful, there will be tangible benefits in terms of a more attractive and diverse environment, which will benefit local people and visitors alike, for example through an improved quality of recreation and lower costs of maintaining the navigation through dredging. Lastly, there are a number of national and international government commitments, which agencies such as the Broads Authority, English Nature and the Environment Agency are obliged to fulfill, such as the U.K. Biodiversity Action Plan and EC Directives mentioned above.

Broads restoration aims

The accepted aim is to restore the rivers and broads to 'a good ecological condition' which will support stable populations of plant and animal life (Broads Authority, 1997). Stability in this sense is meant to mean that the lake is able to buffer itself though internal biological and chemical mechanisms against external influences, which would tend to cause it to switch back to an algal-dominated state. It also implies that once 'restored', it will require minimal management to maintain its integrity and quality.

Nature conservation priorities

Since the early days of research and experimentation on water quality improvement in the Broads, the nature conservation case has been considerably strengthened. Around 20% of the wetland area is recognised as being of international importance for nature conservation, including 15 of the 41 broads, the majority of which are nature reserves. The Broads Natural Area Profile, which assessed the natural resources of Broadland in a national context, drew attention to a number of nationally rare and threatened species limited to Broads (Holve, 1996). The designations in themselves, however, are an unreliable guide to the nature conservation value of individual broads or of their restoration potential, since many were designated as part of a wider wetland complex where other natural features have higher importance.

Ecological targets

Biodiversity and environmental targets and standards, based on pre-eutrophication conditions in the Broads, are not considered appropriate and hence, strictly speaking, the aim is for rehabilitation to an improved state rather than for restoration. True restoration would be frustrated by an incomplete knowledge of the historical nature of the Broad, an acceptance that catchment influences on water quality and water resources can be reduced but not reversed, and that fairly intensive recreational use of many Broads is likely to continue in the medium term. Instead, it has been necessary to develop a vision based on what might be possible and sustainable in the future.

It has become increasingly apparent through research and monitoring work in the Broads and elsewhere that fairly dense beds of submerged macrophytes and a well developed littoral margin are prerequisites to the stability of macrophyte dominated lakes, which also support a diverse assemblage of invertebrates and fish (Timms & Moss, 1984; Perrow et al., 1999). Basic ecological theory suggests that biological diversity is likely to increase with a more complex biological structure within the lake. A well-developed littoral margin, floating and submerged plants together with variation in sediment type, depth and profile would maximise the habitat heterogeneity. Survey and monitoring work on macro-invertebrates (Jackson, 1997) and fish populations in Broads across the full range of ecological conditions (Perrow, 1991; Perrow & Jowitt, 1997), have confirmed this relationship between diversity and structure.

The pursuit of increased overall diversity has to be tempered to some extent by the need to promote conditions appropriate to rare habitats or species. In the Broads, an ideal goal is thought to be the restoration of extensive charophyte beds. This is justified by the previous domination of the broads by charophytes, as proven by reference to sediment cores (Moss, 1988), and the current national rarity of these species assemblages (HMSO, 1995; Holve, 1996). Particular species recovery targets for rare species, such as holly-leaved naiad (*Najas marina*), will be directed to suitable broads and possibly newly excavated peat diggings where the potential for recolonisation is the greatest.

Apart from the direct nature conservation benefits associated with achieving a near pristine condition in Broads, there would be less conflicts with recreational boating due to the low growing nature of charophytes. In Broads which are navigable, restoration attempts could be unpopular unless charophyte dominance is achieved. This is because if tall, dense growth of water plants is achieved instead, there will be disruption to boating and consequent high costs of weedcutting.

Steps in restoration

The research and experimental management work in the Broads has established four essential steps in the restoration process. These are explained by Phillips et al. (1999) and in Madgwick & Phillips (1996). They are in summary:

1. Reduction in external and internal nutrient loading.
2. Biomanipulation to create clear water conditions in the summer.
3. Re-establishment of macrophytes.
4. Stabilisation of the lake ecosystem, including the re-establishment of appropriate fish populations.

Whilst the implementation of a stepwise programme of restoration can be justified based on this simple theoretical framework, the implementation of such a programme is likely to more or less be successful in a particular Broad, depending on a number of catchment and site specific issues. These have been outlined for the Broads Authority in a position statement document (Madgwick, 1996) and are discussed below with reference to their influence on management targets and conservation strategies.

Control of catchment nutrients

The majority of broads are small and distant from the rivers and mainly receive nutrients from the agricultural catchment area. The nutrient inputs to some of these lakes can be significant, as at Alderfen Broad (Moss et al., 1990), and a choice has to be made concerning diversion of nutrient-rich inflows or use of other measures to reduce the nutrient status of the incoming water. As evidence is growing for the importance of water exchange in these moderately eutrophic lakes, there is a growing case for finding means of minimising or intercepting nutrients in supply waters to these broads.

A localised and short-term management option for nutrient control is to isolate some of the broads from the river system by creating an artificial barrier. This was achieved at Cockshoot Broad in 1981, as an early experiment (see case study in Moss et al., 1996). However, due to public navigation rights and a desire to retain the historical landscape and hydrological continuity between rivers, Broads and the surrounding wetland, there is now a general presumption by the Broads Authority and English Nature against using isolation as a management technique (Holve, 1996). There is also increasing evidence from the Broads monitoring data that water exchange or 'flushing' of Broads is important to prevent recycling of nutrients (Pitt et al., 1997).

Since a recovery of aquatic life is desirable ultimately for the whole rivers and broads system, and not just for broads that are naturally or artificially isolated from the rivers, a phased programme of investment for phosphorus reduction has been negotiated between interested parties.

The main source of catchment phosphorus into the northern (Bure, Ant and Thurne) rivers and broads was shown to be from 15 key sewage treatment works (with population equivalents >100). Control of the nutrient inputs to the rivers and connected Broads from major sewage treatment works is ultimately limited by the ability of Anglian Water to invest in phosphate removal plant and its continued maintenance.

At first, this investment was supplemented by the Broads Authority in an experimental phase in the 1980s but more recently, Anglian Water has been able to pass on the costs to its customers. This in turn is limited by OFWAT (the government-appointed water regulator) and the Department of the Environment, who have to agree to any expenditure outside that required by statute. The Environment Agency regulate non-statutory objectives through a consent procedure, agreed with Anglian Water and the Broads Authority. The 1mg l^{-1} total phosphorus standard for the final effluent from 8 sewage treatment works in the northern part of the Broads, was set as an agreed target which is thought to be achievable and reasonable, taking into account of the background diffuse inputs from the catchment and internal loading from rivers and Broads. This negotiation process was considerably assisted in the Broads by the use of demonstration projects which have shown what results may be possible as a result of restorative measures following nutrient reduction.

Although there is no doubt that a phased programme of investment in phosphorus reduction at sewage treatment works is needed, there is recent evidence in the Bure catchment that agricultural inputs of phosphates from arable land and the contribution of a large number of small point source private sewage treatment works, are becoming significant in terms of nutrient loading (Johnes, 1996). Further regulatory mechanisms may be needed to control these diffuse inputs. For agricultural sources this would need to include reductions in application of fertilisers or poultry litter to arable land in certain areas. In time, this kind of measure could be made a condition of continued agricultural subsidies through agri-environment reforms. The use of buffer strips or created wetlands may also be necessary to intercept run off that carries particulate phosphorus (Haycock et al., 1997).

The fundamental objective of phosphorus reduction measures in the Broads has been to achieve nutrient levels in broads that are in the range for a switch to a macrophyte dominated condition to be achievable through biomanipulation. However, from a nature conservation point of view, the greater the reduction in nutrient supply to lakes the better. At the lowest end, with lake water quality below about 50 $\mu g\ l^{-1}$ total phosphorus, a return to charophyte-dominated systems may be achievable and conservation value in terms of the diversity of water plants and the stability of the lake is likely to decrease with increasing nutrient levels (Moss et al., 1996; Hosper, 1997).

Control of internal nutrient loading

Sediment release of phosphates and the control of internal nutrient loading through suction dredging and dosing with iron salts have been investigated in broads through experimental management (Pitt et al., 1996). Iron dosing trials have not justified large scale application but may have a role in broads that remain unconnected to the main river system. In many broads, suction dredging has proved necessary to establish 1–2 metres of water depth and it is currently the only means of reducing the available sediment nutrient supply. Unfortunately, monitoring data on the effectiveness of suction-dredging in the Broads is limited (see Phillips et al., 1999).

One problem with suction dredging concerns the decision about what depth of sediment to remove, and to what extent a nutrient target can be proposed. Analysis of a number of sediment cores in Barton Broad revealed a generalised cut-off point with increased depth in cores from each sector of the broad, beneath which nutrient levels declined dramatically, although there was considerable local variation, presumably influenced by sedimentation patterns and the influence of boat traffic (Figure 2). The final specification in this case incorporated depth requirements for various types of boats (Holzer et al., 1997).

Recent investigations concerning the factors affecting macrophyte recovery (Schutten et al., 1997), conclude that sediment instability is a key factor and in lakes with soft, peaty sediments, seedlings may not survive long as they are so easily uprooted, for example by birds, water or sediment movement. Experience in dredging down to the solid peat in the Broads and in lake Geerplas in the Netherlands has shown that a fluid sediment can reform in one or two years with similar phosphorus release characteristics as before dredging (Pitt et al., 1996). Monitoring of the current large scale dredging project in Barton Broad will help to increase understanding of these effects. It will be necessary to define more precisely the chemical and physical parameters that can serve as management targets for suction dredging, taking into account objectives for nutrient control alongside those for macrophyte recovery and use of the lake.

The main practical constraint of suction dredging, apart from cost, is the local availability of suitable

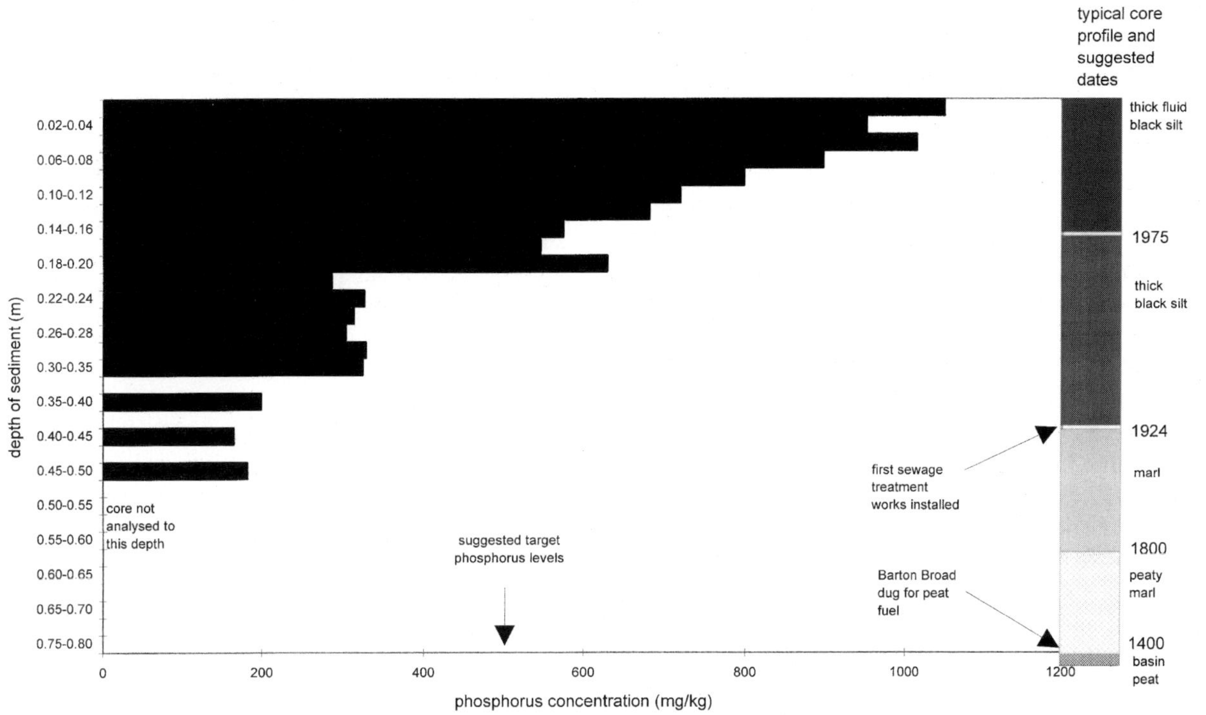

Figure 2. Phosphorus concentration profile in a sediment core from Barton Broads, and its relationship to visible layers, which reflect the history of human influences on the lake.

spoil disposal sites. Use of agricultural land is preferable to wetland of nature conservation importance but this brings further constraints on site preparation and handling of soils (see Barton Broad case study in Moss et al., 1996).

Biomanipulation

'Biomanipulation' or the near complete removal of zooplanktivorous fish to enhance the effects of grazing Cladocera, has been practised in a number of small broads over the last 7 years (Perrow, 1994), and has most recently been successfully applied as a sole restoration technique for one of the largest broads, Ormesby Broad (54 ha), where nutrient levels are moderately low with the mean around 50 μg l^{-1} total phosphorus (see case study in Moss et al., 1996). A range of practical fish removal techniques have been developed to take account of the physical characteristics of different broads and the nature and seasonal behaviour of fish stocks (Holzer et al., 1997).

In the Broads, the obvious practical constraint is the connection of many of the major broads to the river system and hence the need to create temporary, fish barriers to isolate whole or part of each lake. The ability to achieve this is often frustrated by navigation objections and design difficulties for the barriers which need to be able to withstand a tidal regime, allow water exchange, while preventing the passage of fish fry and spawn and resist clogging with epiphytic algae. Sometimes it is also necessary to build in a means for boat passage (Moss et al., 1996; Holzer et al., 1997). Another obvious practical consideration is the use of the lake for angling purposes. The retention of tench (*Tinca tinca*) and pike (*Esox lucius*) in biomanipulated broads has at least enabled some angling to continue in the short term. As successful examples of biomanipulation become available, where the need for fish removal declines and ceases as at Cockshoot Broad (Perrow, 1997), it has become easier to convince passionate angling groups of the future benefits of biomanipulation for the quality of the fishery and hence their sport.

Despite the technical problems, trials in the Broads have shown that even small fish free areas in broads can sustain clear water and water plants over a number of years. Further work is required to establish whether a lake can be restored by the progressive use of exclosures in bays. If this is a practical solution, it will be necessary to establish what proportion of the waterspace should be targeted for biomanipulation to achieve macrophyte dominance.

Re-establishment of macrophytes

In non-biomanipulated broads, the shading effect of dense algal blooms is likely to be the dominant constraint on aquatic macrophyte growth and survival. Even in clear water broads, following biomanipulation, the recovery of macrophytes has been fairly slow and erratic. In several broads it has been observed that many seedlings develop in the first spring clear water period but they have disappeared by the end of the summer, even though the water remains clear. Recently, considerable attention has been given to investigate potential limiting factors for macrophyte recovery (Schutten et al., 1997). A comparative study of a range of environmental variables associated with a limited number of plant species was carried out in 30 shallow lakes in the Netherlands and the U.K. A number of physical and chemical attributes of the sediment, that this survey highlighted as having a strong influence on water plant presence, were further tested in the laboratory. This work has suggested that the instability of peaty sediments in the Broads may be the key limiting factor on seedling survival. Further research and use of modeling will help managers to identify target species or guilds of plants for different lakes or parts of lakes according, for example, to local conditions of wind fetch and sediment variability. Experimentation with ways of treating and stabilising loose sediments in peaty lakes like the Broads, will assist the identification of realistic management techniques to secure water plant recovery.

Grazing by water birds is often cited as an important influence on macrophyte composition and in some situations it is suggested that grazing pressure is such that it may prevent or severely delay the recovery of macrophytes in shallow lakes (Van Donk, 1997). Some detailed survey and experimental work in the Broads showed that losses to grazing birds were negligible in the summer growing period in comparison to potential growth rates (Perrow et al., 1997). However, year to year variations in plant communities may well be determined in part by late summer and autumn grazing.

Further development of models is needed to assist managers in using the data available on existing conditions at a given lake, to predict the likely success of macrophyte recovery and to determine what, if any, mitigation measures might be necessary. It will also be important to carry out autecological research on rare macrophyte species or local, native species that are thought to play a central or 'keystone' role in the plant recovery process. This type of ecological information would greatly add to the ability for managers to determine and prioritise actions in particular lakes and for nature conservation targets to be incorporated into wider lake restoration strategies.

Stabilisation of the lake system

The presence of submerged vegetation and benthic diatoms act as 'buffering mechanisms' to reduce the susceptibility of sediment to wind-induced resuspension, compete with phytoplankton for nutrients and provide refuges for grazing zooplankton (Hosper, 1997). This is more achievable in small lakes (<30 hectares) (Hosper, 1997). The effectiveness of 'top down' control exerted by piscivores, which helps stabilise lake systems is also known to be dependent on the presence of substantial areas of macrophyte beds of varying density, interspersed by patches of water (Grimm & Backx, 1990; Walker, 1994; Perrow & Jowitt, 1997).

Despite good agreement amongst limnologists and fisheries scientists on these general preconditions for lake stability, there is relatively little specific information available for managers, for example concerning the target proportion of submerged macrophyte beds for zooplankton refuge effects to be effective from year to year. Apart from the percentage cover of aquatic macrophytes, targets for the volume, structure and pattern of cover of aquatic and littoral plants may be appropriate. This in turn may influence where biomanipulation works are prioritised within the lake and the scale and type of measures which will be needed for water plant introductions or protection from bird grazing.

Preliminary results from Cockshoot Broad and Ormesby Broad suggest that fish communities can respond fairly rapidly to the presence of submerged macrophyte beds and that diversification of the fish community and a tendency to pike dominance will occur without the need for restocking (Perrow, 1997). However, there is uncertainty about the stability of these lake situations and there is a desire to establish appropriate targets for fish population composition and structure for the range of shallow lake types (Moss et al., 1996). In the Broads, a precautionary approach has been taken in monitoring zooplanktivore populations during the restoration process and continuing to

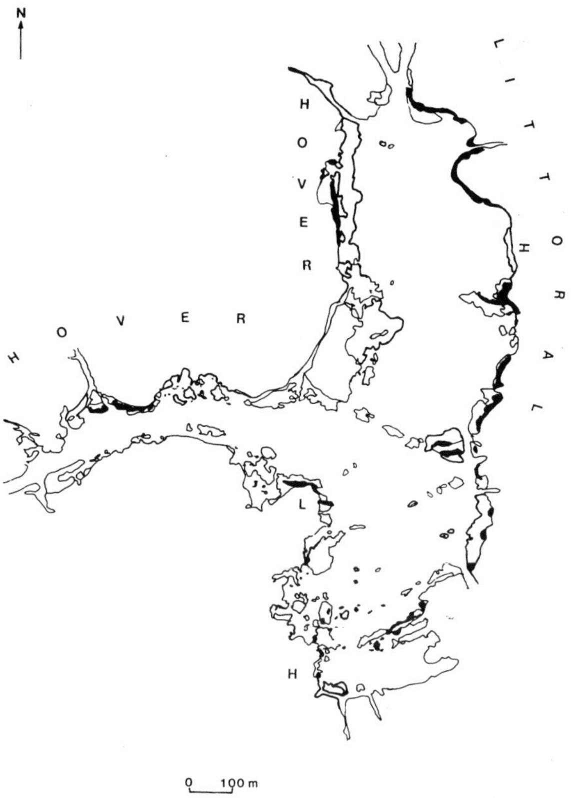

Figure 3. The distribution of hover and littoral growth forms of reedswamp in Barton Broad, showing areas lost to open water (unshaded) between 1946–1977 (from Crook et al., 1983).

keep them low through small scale fish removal operations, until submerged macrophyte beds are seen to occur across the Broad for several years. However, some fairly arbitrary decisions on fish manipulations have needed to be made from year to year. For example, when macrophytes are present, but variable in cover from year to year, it is difficult to know whether it is cost-effective to further reduce the adult bream (*Abramis brama*) population to minimise the predation pressure of young fish in the spring and summer. More information on the habitat requirements for perch (*Perca fluviatilis*), would enable management to encourage this species to play an important piscivore role in a 'restored' lake situation.

The presence of a well developed littoral margin around the lake and a high proportion of reedswamp edge to open water, is thought to be crucial in enabling pike dominance to be re-established quickly vegetation (Jeppeson, pers. comm.; Hosper, 1997). Aerial photographs from the 1930s to the 1950s show a large proportion of the Broad basins covered by floating reedswamp (hover), interspersed with open water. A detailed study of the reedswamp regression revealed that virtually all of this floating hover was lost by the mid 1970s (Crook et al., 1983) and that littoral vegetation was confined to the firmer edges of Broads (see Figures 3 and 6). This floating reedswamp habitat is likely to have played a key role in determining the predator/prey interactions in the fish community and providing sheltered bays for aquatic macrophytes. Hence, restoration of substantial areas of reedswamp may prove to be necessary prior to, or in association with, biomanipulation in many Broads. This, in any case is a desirable objective from a nature conservation point of view, since reedswamp is a valuable transition habitat in terms of plant communities (George, 1992), a wide range of macroinvertebrates (Jackson, 1997) and nationally scarce birds, such as the bittern (*Botaurus stellaris*), (Holve, 1996).

Finally, stabilisation of a macrophyte dominated lake following biomanipulation, may depend on further nutrient controls. This is another important area for future research because most of the lake restoration projects of this kind in western Europe have occurred in the last decade and it is too soon to draw any conclusions. However, there is some evidence that plant communities in the Broads associated with intermediate nutrient levels (around 50–100 μg l^{-1} total phosphorus), typically including tall, leafy plants such as *Ceratophyllum demersum*, *Potamogeton* spp. and *Elodea* spp, are relatively unstable and may show considerable shifts or cycles in plant abundance and composition, due to weather-induced events and other de-stabilising influences. Conversely, charophyte-dominated lake communities which are always associated with low nutrient conditions ($<$50 μg l^{-1} total phosphorus) tend to be relatively stable in terms of the composition and abundance of macrophytes from year to year (Kennison et al., 1998).

Other influences on the success of broads restoration

Apart from the practical constraints, environmental influences and uncertainties mentioned above, which are likely to face most managers of shallow, eutrophic lakes, there are some particular water management issues in the Broads which impose another layer of complexity on decision-making and planning. These are associated with the tidal nature of the Broads waterway system.

Summer freshwater river flow may be a critical factor influencing the success of restoration attempts.

Greater flows reduce algal development by decreasing retention times in Broads connected to the rivers. This reduces the chances of dominance by blue-green algae (in particular *Oscillatoria* blooms) and limits nutrient recycling (Phillips, pers. comm.; Hosper, 1997). Apart from the recreational problems, blue-green algae may cause toxicity to zooplankton grazers. Reduced summer flows can also increase the chance of entry of predators of zooplankton, such as Neomysis integer, which can interrupt the restoration process. Such an event was recorded at Cockshoot Broad in 1992 resulting in cloudy water and poor growth that summer (Moss et al., 1996), although macrophyte recovery accelerated in the following years (Dunsford, 1996). There is also an increased risk of introduction of other toxic algal populations associated with more brackish conditions, such as *Prymnesium parvum*, which can cause dramatic changes in fish communities.

The degree of water exchange is known to influence the type and amount of algae that are produced in broads (see Phillips et al., 1999). The characteristics of those Broads close to the river, at the head of the system, are likely to be the most affected by any future reduction to river flow, which could be caused by increased abstraction and the effects of climate change. Those lowest down the river valley are most influenced by the tidal regime, saline intrusion and nutrient inputs from downstream. Broads connected to the middle reaches of the rivers by long, narrow channels have the longest residence times and hence provide the most stable environment, where blue-green bacterial populations are favoured.

The implication of these water management issues is that broads which are isolated from the river, and those which receive a regular throughput of freshwater, and are furthest from the estuary, are likely to be a safer bet for restoration to clear, freshwater Broads. A case has been made to quantify the river flow requirements that would satisfy Broads restoration objectives for the northern broads system, through hydrological modeling. This would be part of a bigger project to determine an integrated water management strategy for the area (Harper, 1997).

Classification of Broads according to management need

It is possible to group Broads according to their position in the Broads waterway system, the likely influence of river flow and saline intrusion, degree of isolation and ecological condition. An early classification system for the Broads identified three phases of ecological condition, related to a nutrient gradient (Broads Authority, 1987). This linear model was later adapted to reflect the notion of 'alternative stable states' (see Figure 4; Broads Authority, 1997). Recently, a water plant classification system has been derived from a long-term macrophyte monitoring dataset (Kennison et al., 1998). A summary matrix (Figure 5) shows a grouping of broads into three categories, according to the plant and animal communities that are thought to be achievable as a result of restoration management and their relationship to total phosphorus concentrations. The recreational implications of the three alternative scenarios are indicated. The influence of the position of broads in relation to the river system (and hence to tidal effects and salinity) and the degree of current navigational use are also shown. This classification of broads according to restoration potential, taking account of a multiplicity of constraints, will help to inform the Broads restoration strategy and individual site management plans.

Future prospects for Broads restoration

There is uncertainty over the ability of agencies to meet the ultimate target of restoring the whole rivers and broads system. It will require continued large-scale investment in phosphorus removal facilities at all significant sewage treatment works. All parties are committed to this approach in the Broads Plan, 1997. As this progresses, other catchment inputs from agriculture and a large number of septic tanks and private sewage works become more significant. It is only justifiable to tackle these types of sources as part of an integrated, catchment approach will make a significant difference to the water quality and aquatic life of the rivers and broads. New approaches to water quality regulation will be required, along with a joint commitment for investment from the water companies, Environment Agency and Ministry of Agriculture (MAFF). Until the minimisation of nutrient and pesticide losses from agricultural land becomes mandatory, through agri-environment reforms, any schemes to intercept nutrient and particulate run-off from farmland will also require the voluntary cooperation of landowners.

In the medium-term, some temporary restrictions on navigation will be required in the Broads to enable isolation of lakes for biomanipulation purposes. In the longer-term, considerable changes to the intensity and type of boating may be justified in certain Broads and

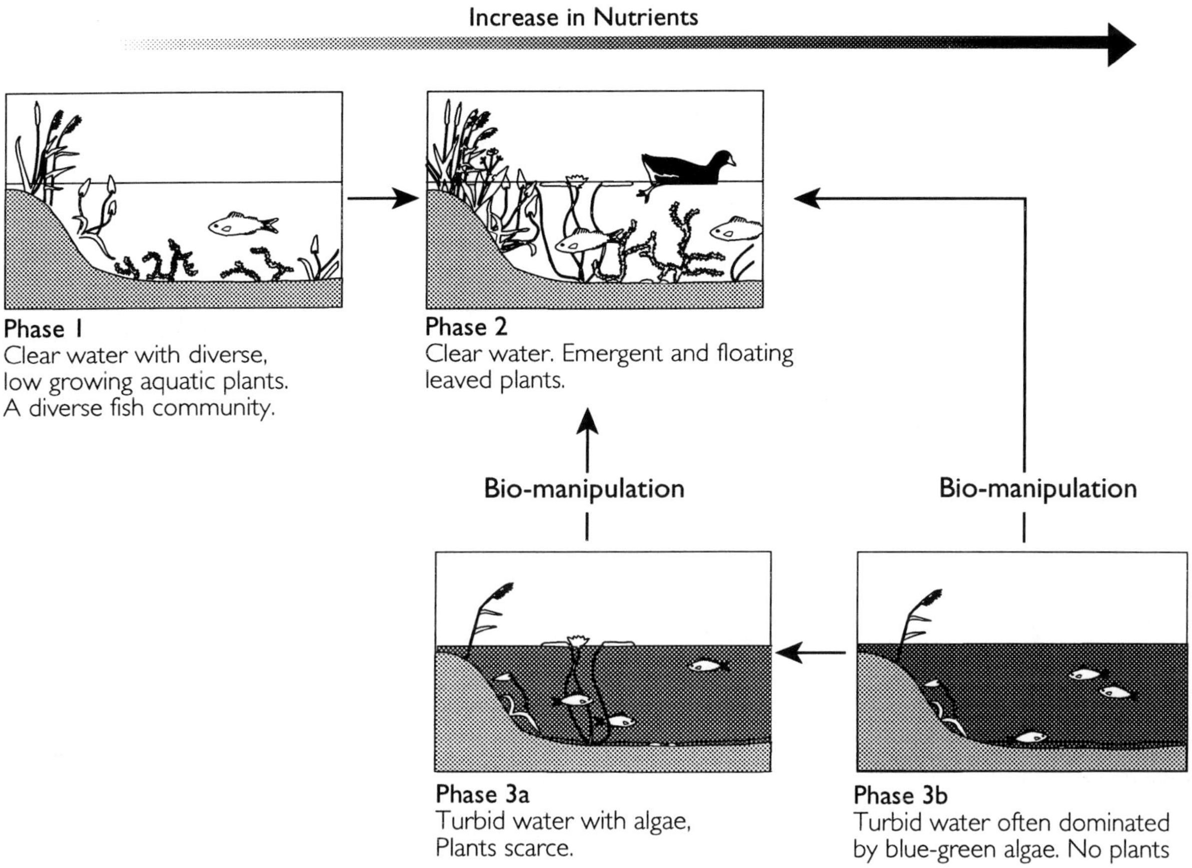

Figure 4. The alternative stable states model for dominance by aquatic plants or phytoplankton in shallow lakes at moderate nutrient levels (50–100 $\mu g\ l^{-1}$ total phosphorus), showing biomanipulation as a 'reverse switch'.

in the rivers themselves. Due to the connectedness of the rivers with the major broads, the rivers would almost certainly need to be rehabilitated to support a diverse fish community (Harper, 1997).

The future prospects for the restoration of the whole rivers and Broads system is likely to also depend on regional and local water resource management. Future trends in agricultural irrigation, population growth and settlement patterns, tourism strategies etc. may place significant and conflicting demands on these resources. The ecological scenarios related to different river flow conditions will need to be fully evaluated. Some broads at the lower end of the river system may inevitably become brackish in the long term due to climate change. For others, some local or strategic changes in water abstraction may be needed to ensure that past restoration efforts are safeguarded and that future freshwater restoration efforts are underpinned. As the legislative mechanisms for such changes are currently unavailable, any water resource obstacles of this kind could prove to be significant.

In the short-term, experimental management will need to continue to focus on small, isolated broads where nutrient reduction has been achieved, and boating and angling influences are small or absent. The outcome of this work is being applied to major broads restoration projects in an integrated manner, accepting that investigatory work is proceeding in tandem with restorative actions.

A bold restoration project has been started at Barton Broad, the largest broad to have suffered from severe eutrophication (see Figure 6). The 67 ha broad straddles the top end of the River Ant, where sufficient phosphorus reduction in the river has now been secured at the sewage treatment works upstream. Since the 1970s, this lake has been almost devoid of aquatic macrophytes and it is dominated by algal blooms in the summer months as a result of high total phos-

(BMT = Broads macrophyte type) References: [1]Kennison et al (1998), [2]Jackson, 1997, [3]Stansfield et al, 1997, [4]Perrow et al, 1997, [5]Phillips et al, in this volume

Achievable plant communities	Other ecological targets	Associated nutrient status	Key 'symptoms' of restored state	Recreational implications	Broads where target applies (__ means target achieved)	Nature reserve & SSSI	Category of broads according to ease of ecological restoration	Tidal #	/Public navigation ~
Charophytes (*Chara* spp) or *Najas marina* dominant (BMT1)[1] Extensive, diverse littoral margin	Diverse macro-invertebrate assemblages[2] Large-bodied Cladocera present throughout plant growing season[3] Pike (*Esox lucius*) and tench (*Tinca tinca*) dominant in fish community[4]	Total P ≤ 50µg/1	• clear water throughout the summer • low risk of blue-green algal blooms • plant community relatively constant - only small perturbations	• high quality pike angling • little conflict with sailing as low growing plants	Martham N Broad Martham S Broad Hickling Broad Horsey Mere Heigham Sound Blackfleet Broad Brundall Garden Upton Broad Calthorpe Broad Ormesby Broad Ormesby Little Broad Rollesby Broad Filby Broad Cockshoot Broad Alderfen Broad Strumpshaw Broad Belaugh Broad Decoy Broad Crome's Broad Barnby Broad Burntfen Broad	* * * * * * * * * * * * * * 	Isolated broads where mainly diffuse agricultural inputs of nutrients or/ broads at top of river system where there are no major point sources of nutrients or/ where benefits of P stripping at sewage treatment works realised and bio-manipulation is a practical option	# # # # # 	 ~ ~
Significant proportion of lake with dense beds of tall leafy submerged plants including, for example, *Potamogeton* spp, *Zanichellia palustris*, Elodea spp; charophytes and *Najas* occur locally (BMT2a)[1] Intact, extensive, diverse littoral margins	Diverse macro-invertebrate assemblages[2] Large bodied Cladocera abundant in plant beds during plant growing season[3] Diverse fish community of cyprinids, percids and pike[4]	Total P 50-100µg/1	• clear water sustained through plant growing season at least in areas of dense plant beds • blue-green algal blooms uncommon • plant and fish communities may show some temporary shifts or cycles due to weather-induced events	• temporary conflicts with boating due to need for bio-manipulation, weedcutting may be needed outside main navigation channels • variety of angling opportunities	Rockland Broad Hoveton Great Broad Hudsons Bay Hoveton Little Broad Pound End Salhouse Little Broad Barton Broad Ranworth Broad Bargate Broad Surlingham Broad Wheatfen Broad Buckenham Broad Hassingham Broad S Walsham Inner Broad	 * * * * * * 	Tidal broads where further investment in nutrient reduction is needed and/or where bio-manipulation will be problematic due to navigation and/or tidal influences	# # # # # # # # # # # # # #	~ ~ ~ ~ ~ ~ ~
Nuphar lutea and *Nymphaea alba* ant lake margins, other submerged plants scarce (BMT2b)[1] Intact littoral margins	Improved macro-invertebrate and fish communities, associated with margins[2,4]	Total P >100µg/l	• turbid water except at margins • water plants restricted to margins • higher risk of blue-green algae	• weed-free water depth for sailing • some angling opportunities	Wroxham Broad Salhouse Broad Bridge Broad Malthouse Broad S Walsham Broad Oulton Broad		Tidal broads where navigational use is probably a dominant constraint but where restoration of marginal macrophyte communities may be possible	# # # # # #	~ ~ ~ ~ ~ ~

Figure 5. Suggested ecological targets for broads grouped according to the ease of ecological restoration, determined by the influences of nutrient sources, geographic posiiton, tidal action and recreational use.

Figure 6. Barton Broad in 1997, the site of a major lake restoration project involves suction dredging, biomanipulation and the regeneration of reedswamp margins, following successful reduction of phosphorus inputs from Stalham sewage treatment works, just upstream.

phorus concentrations, maintained by internal loading. The restoration project will test the whole range of restoration techniques which have been developed in smaller scale experiments and in particular, the effectiveness of suction dredging in permanently removing a major nutrient supply (see case study in Moss et al., 1996). Further nutrient export modeling work will be carried out to investigate the significance of diffuse and small, point-source inputs and depending on the hydrological pathways, some further control measures may be required.

Following suction dredging of the whole lake, a start can be made on biomanipulation activities in some large bays (Figure 6). A greater understanding of the fish populations and their use of the river, broad and surrounding wetland in different seasons, is a prerequisite to planning biomanipulation measures for the whole broad. It may be concluded that a more holistic approach to fish management may be possible and worthwhile, taking into account the whole area affected by fish migration. In the same way, the restoration of flooded marshlands around the Friesland lakes has been proposed to provide optimal nursery areas for pike and hence to help stabilise macrophyte dominated conditions (Grimm, 1994).

Major lake management projects such as this, offer a real opportunity to increase understanding of lake ecology through associated research and monitoring work.

Conclusions

This paper has argued that nature conservation strategies need to operate at various levels. Although the setting of conservation targets can be helpful in directing and prioritising actions, it suggested that ecological targets may need be adjusted as our understanding of what is possible and sustainable increases. There are still gaps in the theoretical framework for shallow lake restoration as well as technical problems to overcome, but this is a fast developing field. It is noted that most progress in developing management techniques has been made through the combination of focused research projects in tandem with large-scale field experiments, against a background of long-term monitoring. The joint programme managed by the Broads Authority and the Environment Agency in the Broads has proved very productive in this respect over the last decade.

Effective conservation and restoration of a representative network of shallow lakes in Europe and in each country will depend on the adoption of a workable, common lake classification system. The use of a 'reference' approach as proposed by Johnes et al. (1994) would enable ecosystem standards to be defined which would reflect appropriate lake conditions for their natural catchments and allow meaningful comparisons of lake condition through time and space. An integrated catchment approach to water management is another pre-requisite to effective lake conservation. Procedures are being formalised to assist this process in the European Union through the proposed Water Framework Directive.

At a more local level, nature conservation priorities need to be considered alongside other uses of the lake or group of lakes in order that common management goals can be agreed between the various stakeholders. The management options will largely depend on a number of factors which influence the ecological potential, such as nutrient loading, morphometry, sediment type, type of recreational use and bird grazing pressure.

As research and experimental management work progresses, more specific advice will become available to managers, for example concerning fish predator/prey targets and the optimal reedswamp edge to lake area ratio. An increased use of multi-lake studies and models is envisaged to analyse complex lake interactions and derive appropriate ecological targets for lake managers. Thus, as the information base grows, the risks, costs and benefits of lake management measures and ecological restoration will become clearer. However, this is not an inevitable process and it will depend on researchers and lake managers working ever more closely.

Acknowledgements

The author wishes to recognise the joint funding and coordination of the programme of research and experimental management in the Broads by the Broads Authority and the Environment Agency. The research work has been assisted greatly by the financial support of the EU Life-Environment programme (1993–1996) and the Soap and Detergent Industry Association Environment Trust (SDIAET) over the last two decades. The support of SDIAET, the Millenium Commission and Anglian Water in the current Barton Broad project is also gratefully acknowledged.

References

The Broads Authority, 1987. The Broads Strategy and Management Plan. Broads Authority, Norwich.

The Broads Authority, 1997. The Broads Plan. Broads Authority, Norwich.

Carvalho, L. & B. Moss, 1995. The current status of a sample of English Sites of Special Scientific Interest subject to eutrophication. Aquatic Conservation 5: 191–204.

Crook, C., R. Boar & B. Moss, 1983. The decline of reedswamp in the Norfolk Broadland: causes, consequences and solutions. BARS 6. Broads Authority, Norwich.

Dunsford, D. S., 1996. Survey of Macrophytes in the Norfolk Broads, Broads Authority, Norwich.

English Nature, 1997. Wildlife and Freshwater – An Agenda for Sustainable Management. English Nature, Peterborough.

European Commission, 1995. Communication from the Commission to the Council and the European Parliament: Wise Use and Conservation of Wetlands COM (5) 189.

European Commission, 1997. Proposal for a Council Directive establishing a framework for Community action in the field of water policy COM (97) 49.

Foy, R. H., R. V. Smith, C. Jordan & S. D. Lennox, 1995. Upward trend in soluble phosphorus loading to Lough Neagh despite phosphorus reduction at sewage treatment works. Wat. Res. 29: 1051–1063.

George, M., 1992. The Land Use, Ecology and Conservation of Broadland. Packard Publishing, Chichester.

Grimm, M. P., 1994. The characteristics of the optimum habitat of northern pike (*Esox lucius*). In I. Cowx (ed.), Rehabilitation of Freshwater Fisheries. 235–243. Blackwell Scientific Publications, Oxford.

Grimm, M. P. & J. J. Backx, 1994. Mass removal of fish from Lake Wolderwijd (2700 ha), The Netherlands. In I. Cowx (ed.), Rehabilitation of Freshwater Fisheries. 390–400. Blackwell Scientific Publications, Oxford.

Harper, D. M., 1997. A Catchment Vision for the Third Millennium – Management Issues and Priorities for the Bure, Thurne and Ant Water Environment. Environment Agency, Broads Authority and English Nature, Norwich.

Haycock, N. E., T. P. Burt, K. W. T. Goulding & G. Pinay (eds), 1997. Buffer zones, their processes and potential in water protection. Quest, Environmental, Harpenden, UK.

HMSO, 1995. Biodiversity – The U.K. Action Plan. 2 volumes. HMSO, London.

Holve, H., 1995. The Broads Natural Area Profile. English Nature and Broads Authority, Norwich.

Holzer, T., M. R. Perrow, F. J. Madgwick & D. S. Dunsford, 1997. Practical methods for Broads restoration. In Restoration of the Norfolk Broads, BARS 14. Broads Authority and Environment Agency, Norwich.

Hosper, H., 1997. Clearing lakes – and ecosystem approach to the restoration and management of lakes in The Netherlands. RIZA, Lelystad, The Netherlands.

Jackson, M. J., 1997. Sampling Methods for Studying Invertebrates in the Littoral Vegetation of Shallow Lakes. BARS 17, Broads Authority, Norwich.

Johnes, P., 1996. Nutrient Export Modelling – River Bure, Norfolk. Operational Investigation OI 581. Environment Agency, Norwich.

Johnes, P., B. Moss & G. L. Phillips, 1994. Lakes – Classification and Monitoring: A Strategy for the Classification of Lakes. National Rivers Authority, Bristol.

Kennison, G. C., D. S. Dunsford & H. Schutten, 1998. Stable or changing lakes? A classification of aquatic macrophyte communities from eutrophic shallow lake systems in the United Kingdom, Aquatic Conservation, 8: 669–684.

Kling, M., M. P. Grimm & S. H. Hosper, 1995. Eutrophication and ecological rehabilitation of Dutch lakes: explanation and prediction by new conceptual framework. Water Science and Technology 31: 207–218.

Madgwick, F. J., 1996. Broads Restoration Strategy – an evaluation of the possibilities for restoring nutrient-enriched Broads to good ecological condition. Internal Report to Broads Authority and EU Life Programme (92-3/UK/031).

Madgwick, F. J. & G. L. Phillips, 1996. Restoration of the Norfolk Broads – Final Report to the EU Life Programme. BARS 14, Broads Authority and Environment Agency, Norwich.

Moss, B., 1988. The palaeolimnology of Hoveton Great Broad, Norfolk: clues to the spoiling and restoration of Broadland. Symposium of the Association for Environmental Archaeology, The Exploitation of Wetlands. 7: 163–191.

Moss, B., 1989. Water pollution and the management of ecosystems: a case study of science and the scientist. In P. J. Grubb & R. H. Whittaker (eds), Towards a More Exact Ecology, 401–422. Blackwells Scientific Publications, Oxford.

Moss, B., J. Stansfield & K. Irvine, 1990. Problems in the restoration of a hypertrophic lake by diversion of a nutrient rich inflow. Verh. int. Ver. Limnol. 24: 568–572.

Moss, B., G. L. Phillips & J. Madgwick, 1996. A guide to the Restoration of Nutrient Enriched Shallow Lakes. Broads Authority, Norwich.

National Rivers Authority, 1995. Yare Catchment Management Plan. NRA Anglian Region, Peterborough.

Nature Conservancy, 1965. Report on Broadland. Nature Conservancy, London.

Palmer, M. A., S. L. Bell & I. Butterfield, 1992. A botanical classification of standing waters in Britain: applications for conservation and monitoring. Aquatic Conservation 2: 125–44.

Perrow, M., 1994. Practical aspects of the biomanipulation of fish populations. In Development of Biomanipulation Techniques and Control of Phosphorus Release from the Sediment. NRA Project Report 475/2/a, Norwich.

Perrow, M., 1997. Status of Fish Populations in Macrophyte-dominated and Biomanipulated Broads. Broads Authority, Norwich.

Perrow, M. & A. J. Jowitt, 1997. Factors affecting water plant recovery – the influence of macrophytes on the structure and function of fish communities. In F. J. Madgwick & G. L. Phillips (eds), Restoration of the Norfolk Broads – Final Report to the EU Life Programme. BARS 14, Broads Authority and Environment Agency, Norwich.

Perrow, M., J. Schutten, J. R. Howes, T. Holzer, F. J. Madgwick & A. J. Jowitt, 1997. Interactions between coot (*Fulica atra*) and submerged macrophytes: the role of birds in the restoration process. Hydrobiologia 342/343: 241–255.

Phillips, G., A. Bramwell, J. Pitt, J. Stansfield & M. Perrow, 1999.Practical application of 25 years' research into the management of shallow lakes. Hydrobiologia 395/396 (Dev. Hydrobiol. 136): 61–76.

Pitt, J., A. Kelly & G. L. Phillips, 1996. Control of nutrient release from sediment. In F. J. Madgwick & G. L. Phillips (eds), Restoration of the Norfolk Broads – Final Report to the EU Life Programme. BARS 14, Broads Authority and Environment Agency, Norwich.

Royal Society, 1983. The Nitrogen Cycle of the United Kingdom. Royal Society, London.

Sharpley, A. N., M. J. Hedley, E. Sibbeson, A. Hillbricht-Ilkowska, W. A. House & L. Ryszkowski, 1995. Phosphorus transfers from terrestrial to aquatic systems. In H. Tiessen (ed.), Phosphorus in the Global Environment. 171–200. John Wiley & Sons, Chichester.

Schutten, J., A. J. Davy, F. J. Madgwick, H. Coops, W. Admiraal, E. H. Lammens & G. L. Phillips, 1997. Factors affecting water plant recovery. In F. J. Madgwick & G. L. Phillips (eds), Restoration of the Norfolk Broads – Final Report to the EU Life Programme. BARS 14, Broads Authority and Environment Agency, Norwich.

Timms, R. M. & B. Moss, 1984. Prevention of growth of potentially dense phytoplankton populations by zooplankton grazing, in the presence of zooplanktivorous fish in a shallow wetlands system. Limnol. Oceanogr. 29: 472–486.

U.K. Round Table on Sustainable Development, 1997. Freshwater. UKRTSD, London.

Van Donk, E., 1997. Switches between the clear water state and turbid state induced by herbivory on macrophytes. In E. Jeppesen, (ed.), The Role of Submerged Macrophytes in Structuring the Biological Community and Bio-geochemical Dynamics in Lakes. NERI, Silkeborg, Denmark.

Van Donk, E. & R. Gulati, 1989. Biomanipulation in The Netherlands, Proceedings of a Symposium. Hydrobiol. Bull. 23: 1–99.

Walker, P. A., 1994. Development of pike and perch populations after biomanipulations of fish stocks, In I. Cowx (ed.), Rehabilitation of Freshwater Fisheries, 376–389. Blackwells Scientific Publishers, Oxford.

Hydrobiologia **395/396:** 325–333, 1999.
D.M. Harper, B. Brierley, A.J.D. Ferguson & G. Phillips (eds), The Ecological Bases for Lake and Reservoir Management.

A reservoir in management-induced transition between ecological states

Peter W. G. Daldorph
Anglian Water Services plc, Endurance House, Vision Park, Histon, Cambs CB4 4ZY, U.K.
Present address: Eastern Hydrology Limited, 7 Hobart Road, Cambridge CB1 3PU, U.K.

Key words: reservoir, phosphorus, algal blooms, water treatment, ferric dosing, macrophytes

Abstract

The impact of phosphorus control on the nutrient dynamics and biological activity of Foxcote Reservoir, a small impounded reservoir operated by Anglian Water are discussed. Phosphorus precipitation using iron salts was commenced in 1981 to improve the treatability of previously algal-laden water and continued until 1994 when the reservoir was closed as a water supply source. A marked decrease in algal biomass was observed after an initial delay, which coincided with an increase in the abundance of aquatic macrophytes. There were also marked shifts in the macrophyte community during the following 10 years, with an increase in *Elodea* and an initial increase then decline in *Chara*. The zooplankton community also changed with an increase in the larger bodied *Daphma pulex*. No overall change in the rate of denitrification was observed, although patterns of silica depletion showed changes in frequency. Trends in wildfowl counts over this period reflect changes in the ecology of the reservoir.

Introduction

The provision of high quality potable water from surface water reservoirs can be compromised by eutrophication and the development of large blooms of phytoplankton (Howard et al., 1987). Shallow reservoirs, in particular, are prone to excessive growths of algae as light limitation is unlikely to develop. In these reservoirs, the reversal of eutrophication normally results in increased abundance of macrophytes. This can, in turn, stabilise the community structure as a result of ecological buffers which, in turn, reduce the tendency to revert to phytoplankton dominance (Ozimek et al., 1990; Hosper & Jagman, 1990). However this shift toward macrophyte dominance can be prevented by recycling of phosphorus, associated with high phytoplankton biomass (Scheffer, 1990; Moss, 1990). In order to improve water quality it may, therefore, be necessary to overcome ecological barriers, as well as reducing nutrient inputs. In many cases biomanipulation has been necessary to overcome these ecological barriers (Søndergaard et al., 1990).

The present paper describes the history of Foxcote reservoir, which has displayed such a shift in ecological state. In this case, the lake has shifted from phytoplankton to macrophyte dominance following the control of phosphorus inputs.

Description of site and history

Foxcote reservoir is a pumped storage reservoir located near Buckingham in the River Great Ouse catchment in Southern England. It was commissioned in 1957 and provided water for treatment for potable supply until 1994 when the reservoir was taken out of service. The reservoir (Figure 1) is 17 ha in surface area with a mean depth of 2.8m and capacity of 550×10^6 l. It received pumped water from the River Great Ouse at a typical rate of 10×10^6 l day^{-1}, resulting in an average retention time of 2 months. Small quantities of water are derived from the local natural catchment (<5%). In the late 1970s severe treatment problems developed because of very large algal populations in the reservoir, making it necessary to switch to direct abstraction from the River Great Ouse on a number of occasions. In order to overcome these problems, a trial was carried out in 1981 to precipitate dissolved phosphorus in the reservoir using ferric sulphate (Hayes et al., 1984). This was considered effective and a permanent dosing system was installed

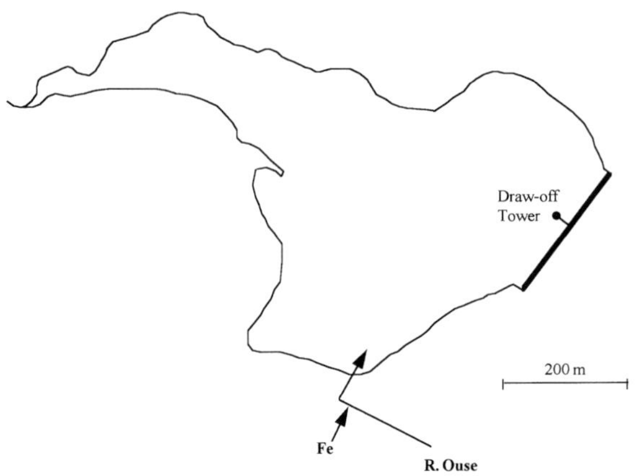

Figure 1. Foxcote Reservoir, showing input from the River Great Ouse.

in November of the same year. The ferric sulphate was dosed directly to the main carrying water from the Ouse to the reservoir, at a rate proportional to the phosphorus loading. The ferric hydroxide floc precipitates phosphorus and settles in the reservoir. This was continued until 1994 when the reservoir came out of service.

Sampling and monitoring

Weekly samples were collected from the reservoir draw-off tower and from the abstraction point on the Great Ouse throughout the period 1976–1995, with some interruption in sampling in 1989. No total phosphorus analysis was carried out before 1982. Algal counts in the abstracted water were carried out using an inverted microscope. Additional buoys in the reservoir were sampled for zooplankton by collecting a vertical haul using a plankton net. Macrophyte surveys were carried out visually over a sampling grid (Young et al., 1991).

Water chemistry, phytoplankton and macrophytes

The changes in annual mean phosphorus concentration in Foxcote reservoir are shown in Figure 2, compared to the influent phosphorus concentration. Although total phosphorus data were not available prior to the commencement of phosphorus control, it is clear that concentrations in the reservoir are low compared to those flowing into the reservoir, and showed no increase when intluent concentrations were particularly high in 1990 and 1991. More typical ratios between in-lake and influent phosphorus based on the Vollenweider model (OECD, 1981) would result in much higher phosphorus concentrations in the reservoir (0.5–0.35 $\mu g\ l^{-1}$ P) than were observed which indicates that most of the phosphorus was precipitated by the ferric sulphate. Dissolved phosphorus values declined after ferric sulphate dosing was commenced in 1981 but increased in the period 1988–1992.

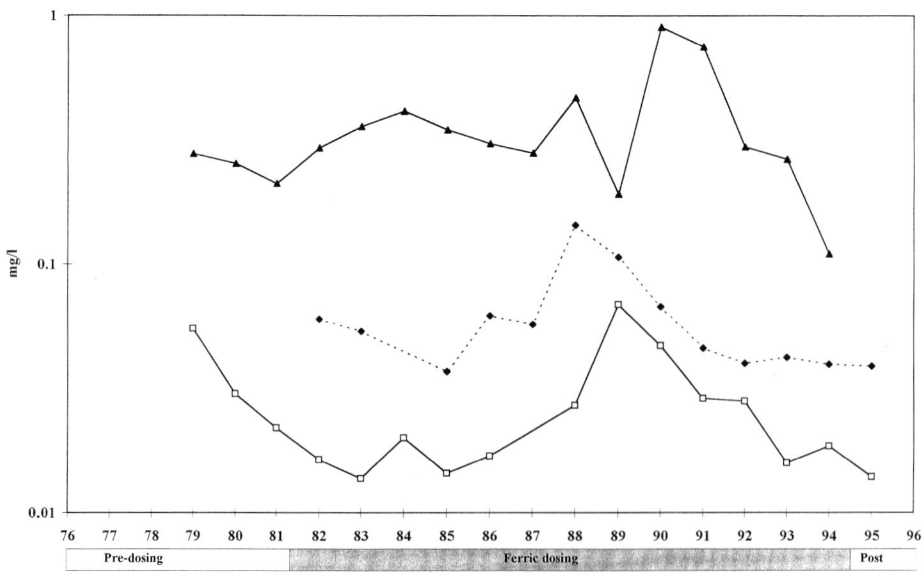

Figure 2. Changes in annual mean values of influent total phosphorus (solid triangles) reservoir total phosphorus (dashed line) and reservoir orthophosphate (open squares).

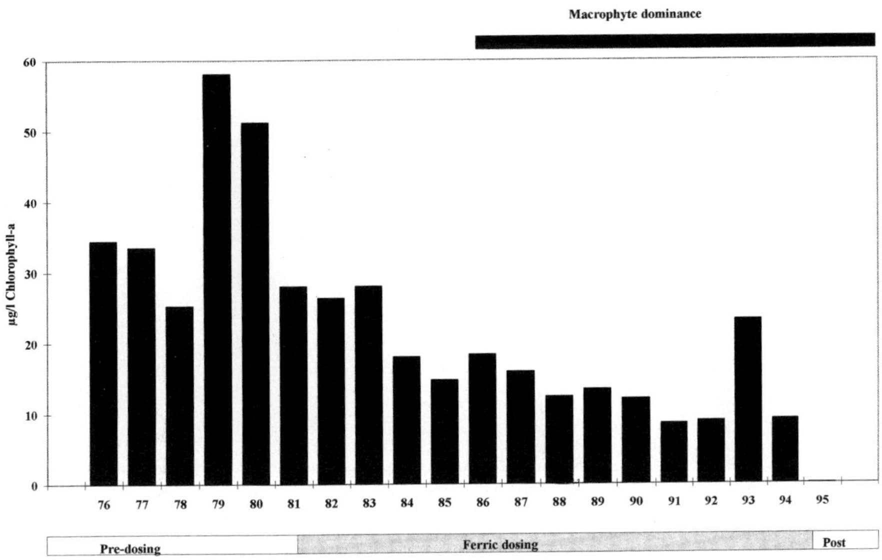

Figure 3. Changes in annual mean chlorophyll-a.

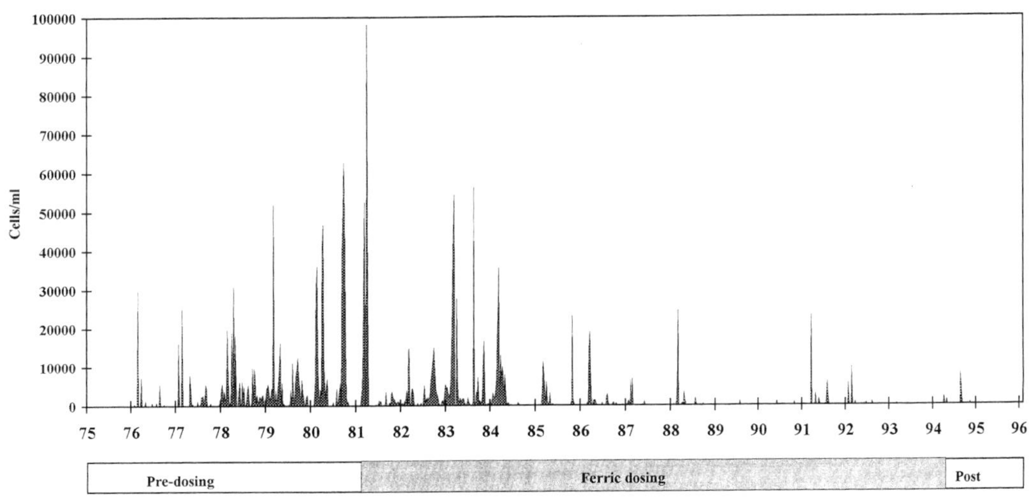

Figure 4. Changes in cell counts of Chlorophyceae.

The changes in annual mean chlorophyll a concentration are shown in Figure 3. Concentrations show a clear downward trend after ferric dosing was commenced with the exception of a large peak in 1993 which was caused by a bloom of *Aphanizomenon*. Counts of all of the dominant groups of algae also declined after dosing was commenced (Figures 4–6) although there was a delay period until 1985. *Microcystis* counts were high immediately after dosing was commenced, reaching counts in excess of 30 000 cell ml^{-1} during the summer of 1981–1984 with counts much lower in other years. Reductions of in the cell counts of Chlorophyceae and Bacillariophyceae were particularly pronounced.

The relative abundance of macrophytes (based on areas of the reservoir covered) are shown in Figure 7. Prior to the commencement of the surveys macrophyte populations were sparse. Beds of filamentous algae covered large areas of the reservoir in 1985 (Figure 8), but these declined in subsequent years. The abund-

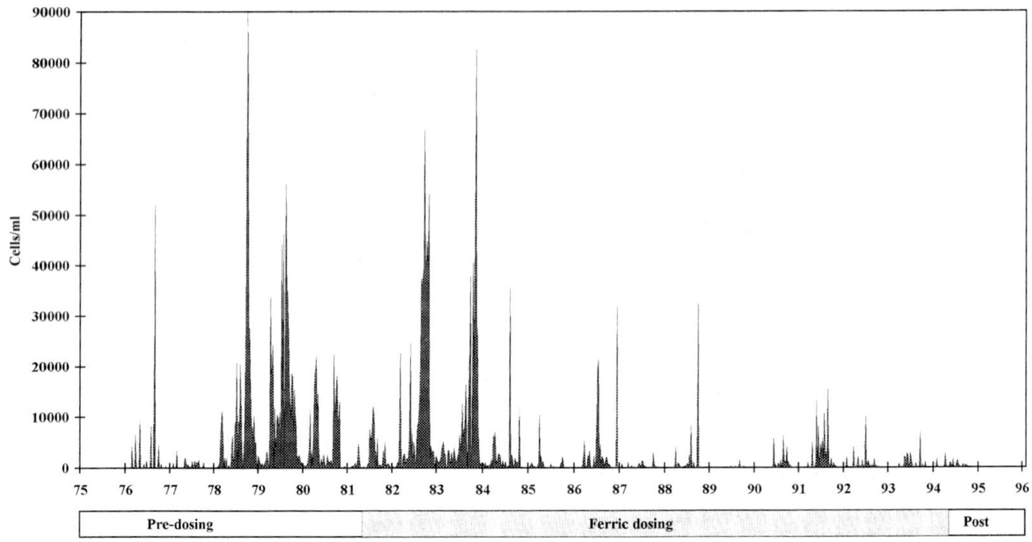

Figure 5. Changes in cell counts of Bacillariophycaea.

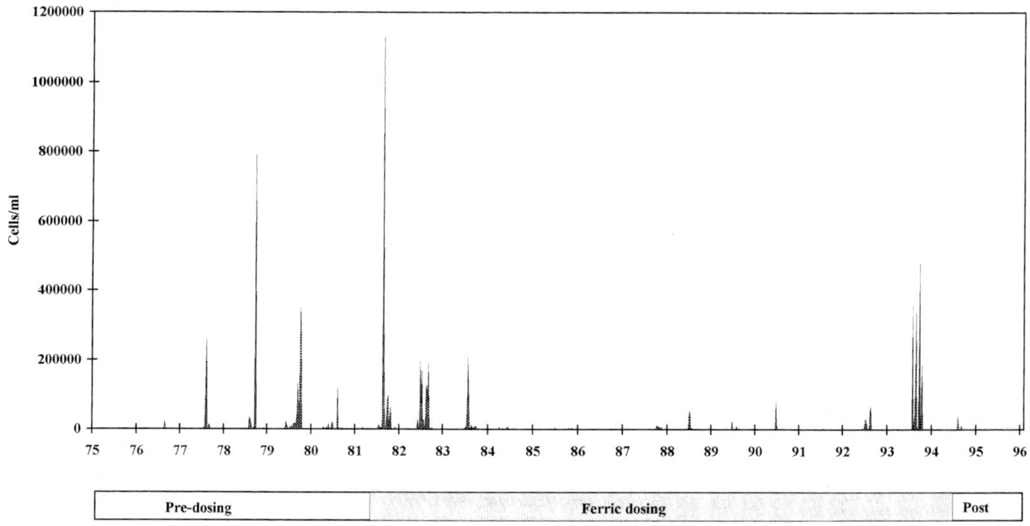

Figure 6. Changes in cell counts of Cyanobacteria.

ance of *Chara* increased until 1991, then declined. *Elodea* became dominant in the 1990s. *Potamogeton pectinatus* abundance was less variable.

Nitrate concentrations in the reservoir and river are shown in Figure 9. The level of denitrification in the reservoir was calculated as the difference between the mass input and the mass output to treatment over a period of one year. Although the populations of algae in the reservoir declined, rates of denitrification showed no clear trend over the period. Silica concentrations declined to below detection (Figure 10), but the number of minima each year decreased. Before phosphorus control commenced, there were two well defined declines in silica concentration each year, but from 1983 onward there was, generally, only one decline.

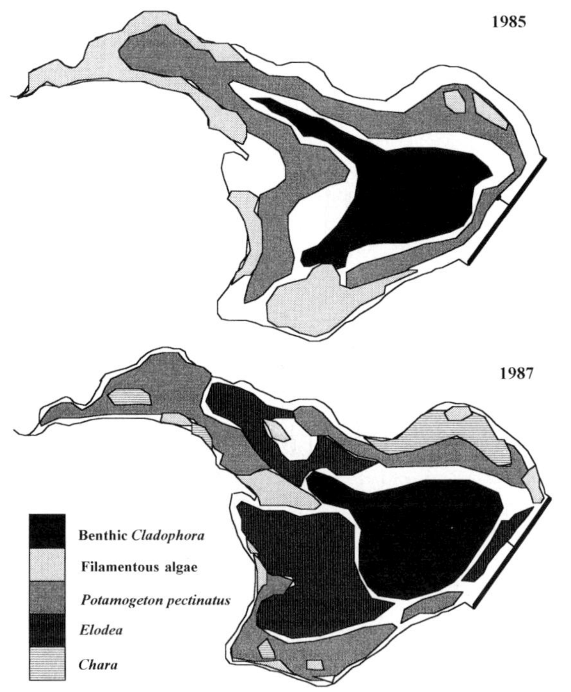

Figure 7. Comparison of macrophyte abundance in 1985 and 1988 (maps recorded in the final week in July in both years).

Wildfowl

The mean monthly wildfowl counts for Foxcote reservoir are shown in Figures 11–12. Species are divided into those commonly associated with water plants as a feeding habitat (pochard, mute swan, coot, gadwall) and fish feeding species (great crested grebe). Cormorants are not included in the latter data, as counts are only available since 1986/87. The numbers of herbivorous species increased over the period. In contrast, numbers of the piscivorous great crested grebe declined.

Discussion

Trophic structure

The change in the ecology of the reservoir was not immediate but took a number of years following the commencement of ferric sulphate dosing. Algal populations remained high for two years and a long-term shift in the macrophyte community occurred. A likely cause of this delayed response is availability of phosphorus from the sediment, which may have sustained high algal biomass (Sas, 1991). As iron spread over the reservoir, this would have increased the phosphorus-binding capacity and eventually reduced the release of phosphorus (Bostrum et al., 1982). Inactivation of sediment phosphorus release by iron may have been a critical process in facilitating the switch toward macrophyte dominance as many shallow lakes have displayed far longer delays in restoration once external loads of phosphorus have

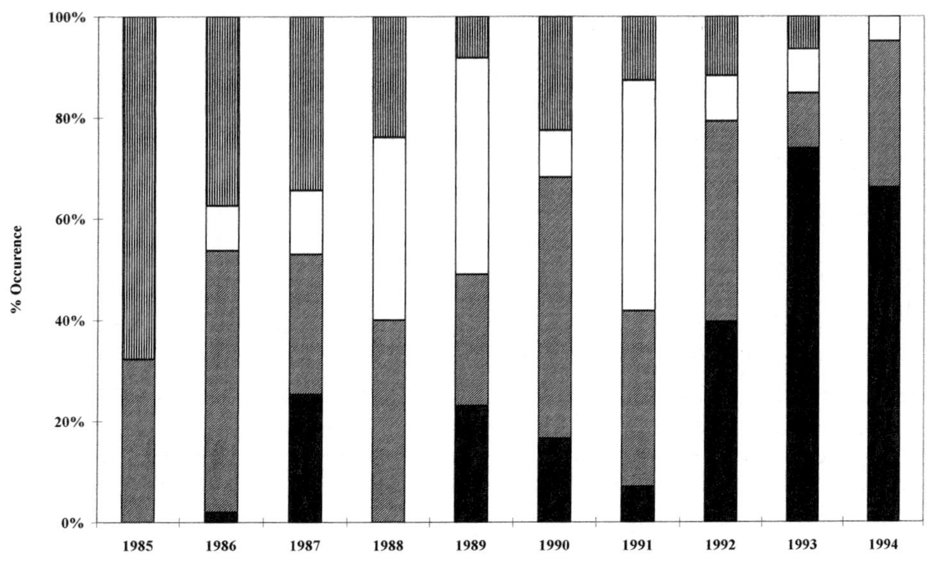

Figure 8. Comparison of the relative abundance of macrophytes (black *Elodea,* white *Chara,* diagonal, stripes *Potamogeton pectinatus,* vertical stripes filamentous algae).

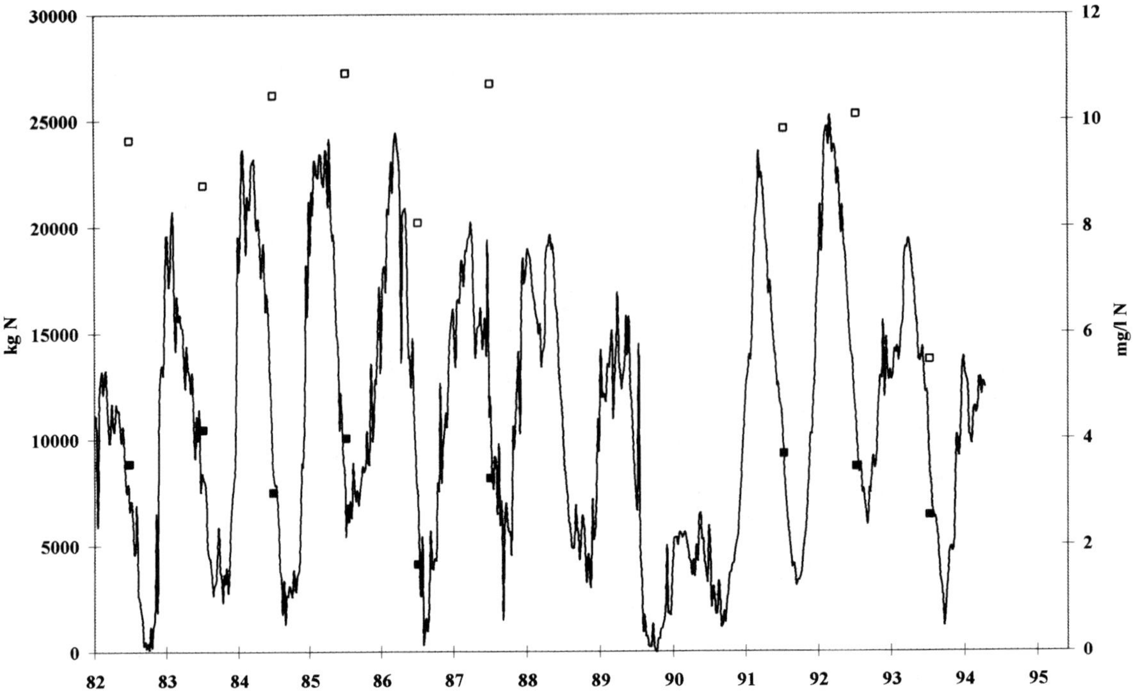

Figure 9. Changes in reservoir oxidised nitrogen, input of nitrogen (open squares) and denitrification (closed squares).

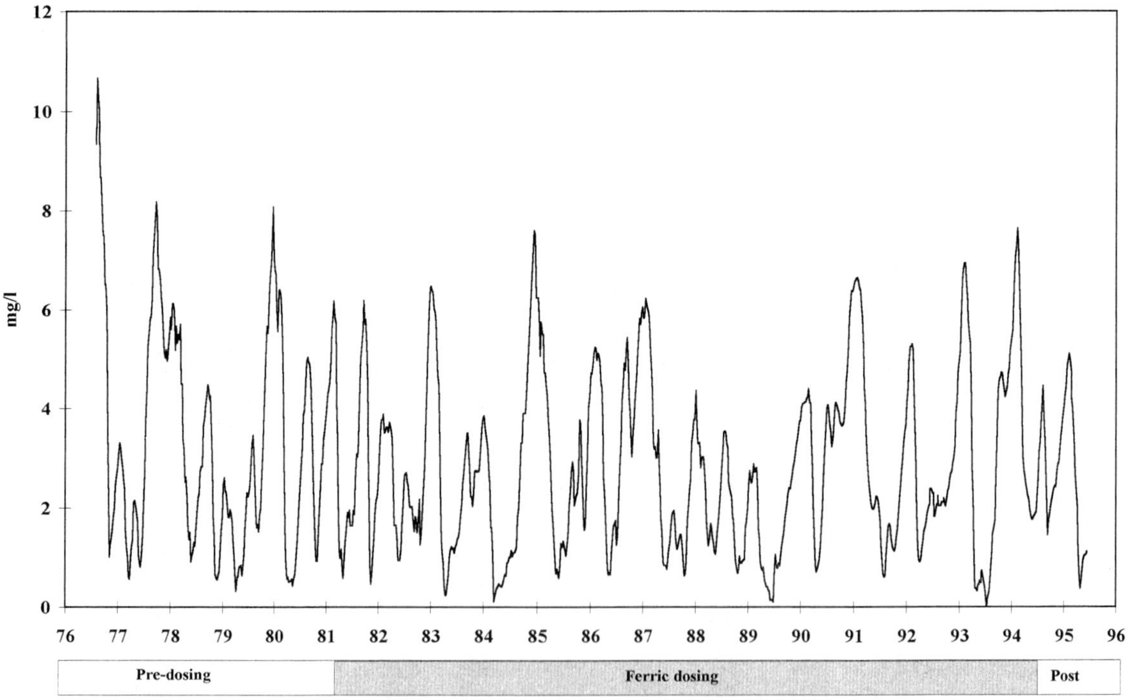

Figure 10. Changes in silica concentration.

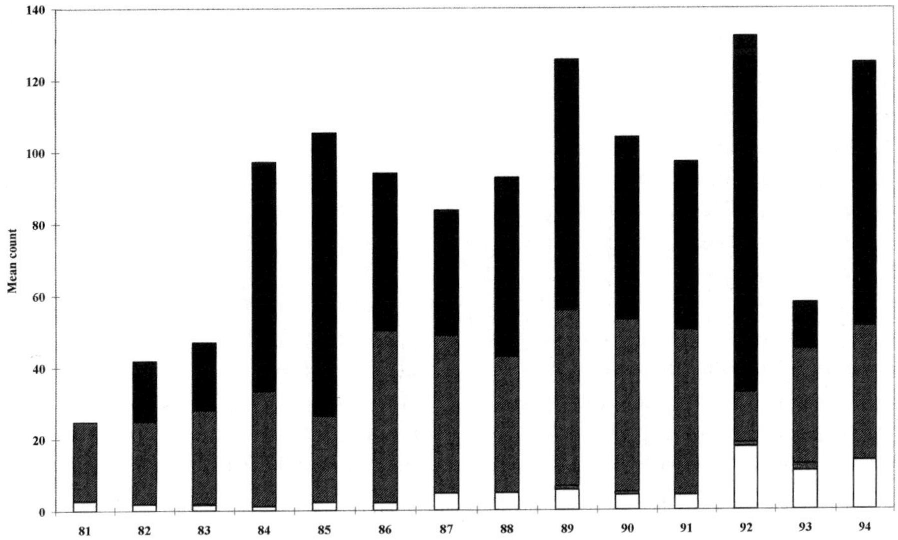

Figure 11. Changes in the mean monthly count herbivorous wildfowl (white – mute swans, horizontal stripes – gadwall, diagonal stripes – pochard, black – coot).

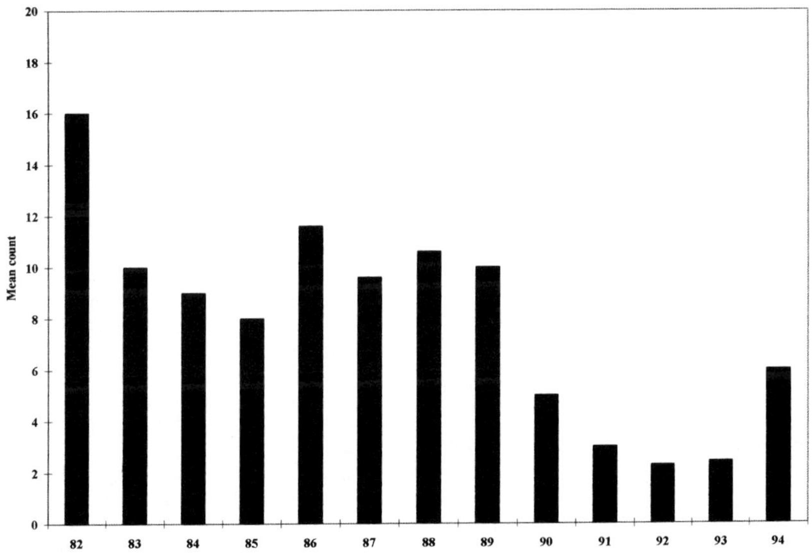

Figure 12. Changes in mean monthly counts of great crested grebe.

been reduced (Scheffer, 1990; Moss, 1990). In many shallow lakes, continued organic loading to sediments resulting from algal blooms is important in causing a release of phosphorus, thereby maintaining a high trophic state. Continuous dosing of iron salts is more likely to be effective than a one-off injection of the sediment (Foy, 1985) which has been tried elsewhere, as most of the phosphorus is potentially bound as it enters the system.

Immediately after ferric dosing commenced, the increase in benthic filamentous algae *Microcystis* observed could have been the result of access to benthic phosphorus released from the sediment. These species have rapid rates of colonisation, compared to higher plants, and so had a competitive advantage before vascular macrophytes colonised the reservoir. Large growths of benthic *Microcystis* have also been observed in Ravensthorpe, another shallow reservoir operated by Anglian Water and similar patterns of extensive growth of benthic filamentous algae have been observed in the Norfolk Broads, following restoration measures. The subsequent increase in rooted submersed macrophytes is a direct result of increased water clarity.

Another key element in the restoration of shallow lakes is the habitat aquatic plant communities provide for zooplankton as a refuge from fish predation. This, in turn, can increase the grazing pressure on algae maintaining clear water conditions. Unfortunately the zooplankton monitoring was not adequate to determine whether these mechanisms were important although a change in the *Daphnia* community was observed with *Daphnia pulex* not recorded before 1986, but present each year after. These processes provide an ecological buffer against the reversion to a phytoplankton dominated state (Gulati, 1990). The fact that no increase of phytoplankton was observed during the period 1988–92, when phosphorus levels in the reservoir showed a substantial increase, provides evidence that some buffering process was in operation. However, in 1993, the stability of the system clearly 'broke down' when a large bloom of *Anabaena* developed in the reservoir. The reasons for this are unclear but a decline in macrophytes was observed unusually early that year.

The wildfowl populations are considered to be of great importance in terms of the conservation value of reservoirs in the U.K., and Foxcote was designated as an SSSI (Site of Special Scientific Interest) as a result of its wildfowl populations. This is also the case for a number of Anglian Water's other reservoirs. The observed changes in wildfowl follow macrophyte dominance. This trend is opposite to the regional pattern for Anglian Water's other reservoirs and the Nene Washes for which there was a downward trend. In contrast, great crested grebes, more completely associated with open water, showed a decline in numbers, although they increased in the region over the same period. In many cases, long data sets for wildfowl are available for waterbodies where limnological data are absent. Comparison of wildfowl trends in relation to regional patterns may therefore be a useful tool in providing evidence for gross long term changes in the ecology.

As a shallow water body with very large inputs of nutrients, Foxcote reservoir provided an ideal environment to support large growths of algae. This caused severe treatment problems in the 1970s, primarily due to elevated iron residuals resulting from interference with coagulation and clogging of the rapid sand filters at the treatment works. As a result of these problems it was necessary to revert to direct river abstraction. This removes the advantages of storage in terms of security of supply, pollution protection and denitrification.

Treatment problems with algae were greatly reduced, following the shift in the reservoir ecology. However, taste and odour problems developed as a result of the growth of filamentous algae and *Chara* in the late 1980s. The latter is well known for producing strong tastes and odours. Reversal of eutrophication can, therefore, result in treatment problems although, usually, not as severe as those caused by large algal blooms.

References

Bostrum B., M., Janssoon & C. Forsberg, 1982. Phosphorus release from lake sediments. Arch. Hydrobiol. 18: 5–59.

Foy, R. H., 1985. Phosphorus inactivation in a eutrophic lake by the direct addition of ferric alumninium sulphate: changes in phytoplankton populations. Freshwat. Biol. 15: 613–629.

Gulati, R. D., 1990. Structural and grazing responses of zooplankton community to biomanipulation of some Dutch water bodies. Hydrobiologia 200/201 (Dev. Hydrobiol. 61): 99–118.

Hayes, C. R., R. G. Clark, R. F. Stent & C. J. Redshaw, 1984. The control of algae by chemical treatment in a eutrophic water supply reservoir. Journal of The Institution of Water Engineers and Scientists 38: 149–162.

Hosper S. H. & E. Jagman, 1990. Biomanipulation additional to nutrient control for restoration of shallow lakes in The Netherlands. Hydrobiologia 200/201 (Dev. Hydrobiol. 61): 523–534.

Hutson R. A., B. S. C. Leadbetter & R. W. Sedgewick, 1987. Algal interference with water treatment processes. Progress in Phycological Research 5: 266–299.

McKinnon, S. L & S. F. Mitchell, 1994. Eutrophication and black swan *Cygnus atratus* Lathum) populations: tests of two simple relationships. Hydrobiologia 279/280 (Dev. Hydrobiol. 96): 163–170.

Moss, B. 1990. Engineering and biological approaches to the restoration from eutrophication of shallow lakes in which aquatic plant communities are important components. Hydrobiologia 200/201 (Dev. Hydrobiol. 61): 367–377.

OECD, 1982. Eutrophication of Water. Monitoring, Assessment and Control. OECD, Paris.

Ozimek T., R. D. Gulati & E. van Donk, 1990. Can macrophytes be usetul in biomanipulation of lakes? The lake Zwemlust example. Hydrobiologia 200/201 (Dev. Hydrobiol. 61): 399–407.

Suter, W. 1994. Overwintering waterfowl on Swiss lakes: how are abundance and species richness influenced by trophic status and lake morphology? Hydrobiologia 279/280 (Dev. Hydrobiol. 96): 1–14.

Sas, H., 1989. Lake restoration by reduction of nutrient loading. Academia Verlag, St Augustin, Germany.

Sondergaard M., E. Jeppesen, E. Mortensen, E. Dali, P. Kristensen & O. Sortkjaer, 1990. Phytoplankton biomass reduction after planktivorous fish reduction in a shallow, eutrophic lake: a combined effect of reduced internal P-loading and increased zooplankton grazing. Hydrobiologia 200/201 (Dev. Hydrobiol. 61): 229–240.

Young, S. N., W. T. Clough, A. J. Thomas & R. Siddall, 1991. Changes in plant community at Foxcote Reservoir following use of ferric sulphate to control nutrient levels. Journal of Institute of Water and Environmental Management 2: 5–12.

Hydrobiologia **395/396:** 335–353, 1999.
D.M. Harper, B. Brierley, A.J.D. Ferguson & G. Phillips (eds), The Ecological Bases for Lake and Reservoir Management.

Ecological principles for management techniques in deeper reservoirs

Bill Brierley[1] & David Harper[2]
[1]*Environment Agency – Anglian Region, Waterside House, Waterside North, Lincoln, LN2 5HA, U.K.*
[2]*Biology Department, University of Leicester, Leicester LE1 7RH, U.K.*

Key words: lowland reservoirs, artificial mixing, ferric dosing, biomanipulation

Abstract

Eutrophic reservoirs produce many management problems for both water supply and recreational management, which are only expensively solved. A range of eco-technical methods have been used over the past 10 years to control phosphorus and nuisance algal bloom levels. Three case studies are used to show that these methods may be succesful but, if used without careful monoitoring or controls, or if not applied in combination with each other or with other techniques, their practical value and benefit-cost ratio may not be apparent.

Introduction

The need to control excessive growths of phytoplankton algae in reservoirs has been well documented (Lund, 1970 & 1981; Steel, 1972; Youngman, 1975; Collingwood, 1977). Algal interference with water treatment includes taste and odour problems (Hayes & Greene, 1984), reduced filtering capacity and filter runs (Youngman, 1975) (Figure 1), disturbance of flocculation process (Collingwood, 1977; Hayes & Greene, 1984), breakthrough of algal cells into water distribution systems providing a carbon source for animals (Smart, 1989; Smart & Harper, 1999) and presence of cyanobacteria toxins (Zalewski, 1999). These problems all contribute to greater water treatment costs although it is very difficult to estimate what this cost must be for individual water treatment works (Harper, 1992). In lakes and reservoirs with a high amenity or conservation value, algal control is also required to reduce the interference with recreational activities and improve the aesthetic nature of the site. At Rutland Water (U.K.), a very severe bloom of the cyanobacteria, *Microsystis aeruginosa* led to the closure of the whole lake to recreational activities in the late summer of 1989.

Ecological principles for the control of phytoplankton and lake restoration

There are many methods used for controlling phytoplankton growth and eutrophication in reservoirs and deep lakes. Sas (1989), Klapper (1991), Harper (1992) and Cooke et al. (1993) review many of these techniques and report on the degree of success. Three case studies are presented here from different reservoirs in central England, which demonstrate different ecological principles, which can be used for control. The techniques are:

1. Artificial mixing, with the aim not only of preventing stratrification, but also of reducing the underwater light available to algal cells, and subsequent disruption to potential maximum algal growth rates, was carried out at Staunton Harold reservoir in Derbyshire, England.

2. Nutrient reduction by direct ferric sulphate dosing into the reservoir to inactivate the phosphorus in the incoming river water, was carried out between 1990 and 1995 at Rutland Water in Leicestershire, England.

3. Biological control has also been investigated at Covenham reservoir, Lincolnshire, England which is designed on the Thames reservoir design (a simple flow through system consisting of a concrete bowl with an inflow and outflow). Both nutrient reduction by ferric dosing (1990–1993) and artificial mixing have also been carried out at Covenham reservoir during the study period.

Artificial destratification to control phytoplankton in Staunton Harold reservoir

Artificial destratification had been successful for improving water quality by preventing deoxygenation since the early 1950s (Hooper et al., 1953) and many new reservoirs are designed with artificial mixing systems (Steel, 1972; Tolland, 1977). The effects of artificial mixing upon phytoplankton have often been contradictory. The success or failure of artificial mixing in altering phytoplankton biomass and succession would appear to depend upon the timing and intensity of mixing energy which Ridley (1971) noted would be different for each lake. Shapiro (1981) suggested that when increases in phytoplankton occurred after mixing, it was probably due to inadequate circulation with the phytoplankton not being thoroughly mixed throughout the water column, but the nutrient concentrations increased in the euphotic zone. In contrast, when a decrease in phytoplankton biomass was recorded, it was due to complete circulation with the algal cells spending less time in the photic zone and thus, their production was reduced. Low algal biomasses were routinely found in the London reservoirs even though the water in the reservoirs had significant concentrations of soluble reactive phosphorus and nitrate (>1mg l^{-1} PO_4-P and 7 mg l^{-1} NO_3-N) (Ridley et al., 1966). The phytoplankton biomass were maintained well below the nutrient carrying capacity (20–60 μg l^{-1} chlorophyll *a*; Duncan, 1990) due to the use of powerful jetted inlet water (Cooley & Harris, 1954; Steel, 1972).

Staunton Harold reservoir forms part of the water supply system for the City of Leicester in the East Midlands. Water was pumped from the River Dove in Derbyshire and used to fill two valley reservoirs – Staunton Harold and Foremark reservoirs. The raw water from the reservoirs was then treated at Melbourne Treatment works before being pumped to Leicester, as part of the potable water supply system (Figure 2). Both reservoirs were considered to be eutrophic and supported summer blooms of cyanobacteria often in excess of 50 μg l^{-1} (Brierley, 1985). The major characteristics of the reservoirs are outlined in Table 1.

Artificial mixing was carried out using a inexpensive and efficient perforated-pipe system (Tolland, 1977; Davis, 1980) in Staunton Harold reservoir. The nearby Foremark reservoir was used as a 'control' – the reservoir had been shown to behave similarly with respect to stratification, water chemistry and phytoplankton activity (Severn Trent Water Authority, unpublished). A 40 mm diameter polythene pipe, perforated for the last 150 m, ran out into the deepest part of the reservoir at right angles to the dam. A compressor was connected to the pipe below the dam. The compressed air left the pipe *via* the small holes and rose to the surface as a curtain of bubbles, water being entrained from the lower levels to the surface. Local destratification around the pipe was then followed by mixing of the whole reservoir. The estimated air flow from the compressor was between 27 and 28.8 l s^{-1} during the study period.

Table 1. The physical characteristics of Staunton Harold and Foremark reservoirs

	Staunton Harold	Foremark
Construction completed	1964	1976
Volume (10^6 m^3)	6.36	13.4
Area (hectares)	85	92
Maximum depth (m)	23	32
Mean depth (m)	7.9	14.3
Catchment area	2590	295
Nominal retention time (days) 1981–1983	108	148

The perforated pipe system was able to fully destratify the reservoir in three days during September 1981, and when used continuously prevented Staunton Harold from stratifying in 1982 and 1983. A maximum temperature difference between the surface and bottom of 3.3 °C occurred in Staunton Harold on 7 June 1982 compared to a difference of 13.25 °C in Foremark reservoir (Brierley, 1985). The stability of the water columns, measured as the Brunt-Vasala frequency – N^2 and defined by Phillips (1966) and Boyce (1974) by the equation:

$$N^2 = -\frac{\mathbf{g}dk}{kdz}\ (\text{sec}^{-2}),$$

where $\mathbf{g}$ is the acceleration due to gravity; k is the density of water at 4 °C; dk/dz is the density gradient over a specified depth interval.

High values of N^2 indicate that density gradients are intense and as a result the resistance to mixing is high whilst a low N^2 value indicates weak or low density gradients and that the water column is unstable or well mixed. The N^2 values were calculated from weekly temperature profiles and the changes in N^2 (0–5 m) and (0–15 m) in Staunton Harold and Foremark reservoir are shown in Figure 3.

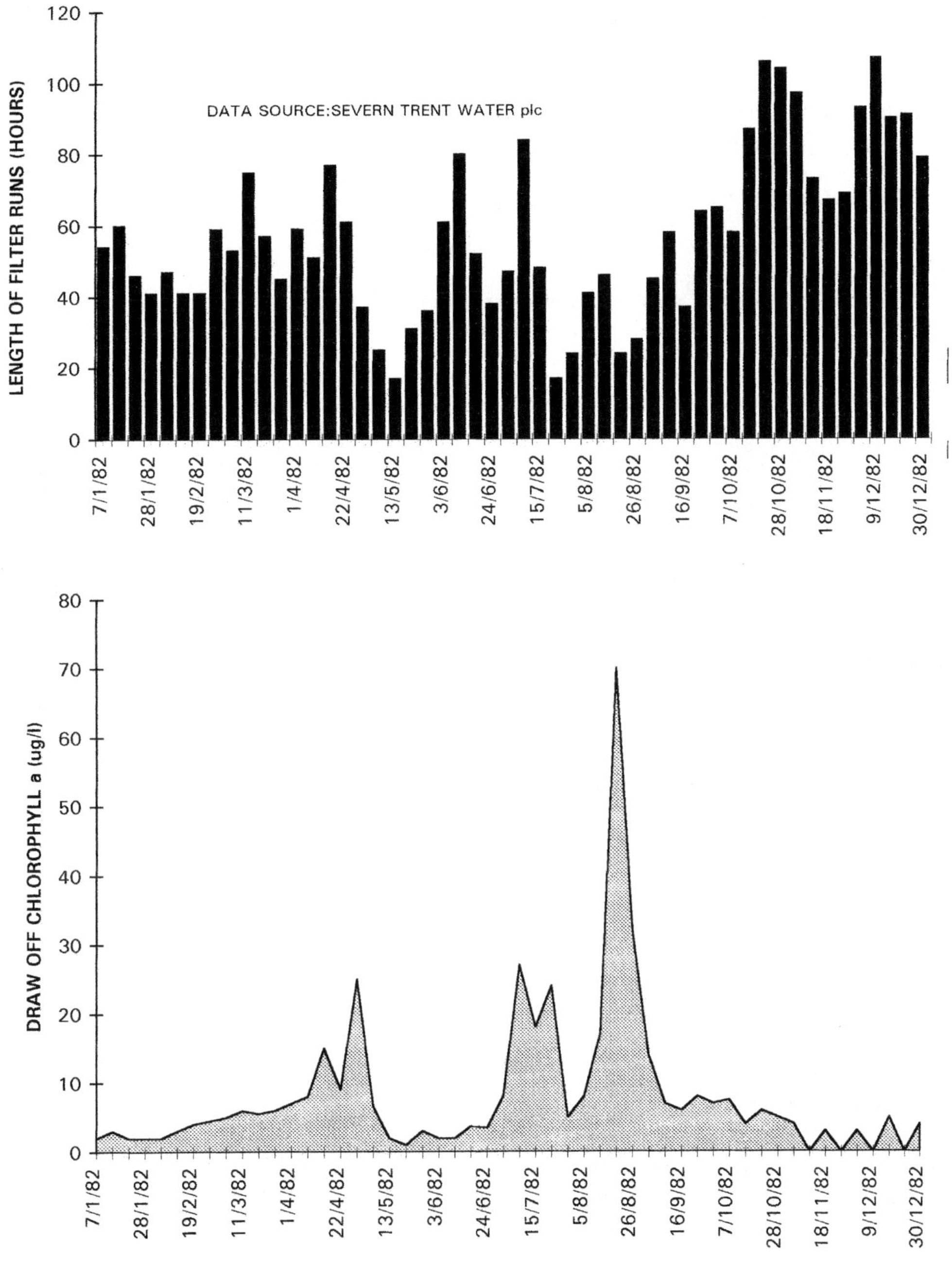

Figure 1. Filter run length at Melbourne Water Treatment works in relation to the development of chlorophyll in Staunton Harold Reservoir.

In 1981, oscillations in both the 0–5 m and 0–15 m N^2 values were similar for both reservoirs. The similarities and low N^2 values were a result of the poor climactic conditions which prevented Staunton Harold from stratifying until late in the season and depressed the metalimnetic layer in Foremark to between 14 and 18 m for most of the summer. The effects of continuous artificial mixing on water column stability in Staunton Harold were very noticeable in 1982 and 1983. N^2 (0–5 m) in Foremark showed large peaks in both years which were related to periods of intense heating of the epilimnion. Generally N^2 (0–5 m) values for Staunton Harold were much lower than those recorded for Foremark. The differences in the

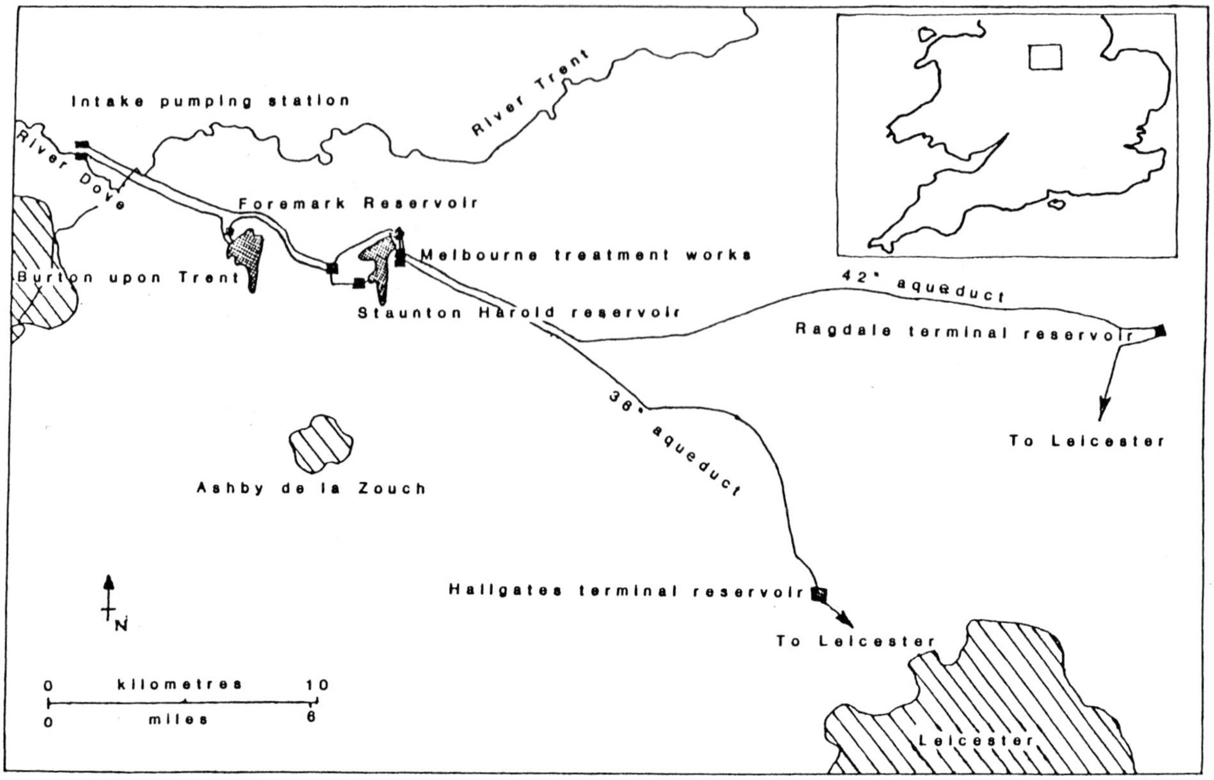

Figure 2. Location of Staunton Harold Reservoir.

water column stability between the reservoirs was further emphasised when using the N^2 (0–15 m) values with Foremark having higher values between May and September in both 1982 and 1983.

The mixing in 1982 and 1983 also prevented any build up of soluble reactive phosphorus and silica in the lower layers, whilst in Foremark typical clinograde distributions of both nutrients occurred for much of the summer period, as the hypolimnion became progressively more anaerobic.

Although the water column remained well mixed throughout most of 1982 and 1983 as a result of the continuous mixing, phytoplankton biomass was observed to stratify (Figure 4 shows chlorophyll *a* isopleths during 1982) when water column stability increased greatly or motile species were present, e.g. *Aphanizomenon flos-aquae* in August 1982. In Foremark, algal stratification occurred in all three summers and the biomass was concentrated in the top 5 m (Figure 5).

Analyses of the individual species' maximal rate of population increases (Table 2) showed that small centric diatoms, *Cryptomonas* spp. and *Rhodomonas* spp. had highest rates during artificially mixed years and lowest rates in stratified years. In contrast, *Aphanizomenon flos-aquae, Anabaena* spp and *Microcystis aeruginosa* all had rates of increases which were lowest in artificially mixed years and highest in the stratified years. The seasonal succession in both reservoirs was typical of eutrophic reservoirs with diatoms in spring, followed by colonial chlorococcales in early summer and cyanobacteria dominating in mid to late summer. The phytoplankton assemblages found in the reservoirs, were identical to those found by Reynolds (1984) in eutrophic reservoirs, and the periodicity of these assemblages appeared to be driven by nutrient availability and water column stability. The only major disturbance from the sequence of assemblages as a result of artificial mixing occurred in 1983 when an assemblage not recorded before and consisting of *Chlamydamonas* sp. and a mixture of small chlorophytes. This assemblage became dominant in July and September and it was proposed that it had resulted from the increased nutrient availability in the euphotic zone as an indirect effect of the mixing.

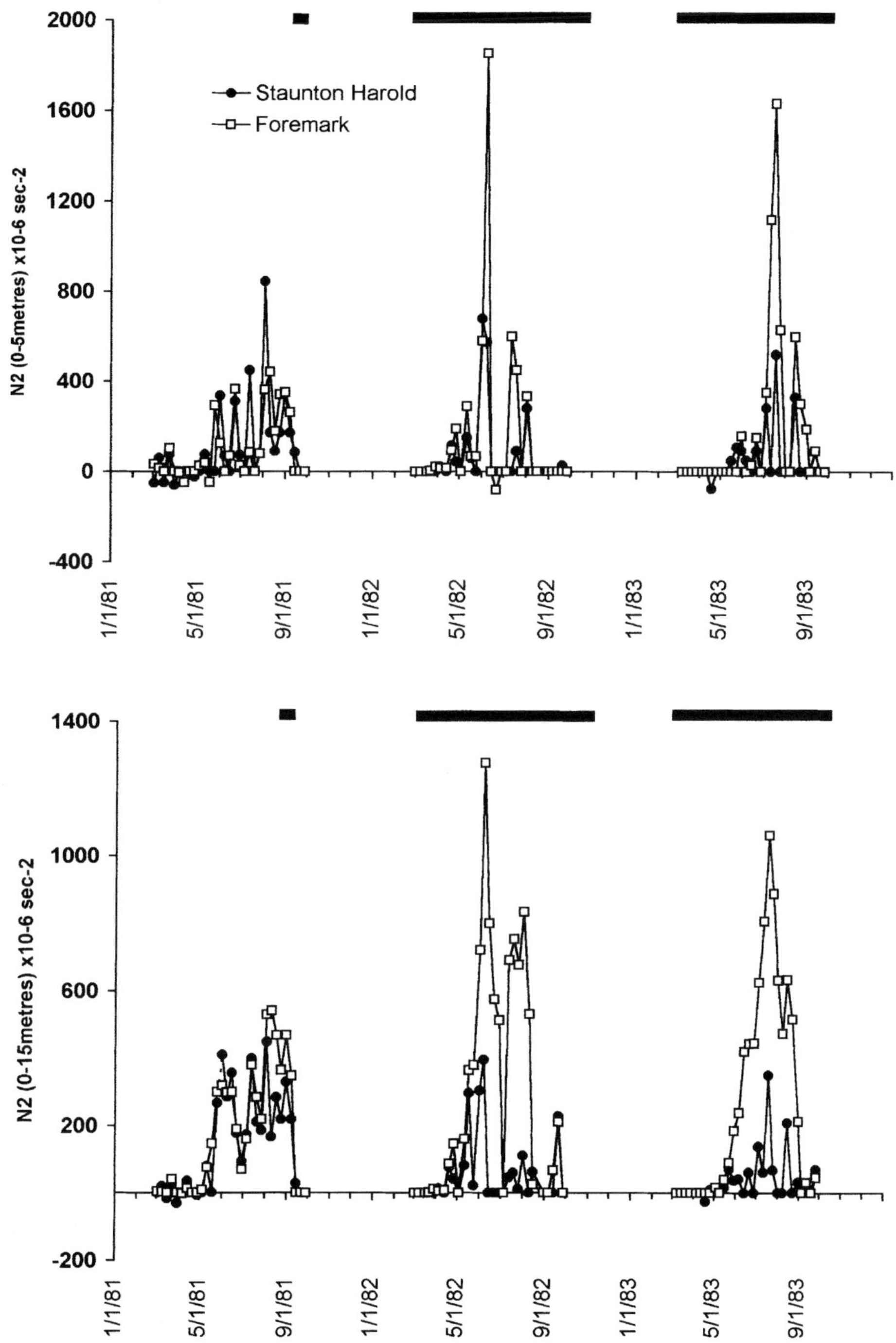

Figure 3. Temporal change in water column stability at 0–5 m and 0–15 m in Staunton Harold Reservoir (closed circles) and Foremark Reservoir (open circles). Solid bars indicate periods of artificial mixing in Staunton Harold.

Mean algal biomass (as chlorophyll *a*) was lower in one year of artificial mixing (1982) and higher in another (1983) when compared to the mean biomass in the naturally mixed year and those in Foremark reservoir which stratified (Table 3), probably due to

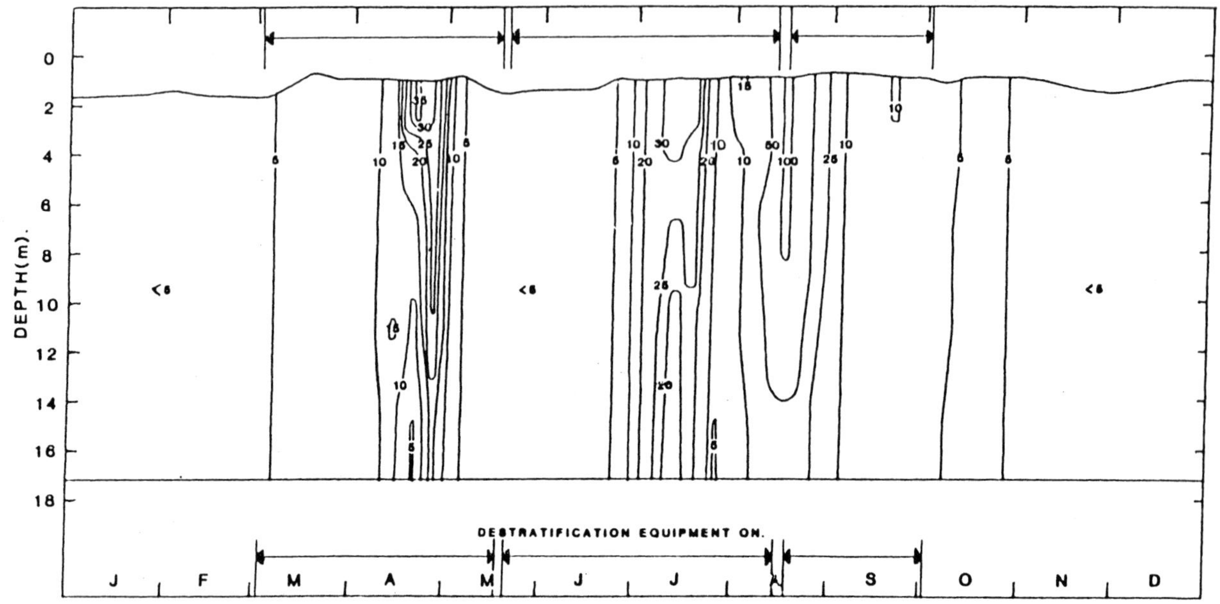

Figure 4. Variations in the vertical distribution of chlorophyll in Staunton Harold. Isopleths in $\mu g\ l^{-1}$.

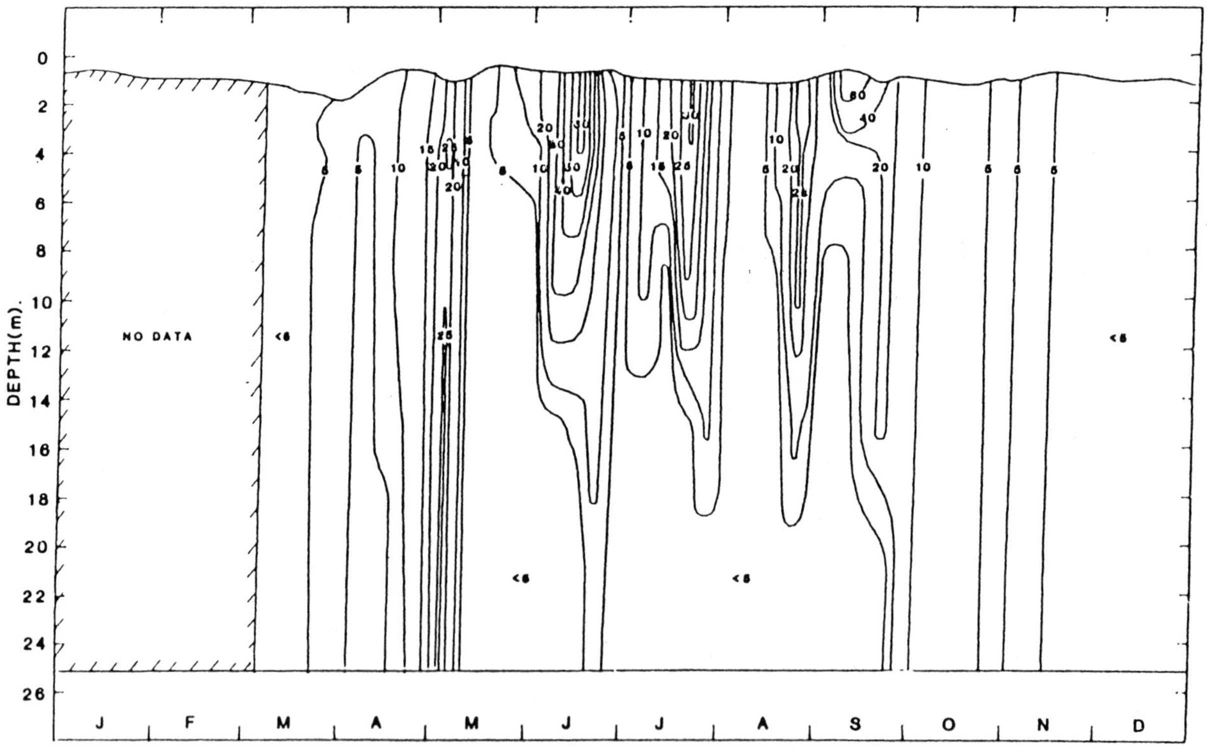

Figure 5. Variations in the vertical distribution of chlorophyll in Foremark. Isopleths in $\mu g\ l^{-1}$.

inadequate mixing and the ability of the phytoplankton to stratify.

The productivity and photo-adaptation of the mixed and stratified phytoplankton populations was

Table 2. The maximum observed summer rates of population increase (kn+ d^{-1}) of selected species of phytoplankton from Staunton Harold and Foremark reservoirs between 1981 and 1983

Reservoir	Artificial mixing		Natural mixing			
	Staunton Harold	Staunton Harold	Staunton Harold	Foremark	Foremark	Foremark
Year	1982	1983	1981	1981	1982	1983
Small centric diatoms	0.38	0.34	0.31	0.3	0.29	0.20
Asterionella	0.27	0.08	-	0.45	0.15	0.15
Cryptomonas	0.35	0.16	0.20	0.06	0.09	0.16
Rhodomonas	0.29	0.42	0.29	0.34	0.28	0.15
Aphanizomenon	0.19	0.13	0.20	0.18	0.31	0.29
Anabaena	0.14	0.18	0.26	0.11	0.23	0.14
Microcystis	0.16	0.19	0.16	0.16	0.17	0.26

Table 3. Annual mean and maximum algal biomass (as chlorophyll *a*) under mixed and stratified conditions in the surface layers of Staunton Harold and Foremark reservoirs

		Chlorophyll *a* (μg l^{-1}) Staunton Harold	Foremark
1981	Maximum	114.0*	213.0
	Mean	17.9*	18.4
1982	Maximum	**69.0**	71.0
	Mean	**10.9**	14.7
1983	Maximum	**43.0**	73.0
	Mean	**11.4**	13.7

Figures in bold indicate artificially mixed years, * 1981 was a naturally (wind mixed) year.

measured in both 1982 and 1983 in both reservoirs by light/dark bottles with titration of changes in oxygen concentration. Depth integrated values for photosynthesis (ΣA) which include variation in underwater light with depth, were significantly lower from mixed populations (Table 4). Much of the variance of integral photosynthesis was reflected by the variation in phytoplankton biomass on the different occasions. This was normalised by integrating the depth profiles of photosynthesis per unit biomass to demonstrate that the integral of the specific rate of photosynthesis per unit chlorophyll (ΣP) was significantly lower for mixed populations also (Table 4).

Productivity experiments showed that phytoplankton cells from different depths in the stratified populations appeared to be light/shade adapted, whilst those from mixed populations were tolerant to a wide range of irradiances (Brierley, 1985). It would appear that this tolerance was a result of a selection of species which have either evolved to favour turbulent conditions, or to control carbon assimilation through excretion of photosynthates and photorespiration, or regulation of their own light climate by vertical movements. These populations also had lower assimilation rates compared to the stratified populations.

Changes in phytoplankton periodicity as a result of mixing have been widely described, although changes in productivity have been rarely investigated. Haynes (1971) found a large decrease in *Aphanizomenon flos-aquae* during mixing was due to the redistribution of filaments throughout the water column, rather than a decline in the population. Reynolds et al. (1983, 1984) imposed an artificially imposed cycle of mixing and restratification on the phytoplankton in large limnetic enclosures (Lund tubes) in Blelham Tarn and that the phytoplankton responded in a predictable way with species of winter/spring phytoplankton (eg. diatoms) were selected for and became dominant during mixing episodes, whilst summer assemblages were favoured by and increased during restratification periods. The growth and dominance of slow growing, late summer specie such as *Microcystis aeruginosa* and *Ceratium hirudinella* were delayed. Reynolds (1980) also predicted that the average biomass would be maintained

Table 4. The mean photosynthetic characteristics (+/- s.e.) of mixed and stratified populations of phytoplankton in Staunton Harold and Foremark reservoirs

Photosynthetic characteristic	Mixed populations	Stratified populations
Integral photosynthesis, $\Sigma A(mgO_2\ m^{-2}\ h^{-1})$	296.5 ($\pm$ 48.8)	596.2 ($\pm$ 72.4)
Integral of specific photosynthesis, $\Sigma P\ (mgO_2\ mg\ chl^{-1}\ m^{-2}\ h^{-1})$	17.52 ($\pm$ 2.09)	26.68 ($\pm$ 3.16)

below the nutrient carrying capacity if mixing was carried out. Ryback (1985) had also found that species more typical of winter/spring assemblages were favoured by mixing.

Steel (1972) concluded from work on the Thames reservoirs that the phytoplankton biomass was lower than expected due to the reduced underwater light that the phytoplankton received as a result of the powerful vertical mixing. The reduced gross photosynthetic rates and maintained respiration rates then led to a lowering of net photosynthesis (Talling, 1957, 1965, 1971) and thus phytoplankton biomass. These changes in rate processes clearly occurred in a full-scale mixing experiment.

Artifical mixing is widespread in lowland reservoirs, but is frequently used in a continuous mode – to maintain well-oxygenated, unstratified water. Its use could, however, not give unimportant control over algal succession and productivity in those reservoirs where water managers have the vision to integrate it with their ecological monitoring results.

Phosphorus inactivation by ferric sulphate dosing and nutrient reduction by changes in the pumping regimes at Rutland Water, 1990–95

The excessive blooms and scums of *Microcystis aeruginosa* at Rutland Water in 1989, with deaths of sheep and dogs (NRA 1990) and the subsequent closure of the reservoir to leisure activities, led to both national and international media attention. In June 1990, Anglian Water Services plc started to directly dose Rutland Water with ferric sulphate. This followed an experiment at Foxcote reservoir (see Daldorph, 1999) where large stands of macrophytes were recorded after ferric dosing successfully reduced the phytoplankton biomass (Young et al., 1991). Initially, dosing at Rutland Water was seen as a short term measure to reduce the soluble reactive phosphorus and potentially lower the peak and mean algal biomasses.

Table 5. The main physical characteristics of Rutland Water

Construction completed	1975
Volume ($m^3 \times 10^6$)	136.9
Area (hectares)	1255
Maximum depth (m)	34
Mean depth (m)	10.9
Nominal retention time (days) – 1993	730
Natural catchment area (hectares)	7400

Rutland Water was the largest man-made, potable, pump-storage reservoir in Europe when constructed in 1976 (Harper, 1982) (and is still the largest in the U.K.). Water is pumped from the eutrophic rivers, Welland and Nene, into the reservoir and enters the reservoir in the south arm via submerged jets. These rivers usually provide between 60 and 90% of the total water input. The main physical characteristics of the site are shown in Table 5 and the location of the reservoir in relation to the catchments is shown in Figure 6.

Water from the reservoir is treated at Wing water treatment works (or it may in emergencies be directly used for river regulation). The water is then put into the water supply system together with two other reservoirs, Grafham and Pitsford, which between them supply up to 1.5 million customers with up to 545 $l \times 10^6\ day^{-1}$ potable water in Northampton, Peterborough, Milton Keynes, Bedford, Kettering and Corby (AWS, unpublished). The reservoir has two multiple depth draw-off towers and the main body is mixed during the summer months (usually May–October) to prevent stratification using Helixor air guns (Harper, 1982).

Recreation and conservation interests were included in the development of the site from early on in the planning and building of the reservoir (Knights,

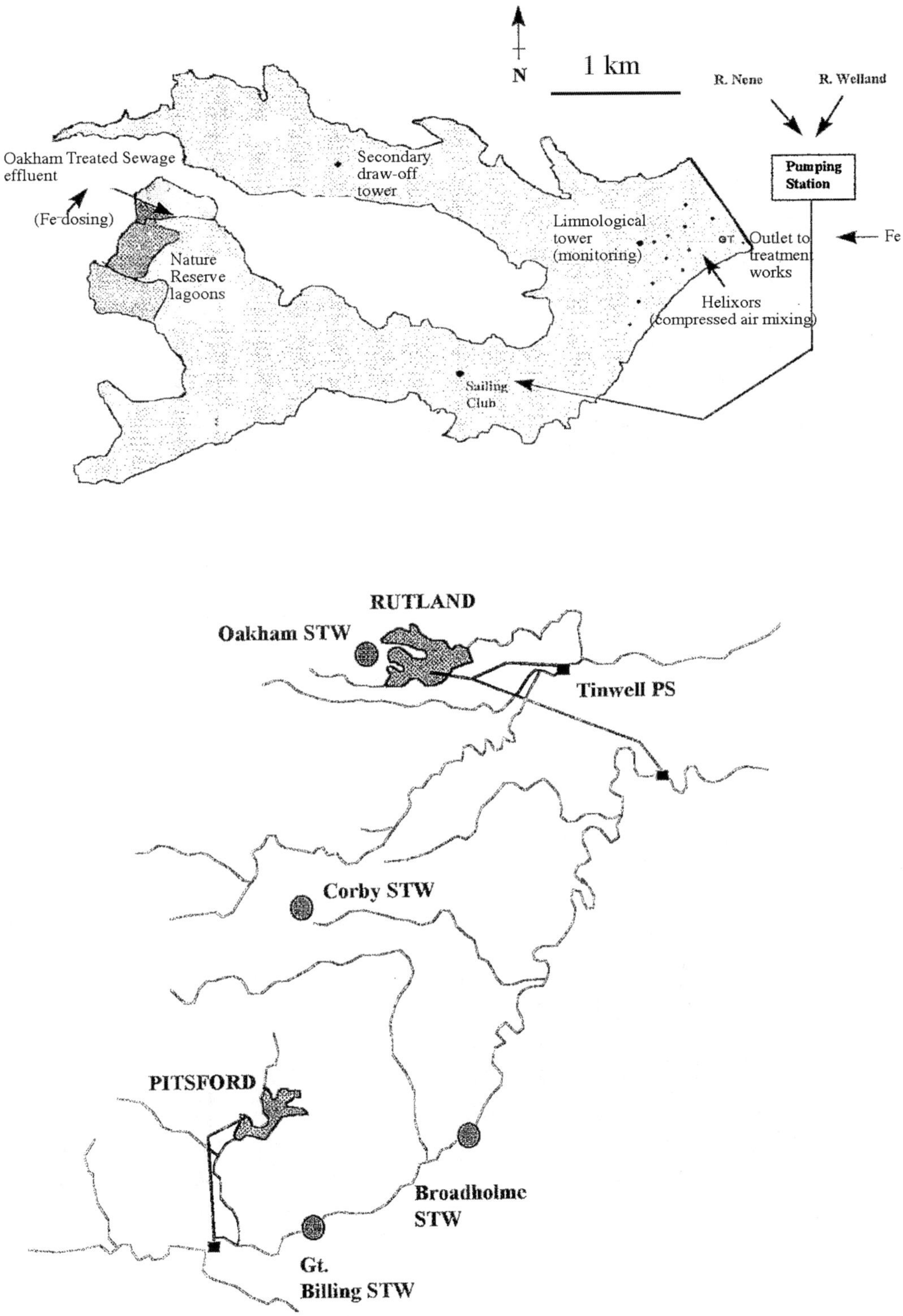

Figure 6. Rutland Water and its catchment area, showing major sewage works location (STW).

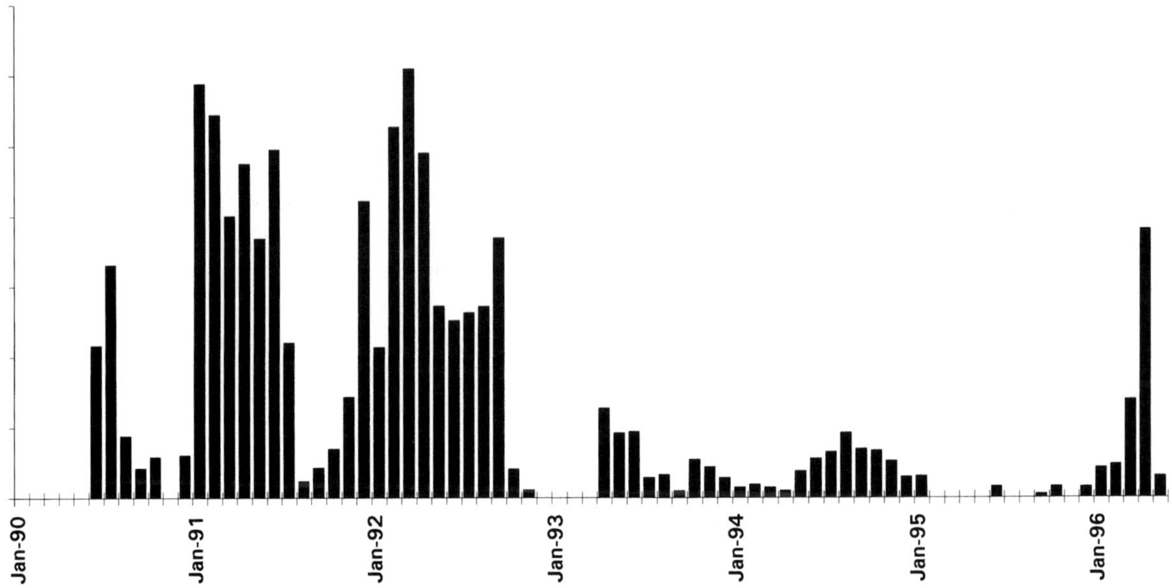

Figure 7. Quantity of ferric sulphate dosed into Rutland Water.

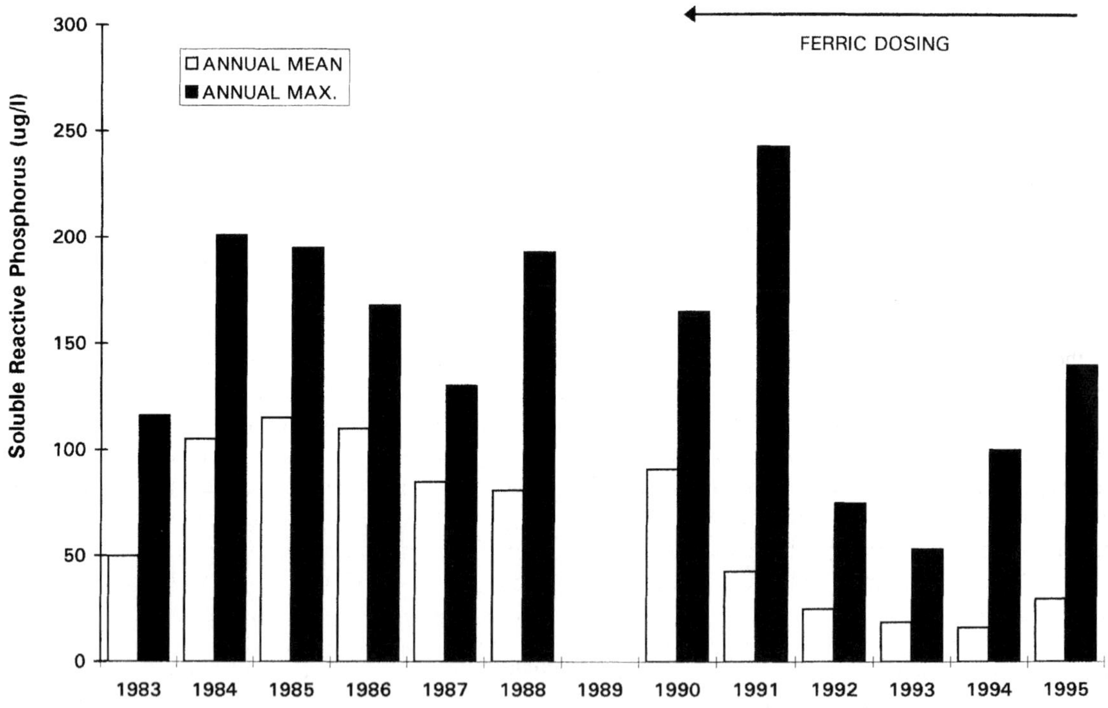

Figure 8. Annual mean and maximum soluble phosphorus in Rutland Water, 1983–85.

1982), with walking, sailing, surfboarding, cycling, picnic areas and conservation centres being some of the attractions. The site was developed as a 'put and take' trout fishery with an international reputation. The site is of national and international importance for resident or visiting birds, and has been designated a Site of Special Scientific Interest, a RAMSAR site and a Special Protection Area under the EU Birds Directive.

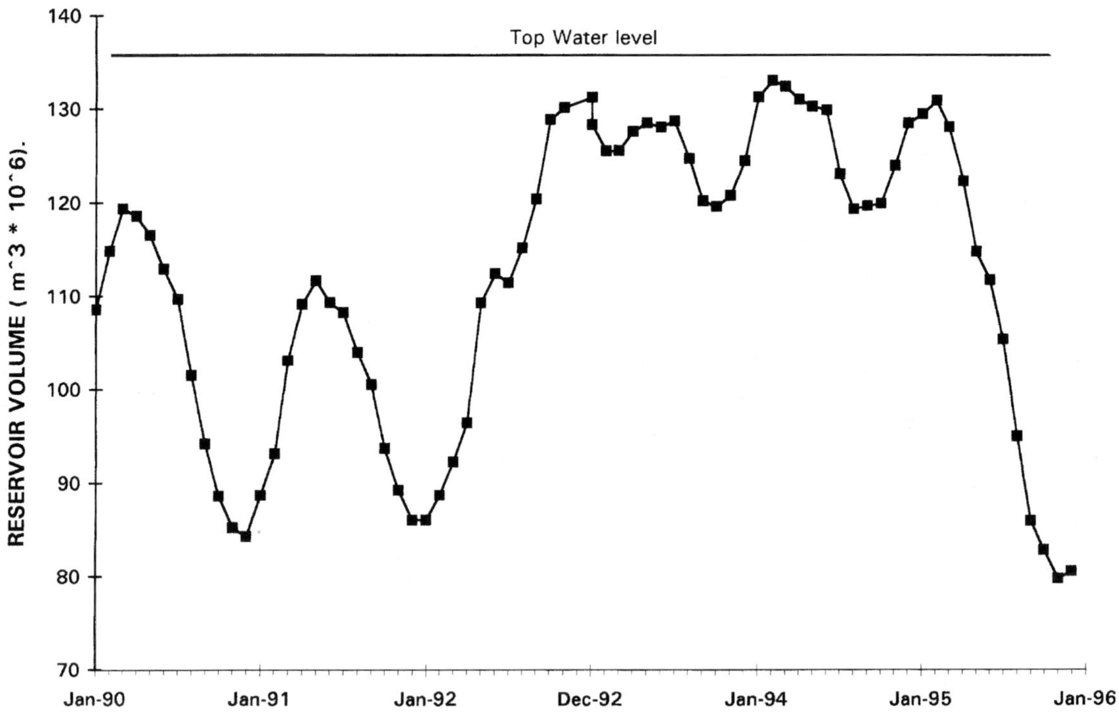

Figure 9. Seasonal changes in reservoir volume, Rutland Water, 1990–95.

Detailed limnological research and monitoring of the reservoir has been carried out since the reservoir started filling in 1975 (Harper & Bullock, 1982). It is eutrophic, with phytoplankton periodicity typical of eutrophic lakes (Ferguson & Harper, 1982). Phytoplankton biomass was initially maintained well below the nutrient carrying capacity, probably as a consequence of inadvertent discontinuous mixing (see above). The continuous operation of the 'Helixors' may be successful in reducing algal biomass as stable conditions persist during anticyclonic weather and full mixing would occur during windy weather. Thus a pattern of intermittent mixing with alternate mixed and stratified water columns results, as predicted by Reynolds et al. (1984).

The Environment Agency and its predecessor, the National Rivers Authority (NRA), have monitored weekly changes in the physical, chemical and biological (phytoplankton and zooplankton) nature at up to five sites at the reservoir since June 1990. Benthic and littoral invertebrates have been monitored, on a less frequent basis, since September 1990. Full results of the invertebrate monitoring will be reported elsewhere. The sites monitored weekly and the main features of the reservoir are shown in Figure 6 .

Ferric sulphate dosing of the reservoir started in June 1990 and continued until 1996. The timing and magnitude of ferric dosed are shown in Figure 7. The annual mean concentration of soluble reactive phosphorus at the Limnological Tower declined from 1990 to 1994 (Figure 8) but increased slightly in 1995. The annual maxima appeared to decrease from 1991 to 1993 but increased in 1994 and 1995. The lower mean and maximum concentrations of soluble reactive phosphorus cannot be attributed solely to the ferric dosing as there were major changes in two other aspects of reservoir management over this period. The dosing period coincided with two major drought periods (1990–1992 and 1995–1997) which led to major changes in the input pumping regimes (because of the minimum flows required in the rivers Welland and Nene). The changes in the reservoir volumes during this period are shown in Figure 9. Also the Helixor air guns used for artificially mixing the reservoir, were upgraded in 1992 (Daldorph, pers. comm.).

There was, thus, a decrease in the quantity of river water pumped into the reservoir (Figure 10a). The phosphorus load entering the reservoir from these main river sources also declined during this period (Figure 10b).

The response of phytoplankton to the decreases in phosphorus as a result of the reduced phosphorus, inputs from the main rivers, and ferric dosing are illustrated by the annual mean and maximum chlorophyll concentrations at the Limnological Tower in Figure 11. The annual mean concentration did not exceed 20 μg l^{-1} from reservoir filling in 1975 to 1995. Since 1990 there were three years when the annual mean was at the lowest recorded levels, this was no doubt a response to the lower soluble reactive phosphorus concentrations. In 1993 a peak of over 120 μg l^{-1} chlorophyll was recorded and in 1994 the highest chlorophyll level measured during routine monitoring was found at buoy S12 and exceeded 220 μg l^{-1}. These high peaks are thought to have been in response to the increased loads of nutrients as the reservoir was being refilled during 1993 after the drought (see Figures 9 and 10) together with an unknown quantity of phosphorus released from the reflooded littoral area. There had been a drawdown greater than 30% in January 1992 which created >370 ha of exposed sediments. Large parts of this exposed area were colonised by terrestrial plants and so when the reservoir refilled there was the potential for release of phosphorus from both plants and sediments. The release of nutrients from exposed and shallow littoral areas is one area of study which needs to be carried out.

Anglian Water Services plc also instigated a longer term remedial measure in 1993. Experimental trials on phosphorus stripping at major sewage works in the River Nene catchment, started prior to the legal requirement to carry out phosphorus stripping under the EC Urban Waste Water Treatment Directive. These trials have been monitored by The Environment Agency (Rose & Balbi, 1997).

In response to the problems encountered at Rutland Water and the complex issues involved in the management of such a site, English Nature, Anglian Water Services and the Environment Agency set up a tripartite group in 1995 which had the task of developing an 'Action Plan' for the cost-effective, sustainable management of the reservoir. To date, the group has developed the framework in which the roles and responsibilities of the three organisations were laid out and both legal and other targets have been proposed. Steps still to be carried out include the determination, appraisal, assessment of costs and benefits and of the impacts of control measures. Then recommendations for the preferred options along with monitoring requirements will then be put out for consultation. The Action Plan is due for completion in late 1998.

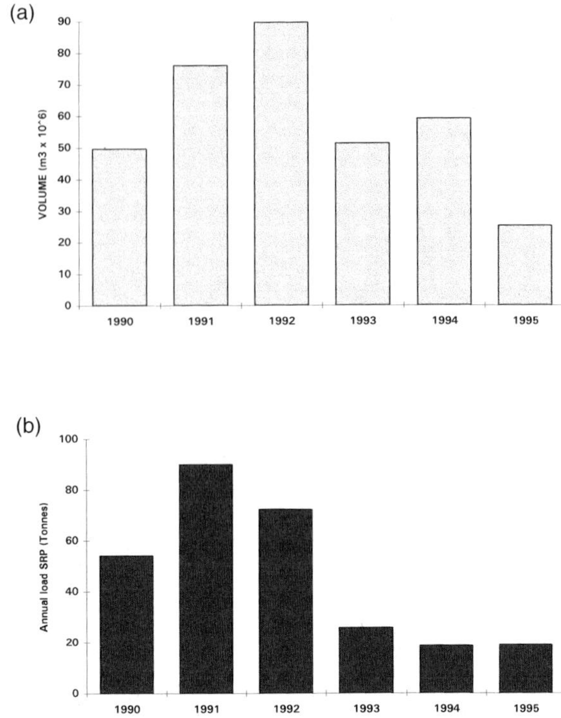

Figure 10. Annual volumes of water (10a) and load of soluble reactive phosphorus (10b) pumped into Rutland Water, 1990–95.

Biological control, direct ferric dosing, reduction of phosphorus loads and artificial mixing at Covenham reservoir in Lincolnshire

The design of Covenham reservoir is based upon that of the London reservoirs – a steep-sided concrete bowl with a jetted inlet. The main physical features are shown in Table 6 and the location in Figure 12. Water is pumped from the Louth Canal into the reservoir *via* the inlet jets, and then is abstracted to the water treatment works which are adjacent to the reservoir for treatment and distribution into the supply network. Direct ferric sulphate dosing was carried out at the reservoir between 1990 and 1993, and dosing of the

Table 6. The main physical characteristics of Covenham reservoir

Construction completed	1970
Volume ($m^3 \times 10^6$)	11.37
Area (hectares)	80
Maximum depth (m)	15.2
Mean depth (m)	14
Nominal retention time (days)	224

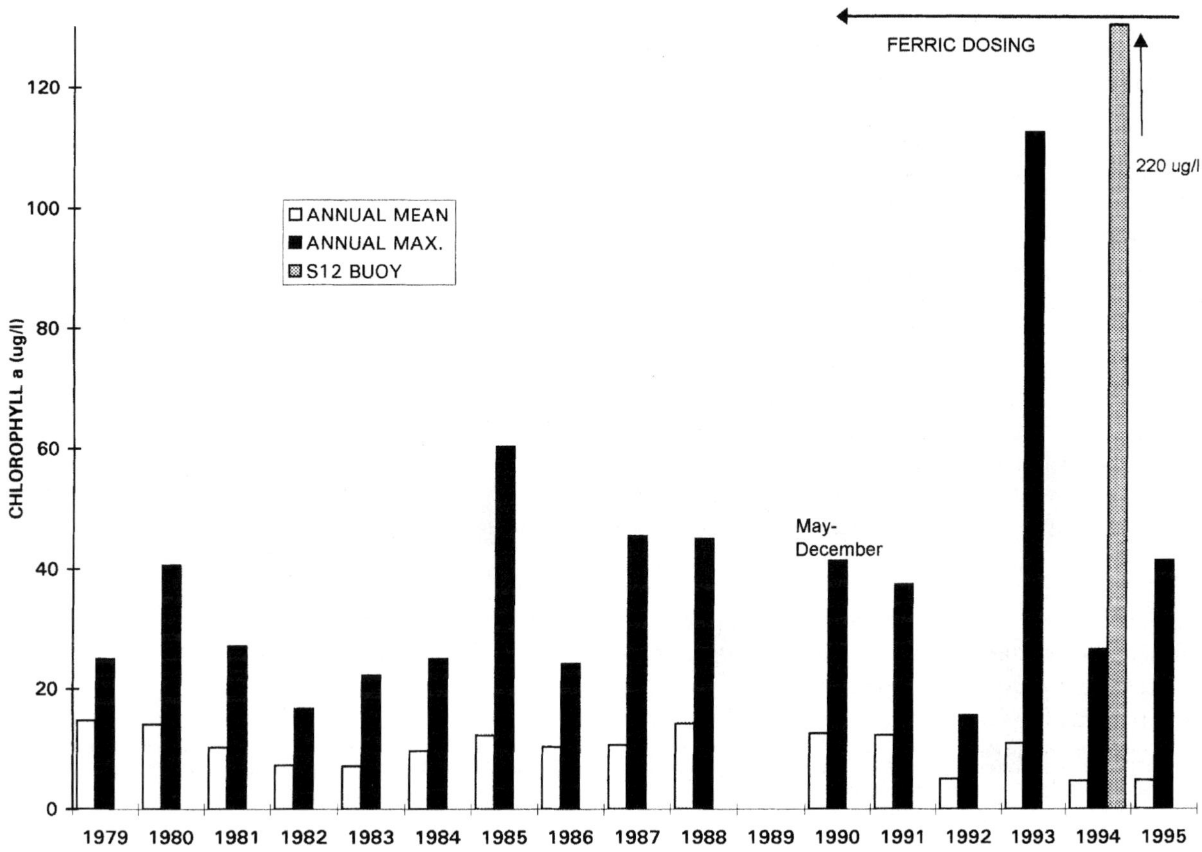

Figure 11. Annual mean and maximum chlorophyll, Rutland Water, 1979–95.

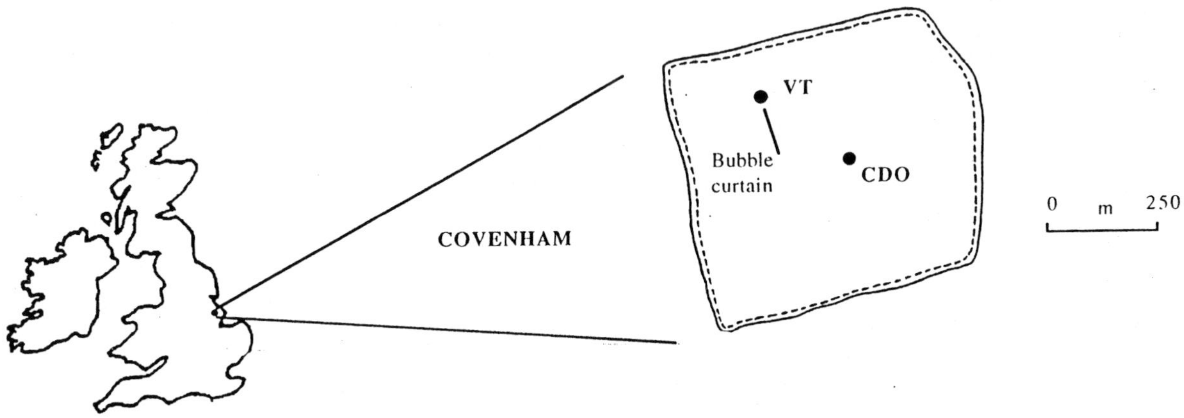

Figure 12. Location and shape of Covenham Reservoir.

effluent from Louth sewage treatment works, which is upstream of the reservoir intake, was undertaken from 1990 onwards. A perforated pipe system was installed in 1990 to aid mixing.

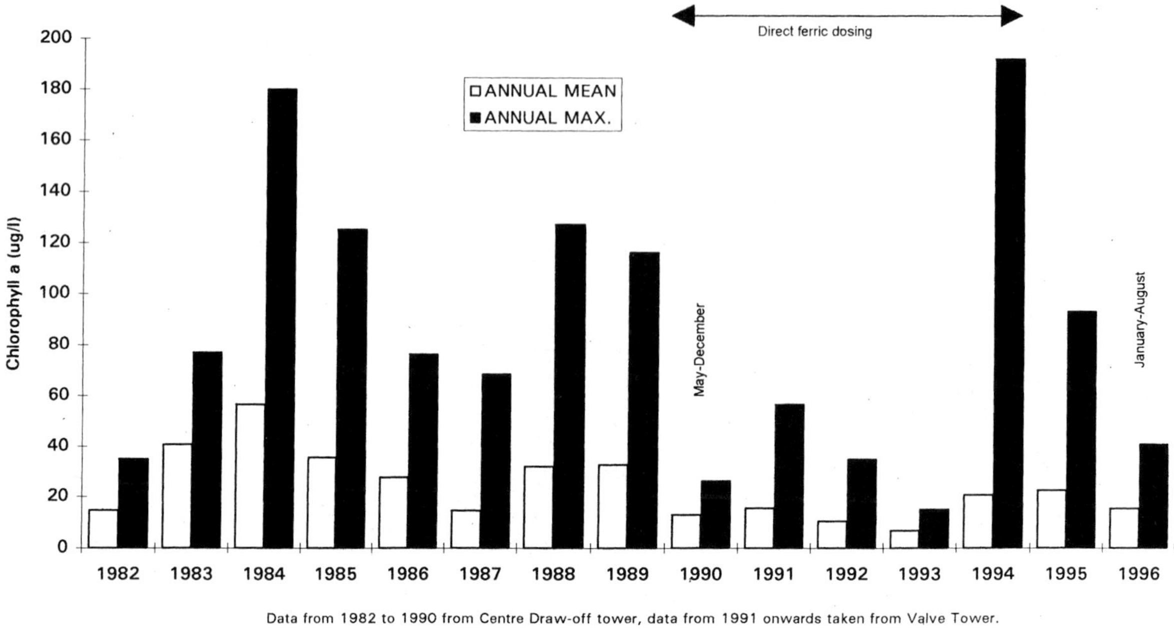

Figure 13. Annual mean and maximum chlorophyll, Covenham Reservoir, 1982–96.

Large phytoplankton biomasses and cyanobacterial blooms and scums have occurred at Covenham reservoir (Anglian Water, unpublished). Figure 13 shows the annual mean and maximum chlorophyll levels. Unlike the London reservoirs (Duncan, 1990), the design alone does not result in low algal biomasses and high water transparencies.

The Environment Agency has been monitoring the limnology of Covenham reservoir since 1990, when Anglian Water started direct ferric dosing into the reservoir. Dosing stopped in October 1994. Artificial mixing using a perforated pipe is used in the summer months to prevent thermal stratification. In 1995, a fish survey to estimate the biomass and community composition was commissioned by Anglain Water and the Environment Agency.

The combination of management control options, which have been undertaken between 1990 and 1995, have led to difficulties in clearly analysing the data to establish which techniques are most successful in reducing phytoplankton. Soluble reactive phosphorus concentrations, expressed as annual mean and maxima, showed a reduction after direct and indirect ferric dosing had been implemented (Figure 14). Both total and soluble reactive phosphorus decreased within one year from the start of dosing and soluble reactive phosphorus remained low (Figure 15). During this period, the phytoplankton did not show such any major reductions in biomass (Figure 13) but there were changes in the successional pattern (Figure 16). The cyanobacterium *Aphanizomenon flos-aquae* was recorded in December 1990 and again in December 1991, together with *Pediastrum, Coelastrum, Oocystis* and *Anabaena.* In 1992, an early bloom of *Aphanizomenon* and *Anabaena* occurred in May, and the late summer biomass peak was dominated by *Cosmarium* and *Ceratium,* rather than the more typical cyanobacterial genera. *Aphanizomenon* was also present with small centric diatoms in March 1993 and March 1994. A large bloom of *Aphanizomenon* occurred in July 1994. The phytoplankton biomass increased slightly but the succession did not return to the pre-dosing pattern once dosing had ceased.

Covenham reservoir, like the London reservoirs, supports three coexisting species of *Daphnia.* The largest of these is *Daphnia magna*, with *D. pulicaria* intermediate and *D. galeata* the smallest. The changes in the dominance of these three species is shown in Figure 17. *D. magna* dominated the cladoceran zooplankton in May and June, 1991, whilst in 1992 *D. galeata* was dominant from May through to October. In 1993, no single species was clearly dominant throughout the summer. *D. pulicaria* peaked in February and early April and remained the dominant species

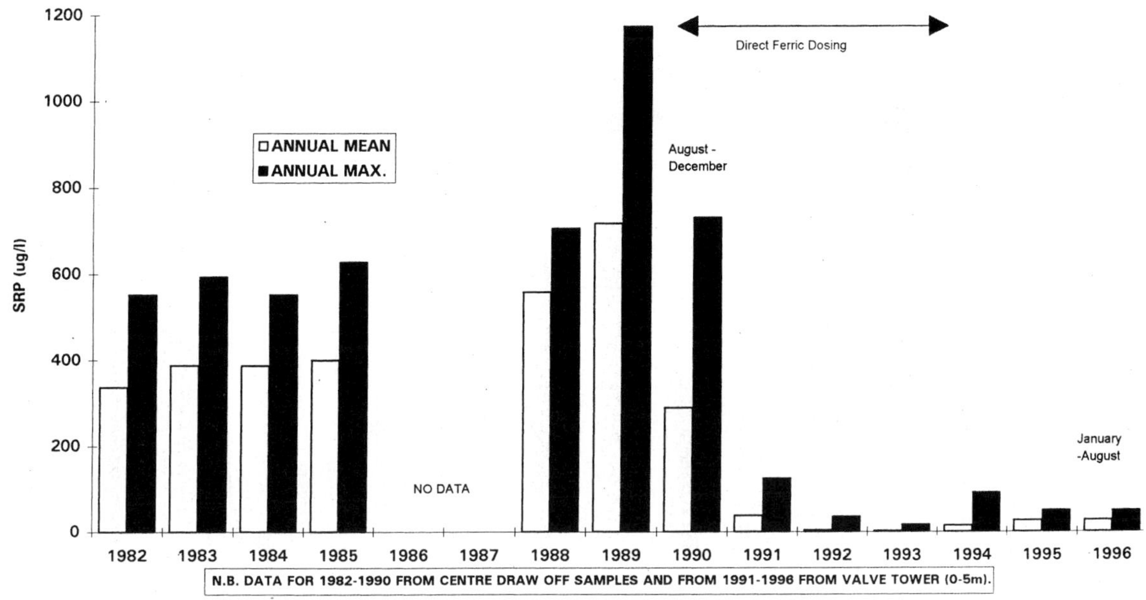

Figure 14. Annual mean and maximum soluble reactive phosphorus, Covenham Reservoir, 1982–96.

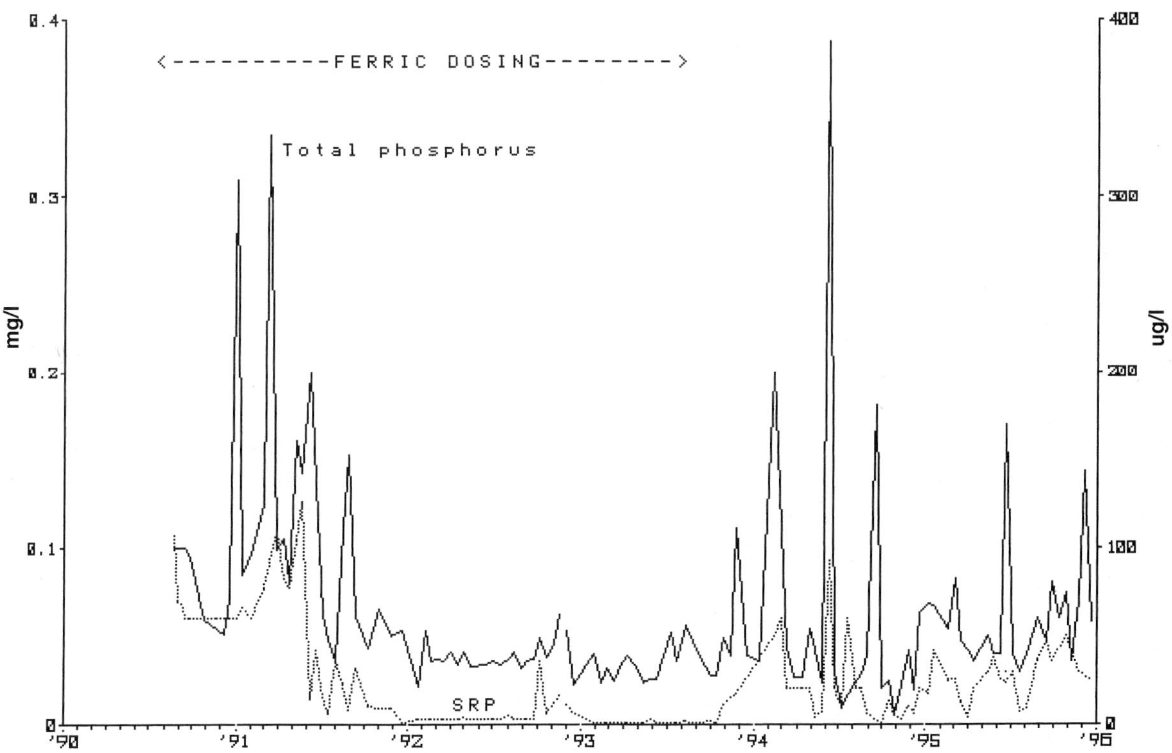

Figure 15. Phosphorus concentrations in relation to ferric sulphate dosing, Covenham Reservoir, 1990–95.

throughout most of the summer. *D. magna* reappeared during August along with *D. galeata*. The difference seen in the zooplankton communities during these years would tend to indicate top-down control by fish

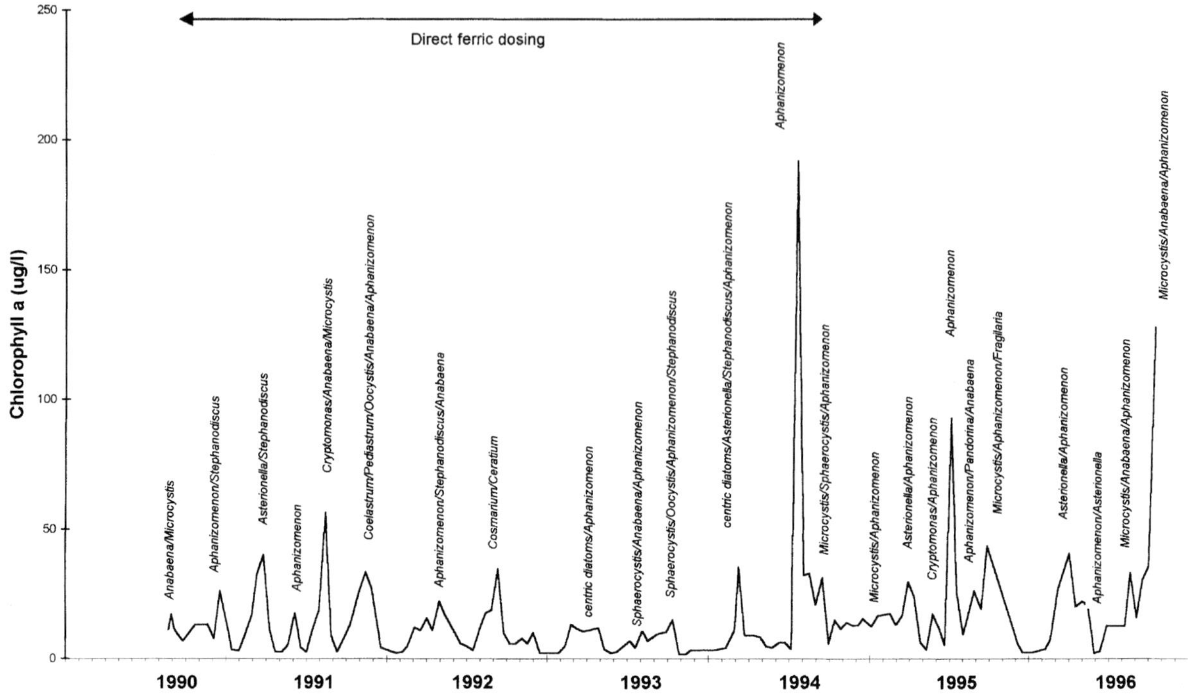

Figure 16. Temporal chlorophyll fluctuations at Covenham reservoir showing the dominant taxa, 1990–96.

predation, possibly with higher recruitment in 1992 of planktivorous fish and fry. The presence of large bodied *Daphnia,* as recorded from Covenham reservoir, would also indicate low fish predation pressure upon the zooplankton. The fish survey, carried out in 1995, showed an extremely low fish biomass of only 4.9 kg ha^{-1}. This is lower than any other biomass recorded for this type of reservoir (Duncan, 1995). The low biomass of fish present, again similar to the London reservoirs, is due to the lack of spawning sites (Duncan, 1990). However, it is unclear as to the reason for the changes in the dominance of the *Daphnia* during this period and it is not certain whether it is a result of top-down control, bottom-up control by mixing and reductions in phosphorus due to ferric dosing, or to a combination of the three.

It is possible that this reservoir is ecologically unstable, with decreasing phosphorus concentrations leading to changes in the phytoplankton which, together with the low fish grazing pressure has prevented a stable annual cycle of zooplankton appearing. Analysis of the zooplankton, phytoplankton and chemical/physical limnology led Harper et al. (1997) in a review of the feasibility of biomanipulation in reservoirs and deeper lakes, to conclude that Covenham should be further studied to ascertain the reason for annual fluctuations, and then whether any biomanipulation is neeeded. In contrast, the natural valley shape of Rutland Water means that cyprinid reproduction will be successful and that any biomanipulation will require tight control of spawning success. Since this cannot be carried out by water level management, due to its pumped storage nature and manipulation of the total fish stocks is impossible because of the economically and recreationally important trout fishery, the use and removal of artificial spawning sites has been recommended on a large-scale trial basis.

Conclusions

These three examples show that, although a variety of methods exist for assisting the management of eutrophic lowland reservoirs, the principal factor of importance is its original design and the second its hydrological regime (cf. Steel & Duncan; Zalewski, this volume). The methods available for reducing algal crops in valley reservoirs with minimal control over inflow and outflows are limited and will only succeeed if their purpose and effects are understood and their

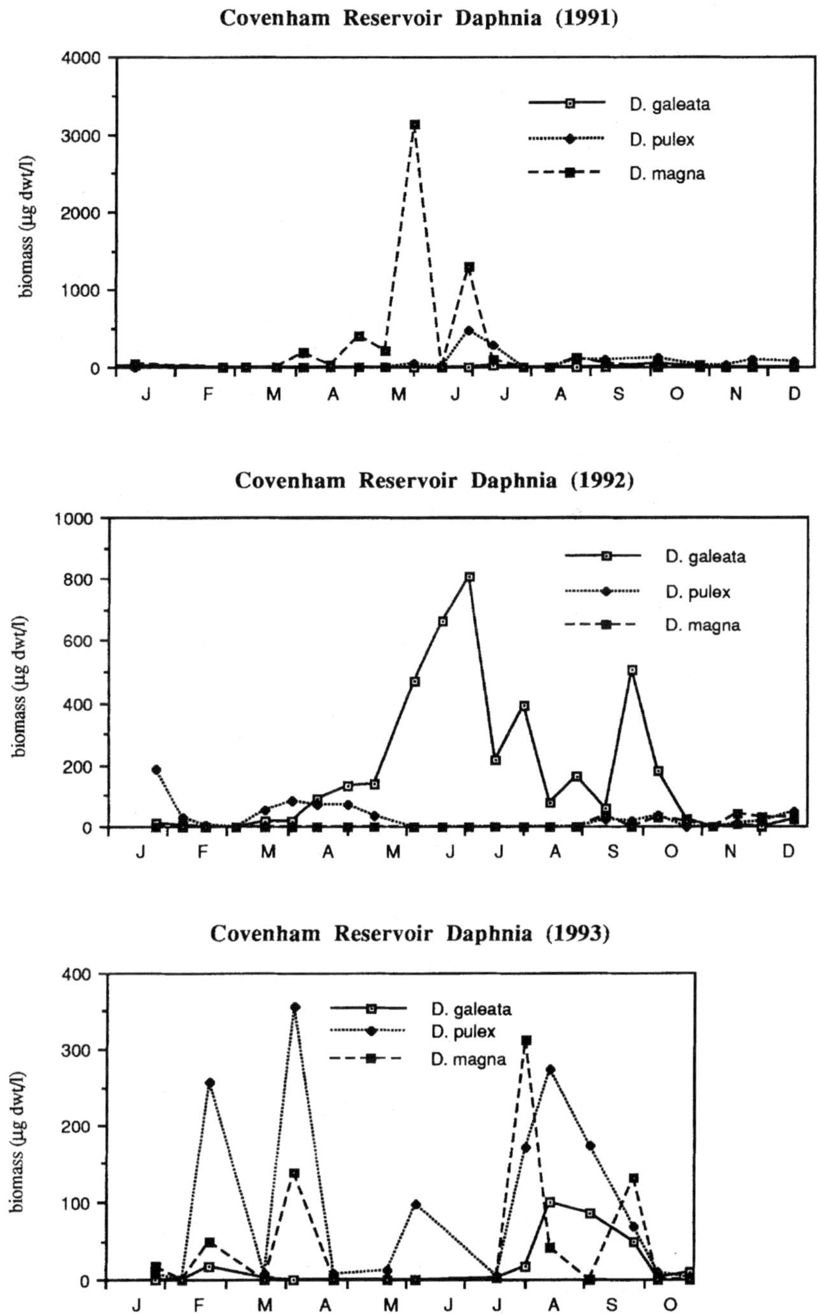

Figure 17. Development of *Daphnia* species in Covenham reservoir, 1991–93.

ecological targets clear. This is a task for the 'sustainable management' to which the Envionment Agency in England and Wales is now committed.

Acknowledgements

The views expressed in this paper are those of the authors. They would like to thank Severn Trent Water for funding the work carried out at Staunton Harold and Foremark reservoirs. Bill Brierley would also like to thank all those colleagues in the National Rivers Authority, especially David Balbi, Sarah Pritchard & Jan Krokowski who have assisted in collecting and preparing data for this paper, and to Chris Extence for his input and encouragement for the work carried out at Rutland Water and Covenham Reservoir.

References

Boyce, F. M., 1974. Some aspects of Great Lakes physics of importance to biological and chemical processes. J. Fish. Res. Bd Can. 31: 689–730.

Brierley, S. J., 1985. The effects of artificial mixing on phytoplankton growth and periodicity. Unpublished Ph.D. Thesis. University of Leicester.

Collingwood, R. W., 1977. A survey of eutrophication in Britain and its effects on water supplies. Technical report, TR40, Water Research Centre, The Centre, Medmenham.

Cooley, P. & S. L. Harris, 1954. The prevention of startification in reservoirs. J. Inst. Water Eng. 8: 517–537.

Cooke, G. D., E. B. Welch, S. A. Peterson & P. R. Newroth, 1993. Restoration and Management of Lakes and Reservoirs. Lewis Publishers.

Davis, J. M., 1980. Destratification of reservoirs – a design approach for perforated-pipe compressed-air systems. Water Serv. 84: 497–504.

Daldorph, P.W.G., 1999. A reservoir in management-induced transition between ecological states. Hydrobiologia 395/396 (Dev. Hydrobiol. 136): 325–333.

Duncan, A., J. Kubecka & J. Seda, 1995. Fish Stock assessment in Covenham Reservoir, June 1995. A report for Anglian Water Services and National Rivers Authority. National Rivers Authority, Peterborough.

Duncan, A., 1990. A review: limnological management and biomanipulation in the London reservoirs. Hydrobiologia 200/201 (Dev. Hydrobiol. 61): 541–548

Ferguson, A. J. D. & D. M. Harper, 1982. Rutland Water phytoplankton: the development of an asset or a nuisance? Hydrobiologia 88 (nos. 1–2) (Dev. Hydrobiol. 8): 117–133.

Harper, D. M., 1982. Introduction – Rutland Water 1970–1981. Hydrobiologia 88 (nos. 1–2) (Dev. Hydrobiol. 8): 1–5.

Harper, D. M., 1992. Eutrophication of Freshwaters: Principles, Problems and Restorations. Chapman & Hall, London.

Harper, D. M. & J. A. Bullock (eds), 1982. Rutland Water – Decade of Change. Developments in Hydrobiology 8. Martinus Nijhoff / Dr W. Junk Publishers, The Hague, 240 pp. Reprinted from Hydrobiologia 88 (nos. 1–2).

Harper, D. M., N. Pacini and R. Sanderson, 1997. The feasibility of biomanipulation in reservoirs and deeper lakes. The Environment Agency. R & D report (01580). Peterborough.

Hayes, C. R. & L. A. Greene, 1984. The evaluation of eutrophication in public water supply reservoirs in East Anglia. Wat. Pollut. Contr. 83: 42–51.

Haynes, R.C., 1971. Some ecological effects of artificial circulation on a small eutrophic New Hampshire Lake. Unpublished Ph.D. Thesis. University of New Hampshire.

Hooper, F. F., R. C. Ball & H. A. Tanner, 1953. An Experiment in the artificial circulation of a small Michigan lake. Trans. am. Fish. Soc. 82: 222–241.

Klapper, H., 1991. Control of Eutrophication in Inland Waters. Ellis Horwood. New York.

Knights, E. A., 1982. Rutland Water – From conception to operation. Hdrobiologia 88 (nos. 1–2) (Dev. Hydrobiol. 8): 7–17.

Lund, J. W. G., 1970. Primary Production. Water Treat. Exam. 19: 332–358.

Lund, J. W. G., 1981. Investigations on phytoplankton with special reference to water usage. Freshwater Biological Association Occaisional Publication 13. Freshwater Biological Association, Ambleside, Cumbria.

N. R. A., 1990. Toxic Blue-green Algae. National Rivers Authority. London.

Phillips, O. M., 1966. The Dynamics of the Upper Ocean. Cambridge University Press, Cambridge.

Reynolds, C. S., 1980. Phytoplankton assemblages and their periodicity in stratifying lake systems. Holarct. Ecol. 3: 141–159.

Reynolds, C. S., 1984. The Ecology of Freshwater Phytoplankton. Cambridge University Press, Cambridge.

Reynolds, C. S., S. W. Wiseman, B. M. Godfrey & C. Butterwick, 1983. Some effects of artificial mixing on the dynamics of phytoplankton populations in large limnetic enclosures. J. Plankton Res. 5: 203–234.

Reynolds, C. S., S. W. Wiseman & M. J. O. Clarke, 1984. Growth – and loss – rate responses of phytoplankton to intermittent artificial mixing and their potential application to the control of planktonic algal biomass. J. appl. Ecol. 21: 11–39.

Ridley, J. E., 1971. Water supply lakes and raw water storage reservoirs. Report from USA Water Quality Office, Environment Protection Agency.

Ridley, J. E., P. Cooley & J. A. Steel, 1966. Control of thermal stratification in Thames Valley reservoirs. Proc. Soc. Water. Treat. Exam. 15. 225–244.

Rose, M. & Balbi, D., 1997. Rivers Nene and Great Ouse eutrophication studies – final report. The Environment Agency, Peterborough.

Ryback, M., 1985. Some ecological effects of artificial circulation on the phytoplankton. Hydrobiologia 122: 89–96.

Sas, H., 1989. Lake Restoration by reduction of nutrient loading. Academia Verlag. Richarz GmbH. St. Augustin.

Shapiro, J., 1981. Effects of artificial circulation in reservoirs on algae. In F. L. Burns & I. J. Powling (eds). Destratification of Lakes and Reservoirs to Improve Water Quality. Australian Water Resources Council, Canberra, Conference Series 2: 376–399.

Smart, A. C., 1989. An Investigation of the ecology of water distribution systems. Unpublished Ph.D.

Smart, A. C. & D. M. Harper, 1999. Life after lakes: the ecology and management of the water distribution network. Hydrobiologia 395/396 (Dev. Hydrobiol. 136): 379–386. Thesis. University of Leicester.

Steel, J. A., 1972. The application of fundamental limnological research in water supply and management. Symp. Zool. Soc. Lond. 29: 41–67.

Talling, J. F., 1957. The phytoplankton population as a compound photosynthetic system. New Phytol. 56: 133–149.

Talling, J. F., 1965. The photosyntheteic activity of phytoplankton in East African lakes. Int. Rev. ges. Hydrobiol. 50: 1–32.

Talling, J. F., 1971. The underwater light climate as a controlling factor in the production ecology of freshwater phytoplankton. Mitt. int. Theor. Agnew. Limnol. 19: 214–243.

Tolland, H. G., 1977. Destratification/ aeration in reservoirs. Water Research Centre, Tech. Report, TR50. Water Research Centre, Medmenham.

Young, S. N., W. T. Clough, A. J. Thomas & R. Siddall, 1991. Changes in plant community, at Foxcote Reservoir following use of ferric sulphate to control nutrient levels. Journal of Institute of Water and Environmental Management 2: 5-12.

Youngman, R. E., 1975. Observations on Farmoor, a eutrophic reservoir in the upper Thames valley, during 1965–1973. In The Effects of Storage on Water Quality. Water Research Centre Symposium Medmenham: 127–164.

Zalewski, M., 1999. Minimising the risk and amplifying the opportunities for restoration of shallow reservoirs. Hydrobiologia 395/396 (Dev. Hydrobiol. 136): 107–114.

Hydrobiologia **395/396:** 355–364, 1999.
D.M. Harper, B. Brierley, A.J.D. Ferguson & G. Phillips (eds), The Ecological Bases for Lake and Reservoir Management.

Ecological and ecophysiological impacts of ferric dosing in reservoirs

Selena Randall[1], David Harper[2] & Bill Brierley[3]
[1]*Environment Agency, Anglian Region, Kingfisher House, Orton Goldhay, Peterborough PE25ZR, U.K.*
[2]*Ecology Unit, Department of Biology, University of Leicester, Leicester LE1 7RH, U.K.*
[3]*Environment Agency, Waterside North, Lincoln LN2 5HA, U.K.*

Key words: eutrophic reservoir, ferric dosing, *Daphnia*, feeding behaviour, thoracic limb

Abstract

The use of iron salts in the control of eutrophication of surface water reservoirs can only be successful if there are no environmental effects. We examined the acute and chronic effects upon *Daphnia longispina*, the dominant member of the zooplankton. Laboratory tests established that, although dissolved iron had no measurable effect, the presence of particulate iron caused mortalities in acute tests (48 hours) and reduced number of broods per female and numbers of young hatching in chronic tests (21 days). At 11.48 mg Fe l^{-1}, 50% mortalities occurred in the population, whilst 4.49 mg Fe l^{-1} caused reductions in reproduction and increased mortality rates over 21 day tests. In inert china clay, similar results were observed, suggesting that it was the particulate nature of the ferric sulphate that caused the effects, rather than any toxic properties. A safe level of 1.69 mg Fe l^{-1} was calculated. The area of the filter of the third thoracic limb of the daphnids, from both the laboratory cultures and the south arm of Rutland Water (which received ferric-dosed inlet water) increased in size in the presence of ferric sulphate and china clay in chronic tests. This is a response to low food concentrations. Abnormal feeding behaviour was also observed in ferric sulphate. Thoracic beat rate, decreased in ferric sulphate but was unaffected in china clay. This suggested that the feeding rate depression was a response to the chemical nature of the ferric sulphate, rather than its particulate nature. The number of times that particles were rejected from the food groove increased in both ferric sulphate and china clay, in response to large clumps of these substances collecting there.

Introduction

Eutrophication, recognised as a problem in freshwaters since the 1940s, results principally from the discharge of phosphate-loaded domestic and industrial sewage effluent to rivers, together with modern farming practices, which use fertilisers rich in nitrates and ammonia. In reservoirs the effects of eutrophication, such as enhanced phytoplankton growth, shading out of macrophytic plants, oxygen depletion and excessive development of cyanobacterial blooms, led to problems at water treatment works and in public water supply systems. Techniques to minimise the effects of eutrophication, such as destratification and biomanipulation, have become widely used. Eutrophication control by removal of nutrients, in particular, phosphorus, has had some success over the past 15 years. One such technique, using iron salts, precipitates available phosphates from the water column to the sediments and prevents internal loading by maintenance of a layer of iron over the sediment. In the lowland east of the U.K., Anglian Water Services began dosing their water supply reservoirs with ferric sulphate in the 1980s. After success in reducing cyanobacterial blooms at Foxcote and following the closure of Rutland Water due to their presence in 1989, ferric dosing began in this, the largest reservoir in the region, in 1990 (see Brierley & Harper, 1999; Daldorph, 1999).

The environmental impact of ferric compounds has been little researched. The effects on filter-feeders is of particular importance, due to their central role in reservoir food chains (Galbraith, 1967; McQueen & Post, 1984; McQueen & Post, 1986; Vaque & Pace, 1992). Direct toxicity of ferric sulphate, or the smothering effect of the floc, may lead to changes in population dynamics with consequences for the algal community. *Daphnia* is the commonest cladoceran zooplankter, and there are established methods for its use in toxicity tests. *Daphnia longispina* O.F. Muller is the dominant daphnid in Rutland Water.

Daphnid feeding morphology and behaviour is complex, and has received much study (Rigler, 1961; Lampert, 1974; Fott et al., 1974; Korinek & Machacek, 1979; Lampert & Schoberl, 1980; Korinek et al., 1985; Phillipova & Postnov, 1988; Urabe, 1991). Environmental standards for iron iron to daphnids do exist (Mance & Campbell, 1988), but direct toxicity may have liitle relevance to field situations where chronic exposure is a more likely occurrence.

This chapter examines the effects of ferric salts upon reservoir and laboratory populations and draws conclusions about the environmental acceptibility of ferric dosing.

Study site and methods

Rutland water is situated approximately midway between Leicester & Peterborough at 50° 40′ N, 0° 37′ W (Figure 1). Its construction began in 1971 and was completed by February 1975. Filling was completed by spring 1977 (see Brierley & Harper, 1999). Raw water from its suply rivers is pumped through a submerged inlet pipe, located at the eastern end of the south arm. This inlet is inclined at 22° to the horizontal to aid mixing of the river and reservoir water. Ferric sulphate was added to the river water at the inlet (Figure 2). Concentrations of total iron fluctuated throughout 1992 between 0–0.5 mg l^{-1}, but through a narrower range in 1993 of 0–0.2 mg l^{-1}. NRA data from 1990 onwards showed iron levels ranged between 0.1–17.5 mg l^{-1}, although values above 0.1 mg l^{-1}were rarely detected, and did not vary significantly between sites in the reservoir. This suggested that iron is rapidly dispersed within the water column, despite the visibility of an orange plume at the inlet site whenever dosing occurred.

Daphnia longispina were collected from the reservoir and cultured in the laboratory in an artificial media containing 0.35 g l^{-1} magnesium sulphate, 0.54 g l^{-1} sodium hydrogen carbonate, 0.01 g l^{-1} potassium chloride and 0.21 g l^{-1} calcium sulphate. The cultures were kept at 20± 2 °C (Enserink et al., 1990; Milbrink & Bengtsson, 1991; Jones et al., 1991), under a light regime of 16 hours light:8 hours dark (OECD, 1981; Vijverberg, 1989) in an environmental cabinet. The cultures were fed the alga *Chlorella* at a concentration of 1.25 × 10^6 cells per ml with yeast extract as an organic additive (Vijverberg, 1989).

Figure 1. Location of Rutland Water.

Acute toxicity tests were carried out over 48 hours, and chronic tests over 21 days in ferric sulphate and china clay (an inert particulate substance which acted as a control). Appropriate china clay concentrations were derived from the dry weight of ferric sulphate in each test concentration to give equivalent amounts of particulate material.

Animals exposed to ferric sulphate or china clay in chronic toxicity tests were collected once mortality occurred or at the end of the test, and preserved in 4% formalin. The filtering area of the third thoracic limb (Figure 3) was calculated from setae length, using the following equation (Egloff & Palmer, 1971; Crittenden, 1981):

$$y = 1.879 \times X1.966,$$

where y = estimated the area of one comb (mm^2); X = mean seta length from 5 measured setae (mm).

The relationships between standard length (measured from the eye to the base of the tail) and filtering area was then compared to determine whether ferric sulphate precipitates had an impact on morphology of *Daphnia*.

Thoracic appendage beat rate was directly observed to determine whether mechanical interference of feeding by the precipitate occurred. *Daphnia* were

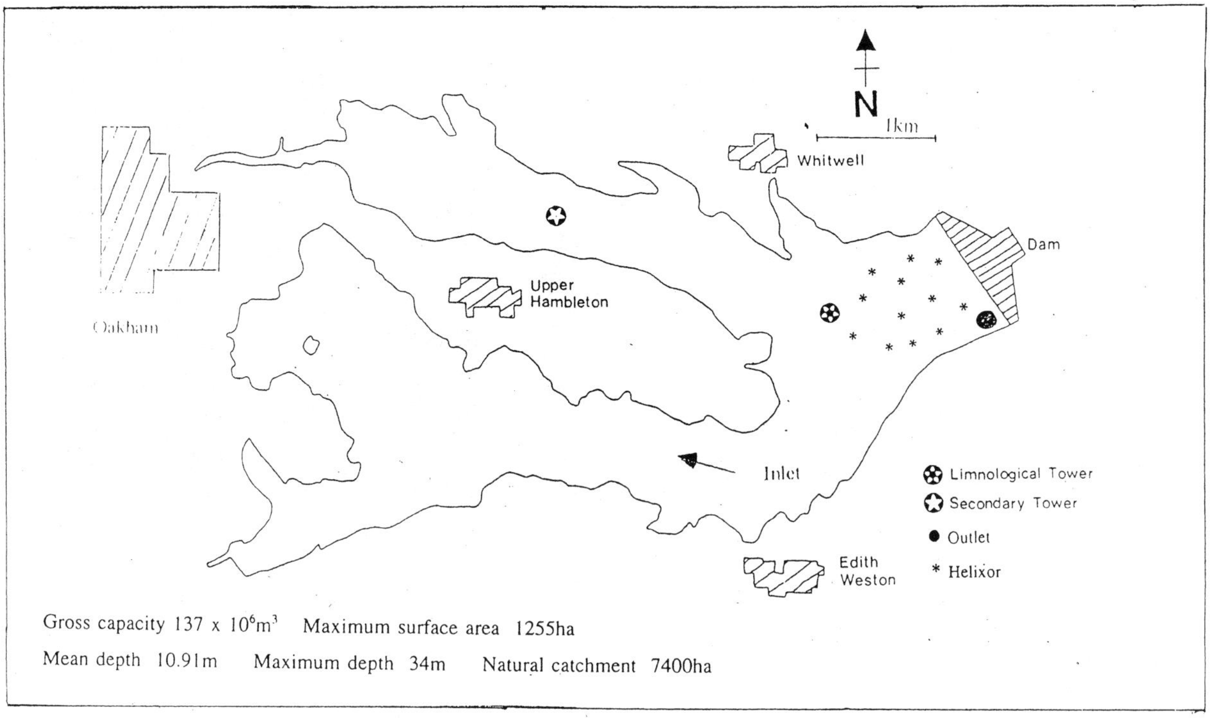

Figure 2. Management features of Rutland Water.

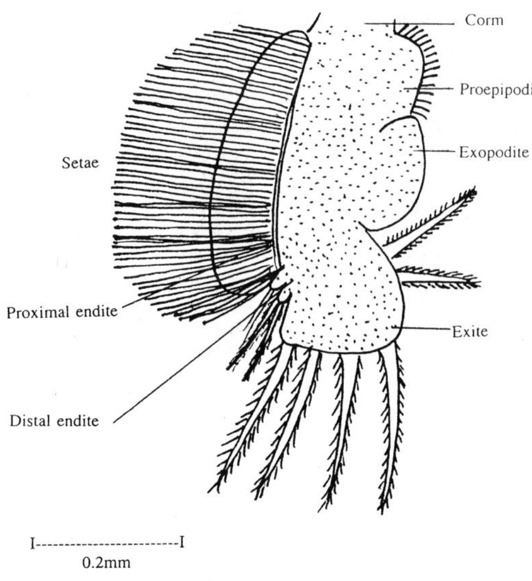

Figure 3. Third thoracic limb of *Daphnia*.

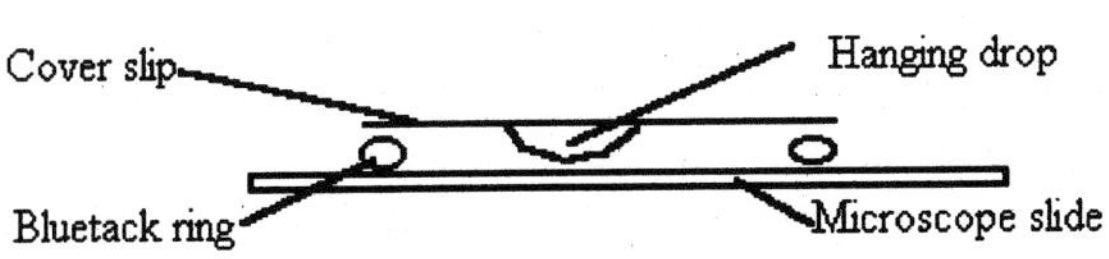

Figure 4. Cross-section through hanging droplet.

exposed to the suspended ferric sulphate in a hanging droplet (Figure 4). Video equipment recorded the thoracic appendage rate, and the number of times food was rejected from the food groove.

Results

There were no significant mortalities (p>0.5) from dissolved iron exposure in acute (48-hour) tests. Percentage mortality increased significantly (p<0.05) with increasing concentration of particulate iron over 8 mg l^{-1} however, suggesting that there was a detrimental effect (Figure 5). The mean effective dose (ED_{50}) was calculated using the method of Litchfield & Wilcoxen (1949), to be 11.48 mg l^{-1} between confidence limits of 12.39 and 10.63 mg l^{-1}.

Mortalities increased significantly in chronic tests (p<0.05) with increasing concentration of particulate iron over 2 mg l^{-1} (Figure 6). The number of broods and the mean clutch size was also affected: no neonates were born in 15.9 mg l^{-1} particulate iron. An ED_{50} of 4.49 mg l^{-1} iron precipitate was calculated as above, between 90% confidence limits of 6.51 and 3.09 mg l^{-1}. The median effective doses

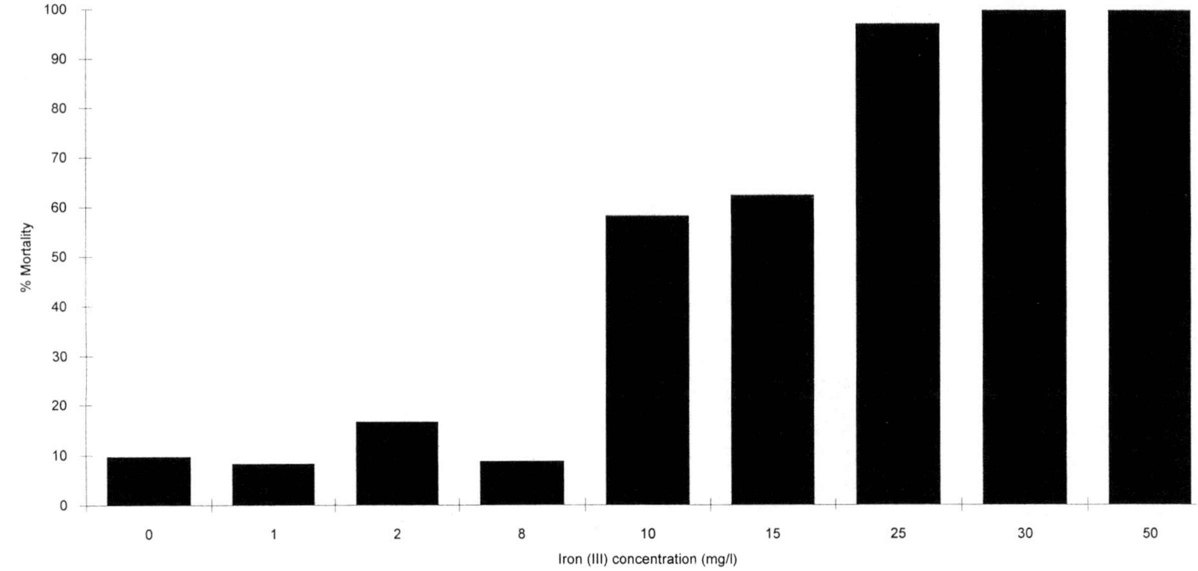

Figure 5. Daphnid mortalities in particulate iron (48 hours).

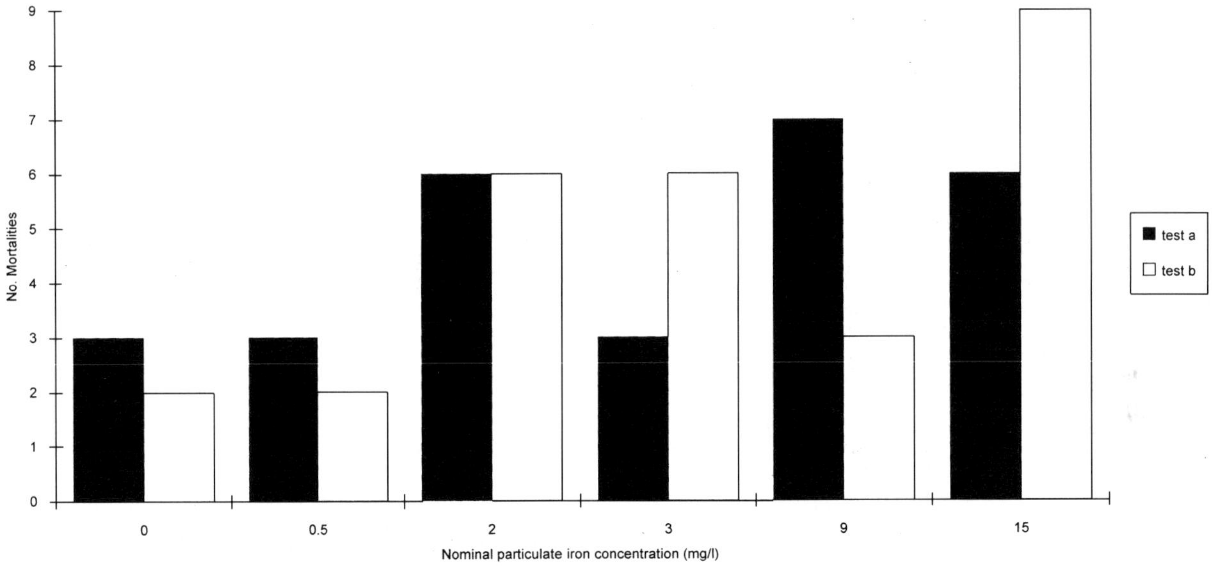

Figure 6. Daphnid mortalities in particulate iron (21 days).

from the acute and chronic tests were used to calculate safe limits of particulate iron of 1.69 mg l^{-1} Fe – that is the concentration of particulate iron below which no harmful effects would be expected in *Daphnia longispina* (Sprague, 1971).

Mortality also increased significantly in china clay (p<0.01) with increasing concentration (Figure 7). These mortalities could not be attributed to toxicity since china clay is inert, and pure grade was used. In the chronic tests mortalities increased significantly as the amount of china clay increased (Figure 8). Time to the first brood was longer and both number of broods and the mean clutch size declined, as china clay concentration increased.

Examination of the individuals which had died during the chronic tests in both ferric sulphate and china clay showed that, with increasing concentration of precipitated iron the increase in filtering area with body size was more rapid (Figure 9). For animals larger than 1.2 mm standard length, a significant difference

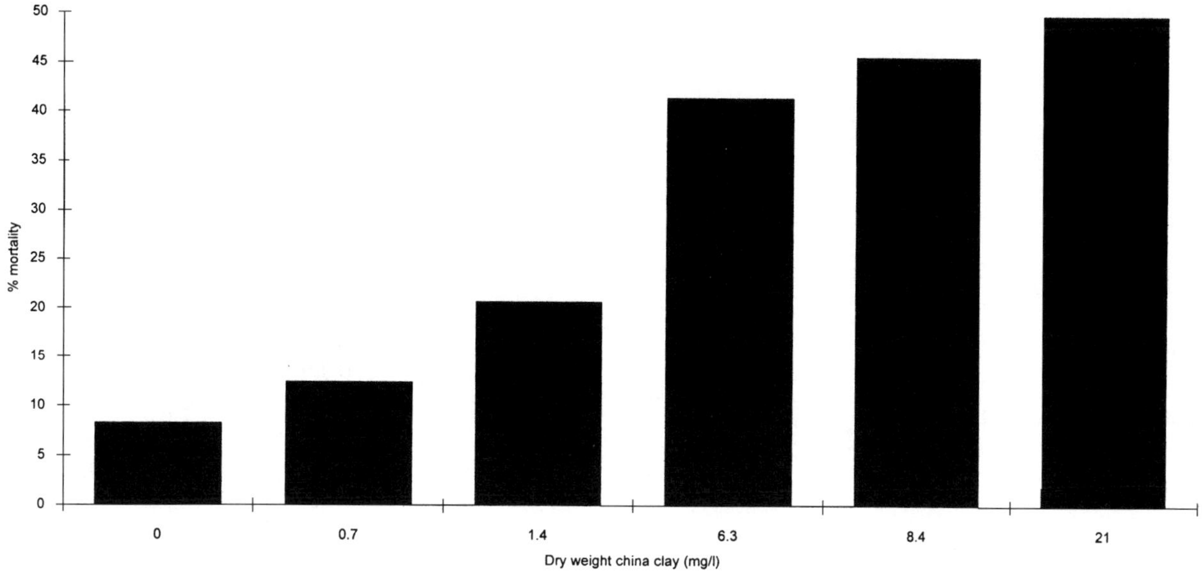

Figure 7. Daphnid mortalities in china clay (48 hours).

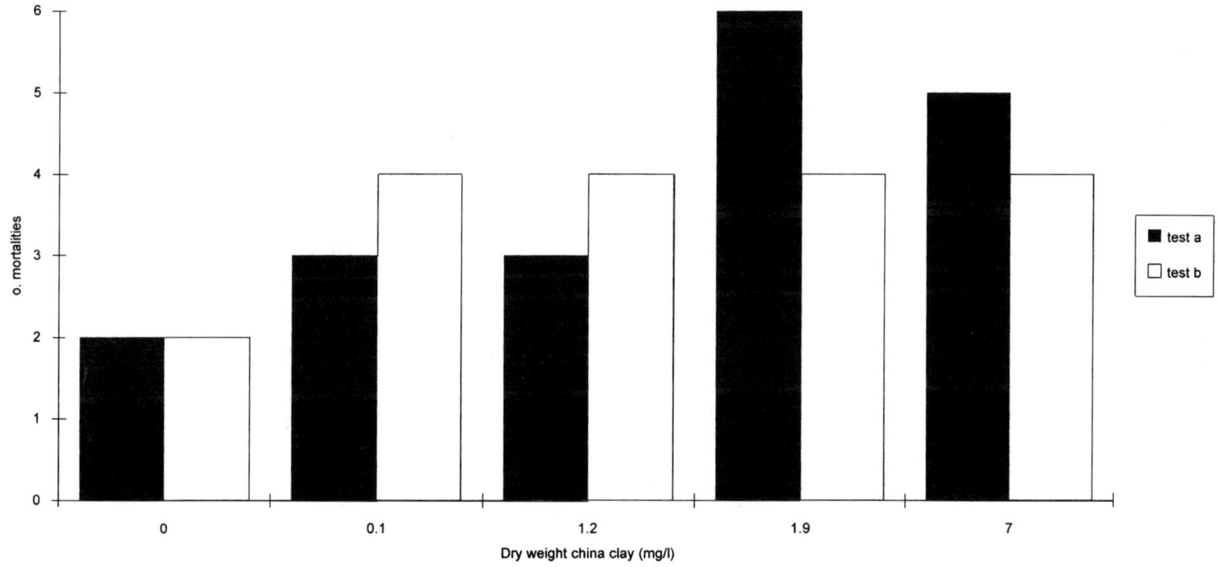

Figure 8. Daphnid mortalities in china clay (21 days).

between filtering area occurred in ferric concentrations of 9 mg $l^{-1}$and above, compared with the control (p<0.01). In china clay as with ferric sulphate, there was a similar increase (Figure 10) significant for the larger animals in china clay concentrations of 7.0 mg l^{-1}dry weight and above compared with a control (p<0.001). There was a significant increase (p<0.001) in the filtering area of individuals above 1.2 mm length collected from the south arm of the reservoir (the arm into which the inlet water is jetted) compared with animals elsewhere in the reservoir (Figures 11, 12).

The *in vivo* counts of the filtering rate showed that mean number of thoracic beats per minute declined as the concentration of iron precipitate increased (Figure 13). Post abdominal rejection rate increased above 0.5mg l^{-1} (Figure 14). There was no significant reduction in thoracic beat rate in china clay compared with the control containing no suspended material (p>0.1). Cessation of the thoracic beat rate did not oc-

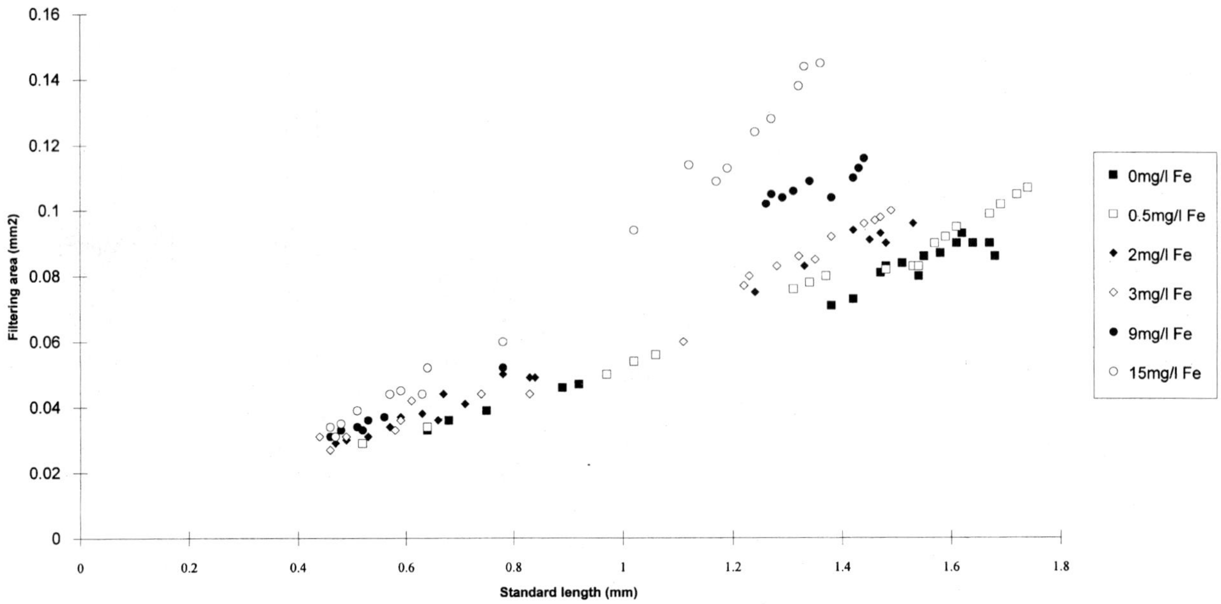

Figure 9. Relationship between standard length and filtering area in ferric sulphate precipitate.

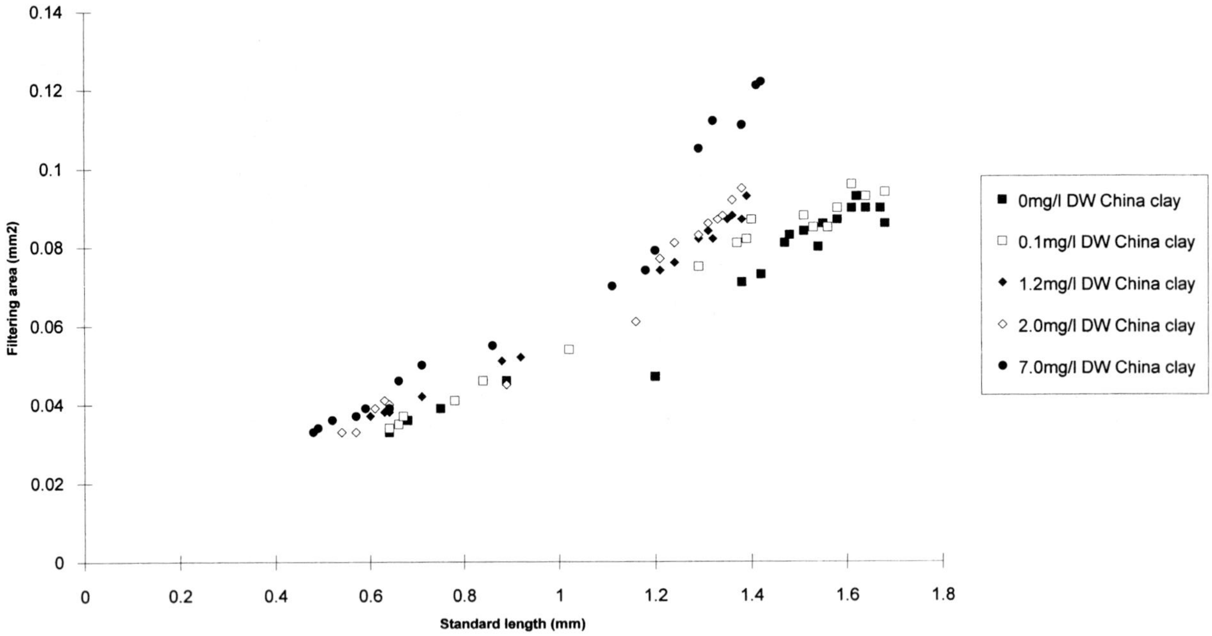

Figure 10. Relationship between standard length and filtering area in china clay.

cur. The post-abdominal rejection rates also increased significantly in china clay (Table 1).

Discussion

In the natural environment ie. the reservoir, this mechanism of rejecting food particles, or stopping feeding altogether in response to the detection of iron (III), would decrease the chances of mortality occurring in the presence of ferric particles. The uptake of ferric particles would not occur at all times during feeding, implying that once the iron was no longer detected normal feeding and growth would resume. If feeding

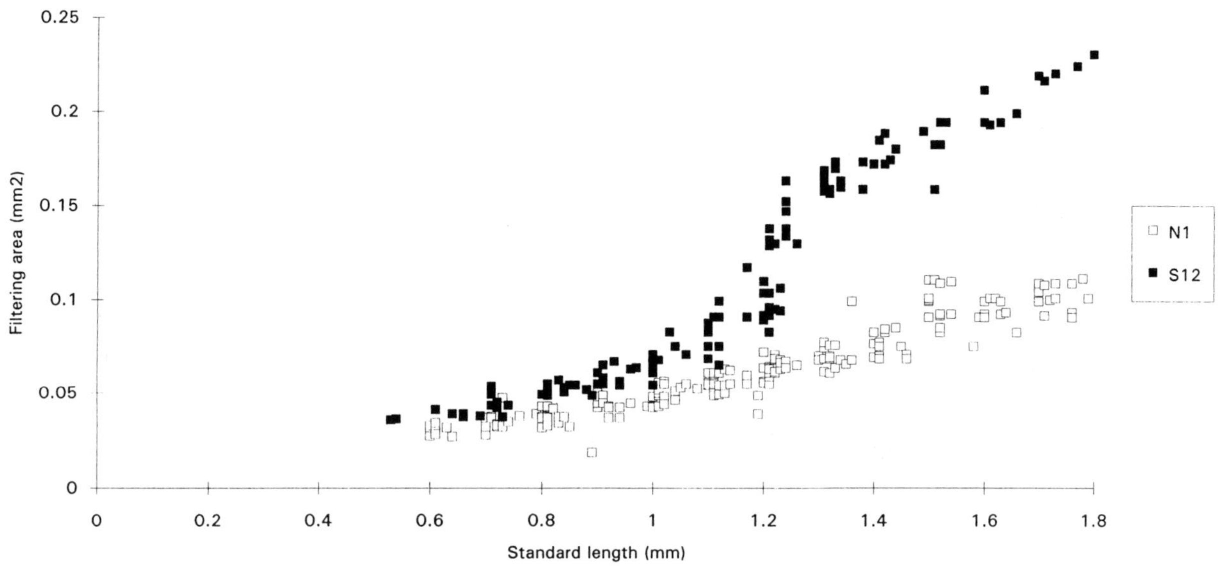

Figure 11. Filter area of third thoracic limb in Daphnids, collected from the north arm of Rutland Water (site N1) and the south arm (site S12), 1992.

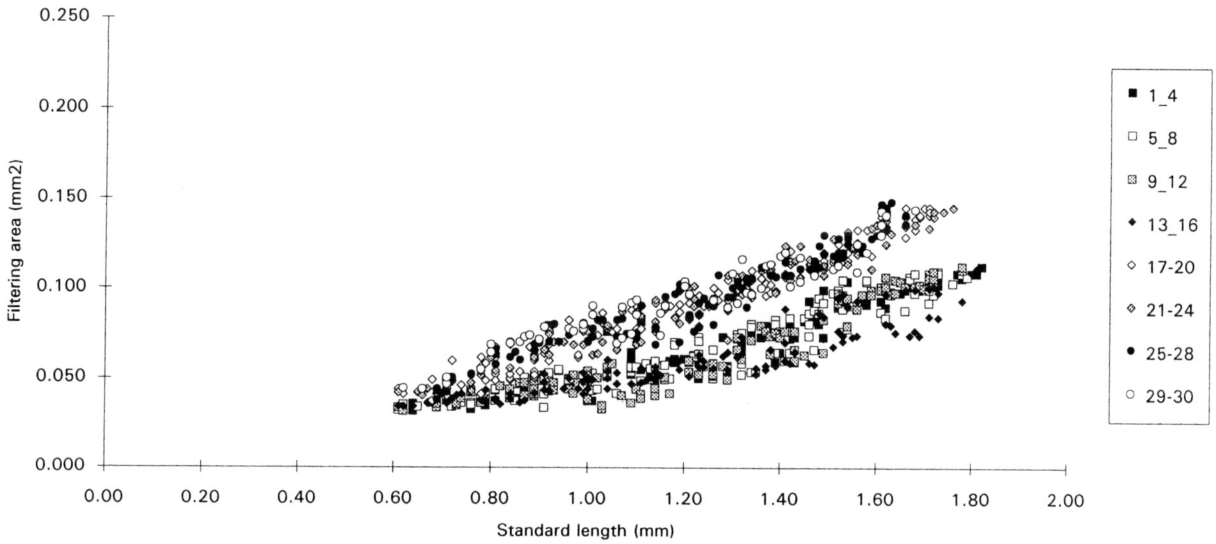

Figure 12. Filter area of third thoracic limb in Daphnids, collected from 30 sites selected at random in Rutland Water, 1993.

Table 1. Post-abdominal rejections in china clay

China clay (mg/l) (dry weight)	Rejections per minute	*n*	Dunnett's value (comparison with 0 mg/l)	Significant
0	7.5	5		
1.5	10	5	3.938	*
25.0	11	5	5.535	*

behaviour ceases altogether, then the daphnid is at risk not only of not taking in enough food to survive, but also of becoming oxygen starved (Lampert, 1987).

The total iron concentrations within the reservoir have rarely been above 5 mg l^{-1} since 1990 and usually below 0.1 mg l^{-1} since July 1992. This latter was well below the safe level of 1.69 mg l^{-1}, determined during toxicity tests. When ferric dosing took place on days when the wind was less than 5 knots, an orange plume was visible in the reservoir around the inlet zone. however, when the winds were stronger, this

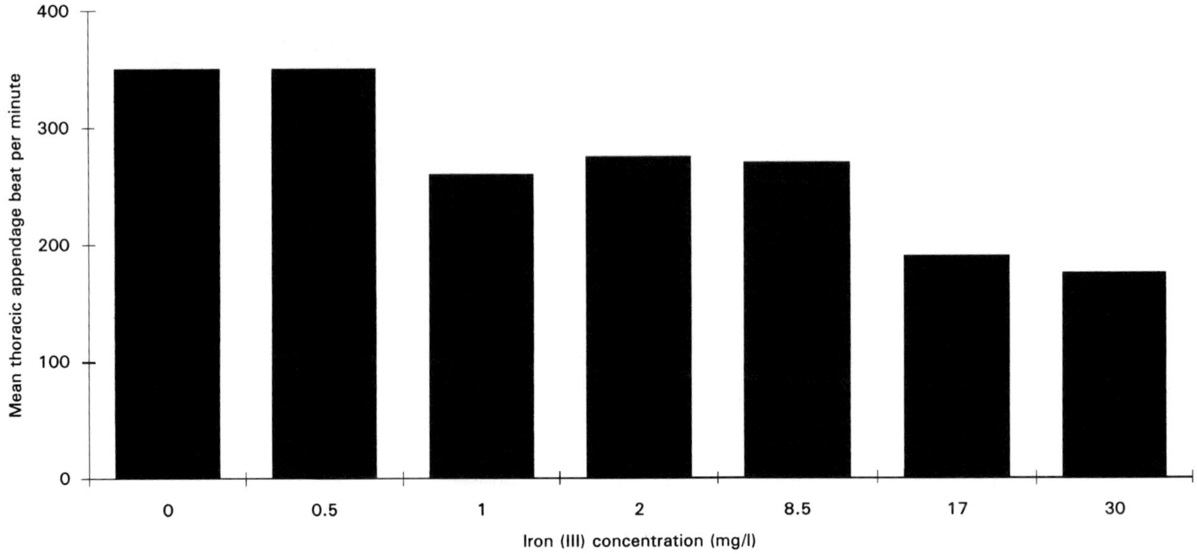

Figure 13. Thoracic appendage beat in the presence of particulate iron.

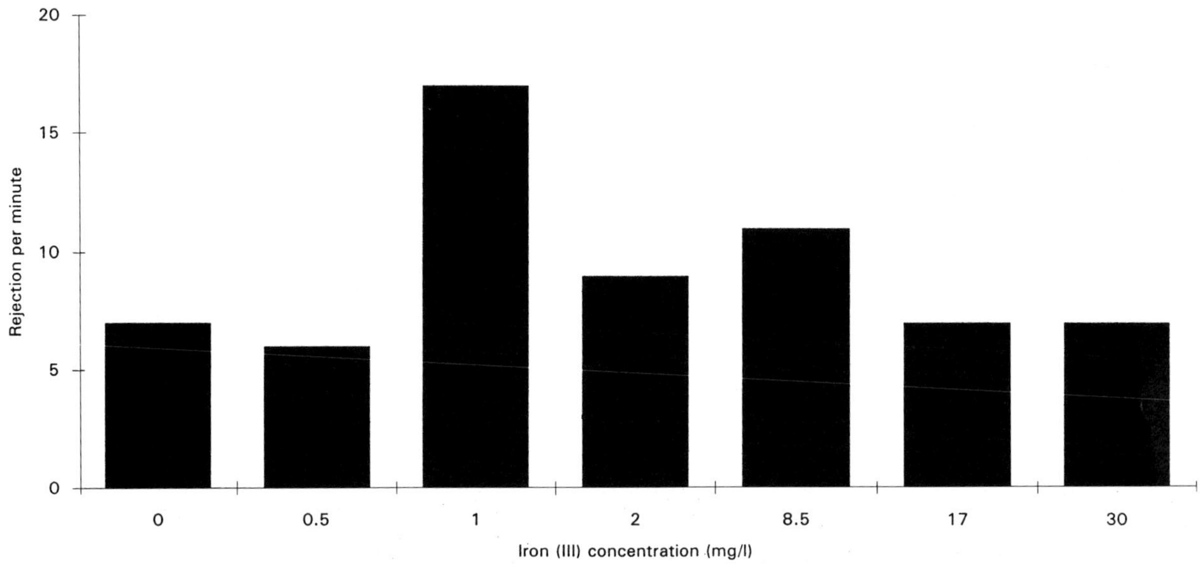

Figure 14. Post-abdominal rejections in particulate iron.

plume was not visible, suggesting that the ferric was circulated by wind-driven currents.

Approximately 10% of the reservoir floor is covered with a depth of between a few centimetres to about 1 m of ferric floc (NRA, unpublished). Most of this floc is unconsolidated, supporting little benthos – chironomids and oligochaetes are the most abundant invertebrates and occur in reduced numbers. Although the majority of the iron is bound to phosphorus in the sediment, the potential for it to mix into the water column is forever present.

At the inlet, pHs of 2–3 have been recorded (NRA unpublished data), associated with the dosant. At these pHs ferric ions are readily soluble (Mance & Campbell, 1988). However, since the reservoir as a whole has a slightly alkaline pH of 8, ferric is precipitated as hydrous ferric oxide ($Fe_2O_3.nH_2O$) within a few days. The NRA rarely recorded dissolved iron concentrations above 0.01 mg l^{-1}. Toxicity studies established that dissolved iron concentrations at pH 7–8 declined rapidly over 48 hours as iron came out of solution. To increase the amount of dissolved iron maintained in the medium required a reduction in pH, which would have its own impact on the daphnids. Dissolved iron did not have an impact on the survivorship of *Daphnia longispina* at pH 7–8.

Many organisms actively take up iron into their tissues. This uptake has been well studied in *Daphnia* (Smaridge, 1956; Hoshi & Kobayashi, 1972; Tazima et al., 1975; Perkins, 1985), and has been noted to occur at accelerated rates in neutral waters (Yan et al., 1989). Any iron that is taken up by daphnids will be recycled in the food chain. Invertebrate and vertebrate predators that ingest iron with the daphnids, as they feed, and bacteria and protozoans will recycle the iron as daphnids and algal cells decompose. Fish and chironomids are also known to store iron in their tissues (Wong et al., 1982). *Daphnia* have regulatory structures to process and excrete iron from their bodies, which would limit any damage that elevated levels of this metal might cause (Hoshi & Kobayashi, 1972). It is not known whether fish and other invertebrates have an iron regulatory system. Iron that is ingested will be changed in form by chemical processes in the body, and once excreted and released into the water column may be in a form more easily taken up by algae, and may promote growth. The concentrations recycled in this way are, however, likely to be small.

Conclusions

The addition of soluble iron as ferric chloride or sulphate to eutrophic reservoirs for the control of phosphorus is unlikely to have direct toxic effects upon planktonic *Daphnia* populations, particularly if the quantity is related to the volume of input water (in this case pumped river water). However, there may be morphological effects, related more to the physical effect of ingested iron particulates, which could reduce zooplankton productivity under extreme conditions and thus affect food web dynamics. Together with the physical effect of iron floc on the benthos, these effects suggest caution in the use of ferric dosing in multi-purpose eutrophic waters.

Acknowledgements

This work was funded by the National Rivers Authority (now Environment Agency). The authors are also grateful for access to unpublished data of the Agency and for permission to publish this paper. All opinions expressed are solely the responsibility of the authors.

References

Brierley, B. & D. M. Harper, 1999. Ecological principals for management techniques in deeper reservoirs. Hydrobiologia 395/396 (Dev. Hydrobiol. 136): 335–353

Crittenden, R. N., 1981. Morphological characteristics and dimensions of the filter structures from three species of *Daphnia* (Cladocera). Crustaceana 41: 233–248.

Daldorph, P. W. G., 1999. A reservoir in management-induced transition between ecological states. Hydrobiologia 395/396 (Dev. Hydrobiol. 136): 325–333.

Davison, W. & E. Tipping, 1984. Treading in Mortimer's footsteps: the geochemical cycling of iron and manganese in Esthwaite Water. Annual Report of the Freshwater Biological Association 52: 91–101.

Egloff, D. A. & D. S. Palmer, 1971. Size relations of the filtering areas of two *Daphnia* species. Limnol. Oceanogr. 16: 900–905.

Enserink, L., W. Luttmer & H. Maas-Diepveen, 1990. Reproductive strategy of *Daphnia magna* affects the sensitivity of its progeny in acute toxicity tests. Aquatic Toxicology 17: 15–26.

Fott, J., V. Korinek, M. Prazakov'a, B. Vodrus & K. Forejt, 1974. Seasonal development of phytoplankton in fish ponds. Int. Rev. ges. Hydrobiol. 59: 629–664.

Galbraith, M. G., 1967. Size-selective predation on *Daphnia* by rainbow trout and yellow perch. Trans. am. Fish. Soc. 96: 1–10.

Harper, D. M. & A. J. D. Ferguson, 1982. Zooplankton and their relationships with water quality and fisheries. Hydrobiologia 88 (Dev. Hydrobiol. 8): 135–145.

Hoshi, T. & K. Kobayashi, 1972. Studies on physiology and ecology of plankton. XXVI Promotion of haemoglobin synthesis by iron in *Daphnia magna* cultured under low oxygen conditions. Scientific Report of Niigata University, Series D. Biology 9: 55–62.

Hrbacek, J., 1987. Systematics and biogeography of *Daphnia* species. Mem. Ist. ital. Idrobiol. 45: 31–35.

Hrbacek, J., B. Desortov'a & J. Popovsky, 1979. Influence of the fish state on the phosphorus–chlorophyll ratio. Verh int. Ver. Limnol. 20: 1624–1636.

Jones, M., C. Folt & S. Guarda, 1991. Characterizing individual, population, and community effects of sublethal aquatic toxicants: An experimental case study using *Daphnia*. Freshwat. Biol. 26: 35–41.

Korinek, V., B. Kreplova-Machackova & J. Machacek, 1985. Filtering structures of Cladocera and their ecological significance. II Relation between the concentration of the seston and their size of filtering combs in some species of *Daphnia* and *Ceriodaphnia*. Vest. Ceskosl. Zool. Spolec. 50: 244–258.

Korinek, V. & J. Machacek, 1979. Filtering structures of Cladocera and their ecological significance I *Daphnia pulicaria*. Vest. Ceskosl. Zool. Spolec. 44: 213–218.

Lampert, W., 1974. A method for determining food selection by zooplankton. Limnol. Oceanogr. 31: 478–490.

Lampert, W., 1987. Laboratory studies on zooplankton–cyanobacteria interactions. New Zealand Journal of Marine and Freshwater Research 21: 283–490.

Lampert, W. & U. Schobel, 1980. The importance of 'threshold food concentrations'. In W. C. Kerfoot (ed.), Evolution and Ecology of Zooplankton Communities. University Press of New England, USA: 264–267.

Litchfield, J. T. & F. Wilcoxen, 1949. A simplified method for evaluating dose-effect experiments. Journal of Pharmacology and Experimental Theory 96: 99–113.

Mance, G. & J. A. Campbell, 1988. Proposed Environmental Standards for List II in Water. Technical Report 258. Water Research Centre, Medmenham, Bucks.

McQueen, D. J. & J. R. Post, 1984. Effects of planktivorous fish on zooplankton, phytoplankton and water chemistry. Proceedings of 4th Annual Conference of NALMS, 35–42.

McQueen, J. & J. R. Post, 1986. Trophic relationships in freshwater pelagic ecosystems. Canadian J. Fish. aquat. Sci. 43: 1571–1581.

Milbrink, G. & J. Bengtsson, 1991. The impact of size-selective predation on competition between two *Daphnia* species: A laboratory study. J. anim. Ecol. 60: 1009–1028.

OECD, 1981. OECD Guidelines for Testing Chemicals. OECD, Paris.

Paloheimo, J. E., 1974. Calculation of instantaneous birth rate. Limnol. Oceanogr. 19, 692–694.

Patalas, K., 1954. Comparative studies on a new type of self-acting water sampler for plankton and hydrochemical investigations. Ecologia Polska 2: 231–242.

Perkins, P. S., 1985. Iron crystals in the attachment organ of the erythrophagous copepod *Cardiodectes medusaeus* Pennellidae. J. Wat. Pollut. Contr. Fed. 48: 591–605.

Phillipova, T. G. & A. L. Postnov, 1988. Effect of food quantity on feeding and metabolic expenditure in Cladocera. Int. Rev. ges. Hydrobiol. 73: 601–615.

Rigler, J., 1961. The relation between concentrations of food and feeding rate of *Daphnia magna* Straus. Can. J. Zool. 39: 857–868.

Smaridge, M. W., 1956. Distribution of iron in *Daphnia* in relation to haemoglobin synthesis and breakdown. Quarterly Journal of Microscopiacal Science 97: 205–214.

Smith, C. D., 1988. Aspects of the Ecology of Crustacean Zooplankton in Rutland Water. PhD. Thesis, University of Leicester.

Sprague, J. B., 1971. Measurement of pollutant toxicity to fish II Sublethal effects and 'safe concentrations'. Wat. Res. 5: 245–266.

Taylor, B. E. & M. Slatkin, 1981. Estimating birth and death rates of zooplankton. Limnol. Oceanogr. 26: 143–158.

Tazima, I., T. Hoshi & S. Inoue, 1975. Histological distribution of ferruginous compounds in *Daphnia*. Zoological Magazine 84: 67–70.

Thompson, J. M., A. J. D. Ferguson & C. S. Reynolds, 1982. Natural filtration rates of zooplankton in a closed system: the derivation of a community grazing index. J. Plankton Res. 4: 545–560.

Urabe, J., 1991. Effect of food concentration on growth, reproduction and survivorship of *Bosmina longirostris* Cladocera: an experimental study. Freshwat. Biol. 25: 1–8.

Vaque, D. & M. L. Pace, 1992. Grazing on bacteria by flagellates and cladocerans in lakes of contrasting food web structures. J. Plankton Res. 14: 307–321.

Vijverberg, J., 1989. Culture techniques for studies on the growth, development and reproduction of copepods and cladocerans under laboratory and in situ conditions: a review. Freshwat. Biol. 21: 317–373.

Wong, P. J. S., Y. K. Chan & D. Patel, 1982. Physiological and biochemical responses of freshwater algae to a mixture of metals. Chemosphere II 4: 367–376.

Yan, N. D., G. I. Mackie & D. Boomer, 1989. Chemical and biological correlates of metal levels in crustacean zooplankton from Canadian Shield lakes: A multivariate analysis. The Science of the Total Environment 87/88: 419–438.

Hydrobiologia **395/396:** 365–377, 1999.
D.M. Harper, B. Brierley, A.J.D. Ferguson & G. Phillips (eds), The Ecological Bases for Lake and Reservoir Management.

Shallow urban lakes: a challenge for lake management

Stephen Birch[1] & Janice McCaskie[2]
[1]SGS Environment, Yorkshire House, Chapel Street, Liverpool, L3 9AG, U.K.
[2]Ecology Officer, Wandsworth Borough Council, London, U.K.

Key words: urban lakes, eutrophication, wildfowl, management, education

Abstract

Urban lakes are very different from other lakes: they are shallow, highly artificial and often hypertrophic yet more people come into contact with them than rural, natural lakes. Our knowledge of their ecology and management is poor. This paper describes a project under the EU Life programme to understand and ecologically manage the most important urban lakes in the London Borough of Wandsworth. One main lake has been evaluated: Battersea Park Lake. The magnitude of anthropogenic impacts are quantified and remedies explored. Computer-based decision trees for urban lake management have been developed and are illustrated.

Introduction

The most common perception of lakes, among the population at large, is probably of a large expanse of water in a mountainous or rural setting, such as occur in Scotland or North Wales. However, if the same exercise then attempted to determine the most frequent contact that people have with lakes, it is likely to be urban lakes which would come at, or near, the top. Given the high population density in urban areas and the year–round accessibility of urban lakes, they provide the most important public contact with lakes. Despite this high level of public interface, and hence importance, of urban lakes, especially in a densely populated country such as England, scientists and managers know less about them and the way they function than any of other lakes. There has been a long-standing concentration of research effort on natural or semi-natural lakes, such as the English Lake District lakes or the Norfolk Broads, but very little attention has been paid to urban lakes. Urban lakes are usually shallow, have many different demands made on them, and as a result are vulnerable to changes in water quality through nutrient enrichment.

Those responsible for managing urban lakes are usually local government officers, very few of whom are likely to have any background in either ecology or freshwater science. Most are from the planning, landscape, leisure or land management professions and usually inherit the lake simply as part of an overall responsibility for parks management. Allied to an often chronic lack of funding, this lack of understanding of the functioning of lakes among those responsible for their management, is a major contributory factor in the decline of many of urban lakes.

Urban lake managers are faced with a variety of practical management problems, some of which are associated either with funding or the people who use the lakes, but many of which are also fundamentally ecological. There is a pressing need to make current ecological knowledge about lakes and the way they function, available to lake managers, so that they can make informed decisions about the management of their urban lakes.

London's public parks make an important contribution to the quality of life of residents and their lakes provide a valuable ecological and recreational resource. Over the past fifty years, the amenity value of park lakes in London has been reduced by poor water quality and the deterioration of the bankside vegetation. Wandsworth Borough Council is responsible for managing six small lakes within its parks. The lakes range in size from under 1 ha to about 8 ha. In 1992, all of the lakes were suffering from eutrophication and the aquatic environment, and its surrounds were impoverished. The problems were:

1. The water was usually green with very low visibility.

2. Algal blooms contained toxic cyanobacteria.
3. The banksides were in need of repair and often devoid of vegetation.
4. There were no higher aquatic plants in the water.
5. The lakes had high populations of waterfowl, particularly Canada geese.
6. Where fish were present, they were often overstocked, and the eutrophic conditions would led to fish deaths in hot weather.

In 1993, Wandsworth Borough Council received funding under the European Union Life Programme for a 3 year demonstration project on the rehabilitation of urban lakes. Three lakes of varying size were chosen to demonstrate an approach to rehabilitation and management, the results of which would be made available throughout Europe. Wandsworth gained European partners in Holland, Denmark and France. In the United Kingdom, the project team comprised:

1. Wandsworth Borough Council – Project management, physical rehabilitation works and dissemination of the results.
2. SGS Environment – Directing water quality improvements, fish management, biological monitoring and the development of a management manual for lake managers.
3. Wetlands Advisory Service – Carrying out a research programme into the waterfowl using the lake in relation to the overall populations in London, and to advise on habitat management techniques.
4. Working Knowledge Transfer – Creating electronic presentations to disseminate the aims and results of the project.
5. South Thames College – Involving the local community and schools in the project and inputting to the training potential of the management manual.

Actions needed for lake rehabilitation

Nutrient sources and biological influences

It is clear that there are a number of issues which need to be tackled to improve the water quality in the lakes. Nutrient inputs and the recycling of nutrients from the sediment need to be reduced. Adequate water clarity to establish macrophytes needs to create and maintain a more diverse water environment. None of the lakes have a natural throughput of water. This means there is no flushing of the system and the lakes have to be maintained by the mains water supply as required. The largest lake at Battersea Park received water from the River Thames to counteract losses from leaks and evaporation. The river water is extremely high in phosphates adding a considerable amount of nutrients to the lake. All the lakes are surrounded by trees, so receive a large amount of leaf litter which adds to the accumulation of sediment on the bottom of the lakes and creates a high oxygen demand. Feeding waterfowl is a popular pastime in parks, which attracts larger numbers of birds to the lakes than would naturally occur and leads to uneaten bread rotting in the water. Canada geese numbers were high in all parks, with their droppings adding a significant amount of phosphorus to the water.

A high nutrient content was available in the sediment, which is released into the water column under low oxygen conditions. Fish such as carp and bream were present in some of the lakes and are known to feed in the benthic layer. They stir up the sediment, so encouraging nutrient release and making the water murky. Any aquatic plants will be uprooted.

The management scenario

A variety of actions were tried out on the three project lakes and Battersea Park, the largest lake, is used here as a case study. Battersea Park is a major London park, containing a lake of 8 ha, with a small connected pool of under 1 ha, called the Ladies Pond (Figure 1). The average depth of the lake is just under 1 m and it was constructed around 1860 of puddled clay. It was, and still is, an important landscape feature within the Park. Old postcards of the lake show water lilies and lush colourful emergent vegetation around the lake edges. The lake was originally used for boating in the summer and skating in the winter. Boating still occurs along with pleasure angling and general enjoyment of the water environment. By 1992, the lake was exhibiting all the problems described earlier and the restoration work was started in 1993.

Water supply

In 1992, the only source of water for topping up the lake was from the River Thames. This water had a phosphate content ranging from 2400–5600 μ l^{-1}. As water was required to cater for a loss in water, through leakages and evaporation, a new water supply was seen as a priority. A new supply was found by drilling a 200 mm borehole through to the chalk strata 125 m below the park. This has supplied low nutrient water ($< 10\,\mu\,l^{-1}$ P) since August 1994, although lake levels

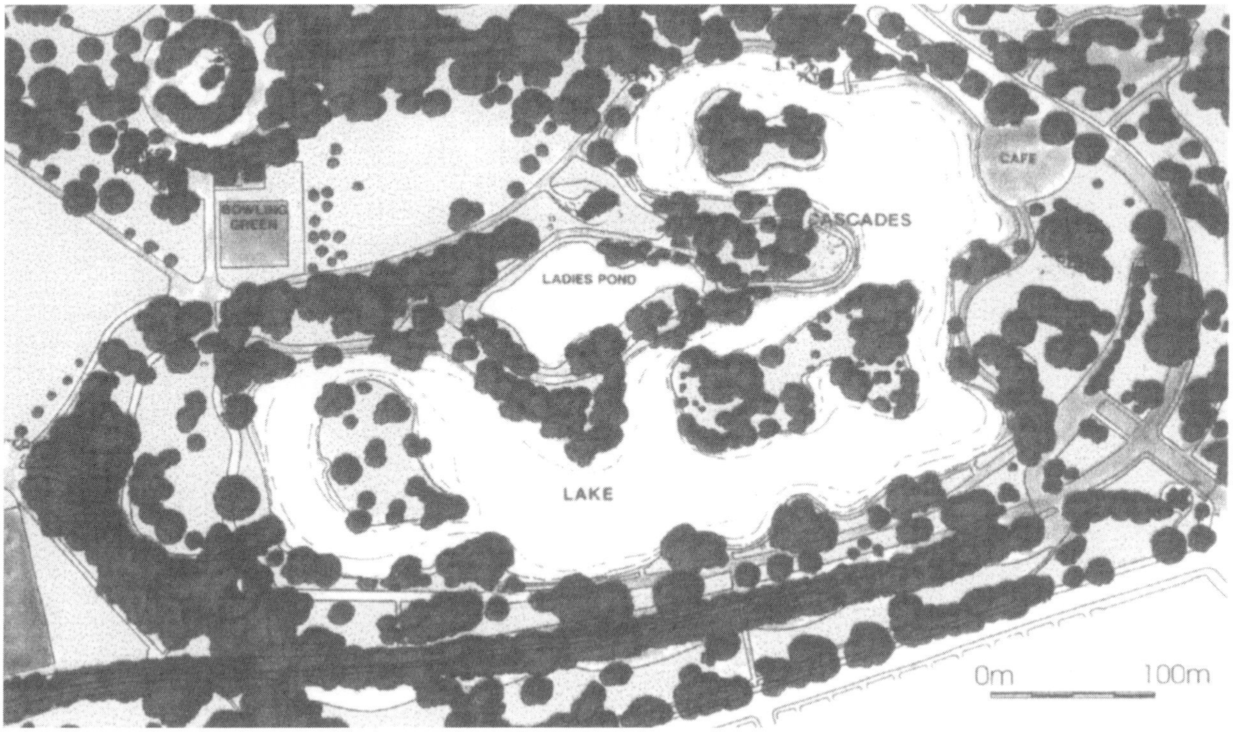

Figure 1. Battersea Park lake, London.

remained higher than this due to sediment phosphorus release (Tables 1 and 2). At present the Council's abstraction licence only allows for enough water to top up the lake, although the borehole could supply more. Options examined include the possibility of increasing the licence to allow the system to be flushed and the water possibly reused for irrigation within the park.

Sediment treatment

The lake had been dredged in the early 1980s at great expense, but with no lasting effect on water clarity. Anoxic mud was present on the lake bed but not at significant depths to warrant dredging again. Instead aerators were installed on the bed of the lake with the dual purpose of keeping dissolved oxygen levels high enough to prevent fish deaths under adverse weather conditions and keeping the sediment layer oxygenated to allow aerobic breakdown of the organic matter.

Low oxygen levels during August 1994 resulted in 200 fish dying before emergency pumps could be installed. The aerators have been operating since December 1994 and no distress to fish was experienced during the very hot summer of 1995. *Ad hoc* reports indicate that the anoxic mud has also disappeared from parts of the lake.

Replanting

At the start of the project, the lake had hard edges and bare banks with compacted soil. There was no vegetation apart from mature shrubs and trees. Certain sections of bank were chosen for replanting and the soil on the banks decompacted and organic matter added. The banks were then replanted with herbaceous plants along the lines of the original planting design.

The establishment of emergent water plants along the edge of the lake was achieved by an organic fibre roll, attached to the edge. The roll acted as a rooting medium and anchor for the plants, which were planted or seeded into it. These have demonstrated a very good rate of establishment and growth. The planting is protected from the public and dogs by new decorative metal railings along the sides of the footpaths. Temporary metal fencing positioned in the water prevents waterfowl from gaining access to the planting.

At the beginning of 1996, the water was considered clear enough to try to establish macrophytes. Circular willow cages, measuring 2.5 m diameter, have been constructed and will be planted up with lilies and submerged species. The willow is covered with mesh to prevent access by fish and the cages have lids to stop waterfowl getting trapped inside or eating the plants.

Table 1. Chemical analysis of Battersea Park Lane

Determainand	Battersea Park Lane												
	Sep–93	Apr–94	Jun–94	Aug–94	Nov–94	Apr–95	Jun–95	Jul–95	Sep–95	Apr–96	Jun–96	Aug–96	Nov–96
pH	–	8.1	7.8	7.7	7.56	8.31	7.81	7.9	8.04	8.3	8.39	8.3	7.77
Ammonia (as NH_4)	0.097	<0.05	0.271	6.27	4.7	0.027	1.69	0.111	<0.026	0.3	2.1	0.59	0.56
Nitrate (as NO_3)	1	1	<0.97	<0.97	1.07	<0.97	39.1?	*	<0.97	<0.89	–	<0.21	<1.15
Total Oxidised Nitrogen (as N)	<1	<1	–	–	–	–	–	–	–	–	<0.35	<0.35	1.17
Nitrate (as NO_2)	0.34	0.24	0.014	0.199	0.049	<0.01	0.129	<0.010	0.031	<0.010	0.4446	<0.066	
Orthophosphate (as P)	0.24	<0.2	0.056	0.76	0.108	0.052	0.2	0.14	0.092	0.036	0.032	0.092	0.047
Total phosphorus (as P)	–	0.322	0.22	0.79	0.148	0.152	0.22	0.32	0.28	0.077	0.068	0.316	0.085
Reactive silicia as (SiO_2)	<1	1.1	2.03	10.8	7.75	0.53	6.18	1.34	0.28	0.8	1.5	7	1.9
Chlorophyll A (meth)	0.058	0.154	0.074	0.319	0.021	0.071	0.039	0.079	0.224	0.036	0.021	0.122	0.016
BOD (5 Day using ATU)	18.6	9.7	4.1	11.9	<2.5	7.6	4.3	*	7.5	4	4.9	4.4	5.4
Iron	0.237	0.552	0.29	0.395	0.163	0.266	0.208	*	0.483	0.119	0.213	0.163	0.106
Chloride (as CL)	33	–	–	–	134	106	105	*	108	87	94	113	100
Aluminium	0.022	0.015	0.07	0.274	0.091	0.142	0.128	0.143	0.138	0.081	0.066	0.195	0.073
Calcium	53	56	52.8	65.8	51.6	61	66.4	*	48.8	67	72	77.5	64
Alkalinity (as HCO_3)	90	159	183	297	232	198	257	278	178	217	243	203	195
Suspended solids	–	–	–	24	18	29	19	*	39	19	48	45	13

All results expressed as mg l^{-1}

Chemical analysis of Ladies' Pond

Determainand	Ladies Pool												
	Sep–93	Apr–94	Jun–94	Aug–94	Nov–94	Apr–95	Jun–95	Jul–95	Sep–95	Apr–96	Jun–96	Aug–96	Nov–96
pH	–	7.6	7.64	7.67	7.55	7.85	7.64	8.03	8.07	8.1	8.09	8	7.91
Ammonia (as NH_4)	0.059	0.56	0.56	2.49	4.49	2	1.35	0.274	0.176	<0.2	<0.2	0.61	0.1
Nitrate (as NO_3)	1	1	<0.97	<0.97	1.4	<0.97	1.34	*	<0.97	<0.89	–	<0.23	<0.28
Total Oxidised Nitrogen (as N)	<1	<1	–	–	–	–	–	–	–	–	<0.35	<0.35	<0.35
Nitrate (as NO_2)	0.018	0.019	0.043	0.047	0.04	0.131	0.037	0.057	<0.010	0.084	0.394	<0.066	
Orthophosphate (as P)	0.68	<0.2	0.068	0.57	0.324	0.056	0.02	0.04	0.028	0.012	0.016	0.132	0.035
Total phosphorus (as P)	–	0.08	0.136	0.6	0.396	0.124	0.1	0.092	0.048	<0.03	<0.03	0.808	0.061
Reactive silicia as (SiO_2)	3.4	1.4	2.29	3.36	4.45	4.26	2.53	1.9	4.12	0.8	1.8	2.9	1
Chlorophyll A (meth)	0.2	0.012	0.063	0.012	0.007	0.033	0.008	0.031	0.072	0.01	0.032	0.04	0.025
BOD (5 Day using ATU)	10.7	1.9	3.1	<2.5	<2.5	4	7.9	*	6.4	2.7	<2.5	2.2	<2.5
Iron	0.094	0.211	0.292	0.459	0.29	0.999	0.116	*	0.253	93.2	0.075	0.255	0.062
Chloride (as CL)	313	–	–	–	156	123	110	*	118	94	105	116	103
Aluminium	<0.015	0.025	0.047	0.267	0.063	0.0874	0.0318	0.135	0.122	0.059	0.0351	0.266	0.0605
Calcium	55	52	56.9	57.6	53.9	125	66.3	*	63.5	68	63	74.7	69.2
Alkalinity (as HCO_3)	162	155	204	249	244	346	239	244	227	193	232	150	184
Suspended solids	–	–	–	5	1	12	4	*	6	14	15	36	11

All results expressed as mg l^{-1}

The lake is surrounded by large numbers of deciduous trees, resulting in considerable amounts of leaf litter entering the lake in the autumn. This can contribute to sediment build up, particularly under anaerobic conditions. New permanent fencing round the lake incorporates discrete leaf traps and, along with the planting itself, should result in less leaves blowing straight into the water. Regular management has also been instituted to remove leaves from the surface of the water during the autumn.

Waterfowl management

Numbers of Canada geese reached a peak of 518 birds in the summer of 1992. This level of geese was perceived as a problem by the Council. How to reduce and maintain lower numbers was not particularly clear and certainly controversial. In London, as a whole, the population of geese had stabilised on the well surveyed sites. However, the overall population was still rising as geese sought out smaller, less favourable sites. The London population of Canada geese is relatively sedentary, with very little movement of geese between London and the surrounding counties. There is considerable seasonal movement between sites as geese try to find the best seasonal conditions – some sites being favoured for breeding, others for moulting and others for winter feeding. Control through management is possible if management is co-ordinated across sites. A range of management techniques has to be used to bring about a decrease in the population of geese and these techniques need to be integrated.

Management of waterfowl before the project began had been *ad hoc*, but based on the need to reduce numbers. Through the project, an Integrated Management Strategy has been developed and successfully implemented. The first clear result obtained from the research was that Battersea Park was favoured as a breeding site for Canada geese. This was due to the presence of large vegetated islands. Birds that had bred at Battersea Park remained for the moult, but it did not provide ideal feeding conditions for them. As soon as they could fly again they would leave the park to find better pre winter feeding. The following management has been undertaken.

- *Physical exclusion*. This was achieved in 1994 by fencing the islands to prevent Canada geese from getting access for nesting. The fence has a gap at the bottom, which allows smaller birds underneath.
- *Habitat modifications*. The habitat is being made less favourable to the geese through the planting schemes and new permanent fencing which prevents easy access to the banks and surrounding grassland.
- *Population management*. Egg pricking has been carried out each year since 1991 resulting in very low recruitment figures. Canada geese are long lived birds so this will only have an effect on the population in the long term. Birds were moved to other sites during the early stages of our work, but this is now seen as inappropriate as it moves a potential problem elsewhere. A small number of birds have been culled, but at a public site, such as a park, this is very difficult. However, it is probably necessary if the current population of birds within an area is too high.

May 1995 saw 35 geese on the lake and there was a very low population all summer. This number of geese is regarded as a sustainable level for good water quality, but how easy it will be to maintain the population at this level is unknown. However, it is clear that ongoing management and monitoring will continue to be necessary.

Fish management

At the start of the project, the fish population was dominated by bream, with all age groups represented. Roach were common, but showing low recruitment. There were low numbers of small perch. Crucian carp were common and large specimens of common carp contributed significantly to the biomass. Battersea Park lake is used by an angling club and the Council's objective was to try to establish and manage a fish population that was satisfactory to anglers, whilst minimising the effect on algal and macrophyte growth. Bream and carp were identified as contributing to water quality problems in the following ways;

1. The fry of bream selectively prey on the large bodied forms of zooplankton. Reduction of these zooplankton reduces grazing pressure on the phytoplankton and so increases algal blooms.
2. Benthic-feeding fish such as carp aid nutrient recycling from the sediment, will uproot macrophytes or prevent their establishment.

The management intention was to reduce the number of bream and carp and restock with tench and crucian carp, managing the fish biomass at an acceptable level. Pike were to be added as a predator when water clarity improved.

The following actions have been undertaken:

1. Eleven different fish netting and removal exercises took place between October 1993 and March 1996. Approximately 2000 kg of carp and 1000 kg of bream were removed from the main lake. No other species of fish were removed. An income was received for the live fish, which will be used to purchase replacement fish.
2. In January 1994, the first restocking of tench occurred, with approximately 200 kg of fish being added to the lake. Subsequent netting exercises revealed evidence of cormorant predation on some of the larger fish and none of the stocked tench were recaptured. Any further stocking of tench would need to be of larger fish, but the cost of buying large tench to stock this 8 ha lake was prohibitive.
3. A fish proof barrier was installed to separate the Ladies pond from the main lake. The pond was drained and as many fish as possible removed. This area will be maintained fish-free at the present time as a comparison with the main lake.

Angling is only permitted within one section of the main lake, so a fish barrier has been erected between this part and the rest of the lake. The barrier allows small fish to pass between the areas but contains large fish within the fishing area. The fishing area is due to be stocked with large tench. The fishing area is 1 ha and the recommended stocking rate is 100 kg of tench, 150 kg of roach and crucian carp relocated from the main part of the lake.

A further netting exercise will take place to remove the few remaining carp from the main lake. Cormorant predation is removing fish of a certain size and age class, so reducing the need for the introduction of pike at this stage.

During 1994, large numbers of *Daphnia* were present in the water at times and water clarity was greatly improved. The phosphate levels in the water had not decreased significantly and the improvements in visibility were most likely to have been as a result of fish removal. As with the waterfowl, it is too early to see how the fish population will stabilise and we do expect to have to continue monitoring and managing the fish stocks as necessary.

Chemical status

In September 1992, the lake was hypereutrophic, measured as total phosphorus of 700 μg l^{-1}, chlorophyll '*a*' 750 μg l^{-1} and Secchi disc extinction of 10 cm. No macrophytes were present and there was a poor invertebrate fauna. The lake is monitored 4 times a year to give annual variability and changes during the project. In 1995 there was better dissolved oxygen levels due to the aeration system. Total phosphate had fallen to 320 μg l^{-1}, chlorophyll *a* to 319 μg l^{-1} and the Secchi disc reading was 50 cm. The establishment of macrophytes is critical and the ability to flush the system with low nutrient water from the borehole is desirable.

A management model

A model of the management process, or system, has been designed to provide a structured approach to facilitate management of these lakes and to be a practical, interactive tool, driven by the needs of the individual lake manager here and elsewhere. It also provides information to lake managers on the important ecological processes in the lake, the way in which management problems arise and the options available for prevention and rehabilitation of such problems. The target user of the model is not the academic community, but those actually responsible for managing lakes on a day to day basis, e.g. local government officers, park managers or management contractors. In addition to this practical application, the model will also have an educational use, particularly in vocational training for park and landscape managers.

The model will be made available in two forms, in document form (as a manual) and as an interactive computer programme, supported by a small manual of operating instructions. The documented version will contain all the information in the model and advice on using the model, but the computer programme will actively direct the user through the model. The user will have to input information and make decisions at key points in order to progress through the model. All input and decisions will be stored in a separate status file which can be printed out as required. A selection of case studies to demonstrate the use of the model and illustrate its main features will also be provided.

Scope of the management model

The first term of reference for the management model was that it should relate to shallow, urban lakes. In order to effectively establish the scope of the model, it was therefore necessary to clearly define what is meant by a 'shallow, urban lake'. For the purposes of the model, the following definition was developed: "Shallow urban lakes are natural or man-made standing freshwaters with a maximum depth of less than 5 m and a mean depth of usually less than 2 m, located in urban situations, mainly within urban parks. They are primarily used for amenity and recreation and do not include reservoirs used either for potable supply, industrial abstraction or irrigation".

Structure and content of the management model

The initial approach taken in developing the model was to adopt a simple decision-tree structure based on the problems and management actions experienced at Battersea Park lake and the other Wandsworth lakes. However, although this approach would have addressed the main problems associated with eutrophication, it was considered that it would not have a sufficiently wide application to the full range of lake management issues. The basis of the model was therefore developed into a use-related (or functional) approach, in which the management of the lake is orientated specifically to one or more uses or functions of the lake. Factors which interfere with the lake function(s) are identified as problems and remedial options are aimed at restoring and maintaining the lake function(s).

The structure of the model is based on the steps explained below.

Setting objectives

The first activity in the management model is to set objectives and the rest of the model is dependent on this activity. At this first stage in the model, we are trying to clearly identify what the lake is being managed for. This is achieved by selecting one or more uses for the lake and adopting objectives related to providing and maintaining conditions suitable for the selected use(s). Where more than one lake use is selected, they must be prioritised, so that subsequent management decisions can be clearly based on the perceived importance of different uses.

A list of potential uses is provided in the management model as a basis for setting the management objectives. This list is based on the most common and most likely uses of shallow, urban lakes. The options provided are listed in Table 3.

Table 3. List of potential uses in the lake model

1.	Fishing
	(a) Pleasure angling
	(b) Competitive angling
	(c) Specimen angling
2.	Water-based activites
	(a) Non-immersion activities
	(b) Immersion activities
3.	Landscape features
4.	Nature conservation
5.	Wildfowl
6.	Education
7.	General amenity

Fundamental requirements

If the lake is to adequately sustain a chosen use or uses, it is reasonable to assume that it must have certain characteristics. These characteristics have been termed 'fundamental requirements' and the maintenance of these requirements provides a basis for the management of the lake. The 'fundamental requirements' for each of the potential uses have been considered in two categories, minimum requirements and additional desirable features. Minimum requirements are essential characteristics without which the lake can not effectively sustain the chosen use. Additional desirable features are those characteristics which would potentially improve the quality of the lake for the chosen use, thereby increasing its value to users. The next activity in the management model is to assess whether the lake has the required characteristics and is adequately fulfilling its selected function. However, this assessment needs to be based on a baseline description of the lake and the model provides a structure for collecting and retaining information about the lake.

Baseline description

This section of the model provides an opportunity for recording any information available about the lake. Within each of the categories listed in Table 4, an extensive series of questions is provided, prompting the user of the model to enter any information held. It is not anticipated that all the 158 questions can, or need to be, answered for every lake. There may only be a few that are relevant, but the extensive list of questions does permit information about all aspects of the lake to be recorded and stored in the model. The description of the lake can also be amended, updated

Table 4. List of categories provided for the management model for baseline descriptions of the lake.

1.	Location and dimensions (physical characteristics)
2.	Landscape and visual appearance
3.	Wind
4.	Water balance
5.	Lake water chemistry
6.	Source water chemistry
7.	Sediment physical characteristics
8.	Sediment chemistry
9.	Microbiology
10.	Algae
11.	Higher plants
12.	Invertebrates
13.	Fish
14.	Birds
15.	Other animals
16.	Facilities ans access
17.	Existing uses and use–related features
18.	User perceptions/conflicts
19.	Existing management

and added to on an ongoing basis, for example by recording monitoring results.

Current status

An evaluation of the current status (or level of performance) of the lake is undertaken using the information provided in the baseline description. For each fundamental requirement of each potential use of the lake, the relevant items of baseline information have been identified. The user of the model can thus match the answers to the relevant questions in the baseline description with each of the fundamental requirements. The model then requires a decision about whether each fundamental requirement is being satisfactorily achieved or not. The decisions on whether the lake is maintaining the characteristics required to sustain the proposed use(s) determine the next step in the model. If the evaluation concludes that the lake is acceptable (i.e. meeting the fundamental requirements), there is no need for immediate remedial action. The user of the model is therefore directed to the section on monitoring. The results of monitoring feed back into the baseline description, which permits a regular re-evaluation of current status on the basis of monitoring results. If, however, the evaluation of current status concludes that the lake is not satisfactorily achieving its fundamental requirements (i.e. is not 'fit for purpose'), the user of the model is directed into the sections of the model dealing with the diagnosis of problems, analysis of causes and identification of remedial options.

Problems and their causes

The problems and remedies sections of the lake management model are based on a core list of 59 problem types, directly related to the potential uses for the lake. The problem types are number coded as a reference for use through the latter stages of the management model. There is some overlap in the problem types, because many of the same problems relate to more than one use and thus receive a different code number for each use of the lake. A summary of the problem types is given in Table 5. The first activity in assessing the 'problem(s)' with the lake is to diagnose the problem(s) and attribute the observed characteristics to one or more of the 59 problem types. An extensive, but probably not exhaustive, list of possible symptoms is provided within the model. For each symptom, the problem types which could cause that symptom are listed. The range of potential symptoms are listed in the following categories:

- General – Facilities and access
- General – Aesthetic conditions
- General – Health and safety
- Fishing
- Water-based activities
- Landscape/aesthetic feature and general amenity
- Nature conservation
- Wildfowl
- Education

If there is a range of symptoms, it is possible that a variety of problem types will be identified and each one can be considered within the structure provided by the model. Having identified a problem type, the user of the model can check the diagnosis by referring to a complete list of the problem types, recorded in order of their reference numbers. For each problem type, a list of all the likely symptoms is given. Once one or more problems are diagnosed, the management model provides a means of evaluating its significance. The assessment of significance can then be used to determine both the need for remedial action and the order of priority for addressing different problem types. The significance of the problem is assessed in terms of its magnitude and its level of interference with the use(s) of the lake. A simple scoring system is used to give a measure of the magnitude of the problem. Scores are allocated within each of six categories and the scores

Table 5. Summary of use-related problem types identified in the model

Category	Problem descriptor	Problem types (number codes)	**Category**	Problem descriptor	Problem types (number codes)
Physical characteristics				Control of users is poor	(33), (56)
	Facilities are inadequate	(1),(14),(47),(53)		Lack of suitable subject matter	(49)
	Inadequate space	(2),(15)			
	Lack of clear water	(3)	**Fish**		
	Insufficient depth of water	(4),(16)		Total fish catch numbers are low	(6)
	Access for users is poor	(43)		Preferred species are absent or low in number	
Aesthetic characteristics				Only small fish are present	(8)
				Fish show lack of condition	(9)
	Lake water appears unattractive	(5),(20),(26),(54)		Fish shows lack of condition	(9)
	Lake margins appear unattractive	(5),(21),(27),(55)			
	Feature is not sufficiently visible	(25)	**Nature conservation/wildfowl**		
	Associated features and structures appear unattractive	(28)		Insufficient area of habitat	(29),(37)
				Quality of habitat is poor	(30),(38)
Health and safety				number of species present is low	(31),(39)
	Safety provisions are inadequate	(17),(48)		Population levels of important species are low	(32),(40)
	Sickness and injury among users of water based activities	(18)		Condition of birds is poor	(41)
	Water quality is substandard	(19)		Deaths of wildfowl occur	(42)
Users					
	Resource is overused	(11),(22),(34),(44),(50), (57)			
	Resource is under–used	(12),(23),(35),(45), (51),(58)			
	Users have poor perception of the quality of the resource	(13),(24),(36), (46),(52),(59)			

Problem types with multiple refernce codes apply to more than one of the potential uses considered in the model, are summed to give a total magnitude score. Above a given threshold score, it is assumed that remedial action is necessary. The six categories used to evaluate the magnitude of the problem are:

- Level of complaint
- Frequency of the problem
- Seasonality of the problem
- Duration of the problem
- Spatial extent of the problem
- Intensity (or severity) of the problem

As well as attributing magnitude scores within each category, the user of the model is also required to make a judgment on the degree of interference with lake use(s) caused by the problem. This is described in terms of:

- Complete prevention
- Substantial interference
- Noticeable interference
- Little interference
- Very little interference

The incidence of noticeable or greater interference, combined with a magnitude score above the threshold is highly indicative of a significant problem, which requires remedial action.

As well as diagnosing problems and evaluating their significance, the management model is also intended to provide background information about the causes of such problems. This information will hopefully provide the users of the model (i.e. lake managers) with a basic understanding of the processes and interactions occurring within their lakes. Current ecological knowledge and understanding of these shallow lake systems has been presented in a technically correct but relatively simple way in the model. The model is intended for managers of urban lakes, most of which are unlikely to have an ecological background, so it has been important to present the ecology of the lake in a clear and understandable way. This is intended to enable the lake managers to understand why problems occur and why certain remedial actions are appropriate, rather than simply provide a 'recipe book' for lake management. The main factors causing the potential range of problems with shallow urban lakes are discussed under the following headings:

- Physical characteristics
- Aesthetic characteristics
- Health and safety
- Users
- Fish
- Nature conservation/wildfowl

Further discussion of the mechanisms underlying the observed problems is also provided in the management model. This is intended to give some explanation of how and why problems are caused and provide an ecological basis for the management of the lake. The list of causal mechanisms discussed in the model is given in Table 6. Some of the categories (e.g. A,B,C,D & E) are not directly related to the ecological processes within the lake and are therefore outside the original scope of the model. They relate mainly to human activities associated with the lake and are discussed in lesser detail than those issues directly related to lake ecology. However, they may still be very important factors in causing management problems with lakes, particularly the planning, design, policy and funding category.

Remedial options

Once the management problem(s) has been identified and described and a basic understanding of its causes has been achieved, the management model provides a range of options for remedial action. These remedial options cover all the potential uses, problem types and causal mechanisms described in the preceding sections of the management model. The letter codes for the causal mechanisms and the number codes for the

Table 6. Categories of causal mechanisms used in the management model

A.	Planning, design, policy, fundung
B.	Level of users/visitors
C.	*Actions of users/visitors*
D.	*Lake maintenance*
E.	*Grounds maintenance*
F.	***Weather conditions***
G.	***Inputs from water supply***
F.	***Sediment characteristics***
J.	*Oxygen concentrations*
K.	Microbiology(excluding algae)
L.	***Algae***
M.	*Vegetation*
N.	*Zooplankton and other invertebrates*
O.	***Fish***
P.	***Wildfowl***
Q.	***Nutrient and organic enrichment***

Categories in italics are relevant to problem type (54), 'Lake water appears unattractive'(see worked example in text),
Categories in bold italics relate specifically to the symptom of coloured water associated with problem type (54),

problem types are maintained throughout this section of the model to provide a clear reference with preceding sections. A total of 40 remedial options are given in the model, some of which have also contain several sub-headings describing different methods of achieving the same remedial action (Table 7). For each problem type, the user of the model is able to see what range of remedial options are available. A general indication of the start-up cost, ongoing cost and timescale of effect is also provided with each option. The computer version of the model enables the user to access the list of problems and list of remedies at the same time, alongside each other on the screen. The provision of detailed advice on the implementation of the appropriate remedial actions is beyond the scope of the management model. The model provides the user (lake manager) with the advice and information he/she needs to be able to make decisions about which actions to take. The development and implementation of an action plan must then be taken forward by the lake manager.

Monitoring performance

The final component of the lake management model is to monitor the performance of the lake. This stage is reached either once the remedial options have been considered or if the current status is perceived to be acceptable. The monitoring has a threefold function:

1. to measure the effectiveness of any remedial measures taken;
2. to provide an early warning of potential problems;
3. to add to the existing baseline information and thus increase the knowledge and understanding of the lake.

The management model provides information on the variety of monitoring which can be undertaken, what parameters can be measured and how they can be measured. Advice is also provided about developing a monitoring programme appropriate to the lake, its uses, the level of management and the availability of resources. The main options for monitoring are considered within the following categories.

- User satisfaction
- Physical parameters (e.g. water level)
- Visual/aesthetic assessment (e.g. standard of maintenance, appearance)
- Microbiology
- Water quality (e.g. dissolved oxygen, BOD, nutrients)
- Sediment quality (e.g. organic content)
- Biology (e.g. algae, higher plants, invertebrates, fish, wildfowl)

Feedback

Information from the monitoring programme can be used to update or add to the baseline description. As the baseline description is revised, it permits a re-evaluation of the current status of the lake. This could be done , for example, on an annual basis as part of an annually reviewed management programme. A very simple measure of the success of the remedial options taken is if the current status of the lake improves and becomes acceptable. There may also be the need to occasionally reconsider the objectives for the lake. For example, if the evaluation of current status consistently indicates that the use-related requirements are either exceeded or fail to be achieved, it may be appropriate to revise the objectives for the lake, either increasing them if the lake is performing well, or reducing them if a particular use cannot be sustained. This revision of objectives is likely to be undertaken on a longer term basis, e.g. every ten years, or in response to particular events or circumstances.

A simple example will be used to demonstrate the way the management model works. Firstly, an objective will be set and, for the purposes of this example, the 'general amenity' use will be chosen. Once the 'general amenity' use is selected, the model identifies the following:

Minimum requirements:

- Suitable public access and facilities, (according to level of use) e.g.
  - access to site (public and private transport)
  - parking and/or toilet facilities
  - access to lake edge (lakeside paths)
  - disabled access
- Aesthetically acceptable water quality, e.g.
  - colour, clarity, odour, litter, foam, oils, biological growth
- Aesthetically attractive lake margins, e.g.
  - absence of litter, foam
  - marginal plants
  - bank stability
  - bankside landscaping (hard or soft edge)
  - pressure of use,

  additional desirable features
- Safety measures, e.g.
  - prevention of entry to water, safety equipment
  - control of dogs, e.g.exclusion of dogs from specified areas to encourage family use.

Table 7. List of remedial options identified by the management model relating to the worked example

F.	Weather conditions				
	(xix)	Aeration and mixing			
G.	Inputs from water supply		(xxiv)	Shading	
				(xxv)	Salinisation
	(xii)	Catchment management		(xxvi)	Filtration of phytoplankton
	(xiii)	Alternative/new water supply		(xxvii)	Algicides
	(xiv)	Diversion or bypass		(xxix)	Barley straw
	(xv)	Change rate of throughput		(xxx)	Planting of aquatic higer plants
	(xvi)	Treatment of inputs		(xxxi)	Biomanipulation
		(a) Pollutant/treatment removal			
		(b) Nutrient reduction/removal	O	Fish	
		(d) Silt trapping			
		(e) Reedbed treatment		(xxxv)	Selection stocking/ removal of fish
				(xxxvi)	Removal of carp
H.	Inputs from discharges				
			P.	Wildfowl	
	(xxi to xvi as above for G)				
				(xxxviii)	Control of wildfowl
I.	Sediment characteristics				(a) Culling
					(b) Treatment of eggs
	(xvii)	Sediment conditioning			(c) Scaring
	(xviii)	Sediment removal(dredging)			(d) Limitation of access
					(e) Habitat modification
L.	Algae				(f) Relocation
	(xxiii)	Nutrient reduction/removal		(xL)	Control of public feeding of wildfowl
		(a) Control of inputs(see G. above)			(a) Limitation of access
		(b) Treatment of lake water			(b) Public education
		(c) Sediment conditioning of dredging			(c) Provision of alternative food
		(see I above)			(d) Enforcement of byelaws
		(d) Clearance of leaf litter			
		(e) Removal of carp			
		(f) Control of wildfowl			
		(g) Reduction in inputs of bread etc.			

Remedial options are referenced by Roman numerals and relate to the range of causal mechanisms (identified by letter code).

If these characteristics can be maintained, the lake will be able to satisfactorily fulfil its general amenity function. Baseline information is required to assess whether these requirements are being met. For the purposes of this example, it has been assumed that water colouration and lack of clarity occurs persistently in the lake. Consequently, it fails to satisfy the requirement for aesthetically acceptable water quality. As the current status of the example lake is not acceptable, the latter stages of the model, describing problems, causes and remedial options can be applied.

The first step in diagnosing problems using the management model is to identify the observed symptoms of the problem. For example, in the 'General - Aesthetic conditions' category, one of the symptoms listed is:

Water is coloured and/or with surface scums. For this symptom, there are four code numbers (relating to problem types) identified:

- (5) Fishing conditions:- Aesthetic conditions are poor
- (20,26,54) Aesthetic conditions:- Lake water appears unattractive

The problem type, 'Aesthetic conditions - Lake water appears unattractive' is common to three different uses, hence the three code numbers. For the general amenity function, problem type (54) would be identified. A complete list of all the possible symptoms associated with this problem type and identified in the management model is as follows.

(54) Lake water appears unattractive

- complaints from visitors
- water is coloured and/or with surface scums
- litter in water
- foam/oil on water
- unpleasant odour from water/sediments
- unsightly mats of aquatic vegetation, especially filamentous algae
- nuisance swarms of midges occur
- eventual reduction in visitor numbers

The most likely causes of this range of symptoms are discussed in the reference text of the management model. For colouration of the water, high concentrations of suspended solids and dense algal phytoplankton are identified as the most likely causes. A brief discussion of the factors contributing to high suspended solid concentrations and dense phytoplankton is also provided. The likely causal mechanisms are described and discussed under the list of categories given in Table 6. Those which are potentially relevant to problem type (54), 'Lake water appears unattractive', (i.e. most categories) are in italics. The causal mechanisms relating specifically to the symptom of coloured water associated with problem type (54) are also in bold type.

On the basis of these causal mechanisms, the list of remedial options provided by the management model relevant to this problem type and, specifically to the symptom of coloured water, is given in Table 7.

Future development of the management model

The management model is currently at a draft stage and is subject to review, revision and refinement. In particular, the computer version is being revised to make it more user-friendly. As described above, the management model will be made publicly available both as a written manual and as a computer programme. The programme will be on CD-ROM and will be available for use in both Macintosh and Windows environments. It is intended to have a European wide application and translated versions of both the text and computer programme will therefore be provided in two other European languages. The nature and style of the accompanying instruction documents has yet to be finalised, but it is anticipated that there will be a simple summary leaflet of the management model and an instruction manual to accompany the computer programme.

In order to aid the user and provide examples of how the model works, a series of case studies, including Battersea Park, will also be incorporated in the model presentation. In addition, a training application, compatible with the U.K.'s National Vocational Qualification programme is being developed. Thus, a complete package will be prepared and this will be available for purchase from Wandsworth Borough Council. Dissemination has been happening at the local level since the beginning of the project. Residents are kept informed of the work of the project through regular coverage in local papers, features in magazines and special exhibitions and events. A computer presentation was produced for the public and schools to explain the problems the lakes have been suffering and the actions needed to improve the situation. Schools have taken part in the project, testing water quality themselves and observing the different works taking place. Their work has formed exhibitions and one local primary school, Swaffield School is producing its own computer presentation about the project.

Hydrobiologia **395/396:** 379–386, 1999.
D.M. Harper, B. Brierley, A.J.D. Ferguson & G. Phillips (eds), The Ecological Bases for Lake and Reservoir Management.

Life after lakes: the ecology and management of the water distribution network

Andrew C. Smart & David M. Harper
Ecology Unit, Department of Biology, University of Leicester, Leicester LE1 7RH, U.K.

Key words: water distribution, treatment, animal penetration, *Asellus*, pyrethrin

Abstract

The fauna of water distribution is rarely studied, and then often only taxonomically. This paper describes an investigation into the community composition of the mains and the extent to which this is dependent upon the source of water. Slow sand filtration results in lower quantities of organic particulate material and organisms passing through into supply, rather than rapid gravity filtration, and the community is smaller and of different structure. A marked change occurred in a system when the supply source was changed.

Introduction

The problem of animal infestations in water supply is widespread but little understood. Water Supply companies accept that animals do exist in the mains but only privately acceed to the scale of the problem. Despite water treatment, in supply areas where the water is from a surface source (river or reservoir derived) rather than from a ground source, there are invariably invertebrate infestations of the distribution system. These infesting invertebrates are known to exit at consumers' taps, sometimes in considerable numbers, leading to complaints which frequently result in bad publicity for the water undertaking concerned. Infesting animals may also cause taste or discolouration problems resulting from stained carapaces or frasse. The community present in the distribution system, is a mixed community dominated by meiobenthos, with a few species of 'surface organisms' that are pre-adapted to survive without light or airspaces. The source of infestations in the distribution system is usually benthic animals penetrating treatment works, surviving and breeding in the distribution system and thus establishing a population of a size which may lead to consumer complaints (Collingwood, 1970).

Available control methods, both physical and chemical, are successful locally, but fail to totally eradicate infesting animals once they are established in a pipe system. Reinfestation, particularly from dead ends (small diameter pipes which have no through flow) at the extremities of a system is common (Mitcham & Shelley, 1980).

There is a lack of correlation between animals penetrating treatment and those in samples from parts of the distribution system, (Greaves & Evins, 1972; Evins & Greaves, 1979; Barham, 1985) and Smart (1989) found evidence suggesting that the survival of animals in the system is influenced by the characteristics of the system rather than by the numbers penetrating treatment. Infestations are almost certainly self-sustaining, with animals feeding on iron bacteria lining the inside of the pipes and organic input through treatment (Collingwood, 1970). The various factors which may affect the presence and numbers of infestations are shown in Figure 1

In the U.K. during the 1950s and 1960s, attention focused on infestations of predominantly oligochaetes and *Asellus aquaticus*, which created serious problems in some systems (Kelly, 1955; Houghton, 1968). These were resolved by treatment using pyrethrin, which enabled the dosing of entire systems without interrupting supply to consumers. More recently, severe infestations involving the parthenogenetic chironomid *Paratanytarsus iniquilinius* Krüger (Williams, 1974), which reproduces as a pharate adult, have been a problem. These animals have also been controlled by pyrethrin (Burfield & Williams, 1975), and more recently permethrin, but infestations rarely appear to be

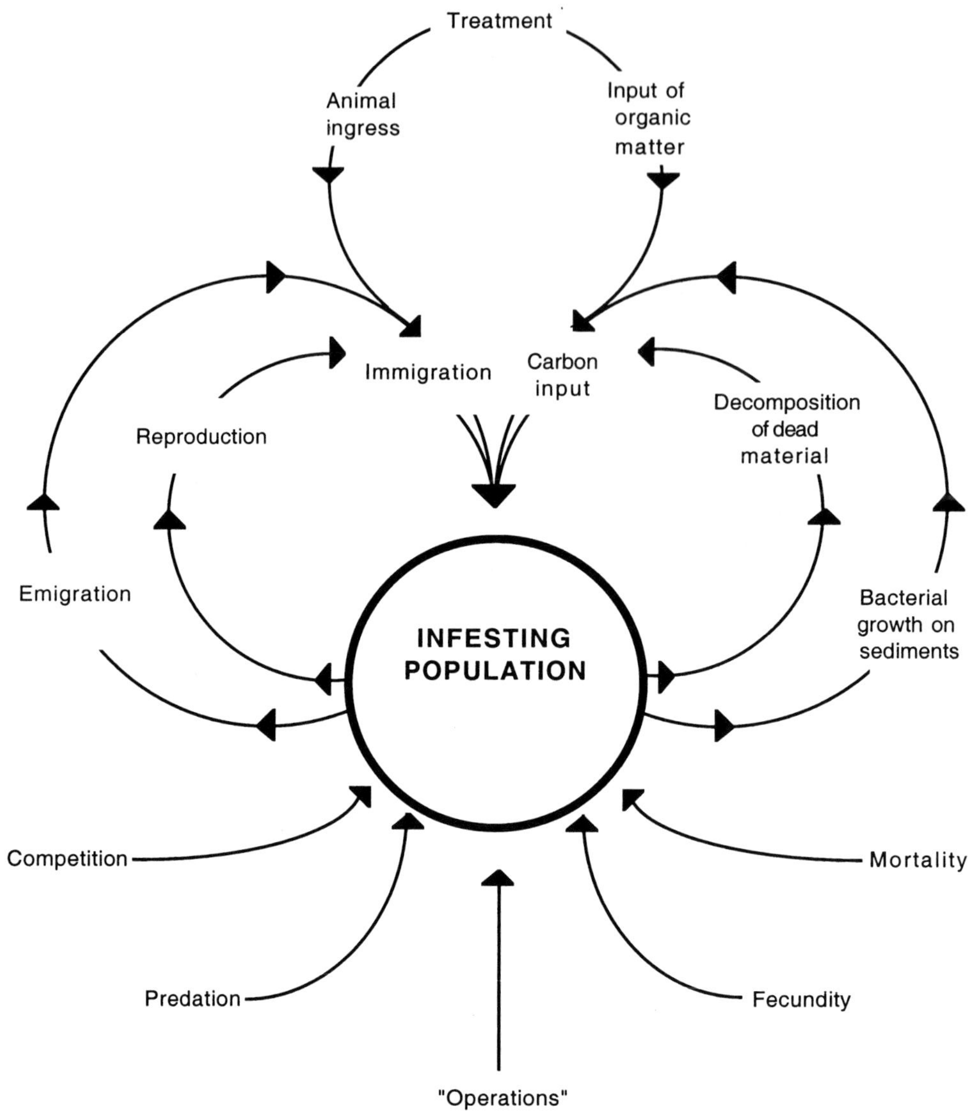

Figure 1. Model of the water distribution pipe ecosystem and factors influencing colonising populations.

cleared. Reinfestation by various taxa is frequent and can be rapid, with animals both surviving in the system and penetrating treatment playing a role.

The community in the study area of this project had moved away from the *Asellus*-dominated community of the 1970s and 1980s following frequent treatment of the water distribution system. What survived and, frequently flourished, was a simple community consisting of Chironomidae (including members of the *Parytanytarsus inquilinius* species complex; Oligochaeta (mainly Naids); Chydoridae; cyclopoid and harpacticoid Copepoda; Hydracarina; Halicaridae and some Mollusca. Many of these were able to penetrate treatment at some stage of the life cycle, to be present in large numbers, in water towers or service reservoirs or to live in low pressure areas of the system or dead-ends.

Unlike ground water (unless mixed with surface derived supplies), potable water in distribution systems from surface water generally has infestations. Penetration of rapid gravity filters (RGF) and slow sand filters (SSF) by invertebrates, algae and organic material provides both the benthic colonisers and a food source for the organisms in the system. Collingwood (1977) suggested RGF is more susceptible to penetration than SSF, and Rowlands (1982) indicated

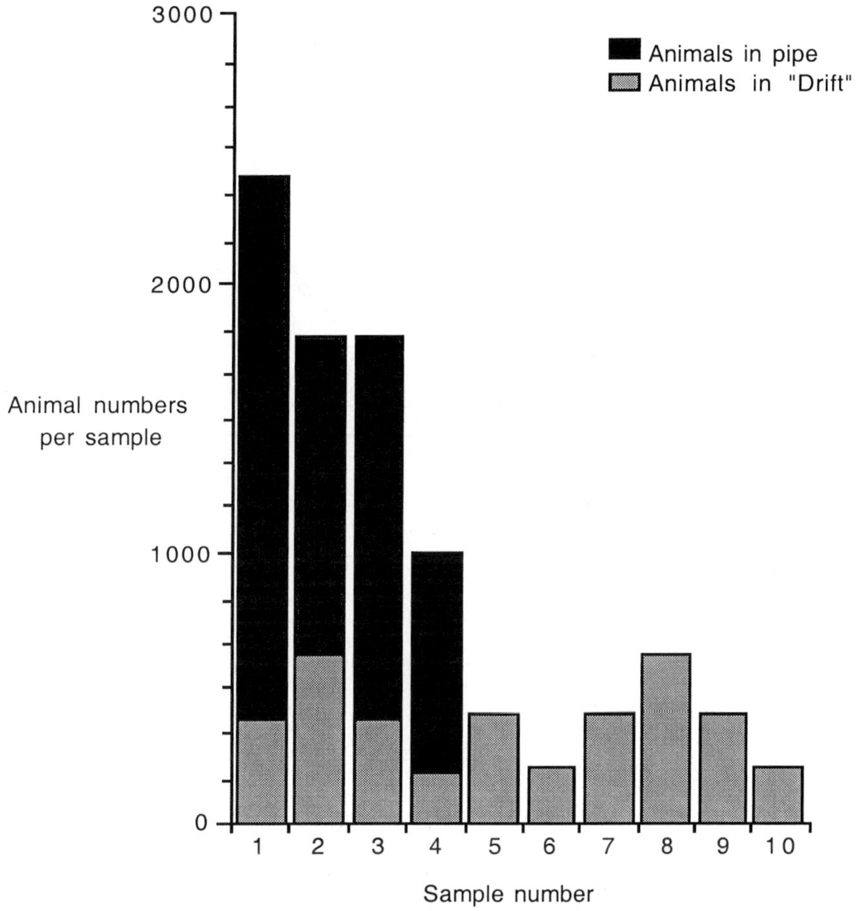

Figure 2. Diagrammatic representation of the assumptions behind the sampling strategy. As samples are removed from the main during a series of flushes, the animals present in the pipe itself are collected along with the 'drift' fauna. As the number of samples collected increases, the animals in the pipe are removed. Further flushing leads to the collection of animals at 'drift' density only.

that RGF final water quality is more variable than that from SSF. This confirms the observation of Houghton (1968) that a change from SSF to RGF led to an increase of available food in the final water, which resulted in a severe infestation of *Asellus aquaticus* in the South Essex system. The quantity of organic material used by animals in the system has not been established, but Collingwood (1970) calculated that at a treatment works experiencing an 'average' algal penetration, there will be sufficient organic material available per year to support a population of 4–5 *A. aquaticus* per metre length of pipe in a 1100 km system.

Sample methods

In the U.K., access to the potable water system is difficult. Access was achieved through fire hydrants situated on small diameter water mains close to the end of the distribution network. These could be opened to flush water through a 1 m-long mesh net pressures of over 7.0 ls^{-1}. Experimental tests showed that between three and five samples at appropriate sites would clear the animal community in the pipe, leaving only a residual number of animals that would have been carried in the water flowing through large diameter feed mains. These residual animals were considered to be 'drift'. Figure 2 shows how the sampling technique operated.

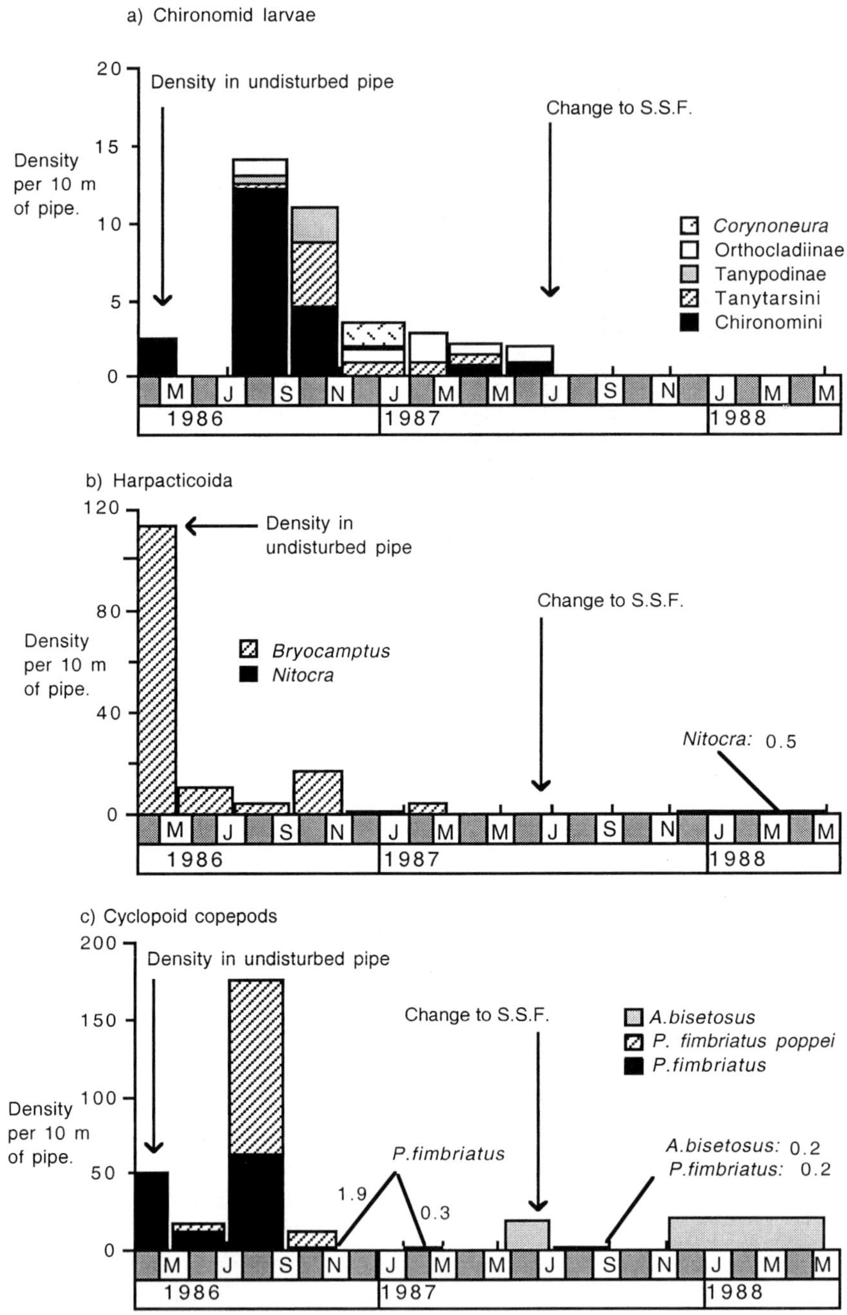

Figure 3. (a–c). Animal density per 10 m length of pipe in a rural site in Northamptonshire experiencing a change in source water treatment works.

Results

Regular sampling of sites at different frequencies over a two year period suggested that what is happening at each site is that a 'competitive lottery' occurs following community removal by flushing, rather than any form of predictable succession. The animals which arrive first are able to effectively exploit a surplus of resouces and continue to do so at the expense of other species. No pipe showed the same pattern of colonisation or community development with time, as any other pipe sampled. This concept of rudereal colonists

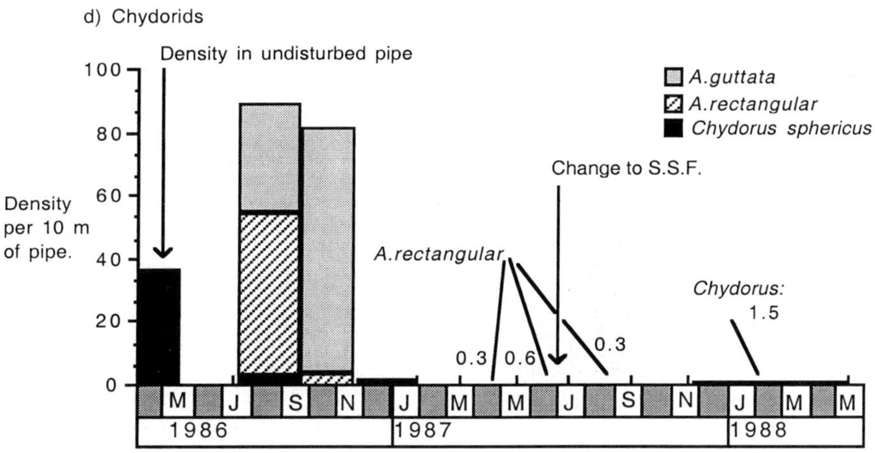

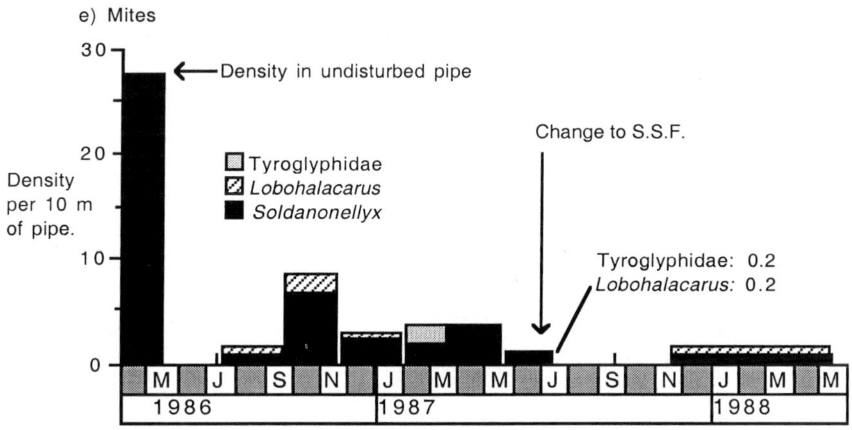

Figure 3. (d–e) Animal density in a 10 m pipe experiencing a change in source water treatment works.

was reinforced by the results at sites where the reservoir feed was switched from RGF to SSF. At these sites (Figure 3) it appeared that, following a switch to SSF, the species involved in recolonisation changed after the supply switch, particularly the copepods (Figure 3c), and, apart from tyroglyphid mites (Figure 3e), no major colonisation or recovery took place over the final six months when each site was left undisturbed.

In the situation where no switch of supply took place, recolonisation did occur in the final six months (Figure 4) and only harpacticoids appeared to be removed by repeated flushing (Figure 4b). Figure 4c shows the large number of *Megacyclops* nauplii, copepodites and adults that penetrated treatment at the rapid gravity filter and completed their life cycle in service reservoirs. This penetration at the end of the summer must have had a major influence on the biomass within the distribution system with dead animals providing a major food source for the ecosystem, and shows the importance of modern methods such as microstrainers to minimise zooplankton passage through treatment works in summer.

Discussion

Animals in the water distribution system will never be eradicated, but must be kept at levels which do not cause undue complaints from the consumer. The target for Water Companies is to manage storage reservoirs and treatment systems in such a way as to minimise the penetration of animals and organic material. The minimisation of the input to the ecosystem will enable costly chemical control methods, or wasteful physical methods such as flushing, to be kept as infrequent as possible.

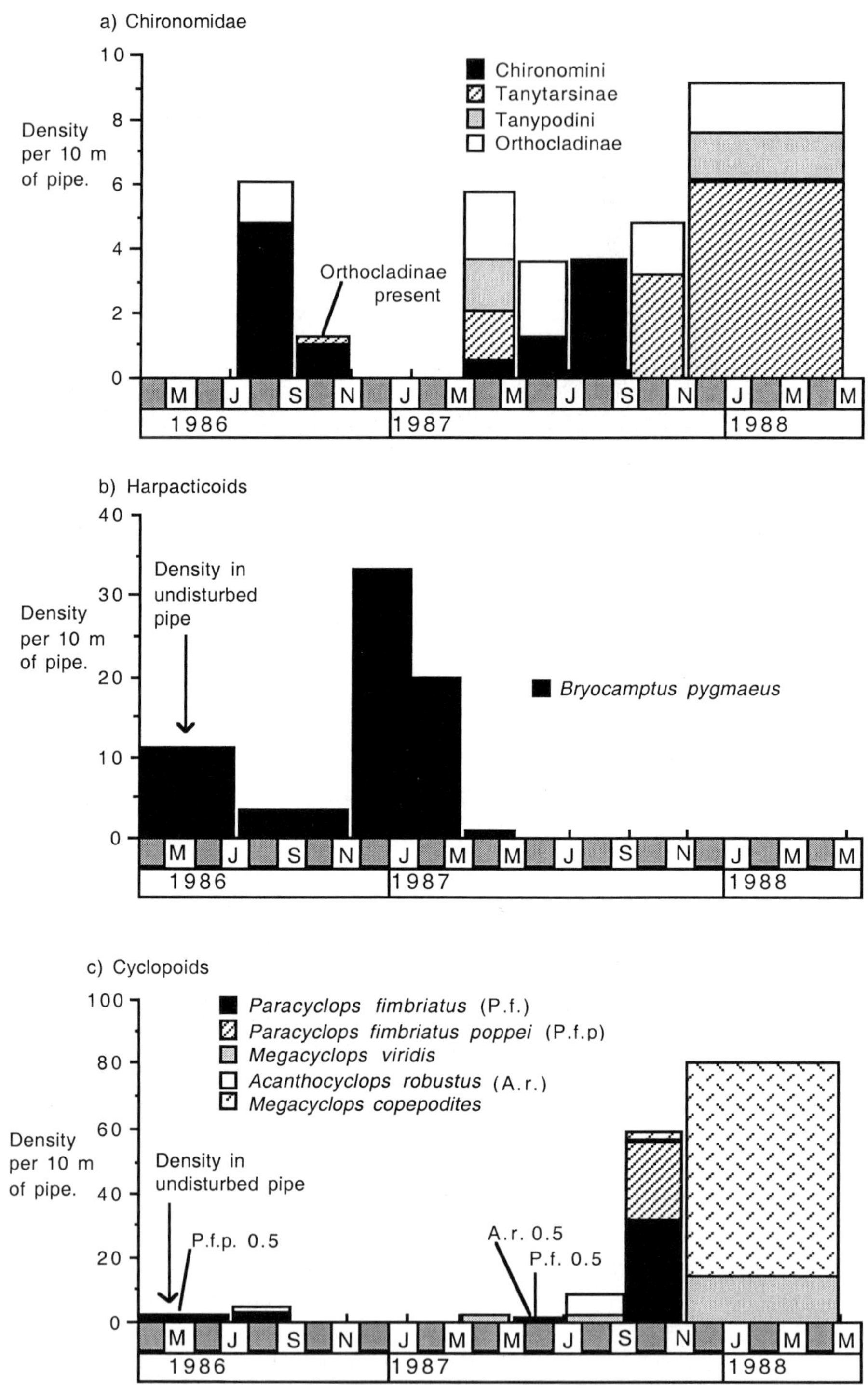

Figure 4. (a–c). Animal density per 10 m length of pipe in a rural site in Northhamptonshire not experiencing a change in source water treatment.

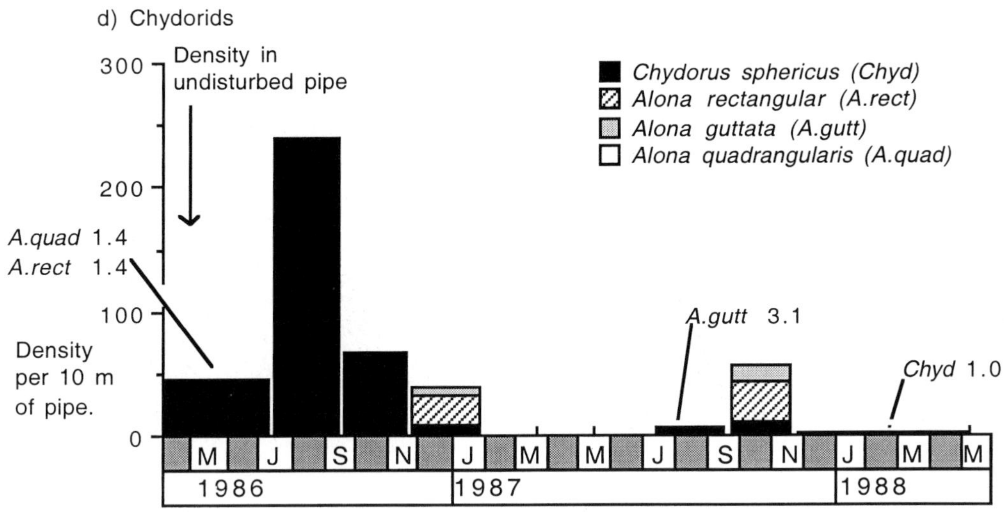

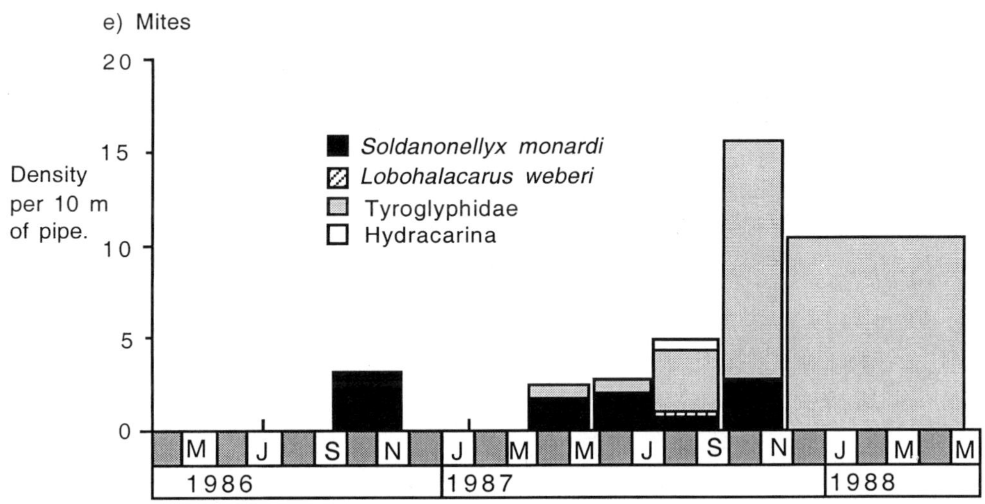

Figure 4. (d–e). Animal density per 10 m length of pipe with no change in source water treatment.

Acknowledgements

This research was funded by the Anglian Water Authority (now Anglian Water plc).The authors are grateful to Bob Markall and Peter Barham who advised the project. The views expressed are the authors' own and do not necessarily relate to views held by Anglian Water plc.

References

Barham, P., 1985. A preliminary account of the biological monitoring of Wing Treatment Works and the primary storage reservoirs at Beanfield. Unpublished Report. Anglian Water, Oundle, Peterborough.

Burfield I. & D N. Williams, 1975. Control of parthenogenetic chironomid with pyrethrins. Wat. Treat. Exam. 24: 57–67.

Collingwood, R W., 1970. Removal of algae and animals, in 'Water Treatment in the 70s' Proceedings of a joint symposium. Society for Water Treatment and Water Research Association, University of Reading, England. Jan 5th–7th 1970.

Collingwood, R W., 1977. Survey of eutrophication in Britain and its effects on water supplies. W.R.C.Technical Report T.R.40. Water Research Centre, Medmenham.

Evins, C. & G F. Greaves, 1979. Penetration of water treatment works by animals. W.R.C. Technical Report T.R.115. Water Research Centre, Medmenham.

Greaves, G F. & C. Evins, 1972. A study of animal penetration through two water treatment plants. I.L.R.170. Water Research Centre, Medmenham.

Houghton, G U., 1968. Observations on *Asellus* problems in South Essex. Wat. Treat. Exam. 17: 127–133.

Kelly, S N. 1955. Infestation of the Norwich, England water system. J. Am. Wat. Wks. Assn. 47: 330–334.

Mitcham, R P. & M W. Shelley, 1980. The control of animals in water mains using permethrin – a synthetic pyrethroid. J. Inst. Wat. Eng. Sci. 34: 474–483.

Rowlands, C L., 1982. Synopsis of characteristics of water supply filters. C.L.R.Rep.613D1, Severn Trent Water Authority (Soar Division), Leicester.

Smart, A C., 1987. Animal Infestation of Water distribution systems. Internal Report. Anglian Water, Oundle, Peterborough.

Smart, A C., 1989. An investigation of the ecology of water distribution systems. Unpublished PhD Thesis, University Of Leicester.

Williams, D N., 1974. Infestation by parthenogenetic chironomid. Wat. Treat. Exam. 23: 215–229.

Section Five

Regional and National Strategies for Lake and Catchment Management

Hydrobiologia **395/396:** 389–401, 1999.
D.M. Harper, B. Brierley, A.J.D. Ferguson & G. Phillips (eds), The Ecological Bases for Lake and Reservoir Management.

Legislative challenges for lake eutrophication control in Europe

Helen Wilson
Department of Environmental Sciences, University of Plymouth, U.K.

Key words: legislation, eutrophication, phosphorus, nitrogen, wildlife conservation, SSSI

Abstract

There is no single piece of legislation which comprehensively addresses the problem of eutrophication of freshwater lakes. An examination is made of the extent to which national and EU laws applicable in England and Wales may be used to reduce phosphorus inputs to eutrophicated lakes. This case study is then used to outline the shortcomings of existing legislation and the future challenges for lake eutrophication control. Applicable legislation may be divided into that relating to nature conservation, to water quality and to agricultural extensification. Nature conservation laws are applicable, in that lakes designated as Sites of Special Scientific Interest (SSSIs), are theoretically safeguarded from degradation. The main limitation of the SSSI system lies in the boundary definition process under which, as a rule, no other category of legally protected land may be designated. Thus, the use of surrounding buffer land to protect water quality at the catchment level is generally precluded. The introduction of consultation areas around SSSIs of international importance under the EU Directive on Habitats is unlikely to alleviate the problem, since the area involved remains minimal compared to the extent of the majority of catchments. Intensive agricultural practices are responsible for significant nutrient enrichment of rivers and lakes, so that legislation aimed at extensification is also relevant. Examples of agricultural extensification include the Environrnentally Sensitive Areas programme and various options available for Set-Aside land. However, the prime purpose of such initiatives is the alleviation of surplus and budgetary problems and, as such, they fall short of a fully integrated approach to the ecological management of farmland. Nutrient enrichment is, essentially, a water quality issue, but policy and legislation in this area are not yet sufficiently developed to address the problem comprehensively. The current regulatory process for water quality carries the potential to work comparatively well for point sources under the system of consents to discharge. This potential is limited, however, by the paucity of information available on ecologically acceptable concentrations of phosphorus in discharges. In addition, the consents system is not constructed to deal effectively with diffuse agricultural losses of phosphorus, since, unlike point sources, these tend to arise from the cumulative effect of many activities. The main legislative challenge for lake eutrophication control lies in the area of diffuse agricultural losses of phosphorus. In this respect, experience in the U.S.A. reveals that the use of comprehensive and catchment-wide 'Best Management Practices' is capable of producing significant water quality improvements, providing that some degree of mandatory compliance is incorporated.

Introduction

Perceptions of the problem of freshwater eutrophication appear to depend on the criteria adopted for assessment. For many people it is associated with major ecological changes, such as the loss of submerged macrophyte communities and the production of algal blooms. These changes are, however, more symptomatic of the eutrophication of naturally productive ecosystems. From the point of view of nature conservation, the enrichment of oligotrophic waters, through the addition of much lower levels of increased nutrient influx, may be equally important (Wilson et al., 1993). Relatively small increases in nutrient supply may cause loss of species, so that by the time more obvious symptoms emerge, community structure may be already severely disrupted. The addition of any level of nutrient which causes a change in species composition and trophic structure is, in ecological terms, eutrophication.

Narrow interpretations of what constitutes eutrophication have led to confusion about the extent of the problem in England. A report by Lund (1980), commissioned by the Soap and Detergent Industry Association, concluded that a eutrophication problem does exist, but that it is highly localised. This idea was perpetuated by the government's interpretation of a further report by Lund & Moss (1990). The Royal Commission on Environmental Pollution (RCEP) Report on Freshwater Quality noted that the government's view on the subject was that "eutrophication is a localised problem in the U.K." (RCEP, 1992, p. 64). This statement was based on the idea that eutrophication is only a problem if it "causes an increase in the growth of algae or other plants and shows a noticeable deleterious effect on the environment" *(ibid.)*.

A comparison of orthophosphate concentrations in rivers with those of natural background levels, detailed in the RCEP Report, revealed that virtually all major rivers in the then NRA regions of Anglian, Thames and Severn–Trent were heavily enriched, as were a high proportion in the North-West, Yorkshire, Southern and Wessex regions. Taking into consideration many other bodies of evidence, the RCEP concluded that "nutrient enrichment would appear to be widespread over large parts of the country" (RCEP, 1992, p. 64). English Nature, in its submission to the House of Lords Select Committee on Municipal Waste Water Treatment, stated that nutrient enrichment is the main environmental problem for inland waters in Great Britain (House of Lords, 1991). A total of 112 river, canal and lake Sites of Special Scientific Interest (SSSIs) were considered by EN to be affected by eutrophication. A survey by Carvalho & Moss (1995) revealed that of 96 SSSIs or groups of SSSIs examined, 84% showed symptoms of eutrophication, and that the conservation status of 72% had been reduced as a result.

If the definition of eutrophication is restricted to the more extreme cases, then it is indeed localised. However, if a wider and scientifically more accurate meaning is used, nutrient enrichment is a widespread and common environmental problem. Its impact requires recognition at government level, via legislative controls on nutrient losses (Wilson et al., 1996). Reductions in external nutrient loadings are a prerequisite for the restoration of eutrophicated lakes and reservoirs (Loehr et al., 1989), even in those cases where internal loadings are sufficient to maintain an elevated trophic state (Sas, 1989). Several comprehensive publications have been produced (e.g. Ryding & Rast, 1989; Cooke et al., 1993), which describe numerous ways in which such objectives may be achieved. They include phosphate stripping at sewage treatment works, diversion of phosphorus-rich effluents, upgrading of stormwater management facilities, reduction in the intensity of agriculture and the establishment of riparian buffer zones. Since there is presently no legislation dealing specifically with eutrophication, it is necessary to examine those Acts and Regulations which address different aspects of the problem. Those most relevant to eutrophication of surface waters may be categorised as relating to nature conservation, to agriculture and to water quality (Figure 1). A summary of the ways in which the various pieces of legislation outlined may be applied to the reduction of external nutrient inputs is illustrated in Figure 2.

Nature conservation

Taking into account the impact of eutrophication on aquatic flora and fauna (Moss, 1988), it is necessary to consider the extent to which current nature conservation legislation protects important aquatic habitats.

Sites of special scientific interest:

The foundation for nature conservation is the system of designating Sites of Special Scientific Interest (SSSIs) under the Wildlife and Countryside Act

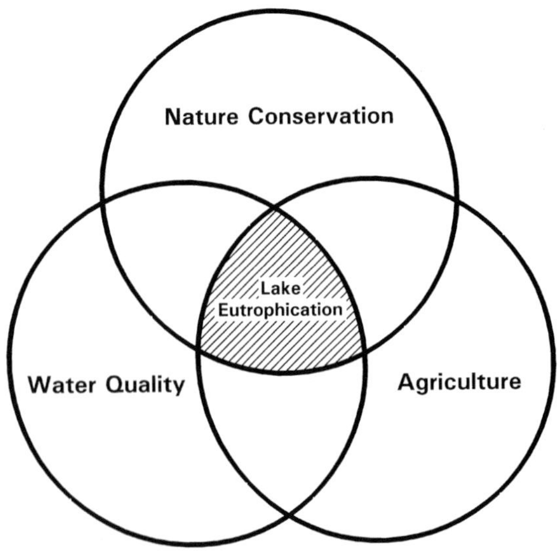

Figure 1. General areas of legislation applicable to the reduction of nutrient loadings on eutrophicated lakes.

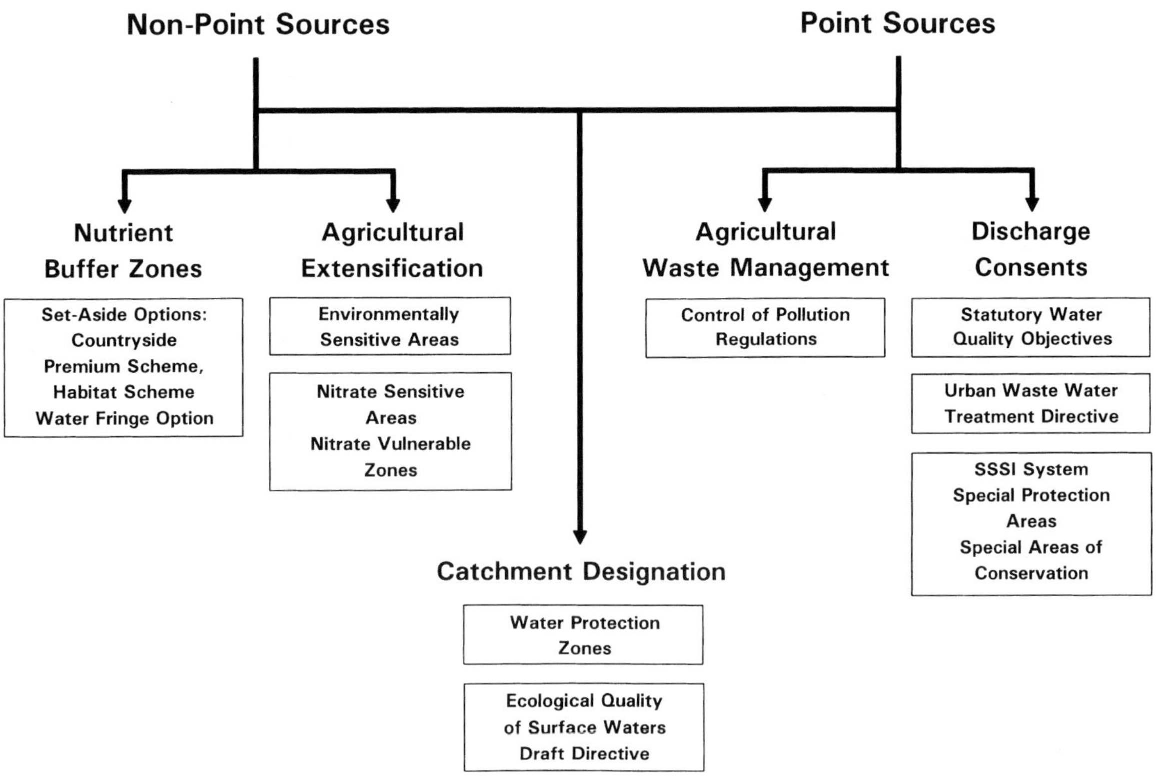

Figure 2. Application of the provisions of various regulations, acts and directives to the reduction of nutrient loadings.

(1981) (as amended). SSSIs are areas of land, or inland water, of special interest by reason of their flora or fauna, or for their geological or physiographical features (Nature Conservancy Council, 1989). Biological SSSIs are selected according to criteria and standards set out in guidelines, primarily designed to enable identification of sufficient habitats to support viable populations of most British wildlife species. For standing waters, site selection first requires classification according both to open water and to emergent vegetation, so that each type may be represented. Once classified, selection is based on further criteria which include species richness, presence of a range of *Potamogeton* species, diversity of physical features, rarity of site type, paleolimnological and geological attributes, presence of an ecological series and naturalness of the catchment. Statutory bodies responsible for the selection and protection of SSSIs in England and Wales are, respectively, English Nature (EN) and the Countryside Council for Wales (CCW).

Questions of SSSI boundary definition often pose problems, particularly for vulnerable sites requiring surrounding 'buffer' land in order to ensure full protection from, in this case, diffuse sources of nutrients. The same guidelines state, however, that there can be no other category of protected land with legal status; surrounding land must be of sufficient interest to be included within a site. It is well recognised that land use changes within the catchments of freshwater SSSIs may affect the special interest of the site. Consequently, where a catchment is still natural or seminatural, and especially where it is small and discrete, the guidelines recommend that the whole of it should be included within the SSSI. Alternatively, a buffer zone around the water may be incorporated which may consist of a grass field's width or the area defined by the first break of slope. A minimum width of 5 m is recommended, or 50 m where there is a threat of afforestation. In practice, therefore, land of lesser intrinsic scientific interest may be designated if it forms part of the same environmental unit as that of interest, but the legal status of surrounding buffer zones is dubious (Ball & Bell, 1994). In relation to eutrophication, the weakness of the boundary definition process lies in the minimal extent of protected area.

Other pieces of legislation impose specific duties in relation to SSSIs. Under the Water Industry Act (1991) and the Environment Act (1995), EN and CCW must notify bodies such as water and sewage undertakers and the Environment Agency of SSSIs which may be affected by their operations. These bodies, once notified, must consult EN or CCW over any intended activity which they think is likely to damage or destroy a site. The Environment Agency is obliged to confer with EN or CCW before granting abstraction licences, and discharge or land drainage consents thought likely to damage an SSSI. These duties relate to all operations and apply to activities in the vicinity of an SS SI. This aspect is particularly pertinent to freshwater SSSIs, since applicable activities include upstream discharges, such as those from sewage treatment, which may contain high levels of nutrients. However, consultation does not necessarily mean that reservations expressed by EN or CCW are taken fully into account. In general, the onus lies with EN or CCW to produce evidence that elevated levels of nutrient inputs would produce detrimental effects on the special interest of sites. In addition, if the relevant bodies fail to consult EN or CCW, there is no recourse unless successful actions for judicial review are brought.

Under the Town and Country Planning (General Development Procedure) Order (1995), where an application for planning permission relates to, or may affect, an SSSI, the local planning authority (LPA) is required to consult EN or CCW. Planning applications for certain developments may also be the subject of Environmental Assessments (EAs), under the Town and Country Planning (Assessment of Environmental Effects) Regulations (1988), implementing EC Directive 85/337. Criteria used to assess the necessity for EAs include proximity to sensitive locations, such as SSSIs. Many activities which damage SSSIs do not, however, require planning permission, because either they are not considered development or they are exempt. Automatic planning permission, or 'permitted development right' (PDR) is granted for many activities, including developments of a minor or temporary nature, those carried out by a wide range of public services, such as water, gas and electricity, and those of favoured activities, especially agriculture and forestry.

In 1991, the Department of the Environment (DoE) issued a consultation paper on proposals for additional planning safeguards for SSSIs. These were required in order to comply with the EC Directives on the Conservation of Wild Birds 79/409/EEC and on the Conservation of Natural Habitats and of Wild Fauna and Flora 92/43/EEC. It is necessary to emphasize that, in order to qualify for national or international designation, sites must already warrant SSSI status. The degree of protection afforded to SSSIs is not, however, sufficient to conform with that required by the above Directives. Certain alterations to planning legislation were therefore necessary, in order to ensure greater protection.

The EU directive on birds

This Directive took full effect in 1981 and requires Member States to adopt measures to maintain sufficient diversity of habitats for all species occurring in their territory, and to designate Special Protection Areas (SPAs) in order to conserve the habitats of certain listed rare, or vulnerable, species. SPAs must be protected so as to avoid pollution, habitat deterioration and disturbance. The relevance of this Directive is that many listed species are wetland birds dependant on aquatic flora and fauna for food. Demise of these sources with eutrophication therefore constitutes habitat deterioration. The provisions of the Wildlife and Countryside Act (1981) achieve broad conformation with the requirements of the Directive, but many SPAs are protected only as SSSIs, without management agreements to ensure maintenance of habitat.

The EU directive on habitats

Although the Directive on Birds has undergone adaptations in terms of species listed in the Annexes, the only amendment to the actual wording comes through the Directive on Habitats, which replaces the Article concerning measures to be taken with regard to SPAs. The Directive represents a broadening of the Directive on Birds to encompass the full range of wild fauna and flora and natural habitats.

The Directive on Habitats was adopted in May 1992 and took effect in May 1994. It aims to conserve habitats of ecological significance, either in themselves or as hosts of threatened species in a European context. Member States are required to establish Special Areas of Conservation (SACs), with the objective of creating by the year 2000 a network (Natura 2000) within Europe. In England, implementation is through the Conservation (Natural Habitats etc.) Regulations (1994).

Once a site is designated, Member States must take appropriate measures to avoid deterioration of the habitat, or disturbance of species for which it was notified.

In the British government's opinion, existing legislation already provides a sound basis for implementation of the Directive, but additional measures apply where conservation and protection were previously achieved through policy. Accordingly, the Habitats Regulations contain amendments to planning legislation and to the Wildlife and Countryside Act. A potentially effective innovation is that the government has requested English Nature and the Countryside Council for Wales to define consultation areas around all SSSIs, giving priority to sites of international importance (DoE, 1994). Consultation areas would not normally reach beyond about 500 m from the boundary of SSSIs, although for wetlands they may extend 2 km upstream. For sites with relatively small catchments, a 2 km consultation area may well be adequate. However, as Figure 3 illustrates, for many catchments such a distance is minimal.

With respect to SPAs and SACs, LPAs must establish whether a proposed development is likely to have a significant effect in terms of the ecological objectives for which the site was designated. Developments may only proceed if the integrity of sites is not affected, or if no alternative sites are available, and the project has to be carried out for imperative reasons of overriding public interest, including those of a social or economic nature. If any development proceeds which damages or destroys a site, the Secretary of State is required to ensure that compensatory measures are taken to ensure the overall coherence of the Natura 2000 network. Government is also reviewing other legislation in order to ensure that, in line with the Directive, activities of statutory bodies do not produce adverse effects on SPAs or SACs.

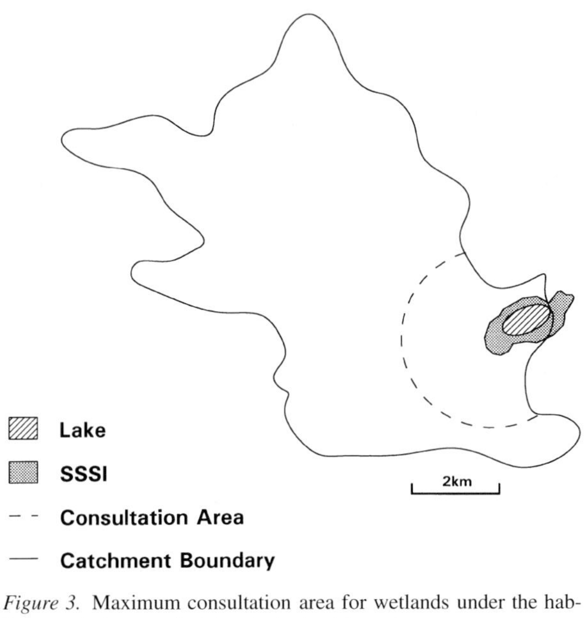

Figure 3. Maximum consultation area for wetlands under the habitats regulations applied to a typical catchment.

The government's position on the issue of presumption against development in SSSIs is stated thus:

"The Government has considered the possibility of introducing a general presumption against development in SSSIs, ... but it does not propose to do so. It considers that the measures proposed ... will ensure that nature conservation interests are taken into account. It is also mindful that notification as an SSSI is a purely scientific statement of the special wildlife and geological interest of the site and is not the only factor to be taken into account in considering planning applications." (DoE, 1991, p. 9).

The legal and political ramifications of a presumption against development in SSSIs would be significant. Nevertheless, the implications of EU legislation to protect internationally important sites may necessitate the adoption of a policy against granting planning permission in these areas (Ball & Bell, 1994). The position of nature conservation in political thinking is that it is seen as but one interest among many which must be considered in disputes over land use. Admittedly, nature conservation is currently granted more recognition as one of these interests than ever before, but in very few circumstances is its status sufficient to warrant absolute priority. This is reflected in the assertion of the DoE, cited above, that notification of land as an SSSI is a *purely scientific* statement. In essence, this implies that nature conservation merits no political, social or economic weighting in being assessed alongside other interests, and that only 'science' will benefit from the protection of sites.

Regulations introduced in order to implement the Directive on Habitats tend to concentrate primarily on protecting sites from adverse developments. However, the deterioration of habitats may not be prevented by this approach. For instance, few provisions have been made for the positive management of sites. In addition, the problem of activities which lie outside the planning system, particularly agriculture, has not been addressed at all. For many eutrophicated sites, measures necessary for the maintenance of habitat may involve reduction of nutrient loadings from wastewater treatment. However, the more limited conservation duty placed on the Environment Agency with respect to its pollution control functions, compared to that of the former National Rivers Authority (NRA), obfuscates responsibility for this aspect of the eutrophication problem. The Directive will provide a rigorous standard against which to test the effectiveness of British nature conservation laws. Designated sites may require stronger protection than is possible even under the amended system.

Agricultural extensification

Agricultural contributions to nutrient loadings on eutrophicated systems are often significant (Wilson, 1995). Elevated levels of nutrient losses are largely the consequence of intensive farming practices encouraged by agricultural policy and legislation. EC/EU policy favouring high levels of production has led to loss of wildlife habitat and landscape through supporting intensification and reclamation of marginal land. The Ministry of Agriculture (MAFF)'s current commitment to conservation is part of a general shift in EU agricultural policy whereby environmental programmes are assimilated because they provide opportunities for alleviation of surplus and budgetary problems. These policy changes promote what is now known as 'extensification' of agriculture, which, regardless of intended objectives, provides opportunities for environmental improvement. The most relevant programmes discussed here are Environmentally Sensitive Areas (ESAs), under which less intensive farming practices are supported, and options available for Set-Aside land which provide opportunities for establishment of nutrient buffer zones.

Environmentally sensitive areas

The relevant Article enabling Member States to introduce ESAs was implemented by the Agriculture Act (1986). Selection of ESAs by MAFF is conducted in consultation with statutory nature conservation advisors and interested parties such as the National Farmers Union. Farmers are offered five year agreements which award them annual payments in return for following a prescribed set of farming practices, including restrictions on fertilizer use and grazing levels. According to MAFF, uptake of the ESA scheme is appreciable, especially in England, where, by the end of the second year, about 87% of total targeted area of land was entered (MAFF, 1989). The attraction of the ESA scheme for farmers, compared to SSSI regulation, is the fact that it is entirely voluntary, less restrictive, simpler to enter, and dealt with by agricultural officials, rather than conservation personnel (Whitby & Lowe, 1994). On the whole, ESAs have gained support from both the farming and the environmental lobbies. The scheme represents a significant divergence for agricultural policy, creating important precedents for the future. The selection procedure, where initial lists of potential ESAs are submitted to MAFF by the nature conservation agencies, ensures that the scheme is targeted at areas which will benefit most. Many of the prescribed practices have a direct impact on reduction of nutrient losses. Nevertheless, ESAs fall short of a fully integrated approach to environmental management, and legitimate criticisms of the scheme exist.

MAFF has rejected calls both from farming and from conservation bodies for a more widespread application of the scheme, which suggests that ESAs remain a well-publicised, but minor component of agricultural policy. The scheme may even be seen as a distraction from a more general reform of agriculture (Whitby & Lowe, 1994). Indeed, amendment of certain aspects of MAFF and EU policy could greatly reduce the need for ESA designation in some sectors. A second criticism relates to the voluntary nature of the scheme. Those farmers who choose not to register, or who enter only part of their farm, may remain eligible for MAFF grants to undertake environmentally damaging improvements. While it is necessary to promote goodwill among farmers, some degree of cross-compliance would not necessarily detract from the appeal of the ESA scheme. For ESAs where monitoring and evaluation are complete, it appears that the scheme has slowed down rates of damage to existing landscapes, habitats and features, but the extent to which conservation interest has been enhanced is limited to a few areas. As EU agricultural policy generates much less pressure for agricultural intensification, and convergence with ESA policy occurs, justification for the latter may be reduced, unless the benefits of positive management become more apparent.

Set-Aside

In 1988, the EC introduced Set-Aside for arable land under Regulation 1272/88, in an attempt to reduce production. This was implemented by Set-Aside Regulations (1988), and later various amendments. The scheme provides for arable land to be removed from production voluntarily by farmers, for periods of five years. Under the basic Scheme, land may be left fallow (permanently or in rotation), used for extensive grazing, or non-agricultural purposes, or converted to woodland. Opportunities exist to exploit Set-Aside for nature conservation, and nutrient export reduction in particular. For example, nutrient buffer zones along streams may be established using fallow or woodland options.

The main disadvantages of the approach adopted under the basic Scheme are that:

1. Uptake is entirely voluntary, and therefore untargeted.
2. Additional management for the benefit of nutrient reduction may not be eligible for compensation.
3. The basic Scheme applies only to arable land.
4. The minimum period for which farmers may enrol land is five years.

It is clear therefore, that further measures are necessary for Set-Aside land to serve eutrophication control objectives. Two Schemes currently exist which aim to enhance nature conservation interest on land taken out of agricultural production.

Countryside premium scheme

In 1989, the Countryside Commission (CC) launched the Countryside Premium Scheme (CPS). The aim is to achieve environmental and recreational objectives through financial incentives, additional to payments under Set-Aside (CC, 1993). Unlike basic Set-Aside, CPS is discretionary, with the CC retaining the right to refuse or alter inappropriate applications. Suitable land may be eligible for grants to establish a number of habitats, including wooded margins, meadowland and wildlife fallow. The creation of these habitats may provide benefits, in terms of reduction in nutrient losses, since no organic or inorganic fertilizers are allowable in their establishment or management. In addition, fallow and wooded margins may act as barriers to soil erosion and nutrient losses. Habitats to be re-instated include wetlands, allowing the application of this option to lake and stream fringes in order to enhance both nature conservation interest and nutrient retention capacity.

Habitat scheme water fringe option

A separate long-term set-aside programme, known as the Habitat Scheme (HS), is currently applied in pilot areas (MAFF, 1994). Authority comes from EC 'Agri-Environment' Regulations 2078/92, which constitute part of CAP reforms. The HS aims to encourage longer term retirement of land from production in selected areas, for creation or improvement of a range of wildlife habitats. The Scheme offers payments per hectare over periods of twenty, rather than five, years, and, unlike basic Set-Aside, applies to permanent grassland as well as arable. Of particular interest is the option to recreate, restore or extend riparian habitats under the Habitat (Water Fringe) Regulations (1994). Selection criteria for pilot areas include river and lake SSSIs suffering from diffuse pollution or fisheries subject to silting. MAFF has so far designated six pilot Water Fringe Areas. The scheme is structured around a number of management options for land adjacent to water courses and lakes, including withdrawal from production of strips of land, and extensive grassland management in whole fields. Any land accepted under these options may also be entered for a raised water level supplement, the purpose of which is to intercept field drainage flows in order to reduce nutrient and sediment export, or to maintain water levels in ditches during winter for the benefit of wildlife.

Since the Water Fringe option is currently at the pilot stage, modifications may yet be applied. Nevertheless, the inclusion of interception of field drainage only as an additional option diminishes its effectiveness to reduce diffuse nutrient loads on eutrophicated rivers and lakes. Some benefit to water quality may accrue from cessation of fertilization and lower stocking densities, but without the interception of underdrainage it is unlikely that improvements will be detectable. Although the scheme is targeted to the extent that sites are carefully chosen by MAFF, in consultation with EN and other appropriate agencies, there remains the usual problems associated with a voluntary programme. Nevertheless, the HS represents an important progression towards integration of environmental objectives into agricultural policy, and recognition of the effects of intensive farming on water quality.

Water quality

Water quality legislation performs an increasingly important role in controlling nutrient concentrations in rivers and lakes. It is here that future policies relating to eutrophication are most likely to be enacted. The Water Resources Act (WRA) (1991), which applies to England and Wales, superseded Part II of the Control of Pollution Act (1974) (COPA). The system of consents for discharges remained much the same, but potentially important changes in relation to eutrophication occurred in other areas. These included establishment of the NRA, introduction of statutory water quality standards and powers to designate water protection zones and nitrate sensitive areas. Sections of the WRA are now superseded by the Environment Act (1995), which transferred the functions of the NRA to the Environment Agency (henceforth referred to as 'the Agency') from 1st April 1996.

The Environment Act requires the Agency to undertake the duties of the former NRA in relation to pollution control, flood defence, water resource planning and management, navigation, freshwater fisheries and water-related conservation and recreation. The primary means by which control of polluting discharges is exercised is through the system of consents to discharge. Conditions are usually attached to a consent, the most significant of which relate to biochemical oxygen demand, levels of toxic or dangerous materials and suspended solids (NRA, 1994a). Generally, nutrient levels are not subject to consent limits.

The water quality implications of a proposed development acts as a 'material consideration' in planning decisions, and as such, the Agency is a statutory consultee for all applications under the General Development Order, although its advice may be disregarded (Ball & Bell, 1994). The Agency also holds a more general power to influence development plans, either directly or through the publication of policy statements. The former NRA's *Guidance Notes for Local Planning Authorities on the Methods of Protecting the Water Environment Through Development Plans* (NRA, 1994b) contains strong advice to resist development which could adversely affect water quality. LPAs are recommended to avoid allowing development on recognised floodplains, and to support initiatives to conserve and enhance river corridors, in recognition of their importance for water quality and nature conservation, of their contribution to the character of the landscape, and as amenity areas.

Additional duties of the Agency relate to conservation and recreation issues. An aspect of direct relevance here is the limited nature conservation duty placed on the Agency in the Environment Act (1995). The duty on the former NRA to "further" nature conservation interests applies to all the Agency's functions except that of pollution control, in the execution of which it need only "have regard to the desirability" of conserving nature. This wording leaves the Agency with a weaker conservation duty towards SSSIs than the water companies and internal drainage boards which it will regulate.

Control of pollution regulations

Under the WRA, the Secretary of State is empowered to make regulations relating to any polluting substances in order to prevent them from entering controlled waters. The only example so far, are the Control of Pollution (Silage, Slurry and Agricultural Fuel Oil) Regulations (1991). These introduce specific controls over storage of these substances, and apply to new or substantially altered facilities. Existing stores may be brought under control if it is believed that there is a significant pollution risk. The Regulations provide the means by which agricultural point sources of nutrients may be addressed.

Nitrate sensitive areas

Provisions for designation of Nitrate Sensitive Areas (NSAs) were included in the WRA. Upon application by the then NRA, MAFF was enabled to make an order designating land as an NSA, in order to prevent or control entry of nitrate into controlled waters as a result of agricultural production. NSAs were designed primarily for protection of potable supplies of groundwaters, in order to reduce nitrate levels to below the 50 mg/l at which human health is thought to be at risk. The introduction of NSAs was in advance of more stringent provisions under the EC Directive on Nitrates (see below). The relevance of NSAs to surface water eutrophication is limited by several factors. First, they relate entirely to nitrates, and the productivity of most rivers and lakes is limited by the availability of phosphorus (Canfield et al., 1989). Second, the nitrate limit of 50 mg/l far exceeds concentrations found in the majority of eutrophicated systems (Johnes et al., 1994). Third, NSAs are targeted at losses to groundwater, rather than surface waters. Fourth, nitrate fertilizer application rates may be adjusted without necessarily changing corresponding amounts of phosphorus (Chalmers et al., 1990). Nevertheless, various land use options available under the scheme, such as converting arable or intensive grassland to extensive grassland, and use of autumn, rather than spring-sown crops, would be beneficial in reducing erosion, and therefore particulate phosphorus transport.

Water protection zones

Under the WRA, areas may be designated as Water Protection Zones (WPZs) by order of the Secretary of State, in response to an application by the former

NRA. Such orders confer the power to prohibit or restrict activities likely to lead to pollution of controlled waters, but no provision exists for compensation payments. The provisions are essentially those contained in COPA, but never brought into force. The main difference is contained in a new provision which states that WPZs may not be used to control nitrate losses from agricultural land, since this function is deemed to be provided by the designation of NSAs. WPZs provide an ideal legal instrument for reducing diffuse losses of phosphorus to surface waters, thereby addressing eutrophication at the catchment level.

The only application by the former NRA for WPZ designation was not, however, directed at diffuse sources, but at industrial pollution in the catchment of the River Dee. The reluctance to utilize WPZs for the prevention of diffuse pollution may reflect unwillingness to press government for implementation of a mandatory scheme for which no compensation is payable, and which focuses on sources that are "less easily controlled . . . [and] cannot be reliably sampled by conventional means" (NRA, 1993, p. 11). With the prospect of compulsory regulation of nitrate losses from farmland under the Directive on Nitrates, it is not surprising that implementation of WPZs to control other diffuse losses is strongly opposed by agricultural interests. The case for mandatory controls on diffuse sources of phosphorus is not easily made under these circumstances. It is likely therefore that, despite their evident applicability, WPZs will not be used to address eutrophication in the near future.

Statutory water quality objectives

Powers to introduce Statutory Water Quality Objectives (SWQOs) under the WRA represent an important innovation, providing potential for establishing a coherent national policy on water quality. SWQOs will be legally binding targets for water quality to be achieved within certain timespans. They will apply to six Use Classifications (UC), one of which, the 'Special Ecosystem' UC relates to sites requiring protection for nature conservation. An additional General Quality Assessment Scheme will facilitate periodic statements to be made on the quality of controlled waters. Once UCs are established by appropriate Regulations, the Secretary of State will set SWQOs for a number of shortlisted pilot catchments, incorporating EU standards where relevant. After SWQOs are formally confirmed for a particular catchment, the Agency and the Secretary of State are obliged to use their powers to achieve required standards, if practicable. For the Agency, such powers include those related to setting consents, as well as those concerned with enforcement, and preventative and remedial actions.

In May 1994, the Surface Waters (River Ecosystem) (Classification) Regulations came into force. These granted considerable leeway to the former NRA in assessing whether results obtained from samples indicate compliance with SWQOs. Where it is believed that high algal levels are responsible for elevated BOD, the readings may be ignored (NRA, 1994c). In other words, even where failures are directly the result of eutrophication, non-compliant waters will appear to meet standards. The DoE stated that the first tranche of SWQOs would be set only "to consolidate existing objectives and not to take big steps forward to higher environmental standards" (ENDS Report, 1993, p. 35). Adoption of 1990 river data as the baseline for maintaining water quality also limits the extent to which implementation of SWQOs will achieve real environmental improvement. In that year, water quality was the worst on record, partly as a result of drought. Acceptance of 1990 as a baseline, therefore ignores the deterioration which occurred during the 1980s.

The initial round of SWQOs applies only to rivers, and not to lakes, estuaries or coastal waters. The development of a classification system for lakes is being addressed through the Agency's national R & D programme. Criteria for other UCs are still under development, and it may be some time before these are addressed, since numerous holdups appear to be delaying the introduction of those for rivers. The accelerated introduction of SWQOs was suggested by the Chief Scientist of the former NRA, Dr Jan Pentreath, who believed it was the only way to fill the credibility gap in government policy on this issue (ENDS Report, 1994).

The impact of the EU on water quality legislation has been highly significant over the last few years. Several Directives contain objectives for nutrients, but only three specifically address eutrophication as part of wider concerns for water quality. These are the Directives on Nitrates and on Urban Waste Water Treatment, and the proposed Directive on the Ecological Quality of Surface Waters.

Directive on nitrates

The Directive Concerning the Protection of Waters Against Pollution Caused by Nitrate From Agricultural Sources 91/676/EEC requires Member States to identify waters affected by nitrate pollution. These are defined as surface freshwaters, or groundwaters intended for drinking, likely to contain more than 50 mg/l nitrate, or any freshwater lakes, estuaries or coastal waters liable to suffer from eutrophication if protective action is not taken. Member States must designate, as vulnerable zones, those areas which drain into such waters and which contribute to pollution.

Proposals by DoE and MAFF to designate around 70 Nitrate Vulnerable Zones (NVZs) were produced in May 1994 and revised in May 1995. However, they appear to fall short of the Directive's requirements. The area covered by the proposed NVZs is a mere 650 000 hectares, and extends only over catchrnents of ground and surface waters used for public supply. The government's narrow interpretation of the Directive will mean that other surface waters subject to eutrophication will remain unprotected.

Although the Directive is specific to nitrate, the strength of measures imposed is likely to elicit an indirect effect on phosphorus losses. Additionally, provided the Directive is implemented in full, identification of nitrogen-limited lakes which are, or may become, eutrophicated, and designation of areas draining into them, may well address this aspect of the problem. But, with respect to the majority of eutrophicated rivers, lakes and reservoirs, which are phosphorus-limited, more specific measures are necessary.

Directive on urban waste water

The Directive Concerning Urban Waste Water Treatment (UWWT) 91/271/EEC was adopted by the EC in May 1991. It sets new criteria for sewage effluents discharged to surface and coastal waters, depending on the size of sewage treatment works (STWs) and the sensitivity of the receiving water. The Directive represents the first attempt to deal directly with nutrient enrichment by sewage effluent and contains a number of measures. The most relevant measure to this discussion is that sewage discharges from STWs affecting designated 'sensitive areas' and serving over 10 000 population equivalent (p.e.) must be treated to remove phosphorus and/or nitrogen by 31st December 1998. Sensitive areas are defined as natural freshwater lakes, estuaries or coastal waters which are, or may become, eutrophicated, or surface waters used for the abstraction of drinking water which could contain more than 50 mg/l nitrate. For STWs discharging into such waters, and serving over 10 000 p.e., effluent must comply with values for concentration or percentage reduction for total phosphorus and/or total nitrogen.

The selection of sites for designation as sensitive areas was left to Member States, as is usually the case. In the U.K., this led to controversy, and served to highlight the disparity between the government's assessment of the extent of eutrophication, and that of EN and the former NRA. The DoE maintained that freshwater eutrophication was restricted to a few sites, whereas EN identified over 100 SSSIs which may be affected, and the NRA stated that 169 sites in England and Wales experienced problems with blue-green algae (House of Lords, 1991). In May 1994, the government announced that 33 sites in England and Wales would be designated under the Directive as sensitive areas. At least 40 STWs will require phosphorus and nitrogen removal by the end of 1998.

All Member States were required to introduce regulations implementing the Directive by 30 June 1993, but it was not until 30 November 1994 that the Urban Waste Water Treatment (England and Wales) Regulations (1994) came into force. Effluent standards will be regulated largely through the Agency's discharge consent system. The requirement to provide tertiary treatment will not apply, however, where the Agency is satisfied, as a result of monitoring, that the minimum percentage reduction of the *overall* load entering the catchment of a sensitive area is at least 75% for total phosphorus and at least 75% for total nitrogen. In other words, where a number of STWs are involved the average percentage reduction will suffice, and at a level lower than that stipulated by the Directive.

In response to a consultation paper on the Regulations, the Institution of Water and Environmental Management (IWEM) considered that an essential element, missing from the UWWT Directive, was reference to water quality objectives on a site-specific basis (IWEM, 1991). The Institution supported nutrient reduction where there might be a risk of eutrophication, but it felt that concentrations in receiving water were more important. The former NRA was concerned about the adoption of the percentage reduction approach for similar reasons (ENDS Report, 1994). Its Chief Scientist, Dr Jan Pentreath, told the DoE that its strategy for implementation may bring about some environmental improvements, but that it could not act as a substitute for uniform discharge limits for STWs, or for the introduction of SWQOs. Nine other sites recommended by the former NRA for designation as sensitive areas were rejected by the DoE following representation from the water companies. These, and a further twenty sites will be monitored and re-evaluated for designation by the end of 1997. EN is concerned that designations will not protect a large number of those SSSIs suffering from eutrophication:

> "The potential benefit of the Directive to nature conservation has been removed by setting a size limit of sewage works where phosphate stripping should be installed ... Small sewage discharges or a series of works along a river cause most of the problem." (House of Lords, 1991)

In addition, the enormous costs involved in implementing the Directive will reduce substantially available funds for addressing eutrophication at other sites.

A draft Directive on the Ecological Quality of Surface Waters was published in August 1994. The proposals aimed to provide a framework for control of pollution from point and diffuse sources, so as to maintain and improve the ecological quality of all surface waters. The draft Directive has since been withdrawn and incorporated into a framework Directive on water resources. It will prepare the ground for the integration of the substantial body of water legislation adopted since 1975. The Directive may provide a more comprehensive framework for water quality objectives than any other piece of EU legislation. In terms of eutrophication, point *and* diffuse sources of nutrients will need to be identified, monitored and addressed. Rather than adopting a sectorial approach, the framework Directive will attempt to deal with all aspects of pollution of surface waters, and, as such, may provide the closest approximation so far to eutrophication legislation.

Discussion

At present, there is no legislation at either the national or EU level which deals directly with the problem of eutrophication in England and Wales in a comprehensive manner. A wide range of laws and policies may be applied to the reduction of phosphorus inputs to eutrophicated surface waters, but, even collectively, these fall far short of addressing the problem. Among several criticisms which may be levelled at the SSSI system of nature conservation, is that it has proved inadequate to prevent eutrophication of freshwater SSSIs. Implementation of the Directive on Habitats is likely to underline inadequacies of the present system, and will require a stronger emphasis on positive management in order to prevent habitat deterioration, and, consequently, better protection for water quality at the catchrnent level. Integration of environmental objectives into agricultural policy is gradually changing the relationship between nature conservation and agriculture. Increasingly, legislation related to extensification of agriculture includes an environmental imperative, a trend which will inevitably extend to nutrient conservation with respect to water quality.

The regulatory system available to the Environment Agency contains the necessary mechanisms to address point sources, but the legislative power to reduce inputs to the majority of eutrophic ated sites is lacking. The system is not constructed to deal effectively with diffuse sources, since these tend to arise from the cumulative effect of many activities and are more difficult to monitor. This aspect of eutrophication control therefore remains a significant challenge, in both practical and legislative terms. There is, however, a wealth of experience and expertise in the U.S.A. in dealing with diffuse agricultural sources of nutrients. This is particularly so in terms of best management practices (BMPs), many of which have been tried and tested for over 60 years. While it is true that farming in the U.S.A. is, in many ways, different from that practiced in England and Wales, there are lessons to be learned.

Agricultural BMPs are based on extensive experience gained during the aftermath of the Dust Bowl in dealing with soil erosion. Methods and procedures have been developed and refined, in order to address the problem of diffuse sources of phosphorus. BMPs may be defined as:

> "those on-farm activities designed to reduce nutrient losses in drainage waters to an environmentally acceptable level, while simultaneously maintaining an economically viable farming operation for the grower" (Bottcher et al., 1995, p. 343).

BMPs for phosphorus control may include the use of minimal tillage, contour ploughing and cultivation, crop rotation, strip cropping, grass waterways, stubble mulches, grazing controls, conversion of cropland to pasture, irrigation and drainage management, fencing of ditches and streams, and establishment of riparian buffer zones (Davidson, 1989; Bottcher et al., 1995). In addition, emphasis is placed on soil fertility management with respect to phosphorus, rather than nitrogen. Particular consideration is given to manure applications, since those designed to provide sufficient nitrogen for crops tend to promote phosphorus accumulation in the soil (Gilliam, 1995).

The 1987 Clean Water Act (CWA) provides the legislative means by which BMPs may be implemented in the U.S.A. Section 319 requires states to develop comprehensive nonpoint source (NPS) Management Programs on a watershed basis, which must then be approved by the Environmental Protection Agency (EPA) and, significantly, provides funds for implementation (Gannon et al., 1996). States may introduce their own water quality objectives, providing they are equal to, or improve upon, Federal standards. The extent to which individual states give priority to NPS control varies enormously. Some, however, have responded to local water quality problems for which Federal laws are inadequate. In several cases, states impose mandatory measures and/or adopt permit systems for agricultural holdings over a certain size, neither of which approaches have been used at the Federal level (M. Ribaudo, USDA, pers. comm.). Several criticisms have been aimed at the CWA as a whole, and at the NPS control provisions it contains. Researchers have described the NPS provisions as "toothless" (Braden et al., 1994, p. 783) and "a start ... far from being dramatic or decisive" (Davidson, 1989, p.43). These criticisms are based mainly on the Act's reliance on voluntary implementation of BMPs, and the lack of sanctions against states which fail to prepare EPA-approved programmes, or which do not implement them.

Since the use of the voluntary approach does not appear to produce sufficient motivation to implement BMPs for eutrophication control, there exists a body of opinion in the U.S.A. which advocates mandatory enforcement. After all, " ... voluntary controls were a dream when applied to point sources; they are no different with respect to nonpoint sources" (Davidson, 1989, p. 48). It seems, therefore, that it is not so much the knowledge and experience of BMPs which is lacking in the U.S.A., but the political will to implement them fully at the national level. The challenge for lake eutrophication control in the European context appears to lie in evolving the necessary legislative and administrative structures to ensure the implementation of measures to reduce diffuse agricultural sources of phosphorus. In the absence of a concerted programme of public education about the amenity, aesthetic and nature conservation value of lakes, and under the present socio-economic system, such structures may need to include measures for the imposition of mandatory controls on agricultural activities.

Acknowledgements

The work on which most of this review is based was funded by the SERC, as part of a CASE award in collaboration with English Nature, Peterborough. The author would like to thank Paddy O'Sullivan and Simon Payne of the University of Plymouth, and Mary Gibson of English Nature, Peterborough, for their assistance and comments, and the Department of Environmental Sciences, University of Plymouth, for financial support to attend the International Symposium on the Ecological Basis for the Management of Lakes and Reservoirs, at which this paper was presented.

References

Ball, S. & S. Bell, 1994. Environmental Law: The Law and Policy Relating to the Protection of the Environment, 2nd edn. Blackstone Press, London.

Bottcher, A. B., T. K. Tremwell & K. L. Campbell, 1995. Best management practices for water quality improvement in the Lake Okeechobee watershed. Ecological Engineering 5: 341–356.

Braden, J. B., N. R. Netusil & R. F. Kosobud, 1994. Incentive-based nonpoint source pollution abatement in a reauthorized Clean Water Act. Water Resources Bulletin 30: 781–791.

Canfield, D. E., J. R. Jones, S.-O. Ryding & D. Uhlmann, 1989. Factors and processes affecting the degree of eutrophication. In: S.-O. Ryding & W. Rast (eds). The Control of Eutrophication of Lakes and Reservoirs, UNESCO Paris and Parthenon Carnforth: 65–84.

Carvalho, L. & B. Moss, 1995. The current status of a sample of English Sites of Special Scientific Interest subject to eutrophication. Aquatic Conservation: Marine and Freshwater Ecosystems 5: 191–204.

Chalmers, A., C. Kershaw & P. Leech, 1990. Fertilizer use on farm crops in Great Britain: Results from the survey of fertilizer practice, 1969–88. Outlook on Agriculture 19: 269–278.

Cooke, G. D., E. B. Welch, S. A. Peterson & P. R. Newroth, 1993. Restoration and Management of Lakes and Reservoirs, 2nd edn. Lewis, Boca Raton, USA.

Countryside Commission, 1993. Countryside Premium for Set-Aside Land – Monitoring and Evaluation 1989–1992, Countryside Commission, Cambridge.

Davidson, J. H., 1989. Thinking about nonpoint sources of water pollution and South Dakota agriculture. South Dakota Law Review 34: 20–62.

Department of the Environment, 1991. Planning controls over Sites of Special Scientific Interest in England and Wales: consultation paper. Journal of Planning and Environmental Law Bulletin May: 6–11.

Department of the Environment, 1994. Planning Policy Guidance: Nature Conservation. HMSO, London.

ENDS Report, 1993. DoE Plans for River Quality Objectives Under Fire. ENDS Report 219: 34–35.

ENDS Report, 1994. Dispute on Sewage Clean-Up Comes to the Boil. ENDS Report 228: 3–4.

Gannon, R. W., D. L. Osmond, F. J. Humenik, J. A. Gale & J. Spooner, 1996. Goal-orientated agricultural water quality legislation. Water Resources Bulletin 32: 437–450.

Gilliam, J. W., 1995. Phosphorus control strategies. Ecological Engineering 5: 409–414.

House of Lords, 1991. Municipal Waste Water Treatment Vol. II, Evidence. HMSO, London.

Institution of Water and Environmental Management, 1991. Comments on the Proposal for a Council Directive Concerning Municipal Wastewater Treatment COM(89) 518 Final. Journal of the Institution of Water and Environmental Management 5: 222–224.

Johnes, P. J., B. Moss & G. L. Phillips, 1994. Lakes – Classification and Monitoring: A Strategy for the Classification of Lakes. National Rivers Authority, Bristol.

Loehr, R. C., S.-O. Ryding & W. C. Sonzogni, 1989. Estimating the nutrient load to a waterbody, In S.-O. Ryding, & W. Rast (eds), The Control of Eutrophication of Lakes and Reservoirs. UNESCO Paris and Parthenon Carnforth: 115–146,

Lund, J. W. G., 1980. Eutrophication in the United Kingdom. The Soap and Detergent Industry Association, Hayes, UK.

Lund, J. W. G. & B. Moss, 1990. Eutrophication in the United Kingdom – Trends in the 1980s. The Soap and Detergent Industry Association, Hayes, UK.

Ministry of Agriculture, Fisheries and Food, 1989. Environmentally Sensitive Areas: First Report. HMSO, London.

Ministry of Agriculture, Fisheries and Food, 1994. The Habitats Scheme Water Fringe Areas. Explanatory Booklet. MAFF, London.

Moss, B., 1988. Ecology of Fresh Waters. Blackwell, Oxford.

National Rivers Authority, 1993. NRA Water Quality Strategy. National Rivers Authority, Bristol.

National Rivers Authority, 1994a. Discharge Consents and Compliance: The NRA's Approach to Control of Discharges to Water. Water Quality Series No.17. HMSO, London.

National Rivers Authority, 1994b. Guidance Notes for Local Planning Authorities on the Methods of Protecting the Water Environment Through Development Plans. National Rivers Authority, Bristol.

National Rivers Authority, 1994c. Water Quality Objectives: Procedures used by the National Rivers Authority for the Purpose of the Surface Waters (River Ecosystem) (Classification) Regulations 1994. National Rivers Authority, Bristol.

Nature Conservancy Council, 1989. Guidelines for Selection of Biological SSSIs. Nature Conservancy Council, Peterborough.

Royal Commission on Environmental Pollution, 1992. Freshwater Quality. HMSO, London.

Ryding, S.-O. & W. Rast, (eds.), 1989. The Control of Eutrophication of Lakes and Reservoirs. UNESCO, Paris and Parthenon, Carnforth.

Sas, H., 1989. Lake Restoration by Reduction of Nutrient Loading: Expectations, Experiences, Extrapolations. Academia Verlag, Richarz, Sankt Augustin.

Whitby, M. & P. Lowe, 1994. The Political and Economic Roots of Environmental Policy in Agriculture, In M. Whitby, (ed.), Incentives for Countryside Management: The Case of Environmentally Sensitive Areas. CAB International, Wallingford: 1–24.

Wilson, H. M., M. T. Gibson, & P. E. O'Sullivan, 1993. Analysis of current policies and alternative strategies for the reduction of nutrient loads on eutrophicated lakes: the example of Slapton Ley, Devon. Aquatic Conservation: Marine and Freshwater Ecosystems 3: 239–251.

Wilson, H., S. Payne, P. O'Sullivan, & M. Gibson 1996. Policy and legislation relevant to the conservation of freshwater SSSIs subject to eutrophication. Journal of the Institution of Water and Environmental Management 10: 348–354.

Wilson, H. M., 1995. An Evaluation of Alternative Management Strategies for Shallow Eutrophicated Lakes and Reservoirs. Unpublished PhD Thesis, University of Plymouth.

Hydrobiologia **395/396:** 403–409, 1999.
D.M. Harper, B. Brierley, A.J.D. Ferguson & G. Phillips (eds), The Ecological Bases for Lake and Reservoir Management.

Eutrophication control in the Netherlands

Diederik T. van der Molen & Paul C. M. Boers
Institute for Inland Water Management and Waste Water Treatment, P.O. Box 17, 8000 AA Lelystad, The Netherlands

Key words: eutrophication, shallow lakes, Netherlands

Abstract

Due to the limited perspectives of the Dutch national policy and due to the unique conditions and constraints of individual lake systems, tailor-made programs are needed for the restoration of freshwater lakes. Based on an evaluation of several restoration projects, a guideline is provided to enable water managers to develop such a restoration program. This guide-line is summarized in the paper and important topics, such as the definition of targets and standards, system identification and measures, are highlighted. Targets should include a natural reference, but also the constraints imposed by other functions of the lake. Monitoring and system identification are necessary to determine the extent of remedial measures and whether they should focus on external nutrient sources, internal sources or other factors, such as the food web structure. It is advocated that for lakes which have not degraded to a green, algal-dominated, turbid state, prevention or re-establishment of its ability to self-restore are preferred to 'intensive surgery' on the system. Lake re-creation can be applied if the system is heavily perturbed. In that case the re-creation of a more beneficial system is a more fair-minded approach than restoration of the natural state.

Introduction

Since the second half of the 1980s, nutrient concentrations have decreased in the main freshwater systems of the Netherlands (Portielje & van der Molen, 1998). The decrease in phosphorus concentration is mainly due to phosphorus removal at wastewater treatment plants, the use of phosphate-free detergents and industrial efforts. The contribution of the agricultural sector in the reduction of nutrient emission is negligible. Because this sector is responsible for the major part of nitrogen emissions, the decrease in nitrogen concentrations has been small compared to the decrease in phosphorus concentrations (van der Molen et al., 1998). Similar developments to those of the Netherlands have taken place in the countries that are also part of the catchment of the main river in the Netherlands, the river Rhine.

Based on the efforts and results of the last decade, it is expected that, for the next decade, only a little further improvement can be made with national, general policy. In many water systems the phosphorus concentration will reach levels of about 150 μg P l^{-1}, the national standard for basic water quality in the Netherlands, but in only a few cases are lower concentrations expected. Therefore, for lake restoration, additional measures have to be carried out on a regional scale. A guideline, based on experiences of several projects in which the external phosphorus loading was decreased, is set-up to facilitate water managers to develop tailor made restoration programs for specific lakes. In this paper, the leading thread running through the guideline is presented (Figure 1) and some of the topics will be emphasised. The guide-line is set up for large (>10 ha), non-stratifying lakes. In addition to existing manuals (e.g. Olem & Flock, 1990; Cooke et al., 1993; Moss et al., 1996), this guide-line is focused on shallow, eutrophic lakes in north-western Europe. It is developed to serve regional water managers in the Netherlands, although the topics mentioned in this paper do also apply for water managers in several other countries.

Shallow lakes have two alternative stable states, a macrophyte-dominated, clear state and a phytoplankton-dominated, turbid state (Scheffer, 1989). Both states are able to persist over a wide range of nutrient conditions and both states are buffered by a variety of mechanisms to prevent a switch to

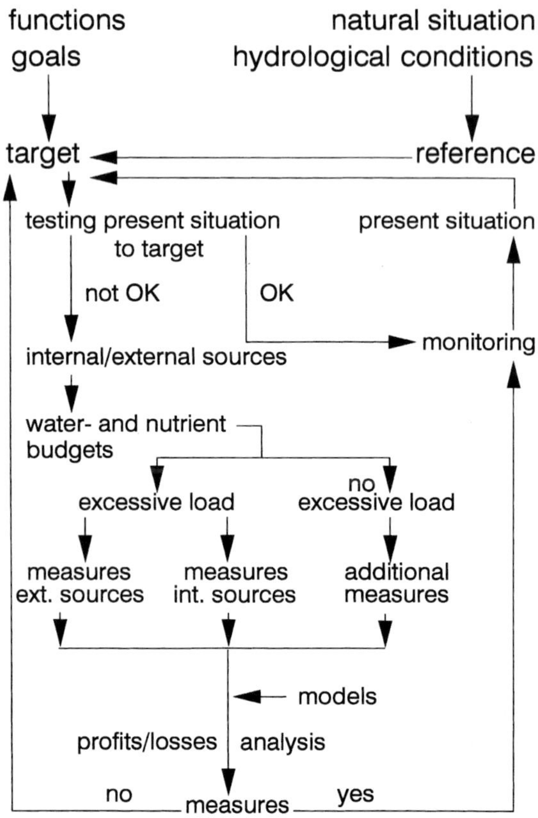

Figure 1. Schematic presentation of a guideline for the restoration of shallow lakes.

the other state. The macrophyte-dominated state is able to resist a high nutrient loading, while the phytoplankton-state can be maintained despite of low nutrient levels (Jeppesen et al., 1990). A phosphorus concentration of about 50 μg P l^{-1} will result in a macrophyte–dominated system, beyond this concentration either a macrophyte–dominated or a phytoplankton-dominated state can exist (Jeppesen et al., 1990; Schreurs, 1992). The corresponding concentration for nitrogen will be about 1000 μg N l^{-1}. However, the higher the nutrient level, the more the system is at risk of switching to the phytoplankton-dominated state. Therefore, in the guide-line, nutrient load reduction is always the first issue to consider when recovering shallow, eutrophicated lakes.

Targets and standards

A restoration program for a specific lake starts with a definition of the targets or objectives of the lake. Possible targets range from the control of symptoms, like avoiding (toxic) cyanobacteria, to the restoration of the pristine state of a lake. However, in general, water managers pay little attention to a further definition of the targets, and an analysis to the feasibility of the targets is often not made.

Some natural reference state is often included in the objective for lake restoration. The natural reference state of a lake can be derived from historic data or paleolimnology. It appears that some lakes have had turbid, cyanobacterial-dominated periods over the past centuries, although it is not clear whether this is a natural or a cultural phenomenon (Andersen, 1995). Due to man's influence, in the Netherlands there are hardly any pristine lakes that can serve as a natural reference. Most of the lakes were created by peat digging or land reclamation. For example, lake Veluwe (3280 ha) was created with the reclamation of the Flevopolder to protect the old land from major ground water losses. The natural reference thus has to be replaced by some historical reference. Even if the natural reference state is known and maybe even achievable, yet another reference state might be preferred. Man-made polders in the Danube Delta, Romania, are inhabited by many birds. Taken with the remaining reed–areas, these have increased the number of habitats available. Therefore, the original natural state is not always preferred, even from a 'natural' reference point of view.

Several factors affect the feasibility of a natural reference and derived water quality targets: the present and future function of the lake, and, often related to these functions, hydrological and nutrient constraints. Functions like recreation and commercial fisheries limit the ecological possibilities of a lake. Disturbance is detrimental to several bird species and mammals. Recreation and other water transport can seriously affect the shoreline of waterbodies and their habitats and will have an impact on the resuspension of sediments. Restricted water level fluctuation can serve as an example for hydraulic constraints. To meet agricultural needs in the Netherlands, water level fluctuations are limited to only a few centimetres (Figure 2). This affects recolonization of macrophytes and spawning places for pike. It is clear that most functions of a waterbody other than 'nature' conflict with ecological water quality targets. Nutrient loadings can be dominated by high concentrations in seepage water or precipitation. In deep polders in the western part of the Netherlands, the nutrient concentration in seepage reaches levels above 1000 μg P l^{-1}. In some areas, phosphorus concentrations in precipitation up to 90 μg P l^{-1} are reported. These nutrient flows are

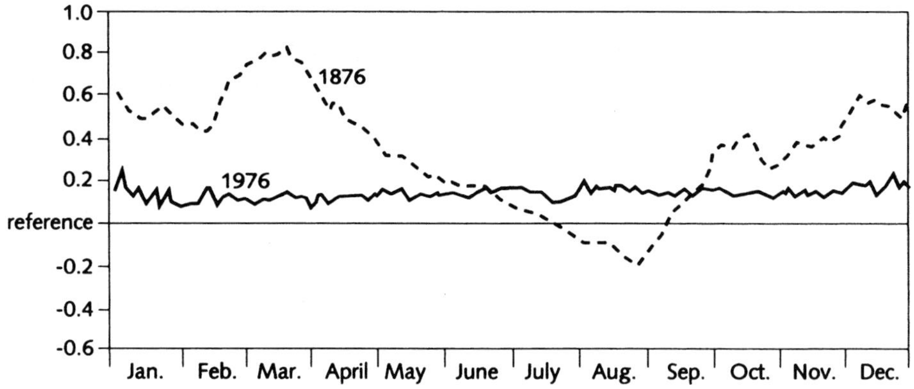

Figure 2. Water level fluctuation of the Frisian lakes, the Netherlands, in 1876 and 1976 (data from province of Friesland).

difficult to manage and limit the feasibility of stringent nutrient targets.

A qualitative description of targets has to be given in measurable standards. This can be a level for nutrient concentrations, a critical transparency, the numbers of species and occurrence or absence of a species or group. This suggests that the development of a lake is predictable. If the aim is to restore the conditions for self-recovery, so the lake will look after itself, standards should be defined in accordance with that aim.

System identification

After the targets and standards are set, they have to be compared with the present situation (Figure 1). A prerequisite is that the lake has been monitored. If the deviation between the target and the present situation is not acceptable, the lake has to be studied in more detail. The first step is the identification and quantification of external and internal sources for water and nutrients. Although the guideline is focused on nutrients, also other relevant factors should be taken into account. For example hydrological conditions, toxic contamination and the food web structure. It is often necessary to have a more detailed look to the hydrological conditions as compared to the definition of the target. Apart from the water level and the water level fluctuation, the hydraulic residence time may be important. In general, a lower critical nutrient loading is allowable for systems with a longer residence time. Toxic contamination may affect the food web structure. Larger zooplankton are most vulnerable to a wide variety of chemicals and therefore a reduction in top down control may result in elevated phytoplankton levels. Insight into the food web structure is especially important when biomanipulation is considered as a restorative measure. Shallow lakes that have been stocked with large amounts of carp will not likely clear up, even if the nutrient loading is sufficiently reduced.

With the information on nutrient and hydrological loading, mass budgets can be made (Figure 3), which in turn can be used as inputs for models to determine whether the nutrient loading is too high and which (combination of) measures will be most promising for its reduction. Related to nutrient budgets, even for a 'well mixed' lake, several choices have to be made considering the temporal scale of the nutrient budgets, the bioavailability of nutrients and the use of nutrient loading or the nutrient concentration in the input.

The temporal scale of mass budgets is important over several years and within a year. Meteorological conditions have a large impact on water- and nutrient flows. A loading can easily vary by a factor of two or more, over different years. So a range of meteorological conditions should always be analysed. The summer is the most important period for eutrophication effects. Some water managers are therefore only interested in the nutrient loading in summer. However,

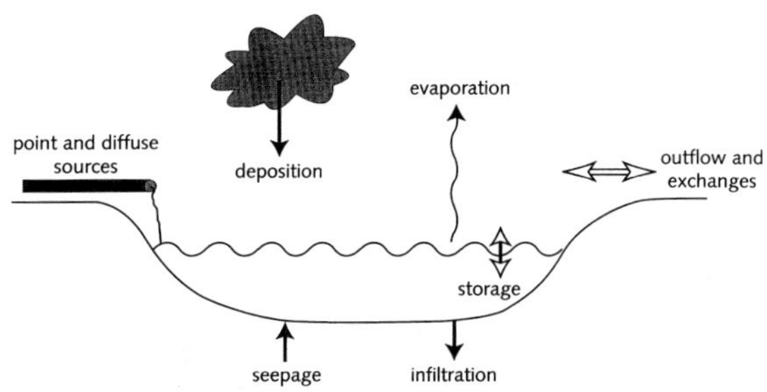

Figure 3. Schematic presentation of a mass budget of a lake.

in many lakes, summer phosphorus input is dominated by internal loading from the sediments and the phosphorus concentration in summer is only weakly correlated with the external loading (van der Molen & Boers, 1994). In summer a significant percentage of the lake water can still be water that entered the system during winter months. Figure 4 shows the phosphorus loading of lake Volkerak, the Netherlands, and that the share from the river Dintel is highest from December until April. Due to the large volume of the lake compared to the hydraulic loading, the percentage of the water that originates from the river Dintel remains more or less constant over the year. Therefore, in most cases nutrient loadings have to be determined on a yearly basis.

In several studies, the bioavailability of nutrients in the external loading is analysed (Ekholm, 1994). Bioavailability, measured with bio-assays, gives a result that is limited by the method used. In the lake, the material meets other conditions, with regard to redox and pH, and the availability of the nutrient might change significantly. Further, the bioavailability of nutrients within the lake is determined by the loading of both nutrients and adsorbens. A loading of so-called non-bioavailable phosphorus might have replaced a loading of only adsorbens in the past. Therefore, even if the loading itself is not available, than the buffer capacity of the lake towards other phosphorus loadings might still decrease.

Water managers often focus on the nutrient loading as a steering factor for eutrophication. In fact, the average nutrient concentration in the lake input or nutrient loading, corrected for the hydraulic residence time, is a better measure than the total nutrient loading (Vollenweider & Kerekes, 1982; van der Molen & Boers, 1994). In lake Veluwe, the Netherlands, flushing was used as a additional measure for phosphorus removal at the water treatment plant in 1979 to combat eutrophication effects (Jagtman et al., 1992). This flushing increased the phosphorus loading, but decreased the average phosphorus concentration in the input and also decreased the phosphorus concentration in the lake. Figure 5 illustrates that the yearly average lake concentration shows a higher correlation with the average phosphorus concentration of the inlet compared to the external loading. In the regression analysis, data of the year 1979 are not used, because the lake concentration of this year was determined by the internal loading.

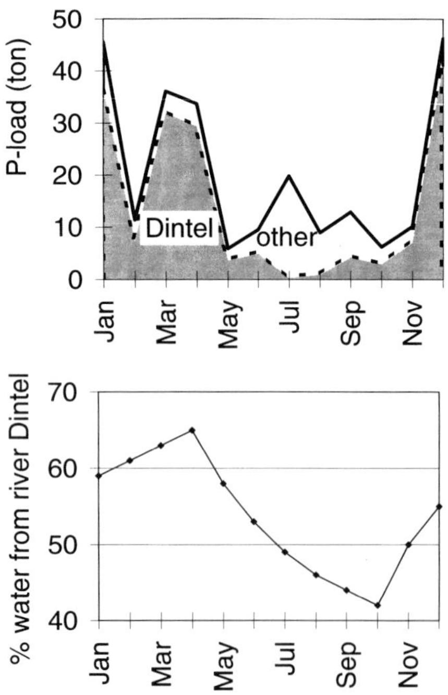

Figure 4. Phosphorus loading (top) of lake Volkerak in 1994 from the river Dintel and other sources, and the share of water from the river Dintel in the lake (bottom).

Measures

Water- and nutrient budgets are a necessity to determine whether measures should focus on external nutrient loading, internal nutrient loading, the most promising nutrient to combat, or if other aspects other than nutrients are important (Figure 1). The tailor-made measures can be divided into:

- Further reduction of the external nutrient loading.
- Physical measures.

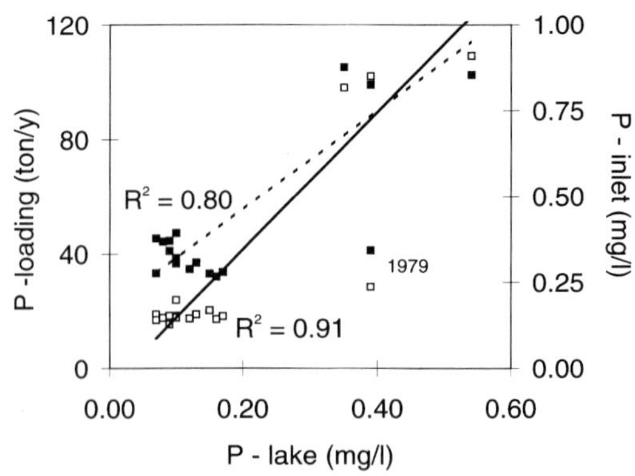

Figure 5. The yearly averaged phosphorus concentration of lake Veluwe explained by the external phosphorus loading (dashed line) and the average concentration in the lake input (solid line).

- Chemical measures.
- Biological measures.

An extra reduction of the external nutrient loading, in addition to measures based on a national policy, is often necessary to approximate concentrations that assure a switch to the stable, macrophyte-dominated state. After the reduction of nutrient emissions from point sources, diffuse sources become relatively more important. However, 'end-of-the-pipe' treatment is very ineffective for diffuse sources. Buffer zones are successful for nitrogen, due to the denitrification process. However, for phosphorus, positive results are generally only temporary (Osborne & Kovacic, 1993). If the total external loading is decreased to an acceptable level, lake resistance by internal loading of phosphorus can frustrate any effect on the lake concentration (Jeppesen et al., 1991; van der Molen & Boers, 1994). A new equilibrium with the external loading in shallow lakes can be accelerated by dredging of sediments that formerly stored phosphorus (Ryding, 1982) or by chemical inactivation of phosphorus (Boers et al., 1994). Both successful and disappointing experiences have been reported (see for example Cooke et al., 1993). These types of measures are mainly focused on nutrients and the consequences for further lake restoration, for example the recolonization of macrophytes, are often not studied.

If the nutrient loading and primary production are reduced, an improvement of the transparency and eventually a switch to another stable state can be accelerated by measures to reduce suspended matter and by food web measures. Improvement of the underwater light climate by dams or deep holes in the lake can be applied, to increase the net sedimentation of particulate matter. Detritus and inorganic suspended matter can determine transparency to a large extent (Blom et al., 1994). Stimulation of the top–down grazing effect on phytoplankton and detritus can also contribute to the water transparency and trigger the change from a turbid to a clear water state (Benndorf, 1987; Gulati et al., 1990). Above phosphorus concentrations of 50–100 $\mu g\ l^{-1}$, the effect of fish removal in large shallow lakes is small or temporary at its best (Jeppesen, 1999). Fish removal in lake Wolderwijd (2615 ha), the Netherlands, at a phosphorus concentration of less than 80 $\mu g\ l^{-1}$ resulted in only six weeks of clear water (Meyer et al., 1994). However, this might have stimulated the recolonisation of macrophytes in the most shallow areas of the lake.

Models can be helpful to selection of measures. There is a wide variety of models. They range from simple regression equations to complex, deterministic models (e.g. Vollenweider & Kerekes, 1982; van der Molen & Boers, 1994; van der Molen et al., 1994). The type of model that is needed depends on the questions that have to be answered, the required precision of the forecasts and the availability of input data. In general conceptual, quantitative models based on closed mass budgets and adjusted for the specific system are the best compromise. If nutrients (phosphorus and nitrogen), phytoplankton species composition, transparency, top-down effects and long term effects have to be included, a single 'complete' model will not be the optimum. Specific models for specific questions and variables are often preferred, especially when higher trophic levels are concerned.

An analysis of measures and the expected effects should result in several scenarios with an expectation of the results, with respect to the target. A cost-benefit analysis is needed to determine whether measures have to be taken and if so, which scenario is the most promising one. If no (sufficient) measures are taken, another target should be chosen (Figure 1). If measures are taken, the lake has to be monitored, and after a period, the measures should be evaluated by comparing the lake condition with the target.

Discussion

Eutrophication research originated in the 1960s and 1970s. Therefore, the relation between nutrients and lake eutrophication is reasonably well understood. In the past 5 years it is recognized that many more factors are affecting the eutrophication process in addition to, or maybe even instead of, the nutrients. However, not much progress has yet been made in the quantification of these factors. Because we are unacquainted with the influence of the disturbances on water quality, a lake restoration target can not be definite. Targets should be realistic and achievable in a surveyable period, and should be inspiring. A target should not be seen as a fixed value, but as a tool for a step by step approach of descriptive, long-term goals.

The effects on lake management are related to the condition of the system (Figure 6). Based on the perspectives and costs of restoration, protection of lakes, not affected by pollution, needs the highest priority. Minor influences on lakes can be combatted by re-establishing the ability for self-restoration. When the conditions can be created for a natural recovery, patience is preferred over a hasty approach with addi-

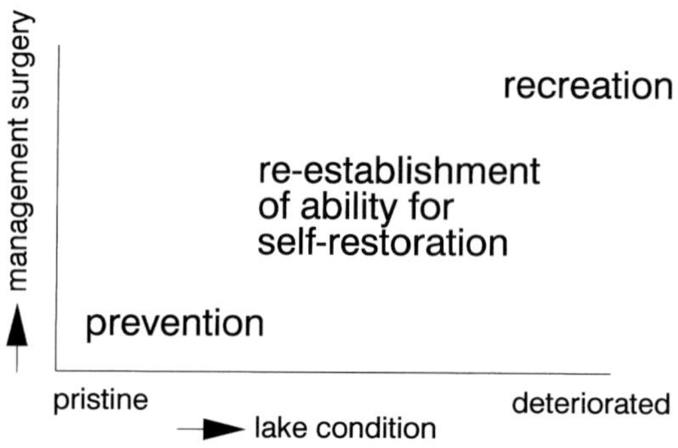

Figure 6. Relation between lake condition and lake management.

tional measures to achieve the target. For most Dutch lakes, however, lake re-creation is necessary. Due to many factors affecting the lake water quality, water managers should be aware that the re-creation of a more beneficial system is a more fair-minded approach than to re-create the pristine state.

In many cases, tailor-made projects for specific lakes are indispensable in addition to the general policy. It is very important to involve people in the region with all stages of the lake restoration projects. Local people have to support the targets if they are involved in the measures. Their knowledge can be used in system identification and their devotion can optimise the effect of measures. However, some conflicts cannot be solved on a regional scale. For example, agricultural policy in the Netherlands limits the possibilities for individual farmers to adjust their agricultural practice. In several other cases, environmental improvements are in conflict with the economic position of a group of people. Also, problems on a regional scale can be expected when pollution and observable adverse effects do not take place at the same time. It is very difficult to convince farmers on a local scale that phosphorus accumulation in soils will affect water quality in a few decades, or that, if it does, it is important.

Conclusions

Eutrophication effects are caused by an elevated loading of nutrients and several other factors, related to the hydrology, chemistry and ecology of the system. Reduced water level fluctuations, toxic contamination and intensive fisheries may serve as examples. Only little progress has yet been made in quantifying the effect of these factors on the eutrophication process. These factors, often related to other functions of the lake than a ecological sound system, imply that the targets for nutrient reduction programmes cannot be equated with a natural reference. Targets for lake restoration programmes should take into account the constraints imposed by the functions of the lake and non-manageable conditions. Therefore, lake restoration is often the re-creation of a more beneficial system, instead of the return of an ecosystem to a former natural condition. If a lake has not deteriorated to the phytoplankton-dominated, turbid state, prevention or re-establishment of its ability for self-restoration are preferred to 'intensive surgery' on the system.

Because national regulations are insufficient in the Netherlands, to reduce nutrient loadings to acceptable levels, and because the causes of eutrophication originate in both nutrient loading and other factors, tailor-made lake restoration programs are needed in addition to a national policy. A guide-line for lake restoration, of which the leading thread is elaborated in this paper, can be a helpful tool to develop a restoration program. Firstly, a target has to be defined, which outlines the system that is aimed for. Next, all relevant information of the system has to be gathered to facilitate the choice of an adequate set of measures. Hydrological and nutrient mass budgets are an essential to determine whether measures should focus on external nutrient loadings, internal nutrient loadings or other factors, like food web manipulation. Finally, after a cost-benefit analysis should be decided if an attempt is made to achieve the targets.

References

Anderson, N. J., 1995. Naturally eutrophic lakes: reality, myth or myopia. Trends in Ecology & Evolution 10: 137–138.

Benndorf, J., 1987. Food web manipulation without nutrient control: a useful strategy in lake restoration. Schweiz. Z. Hydrol. 49: 237–248.

Blom, G., E. H. S. van Duin & L. Lijklema, 1994. Sediment resuspension and light conditions in some shallow Dutch lakes. Wat. Sci. Tech. 30: 243–252.

Boers, P. C. M., J. van der Does, M. Quaak & J. van der Vlugt, 1994. Phosphorus fixation with iron(III) chloride: a new method to combat internal phosphorus loading in shallow lakes? Arch. Hydrobiol. 3: 339–351.

Cooke, G. D., E. B. Welch, S. A. Peterson & P. R. Newroth, 1993. Restoration and Management of Lakes and Reservoirs. Lewis Publishers, Boca Raton, 548 pp.

Ekholm, P., 1994. Bioavailability of phosphorus in agriculturally loaded rivers in southern Finland. Hydrobiologia 287: 179–194.

Gulati, R. D., E. H. R. R. Lammens, M.-L. Meijer & E. van Donk, (eds.), 1990). Biomanipulation – Tool for Water Management. Developments in Hydrobiology 61. Reprinted from Hydrobiologia 200/201. Kluwer Academic Publishers, Dordrecht, 628 pp.

Jagtman, E., D. T. van der Molen & S. Vermij, 1992. The influence of flushing on nutrient dynamics, composition and densities of algae and transparency in Veluwemeer, The Netherlands. In L. Van Liere & R. D. Gulati (eds.), Restoration and Recovery of Shallow Lake Ecosystems in The Netherlands. Hydrobiologia 233: 187–196.

Jeppesen, E., J. P. Jensen, P. Kristensen, M. Søndergaard, E. Mortensen, O. Sortkjaer & K. Olrik, 1990. Fish manipulation as a lake restoration tool in shallow, eutrophic, temperate lakes 2: threshold levels, long-term stability and conclusions. Hydrobiologia 200/201 (Dev. Hydrobiol. 61): 219–227.

Jeppesen, E., P. Kristensen, J. P. Jensen, M. Søndergaard, E. Mortensen & T. Lauridsen, 1991. Recovery resilience following a reduction in external phosphorus loading of shallow, eutrophic danish lakes: duration, regulating factors and methods for overcoming resilience. Mem. Ist. ital. Idrobiol. 48: 127–148.

Jeppesen, E., M. Søndergaard, B. Kronvang, J. P. Jensen, L. M. Svendsen & T. L. Lauridsen, 1999. Lake and catchment management in Denmark. Hydrobiologia 395/396 (Dev. Hydrobiol. 136): 419–432.

Meijer, M.-L., E. H. van Nes, E. H. R. R. Lammens, R. D. Gulati, M. P. Grimm, J. Backx, P. Hollebeek, E. M. Blaauw & A. W. Breukelaar, 1994. The consequences of a drastic fish stock reduction in the large and shallow Lake Wolderwijd, The Netherlands. Can we understand what happened? Hydrobiologia 275/276 (Dev. Hydrobiol. 94): 31–42.

Moss. B., J. Madgwick & G. Phillips, 1996. A guide to the restoration of nutrient-enriched shallow lakes. W. W. Hawes, UK, 180 pp.

Olem, H. & G. Flock, (eds.), 1990. Lake and Reservoir Restoration Guidance Manual. 2nd edn. EPA 440/4-90-006. U.S. Environmental Protection Agency, Washington, DC.

Osborne, L. L. & D. A. Kovacic, 1993. Riparian vegetated buffer strips in water-quality restoration and stream management. Freshwat. Biol. 29: 243–258.

Portielje, R. & D. T. van der Molen, 1998. Trend-analysis of eutrophication variables in lakes in The Netherlands. Wat. Sci. Tech. 37: 235–240.

Ryding, S-O., 1982. Lake Trehornigen restoration project. Changes in water quality after sediment dredging. Hydrobiologia 91–92 (Dev. Hydrobiol. 9): 549–588.

Scheffer, M., 1989. Alternative stable states in eutrophic shallow freshwater ecosystems, a minimal model. Hydrobiol. Bull. 23: 73–83.

Schreurs, H., 1992. Cyanobacterial dominance, relations to eutrophication and lake morphology. Unpublished Ph.D.Thesis University of Amsterdam.

van der Molen, D. T. & P. C. M. Boers, 1994. Influence of internal loading on phosphorus concentration in shallow lakes before and after reduction of the external loading. Hydrobiologia 275/276 (Dev. Hydrobiol. 94): 379–389.

van der Molen, D. T., F. J. Los, L. Van Ballegooijen & M. P. van der Vat, 1994. Mathematical modelling as a tool for management in eutrophication control of shallow lakes. Hydrobiologia 275/276 (Dev. Hydrobiol. 94): 479–492.

van der Molen, D. T., R. Portielje, P. M. De Nobel & P. C. M. Boers, 1998. Nitrogen in Dutch freshwater lakes: trends and targets. Env. Poll. 102 (S1): 553–558.

Vollenweider, R. & J. Kerekes, 1982. Eutrophication of waters, monitoring, assessment and control. OECD, Paris.

Hydrobiologia **395/396:** 411–417, 1999.
D.M. Harper, B. Brierley, A.J.D. Ferguson & G. Phillips (eds), The Ecological Bases for Lake and Reservoir Management.

A strategy for managing still waters in the United Kingdom

Mark Everard
The Natural Step UK, 9 Imperial Square, Cheltenham, Gloucestershire GL50 1QB, U.K.

Key words: still waters, ponds, conservation, management

Abstract

The ecological and amenity values of fresh still waters do not depend primarily upon the size or permanence of the water body. The ecological importance of still waters may be recognised by global, European, national or local legislation or agreements. However, non-designated still waters may also be of great importance, both to their local human population and through their contribution to ecological processes. In particular, small still waters, whilst individually of low importance at the national scale, may nevertheless have a disproportionately high ecological and/or cultural value at the local scale. The uses to which we put our still water resources, and the land within their upstream catchments, exert pressure on both ecological and utility values. Historically, still-water protection has been addressed on a reactive basis. Although a considerable, and largely under-utilised, armoury of powers exist for their protection, resource limitations place real constraints upon the effort that regulatory Authorities can expend upon still waters. Broader public awareness and ownership of the key issues and values, and of the underpinning legislation, must therefore form part of protective strategies for the future. The Environment Agency has a key role to play in this education process, which may represent a cost-effective means for it to discharge its responsibilities to the water environment.

Introduction

The diverse geomorphology and predominantly wet atlantic climate of the United Kingdom gives rise to an abundance of fresh water bodies. In England and Wales, the natural still water resource is augmented by significant numbers of man-made water bodies of many types.

At a global scale, the still waters of England and Wales are relatively small. There is also a marked regional variation in opinion across the country as to how big a still water must be to quality as a lake (Mockerkin Tarn and Red Tarn in Cumbria, though not qualifying by local definition as lakes, are appreciably bigger than the Heath Lake SSSI in the drier southern county of Berkshire). The distinction between lakes and reservoirs is equally vague; some natural lakes serve as a source for potable supply, many man-made still waters are not utilised as sources for water abstraction, whereas lakes such as Cumbria's Seathwaite Tarn are in fact natural lakes enlarged by dams to serve as a source of drinking water. In this paper, the national resource of still waters is considered as a continuum, avoiding unhelpful distinctions between lakes, reservoirs, tarns and ponds.

The ecological processes and utility values performed by these still water bodies are diverse, as are the threats that they face. The purpose of this chapter is to examine the reasons why we should pay attention to the needs of our still water resource, and to recommend an affordable strategy for its future management. This will be addressed in three parts: an overview of our still water resource; a summary of priorities accorded to still waters in recent years; and recommendations for the future.

The still water resource of England and Wales

The question we must first address in considering the management of our still waters is the quantity and nature of the present resource, the trends observed in this resource, its importance to nature and to the human population, and the threats it faces.

The extent of the resource

Johnes et al. (1994) estimate that there are some 1,700 fresh still water bodies with an area of 4 hectares or more in England and Wales, 12,500 of 2 hectares or more, and 50,000 of 1 hectare or more. Throughout this century, there has been a net increase in the numbers of large still waters due to the creation of reservoirs, mineral extraction pits (predominantly clay, gravel, sand and peat) and balancing ponds. In the Environment Agency's Thames Region alone, the two natural lakes (Frensham Great and Little Ponds near Godalming, Surrey) represent only a small proportion of the still water resource compared to the 25 square miles of reservoirs and numerous gravel extraction pits along the Thames Valley (Sweeting, personal communication). The net trend in smaller still waters is, however, one of sharp decline. In Britain as a whole, 90% of fresh water bodies are less than 1 hectare in area (DoE, 1993). There are probably some 400,000 remaining ponds in England and Wales, the remnants of a 65% loss over the last century, with an estimated continuing loss of 1% (around 9000 ponds) per annum (Sansom, 1993).

Very few of these water bodies are truly 'natural'. Many are modified to some extent from their original state; usually enlargement by dams or dredging, although groundwater abstraction can affect the water regime in many water bodies. The larger water bodies are almost exclusively permanent, although seasonal drawdown at the lake margins may be ecologically important. However, a large number of smaller water bodies are temporary, drying on a seasonal basis or during extended dry periods.

Scales of importance

Our still waters are important at a range of scales from the global to the local.

At the **global scale**, a significant number are designated as RAMSAR sites, or may otherwise make a significant contribution to the maintenance of biodiversity (The UK became a signatory of the Biodiversity Convention at the 'Earth Summit', Rio de Janiero, in 1992.)

At the **European scale**, a large number of still waters are contained within designations under various EC/EU Directives including:

- SACs (Special Areas for Conservation) under the EC Habitats Directive (CEC, 1992).
- SPAs (Special Protection Areas) under the EC Birds Directive (CEC, 1979).
- SA[E]s (Sensitive Areas [Eutrophic]) under the EC Urban Waste Water Treatment Directive (CEC, 1991a).
- NVZs (Nitrate Vulnerable Zones) under the EC Nitrates Directive (CEC, 1991b).
- Stretches or areas designated as fisheries under the EC Freshwater Fish Directive (CEC, 1978).
- Abstraction points designated under the EC Surface Water Abstraction Directive (CEC, 1975).

In addition, the EC Bathing Waters Directive (CEC, 1976) does not preclude the designation of freshwater bathing beaches (there are many instances in continental Europe), although the UK Government has chosen not to designate any such freshwater beaches. The proposed EU Directive on the Ecological Quality of Surface Waters, and the prospective wetland directive that may stem from the recent EC Communication on Wetlands (EC, 1995) may also relate to the still fresh waters of England and Wales.

At the **national scale**, various still fresh waters fall within, or are either partially or wholly covered by, designated NNRs (National Nature Reserves) or SSSIs (Sites of Special Scientific Interest) under the Wildlife and Countryside Act 1981, or its predecessor 1949 Act. Still waters form a significant component of the ecological and landscape interest of various National Parks, most significantly the Broads of East Anglia and the Lake District of Cumbria, as well as of AONBs (Areas of Outstanding Natural Beauty) throughout the country. Mesotrophic lakes, and wetland habitats skirting lakes of all types, are noted as of particular ecological value in *Biodiversity: The UK Action Plan* (DoE, 1994), and many other types of still waters may be important habitats for species scheduled in the Plan.

At the **local scale**, smaller still waters are frequently important features of LNRs (Local Nature Reserves), and may also be important components of sites of heritage value (for example, the famous concert venue at Kenwood House in North London, or the Estate Lakes of Blenheim Palace and other such examples of traditional English landscaping).

The values of still waters

It is important to note that all still waters, whether designated or not under domestic or international legislation, or forming part of a high profile site, may nevertheless be of importance for any or all of a wide range of reasons summarised in Table 1. A village pond, though individually of low importance at the larger scale, may act as an important focal point for village life. In addition, these small water bodies may provide an important education resource, provide valued fisheries and leisure opportunities, and act as

Table 1. The Potential Importance of Still Water Bodies

Ecological values:
Habitats of importance to aquatic species
Habitats of importance to migratory species
Source of primary and secondary productivity
Amenity values:
Fisheries
Field sports (e.g. wildfowling)
Leisure opportunities
Water sports (e.g. diving, canoeing, swimming, paddling, etc)
Utility values:
Abstraction for potable supply
Provision of water to industry
Provision of water for livestock watering and spray irrigation
Balancing ponds for canal water
Water management values:
Balancing ponds and floodwater detention/desynchronisation
Water Quality improvements/waste treatment
Long-term environmental/heritage records:
Diatom frustules; chrysophyte scales; photosynthetic pigments
Pollen grains
Cladoceran and Ostracod carapaces
Chironomid head capsules
Carbonaceous granules from coal-burning
Archaeological remains
Other values:
Component of landscape, heritage and natural beauty
Educational resource
Shellfish farming

important corridors or 'islands' for wildlife. Small still waters then, whilst perhaps of low significance individually at the national scale, nevertheless collectively represent a highly important and largely overlooked natural resource. As previously observed, this resource of small still water bodies is inadequately protected.

A surprising number of Red Data Book species (our most endangered species of plants and animals) occur in temporary water bodies (Biggs et al., 1994; Collinson et al., 1995), and Pond Action (1993) conclude that the wildlife value of ponds is likely to be high where the surrounding land has been sympathetically managed. The importance of a water body does not therefore depend only upon its size or permanence. Even the humble, yet totally artificial, garden pond may be important for many of the reasons outlined above, and may provide important sites for threatened aquatic species (for example, the great crested newt *Triturus cristatus*) which feature in Red Data Books or are scheduled under the Wildlife and Countryside Act 1981, the EC Habitats Directive (CEC, 1992) or the UK Biodiversity Action Plan (DoE, 1994).

The current basis for still water management

Greater attention is paid to rivers, as opposed to still fresh waters, in the UK compared to most other countries in northern Europe. The reasons most commonly given for this difference in emphasis are that the British still water resource comprises a much higher proportion of smaller water bodies and that the coast provides for a larger amount of our leisure needs.

A 'straw poll' of water quality managers in all eight of its regions revealed that the Environment Agency has generally lacked a cohesive strategy for addressing the needs of the still waters under its control. Whereas some still waters, and particularly larger water bodies and those with a higher political profile (owing to designations under EC/EU Directives relating to human uses, or heritage considerations), were the subject of detailed investigations and/or management plans, or were included in Catchment Management Plans, the consensus amongst the managers polled was that the Agency dealt with still waters on a largely reactive basis. Where still waters were given attention, the reason for this attention was generally related to human uses of the water body rather than to innate ecological values. Despite this *ad hoc* approach, a range of threats (summarised in Table 2) face our still waters and particularly our smaller still water bodies.

Priorities for the future

The challenge for the future is to find cost-effective means of sustaining the nation's resource of fresh still waters, managing them in a way that strikes a balance between their ecological needs and the uses to which we put them and their upstream catchments. The evidence presented above suggests that, as a nation, we are failing in our obligations to protect our resource

Table 2. Significant Threats to Fresh Still Water Bodies

Organic pollution:	Domestic sewage, Agricultural activity, Food processing industries
Eutrophication:	Point sources, Diffuse sources
Acidification:	Aerial Deposition, Afforestation
Siltation:	Land use changes, Deforestation
Drainage/Infilling:	Agricultural intensification, Urban development
Dumping:	Cheap disposal sites for land-owners, Fly-tipping
Introduction of alien species of plants and animals	
Complex pollutants:	Organophosphates and other synthetics, Heavy metals, Oestrogenic substances

of small still waters, which may have a disproportionately large cumulative value. In this section, the constraints under which regulatory authorities operate are considered prior to making final recommendations about future priorities for action to protect the ecological and utility values of our still waters.

Resource constraints

Regulatory authorities by necessity operate under financial and human resource constraints. This factor alone is sufficient to explain why focus has been given to so few of our water bodies. Resource constraints require us to limit ourselves to larger and higher profile lakes, and one has to weigh up the desirability of increasing this level of investment against other national priorities such as the provision of hospital beds or educational services. Assuming 200 working days per person per year, and an average of four still waters per working day, it would take in excess of 560 person years of resource per annum to visit each of our 450 000 still water bodies just once per year. (This figure will be an under-estimate due to periodic unsuitable weather conditions.) A single visit in any one year would be inadequate to take account of the considerable spatial and temporal variation that occurs within a still water body, and so would not represent a cost-effective deployment of the substantial amount of human resource.

Although the National Rivers Authority before and the Environment Agency now have invested significantly in technology that may ultimately provide a more cost-effective means of assessing still waters (remote sensing, test kits, biomarkers, etc), these technologies are as yet immature or under used. Technology alone can not necessarily be relied upon to surmount the substantial resource constraint, which is currently in shortfall by some three orders of magnitude, were we to need to deal with every still water. The issue then is not to demand ever more funding but, particularly in the light of the new duty upon the Environment Agency to have regard for costs and benefits, to find more efficient ways of utilising our available resources.

Legal constraints

The existing legal framework under which regulatory authorities operate in England and Wales provides us with a range of powers and duties, in addition to some conspicuous oversights. The legislation available to the Environment Agency can be broken down into three major categories: legislation inherited from constituent organisations; new legislation introduced by the Environment Act 1995; and other strands of legislation that provide the Agency with additional powers.

The legislation inherited from constituent organisations (NRA, Her Majesty's Inspectorate of Pollution, and the Waste Regulation Authorities) has been rolled forward into the Environment Agency and consolidated under the Environment Act 1995. Although strands of former HMIP powers are directly relevant in controlling acidic aerial deposition and the discharge of dangerous substances to water, the bulk of powers and duties relating to the water environment are inherited from the NRA through consolidation of the Water Resources Act 1991. Most of the powers and duties contained in the Water Resources Act are limited to 'Controlled Waters', a definition from which still water bodies that are not connected directly to surface water streams or to groundwater are excluded. It is therefore precisely those smaller still waters that appear to be most at risk that are excluded from the legal armoury inherited from the NRA.

Although perhaps not their primary intention, various pieces of new legislation introduced by the Environment Act 1995 may add considerably to the powers of the Agency to deal with matters affecting our still waters. These new pieces of legislation include:

- Section 4 of the Act introduces a new 'umbrella' duty to the achievement of sustainable development. Sustainability relates to the protection of both ecological and societal values, and the protection of biological and heritage resources therefore form central aspects of the duty. It is reasonable to presume that the Agency will target its efforts to deliver the maximum benefits to the water environment.
- The new requirement under Section 5(2)(b) of the Act to "...form an opinion of the state of pollution of the environment..." of England and Wales exists without reference to any particular legal restrictions, and so the Agency is therefore at liberty to offer objective opinions of the state of our national freshwater resource and upon the actions that may be required to safeguard and/or restore it.
- The Agency is required to take account of the costs and benefits of its decisions. Given the unfortunate trend in loss of small still waters, and their disproportionately large ecological and societal values, it is possible that the greatest ecological and societal benefits per unit investment may be achieved by targeting these largely neglected water bodies. It is also incumbent upon the Agency to investigate innovative and cost-effective means to better utilise its limited resources to deliver the best value-for-money.
- Under arrangements with local authorities, health and safety registers of manmade still waters greater than 25,000m^2 will transfer across to the Agency in due course, giving the Agency new powers to control aspects of these still water bodies.

The United Kingdom has made various other commitments to ecological protection, which provide additional powers for regulatory authorities. These include:

- The UK Biodiversity Action Plan, produced in compliance with the requirements of the Biodiversity Convention.
- *The Conservation (Natural Habitats etc) Regulations 1994*, which require all regulatory authorities to use their powers to promote nature conservation in SACs designated under the EC Habitats Directive (CEC 1992) and SPAs designated under the EC Birds Directive (CEC 1979).
- The *Convention on Wetlands of International Importance Especially as Waterfowl Habitat* (The "Ramsar Convention"), which aims to stem progressive encroachment on and loss of wetlands, and ratified by the UK in 1976. Under the terms of the convention, the definition of wetlands includes fresh and saline waters up to a depth of 6 metres (i.e. including most of the smaller still waters and much of the littoral zone of many of our larger water bodies), and signatories are required to develop a national strategy for their protection. The UK has not to date developed such a strategy, however, the Ramsar Commission has recently recognised that Water Management will be a crucial issue in the 21st century. Although not directly providing additional powers to the Agency, the Ramsar Convention provides additional justification for undertaking actions affecting aquatic systems.

As part of its ongoing activities, it is necessary for the Agency to influence partner organisations to use the powers and duties at their disposal to provide protection for the freshwater environment. These partners include both statutory agencies (Ministry of Agriculture, Fisheries and Food, English Nature, Countryside Council for Wales, etc) as well as Non-Governmental Organisations (e.g. Royal Society for the Protection of Birds, Wildfowl and Wetlands Trust). In addition, the Agency seeks to influence land owners and land users according to best practice.

In concluding this subsection on legal constraints, it is worth noting that the legislation available to protect small still water bodies is weak. Prior notification to Local Authorities is required before building a pond, other than a garden pond, and planning permission may be required if the pond is put to non-agricultural uses and is within 25 m of a classified road or is created by exporting soil or gravel from the site. However, no such notification and permission is required for filling in a pond. In the light both of the declining trend and the Government's commitments to maintaining biodiversity, this omission in the legislation is in urgent need of review.

Knowledge constraints

The conference and these proceedings have gone a long way towards bringing together the state of the art in current understanding of lake and reservoir management. The Environment Agency also has a programme of continuing research including participation in the 1996–8 Pond Survey (joint funding with DoE); a project on lake classification and monitoring (linked to a closely-related sediment profiling study funded by English Nature); continued development of lake eutrophication models; and studies on novel techniques for the biological assessment of still waters. In addition to its own programme of research, the Environment Agency also participates in research led by partner organisations.

A body of research into the functional analysis of wetland (e.g. Maltby, 1991; Dugan, 1993; Maltby et al., 1994) and riverine (Harper et al., 1992; 1995) systems has been steadily growing over the past decade. The essence of functional analysis is to determine, where possible on a semi-quantitative basis, the *processes* that habitat features perform within ecosystems. In aquatic systems, these processes are hydrological (for example, desynchronisation and detention of floodwater), physico–chemical (for example, nutrient stripping, settlement of suspended matter, etc), ecological (for example, primary productivity or provision of diverse habitats), and heritage (for example, the preservation of archaeological remains). In recent years, various workers have sought to link functional analysis with environmental economics (e.g. Denny, 1994; Everard, 1998a,b) as a justification to protect, restore or recreate wetland habitats, and to integrate conservation and management. It is probable that the extension of functional analysis techniques to still waters will provide a fruitful line of research.

Environment Agency policy should be led by the best available science. However, having briefly considered present and potential future lines of investigation, and acknowledged that better scientific understanding will always improve management decisions, the clear trends outlined previously in this paper suggest that lack of knowledge is not one of the major constraints to action. In light of the serious decline of our smaller still waters, lack of knowledge should not be used as an excuse for lack of action.

A recommended strategy for the management of still water bodies

Whilst it may be desirable to include all of our still waters in Catchment Management Plans (or their successor Local Environment Agency Plans), or to formulate Action Plans or utilise elegant models such as those addressed elsewhere in this book, to address their specific needs, we must also acknowledge the resource constraints under which we operate. For this reason, the present focus of intensive activity on just a few larger and/or high profile water bodies is unlikely to change in the foreseeable future. Nor is it likely that technological breakthroughs will significantly enhance our ability to deal with the key issues. It is hoped that, through consultation on Catchment Management Plans/Local Environment Agency Plans), the concerned public and external organisations can notify us of perceived problems in additional still waters, which we can then prioritise for detailed study and action.

What then of the remainder of the 450,000 water bodies that fall outside of this safety net? From the above discussion, it is apparent that, although individually they may be of low significance at the national scale, our smaller still waters collectively represent a highly valuable resource, the needs of which are only poorly considered or recognised in existing legislation. This situation begs two questions:

- Financially, can we afford to consider them?
- Ecologically, can we afford *not* to consider them?

There is an innate tension between these two questions, and it is unlikely that Government will increase public expenditure on our still waters in the face of other national priorities. It is therefore up to the Environment Agency, and particularly in the light of its new duties, to investigate innovative, cost-effective methods to better account for the needs of the nation's still water resources.

Promotion of a wider public ownership of the problems is one of the options available to the Agency to maximise its effect per unit investment. This recommendation is not offered merely as a platitude; rather it recognises that the Agency has a central pro-active role to play in promoting wider awareness and ownership of problems of common concern. The terms of this pro-active role may include:

- Promoting a wider awareness of the importance and values of still waters (for example, through RSNC or the schools national curriculum, defining the role of still water protection within Agenda 21, liaison with the media for public education, etc), ensuring that the relevant information is presented in a form comprehensible to a form comprehensible to and usable by the general public.
- Notifying partner organisations of particular problems observed during routine Agency investigative work.
- Promoting a wider awareness of the powers available to the Agency, to its partners and to the wider public to utilise for the purpose of still water protection.
- Targeted support, both technical and financial, for specific projects.
- Applying pressure upon Government to close loopholes in the law such as those preventing the loss of ponds and to comply with the requirements of the Ramsar Convention.

Public awareness and ownership of problems are frequently cited as being central to sustainable solutions. This principle is clearly applicable to the protection of our diverse national still water resource. The Environment Agency has a key role to play in the protection of natural resources, yet protection of these resources frequently relies upon action by those over which we have no direct regulatory powers. Active promotion of public participation and ownership of problems affecting still waters by the Agency may therefore not only represent a cost-effective and innovative solution, but would appear to offer the best prospect for sustained protection.

Acknowledgements

The author wishes to thank the conference organisers for inviting this contribution to the proceedings, and for the production of a 'state of the art' volume on the ecological management of lakes and reservoirs. The opinions expressed here are the author's own and do not necessarily represent any official view.

References

Biggs, J., A. Corfield, D. Walker, M. Whitfield & P. J. Williams, 1994. New approaches to the management of ponds, British Wildlife 5: 273–287.

CEC, 1975. EC Directive on the Quality of Surface Water Intended for the abstraction of drinking water. 75/440/EEC, CEC Brussels.

CEC, 1976. EC Directive on the quality of bathing water. 76/160/EEC, CEC Brussels.

CEC, 1978. EC Directive on the quality of freshwaters meeting protection or improvement in order to support fish life. 78/659/EEC, CEC Brussels.

CEC, 1979. Directive on the conservation of wild birds. 79/409/EEC, CEC Brussels.

CEC, 1991a. EC Directive on urban waste water treatment. 91/271/EEC, CEC Brussels.

CEC, 1991b. Council Directive concerning the protection of waters against pollution caused by nitrates from agricultural sources. 91/676/EEC, CEC Brussels.

CEC 1992. EC directive on the conservation of natural habitats and of wild fauna and flora. 92/43/EEC, CEC Brussels.

Collinson, N. H., J. Biggs, A. Corfield, M. J. Hodson, D. Walker, M. Whitfield & P. J. Williams, 1995. Temporary and permanent ponds: an assessment of the effects of drying out on the conservation value of aquatic macroinvertebrate communities, Biological Conservation 74: 125–133.

Denny, P., 1994. Biodiversity and wetlands. Wetlands Ecology and Management 3: 55–61

DoE, 1993. Countryside Survey 1990: Main Report. Department of the Environment, London, 174 pp

DoE, 1994. Biodiversity: The UK Action Plan. cm2428, HMSO, London.

Dugan, P. J., 1993. Wetlands in Danger. Mitchell Beazley, London, 187 pp

EC, 1995. European Commission: Communication on conservation of wetlands. Europe Environment Document No. 456, June 13, 1995.

Everard, M., 1998a. The contribution of reedbeds to floodplain and river systems. In: R. G. Bailey, P. V. José & B. R. Sherwood (eds), United Kingdom Floodplains. Westbury Academic & Scientific Publishing, Otley: 117–142.

Everard, M., 1998b. Floodplain protection: challenges for the next millennium. In: R. G. Bailey, P. V. José & B. R. Sherwood (eds), United Kingdom Floodplains. Westbury Academic & Scientific Publishing, Otley: 477–485.

Harper, D., C. Smith & P. Barham, 1992. Habitats as the building blocks of river conservation. In Boon, P., P. Calow & G. Petts (eds), River Conservation and Management. John Wiley & Sons, Chichester.

Harper, D., C. Smith P. Barham & R. Howell, 1995. Ecological basis for managing the natural environment. In Harper D. M. & A. J. D. Ferguson, (eds), The Ecological Basis for River Management. John Wiley & Sons Ltd, Chichester.

Johnes, P., B. Moss & G. Phillips, 1994. Lakes Classification and Monitoring: A Strategy for the Classification of Lakes. National Rivers Authority R&D Note 253, NRA, Bristol.

Maltby, E., 1991. Wetlands and their values. In M. Finlayson & M. Moser (eds), Wetlands, 8–26. International Waterfowl and Wetlands Research Bureau, Slimbridge, Gloucestershire.

Maltby, E., D. V. Hogan, C. P. Immirzi, J. H. Tellam & M.J. van der Peijl, 1994. Building a new approach to the investigation and assessment of wetland ecosystem functioning. In Mitsch, W. J., (ed), Global Wetlands: Old World and New, 637–658. Elsevier, Amsterdam.

Pond Action, 1993. The Oxfordshire Pond Survey: A Report to the World Wide Fund for Nature. Pond Action, Oxford Brookes University.

Sansom, A, 1993. Ponds and Conservation. NRA Northumbria and Yorkshire Region, Leeds.

Hydrobiologia **395/396:** 419–432, 1999.
D.M. Harper, B. Brierley, A.J.D. Ferguson & G. Phillips (eds), The Ecological Bases for Lake and Reservoir Management.

Lake and catchment management in Denmark

Erik Jeppesen, Martin Søndergaard, Brian Kronvang, Jens P. Jensen, Lars M. Svendsen & Torben L. Lauridsen
National Environmental Research Institute, Vejlsøvej 25, P.O. Box 314, DK-8600 Silkeborg, Denmark

Key words: Integrated lake management, biomanipulation, lake recovery, phosphorus, nitrogen, eutrophication, restoration

Abstract

The majority of Danish lakes are highly eutrophic due to high nutrient input from domestic sources and agricultural activities. Reduced nutrient retention, and more rapid removal, in catchments as a result of agricultural drainage of wetlands and lakes and channelisation or culverting of streams also play a role. Attempts have recently been made to reduce nutrient loading on lakes by intervening at the source level and by improving the retention capacity of catchment areas. The former measures include phosphorus stripping and nitrogen removal at sewage works, increased use of phosphate-free detergents, and regulations concerning animal fertiliser storage capacity, fertiliser application practices, fertilisation plans and green cover in winter. In order to improve nutrient retention capacity of catchments, wetlands and lakes have been re-established and channelised streams have been remeandered. In addition, cultivation-free buffer strips have been established alongside natural streams and there has been a switch to manual weed control. These measures have resulted in a 73% reduction of the mean total phosphorus concentration of point-source polluted streams since 1978; in contrast, there has been no significant change in the total nitrogen concentration. Despite the major reduction in stream phosphorus concentrations, lake water quality has often not improved. This may reflect a too high external or internal phosphorus loading or biological resistance. Various physico-chemical restoration measures have been used, including dredging and oxidation of the hypolimnion with nitrate and oxygen. Biological restoration measures have been employed in 17 Danish lakes. The methods include reducing the abundance of cyprinids, stocking with 0^+ pike (*Esox lucius*) to control 0^+ cyprinids, and promoting macrophyte recolonization by protecting germinal submerged macrophyte beds against grazing waterfowl and transplanting out macrophyte shoots. In several lakes, marked and long-lasting improvements have been obtained. The findings to date indicate that fish manipulation has a long-term effect in shallow lakes, providing nutrient loading is reduced to a level so low as to ensure an equilibrium lake water phosphorus concentration of less than 0.05–0.1 mg phosphorus l^{-1}. If nitrogen loading is very low, however, positive results may be obtained at higher phosphorus concentrations. Macrophyte refuges and transplantation seem to be the most successful as restoration measures in the same nutrient-phosphorus regime as fish manipulation.

Action plans to combat eutrophication

Danish lakes are typically small, shallow, rapidly-flushed and highly eutrophic (Table 1). Eutrophication is largely attributable to high point-source loading in the past, and high nutrient input from arable land at present. In order to combat eutrophication, the Danish government has implemented a number of action plans. These have two objectives:

1. To reduce external nutrient loading (Table 2).
2. To improve nutrient retention and removal by wetlands (Table 3).

The reason for the latter is that wetland area has reduced considerably over the past 150 years due to land reclamation and drainage.

The major environmental plans are the 1987 'Action Plan on the Aquatic Environment', which is a national plan aimed at reducing land-based nitrogen and phosphorus loading of the aquatic environment by 50% and 80%, respectively. The reduction target for nitrogen was reiterated in the 1991 'Action Plan on

Table 1. Morphometric and environmental parameters for 200–283 Danish lakes

	Mean	Median	Minimum	Maximum
Catchment area (km^2)	89	11	0.11	1500
Surface area	0.88	0.14	0.00031	42
Maximum depth (m)	5.7	3.7	0.3	37.7
Mean depth (m)	2.9	2.0	0.3	16.5
Hydraulic retention time (years)	1.1	0.27	0.0010	21
Total phosphorus (mg P l^{-1})[a]	0.37	0.21	0.017	2.4
Total nitrogen (mg N l^{-1})[a]	2.4	2.1	0.21	9.0
Chlorophyll *a* (μg chl*a* l^{-1})[a]	73	57	2	445
Secchi depth (m)[a]	1.2	1.4	0.2	6.1

[a] Summer mean values (May 1 – Oct 1).

Table 2. Major measures implemented in Denmark to reduce nutrient loading of the aquatic environment

Incorporation of P-stripping at sewage works (>500–5000 P.E.)
Incorporation of N-removal at sewage works (>15 000 P.E.)
Increased use of phosphate-free detergents
Regulations on the amount of fodder used in fish farms
Regulations on livestock density (1.7 – 2.3 DE ha^{-1})
Regulations on utilization of liquid and solid manure (30–45% fertilization efficiency depending on type)
Increased storage capacity for animal manure
Preparation of fertilization budgets/plans and regulations concerning application practices for manure (no use of liquid manure from harvest to spring ploughing)
Mandatory regulations on set-aside and green cover in winter
Increased afforestation
Cessation of subsidies for drainage
Establishment of 2 m wide cultivation-free riparian buffer strips along natural streams and rivers

Sustainable Agricultural Development'. These plans are supplemented by recipient quality objectives stipulated in each Danish County's 'Region Plan'. The recipient quality plans are differentiated plans, adapting catchment loading reductions to the specific water quality objectives, set for each water body and to the sensitivity of the environment. These plans were first elaborated by the Counties in the 1980s and have subsequently been revised every five years. Several more recent plans may also help reduce nutrient loading of the aquatic environment, e.g. the afforestation plan, aimed at doubling the forest area in Denmark during the next 60–80 years (5 000 ha ann^{-1}), the 'Strategy, on Marginal Lands', aimed at restoring 20 000 ha of former wetlands (equivalent to 10% of the drained area) over the next 10–20 years, and finally, the EU Set-Aside scheme, which required that 9% of Danish farmland was removed from production in 1994 (Skotte-Møller, 1995).

Table 3. Major measures implemented in Denmark to improve nutrient retention and removal by wetlands

A switch to gentle weed clearance and reduced dredging in many public streams
Remeandering of channelized streams
Re-establishment of wetlands, streams and lakes
Biomanipulation of lakes

Several attempts have been or are currently being made to reduce external loading of nutrients (Table 2). The Action Plan on the Aquatic Environment requires that phosphorus stripping be incorporated at sewage works with a capacity exceeding 5000 person equivalents (PE), and nitrogen removal at works with a capacity exceeding 15 000 PE. In many of the recipient quality plans, however, phosphorus stripping is required at works with a capacity exceeding 500

Table 4. National tally of re-establishment of wetlands and other natural areas over the period 1989–94 (Skotte-Møller, 1995)

Lakes	1700 ha
Small ponds	1600
Channelized or culverted streams	60 km
Salt marshes	2200 ha
Freshwater meadows	2300 ha
Bogs	2000 ha
Heathland	4800 ha
Grassland	1300 ha
Natural forest	100 ha

PE. As sewage works are usually centralized, sewage from many communities of less than 500 PE will also be subjected to phosphorus stripping and nitrogen removal. These measures, and increased use of phosphate-free detergents on a voluntary basis, have led to a 75% reduction in total phosphorus loading from point sources to freshwater systems over the period 1989–94 (Madsen, 1995). Multiple measures have been implemented to reduce nutrient loading from arable soils, including requirements on the following:

1. Storage capacity for animal fertiliser on farms with more than 31 livestock units and application practices.
2. The preparation of fertiliser budgets/plans.
3. The establishment of uncultivated buffer strips along all natural streams.
4. The maintenance of green cover of fields in winter.
5. The set-aside of agricultural land (EU-directive) (Table 2). Set-Aside will hopefully be concentrated in environmentally sensitive areas in future.

During the last 150 years, a substantial percentage of Danish wetlands, lakes and streams have disappeared due to land reclamation (Iversen et al., 1993). Moreover, to enhance drainage capacity most of the remaining streams were culverted or channelized and 'hard-handed' maintenance was introduced (dredging and mechanical weed control). In addition, stream beds were lowered and stream width enhanced to diminish the risk of flooding (Iversen et al., 1993). These steps have resulted in a significant reduction in hydraulic retention time, and the overall capacity, for temporary or more permanent retention, and removal of nutrients by wetlands has consequently decreased. On the basis of the 1989 Nature Conservation Act, a number of projects have been initiated aimed at re-establishing wetlands, streams, ponds and lakes, at remeandering channelized streams (Table 4), and at speeding up the restoration of lakes showing resilience following a reduction in external nutrient loading. In addition, there has been a shift towards more environmentally sound maintenance practices, weed control now being undertaken manually with scythes in many public streams (Madsen, 1995). The latter significantly increases the nutrient retention capacity of streams during summer: an investigation undertaken in the river Gjern showed that net retention of total phosphorus was, highly positively, related to the size of the riparian zone and macrophyte coverage as well as being positively related to the concentration and negatively to the water velocity (Box 1 and Svendsen et al., 1998). While temporary retention in streams does not affect annual total phosphorus transport to lakes, it is of significance for their overall environmental state during summer, especially in the case of rapidly-flushed lakes (the typical Danish lake type) that are heavily influenced by the immediate nutrient input from catchment streams (Kristensen et al., 1990).

The new more environmentally sound weed-clearance practices together with the remeandering of streams and the raising of stream beds have increased the frequency of flooding, thereby increasing hydraulic retention time, and hence nitrogen loss through denitrification. Flooding also increases retention of the particulate phosphorus being transported by the streams (Table 5), thereby significantly affecting phosphorus transport since particulate phosphorus typically constitutes 35–77% of the total phosphorus transport in Danish streams (Kronvang, 1992; Svendsen et al., 1995). The potential retention capacity of flooded riparian zones is illustrated by the river Gjern study, where a 0.5 ha riparian area in the lower part of the 114 km^2 catchment retained 94 kg phosphorus ha^{-1}in the flooded area from November 1992 to March 1993, which corresponds to a phosphorus loss from a 300-fold greater area of agricultural land (Iversen et al., 1995).

Phosphorus loading of freshwaters has decreased considerably since the implementation of the various Action Plans. Thus, stream total phosphorus concentrations have fallen 73% from 1978 to 1993, when the annual discharge-weighted mean was 0.14 mg phosphorus l^{-1} (Figure 1). The decline is mainly attributable to improved sewage treatment. Thus the percentage of total phosphorus input to freshwaters

Table 5. Total nitrogen and total phosphorus removal and retention in various Danish aquatic ecosystems

	Denitrification (kg N $ha^{-1}ann^{-1}$)	Sedimentation adsorption/desorption (kg N $ha^{-1}ann^{-1}$)	Sedimentation adsorption/desorption (kg P ha^{-1} ann^{-1})
Streams (annual)	5–200[1]	3–380[2,5]	0
Streams (summer)	5–1,460[1]	10–500[2,5]	2–300[2,5]
Riparian zones (summer)	?	20–200[2,5]	8–450[2,5]
Flooded riparian areas	?	?	15–94[10]
Irrigated riparian areas	30–2,725[7,8,9,12,13]	?	−30–60[6,7,8,10]
Meadows	42–2,180[7,8,9,12]	?	0–3[7,8,10,12]
Lakes	65–220 [3,4]		−2.7–5[11]

[1]Christensen et al., 1991; [2]Svendsen & Kronvang, 1993; [3]Jensen, J. P. et al., 1991; [4]Windolf et al., 1996; [5]Svendsen et al., in press; [6]Hoffmann 1996; [7]Rebsdorf et al., 1994; [8]Hoffmann et al., 1993; [9]Hoffmann, 1998a; [10]Iversen et al., 1995; [11]Windolf et al., 1993, [12]Hoffman, 1998b; [13]Ambus & Hoffman, 1990.

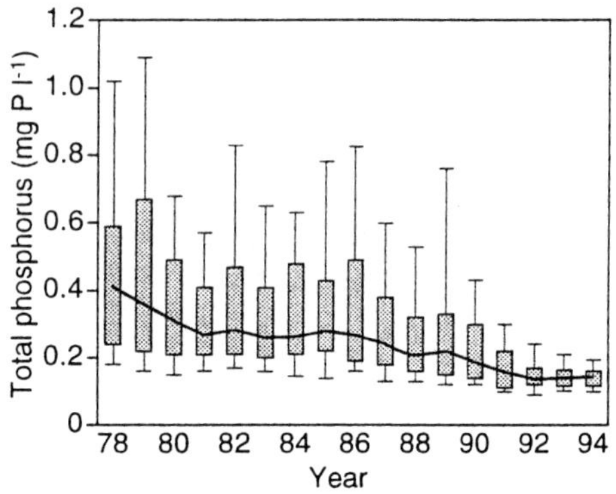

Figure 1. Trend in annual mean concentration of total phosphorus in Danish streams over the period 1978–92. Mean quartiles and 5 and 95 percentiles (5% and 95%) are also shown.

accounted for by point sources has fallen from 68% in 1989 to only 26% in 1994; conversely, the percentage accounted for by arable land has increased from 15% to 49% (Figure 2). Even though point-source nitrogen loading of freshwaters has been reduced by 42% since 1981–88 (Larsen et al., 1995), there has been no significant change in stream nitrogen transport, which is completely dominated by the input from arable land (88% in 1989 and 95% in 1994) (Kronvang et al., 1995). Export coefficient models for diffuse nutrient losses have been developed based on the 1989–94 data from approximately 80 small, representative catchments encompassed by the Danish Nationwide Monitoring Programme under the Action Plan on the Aquatic Environment (Kronvang et al., 1995). The models estimate catchment losses of total nitrogen and total phosphorus by use of simple predictor variables (run-off, proportion of arable land, soil type). The models can be used to estimate diffuse nutrient loading of lakes from otherwise non-monitored areas, as well as to assess measures taken to diminish diffuse nutrient loading by reducing the area of arable land under cultivation.

The impact of the reduction in external loading on lake environmental conditions can be evaluated by using loading response models originally developed by Vollenweider et al. (1976), and subsequently adapted for Danish lakes by Kristensen et al. (1990, 1991). Such calculations have been made for the 37 lakes included in the Nationwide Monitoring Programme (Figure 3). The lakes are considered representative of Danish lakes with respect to loading, size and mean depth. The scenarios, run over 1989–94, show that fulfilment of sewage plant treatment requirements is insufficient to bring many of the lakes into the clearwater state that is the aim of most Region Plans (Figure 3). Not even a cut-off of the sewage contribution will suffice. The aim can only be achieved if efficient measures are also taken against diffuse sources.

Lake restoration measures

External phosphorus loading of some Danish lakes is now so low that a shift to the clearwater state ought to have taken place. However, resilience to recovery is sometimes observed due to both chemical and biological factors. Phosphorus release (internal loading) from the phosphorus pool accumulated in lake sediments during the time when loading was high may counteract the reduction of external loading. The duration of resilience depends on conditions such as the magnitude and duration of loading, hydrological re-

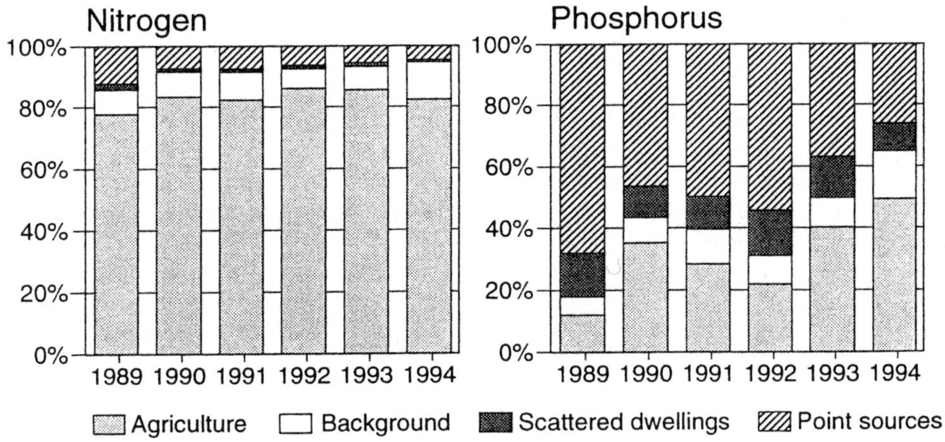

Figure 2. Trend in source apportionment of total phosphorus and total nitrogen inputs to Danish streams over the period 1989–94.

tention time and iron input (Cullen & Forsberg, 1988; Sas, 1989; Jeppesen et al., 1991; Jensen, et al., 1991). The period of resilience may be long-lasting, even in rapidly-flushed lakes. A good example is the shallow Lake Søbygaard (retention time 15–25 days) in which internal loading was still so high thirteen years after a 90% reduction in external loading that it exceeded external phosphorus loading (Figure 4). Phosphorus is now released from a sediment depth of 20–25 cm, and it is estimated that a further 10–20 years will pass before the lake reaches equilibrium at the present level of external phosphorus loading (Søndergaard et al., 1993).

In Denmark, various physical and chemical methods have been employed to combat internal phosphorus loading in order to speed up the recovery of lakes following a reduction in external loading. The methods include sediment dredging and oxygenation of the hypolimnion by pure oxygen or nitrate. The most extensive dredging restoration project so far undertaken in Denmark was that at shallow Lake Brabrand (surface area 1.5 km^2, mean depth 1 m). Approx. 500,000 m^3 of phosphorus-rich sediment were removed to prevent the lake silting up and to reduce internal phosphorus loading. Although this significantly reduced internal loading, the quality of the lake water is still poor due to high external loading and high density of planktivorous fish (particularly bream) (Jørgensen, 1998).

Hypolimnic oxidation with pure oxygen has been undertaken for the last ten years in Lake Hald (surface area 3.4 km^2, mean depth 13.1 m and max. depth 31 m) to supplement the reduction in external phosphorus loading, which has been reduced to 0.7 g phosphorus m^2 ann^{-1} from approx. 1.3 g phosphorus m^2 ann^{-1} before restoration (Rasmussen, 1998). The main objective was to ensure the survival of an endangered chironomid (*Chironomus anthracinus*) and to reduce internal nutrient loading. Hypolimnic oxidation markedly reduced internal phosphorus loading which, together with the reduction in external loading, significantly lowered the average summer epilimnion phosphorus concentration from 130 μg phosphorus l^{-1} to 30 μg P l^{-1}. As a consequence, algal biomass has decreased and transparency has improved considerably (Rasmussen, 1998). Oxidation is now gradually being phased out.

Nitrate treatment of lake sediment (Ripl, 1976) or the hypolimnion (Erlandsen et al., 1988) may be a valuable alternative oxidation method to the use of pure oxygen. An experiment recently initiated in Lake Lyng (10 ha, max. depth 7.6 m) has confirmed earlier observations that nitrate treatment of the hypolimnion may significantly reduce phosphorus release from iron-rich sediment (M. Søndergaard & E. Jeppesen, unpublished observation). The findings, together with those of Erlandsen et al. (1988), indicate that nitrate treatment of the hypolimnion is a feasible alternative to the far more expensive and technically demanding sediment treatment method developed by Ripl (1976). Like oxygen, the nitrate serves two purposes. In the short-term it improves sediment redox conditions, thereby diminishing the phosphorus release in the transitional phase following a reduction in external loading. In the long-term, it stimulates aerobic mineralization of organic matter, thereby enhancing the sediment retention capacity for phosphorus.

Box 1. Nutrient retention in the river Gjern

Temporal and more permanent retention of particulate matter and associated nutrients has been measured by *in situ* sampling techniques in different stream environments and in riparian zones at 35 stream reaches in the 114 km^2 river Gjern catchment (Svendsen et al., 1998). Phosphorus and nitrogen retention was low on coarse stream bed substrates, but high within macrophyte patches and in vegetated riparian zones (Table 6). Dividing the Gjern river system into four stream orders revealed that median total sediment phosphorus (TSP) retention during the low-flow period mid-March to mid-August 1994 increased significantly with increasing stream order ($p>0.0001$, Wilcoxon rank-sum test) (Table 7). Net catchment TSP retention in stream beds and riparian zones during the same period amounted to 850±20 kg P, corresponding to 25% of gross total phosphorus export from the watershed during the period and 12% of 1994 annual total phosphorus export (or 0.65 kg P ha^{-1}) (Svendsen et al., 1998). Phosphorus retention mainly results from the sedimentation of particulate matter-associated phosphorus, although dissolved phosphorus adsorbed to sediment or assimilated by macrophytes, benthic algae, etc. may also contribute.

In the main channel of the river Gjern, up to 80% of dissolved reactive phosphorus (DRP) was adsorbed or assimilated during spring (Svendsen & Kronvang, 1993). The pools of particulate nitrogen and phosphorus retained during summer are flushed out of the stream system following weed clearance (usually during August and September) or during the first major autumn storm events (Svendsen & Kronvang, 1993 and Svendsen et al., 1995). Thus, net TSP resuspension in September 1987 and October/November 1987 amounted to 77% and 20%, respectively, of particulate phosphorus export from the river Gjern system during these two periods. The corresponding figures for total suspended nitrogen were 20% and 15%, respectively. Reduced stream maintenance, remeandering of stream channels and cutting of drains before entering the riparian zone will increase water residence time within stream systems, thereby increasing temporary retention of particulate matter and associated substances (nutrients, pesticides and heavy metals), prolonging the period with positive net retention within the stream system and increasing the frequency of flooding of riparian areas (thus enhancing permanent retention of particulate matter).

Table 6. Mean phosphorus and nitrogen retention measured during summer (June–August) in different types of stream bed and in riparian zones. Values are the average of five different stream reaches in the Gjern river system. Mean accumulation during winter was 3.4 g m^{-2} (after Svendsen & Kronvang, 1993 and Svendsen, unpublished)

Retention g m^{-2}	Type of stream bed		Riparian zone		
	Gravel	Sand	*Batrachium pel.*	*Sparganium emers.*	*Sparganium erectum*
Phosphorus	4	8	18	18	29
Nitrogen	4	9	20	20	32

Table 7. Median and net watershed total sediment phosphorus (TSP) retention in the river Gjern during mid-March 1994 to mid-August 1994 apportioned by stream order (sensu Strahler, 1957). For median TSP retention the 25% and 75% percentiles are given in parentheses. Modified from Svendsen et al. (1998)

Stream order	No. of reaches	Stream bed and riparian zone area (1000 m^{-2})	Median TSP retention (kg P km^{-1})	Net TSP (kg ±SD)
1st order	12	59	3.9 (2.1–7.8)	220 ±74
2nd order	11	38	8.3 (5.3–12)	180 ±5.7
3rd order	5	19	29 (23–30)	150 ±8.7
4th order	7	48	31 (21–35)	300 ±8.3
River system	35	164	–	850 ±20

Box 2. Impact of remeandering the river Gelså on nutrient retention

Numerous formerly channelised Danish streams and rivers have been remeandered by regional authorities over the last 5–10 years. Habitat diversity and hence biodiversity have been documented to increase following remeandering, an example being the remeandering of a 1850 m reach of the river Gelså in 1989 (Kronvang et al., 1994). Following an initially unsteady period with excessive erosion of the newly excavated watercourse, macroinvertebrate density and species composition increased, as did the number of submerged and emergent macrophyte species (Friberg et al., 1994; Kronvang et al., 1998). Remeandering of rivers also influences their sediment and nutrient retention capacity, which increases due to an increase in the total area of riparian zones and restoration of hydrological contact between the river and its riparian area. Re-establishment of temporarily flooded wet meadows thus results in the removal of large quantities of nitrate nitrogen by denitrification, as well as the retention of sediment and sediment-associated phosphorus during flooding. In the case of the river Gelså, the mass balance for the restored reach and an upstream reference reach revealed a decrease in nitrate loss to the river of 80 kg N ha^{-1} wet meadow (Kronvang et al., 1994). The restored reach of river Gelså also acted as a sink for suspended sediment and phosphorus, during the second year following remeandering, probably due to the significant reduction in discharge capacity and hence increased flooding of riparian zones and riparian areas (Kronvang et al., 1998). Excessive loss of sediment and phosphorus during the first year following restoration of the river channel, was succeeded by a net retention of 42 kg suspended sediment and 67 kg phosphorus km^{-1} restored river channel during the second study year.

Biological homeostasis is another factor affecting internal phosphorus loading. Thus planktivorous and benthivorous fish seem to contribute significantly to biological resilience in shallow eutrophic lakes. By feeding on large zooplankton and stirring up sediment when foraging on benthic invertebrates, they prevent efficient grazer control of phytoplankton, thereby keeping the lake in the turbid state despite a reduction in external loading. A shift to the clearwater state may therefore be prevented or delayed (Jeppesen et al., 1990; Hosper & Jagtman, 1990; Scheffer, 1990; Scheffer et al., 1993). To overcome such biological resilience, various fish manipulation methods have been and are currently being developed. One such method is the enhancement of top-down control of phytoplankton by the selective removal of planktivorous fish, a method that has been employed with varying degrees of success in many parts of the world (Shapiro & Wright, 1984; Benndorf, 1990; De Melo et al., 1992; Meijer et al., 1994; Gulati et al., 1995; Phillips et al., 1996).

An alternative or supplementary method to fish removal is to stock high numbers of 0^+ pike (1,500–3,000 ha^{-1}) to control newly hatched planktivorous fish such as roach, bream and bleak (Prejs, 1994; Søndergaard et al., 1997; Berg et al., 1997) (see Box 3). However, stocking has to be repeated every year until macrophytes have become established, it not usually being possible to maintain a sufficiently high pike density to control the young-of-the-year (YOY) due to cannibalism among the pike.

The first Danish full-scale experiments involving fish manipulation were initiated in 1986, and fourteen additional experiments have since been undertaken. Lake Væng is an example of a successful restoration. The findings to date indicate that the method will have a long-term effect in shallow lakes if nutrient loading has been reduced to a level such that the TP concentration in the future state of equilibrium will be below 0.05–0.1 mg P l^{-1} (E. Jeppesen & M. Søndergaard, unpublished observations), a level somewhat lower than the threshold of 0.08–0.15 mg P l^{-1} given by Jeppesen et al. (1990). However, if nitrogen loading is low, stocking may have a positive impact at higher phosphorus concentrations. The threshold of 0.05–0.1 mg P l^{-1} is in accordance with the results of a survey of 65 Danish lakes (Figure 7) as it is within this range that major changes in lake biological structure are usually observed: The biomass of planktivorous fish decreases and the percentage of predator fish increases, thus indicating greater control of prey fish by the carnivorous fish. Moreover, the zooplankton: phytoplankton ratio increases, thus indicating greater grazer control of phytoplankton, and chlorophyll *a* declines and transparency increases. Fish manipulation may cascade to the nutrient levels as well. A 30–50%

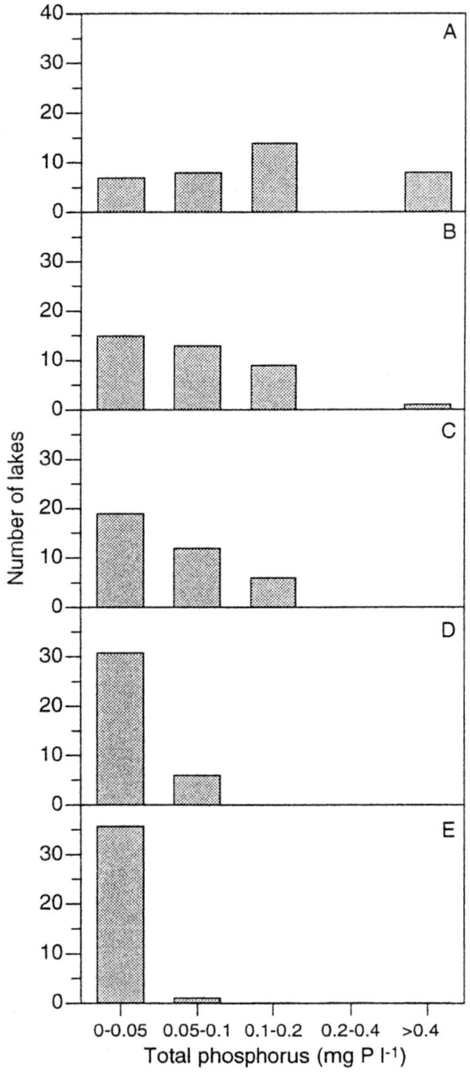

Figure 3. Effects of various measures aimed at lowering annual average phosphorus loading of 37 Danish lakes (all Danish lake types are represented): (a) recorded averages for the period 1989–95, (b) when the lakes are in equilibria at the current level of loading according to Vollenweider's retention models (Vollenweider, 1976), (c) cut-off of all loading point-sources, excluding the contribution from scattered dwellings, (d) cut-off of point-sources and reduction of diffuse loading by 50%, (e) as (d), but a 70% reduction of diffuse loading. The results show that effective measures need to be taken against diffuse loading, if the phosphorus level of Danish lakes is to be reduced to a level so low that a shift to the clearwater state may take place (see Figure 7).

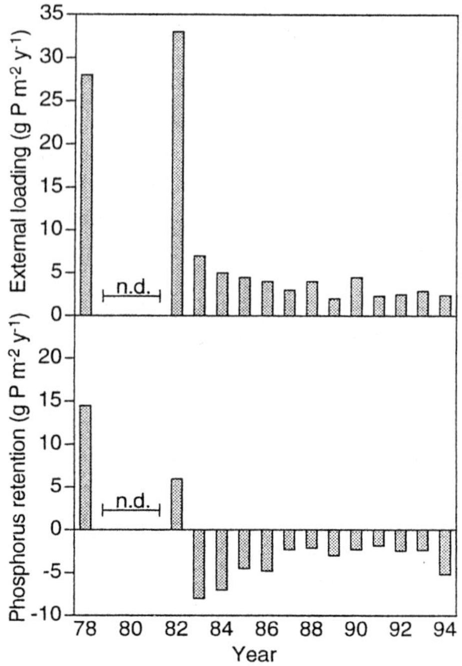

Figure 4. Trend in phosphorus loading and net retention in Lake Søbygaard over the period 1978–94.

reduction in lake concentrations of total nitrogen and total phosphorus has been found in the most successful fish manipulation experiments in Denmark, even in cases when macrophytes were absent (Jeppesen et al., 1998; E. Jeppesen & M. Søndergaard, unpublished). Consequently, lake nutrient retention capacity increases and loading of the marine areas decreases. Fish manipulation methods are still being developed, but there is little doubt that they will become an important tool to promote the restoration of eutrophic lakes once external nutrient loading reduction has been reduced.

A further factor contributing to resilience following a reduction in nutrient loading is grazing by waterfowl, in particular coot (*Fulica atra*) and mute swan (*Cygnus olor*), which has been reported to hinder successful recolonization by submerged macrophytes (Lauridsen et al., 1993, 1994; van Donk et al., 1994; Søndergaard et al., 1996). The construction of refuges to protect macrophytes against grazing has therefore been employed as an alternative or supplementary restoration tool. The refuges need only be simple, e.g. wooden poles surrounded by garden fencing or chicken wire. The refuges enable the macrophytes to grow in a grazer-free environment from where they may spread and colonize the remainder of the lake. In addition, macrophyte refuges serve as a daytime refuge for pelagic zooplankton (*Daphnia*, etc.), thereby enabling them to avoid fish predation (Timms & Moss, 1984; Phillips et al., 1996; Schriver et al., 1995; Lauridsen & Lodge, 1996). At night,

Box 3. Fish manipulation by stocking of 0^+ pike (*Esox lucius*) in Lake Lyng

Lake Lyng has a surface area of 10 ha, a mean depth of 2.4 m and a maximum depth of 7.6 m. Approximately 10% of the lake area has a depth above 5 m and stratifies during summer. The lake has no surface inlets except during periods of heavy precipitation, and the hydraulic retention time is estimated to be more than one year. It is eutrophic and in 1989, before pike stocking began, mean summer Secchi depth was 0.72 m and total phosphorus 0.79 mg P l^{-1} (epilimnion). In spring 1990–93, stocking of juvenile pike (mean length: 2–6 cm) was undertaken in densities ranging from 500 to 3,600 ind. ha^{-1} yr^{-1}. Catch per unit effort of the dominant species, roach, varied from 17 to 272 during the period 1989–94. YOY (young of the year roach) density was high in years with little or no stocking, but was low in years with high stocking. There were corresponding marked changes in the abundance and relative composition of zooplankton, and the percentage contribution of *Daphnia* to total cladoceran density was highest and rotifer density lowest in years with low YOY density (Figure 5). Mean summer chlorophyll *a* correlated significantly with YOY density, suggesting an impact of stocking at the phytoplankton level. Finally, total phosphorus concentration was positively related to YOY density, irrespective of the fact that no changes were observed in the external phosphorus loading. It is concluded that pike stocking can be used as a lake restoration tool to increase lake water transparency; however, the effect only lasts for the season in which stocking is undertaken. The method is therefore considered to be most useful in shallow, turbid lakes in which nutrient loading has been reduced sufficiently to permit substantial and permanent macrophyte coverage if clearwater conditions are established. Further information about the pike stocking trials is given in Berg et al. (1997) and Søndergaard et al. (1997).

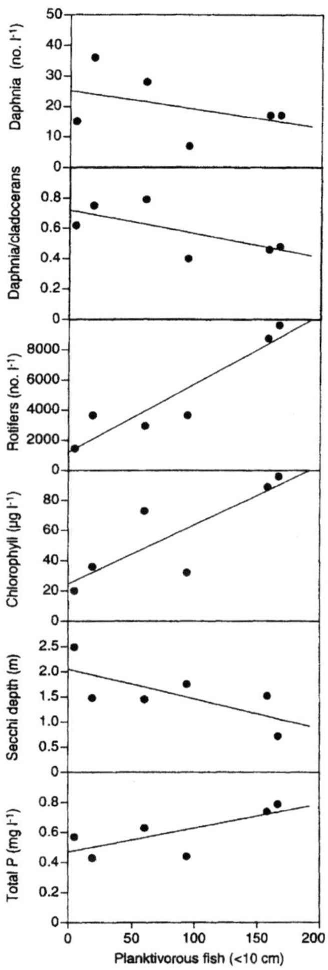

Figure 5. *Daphnia* abundance, percentage contribution of *Daphnia* to total cladoceran density, rotifer abundance, chlorophyll *a*, Secchi depth and total phosphorus versus planktivorous fish (< 10 cm) catch in multiple mesh size gill nets (one measurement per year) in Lake Lyng in August during 6 years with different levels of stocking of 0^+pike (0–3,620 indiv. ha^{-1}).

when predation risk is lower, the zooplankton migrate into open water in search of food. Macrophyte refuges thus help augment grazing pressure on phytoplankton, thereby enhancing water transparency and further improving growth conditions for the macrophytes. Experiments undertaken in Denmark show that zooplankton density during day is highest in areas of high macrophyte density (Jeppesen et al., 1997), this being in accordance with the fact that fish tend to avoid such areas (Winfield, 1986; Engel, 1988; Phillips et al., 1996; Jeppesen 1997). Moreover, the proportion of zooplankton migrating into open water at night was highest from dense macrophyte refuges (Jeppesen et al., 1997), just as daytime concentration and night-time migration of zooplankton per unit area at fixed macrophyte density were highest from small refuges (Lauridsen et al., 1996). The latter may be attributable to the fact that migratory pelagic zooplankton

Box 4. Fish manipulation by removal of planktivorous fish in Lake Væng

Lake Væng covers 15 ha, has a mean depth of 1.2 m, a maximum depth of 1.8 m and a hydraulic retention time of 15–25 days. Groundwater contributes 95% of the inflow, and biologically treated sewage effluent from a small village was discharged into the lake until 1981. No changes in biological structure and Secchi depth were observed during the first five years following cessation of sewage discharge. The fish stock was dominated by plankti-benthivorous fish and the zooplankton biomass was low, while the phytoplankton biomass was high and dominated by cyanobacteria.

From October 1986 to July 1988, 50% of the plankti-benthivorous fish biomass was removed, thereby causing a trophic cascade resulting in low phytoplankton biomass and low percentage of cyanobacteria. Secchi depth increased significantly from a summer mean of approx. 0.6 m and peaked in 1989–90 (Figure 6). After an initial delay that is probably an attributable to grazing by waterfowl, submerged macrophytes colonized the lake and the concentration of nitrogen, and in most years also of phosphorus, decreased. The piscivore: plankti-benthivore ratio increased to 50%, thereafter fluctuating between 30% and 60%. The macrophytes almost disappeared during winter 1991–92. However, except for a short-term deterioration in late summer (a few weeks), the lake stayed clear, supposedly due to the high density of piscivorous fish. The macrophytes reappeared in 1993, and have since increased their coverage. Thus the lake has stayed clear for seven years following the fish manipulation initiative. The Lake Væng case study is a good illustration of biological resilience caused by the plankti-benthivorous fish, as well as of the efficacy of fish removal in cases where the conditions for such an intervention are fulfilled (see main text).

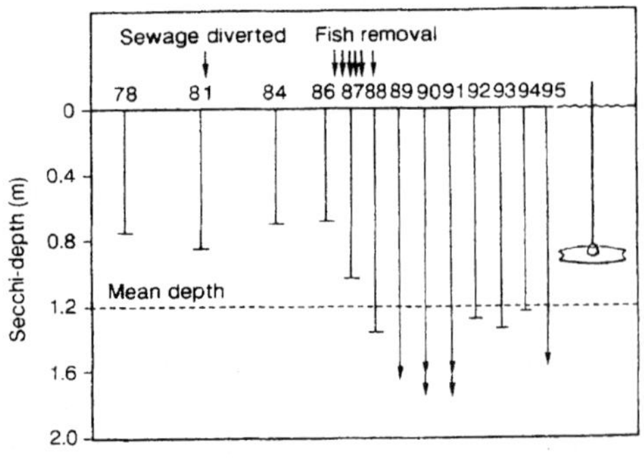

Figure 6. (a) Trend in the summer mean Secchi depth (May–October) in Lake Væng following the cessation of sewage discharge in 1981 and fish manipulation in 1986–88 involving removal of 50% of the planktivorous fish biomass. Arrows indicate that Secchi depth reached the bottom at some of the sampling dates; the mean values are therefore higher.

It also shows a cascading effect on the nitrogen and phosphorus levels in the lake, in particular following the appearance of submerged macrophytes. Further information about the study can be found in Søndergaard et al. (1990), Jeppesen et al. (1990, 1991) and Lauridsen et al. (1993, 1994).

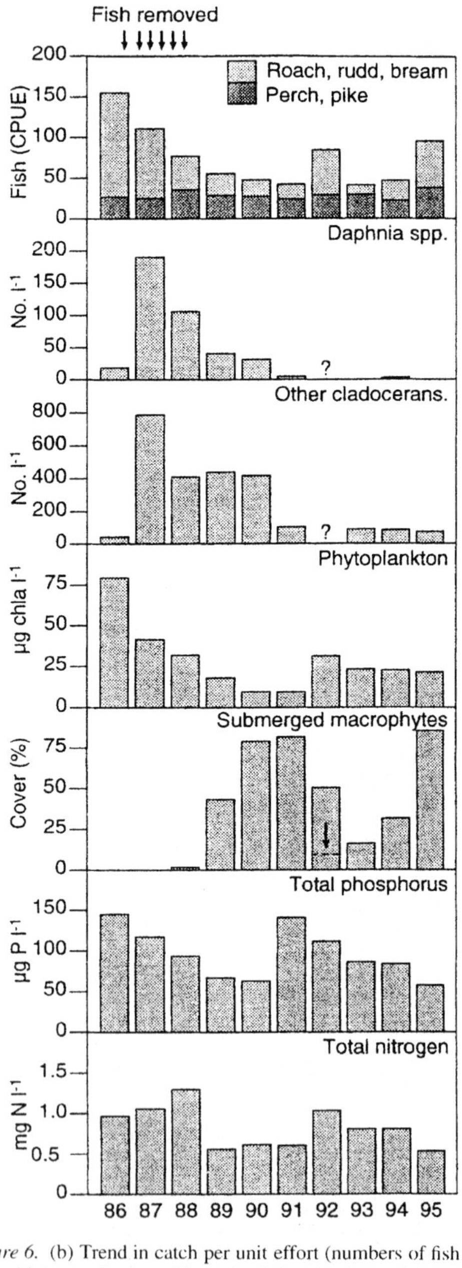

Figure 6. (b) Trend in catch per unit effort (numbers of fish caught in multiple mesh size gill nets) of the dominant fish species in Lake Væng following fish manipulation. Also shown are summer mean *Daphnia* and *Bosmina* abundance, chlorophyll *a*, lake water total phosphorus and total nitrogen concentrations, and maximum submerged macrophyte coverage during summer.

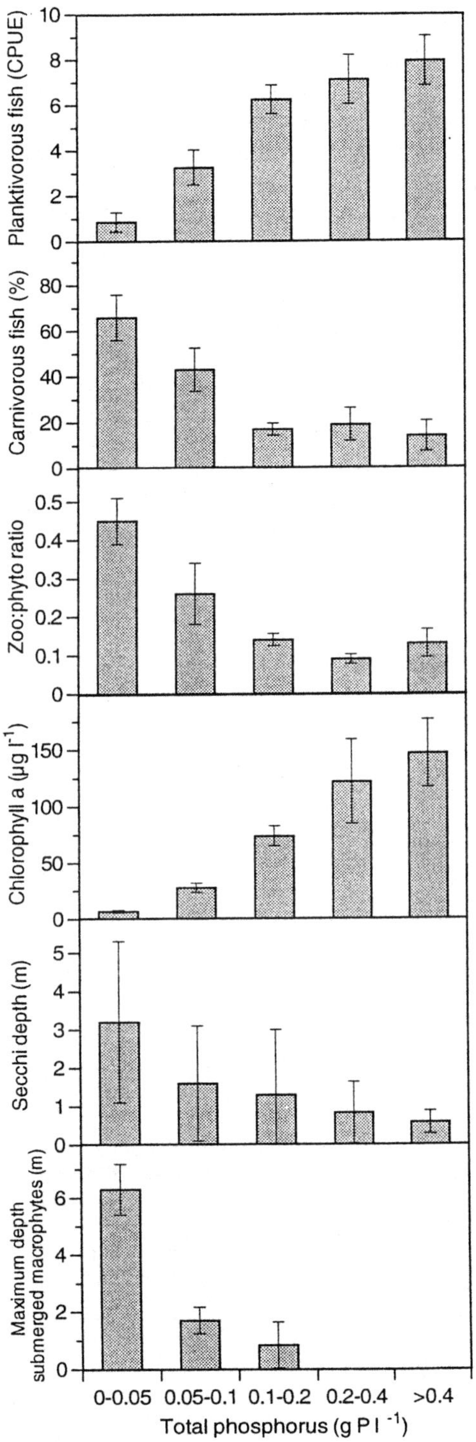

Figure 7. CPUE in terms of biomass of plankti-benthivorous fish in multiple mesh-sized gill nets (14 different mesh sizes 6.25–75 mm) sampled in August for all Danish lakes on which information about fish CPUE and zooplankton is available (n = 65). Also shown are the percentage of carnivorous fish, summer mean of zooplankton:phytoplankton biomass ratio, epilimnion chlorophyll a concentration, Secchi depth and the maximum depth of submerged angiosperm macrophytes versus total phosphorus. Mean ± SD of the five total phosphorus groups is shown.

seem to prefer the transitional zone between plants and open water, for which reason their density is greatest in refuges having a high perimeter:area ratio (Lauridsen & Buenk, 1996). Small, dense macrophyte refuges thus seem to have the highest flux of migrating zooplankton per unit area, and hence probably have the greatest grazing impact on the open water phytoplankton. The usefulness of plant refuges as a restoration tool is probably limited to the same nutrient level as the fish manipulation methods (Jeppesen et al., 1997), and their effect is probably greatest in small lakes (Jeppesen et al., 1997; Søndergaard et al., 1996) where coot density in the absence of plants is highest (Brøgger-Jensen & Jørgensen, 1992).

Conclusion

The Danish findings indicate that, even if population density is high and agricultural activity intensive, then lake eutrophication problems can still be remedied. This demands a multiple approach encompassing measures aimed at reducing point source and diffuse loading, as well as enhancing nutrient retention and removal in the catchment, e.g. by re-establishing lost wetlands, remeandering streams and changing stream maintenance practice. To speed up lake recovery following a reduction in nutrient loading, lake restoration measures may be used. At relatively low nutrient concentrations, particularly in shallow lakes, recovery may be enhanced by biomanipulation. At high nutrient concentrations, however, sediment dredging and hypolimnic oxidation with pure oxygen or nitrate may be more valuable. Danish experience also indicates that cooperation between state and regional authorities on the exchange of know-how, data and models can ensure an efficient and scientifically sound approach to the lake eutrophication problem.

Acknowledgements

We thank the Danish Counties for access to some of the data used in the analyses. The assistance of the technical staff of the National Environmental Research Institute, Silkeborg, is gratefully acknowledged. Field and laboratory assistance was provided by L. Hansen, J. Stougaard, B. Laustsen, J. Glargaard, K. Jensen & L. Nørgaard. Technical assistance was provided by K. Møgelvang & A. M. Poulsen. D. I. Barry provided valuable editorial comments. The study was supported by the Centre for Freshwater Environmental Research.

References

Ambus, R. & Hoffman, C.C., 1990. Kvcelstofomscetning of stofbalance i åncere områder [Nitrogen turnover and mass balance in riparian areas] NPO research from the Danish Environmental Protection Agency No. C13. 68 pp. (In Danish)

Benndorf, J., 1990. Condition for effective biomanipulation: Conclusion derived from whole-lake experiments in Europe. Hydrobiologia 200/201 (Dev. Hydrobiol. 61): 187–204.

Berg, S., E. Jeppesen & M. Søndergaard, 1997. Pike (*Esox lucius* L.) stocking as a biomanipulation tool. 1. Effects on the fish population in Lake Lyng (Denmark). Hydrobiologia 342/343: 311–318.

Brøgger-Jensen, S. & H. E. Jørgensen, 1992. Vandfugle og søers miljøtilstand. Miljøprojekt nr. 200 [Waterfowl versus the environmental state of lakes. Environmental Project no. 200], Danish Environmental Protection Agency, Copenhagen, 64 pp. (In Danish).

Christensen, P. B., L. P Nielsen & N. P. Revsbech, 1991. Denitrification i våd- og vandområder [Denitrification in wetlands, lakes, rivers and fjords]. In J.-O. Frier & J. R. Christensen (Eds). Kvælstof, Fosfor Og Organisk Stof i Jord- Og Vandmiljøet [Nitrogen, Phosphorus and Organic Matter in Land and Water Environment]. Rapport fra konsensuskonference 31. januar, 1. og 4. februar 1991, Danish Environmental Protection Agency, 10–24. (In Danish).

Cullen, P. & C. Forsberg, 1988. Experience with reducing point sources of phosphorus to lakes. Hydrobiologia 170: 321–336.

De Melo, R., R. France & D. J. McQueen, 1992. Biomanipulation: Hit or myth? Limnol. Oceanogr. 37: 192–207.

Engel, S., 1988. The role and interactions of submersed macrophytes in a shallow Wisconsin lake. J. Freshwat. Ecol. 4. 329–340.

Erlandsen, H., P. Breetum, J. E. Løvik, S. Markager & T. Källqvist, 1988. Kolbotnvatnet. Sammenstilling av resultater fra perioden 1984–89. Report from Norsk Institut for Vannforskning, NIVA, Oslo, Norge, 118 pp. (In Norwegian).

Friberg, N., B. Kronvang, L. M. Svendsen, H. O. Hansen & M. B. Nielsen, 1994. Restoration of a channelized reach of the River Gelså, Denmark: Effects on the macroinvertebrate community. Aquatic Conservation: Marine and Freshwater Ecosystems 3: 73–92.

Gulati, R., 1995. Manipulation of fish population for lake recovery from eutrophication in the temperate regions (Chapter 4). In R. De Bernardi & G. Guissani (Eds), Biomanipulation in Lakes and Reservoirs Management. Guidelines of Lake Management 7: 53–79.

Hoffmannn, C. C., M. Dahl, L. Kamp-Nielsen & H. Stryhn, 1993. Vand- og stofbalance i en natureng [Water and nutrient mass balance in a meadow]. Danish Environmental Protection Agency, Copenhagen, 152 pp. (In Danish).

Hoffmannn, C. C., 1996. Fate of phosphate, nitrate and other elements during short-term flooding of a riparian meadow. In: B. Kronvang, S. E. Sibbesen & L. M. Svendsen (Eds), Proceedings From An International Workshop "Sediment and phosphorus. Erosion and delivery, transport and fate of sediment-associated nutrients in watersheds" held October 9-12 1995 in Silkeborg Denmark. Ministry of Environment and Energy – NEI Technical Report No. 178: 135–142.

Hoffmannn, C. C., 1998a. Nitrate removal in a regularly flooded riparian meadow. Verh. Int. Ver Limnol. 26: 1352–1358.

Hoffmann, C.C., 1998b. Nutrient retention in wet meadows and fens. PhD thesis National Environmental Research Institute, Silkeborg, Denmark. 134 pp.

Hosper, S. H. & E. Jagtman, 1990. Biomanipulation additional to nutrient control for restoration of shallow lakes in the Netherlands. Hydrobiologia 200/201 (Dev. Hydrobiol. 61): 523–534.

Iversen, T. M., B. Kronvang, B. L. Lauge, P. N. Markmann & M. B. Nielsen, 1993. Re-establishment of Danish streams: restoration and maintenance measures. Aquatic Conservation: Marine and Freshwater Ecosystems 3: 73–92.

Iversen, T. M., B. Kronvang, C. C. Hoffmannn, M. Søndergaard & H. O. Hansen, 1995. Restoration of aquatic ecosystems and water quality. In H. Skotte-Møller (Ed.), Nature Restoration in the European Union. Proceedings from a seminar in Denmark 29-31 May 1995. Ministry of Environment and Energy, The National Forest and Nature Agency Copenhagen: 63–69.

Jensen, J. P., P. Kristensen & E. Jeppesen, 1991. Relationships between nitrogen loading and in-lake nitrogen concentrations in shallow Danish lakes. Verh. Int. Ver. Limnol. 24: 201–204.

Jensen, H. S., P. Kristensen, E. Jeppesen & A. Skytthe, 1991. Iron:phosphorus ratio in surface sediment as an indicator of phosphorus release from aerobic sediments in shallow lakes. Hydrobiologia 235/236 (Dev. Hydrobiol. 75): 731–743.

Jeppesen, E., J. P. Jensen, P. Kristensen, M. Søndergaard, E. Mortensen, O. Sortkjær & K. Olrik, 1990. Fish manipulation as a lake restoration tool in shallow, eutrophic, temperate lakes 2: Threshold levels, long-term stability and conclusions. Hydrobiologia 200/201 (Dev. Hydrobiol. 61): 219–227.

Jeppesen, E., P. Kristensen, J. P. Jensen, M. Søndergaard, E. Mortensen & T. Lauridsen, 1991. Recovery resilience following a reduction in external phosphorus loading of shallow, eutrophic Danish lakes: Duration, regulating factors and methods for overcoming resilience. Mem. Ist. ital. Idrobiol. 48: 127–148.

Jeppesen, E., J. P. Jensen, M. Søndergaard, T. Lauridsen, L. J. Pedersen & L. Jensen, 1997. Top-down control in freshwater lakes: the role of nutrient state, submerged macrophytes and water depth. Hydrobiologia 342/343 (Dev. Hydrobiol. 119): 151–164.

Jeppesen, E., J. Windolf, J. P. Jensen, T. L. Lauridsen, M. Søndergaard, K. Sandby & P. H. Møller, 1998. Changes in nitrogen retention in four shallow eutrophic lakes following a decline in density of cyprinids. Arch. Hydrobiol. 142: 129–152.

Jørgensen, T.B. (1998) Braband Sø In: M. Søndergaard, E. Jeppesen & J.P. Jensen (eds), Sørestaurering i Danmark [Lake restoration in Denmark]. Environmental News no. 28, Danish Environmental Protection Agency, Copenhagen (In Danish).

Kristensen, P., J. P. Jensen & E. Jeppesen, 1990. Eutrofieringsmodeller for søer. [Eutrophication models for lakes]. NPo-forskning fra Miljøstyrelsen C9, C-abstracts. Danish Environmental Protection Agency, Copenhagen, 120 pp. (In Danish).

Kristensen, P., J. P. Jensen & E. Jeppesen, 1991.Simple empirical lake models. Nitrogen and phosphorus in fresh and marine waters. NPo-forskning fra Miljøstyrelsen, C-abstracts. Danish Environmental Protection Agency, Copenhagen: 125–145.

Kronvang, B., 1992. The export of particulate matter, particulate phosphorus and dissolved phosphorus from two agricultural river basins: Implications for estimating the non-point phosphorus load. Wat. Res. 26: 1347–1358.

Kronvang, B., P. Græsbøll, L. M. Svendsen, N. Friberg, A. B. Hald, G. Kjellson, M. B. Nielsen, B. D. Petersen & O. Ottosen, 1994. Restaurering af Gelså ved Bevtoft – miljømæssig effekt i vandløb og de vandløbsnære arealer. [Restoration of the river Gelså at Bevtoft – environmental effect in streams and riparian zones.

National Environmental Research Institute. Scientific Report no. 110]. Faglig rapport fra DMU nr. 110, 87 pp. (In Danish).

Kronvang, B., R. Grant, S. E. Larsen, L. M. Svendsen & P. Kristensen, 1995. Non-point-source nutrient losses to the aquatic environment in Denmark: Impact of agriculture. Marine & Freshwater Research 46: 167–77.

Kronvang, B., L. M. Svendsen, O. Ottosen, M. B. Nielsen & L. Johannesen, 1998. Remeandering of rivers: short-term implications for sediment and nutrient transport. Ver. Int. Ver. Limnol. 26: 929–934.

Larsen, S. E., J. Erfurt, P. Græsbøll, B. Kronvang, E. Mortensen, C. A. Nielsen, N. B. Ovesen, C. Paludan, Aa. Rebsdorf, L. M. Svendsen & P. Nyegaard, 1995. Ferske vandområder – Vandløb og kilder. Vandmiljøplanens Overvågningsprogram 1994. [Freshwater areas – streams and springs. The Aquatic Monitoring Programme 1994. National Environmental Research Institute. Scientific Report no. 140]. Danmarks Miljøundersøgelser. Faglig rapport fra DMU nr. 140, 196 pp. (In Danish).

Lauridsen, T. L., E. Jeppesen & F. Ø. Andersen, 1993. Colonization of submerged macrophytes in shallow fish manipulated Lake Væng: Impact of sediment composition and water fowl grazing. Aquat. Bot. 46: 1–15.

Lauridsen, T. L., E. Jeppesen & M. Søndergaard (1994). Colonization and succession of submerged macrophytes in shallow Lake Væng during the first five years following fish-manipulation. Hydrobiologia 275/276 (Dev. Hydrobiol. 94): 233–242.

Lauridsen, T. & I. Buenk, 1996. Diel changes in the horizontal distribution of zooplankton in the littoral zone of two eutrophic shallow lakes. Arch. Hydrobiol. 137: 161–176.

Lauridsen T. L. & D. Lodge, 1996. Avoidance by *Daphnia magna* Strauss of fish and macrophytes: Chemical cues and predator-mediated use of macrophyte habitat. Limnol. & Oceanogr. 41: 794–798.

Lauridsen, T. L., L. Junge Pedersen, E. Jeppesen & M. Søndergaard, 1996. The importance of macrophyte bed size for composition and horizontal migration of cladocerans in a shallow lake. J. Plankton Res. 18: 2283–2294.

Madsen, B. L., 1995. Danish watercourses – ten years with the new Watercourse Act. Miljønyt no. 11, Danish Environmental Protection Agency, Copenhagen, 206 pp.

Meijer, M-L., E. Jeppesen, E. van Donk., B. Moss, M. Scheffer, E. Lammens, E. Van Nes, J. A. Berkum, G. J. de Jong, B. A. Faafeng & J. P. Jensen, 1994. Long-term responses to fish-stock reduction in small shallow lakes: Interpretation of five year results of four biomanipulation cases in the Netherlands and Denmark. Hydrobiologia 275/276 (Dev. Hydrobiol. 94): 457–466.

Phillips, G. L., M. Perrow & J. Stansfield, 1996. Manipulating the fish-zooplankton interaction in shallow lakes: a tool for restoration.In S. P. R. Greenstreet & M. L. Tasker (eds), Aquatic Predators and Their Prey. Blackwell Scientific Publications, Oxford: 174–183.

Prejs, A., A. Martyniak, S. Born, P. Hliwa & P. Koperski, 1994. Food web manipulation in a small eutrophic Lake Wirbel, Poland: Effect of stocking with juvenile pike on planktivorous fish. Hydrobiologia 275/276. (Dev. Hydrobiol. 94): 65–70.

Rasmussen, K., (1998). Hald Sø. In M. Søndergaard, E. Jeppesen & J.P. Jensen (eds), Sørestaurering i Danmark [Lake Restoration in Denmark]. Environmental News no. 28, Danish Environmental Protection Agency, Copenhagen (In Danish). 259–268.

Rebsdorf, A. A., N. Friberg, C. C. Hoffmannn & B. Kronvang, 1994. Ånære arealers samspil med vandløb. [The interaction of riparian zones with streams]. Danish Environmental Protection Agency, Copenhagen, 140 pp. (In Danish).

Ripl, W., 1976. Biochemical oxidation of polluted lake sediment. A new restoration method. Ambio 5: 409–452.

Sas, H. (ed.), 1989. Lake Restoration by Reduction of Nutrient Loading. Expectation, Experiences, Extrapolation. Acad. Ver. Richardz Gmbh., 497 pp.

Scheffer, M., 1990. Multiplicity of stable states in freshwater systems. Hydrobiologia 200/201 (Dev. Hydrobiol. 61): 475–486.

Scheffer, M., S. H. Hosper, M.-L. Meijer, B. Moss & E. Jeppesen, 1993. Alternative equilibria in shallow lakes. Trends in Ecology and Evolution 8: 275–279.

Schriver, P., J. Bøgestrand, E. Jeppesen & M. Søndergaard, 1995. Impact of submerged macrophytes on the interactions between fish, zooplankton and phytoplankton: Large-scale enclosure experiments in a shallow lake. Freshwat. Biol. 33: 255–270.

Shapiro, J. & D. I. Wright, 1984. Lake restoration by biomanipulation. Round Lake, Minnesota – the first two years. Freshwat. Biol. 14: 371–383.

Skotte-Møller, H., 1995. Nature restoratrion in Denmark – status and perspectives. In H. Skotte-Møller, (Ed.), Nature Restoration in the European Union. Proceedings from a seminar in Denmark 29-31 May 1995. Ministry of Environment and Energy, The National Forest and Nature Agency: 63–69.

Strahler, A. N., 1957. Quantitative analysis of watershed geomorphology. American Geophysical Union Transactions 38: 913–920.

Svendsen, L. M. & B. Kronvang, 1993. Retention of nitrogen and phosphorus in a Danish lowland river system: Implications for the export from the watershed. Hydrobiologia 251: 123–135.

Svendsen, L. M., B. Kronvang, P. Kristensen & P. Græsbøll, 1995. Dynamics of phosphorus compounds in a lowland river system: Importance of retention and non-point sources. Hydrobiological Processes 9: 119–142.

Svendsen, L. M., B. Kronvang, A. R. Laubel, S. E. Larsen & B. Andersen, 1998. Phosphorus retention in a Danish lowland river system. Verh. Int. Ver. Limnol. 26: 956–962.

Søndergaard, M., E. Jeppesen, E. Mortensen, E. Dall, P. Kristensen & O. Sortkjær, 1990. Phytoplankton biomass reduction after planktivorous fish reduction in a shallow, eutrophic lake: A combined effect of reduced internal P-loading and increased zooplankton grazing. Hydrobiologia 200/201 (Dev. Hydrobiol. 61): 229–240.

Søndergaard, M., P. Kristensen & E. Jeppesen, 1993. Eight years of internal phosphorus loading and changes in the sediment phosphorus profile of Lake Søbygaard, Denmark. Hydrobiologia 253: 345–356.

Søndergaard, M., E. Jeppesen & S. Berg, 1997. Pike *(Esox lucius)* stocking as a biomanipulation tool, 2. Effects on lower trophic levels in Lake Lyng, Denmark. Hydrobiologia. 342/343 (Dev. Hydrobiol. 119): 319–325.

Søndergaard, M., L. Olufsen, T. Lauridsen, E. Jeppesen & T. Vindbæk Madsen, 1996d. The impact of grazing waterfowl on submerged macrophytes: In situ experiments in a shallow eutrophic lake. Aquat. Bot. 53: 73–84.

Timms, R. M. & B. Moss, 1984. Prevention of growth of potentially dense phytoplankton populations by zooplankton grazing in the presence of zooplanktivorous fish, in a shallow wetland ecosystem. Limnol. Oceanogr. 29: 472–486.

Van Donk, E., E. De Deckere, J. G. P. Klein Breteler & J. Meulemans, 1994. Herbivory by waterfowl and fish on macrophytes in a biomanipulated lake: Effects on long-term recovery. Verh. Int. Ver. Limnol. 25: 2139–2143.

Vollenweider, R. A., 1976. Advance in defining critical loading levels for phosphorus in lake eutrophication. Mem. Ist. Ital. Idrobiol. 33: 53–83.

Windolf, J., E. Jeppesen, M. Søndergaard, J. P. Jensen & L. Sortkjær, 1993. Ferske vandområder – søer. Vandmiljøplanens Overvågningsprogram 1992. [Freshwater areas – lakes. The Aquatic Monitoring Programme 1992. National Environmental Research Institute. Scientific Report no. 90]. Danmarks Miljøundersøgelser. Faglig rapport fra DMU, nr. 90, 130 pp. (In Danish)

Windolf, J., E. Jeppesen, J. P. Jensen & P. Kristensen, 1996. Modelling the seasonal variation in in-lake concentration and retention of nitrogen in 16 shallow lakes: a four-year mass balance study. Biogeochemistry 33: 25–44.

Winfield, I. J., 1986. The influence of simulated aquatic macrophytes on the zooplankton consumption rate of juvenile roach, *Rutilus rutilus,* rudd, *Scardinius erythrophthalmus,* and perch, *Perca fluviatilis.* J. Fish Biol. 29: 37–48.

Hydrobiologia **395/396:** 433–453, 1999.
D.M. Harper, B. Brierley, A.J.D. Ferguson & G. Phillips (eds), The Ecological Bases for Lake and Reservoir Management.

A quality classification for management of Scottish standing waters

Ian Fozzard[1], Ross Doughty[2], Robert C. Ferrier[3], Tom Leatherland[1] & Roger Owen[4]
[1]*Scottish Environment Protection Agency, East Region E-mail: ian.fozzard@sepa.org.uk*
[2]*Scottish Environment Protection Agency, West Region*
[3]*Macauley Land Use Research Institute, Aberdeen*
[4]*Scottish Environment Protection Agency, North Region*

Key words: Scotland, lochs, eutrophication, acidification, toxic substances, classification

Abstract

The development of a national scheme for the standing waters of Scotland is described. It classifies waters along three gradients – phosphorus, loss of acid neutralising capacity and degree of contamination with toxic suubstances. A series of case studies provide examples of the application of the classification in lake management.

Introduction

The major problems affecting standing waters have been recognised for at least two decades, but their quantification and classification for environmental managers has proved elusive. The European Environment Agency has recently described the condition of European freshwater resources and their management as a prominent environmental problem with nutrient enrichment, acidification, and contamination by toxic substances identified as major impacts (Stanners & Bourdeau, 1995). The classification of the quality of standing waters, that is the ranking of the degree of impact of man's activities, is a necessary first step in the development of any management strategy. The Scottish classification scheme has now been developed to inform policy makers and the public of the state of the environment, although it may not be the most useful product of the process. Other important products of a science-based scheme should be data on current status, information on the distance from the desired state (setting the water in context where change and sources of change can be identified) and some indication of the actions necessary for improvement.

Classification schemes

Trophic status

Classification schemes for standing waters have been the subject of academic research for over a century, initially to categorise the natural range of lake types in an effort to gain understanding of their ecology, and latterly to investigate the impact of man's activities on the range of natural variability observed. The major focus of most research has been trophic status. In standing waters inorganic plant nutrients, principally phosphorus, nitrogen and silica play an important role. Standing waters in Britain have naturally varying concentrations of nutrients determined by the physical, chemical and biological character of their catchments. This results in a natural range of trophic status and ecological character. The location of the reserves of these nutrients within the standing water ecosystem, the transformation pathways and the specific requirements of the biota have resulted in significant difficulties in sampling, measurement, evaluation and classification.

Classifications using various taxonomic groups, or assemblages from specific habitats within the standing water, have also encountered difficulties. Harper (1986) related increasing degrees of anthropogenic enrichment to changes in the littoral, planktonic and benthic fauna in a series of three very different Angus and Tayside lochs. However, littoral habitats often reflect local conditions, for example substratum particle size or hydraulic energy inputs, rather than the quality of the water as a whole. Conservation agencies in the United Kingdom have focused their efforts on the classification of standing waters using macrophytes and the empirical relationship between species and measures of trophic status (Palmer, 1989). However phosphorus data has not been regularly collected during these surveys and the relationship between the macrophyte-based indices and water column phosphorus concentrations is unclear. Macrophytes may of course, obtain their phosphorus from the centuries of accumulation in the sediments of standing waters and therefore be largely independent of current water column concentrations. The profundal zoobenthos promised a simple scheme of classification based upon the relationship of this assemblage to the productivity of the phytoplankton in the overlying water column, thus integrating the whole loch water column character (Brinkhurst, 1974; Saether, 1979; Wiederholm, 1980). The difficulties of separating natural variation in productivity and sediment character from perturbation effects again limits its usefulness. Additionally, in some Scottish lochs, the influence of natural allochthonous organic matter (principally the leaves of deciduous trees) supplement the organic matter derived from the plankton and confound the relationship between open water nutrient status and profundal zoobenthos (Weerekoon, 1956; Fozzard et al., 1994).

Classification schemes based upon plankton offer perhaps the most directly relevant information to water managers, assuming that the plankton have a strong relationship with water column chemistry. Both phytoplankton and zooplankton have been used for classification of standing waters in Scotland (Fozzard & Marsden 1990, Forth River Purification Board, unpublished), but the low species richness of the zooplankton (Maitland et al., 1981) and the difficulties of species level identification of phytoplankton and the complexity of interpretation (Bailey-Watts & Duncan 1981b), have limited their routine application.

Acidification

Surface water acidification is a relatively recently-identified phenomenon affecting standing waters. As such it has not received the same degree of attention as nutrient enrichment. In Scotland, many surface waters are acidic, but in most cases this may be regarded as their natural condition. Uncontaminated rainfall is naturally acidic and the many 'brown water' acid lochs are also regarded as naturally acid, their acidity being derived from the organic acids of peat and from acids produced in biological processes. Surface waters acidified by man's activities are the product of interactions between atmospheric contamination, climate, vegetation and soils. Acidified surface waters may be defined as those which have lost some of their alkalinity due to the input of strong acids from the atmosphere. Biota indicating acid conditions and which could be used to classify standing waters have been identified (Raddum & Saether, 1981; Raddum & Fjellheim, 1984; Raddum et al., 1988; Doughty, 1990) but the difficulties of separating acidified from naturally acidic sites limit the value of this method.

Toxic substances

The presence of toxic substances in standing waters is perhaps a simpler problem to address in relation to quality classification. Many trace elements and derivatives, such as ammonia, are naturally present at low concentrations, often below nationally agreed environmental quality standards. As concentrations increase, then the likelihood of this being due to man's activities increases. Concentrations above environmental quality standards are actually, or potentially, damaging and should result in a classification downgrading.

Classification schemes: the solution

Environmental managers need some measure of human-mediated environmental change, which spatial classification schemes cannot provide, for those impacts which extend the range of natural variability. For Scottish standing waters, naturalness is an important concept, and an understanding of context is vital to management. The concept of 'changed state' or temporal (rather than spatial) classification schemes for standing waters was introduced to many in Scotland by Moss (Moss et al., 1994). The central theme was of 'hindcasting' of key environmental parameters, notably phosphorus concentration, on the basis of empirical relationships between measured and estimated factors. A more direct examination of the ecological history of standing waters was also underway as part of the research effort to understand the acidification of surface waters.

Battarbee and his team from University College London demonstrated the degree and timing of acidification of lakes based upon the identification of sub fossil diatom remains in lake sediments and the species level relationships with pH value (Battarbee, 1990). These new approaches to quality classification were seen to have potential value. The chemical environment in standing waters may be considered the principal determinant influencing the biota in contrast to the importance of the physical structure in flowing waters. In natural systems, lakes come into chemical balance with their catchments, man's activities modify chemical fluxes and initiate ecological change. Changed state classification schemes necessitate the use of 'hindcastable' parameters. These are most conveniently chemical determinands and as such, appropriate to the classification of standing waters.

Development of the Scottish scheme

For many years prior to the development of a quality classification scheme for Scottish standing waters, it had been clear to River Purification Board scientists that the existing scheme for the classification of running waters was not appropriate for standing waters (lochs). Further incentive to correct this situation came with the publication in 1992 of the 16th Report of the Royal Commission on Environmental Pollution. Recommendation number 6 of that report stated that; "U.K. waters of 0.4 hectares and above should be surveyed and classified on a regular basis and that as far as possible the classification should adopt an internationally recognised system such as the OECD one". (Royal Commission on Environmental Pollution, 1992).

At the end of 1993 the Directors of the Purification Boards sanctioned the establishment of a group to devise a scheme appropriate to the classification of standing waters. The objectives were that this scheme should be acceptable to the Scottish Office, and that the output would be included in the 1995 quinquennial survey of water quality published by the Scottish Office. This required that the scheme be effectively finalised and agreed before the end of 1994, so that any additional environmental monitoring required could be planned, resourced and carried out during 1995. These objectives were met. The classification group first met in January 1994 and their proposed scheme was agreed by the Directors in August 1994. The scheme which has been adopted places standing waters into four classes in relation to the major impacts on water quality. The other quality classification schemes in use in Scotland for rivers, estuaries and coastal waters, also have just four classes.

The imposed timetable and the limited resources available demanded a pragmatic approach be adopted. Five stages were identified; these were to:

1. Decide the impacts of relevance to Scottish lochs.
2. Identify what methods were available for making a 'changed state' assessment of these impacts.
3. Assemble these methods into a simple scheme based around a decision tree.
4. Commission work to address perceived weaknesses.
5. Seek validation of the outputs.

Clearly the commissioned research and the validation phase were unlikely to be completed in their entirety before implementation. The scheme was tested on a range of lochs in 1994. The lochs chosen included some deemed unimpacted and others impacted to varying degrees by nutrient enrichment, acidification and toxic substances. The testing indicated that the scheme had the potential to separate standing waters into intuitively meaningful quality classes. Copies of the classification scheme (Association of Directors and River Inspectors of Scotland 1995) are available from the authors.

Scheme outline

The scheme classifies waters along three gradients: increases in phosphorus concentration as an indication of the potential for eutrophication; loss of acid neutralising capacity as an indication of acidification; and the degree of contamination with toxic substances. Standing waters are placed into one of four quality bands along each of these gradients, the overall quality class is that of the lowest quality band. The user is guided through the classification process by a decision tree which provides all the necessary stages. The scheme incorporates the use of any available biological data to validate or modify the assessments made. The output from the scheme is an overall classification into four classes, each of which has been given a simple description to aid interpretation by non-specialists.

Phosphorus banding

Banding is determined on the basis of annual average total phosphorus concentrations. In northern temperate lakes, phosphorus is generally regarded as being the factor limiting the biomass development of planktonic algae in most situations (Vollenweider, 1968), furthermore it is the factor most modified by human activities and most amenable (technically and legally) to control. Although it is acknowledged that phosphorus concentration cannot be directly equated with the symptoms of eutrophication, there was little choice of suitable surrogates which could be subjected to hindcasting techniques. Problems with the use of phosphorus as a surrogate for eutrophication in Scottish waters include the mitigating effects of both dystrophic peaty waters and highly flushed waters, where elevated phosphorus levels do not necessarily result in increases in phytoplankton biomass. Nevertheless, whilst it can be argued that elevated phosphorus does not invariably lead to elevated phytoplankton biomass, the potential would remain for enhanced algal or macrophyte production in shallow littoral areas and downstream rivers.

The banding is determined by the ratio of current total phosphorus concentration to a hindcast 'original' total phosphorus concentration;

$$\text{Ratio}\frac{[TP]_{\text{current}}}{[TP]_{\text{original}}},$$

where:

$[TP]_{current}$; current total phosphorus concentration measured by annual mean value or assigned surrogate value derived from available physical, chemical, or biological data.

$[TP]_{original}$; 'original' total phosphorus concentration estimated by land use/loss co-efficient method. (The nominal date for this hindcast value is *circa* 1850, *"before the import into the catchment of substantial quantities of phosphorus in the form of plant fertiliser or animal foodstuffs")*.

The ratio is then banded as shown in Table 1.

There are approximately three fold increases between each band, these being derived originally from the OECD (Organisation for Economic Co-operation and Development, 1982) phosphorus values representing the different trophic states of a large population of northern temperate lakes. The Scottish classification is not an attempt to describe such trophic states however, but a method of assessing significant movement along the gradient of phosphorus enrichment.

The hindcast total phosphorus concentration, $[TP]_{original}$, is derived according to a protocol produced under contract by The Macaulay Land Use Research Institute (MLURI) (Ferrier et al., 1996). The basic methodology is to determine the diffuse loadings of phosphorus from land cover classification before the agrarian revolution in Scotland. MLURI hold a variety of unique spatial datasets covering the whole of Scotland. The Land Cover of Scotland (LCS88) contains more than 120 different discrete units along with 12 000 mosaics representing combinations of individual cover types. These have been summarised into 30 categories with which are associated a range of phosphorus loss coefficients. This has formed the basis of a predictive model called PLUS (*P*hosphorus/*L*and *U*se/*S*lope) which determines the spatial loading of phosphorus, within a catchment, within a range of predicted maxima and minima.

Table 1. Phosphorus banding based on the ratio of present to hindcast phosphorus concentrations

$\frac{TP(\text{current})}{TP(\text{hindcast})}$	≤2	≤6	≤20	>20
Allocated Band	Band 1	Band 2	Band 3	Band 4

To develop an historical land cover set for the whole of Scotland, relationships between the LCS88 and the distribution of soil and vegetation types contained in the MLURI Soils' database were examined. During the classification of soils in Scotland, other attributes such as soil association, soil type, land form, parent material and vegetation cover were determined and are held in a relational database. The allocation of vegetation cover type was carried out by relating the soil survey to the Common Plant Communities of Scotland (Robertson, 1984), representing 58 natural and semi-natural communities. Commonalities between the overlays of these two datasets were determined in three bio-geomorphological zones covering Scotland and a rule-based hindcasting scheme, based on the occurrence of linked classification units determined. The distribution of ancient woodlands was incorporated into the historical land cover by utilising the Ancient Woodlands Inventory (dataset held by Scottish Natural Heritage and used with permission).

Calibration of the phosphorus loss coefficient model (PLUS) was performed on a site by site basis, where adequate data existed. The basic calibration steps were to; determine a theoretical phosphorus load using $[TP]_{current}$ and OECD regression equations (Organisation for Economic Co-operation and Development, 1982); apportion this total load between point sources and diffuse sources within the catchment; assign phosphorus loss coefficients from within an agreed range to each land use within the catchment to account for the theoretical diffuse load; determine $[TP]_{derived}$ and compare with $[TP]_{current}$. When appropriate loss coefficients had been determined within each catchment, land cover hindcasting was undertaken and the catchment specific coefficients were applied to the historical cover. For hindcasted arable land, a new unit was formed to represent low intensity agriculture. This was ascribed a loss coefficient corresponding to the lowest level of the next most highly managed unit of land cover, namely improved grassland. The total phosphorus load derived from these loss coefficients and historic land use areas was finally converted into an hindcast in-loch phosphorus concentration ($[TP]_{original}$) using the appropriate OECD regression equation.

Acid neutralising capacity banding

Acid neutralising capacity (ANC) is a measure of the sensitivity of water to acidification. ANC represents the capacity of a chemical system to react with added strong acid and can be represented by: ANC = (sum of buffering components) + $[OH^-]$ - $[H^+]$ where (sum of buffering components) includes weak Bronsted bases such as HCO_3^-, CO_3^{2-}, $B(OH)_4^-$, $Al(OH)_2^+$, $Al(OH)_4^-$, NH_3 and organic acid anions (Cantrell et al., 1990).

Acidification status is assessed by comparing estimates of present day ANC (ANC_t) with estimates of baseline ANC (ANC_o). ANC_t is calculated from an empirical relationship between alkalinity and dissolved organic carbon (DOC) to take account of the additional buffering provided by humic substances (Cantrell et al., 1990).

$$ANC_t(\mu_{eql}^{-1}) = \text{alkalinity}(\mu \text{eql}^{-1}) + 4.5\text{DOC}(\text{mg l}^{-1})$$

ANC_o may be estimated by two methods, based on the observation that in non acidified waters, bicarbonate alkalinity is approximately equivalent to the total base cation concentration (excluding that derived from sea salts).

The first uses methodology developed for critical loads calculations employing base cations, chloride and sulphate.

$$ANC_o = [BC_t]^* - F([SO_4{}^{2-}{}_t]^* - [SO_4{}^{2-}{}_o]^*) - [SO_4{}^{2-}{}_o]^*,$$

Where $[BC_t]$ is total base cations.

* denotes non-marine ions

$$F = \sin(90[BC_t]^*/S \text{ and } S = 400$$
$$[SO_4{}^{2-}{}_o]^* = 15 + 0.16[BC_t]^*.$$

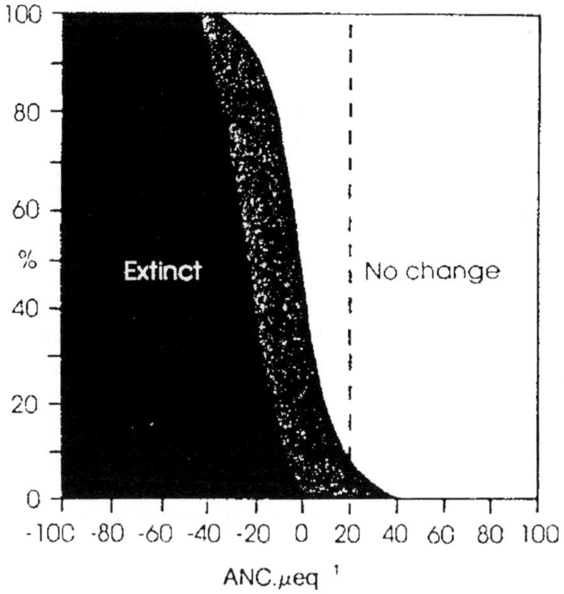

Figure 1. Relationship between fish populations and acid neutralising capacity in Norweigian lakes. (Henriksen et al., 1992) (by permission)

Apart from the validity of the assumptions implicit in the use of such models, the greatest sources of uncertainty in calculating ANC_o by this method are likely to be in estimating F (a measure of the increase in base cation leaching caused by acidification). In addition, the relationship between ANC, alkalinity and DOC has not been widely tested in U.K. waters. These relationships will be the subject of commissioned research.

The second method is based on the simplest version of the Henriksen model (Wright & Henriksen, 1983) and is provided as an alternative method of estimating ANC_o where available data is limited.

$$ANC_o = 1.23[Ca^{2+}]^* - 10 + 4.5\text{DOC},$$

where DOC is measured in mg l^{-1}.

Enhanced base cation leaching and background sulphate are ignored, but organic anions are taken into account. It is assumed that DOC does not change with acidification.

The banding of ANC_t against ANC_o is derived from a look up table (Table 2).

The range and scaling of the ANC values chosen were based on the relationship between the probabilities of damage to fish populations and ANC values derived from studies on Norwegian lakes (Henriksen et al., 1992). Figure 1 illustrates this relationship, it suggests that bands between $+40\mu$eq l^{-1}, at which the

Table 2. Banding of acid neutralising capacity

Baseline ANC_0	Present ANC_t ≥40	39 to 20	19 to 0	-1 to -20	-21 to -40	<-40
≥40	Band 1	Band 2	Band 2	Band 3	Band 3	Band 4
20–39		1	2	3	3	4
0–19			1	2	3	4

probability of damage to fish populations is zero, and -40μeq l^{-1}, below which the probability of extinction is 100%, would be appropriate.

Toxic substances banding

Whilst nutrient enrichment and acidification are the main threats to the quality of Scottish standing waters, a comprehensive scheme must consider all potential impacts. The third element of the classification scheme therefore embraces all potentially toxic substances. The quality banding for toxic substances is determined by both the concentration and the nature of the substances present;

100% EQS compliance of all samples Band 1
Annual EQS compliance of samples Band 2
Exceedence of any List II EQS Band 3
Exceedence of any List I EQS Band 4

100% compliance with an environmental quality standard (EQS) means that all samples must be at or below the EQS, annual EQS compliance means that the annual average must be at or below the EQS. Exceedence of local, national and international (EC) environmental quality standards for List I, List II or other hazardous substances would generally indicate a substantial change from the natural state and result in downgrading (Official Journal of the European Communities, 1976; Department of the Environment, 1990; North Sea Task Force, 1990). Edwards (1992) provides a useful guide to dangerous substances in water, including an EQS summary.

The decision tree

A simplified version of the decision tree is shown in Figure 2. The tree was drawn up specifically for use in the 1995 quinquennial water quality survey of Scotland and includes, as a first step, a check on the size criterion implemented for that survey (see below).

The decision tree incorporates a further screening step.The scheme is default based; lochs which are currently deemed oligotrophic ([TP] $< 10\mu g\ l^{-1}$), have an alkalinity greater than 2 mg l^{-1}, and whose waters comply with local, national and international environmental quality standards for toxic substances, default to class 1. This makes application of the scheme cost-effective in identifying those waters which have been significantly impacted, permitting targeting of monitoring resources. Where data indicate that a waterbody may not meet these primary criteria, then further data collection and hindcasting are used as described to determine the quality banding along the appropriate gradient. This does not imply that the water will be less than class 1, as further analysis may still result in a band 1 designation. After this process, the waterbody is banded for the three quality gradients, the overall class being that of the lowest derived band.

Quality class descriptors

Output from the scheme is an overall classification into four classes, based upon the lowest of the three bands for phosphorus enrichment, acidification, and toxic substances. The general character of the lochs within each of the classes is described below:

Class 1. Lochs whose water quality and aquatic ecology are not significantly altered by human activity.

Class 2. Lochs whose water quality and/or ecology are significantly altered by human activity.

Class 3. Lochs whose water quality and ecology are seriously downgraded by human activity.

Class 4. Severely polluted lochs incapable of supporting fisheries due to gross enrichment, acidification, presence of toxins or deoxygenation.

Simple descriptors have been applied to the four quality classes to help interpretation of the outputs by the public and non-specialists. This is not without problems of course, and the meaning imparted by phrases such as ‘significantly altered’ and ‘seriously downgraded’ depend very much on the recipients perspective. Nevertheless, a major role of classification schemes is to disseminate information to the widest

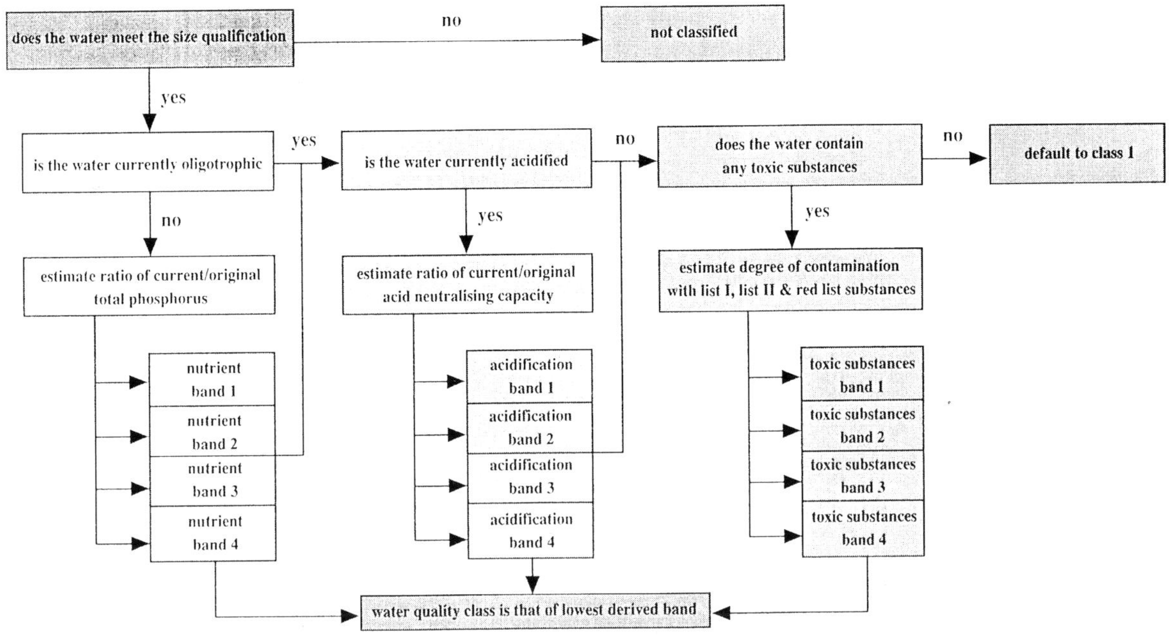

Figure 2. Decision tree for the classification of standing waters.

audience and we felt it worthwhile to include these descriptions as a general guide to quality.

Commissioned work

In addition to the phosphorus hindcasting contract carried out by MLURI and discussed above, three other areas of work have been identified during the development of the classification scheme. These are:

1. An investigation into sampling design protocols for standing waters.
2. A review of the methods for the analysis and interpretation of alkalinity and acid neutralising capacity, of surface waters paying particular attention to the methodology adopted for the standing waters scheme.
3. A validation (or alternative assessment), by the use of palaeolimnological techniques, of the hindcast estimates for both phosphorus and acid neutralising capacity in a suite of lochs.

These projects will be funded by the Scottish and Northern Ireland Forum for Environmental Research (SNIFFER) or directly by SEPA. Reports should be available from SEPA.

The 1995 survey of Scottish standing waters

It is estimated that there are 3788 freshwater lochs of over 4 ha in Scotland (Smith & Lyle, 1979). In terms of the Royal Commission recommendation on the lower size limit for classification (Royal Commission on Environmental Pollution, 1992), there are estimated to be over 30 000 waters of over 0.4 ha. Resource constraints prevented classifying all of these in one year, therefore in 1995 the classification was restricted to those lochs over 1 km^2 and those smaller lochs considered to be of local importance or concern. There is a risk that including such smaller, possibly polluted lochs would result in an apparent downgrading of the overall quality of Scottish standing waters. The results of the classification are therefore to be reported as both numbers and as areas in each class. For the purposes of this paper, numbers only are given in Table 3.

174 standing waters were classified, the overall classification resulting in 147 in class 1 (85%), 24 in class 2 (14%), two in class 3 (1%) and one in class 4 (0.5%). Enrichment with phosphorus was assessed as the major cause of downgrading (11%), loss of acid neutralising capacity accounting for 4.6% of downgrading, with ammonia at one site being the only toxic substance identified as exceeding an en-

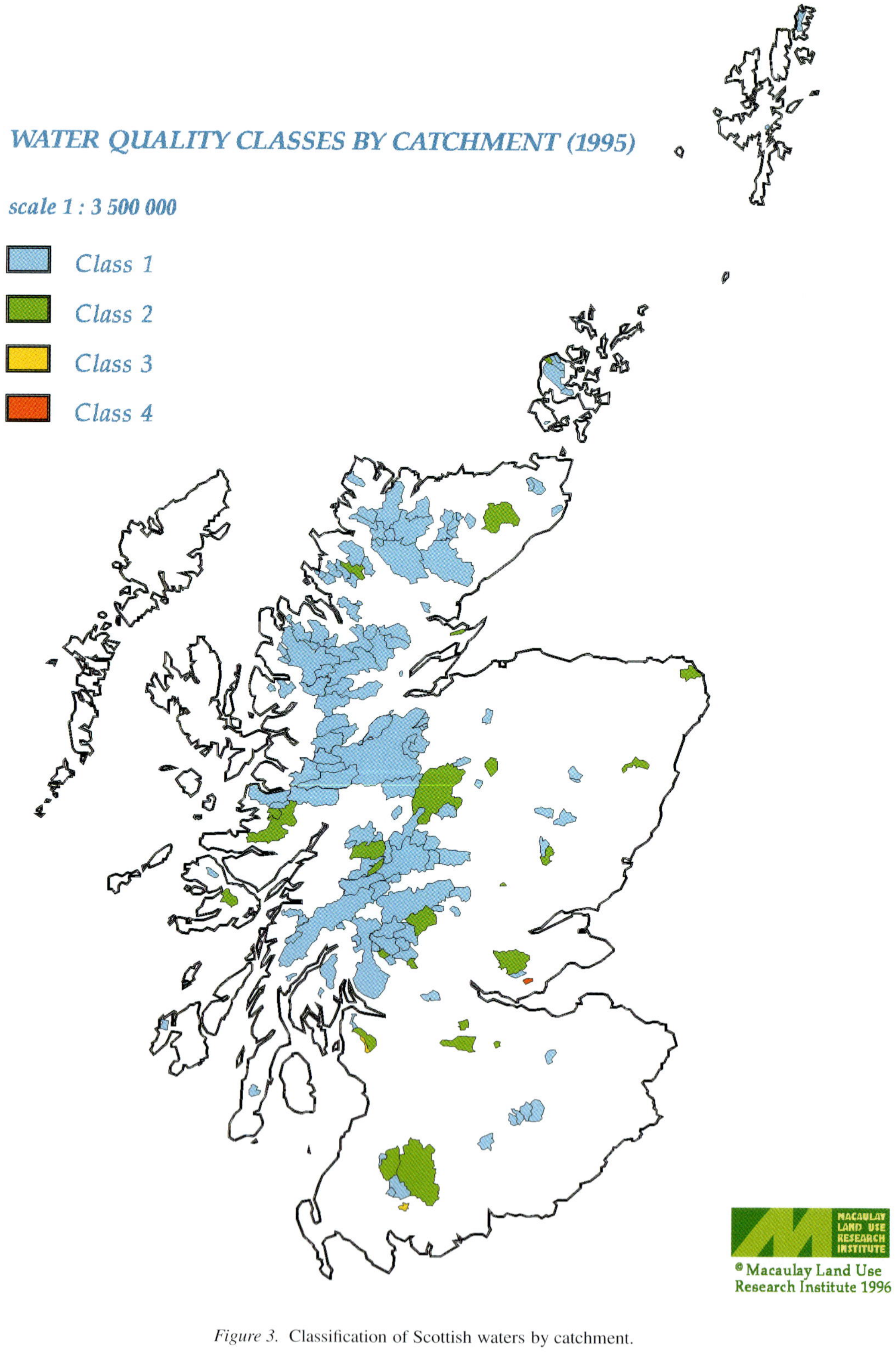

Figure 3. Classification of Scottish waters by catchment.

Table 3. Summary of the loch classification from the 1995 Scottish survey

Standing Water Name	rpa area	km^2	TP band	ANC band	Toxic sub band	Quality class
Loch Achall #	Highland	1.33	1*	1*	1*	1
Loch Affric	Highland	2.48	1*	1	1*	1
Loch Ailsh	Highland	1.07	1*	1*	1*	1
Loch Alvie	North East	0.61	1*	1*	1*	1
Loch Ard	Forth	2.22	1*	1	1*	1
Loch Arkaig	Highland	16.11	1*	1	1*	1
Loch Arklet	Clyde	2.24	1*	1	1*	1
Loch Ashie	Highland	1.40	1*	1*	1*	1
Loch Assynt	Highland	8.01	1*	1*	1*	1
Loch Avich	Clyde	3.49	1*	1*	1*	1
Loch Awe	Clyde	38.5	1*	1*	1*	1
Loch Ba	Clyde	3.15	1*	2	1*	2
Loch Ba	Tay	2.95	1*	1	1*	1
Backwater Reservoir	Tay	1.78	1*	1*	1*	1
Loch Badanloch	Highland	3.69	1*	1*	1*	1
Bea Loch	Orkney	0.38	1	1*	1*	1
Loch a Bhealaich #	Highland	1.08	1*	1*	1*	1
Loch Beannacharan	Highland	1.03	1*	1	1*	1
Loch Beoraid #	Highland	1.53	1*	1*	1*	1
Loch a Bhraoin	Highland	1.65	1*	1	1*	1
Blackwater Reservoir #	Highland	11.01	1*	2	1*	2
Loch of Boardhouse	Orkney	2.35	2	1*	1*	2
Loch Bradan	Clyde	2.14	1*	1	1*	1
Loch Brora	Highland	2.25	1	1*	1*	1
Loch Bruicheach	Highland	0.96	1*	1*	1*	1
Loch of Burness	Orkney	0.13	1	1*	1*	1
Loch Calder	Highland	3.8	1*	1*	1*	1
Cam Loch	Highland	2.57	2	1*	1*	2
Carron Valley Reservoir	Forth	3.81	1	1*	1*	1
Loch Chon	Forth	1.09	1*	2	1*	2
Loch a Chroisg	Highland	2.60	1*	1	1*	1
Loch nan Clar	Highland	3.20	1*	1*	1*	1
Clatteringshaws Loch	Solway	4.96	1*	1	1*	1
Loch Cliff	Shetland	1.04	1*	1*	1*	1
Loch Cluanie	Highland	10.76	1*	1	1*	1
Cobbinshaw Reservoir	Forth	1.28	2	1*	1*	2
Daer Reservoir	Clyde	2.02	1*	1*	1*	1
Loch an Daimh	Highland	0.69	1*	1*	1*	1
Loch an Daimh	Tay	2.83	1*	1*	1*	1
Loch Damh	Highland	3.32	1*	1*	1*	1
Loch Davan	North East	0.31	1	1*	1*	1
Loch Dee	Solway	1.01	1*	1	1*	1
Loch Doon	Clyde	10.63	1*	2	1*	2
Lochindorb	Highland	2.21	1*	1*	1*	1
Loch Druim a Chliabhain #	Highland	1.41	1*	1*	1*	1

Continued on p. 442

Table 3. Continued

Standing Water Name	rpa area	km^2	TP band	ANC band	Toxic sub band	Quality class
Loch Dughaill	Highland	0.98	1*	1*	1*	1
Dunalastair Water	Tay	1.47	1*	1	1*	1
Loch Duntelchaig	Highland	5.61	1	1*	1*	1
Lochan Na H-Earba #	Highland	1.67	1*	1*	1*	1
Loch Earn	Tay	9.53	2	1*	1*	2
Loch Eck	Clyde	4.33	1*	1*	1*	1
Loch Eigheach	Tay	1.36	1*	1*	1*	1
Loch Eilde Mor #	Highland	0.96	1*	1*	1*	1
Loch Eilt	Highland	1.82	1*	2	1*	2
Loch Einich	North East	0.75	1*	1	1*	1
Loch Ericht	Tay	22.62	1*	1*	1*	1
Loch Errochty	Tay	3.03	1*	1*	1*	1
Loch Eye	Highland	1.56	2	1*	1*	2
Lochan Fada #	Highland	3.73	1*	1*	1*	1
Loch Fannich #	Highland	10.87	1*	1*	1*	1
Loch Fiag	Highland	1.50	1*	1	1*	1
Glen Finglas Reservoir	Forth	1.41	1*	1*	1*	1
Fionn Loch #	Highland	10.18	1*	1*	1*	1
Loch Freuchie	Tay	1.40	1*	1*	1*	1
Loch Frisa	Clyde	4.44	1*	1*	1*	1
Fruid Reservoir	Tweed	1.10	1*	1*	1*	1
Loch Garry	Highland	7.80	1*	1	1*	1
Loch Garry	Tay	1.66	1*	1*	1*	1
Loch Garve	Highland	1.51	1*	1*	1*	1
Loch Gelly	Forth	0.59	4	1*	3	4
Loch Bad a Ghaill	Highland	2.86	1*	1*	1*	1
Loch a Ghriama	Highland	1.12	1*	1*	1*	1
Loch Girlsta	Shetland	0.99	1*	1*	1*	1
Gladhouse Reservoir	Forth	1.59	1	1*	1*	1
Loch Glascarnoch	Highland	5.28	1*	1	1*	1
Loch Glashan	Clyde	1.95	1*	1*	1*	1
Loch Glass	Highland	4.83	1*	1	1*	1
Loch Gorm	Clyde	2.72	1*	1*	1*	1
Loch Grannoch	Solway	1.11	1*	3	1*	3
Loch of Harray	Orkney	9.95	1	1*	1*	1
Loch Hempriggs	Highland	0.84	1	1*	1*	1
Hillend Reservoir	Clyde	1.36	2	1*	1*	2
Loch Hope	Highland	6.41	1*	1	1*	1
Loch of Hundland	Orkney	1.00	1	1*	1*	1
Loch Insh	North East	1.79	2	1*	1*	2
Loch Katrine	Forth	13.27	1*	1*	1*	1
Loch Ken	Solway	4.40	2	1*	1*	2
Kilbirnie Loch	Clyde	0.83	3	1*	1*	3
Loch Kinord	North East	0.71	1	1*	1*	1

Continued on p. 443

Table 3. Continued

Standing Water Name	rpa area	km^2	TP band	ANC band	Toxic sub band	Quality class
Loch of Kirkbister	Orkney	1.02	1	1*	1*	1
Loch Laidon	Tay	4.95	1*	2	1*	2
Loch Laggan	Highland	9.53	1	1*	1*	1
Loch Lednock	Tay	1.73	1*	1*	1*	1
Loch Lee	Tay	0.92	1*	1	1*	1
Loch Leven	Forth	13.77	2	1*	1	2
Loch of Lintrathen	Tay	1.75	2	1*	1*	2
Loch Lochy	Highland	16.30	1	1*	1*	1
Loch Lomond	Clyde	71.1	1*	1*	1*	1
Loch of the Lowes	Tay	0.87	2	1*	1*	2
Loch Loyal	Highland	6.55	1*	1*	1*	1
Loch Loyne	Highland	6.27	1*	1*	1*	1
Loch a Bhaid Iuachraich	Highland	1.30	1*	1*	1*	1
Loch Lubnaig	Forth	2.39	1*	1*	1*	1
Loch Luichart	Highland	7.49	1*	1	1*	1
Loch Lundie #	Highland	1.34	1*	1*	1*	1
Loch Lurgainn	Highland	3.25	1*	1	1*	1
Lussa Loch	Clyde	1.32	1*	1	1*	1
Loch Lyon	Tay	4.99	1*	1*	1*	1
Loch Maree	Highland	28.01	1*	1*	1*	1
St.Marys Loch	Tweed	2.53	1*	1*	1*	1
Megget Water	Tweed	2.48	1*	1*	1*	1
Loch Meiklie	Highland	0.78	1	1*	1*	1
Lake of Menteith	Forth	2.45	2	1*	1*	2
Loch Merkland	Highland	1.77	1*	1	1*	1
Loch Beinn a Mheadhoin	Highland	4.01	1*	1	1*	1
Loch Mhor	Highland	3.88	1	1*	1*	1
Loch Migdale	Highland	1.05	1	1*	1*	1
Loch Monar	Highland	10.86	1*	1	1*	1
Loch Morar	Highland	26.81	1*	1*	1*	1
Loch More I (Sutherland)	Highland	3.81	1*	1*	1*	1
Loch More II (Caithness)	Highland	1.93	2	1*	1*	2
Loch Morie	Highland	2.34	1*	1*	1*	1
Loch Morlich	North East	1.24	2	1*	1*	2
Loch Muick	North East	2.21	1*	1	1*	1
Loch Mullardoch	Highland	9.53	1*	1	1*	1
Loch Nant	Clyde	2.05	1*	1*	1*	1
Loch Naver	Highland	7.86	1*	1	1*	1
Loch Nell	Clyde	1.29	1*	1*	1*	1
Loch Ness	Highland	56.4	1*	1*	1*	1
Loch na h-Oidnche	Highland	1.39	1*	1*	1*	1
Loch Oich	Highland	1.97	1*	1*	1*	1
Loch Ore	Forth	1.02	1	1*	1*	1
Orrin Reservoir	Highland	3.19	1*	1	1*	1
Loch Oscaig	Highland	1.68	1*	1	1*	1

Continued on p. 444

Table 3. Continued

Standing Water Name	rpa area	km^2	TP band	ANC band	Toxic sub band	Quality class
Loch Ossian	Highland	2.57	1*	1*	1*	1
Loch Quoich	Highland	17.55	1*	1	1*	1
Loch Rannoch	Tay	18.96	1*	1	1*	1
Loch Rimsdale	Highland	4.03	1*	1*	1*	1
Loch an Ruathair	Highland	2.31	1	1*	1*	1
Loch Ruthven	Highland	1.55	1*	1*	1*	1
Loch Saintear	Orkney	0.16	1	1*	1*	1
Sandwood Loch	Highland	0.99	1	1*	1*	1
Loch Scarmclate	Highland	0.75	1	1*	1*	1
Loch na Sealga #	Highland	3.71	1*	1*	1*	1
Loch an t-Seilich	North East	1.14	1*	1*	1*	1
Castle Semple Loch	Clyde	1.65	2	1*	1*	2
Loch Bad an Sgalaig	Highland	1.34	1*	1*	1*	1
Loch Shiel	Highland	20.02	1*	2	1*	2
Loch Shin	Highland	32.65	1*	1	1*	1
Lochan Shira	Clyde	1.25	1*	1*	1*	1
Loch Sionascaig	Highland	6.25	1*	1*	1*	1
Loch of Skene	North East	1.13	2	1*	1*	2
Loch Sloy	Clyde	1.28	1*	1	1*	1
Spey Dam Loch	North East	0.87	1*	1*	1*	1
Loch Ma Stac #	Highland	0.67	1*	1*	1*	1
Loch Stack	Highland	2.59	1*	1*	1*	1
Loch of Strathbeg	North East	2.03	2	1*	1*	2
Strathclyde Park Loch	Clyde	0.82	2	1*	1*	2
Loch Swannay	Orkney	2.34	1	1*	1*	1
Talla Reservoir	Tweed	1.22	1*	1*	1*	1
Loch Tarsan	Clyde	1.05	1*	1	1*	1
Loch Tay	Tay	27.30	1*	1	1*	1
Loch Thom	Clyde	1.48	1*	1*	1*	1
Loch Treig	Highland	7.39	1*	1*	1*	1
Loch Tulla	Clyde	2.69	1*	1	1*	1
Loch Tummel	Tay	5.60	1*	1*	1*	1
Glen Turret Reservoir	Tay	1.76	1	1*	1*	1
Loch Ussie	Highland	0.86	1	1*	1*	1
Loch Urigill	Highland	2.01	1	1*	1*	1
Loch Vaich	Highland	1.64	1*	1	1*	1
Loch Venachar	Forth	5.2	1*	1*	1*	1
Loch Veyatie	Highland	2.60	1	1*	1*	1
Loch Voil	Forth	2.7	1*	1*	1*	1
Loch Watten	Highland	4.54	1	1*	1*	1

Note; # indicates remote, not sampled * indicates default to band 1 (see decision tree)

vironmental quality standard. Although 85% of this sample of Scotland's standing waters was deemed to be of class 1 quality in 1995, there are two provisos which must be considered when evaluating these results. Firstly these were, by and large, Scotland's largest waterbodies and therefore least susceptible to pollution effects; secondly, the assessment for almost all waters indicated some change over the past cen-

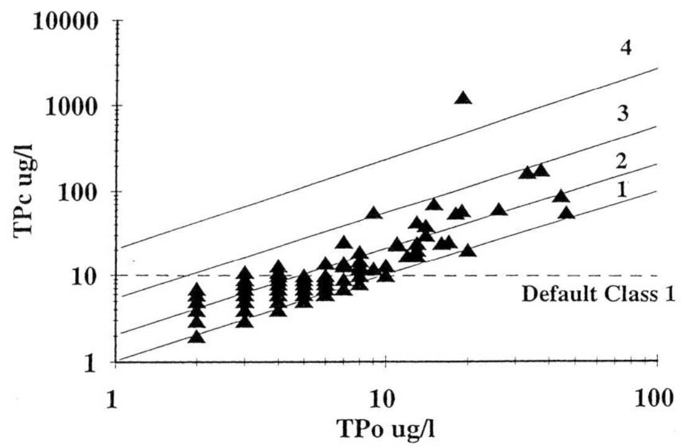

Figure 4. The relationship between $[TP]_{original}$ and $[TP]_{current}$ for the classified waters.

tury or so, but not necessarily of sufficient magnitude to cause downgrading. These qualifications when applied to the classification results do confirm that both nutrient enrichment with phosphorus and acidification should be considered as serious threats to the quality of a significant proportion of Scotland's water resources.

The geographical distribution of classified waters

Figure 3 illustrates the catchments of the standing waters classified and indicates the overall quality class assigned to the standing water in each catchment in 1995. A fundamental principle of any quality classification scheme is that the distribution between classes should correspond with the identified impacts. For phosphorus enrichment, this is evident in relation to the distribution of agricultural land, especially in the south and east, and in the central belt in relation to urbanisation. In the north and west of Scotland, where natural and semi-natural land cover are more prevalent, identified impacts are related to commercial afforestation and point sources such as fish farms. For loss of acid neutralising capacity, interactions between deposition rates, land use and soils are of proven importance. The distribution of these waters follows the distribution identified by palaeolimnological techniques, and includes waters in Galloway, the Trossachs and the Grampians.

Total phosphorus

Figure 4 illustrates the relationship between $[TP]_{original}$ and $[TP]_{current}$ for the classified waters. Hindcast total phosphorus concentrations varied between 2 $\mu g\ l^{-1}$ and 50 $\mu g\ l^{-1}$. Current measured total phosphorus concentrations were generally within this range but

Table 4. Land Cover Factor in the catchments of waters which are downgraded by phosphorus banding

Standing water name	Land Cover Factor	Major land cover changes
Loch Earn	0.11	Afforestation 6%
Strathclyde Park Loch	0.12	Urbanisation 20% Afforestation 13% Quarries 5%
Cobbinshaw Loch	0.14	Afforestation 25%
Loch Gelly	0.16	Urbanisation 25%
Loch Insh	0.20	Afforestation 7%
Castle Semple Loch	0.37	Urbanisation 5%
Hillend Reservoir	0.46	Afforestation 13%
Loch Leven	0.73	Afforestation 8% Urbanisation 2%
Lake of Menteith	0.79	Afforestation 13% Arable 3%
Kilbirnie Loch	0.95	Urbanisation 14%
Loch of Skene	0.95	Afforestation 13%
Loch of Strathbeg	0.96	Arable 29%
Loch More II (Caithness)	1.00	Afforestation 15%
Cam Loch	1.00	Afforestation 14%
Loch Eye	1.00	Afforestation 6%

with a few waters between 100 and 200 $\mu g\ l^{-1}$ and one water having an annual mean concentration of 1246 $\mu g\ l^{-1}$ for 1995. The scheme has identified particular waters which have undergone significant change in trophic status, whilst not undergoing clearly identifiable change in either land use or point sources of phosphorus. These are waters where the total area of land under cultivation has not increased, but the management of that land in terms of the intensity of use has altered.

The scheme also identifies the relative importance of point sources *versus* diffuse sources in those catchments which have generated enhanced phosphorus concentrations in the receiving water. This has important consequences for the future management of these systems in relation to the relative ease of control of the sources of this nutrient. Land Cover Factors (LCF) have been assessed for non band 1 (phosphorus) waters and are presented in Table 4. Land Cover Factor indicates the proportion of change in total loads which are judged to have arisen from land cover change. For example, if change is solely due to point sources then LCF would be 0, if there are no point sources and change is due entirely to land cover changes then LCF would be 1. Table 4 indicates that changes in land use are a significant cause of phosphorus enrichment in downgraded waters, and that afforestation is the major land use change responsible.

It is also apparent that with some waters which have a high $[TP]_{current}$, the scheme does not adequately account for all the influences determining the observed phosphorus concentration. The recycling of phosphorus within these waters is proposed as the major mechanism resulting in poor calibration ($[TP]_{current}$:$[TP]_{derived}$).

Some standing waters were considered too remote to sample adequately. These waters were banded for phosphorus enrichment employing $[TP]_{original}$ and $[TP]_{derived}$. Examples of this method of banding are presented in Table 5. This illustrates a further aspect of the phosphorus hindcasting scheme, an estimate of change can be made in the absence of an estimate of $[TP]_{current}$. As would be expected in these remote waters, there appears to be little human influence on phosphorus concentrations.

The large lochs of Scotland, Lomond, Awe, Ness, Morar and Shiel were all band 1 for phosphorus enrichment. Nevertheless, all have apparently undergone considerable enrichment with phosphorus, leading typically to an approximate doubling of annual mean phosphorus concentrations (e.g. Loch Morar $[TP]_{current}$ 6 $\mu g\ l^{-1}$, $[TP]_{original}$ 2 $\mu g\ l^{-1}$, Loch Shiel $[TP]_{current}$ 8 $\mu g\ l^{-1}$, $[TP]_{original}$ 4 $\mu g\ l^{-1}$). These lochs have such a large volume and surface area that local eutrophication problems may occur without causing a shift in overall classification. This has to be addressed by site-specific investigations, targeting known sources which should be indicated by the classification analysis. In these large lochs, very large loads of phosphorus and volumes of water are modelled and result in small changes in low concentrations of water column phosphorus. To investigate the sensitivity of these outputs to changes in model parameters, maxima and minima modelled phosphorus concentrations were calculated for both historical and present day conditions in Lochs Awe, Lomond and Ness. Table 6 presents the modelled diffuse and estimated point phosphorus loads, Figure 5 compares the modelled ranges of historical and present in-lake concentrations and indicates the current measured total phosphorus concentration. The analysis does suggest that an increase in phosphorus concentration is most likely to have occurred in Lochs Lomond and Ness, but the indications are less clear for Loch Awe. Loch Lomond has substantial point sources of phosphorus, which may contribute more to the loch than can be accounted for from diffuse sources, even by employing maximum loss coefficients. The estimate of point sources in the Lomond catchment should be regarded as a maximum, however, since it has been derived by assuming that sewage treatment works operate continuously at design capacity. In practice, sewage treatment works serving tourist facilities operate below this level for much of the year. Both Loch Awe and Loch Ness show enhanced phosphorus inputs from diffuse sources.

In general, standing waters identified as having significant increases in phosphorus concentration by the classification scheme are those waters which are known to have eutrophication problems. Remedial measures are planned or underway at many of these sites, some of which are described in a later section of this paper. Some of the outputs from the scheme will undoubtedly stimulate further investigatory work, or lead to a precautionary approach when reviewing or adopting policies towards phosphorus sources.

Acid neutralising capacity banding

In Scotland, smaller waterbodies tend to be the most strongly acidified. This is due principally to the heterogeneity of the underlying geology of much of Scotland, and the consequent frequent occurrence of strata and soils having the ability to neutralise these acidic inputs. It is not surprising, therefore, that over 95% of large waterbodies with correspondingly large catchments were deemed to be band 1 for acid neutralising capacity in the 1995 survey. Although only some 4.6% of standing waters were of less than class 1 quality due to loss of ANC, around 25% had alkalinities close to or less than the 2mg l^{-1} default criterion, suggesting vulnerability to acid episodes.

The eight lochs downgraded due to loss of ANC were Loch Ba (Clyde), Blackwater Reservoir, Loch Chon, Loch Doon, Loch Eilt, Loch Laidon and Loch Shiel, all Band 2, and Loch Grannoch, Band 3. The acidified status of several of these lochs is reported in the literature, for example, Loch Chon (Kreiser et al.,1990) and Loch Grannoch (Patrick et al., 1995). Studies of the sub fossil diatom record in the sediments of these two lochs indicate contrasting patterns of acidification. Loch Chon in the headwaters of the River Forth in central Scotland has undergone a marked acceleration in acidification after 1960, due to forestry plantings increasing the flux of atmospheric acid forming pollutants. Loch Grannoch in Galloway,

Table 5. Phosphorus banding for remote standing waters

Standing water name	$[TP]_{original}$ μgl^{-1}		$[TP]_{derived}$ μgl^{-1}		Phosphorus banding ($[TP]_d$: $[TP]_o$)
	minimum	maximum	minimum	maximum	
Loch Achall	2	6	2	6	1
Loch a Bhealaich	1	4	1	4	1
Loch Beoraid	1	4	1	4	1
Loch Druim a Chliabhain	2	6	2	6	1
Loch na Earba	1	5	1	5	1
Loch Eilde Mor	1	4	1	4	1
Lochan Fada	1	1	1	1	1
Fionn Loch	1	4	1	4	1
Loch na Sealga	1	4	1	4	1
Loch ma Stac	1	4	1	4	1

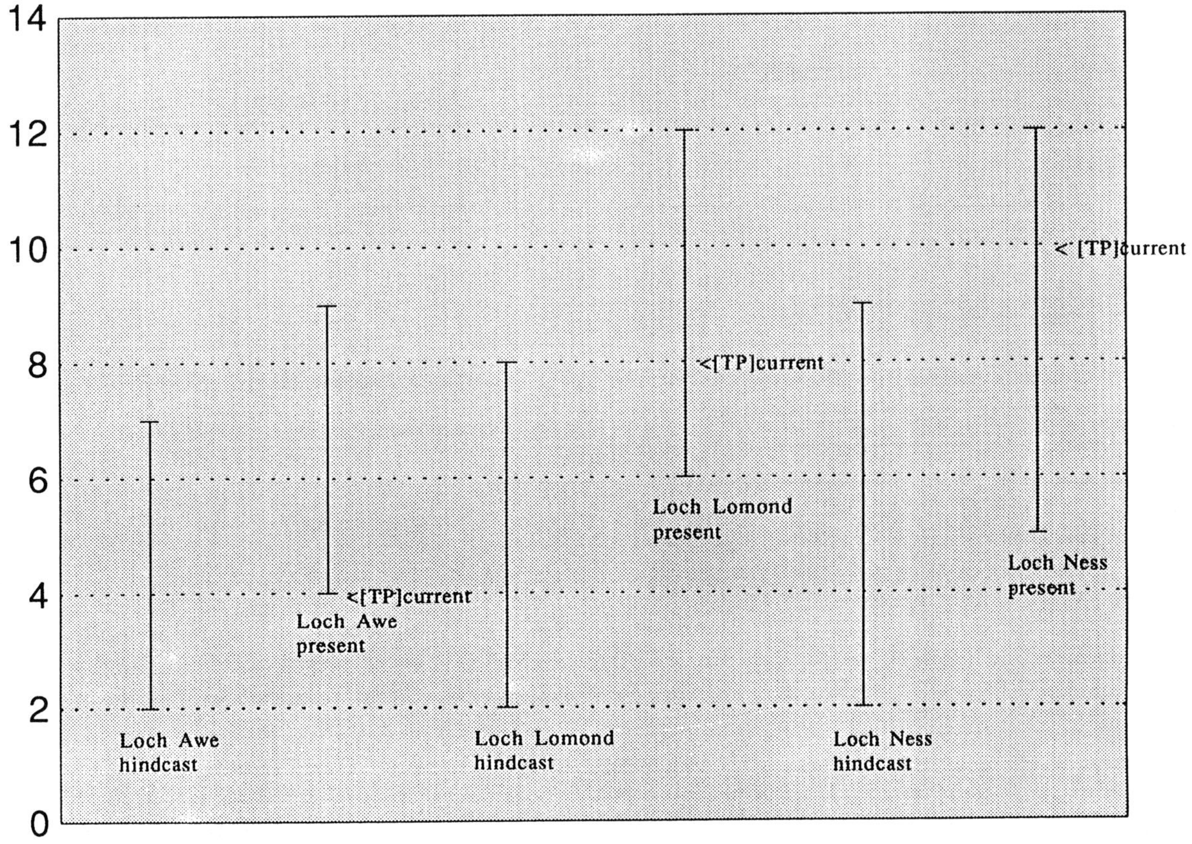

Figure 5. Modelled mean total phosphorus concentrations, present day and hindcast.

Table 6. Comparison of maximum and minimum modelled loads (total phosphorus, kg yr^{-1}) for Scotland's largest lochs. Current measured mean total phosphorus also shown (see also Figure 4)

	Loch Awe	Loch Lomond	Loch Ness
Diffuse hindcast load, minimum	3 700	2 700	8 100
Diffuse hindcast load, maximum	13 800	10 000	30 500
Diffuse present load, minimum	4 500	3 200	9 900
Diffuse present load, maximum	16 400	11 600	36 700
Estimated point source loads	7 815	15 604 (maximum based on STW design capacity)	3 699
$[TP]_{current}$, north basin ($\mu g l^{-1}$)	4	8.2	10 (whole loch mean)
$[TP]_{current}$, south basin ($\mu g l^{-1}$)	4	8.8	

south west Scotland, also has an afforested catchment but has shown a steady decline in pH value since the middle of last century.

Of the large lochs of Scotland, Loch Shiel (surface area 20.02 km^2), was identified as acidified. Maitland (1981) reports that none of the four sub-catchments of Loch Shiel have any base rich rocks, and furthermore, settlements and arable land are concentrated at the western, outflow, end of the loch thus minimising any human enrichment. Bailey-Watts & Duncan (1981a) in the same volume compare Loch Shiel with the other large Scottish Lochs (Lomond, Awe, Ness and Morar). Loch Shiel in 1977/78 had the lowest conductivity (29 $\mu S\ cm^{-1}$), lowest alkalinity (2.10mg l^{-1}), and lowest pH value (6.14), of these five lochs.

Toxic substances banding

Only one loch, Loch Gelly in Fife, central Scotland, was downgraded due to toxic substances. Loch Gelly is just less than 1 km^2 in surface area but was included in the classification since it is a high priority loch for environmental improvement. The loch receives treated sewage effluent from a medium sized town and was downgraded on the basis of both ammonia (band 3) and phosphorus (band 4). It is described in more detail in the next section.

Other standing waters in Scotland undoubtedly contain traces of toxic substances, both in the water column and in the sediment. However, the cost and complexity of analysis of these often novel substances precludes their routine determination on a large scale. The policy generally adopted is to target analysis at locations where either a compound is known to be present from knowledge of effluent discharges, or where biological effects are recorded symptomatic of significant contamination of a toxic substance. Biological monitoring is established in Scotland as an extremely cost effective method of environmental audit. Typical problems in Scottish freshwaters associated with toxic substances include sheep dip chemicals, mothproofers from wool industries and other agricultural pesticides such as those used for parasite control in intensive chicken rearing. Loch Leven in central Scotland has a wool and cashmere spinning mill in its catchment, which was the source of several persistent mothproofing compounds up to 1988 (Fozzard, 1994). Dieldrin (to 1964), chlorphenylid (1964–1980) and flucofenuron (1980–1988) were used at the mill and were discharged to the loch *via* an inflow stream. Residues of these mothproofers have been identified in trout tissue and heron eggs at Loch Leven but the only identified harmful impacts were a reduction in benthic invertebrate diversity in the influent stream receiving the effluent and possibly the absence of *Daphnia hyalina* for a period in the 1960s and early 1970s. The mill ceased using mothproofers in 1988.

The management of standing waters in Scotland

Quality classification schemes provide one mean of determining quality objectives for waters. They also enable overall long term quality trends to be monitored. The Scottish River Purification Boards (now part of the Scottish Environment Protection Agency) have, for 40 years, sought to maintain and where necessary improve water quality in controlled waters. Much has been achieved, but standing waters have, until recently, received relatively little attention. The exceptions to this being drinking water reservoirs, where strict catchment controls have been widely employed to maintain water quality. Loch Katrine for example, Glasgow's main drinking water supply, has a catchment which is wholly managed by the water supply authority, the complete catchment (110.5 km^2) being purchased in 1920 for £107 000.

The classification, the data collected and the information gained from the analyses undertaken, will be used within the Scottish Environment Protection Agency to manage standing waters both directly by catchment based actions and indirectly in the case of acidic inputs from the atmosphere. The formation of the Agency will permit a national perspective and an equitable allocation of resources to apply to rehabilitation of the environment. The river purification boards had disparate resources related in part to population densities within their boundaries. As this survey has shown, this is appropriate to the control of some impacts, but others, for example forestry, fish farming and atmospheric mediated inputs to freshwater, are often inversely related to population density and the environmental impacts caused were dealt with by authorities with restricted resources. The Forth River Purification Board within the central belt of Scotland and with a headquarters in Edinburgh, was relatively well resourced and was closely involved with initiatives in the management of standing waters. The Forth catchment provides examples of the management of standing waters across the spectrum of quality found in central Scotland, these are described below to illustrate the variety of solutions needed for restoration of damage and protection of existing good quality.

Loch Gelly

Loch Gelly has been classified as class 4 in 1995. The loch was banded 4 for phosphorus and 3 for toxic substances (ammonia). Loch Gelly is situated in the Fife coalfield in central Scotland, it is small (0.59 km^2), shallow (mean depth 1.8 m) and rapidly flushed (water residence time 0.16 year). Loch Gelly is characterised in summer by extensive floating mats of *Enteromorpha* and *Cladophora*. Annual mean total phosphorus concentration is typically over 1000 $\mu g\ l^{-1}$ and was 1246 $\mu g\ l^{-1}$ in 1995. The cause is the treated sewage effluent from a population of 10 600 and industrial sources, which enters the loch *via* the main inflow. The sewage effluent forms the bulk of the flow in what would otherwise be a small stream. The Land Cover Factor (LCF) was assessed as 0.16 (Table 4), indicating the dominance of this point source relative to diffuse sources of phosphorus. As well as the ecological impact on the loch and the outflow stream, use of the loch as a waterski venue was impaired due to surface accumulations of algae and *Potamogeton pectinatus*.

The loch is clearly identified by the classification scheme and diversion of the treated sewage effluent to a nearby trunk sewer is nearing completion. In this particular case, upgrading of the sewage treatment facility was assessed as not a viable option. Diversion of the effluent may of course be only the initial stage of restoration. Further actions will depend upon the quality objectives set for the loch and possibly cost-benefit analysis of the achievable options.

Loch Leven

Loch Leven has been classified as class 2 in 1995, downgrading being due to a band 2 phosphorus designation. Loch Leven is a high profile Scottish loch, which has suffered the effects of eutrophication for much of this century. Extensive macrophyte beds have been lost, notably in the late 1940s, and blooms of cyanobacteria occur every year. One bloom of *Aphanizomenon flos-aquae* began in 1963 and persisted for 14 months. (Morgan, 1970; Morgan, 1974; Fozzard, 1994). Loch Leven has a statutory designation as both a Site of Special Scientific Interest and a National Nature Reserve. It supports a good brown trout (*Salmo trutta*) sport fishery and its outflow is regulated to provide cooling and process water for paper mills on the river downstream. The financial value to the local economy was estimated to be over £1.5m in 1987 (Forth River Purification Board, unpublished). The algal blooms interfere with these various uses of the loch and its water and are a cause for concern. A severe (but not atypical) bloom of *Anabaena flos-aquae* in summer of 1992 coincided with a major lochside media-orientated event. The resultant adverse publicity and public reaction has been estimated to have cost the local economy up to £673 000 in tourism losses, the fishery losses were estimated at £110 000 and the bloom increased costs to the paper mills to the order of £160 000 (Loch Leven Area Management Advisory Group, 1993).

The Forth River Purification Board has achieved a progressive reduction in sources of phosphorus since the late 1960s. A woollen mill was the largest point source of phosphorus, contributing up to an estimated

Table 7. Appraisal of principal remedial options for Loch Leven, Kinross (Fozzard, 1994)

Option	Costs	Phosphorus removal	Environmental costs	Conclusions
Phosphorus reduction at sewage treatment works	£2m	9.2kg/day (21% external load)	None	Approved and actioned
Increased loch flushing	Not costed	Not significant	Potential flooding	Not viable
Diversion of sewage treatment works effluent	£3.2m	14kg/day (32% external load)	Downstream river quality	Superseded by option 1
Inflow treatment	£60,000	5.0kg/day (12%	Loss of farmland	Ongoing
Iron salt treatment of loch water/sediment	Not costed	None	Toxic/damage to valuable habitat	Rejected
Sediment dredging	£10m+	50 tonnes	Massive habitat damage	Rejected
Algal harvesting	£150,000	Not significant	Disposal problems	Ongoing

150 kg every week (Holden & Caines, 1974), although some estimates put the figure as high as 500 kg each week. This was progressively reduced, finally the mill ceasing usage in 1988. New sewage treatment works serve the two main settlements in the catchment. The first of these serving Kinross was commissioned in 1981 and removed an estimated 45% of phosphorus due to the inherent properties of its design (extended aeration system). Both works now have tertiary phosphorus removal. A third smaller settlement discharges to the loch and this effluent will be diverted out of the loch catchment by the end of 1997. The catchment of Loch Leven is good arable land with a highly erodible sandy soil. By the late 1980s phosphorus derived from diffuse sources within the catchment (primarily agricultural) was estimated to be supplying around 2/3 of the total load. The LCS for Loch Leven was estimated at 0.73 in 1995, confirming the importance of land use as an influence. The iterative approach to phosphorus reduction, adopted due to the uncertainties involved, was no longer deemed appropriate after the events of 1992. A working group was set up, the Loch Leven Area Management Advisory Group, and examined available remedial options. Their appraisal of the principal remedial options is outlined in Table 7. Seven main options centred around the control of external and internal loadings of phosphorus, and the control of the algal blooms by removal were investigated. Massive intervention was rejected as inappropriate on the grounds that it would be too damaging to the ecology of the loch and downstream uses.

The long term environmental target for the loch is to achieve a standard of water quality which will permit sustained increases in diversity and abundance of aquatic macrophytes and invertebrates. There have been encouraging signs of ecological change within the loch, lower algal abundances, increased water clarity, increases in the abundance of macrophytes in some years, increased numbers of herbivorous water birds (mute swan, pochard) and increased diversity and abundance of invertebrates such as mayflies, caddis flies, stoneflies and damselflies (Fozzard, 1994). Nevertheless, weather conditions continue to be a dominant influence in determining the size and duration of algal blooms, and severe blooms are still a regular feature during periods of warm weather and low flushing (Armstrong et al., 1994). Further reductions in the external load of phosphorus would almost certainly be of benefit to the recovery of the loch, but it seems likely that habitat management within the loch and its catchment would accelerate this process and may be necessary for the full achievement of objectives. Such habitat management might include measures in the loch to protect macrophytes from wave action and waterfowl grazing until re-established. The importance of these other factors was identified almost 20 years ago by Jupp & Spence (1977) who found that less than 5% of the shallow water available for macrophyte growth at Loch Leven was actually colonised (growth in deeper water was found to be limited by phytoplankton shading). The PLUS model has been applied to the Loch Leven catchment to help identify areas potentially contributing the largest quantities of phosphorus.

Lake of Menteith

Lake of Menteith lies 22 km west of Stirling in central Scotland, it is south of the Highland boundary fault and lies on superficial deposits in the catchment of the River Forth. The mean total phosphorus concentration in 1995 was 19 μg l^{-1} and the lake supports a typically mesotrophic assemblage (Fozzard & Marsden, 1990). Lake of Menteith is designated a Site of Special Scientific Interest for its macrophyte interest. The slender naiad (*Najas flexilis*) is found in Lake of Menteith. Both mesotrophic lakes and the slender naiad are listed in the Biodiversity Action Plan Steering Group Report as requiring special measures for conservation. The lake suffers annual cyanobacteria blooms, particularly in late autumn, but has not so far been regarded as highly impacted by eutrophication.

Lake of Menteith was classified as class 2 in the 1995 survey, on the grounds of phosphorus enrichment. The phosphorus banding was fairly marginal, the $[TP]_{current}$ to $[TP]_{original}$ ratio being 2.38. The LCS was estimated at 0.79. Point sources of phosphorus include two land-based fish rearing units and fish cages in the lake, diffuse sources are arable and improved grassland and forestry. The Control of Pollution Act (1974) as amended by Schedule 23 of the Water Act (1989) requires application for consent to be made for discharges of sewage effluent, trade effluent or any other matter. Section 56 of Schedule 23 clearly defines fish farms as trade premises, bringing them within the consenting procedure. The discharge consent is a license permitting a discharge to be made subject to certain conditions, including volume and quality. Other factors such as the maximum size of development or the annual quantity of fish food permitted can be specified.

Sources of phosphorus have been investigated within the catchment utilising an early version of PLUS and flow and concentration measurements (Marsden et al., 1995). The phosphorus budget for Lake of Menteith demonstrates the importance of diffuse sources, accounting for 56% of the annual load. The analysis indicated that action was necessary on a broad front. Consents limiting phosphorus discharge have been issued to the three fish rearing facilities at Lake of Menteith, including a limit on tonnage and mean phosphorus content of fish food used at the in-lake cage units. This is necessary since effluent sampling at such sites is not possible. The fish farmer keeps detailed purchase and use records which are available for inspection, and samples of food may be analysed to confirm phosphorus content (0.96% phosphorus in 1994). On an annual basis, the four large consented point sources accounted for only 19% of the total load, whereas in the sensitive summer period (May–September inclusive), these point sources contributed an estimated 65% of the load. It is therefore important to appreciate the implications of seasonality (and bioavailability, which is likely to be high from such point sources) on algal production. One of the four point sources, tertiary treated hotel sewage effluent, has since been diverted (1995) out of the lake catchment.

Limitations in the scope of current legislation prevent the extension of controls over all diffuse sources of phosphorus, effective cooperation with other agencies whose powers can influence land use and land management practice is therefore vital. The importance of inter-agency cooperation is illustrated by the understanding between the Forestry Authority and the Forth River Purification Board which defines acceptable ground preparation practices and ensures that fertilisation proposals are given close scrutiny.

The Trossachs Lochs

The Forth River Purification Board extended phosphorus control policies to standing waters designated as oligotrophic. In 1988 the Board adopted a policy to limit further inputs of phosphorus to lochs in the Trossachs region of central Scotland. This is an area of outstanding importance for tourism and nature conservation. To protect the trophic status of these waters it was resolved that phosphorus concentrations should not be permitted to rise by more than 20% of their 1988 baseline value. For the most vulnerable bodies, such as Lake of Menteith, a 5% limit was imposed. An upper management threshold was also set which defined the maximum annual mean total phosphorus concentration for oligotrophic (8 μg l^{-1}) and mesotrophic (25 μg l^{-1}) lakes. These standards demanded an understanding of nutrient loading over the whole catchment, necessitating an assessment of the relative importance of all phosphorus sources, which would permit the development of control strategies. Loch Ard (1995 mean [TP] 6 μg l^{-1}) and Loch Achray (1995 mean [TP] 5 μg l^{-1}) both have phosphorus control measures in their catchments including phosphorus removal at sewage treatment works serving timeshare developments. In general, residents and local businesses recognise that their livelihood and quality of life depends on high environmental quality, and therefore support the policies adopted.

Loch Lubnaig (1995 [TP] 8 μg l^{-1}) in the Trossachs has a statutory Site of Special Scientific Interest at its inlet, designated for its macrophyte interest. Like other oligotrophic (mean [TP] $<10\ \mu g\ l^{-1}$) lochs it is designated band 1 for phosphorus and in this case class 1 overall. Phosphorus is largely derived from diffuse sources, there is no arable land in the catchment and forestry is the single largest controllable source. In the summer of 1994, 2.8 km^2 of forestry plantation in the Loch Lubnaig catchment was scheduled for aerial fertilisation by rock phosphate. This amounted to 15 tonnes of phosphorus. Concerns arose over the probability of a summer bloom of algae and negotiation was initiated with the forestry company. An agreement was reached whereby the aerial application was restricted to 0.9 km^2 with a further 0.5 km^2 being fertilised by hand application. Monitoring of the site indicated some increase in the phosphorus content in the small streams draining the fertilised area, but this was small in relation to the overall input to the loch, and no subsequent algal bloom was recorded.

Loch Chon

Loch Chon is also in the Trossachs region of central Scotland and has a catchment which is 51% afforested (Kreiser et al., 1990). Nutrient enrichment is not perceived as the main threat to environmental quality however, Loch Chon being an Acid Waters Monitoring Network site, identified as having undergone acidification after 1960 associated with afforestation (Kreiser et al., 1990; Patrick et al., 1995). Loch Chon was banded band 2 for loss of ANC in the 1995 survey. Management options in this situation are limited to lobbying Central Government in relation to National and International agreements on atmospheric emissions, and influencing land use and land management which might affect the scavenging of atmospheric pollutants within the catchment.

An opportunity to influence land use came in 1995, with consultation from the Forestry Authority on the Strategic Forest plans for the Trossachs area. In addition to compliance with the Forest and Water Guidelines, the Board recommended a reduction in total planted area at restocking, the area to be decided after a joint study involving both agencies. The objective being to reduce atmospheric scavenging of acid forming pollutants to a level which would permit recovery in Loch Chon.

Conclusions

The maintenance or restoration of the appropriate water quality in standing waters is a prerequisite for the protection of the ecosystem. There are, nevertheless, some sites in Scotland where improved water quality alone may be insufficient to ensure restoration, but in general, management of the chemical environment through effective input control should ensure protection of established biota. Water quality management must be on a site by site basis taking account of context, related to the natural condition of the site.

The 1995 classification of standing waters, for which many highland lochs have been monitored for the first time, will provide a baseline against which to judge the success of the Scottish Environment Protection Agency (established in April, 1996) in achieving its principal aim; *"to provide an efficient and integrated environmental protection system for Scotland which will both improve the environment and contribute to the Government's goal of sustainable development"*.

Acknowledgements

The Standing Waters Classification Scheme is the product of an ADRIS sub group comprising the SEPA authors and Dr Julian Hunter. Thanks are due to Ron Harriman of the Freshwater Fisheries Laboratory, Pitlochry, for comments and advice on the acidification aspects of the scheme.

References

Armstrong, J.A., I. R. Fozzard & R. J. Sargent, 1994. Development and collapse of an *Oscillatoria* bloom in Loch Leven during July 1994. Freshwater Forum 4: 203–210.

Association of Directors and River Inspectors of Scotland, 1995. Quality Classification of Scottish Standing Waters. Unpublished Report (fourth revision), ADRIS sub group on Classification of Standing Waters, May 1995.

Bailey-Watts, A.E. & P. Duncan, 1981a. Chemical characterisation – a one year comparative study. In P.S. Maitland (ed.), The Ecology of Scotland's Largest Lochs: Lomond, Awe, Ness, Morar and Shiel, Dr W. Junk Publishers, The Hague.

Bailey-Watts, A. E. & P. Duncan, 1981b. The phytoplankton. *In* P. S. Maitland (ed.), The Ecology of Scotland's Largest Lochs: Lomond, Awe, Ness, Morar and Shiel, Dr W. Junk Publishers, The Hague.

Battarbee, R. W., 1990. The causes of lake acidification, with special reference to the role of acid deposition, Phil. Trans. R. Soc. Lond. B 327: 339–347.

Brinkhurst, R. O., 1974. The Benthos of Lakes. Macmillan, London.

Cantrell, K. J., S. M., Serkiz & E. M. Perdue, 1990. Evaluation of acid neutralising capacity data for solutions containing natural organic acids. Geochimica et Cosmochimica Acta 54: 1247–1254.

Department of the Environment, 1990. "United Kingdom North Sea Action Plan 1985 – 1995" DoE, London.

Doughty, C. R., 1990. Acidity in Scottish rivers. A Chemical and Biological Baseline Survey, Scottish River Purification Boards and the Department of the Environment, East Kilbride, Scotland.

Edwards, P., 1992. Dangerous substances in water: a practical guide. Environmental Data Services Ltd., London.

Ferrier, R. C., A. Malcolm, E. McAlister, A. Edwards & J. Morrice, 1996. Hindcasting of in-loch phosphorus concentrations based on Land Cover Classification. Report for Scotland and Northern Ireland Forum for Environmental Research, Edinburgh.

Fozzard, I. R., 1994. Controlling water quality. In Hutchinson P. & A. Walker (eds), The Loch Leven Trout Fishery – its Future, Proceedings of conference organised by the Institute of Fisheries Management and Scottish Natural Heritage, Kinross.

Fozzard, I. R., C. R. Doughty, B. E. Clelland, 1994. Invertebrates. *In* P. S. Maitland, P. J. Boon & D. S. McLusky (eds), The Freshwaters of Scotland. John Wiley & Sons, Chichester.

Fozzard, I. R. & M. W. Marsden, 1990. The Lake of Menteith: some aspects of its ecology. The Scottish Naturalist 102: 97–129.

Harper, D. M., 1986. The effects of artificial enrichment upon the planktonic and benthic communities in a mesotrophic to hypertrophic loch series in lowland Scotland. Hydrobiologia 137: 9–19.

Henriksen, A., J. Kamari, M. Posch & A. Wilander, 1992. Critical loads of acidity: Nordic surface waters. Ambio 21: 356–363.

Holden, A. V. & L. A. Caines, 1974. Nutrient chemistry of Loch Leven, Kinross. Proceedings of the Royal Society of Edinburgh B 74: 101–121.

Jupp, B. P. & D. H. N. Spence, 1977. Limitations of macrophytes in a eutrophic lake, Loch Leven. II, Wave action, sediments, and waterfowl grazing. J. Ecol. 65: 431–446.

Kreiser, A. M., P. G. Appleby, J. Natkanski, B. Rippey & R. W. Battarbee, 1990. Afforestation and lake acidification: a comparison of four sites in Scotland. Phil. Trans. R. Soc., Lond. B 327: 377–383.

Loch Leven Area Management Advisory Group, 1993. Loch Leven, the report of the Loch Leven Area Management Advisory Group, Forth River Purification Board, Edinburgh.

Maitland, P. S., B. D. Smith, & G. M. Dennis, 1981. The crustacean zooplankton, In Maitland P. S. (ed.). The Ecology of Scotland's Largest Lochs: Lomond, Awe, Ness, Morar and Shiel. Dr W. Junk Publishers, The Hague.

Marsden, M. W., I. R. Fozzard, D. Clark, N. McLean, & M. R. Smith, 1995. Control of phosphorus inputs to a freshwater lake: a case study. Aquaculture Research 26: 527–538.

Morgan, N. C., 1970. Changes in the fauna and flora of a nutrient enriched lake. Hydrobiologia 35: 545–553.

Morgan, N. C., 1974. Historical background to the International Biological Project at Loch Leven, Kinross. Proceedings of the Royal Society of Edinburgh B 74: 45–55.

Moss, B., P. J. Johnes & G. L. Phillips, 1994. August Thienemann and Loch Lomond – an approach to the design of a system for monitoring the state of north-temperate standing waters. In Murphy K. J., M. C. M. Beveridge, R. Tippett (eds), The Ecology of Loch Lomond, Kluwer Academic Publishers, Dordrecht.

North Sea Task Force, 1990. Final Declaration of the Third International Conference on the Protection of the North Sea. North Sea Task Force, The Hague.

Official Journal of the European Communities, 1976. Council Directive on pollution caused by certain dangerous substances discharged into the aquatic environment of the community. (76/464/EEC) No.L129, 18.5.1976, 23 pp.

Organisation for Economic Co-operation and Development, 1982. Eutrophication of Waters, Monitoring Assessment and Control, OECD, Paris.

Patrick, S., D. T. Monteith & A. Jenkins (eds), 1995. UK Acid Waters Monitoring Network: The First Five Years, Department of the Environment and Department of the Environment (Northern Ireland), ENSIS Publishing, London.

Raddum, G. G. & O. A. Saether, 1981. Chironomid communities in Norwegian lakes with different degrees of acidification, Verhandlungen der Internationalen Vereinigung für Theoretische und Angewandte Limnologie 21: 399–405.

Raddum, G. G. & A. Fjellheim, 1984. Acidification and early warning organisms in freshwater in Western Norway, Verh. Int. Ver. Limnol. 22: 1973–1980.

Raddum, G. G., A. Fjellheim & T. Hesthagen, 1988. Monitoring acidification by the use of aquatic organisms, Verh. Int. Ver. Limnol. 23: 2291–2297.

Robertson, J. S., 1984. A key to the common plant communities of Scotland. Macauley Institute for Soil Research, Aberdeen.

Royal Commission on Environmental Pollution, 1992. Sixteenth Report, Freshwater Quality. HMSO, London.

Saether, O. A., 1979. Chironomid communities as water quality indicators, Holarct. Ecol. 2: 65–74.

Smith, I. R. & A. Lyle, 1979. Distribution of Freshwaters in Great Britain. Institute of Terrestrial Ecology, Cambridge.

Stanners, D. & P. Bourdeau (eds), 1995. Europe's Environment: The Dobris Assessment, European Environment Agency. Office for the Official Publications of the European Communities, Luxembourg.

Vollenweider, R. A., 1968. Scientific fundamentals of the eutrophication of lakes and flowing water, with particular reference to phosphorus and nitrogen as factors in eutrophication. Organisation for Economic Cooperation and Development Technical Report. PAS/CS1/68, Paris.

Weerekoon, A. C. J., 1956. Studies on the biology of Loch Lomond 1. The benthos of Auchentullich Bay, Ceylon J. Sci. (C) 7: 1-94.

Wiederholm, T., 1980. Use of benthos in lake monitoring, J. Wat. Pollut. Cont. Fed. 52: 537–547.

Wright, R. F. & A. Henriksen, 1983. Restoration of Norwegian lakes by reduction in sulphur deposition. Nature 305: 422–424.

Hydrobiologia 395/396: 455–469, 1999.
D.M. Harper, B. Brierley, A.J.D. Ferguson & G. Phillips (eds), The Ecological Bases for Lake and Reservoir Management.

Keeping pristine lakes clean: Loughs Conn and Mask, western Ireland

M. L. McGarrigle[1] & W. S. T. Champ[2]
[1]*Environmental Protection Agency, Castlebar, Co. Mayo, Ireland*
[2]*Central Fisheries Board, Glasnevin, Dublin 9, Ireland*

Key words: Ireland, oligotrophic, charr, Lough Conn, Lough Mask

Abstract

The large western Irish lakes are natural salmonid lakes and a unique ecological resource. Lough Mask (82 km^2) and Lough Conn (50 km^2), Co. Mayo, are among the finest, natural, wild brown trout fisheries in Europe. Recently, however, threats have arisen to water quality in Lough Conn as a result of agricultural intensification. A doubling of phosphorus (P) inputs to Lough Conn for the period 1980 to 1990 caused a number of ecological changes such as an apparent increase in littoral algal production; while arctic charr (*Salvelinus alpinus* L.) have also disappeared from the lake over the same period of time. The changes in Lough Conn appear to represent an early phase in the eutrophication process. The western Irish lakes represent a super-sensitive ecological category requiring particularly strict catchment controls in order to protect the diversity of native flora and fauna including unusual glacial relict species and the unique genetic strains of fish which are present. Catchment management committees were established in order to study and reverse the perceived early-stage eutrophication problem in Lough Conn and in order to prevent the occurrence of similar problems in the case of Lough Mask. These committees have a sectoral basis covering: agriculture, local authority functions (sewage, rural septic tanks, planning and environmental control), forestry, peat harvesting and drainage works, as well as having representation from the fisheries boards and the Environmental Protection Agency (EPA). Management of these relatively pristine lakes is perhaps a more difficult problem than that of cleaning up an obviously degraded resource. This is especially so since they are located in regions which are economically underdeveloped and where a real need exists to improve economic returns from farming, tourism and industry in the catchments. The importance of developing a sustainable, ecologically-based management strategy which will maintain existing high quality lakes into the next century with their irreplaceable genetic and ecological resources intact cannot be over-emphasised. The challenge is to devise refined management strategies which will eliminate nutrient wastage to waters and yet allow an appropriate level of sustainable development in the area. The management of Lough Conn and Lough Mask is described in some detail concentrating in particular on the efforts to reduce phosphorus loss from agriculture in the catchment but also emphasising the importance of a multi-sectoral approach to phosphorus load reduction.

Introduction

The large western lakes in Ireland are among the most important natural wild brown trout fisheries in Europe. In addition to their importance as angling lakes, they hold important populations of glacial relict species such as *Mysis relicta* (Lovén). Some of the more important lakes are characterised in Table 1. Lough Corrib in County Galway together with Lough Conn, and Lough Mask in County Mayo are the main western lakes which with their sister lakes, Cullin, Carra and Nafooey cover a surface area in total of some 330 km^2. Lough Melvin in Leitrim, Lough Arrow in Sligo and the Killarney Lakes, in particular Lough Leane, are also grouped under the general category of 'western lakes' for the purposes of this paper. In total these lakes comprise a large proportion of standing waters in the Republic of Ireland. Loughs Conn, Mask and Corrib are situated where the Carboniferous limestone of the Central Plain meets the harder more resistant siliceous rocks (granite, gneiss and schist) of the Connaught Uplands. Lough Leane in Kerry has a similar meeting of old red sandstone and granite. Lough Melvin in Leitrim/Donegal/Fermanagh is located on limestone bedrock. Lough Arrow, Co. Sligo is also situated on limestone with significant karst areas in the catchment. Limestone solution may have played some part in the formation of Lough Conn, Mask and Corrib, but their relatively great depth and position astride the major geological boundary suggests that selective glacial erosion by the Connemara ice-sheet may be the principal factor in the formation of these lakes.

This paper concentrates in particular on the management of Lough Mask (82 km^2) and Lough Conn (50 km^2), two of the largest lakes in this western group of lakes (Figures 1a, b). Recent, quite subtle water quality problems in Lough Conn are outlined and the approach taken to management of these problems is discussed. Preventative management has been initiated in the case of Lough Mask which is the most pristine of the western lakes group. Ecological management of fish populations is undertaken on an ongoing basis by the Central and Regional Fisheries Boards since the early 1960s. Both lakes are also important as a source of high quality drinking water for towns and rural communities in their catchments. The most pressing water management issue in respect of these lakes is that of controlling phosphorus inputs. Approaches to reducing the phosphorus load to the lakes are discussed and prospects for the future management of the lakes based on ecological principles are considered.

Lough Conn discharges southward through Lough Cullin to the River Moy. Typical runoff from Lough Conn's total catchment area of 414 km^2 averages approximately 12 m^3 s^{-1}. The River Moy/Lough Conn system was subjected to arterial drainage in the 1960s with a significant lowering in the level of Lough Conn such that the original shoreline and littoral regions are now quite different to that which evolved naturally following the lake's formation. Occasionally the flow at Pontoon bridge between the two lakes is reversed and enriched water of higher ionic content moves from Lough Cullin into Lough Conn – generally following periods of heavy rainfall in the southern part of the catchment but perhaps also during periods of strong southerly winds when a wind-driven current occurs.

Lough Mask, situated upstream of the Lough Corrib catchment, receives drainage from Lough Carra (a highly calcareous system) and discharges principally through underground channels, through karst limestone, to Lough Corrib. A canal at Cong was excavated between 1848 and 1854 to link Loughs Corrib and Mask. The canal was never completed, however, due to a combination of economic and engineering circumstances. Today the channel is dry over much of its length during the summer months as the canal was never staunched to prevent water flowing into the underlying caverns (Delany, 1988). It is necessary to rescue trout from isolated pools along the 6.5 km channel as levels fall each year. Lough Corrib tributaries were subject to modern arterial drainage – the River Clare and tributaries in the period 1954–1962 and the Black and Headford region between 1967 and 1973. Arterial drainage was also carried out in the Carra/Mask catchment between 1979 and 1985.

Irish water quality management generally is guided by the principle that Irish waters are inherently salmonid due to climatic and biogeographical factors. Thus, salmonid criteria are generally applied to rivers and lakes. While the fish and associated ecological communities of salmonid waters are intrinsically important, the use of salmonid criteria which imply strict standards and highest quality waters has a range of added benefits in that:

1. all amenity uses are ensured if the salmonid criteria are met; and
2. the treatment requirements for drinking water and industrial abstraction are significantly less than they are for water from eutrophic sources.

Fish populations

Table 1 lists the fish species found in Lough Mask and Lough Conn, and compares them with a number of the other large western Irish lakes. These lakes are all important salmonid lakes and continue to support natural indigenous fish stocks dating from the Ice Age. These systems are ideally suited to the proliferation and propagation of salmonid species. The lakes are well supplied with natural, fast-flowing tributaries which for the most part are clean and unpolluted providing ample spawning and nursery areas for salmonids. Such nursery and spawning rivers yield up to 0.1 salmon parr per m^2 (O'Grady, pers. comm.). Because of the insular location and oceanic influence on the climate, water temperatures normally fluctuate between 4 and 21°C. Ice cover is rare, except for Lough Carra adjoining Lough Mask where substantial ice cover can occur occasionally, e.g., the winter of 1982/1983. The lakes are relatively shallow, expansive sheets of water exposed to turbulent mixing with mainly erosive stony shores.

Both Lough Conn and Lough Mask hold large stocks of wild brown trout (*Salmo trutta* L.). The latter

Table 1. Location, morphology and fish species present in some large western Irish lakes

Lake	OS Reference	Area (ha)	Max Depth(m)	Altitude (m)	Trophic status	Fish species
Arrow	G 79 13	1264	28	55.2	Mesotrophic	3,5,6,9
Carra	G 17 74	1620	20	20.4	Mesotrophic	3,5,6
Conn	G 18 13	5211	40	12.5	Mesotrophic	1,3,[4],5,6,8*,
Corrib	M 27 33	16841	50	8.8	Oligotrophic/ Mesotrophic	1, 2, 3, [4], 5, 6, +7, 8,9
Leane	V 93 88	2000	67	19.8	Mesotrophic	1, 2∗, 3, 4, 6, 7*, 10, 11
Mask	M 11 62	8294	58	20.7	Oligotrophic/ Mesotrophic	3, 4, 5, 6, +7*
Melvin	G 90 52	2125	45	21.3	Mesotrophic	1, 2, 3, 4∗, 6, 8

Key to fish species:
1 = Salmon, *Salmo salar* L.; **(2)** = Sea Trout,*Salmo trutta* L.;
(3) = Brown Trout, *Salmo trutta* L.; **(4)** = Arctic Charr, *Salvelinus alpinus* L.;
(5) = Pike, *Esox lucius* L.; **(6)** = Perch, *Perca fluviatilis* L.; **(7)** = Roach,*Rutilus rutilus* L.;
(8) = Rudd,*Scardinius erythrophthalmus* L.; **(9)** = Bream,*Abramis brama* L.;
(10) = Tench, *Tinca tinca* L.; **11** = Shad, *Alosa fallax killarnensis* (Lacepede).
+7 = present since 1977. **[4]** = no longer present. * occasional encounter only.
Note: Eels, Minnows, Three- and Ten-spined Sticklebacks are present in all these lakes.

water contains several brown trout sub-populations: including the large ferox trout which feed on other fish and 'Gillaroo' which feed primarily on molluscs. These varieties also occur in Lough Melvin where they have been shown to be genetically distinct (Ferguson & Mason, 1981). The relatively rich feeding conditions in these predominantly limestone lakes allows trout to grow to substantial sizes in comparison with the stunted fish characteristic of oligotrophic mountain lakes in regions of igneous bedrock geology (Kennedy & Fitzmaurice, 1971). Lough Conn has an Atlantic salmon run (*Salmo salar* L.) but L. Mask is isolated by its underground outflow which links it to Lough Corrib and, therefore, does not have a migratory salmon run. In the past both lakes have also had significant arctic charr populations (*Salvelinus alpinus* L.). Maitland & Campbell (1992) note that it is unusual for charr to occur in relatively productive lakes such as the large Irish lakes. Lough Conn, however, suffered a dramatic collapse of its charr population in the late 1980s (O'Grady, 1992) due perhaps to early-stage eutrophication impacts (McGarrigle et al., 1993). Similarly, recent stock surveys of Lough Corrib suggest that the charr population of Lough Corrib has suffered a similar decline within the past decade (O'Grady, pers. comm.). Lough Mask, however, still holds a large stock of charr (O'Grady, pers. comm.).

Kennedy & Fitzmaurice (1971) reported ferox trout eating other trout in Lough Conn in the winter of 1968/1969 but their occurence is thought to have been related to a shortage of food following the lowering of the lake. Lough Conn is regularly fished for salmon using trolled baits; and ferox trout, if present, would be captured using this method also (Tolan, pers. comm.).

Both Lough Conn and Lough Mask also have populations of introduced pike (*Esox lucius* L.), which in Irish is called *an gall iasc* or 'the foreign fish'. Coarse fish species such as perch (*Perca fluviatilis* L.) and rudd (*Scardinius erythrophthalmus* L.) occur in some of the lakes listed in Table 1 and roach (*Rutilus rutilus* L.) were recently introduced into the Lough Corrib and Lough Mask catchments. The unique combination of species makes these lakes particularly important in European terms.

Other ecological features of the western Irish lakes

The open water of these lakes is moderately hard with relatively low colour and good transparency (Flanagan & Toner, 1975). The isolated projection to the southwest of Lough Mask (known locally as 'Upper Mask') and a similar north western extension of Lough Corrib contain low ionic water draining from the siliceous mountainous formations adjoining this area of the lakes. This part of Lough Corrib is an area of special scientific interest because the aquatic floral assemblages exhibit special features which are pecu-

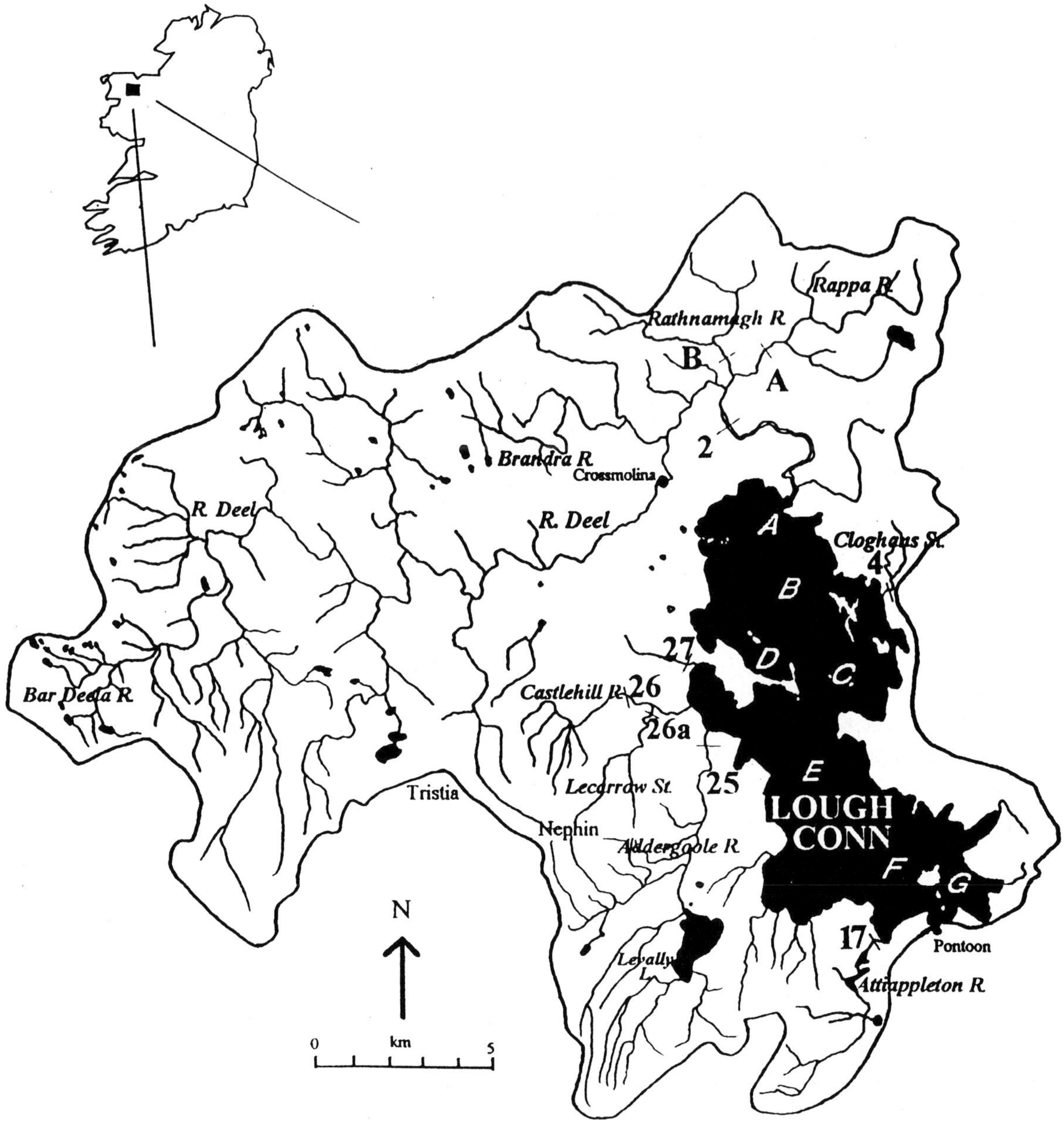

Figure 1. (a) Map of Lough Conn showing main tributaries and phosphorus loading sample stations.

liarly Irish: in particular, the occurrence of *Eriocaulon septangulare* With. and *Najas flexilis* Rostk & Schmidt (Krause & King, 1994). These authors describe these species as being widely distributed in North America and *Najas* is also widely distributed in Ireland, but usually in more enriched waters, while *Eriocaulon* is well distributed but largely confined to Connemara. Elsewhere in Europe, both species have a limited occurrence in Scotland, with *Najas* also recorded from Russia. *Eriocaulan septangulare* also occurs in 'Upper Mask' (King, unpublished data).

Stoneworts (charophytes) are widely distributed in Loughs Corrib and Mask with the densest communities occurring on the eastern embayments where they extend to depths of 6.0 m. One species, *Chara desmacantha* J. Gr. & B.W., grows to 20 to 35 cm and overwinters successfully as green plants. The species is largely confined to Ireland, Scotland and England. In locations where it occurs on the Continent the plants

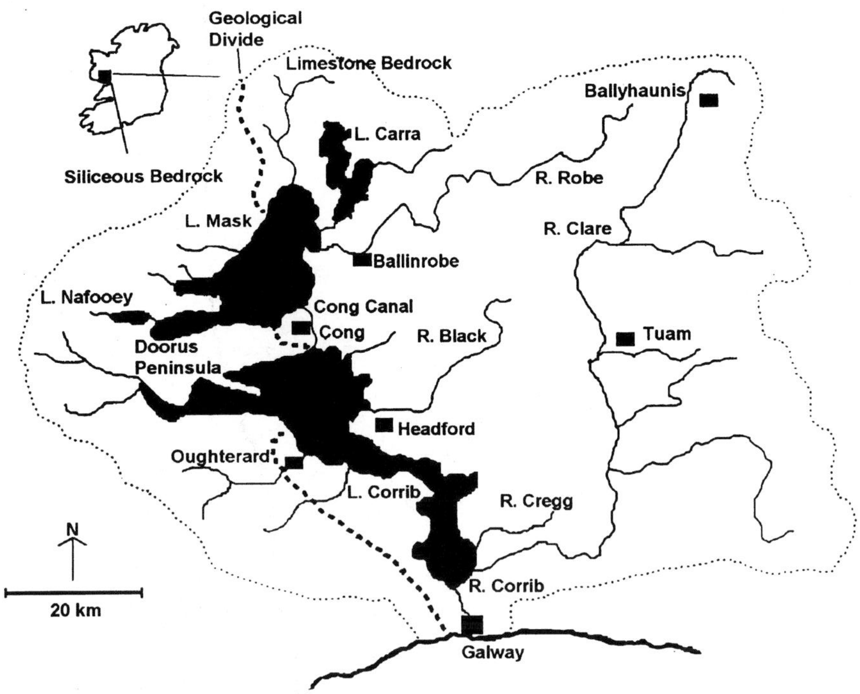

Figure 1. (b) Map of Lough Mask and Lough Corrib system showing main tributaries and east-west geological divide.

are stunted due to climatic factors there preventing overwintering of green apical buds (Krause & King, 1994). This species also occurs in Lough Conn, but the charophytes are more localised there, being confined to sheltered bays (Cloghans Bay and Pratt's Bay) with sparse cover in sheltered locations near Pontoon at the southern end of Lough Conn (Champ & King, 1988).

The western lakes are also noteworthy in that they have a number of unusual macroinvertebrate species; the glacial relict *Mysis relicta* Lovén occurs in Lough Corrib, *Niphargus kochianus irlandicus* Schellenberg, a sightless cave dwelling amphipod has been reported from deep water in Lough Mask (Gledhill et al., 1993) and *Gammarus tigrinus* Sexton has been recorded in Lough Conn O'Grady (pers. comm.). The interesting glacial relict chironomid species, *Corynocera ambigua* (Zetterstedt), has been recorded in Lough Corrib and perhaps may occur in others of the western lakes also (Murray & Ashe, 1983).

The large lakes along the western seaboard, therefore, possess unique ecological communities as evidenced by the flora, macroinvertebrate fauna and fish. Ferguson (1986) dealing with the genetic divergence of genetically isolated 'island' populations of brown trout in Lough Melvin states "The salmonid fishes are amongst the most native and unique components of the Irish fauna. They have been around for 10 000 years and in some cases are found nowhere else in Western Europe".

Water quality status of Lough Conn and Lough Mask

In terms of trophic status, Lough Conn is mesotrophic but with eutrophic tendencies in its North Basin in recent years; while Lough Mask remains on the oligotrophic/mesotrophic boundary. Table 2 gives the typical range of annual mean values for total phosphorus (TP), chlorophyll and Secchi disk. Figure 2 illustrates annual mean TP and chlorophyll values in Lough Conn over the 20 year period 1975–1995 and Figure 3 illustrates annual mean TP and chlorophyll levels for Lough Mask for the period 1976–1995.

Lough Conn

The arctic charr population of Lough Conn suffered a dramatic crash in numbers in the late 1980s and the species is now believed to be virtually extinct in the lake (O'Grady, 1992; McGarrigle et al., 1993). As stratification rarely lasts for more than a few weeks

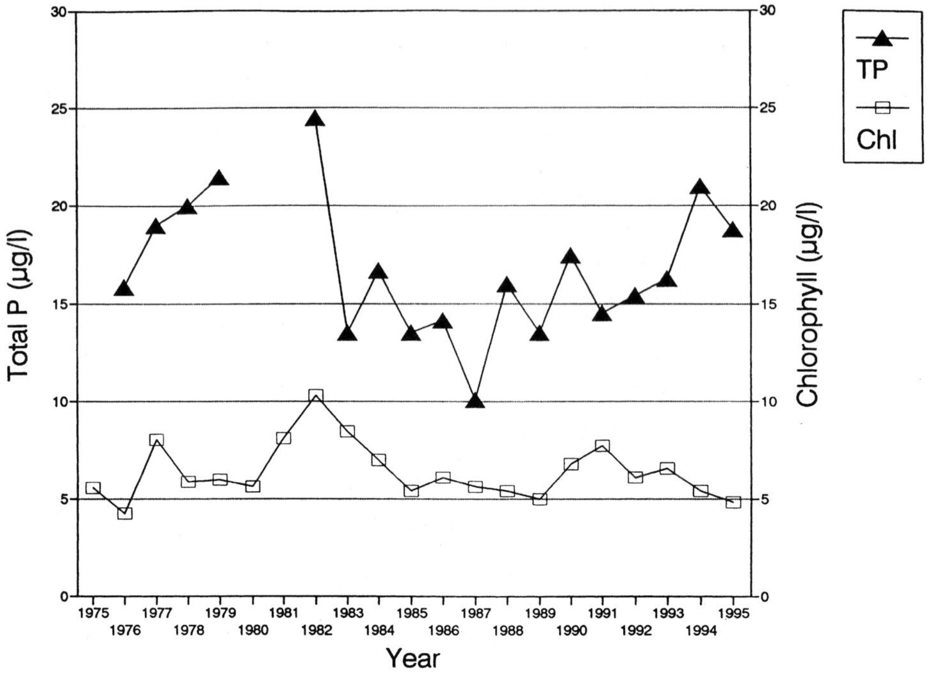

Figure 2. Lough Conn mean annual chlorophyll and total phosphorus 1975–1995

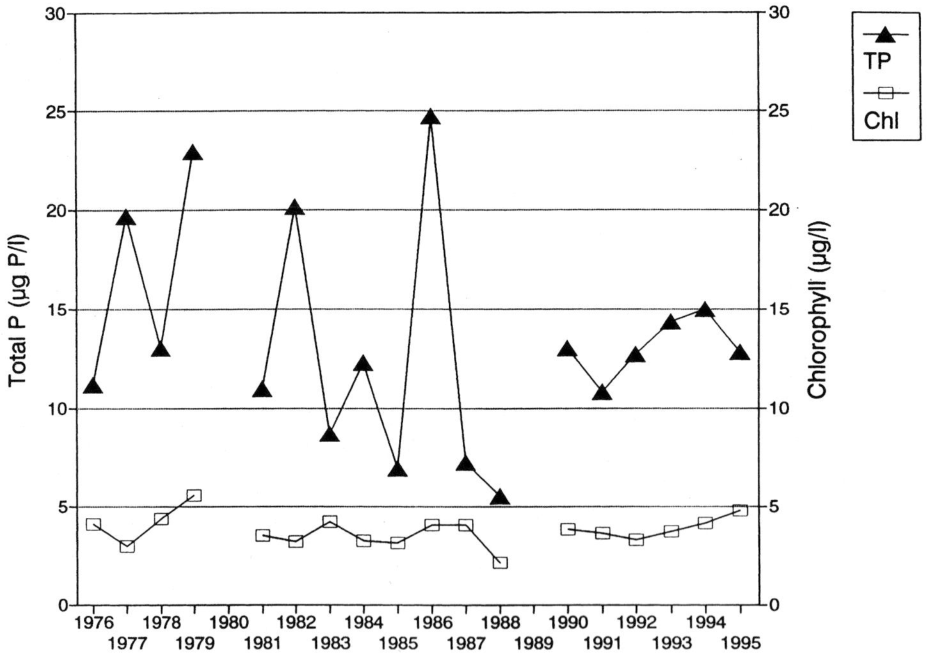

Figure 3. Lough Mask annual mean chlorophyll and total phosphorus 1976–1995

at most, even in the calmest of years, classic hypolimnetic oxygen depletion does not occur. Thus, the loss of charr could not be attributed to hypolimnetic oxygen depletion. Nevertheless, the possibility that eutrophication was responsible for the loss of arctic charr in Lough Conn led to an intensive study to quantify the phosphorus (P) loading to the lake. Monthly open lake monitoring carried over the period 1975–1992 by the Central Fisheries Board did not at the time show any statistically significant trends in chlorophyll, Secchi disk or in the nutrient chemistry of Lough Conn (see Figure 2 for annual mean total phosphorus and chlorophyll). There did, however, appear to be a significant increase in the frequency and intensity of littoral algal growths and accumulations of *Anabaena* (cyanobacteria). These signs, together with the fact that in the Spring of 1990 gill nets placed in the lake overnight became clogged with algae were taken to indicate a change in the ecology of the lake. Significant deposits of senescent *Oscillatoria* combined with filaments of *Mougeotia* were noted on stony littoral areas which would have previously been suitable for charr spawning. It is hypothesised that this may have been implicated in the charr's demise as Baroudy (1995) points out that charr eggs may be particularly sensitive to adverse conditions.

Lough Conn had been reported as mesotrophic since the 1950s, with algal types and abundance indicating mesotrophic conditions e.g. Macan & Lund (1954), Round & Brook (1959), Flanagan & Toner (1975), Champ (1977), Toner et al. (1986) and Clabby et al. (1992). West & West (1906), however, surveying the algae in late August 1904, found dominant algal types which suggested that Lough Conn was oligotrophic or at least oligo-mesotrophic at the time. The suggestion of a change in trophic status is also supported by Murray (1980) who, on the basis of a paleolimnological investigation, suggested that the lake had undergone accelerated eutrophication particularly since the 1960s as evidenced by increasing level of algal pigments in the upper sedimentary layers deposited since the 1960s. In these studies the arterial drainage programme of the 1960s is clearly evident in the sedimentary record as sharp peaks in sodium, iron and manganese levels in cores from the lake. A more detailed study of phosphorus inputs to Lough Conn had been carried out over the period 1979–1982

Table 2. Trophic characteristics of Lough Conn and Lough Mask

	Mask	Conn
Total P (μg P l^{-1})	8–25	9–44
Chlorophyll (μg l^{-1})	0.9–10.3	2.0–13.6
Secchi Disk (m)	3.7–6.1	1.2–3.6

covering all the major inflowing streams to the lake (An Foras Forbartha, unpublished data). This study was one of a series of studies on the trophic status of Irish lakes and it provided a good baseline for a repeat survey carried out in the early 1990s (McGarrigle et al., 1993). This latter survey revealed a striking change in the phosphorus runoff pattern in some of the larger tributaries of Lough Conn. In the 1979–1982 survey in-stream phosphorus concentration was not correlated with rainfall or streamflow in any of Lough Conn's tributaries. In the 1990–1991 survey, however, a strong positive correlation was found between rainfall or streamflow and in-stream unfiltered molybdate reactive phosphate (MRP) in two of the lake's largest tributaries: the River Deel (246 km^2) and in the Addergoole River (40 km^2) (Figure 1). Not all streams exhibited this change suggesting that a real change had occurred in some areas of the catchment and that the difference in runoff pattern was not simply an artefact due to methodological or climatic differences.

The pattern of increasing phosphorus concentration in rivers was taken as an indication of non-point source runoff from agricultural sources which is quite episodic in nature. The implications of this type of episodic increase in phosphorus concentrations during rainfall events is quite important when calculating phosphorus loading to the lake, as there is a multiplier effect between increasing hydraulic load and simultaneously increasing phosphorus concentration. Thus, in the 1990–1991 period the highest concentrations occurred at times of highest flow; whereas 10 years previously there would not have been any increase in concentration during a wet weather event. This meant that the pattern of phosphorus loading to Lough Conn had undergone a remarkable change in quite a short period. In fact the investigation suggested that the loading of TP to Lough Conn had approximately doubled from 18 to 35 t P ann^{-1} in the period between 1980 and 1990 (McGarrigle et al., 1993). The 'natural' background loading of phosphorus to the lake was expected to be of the order to 9 t P assuming a reasonably high specific runoff value of 0.25 kg P ha^{-1} ann^{-1}. This is a typical export figure for some of the cleaner stream sub-catchments of Lough Conn but nevertheless, still significantly greater than the export coefficients noted in the western Lough Mask streams, for example. Thus, the load to the lake was already quite high in 1980 at double this estimated background figure. Follow-up studies since the initial Lough Conn Report in 1993 (McGarrigle et al., 1993) confirm the strong positive correlation

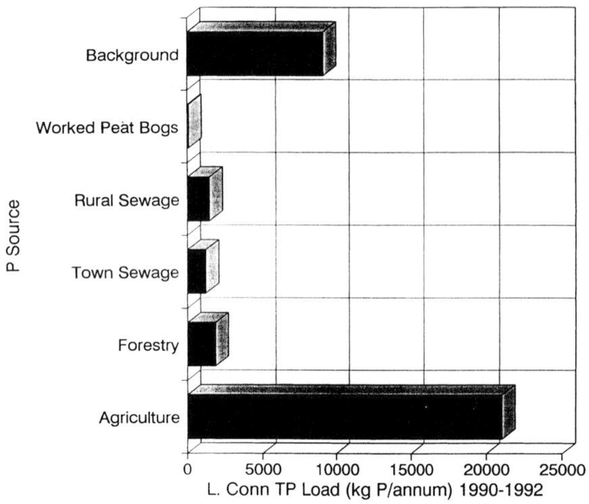

Figure 4. Sectoral breakdown of annual phosphorus load to Lough Conn in the period 1990–1993.

between in-stream phosphorus concentration and rainfall or streamflow. Using autosamplers, taking up to 8 samples per day in selected sub-catchments, it has been possible to demonstrate that there is an underlying, highly predictable, component of phosphorus loading which is dependent on rainfall amounts or streamflow. This predictable proportion can be as high as 90% of the total export from some small catchments (McGarrigle et al., 1997).

Total phosphorus loading from each of the major phosphorus sources in the Lough Conn catchment is illustrated in Figure 4. The manner in which this budget was established is discussed in detail in McGarrigle et al. (1993). The major component of the phosphorus budget is agricultural and as many of the other phosphorus sources, such as town sewage, have not changed very much over the 10-year period, the increase in phosphorus load was believed to be largely agricultural in origin. This is not unexpected, however, as major changes in agricultural practices have occurred in the west of Ireland consequent on European Union Common Agricultural Policy and linked Irish Government policies. This has resulted in a move away from traditional extensive type farming (i.e. grazing, hay making and outwintering of cattle) towards intensive fertilisation and silage making, with winter housing of cattle and spreading of animal manures.

The increased phosphorus load may now be affecting in-lake TP as there is an apparent increase from 1985 to 1995 whereas in the early 1990s a statistically significant pattern was not evident (Figure 2). Open lake chlorophyll values, however, do not yet show an upward trend. The lake sampling programme was primarily a monthly, two location mid-lake sampling programme. In view of the episodic nature of the phosphorus inputs it is possible that this programme may miss such peak values. The inflowing streams were sampled on a weekly basis and even then, many of the peak-flow, high concentrations were missed due to the episodic nature of the inputs. Unweighted MRP concentrations in the Deel River, for example, increased only slightly from 19 to 25 $\mu g\ l^{-1}$ P ($P<0.05$) over the period 1980 to 1990; whereas the flow-weighted concentration doubled in the same period. Thus, it is not unexpected that the monthly open lake sampling programme could miss many of the high TP values. Rapid sedimentation may also be partly responsible for the lack of an apparent increase in mid-lake concentration. The upper, North Basin of Lough Conn is large but quite shallow (2161 ha and 4.4 m mean depth) such that there is plenty of contact between water and sediments and ample opportunity for sedimentation processes to rapidly reduce the inflowing phosphorus concentration. Occasionally, however, very high TP values have been observed following heavy rainfall. A period of intense rainfall occurred in late May and early June in 1993. Following this on 10 June 1993, a TP value of 135 μgl^{-1} P was observed in the lower Deel River and a corresponding mid-lake TP value of 53.5μgl^{-1} P in the North Basin. A chlorophyll peak of 17.8 μgl^{-1} was also measured on the same day in the North Basin. In comparison, a South Basin TP value of 14.6 μgl^{-1} P was recorded on the same day. It is possible that these events which are short-lived, but nevertheless accounting for a disproportionately high percentage of the annual phosphorus load, are not adequately represented in the normal monthly sampling programme. What was quite noticeable (and perhaps more so in retrospect) was the occurrence of winter shoreline blooms or, more correctly, algal accumulations following wet weather events. In Lough Conn these accumulations along shorelines are almost invariably dominated by the cyanobacterial genus *Anabaena* which is positively buoyant and cannot hold its position during calm conditions. In the autumn of 1995, for example, following a long dry summer the first heavy rains brought a series of shoreline blooms during calm weather following the initial wet spell. These shoreline accumulations can occur throughout the winter months. The frequency and intensity of cyanobacterial blooms (*Anabaena* and *Oscillatoria*) may be quite an important phenomenon in these large lakes, perhaps even with regard to their impact on the early life stages of the *Salvelinus alpinus* which spawn in the gravel areas of the shallows.

Lough Mask

Figure 3 illustrates the annual mean TP and chlorophyll levels in Lough Mask over the period 1976–1995. It can be seen that Lough Mask is on the oligo-mesotrophic boundary on the basis of the OECD fixed boundary system (OECD, 1982). Recent fish stock surveys show that it still has a large arctic charr population and while brown trout numbers have dropped, this is believed to be due to a combination of pike predation and recruitment problems due to pollution of some of the smaller streams in the River Robe catchment especially (O'Grady, pers. comm.). In general the ecological quality of the lake is of a very high standard. In view of the problems with Lough Conn, however, a very conservative approach is being taken to the issue of ecological quality in Lough Mask. Lough Mask is the 'jewel in the crown' of the western Irish lakes and it is important that it is maintained in top quality condition.

A TP load measurement programme was put in place in December 1995 with daily autosampling of two of the major tributaries (River Robe and River Owenbrin) supplemented by a weekly grab sample from each of the other important tributaries. Only preliminary results are available from this survey but it has demonstrated that the River Robe contributes a disproportionately high fraction of the TP input to the lake (preliminary findings suggest 85% of the total load). A new phosphorus removal plant came into operation at Ballinrobe Town (population equivalent 5000) near the mouth of the River Robe in early 1996 and this will reduce the annual phosphorus loading to the lake by at least 3 t P per annum. Winter loading of agricultural phosphorus is also a problem in this catchment and a diffuse runoff pattern similar to some of the Lough Conn streams is evident in the upper Robe catchment where there are no industrial or municipal discharges. In national surveys the River Robe has been classified as eutrophic in recent years on the basis of results of routine ecological assessments and physico-chemical surveys (e.g. Bowman et al., 1996; McGarrigle et al., 1996). On the western side of the catchment there has been a problem with overgrazing by sheep on the upland bog and moorland regions of the Partry Mountains. The Owenbrin catchment is affected in this manner, but the overgrazing does not appear to have had any significant effect on the total phosphorus concentrations in the river, as total phosphorus levels are very low throughout the year ($<10\ \mu g l^{-1}$ P). Some afforestation has taken place in the catchment particularly in the last 15 years but forest cover accounted for less than 3% of the total catchment area in 1990.

Lake catchment management strategies

Lough Conn

A Lough Conn Management Committee was convened by Mayo County Council in 1992 as a result of the perceived water quality problem in the lake. The Committee has representatives from all the major bodies which are likely to have an impact on the lake as well as the regulatory agencies. Included are the local authority, Mayo County Council, the convenor of the Committee; Coillte, the state forestry company; Bord na Móna, the state peat harvesting company; the state's farm advisory service, Teagasc; the Farm Development Service of the Department of Agriculture, Food and Forestry; the Central and North Western Regional Fisheries Boards, the fisheries managers who also represent the anglers; and the Environmental Protection Agency. While there is potential for conflict due to the diverse interests of the bodies involved, the Committee has been highly successful in obtaining agreement among the participants regarding management strategies and tactics. The fundamental guideline was that all sectors were required to reduce their phosphorus inputs to the lake. Thus, even those responsible for relatively small proportions of the lake's overall phosphorus loading had to demonstrate some progress was being made towards reducing this loading. This approach makes it more acceptable for the larger contributors, especially the agricultural sector, to introduce more wide-ranging programmes for phosphorus reduction. It was felt to be important that the local authority, in particular, should be seen to be reducing the municipal sewage phosphorus loading as a gesture of 'good faith', so to speak, in order to encourage farmers to participate in the phosphorus reduction programme. The Committee's technical report which outlined the nature of the Lough Conn problem, also agreed a wide ranging series of management recommendations aimed at halving the phosphorus load to the lake (McGarrigle et al., 1993). These were divided up on a sectoral basis covering agriculture, forestry, town sewage, rural sewage and peat harvesting. A series of recommendations for further research into the particular problems of these relatively pristine western lakes was also drawn up.

The following sections outline the principal recommendations of the Lough Conn Report in each of the sectors relevant in the Lough Conn catchment.

Agriculture

In the mid- to late 1980s, EU-funded agricultural improvement schemes in the catchment led to the installation of many new cattle housing units in the Lough Conn Catchment with concomitant slurry storage facilities. Much of this was aimed at general farmyard pollution problems. However, new problems arose due to the consequent increase in land application of slurry in the autumn and a build up of soil phosphorus levels caused by increased use of phosphorus fertiliser. The recommendations, therefore, emphasised the need for further environmental improvements and farm nutrient management measures.

Advice and education

A specific educational campaign for farmers in the catchment was recommended to increase their awareness of the water quality problems caused in the lake by agriculture and to make them more aware of the practical solutions that are available. This includes the commissioning of videos and booklets/leaflets, evening classes, demonstrations and public meetings, and covering topics such as nutrient management planning (NMP) concepts, timing and rate of nutrient applications, slurry spreader calibration to ensure target applications are achieved and quality control of slurry storage systems to minimise dilution with waters. Significant progress has been made in raising local awareness of these topics.

Nutrient Management Planning (NMP)

An important recommendation was for the introduction of an NMP service for farmers in the catchment which will assist them in 'fine tuning' their nutrient application rates on a field-by-field basis. The objective of NMP on the farm is to maximise the efficiency of crop uptake of applied nutrients from either artificial fertilisers or animal manures. This can be achieved by applying the nutrients at the required rate and at the correct time. This simultaneously achieves a reduction in nutrient wastage to waters, effecting an improvement in water quality.

The rate of nutrient application is determined by the crop to be grown (e.g. grass for silage or grazing) and soil fertility levels (i.e. the lower the soil fertility the higher the nutrient requirement). NMP focuses attention on the importance of soil and slurry testing in determining the crop's nutrient requirements and in establishing the nutrients present in organic 'wastes'. The fertiliser value of the nutrients in animal manures is often discounted by farmers. As a consequence the full complement of artificial fertilisers may be applied following slurry applications – resulting in excessive nutrients being applied. The fertiliser value of the applied animal manures must be emphasised and due account taken of this contribution to the crop's requirement. NMP emphasises this concept by recommending appropriate application rates for slurries and 'top up' fertiliser application rates on a field-by-field basis, taking soil fertility, slurry nutrient content and crop characteristics into consideration. The objective is to reduce to a minimum the quantity of nutrients which might be washed off the land. NMP also reminds farmers of the savings that can be made by putting a monetary value on the nutrient content of animal manures. NMP has since been incorporated into national policy measures as part of the REPS programme and the Waste Act.

Slurry application

Recommendations were made for calibration of vacuum tankers to ensure that the target slurry application rates are achieved. Similarly, slurry application equipment and methods were recommended including buffer zones near watercourses and ensuring that sufficient vacuum tankers were available to farmers in the catchment in order to provide maximum flexibility with respect to weather conditions.

Nutrient applications should be made during the growing season and not in late autumn/winter – e.g. slurry should be spread on silage ground, ideally after the first and second cut silage. A co-ordinated and targeted campaign was recommended to convince farmers that fertiliser and slurry applications prior to a heavy rainfall event, increases the risk of nutrient runoff and results in an increased pollution risk and a loss of expensive nutrients. In general, there is a strongly held belief among farmers in the region that rain is required following nutrient applications to 'wash' them into the soil. In a high rainfall area, like the Lough Conn Catchment, this practice of waiting for rain prior to the animal manure application may be a very important factor which increases the risk of phosphorus runoff. Changes have since occurred in the slurry spreading practices used in the catchment with, for example, slurry spreading immediately after silage cutting now being quite a common practice.

Control of dirty water

Dirty or soiled water from farmyards is a significant source of nutrient loss from agriculture. Control of dirty water from farmyards can only be achieved on an individual farm basis as the appropriate controls (volume reduction, storage and land spreading options) require farm-specific data. The Committee recommended the provision of grants to assist farmers to install efficient, cost-effective control systems for dirty water.

Cattle access to streams

The fencing of stream and river banks was recommended in order to control animal access, to create buffer strips and riparian corridors. A small number of cattle standing in a stream can cause severe disruption. Deposition of dung and urine can result in an increased nutrient load to the lake as well as a significant BOD loading to the stream itself. In addition, grazing and trampling of bankside vegetation causes destabilisation of river banks.

EU Agri-environmental programme

In advance of the introduction of the EU Agri-Environmental Programme a series of suggestions were made which the Committee hoped would be incorporated into the Irish Government's Rural Environmental Protection Scheme (REPS). These recommendations were again aimed at reducing phosphorus losses from agriculture to water and they covered grant-aid for machinery, for NMP, soil testing and reduction in the use of artificial fertiliser. Assistance for streambank fencing was also recommended and for the creation of buffer strips along rivers and streams. Programmes were also recommended for the redirection of farm building roof water away from cattle yards, thereby minimising slurry dilution, controlling dirty water and for general improvement of slurry storage facilities. Many of these recommendations were actually incorporated into the REPS programme for Irish farmers which was introduced in 1994. After a slow uptake initially, approximately 30% of farms in certain key catchments are now participating in the REPS programme and obtaining financial support for the type of nutrient loss reduction practices outlined above. The nature of the REPs programme means, however, that it is more attractive to less intensive farmers. Intensive farmers who would be required to reduce their nutrient inputs to participate in REPS are slower to become involved.

Research needs

Recommendations were also made for research programmes related to agricultural impacts in particular: research into the precise forms of phosphate runoff which are causing problems – e.g. to quantify the relative importance of runoff from fields vs. farm- yards and of slurries vs. artificial fertiliser. Research was also recommended into the use of vegetated buffer strips and silt/nutrient traps as a means of reducing nutrient runoff from diffuse agricultural sources under Irish conditions. Similarly, areas such as research aimed at further refinement of the timing and methodology of slurry and fertiliser application in the Lough Conn, catchment were mentioned in the context of research into diffuse agricultural runoff.

Forestry

A number of recommendations were made aimed at controlling the loss of phosphorus from forestry. While previously most attention in Ireland was directed towards the acidification impacts of forestry in poorly buffered regions, the western lakes committees were concerned at the possibility of phosphorus leaching from forestry in upland bog areas. Thus, recommendations covered areas such as minimisation of phosphate leaching by following correct procedures in ground preparation and fertilisation, e.g. Fishery and Forestry Guidelines (Coillte/CFB, 1990). Ground preparation in all circumstances within the catchment should be by mounding and it was recommended that double mould-board mole ploughing should be discontinued. Fertiliser should be spread before turning the sod so that the fertiliser is trapped beneath the sod at the beginning in order to reduce the likelihood of leaching. Aerial fertilisation of trees should not be allowed within the catchment. Reduction of the likelihood of phosphate leaching from peaty soil should be an important management criterion. It was also recommended that large-scale planting of conifers in blanket peat areas should not take place. It was recommended that the lower size threshold of 200 ha for mandatory forestry Environmental Impact Statements should be eliminated altogether such that all forestry developments would require EISs. (In fact, this size limit was reduced to 70 ha by the Department of Agriculture, Forestry and Food in 1996). An emphasis was also placed on the importance of monitoring of phosphorus leachate from coniferous plantations within the catchment – especially in the 4 to 5-year period immediately following planting and fertilisation. Similarly, it was felt that the impact of clear-felling on erosion and leaching and phosphorus transport should be carefully monitored in the catchment. The hydraulic implications of large-scale drainage of blanket and raised bog areas for forestry (and peat production) also requires study. On a related matter it was felt that an assessment should be made of the potential long–term hydrological impacts of mature canopies – evapotranspiration, water retention (possibly with a reduced or rainfall regime under a global warming scenario).

Municipal sewage treatment

Upgrading of the secondary treatment capacity of the Crossmolina works was regarded as an important step as the existing secondary treatment plant is overloaded and phosphate removal is required in order to reduce the output of phosphorus from the town to the lake (population equivalent 1100). In the summer months phosphate values are routinely measured downstream of the STW which are some 10–100 times the level required to limit eutrophication. Phosphate stripping at a 90% removal rate would reduce the contribution from Crossmolina to approximately 125 kg phosphorus per annum and make a substantial contribution to reducing the overall load to Lough Conn. It was also recommended that an educational programme to encourage the use of phosphate-free and low phosphate detergents in the Crossmolina area be undertaken as inevitably it will take some time before a new sewage treatment plant could be constructed and brought on line. Low phosphate detergents have become quite popular with a market share of approximately 40% even in the absence of bye-laws to prohibit the sale of high phosphate detergents.

Hotels and institutions

With respect to point sources, such as hotels and institutions located near Lough Conn, it was recommended that the phosphorus loading to the lake from these institutions should be quantified, and targets set for reducing their individual phosphorus contributions. Phosphate-free detergents should be encouraged for dishwashers and laundry in these hotels and institutions.

Rural septic tanks

Domestic septic tanks are known to cause significant problems to surface waters in waterlogged areas or to groundwater in areas with thin soil overburden layers. A series of recommendations were made aimed at minimising the contribution from this source. Strict planning controls are required for individual houses adjacent to the lake. Groups of houses with septic tanks should be actively discouraged within 400 m of the lake. A number of other recommendations such as improved maintenance of septic tanks were also included in the Committee's recommendations. Mayo County Council subsequently made a number of changes in their planning regulations implementing many of these suggestions.

Peat harvesting

In general terms it is recommended that every effort should be made to minimise the release of both particulate and dissolved humic substances from the commercial bogs in the catchment. Improved design and location and maintenance of settling ponds may be required particularly in order to avoid the flushing out of already settled material in wet weather. The area of silt ponds in use was doubled following the Committee's recommendations.

General research and monitoring recommendations

The Lough Conn report demonstrated the value of baseline surveys and ongoing lake water quality surveys. Without the previous survey in 1979–1982 it would not have been possible to gauge the extent of the change which occurred in the late 1980s and early 1990s. A number of gaps are apparent, however, in our knowledge of the lake. In particular it appears that the mid-lake sampling programme may need to be supplemented or intensified in order to detect the subtle early-stage eutrophication changes which the lake was undergoing. In the case of Lough Conn it was recommended that greater emphasis be placed on the assessment of the shallow littoral areas of the lake in order to obtain a better understanding of the nature of the algal blooms that occurred in these areas. Regular mapping of the lake's rooted vegetation was also required. Intensive lake sampling should be undertaken following rainfall events. Modelling of the phosphorus pathways through the lake itself should be undertaken – with particular emphasis on phosphorus sedimentation processes, the uptake of phosphorus by phytoplankton and rooted vegetation and the role of humic substances in transporting P.

The Lough Conn investigations highlighted the need for a broadly-based ecological approach to the assessment of lakes and in particular the western lakes which are relatively pristine. This would include an assessment scheme which would include surveys of the extent of littoral vegetation, frequency of shoreline blooms and algal accumulations, and surveys of benthic invertebrates, zooplankton and phytoplankton species and assessment of fish populations. While Irish rivers are regularly assessed both in terms of their water chemistry and biology, Irish lakes have not received adequate coverage in the past – due to the large numbers of lakes (over 5500) and the limited resources available.

Since the publication of the Lough Conn Report, however, a range of lake-related research projects have been supported by the Irish Environmental Protection Agency with European Regional Development Fund (ERDF) funding. The projects included research to develop an ecological 'early warning' system to detect early stage eutrophication problems in the western lakes concentrating particularly on the littoral zone of the large western lakes. Another project is developing an ecological assessment methodology for lakes more generally which will be capable of filling the requirements of the Draft Directive on the Ecological Quality of Surface Waters as well as predicting state change within individual lakes. A third project is refining remote sensing chlorophyll detection algorithms for Irish lakes and developing techniques for aquatic vegetation mapping and catchment land use mapping. A fourth research project is developing catchment models predicting phosphorus losses to water from soils based on soil phosphorus burden, soil type, rainfall amounts, slope, etc. These projects are also charged with producing a common geographical information system (GIS) for Irish lakes which will combine data sets from all of the projects. Many hundreds of lakes and their catchments have been surveyed and assessed in the course of these and other lake research projects which are currently being funded.

Management of Lough Mask

Following the experience with Lough Conn a catchment management committee was set up for Lough Mask and more intensive monitoring programme was started in order to provide a solid basis for management recommendations. Rather than waiting for the results of such monitoring to be completed, however, it was decided that relevant recommendations from the Lough Conn catchment should be put in place for the Lough Mask catchment immediately. This approach was agreed in view of the perceived urgency of the need to protect these important lakes, and in view of the results obtained for Lough Conn which is a neighbouring catchment with similar land use practices. In catchment management terms the Lough Conn recommendations were generally desirable and uncontroversial; such that they should be more widely implemented in any case.

In addition to the Lough Mask Committee, a separate programme under the EU Tourism Angling Measure for Ireland is in place aimed at improving and maintaining the fish stocks of the Lough Mask and Lough Corrib catchments. This programme also has a water quality element which is carried out in liaison with the Lough Mask Committee. Ecological management of fish populations is part of this programme and it includes stream habitat rehabilitation (especially following arterial drainage programmes) in order to maximise salmonid spawning, nursery and holding areas over the full life cycle of salmonid fish. Management of introduced species such as pike, which are salmonid predators, also forms part of this programme and fish population assessments have recently been completed for Lough Mask and Lough Corrib (O'Grady, *pers. comm.*). Measures to prevent the introduction of alien species are also in place nationally, as well as controls on the movement of fish between water bodies.

Farm surveys

In both the Lough Conn and the Lough Mask catchments the intensive phosphorus loading surveys carried out identified regions in the catchments which were responsible for a large proportion of the phosphorus exported. These were predominantly the more intensive agricultural regions with the main activities being grassland dairy and beef farming. Farms in these areas were targeted for a blanket survey in order to build up a picture of the phosphorus budget for the agricultural sector in defined sub-catchments. Voluntary co-operation was received from virtually all farms approached and detailed individual farm questionnaires were completed covering details such as animal numbers, artificial fertiliser usage, foodstuffs usage, the area devoted to growing silage on the farm, the area used for grazing of animals, the volume of tanks used for slurry and soiled water containment. Soil phosphorus levels are also recorded where available. The general condition of the farmyards and of their overall pollution risk is also assessed. The data gathered were incorporated into a catchment GIS.

Large catchments such as Lough Conn and Lough Mask (combined area of almost 1500 km^2) at first sight present a major logistics problem in carrying out farm surveys and in pinpointing pollution sources, especially diffuse sources, as there are many thousands of farms in the catchments. The detailed surveys of phosphorus loading from subcatchments demonstrated that there are localised 'hot spots' which are responsible for a large fraction of the phosphorus exported from the catchment. This concept leads to the possibility that limited subcatchments can be targeted for intensive surveying and pollution control measures in order to maximise the return on effort invested. Two approaches have proved useful in the case of these western lakes. First, the use of the GIS-based CORINE land use map for Ireland enables all areas of more intensive grassland agriculture to be pinpointed on a map. Areas of 'improved pasture' in particular tend to coincide with the more intensively farmed areas with higher soil phosphorus burdens, for example. The second technique in use is the rapid ecological screening of streams in a catchment using the national Quality Rating or 'Q-Value' System (Flanagan & Toner, 1972). This entails surveying streams which are smaller than those normally included in national river surveys. Rapid surveys allow colour-coded GIS maps to be prepared which highlight the polluted streams in the catchment. Overlaying the two sets of maps and placing the polluted streams over the intensively farmed regions gives an initial set of priority regions for immediate attention. This region of overlap will comprise quite a small proportion of the total catchment area and thus, allow surveys which are more manageable in that they include hundreds of farms rather than many thousands. It is hoped to refine this system further by producing a more detailed phosphorus loss model which takes a wider range of driving forces into account – this is currently the subject of an EPA funded research programme as mentioned above. This should also help to further narrow down the main sources of diffuse agricultural phosphorus loss in a catchment. Significant reductions in phosphorus losses can be achieved by the use of farm surveys followed up by advice and encouragement to use NMP on farms and, if necessary, legal action to prevent unnecessary phosphorus losses from farmyards and fields. It is concluded that reducing agricultural phosphorus losses to water can be a manageable proposition by pinpointing the catchment's most likely 'hot spots' for phosphorus loss and then targeting the farms in these relatively limited regions.

The follow-up programme deals with farms which were obviously causing water quality problems due to either direct farmyard losses or to field surface and subsurface runoff. Initially general pollution control advice is given and warnings regarding the likelihood of legal action if a major improvement is not apparent in subsequent inspections. Failure to make the required improvements results in the farmer being issued with a 'Section 12' Notice under the Water Pollution Act which requires specified changes or improvements to be made on the farm. Legal proceedings follow subsequently if this formal legal notice is not complied with. Another recent introduction in Irish environmental law (Waste Act) is the facility to allow local authorities to require farmers to prepare nutrient management plans in order to control the nutrient fluxes on farms and thus to reduce nutrient losses to water. The Fisheries Boards may also prosecute polluters under the Irish Fisheries Acts.

Conclusion

The large western Irish lakes are important in the wider European context and not solely in the Irish context. Ecological management is crucial for the future well being and ecological health of these lakes. The current management strategy has placed strong emphasis on the need to reduce the phosphorus loadings to these lakes. The management teams now in place for Lough Conn and Lough Mask have worked well in this context and in future the same committees should be capable of tackling other ecological problems such as overgrazing which is a major issue in parts of these catchments at present. Similarly, problems which are less pressing at the moment but which may become more important in future – climate change impacts, shoreline aesthetics, alien species introductions and the like – can be tackled by multi-agency committees of the type now in place for Lough Conn and Lough Mask.

References

Baroudy, E., 1995. Arctic charr (*Salvelinus alpinus*) in Windermere (Cumbria). Freshwater Forum 5: 185–192.

Bowman, J J., K J. Clabby, J. Lucey, M L. McGarrigle & P F. Toner, 1996. Water Quality in Ireland 1991–1994. Environmental Protection Agency, Ardcavan, Wexford.

Champ, W S T., 1977. Trophic status of fishery lakes. In Downey, W. & G. Ni Uid (eds), Lake Pollution Eutrophication Control. Stationery Office, Dublin: 65–78.

Champ, W S T. & J J. King, 1988. The Trophic Status of Lough Conn: An Extended Study. Central Fisheries Board, Dublin.

Clabby, K J., J. Lucey, M L. McGarrigle, J J. Bowman, P J. Flanagan & P F. Toner, 1992. Water Quality in Ireland 1987–1990, Part One: General Assessment. Water Resources Division, An Foras Forbartha, Dublin. 189 pp.

Coillte and Central Fisheries Board, 1990. Forestry and Fisheries Guidelines. Department of Agriculture Forestry & Food, Central Fisheries Board, Dublin.

Delany, R., 1988. Ireland's Inland Waterways. Appletree Press., Belfast, ISBN 0-86281-200-3. 200 pp.

Ferguson, A. & F. M. Mason, 1981. Allozyme evidence for reproductively isolated sympatric populations of Brown Trout *Salmo trutta* L. In Lough Melvin, Ireland. J. Fish. Biol. 18: 629–642.

Ferguson, A., 1986. Lough Melvin, a unique fish community. Went Memorial Lecture, 1985. Occasional Papers in Irish Science and Technology, Royal Dublin Society, Dublin.

Flanagan, P F. & P T. Toner, 1972. The national survey of Irish rivers: A report on water quality. An Foras Forbartha, Water Resources Division, Dublin. 213 pp.

Flanagan, P J. & P F. Toner, 1975. A Preliminary Survey of Irish Lakes. An Foras Forbartha. Dublin.

Gledhill, T., D W. Sutcliffe, & W D. Williams, 1993. British freshwater Crustacea Malacostraca: A key with ecological notes. Freshwater Biological Association, Scientific Publication No. 52. 173 pp.

Kennedy, M. & P. Fitzmaurice, 1971. The growth and food of brown trout *Salmo trutta* L. in Irish waters. Proc. R. Ir. Acad. 71B 18: 269–352.

Krause, W. & J J. King, 1994. The ecological status of Lough Corrib, Ireland, as indicated by physiographic factors, water chemistry and macrophytic flora. Vegetatio 110: 149–161.

Macan, T T. & J W G. Lund, 1954. Records from some Irish Lakes. Proceedings of the Royal Irish Academy, 56B: 135.

Maitland P S. & R N. Campbell, 1992. Freshwater Fishes. Harper Collins, 368 pp.

McGarrigle, M L., 1993. Aspects of river eutrophication in Ireland. Ann. Limnol. 29: 355–364.

McGarrigle, M L., W S T., Champ, R. Norton, M. Moore & P. Larkin, 1993. The Trophic Status of Lough Conn. 89pp. Mayo County Council, Ireland.

McGarrigle, M L., K J. Clabby, P F. Toner, M. Neill, J. Lucey, M. Flanagan & M B. Quinn, 1996. Water Quality in Ireland 1991–1994, Statistical Compendium of River Quality Data. Environmental Protection Agency, Ardcavan, Wexford, 904 pp.

McGarrigle, M L., L. Kilmartin & R. Hallissey, 1997. Phosphorus loss from agriculture to water - a twin-catchment analysis in the Lough Conn catchment. In Collins, J. (ed.), Proceedings of the Irish Soil Science Association, University College Dublin, April 1977.

Murray, D A., 1980. Paleolimnological studies on lakes of known differing trophic status to establish eutrophication rates and standards. Report to the European Commission, contract No. 217-77-1 ENV. EIR.

Murray, D A. & P. Ashe, 1983. An inventory of the Irish Chironomidae (Diptera). Memoirs of the American Entomological Society 34: 223–233.

OECD, 1982. Eutrophication of waters, monitoring assessment and control. OECD, Paris.

O'Grady, M., 1992. A review of fish stocks in Lough Conn 1978-1990. In; The Future of the Moy Fisheries, Integrated Management and Development Plan for the River Moy System. North Western Fisheries Board, Ballina, Ireland.

Round, F E. & A J. Brook, 1959. The phytoplankton of some Irish Loughs and an assessment of their trophic status. Proceedings of the Royal Irish Academy, 60B: 67.

Toner, P F., K J. Clabby, J J. Bowman & M L. McGarrigle, 1986. Water quality in Ireland, The current position, Part One: General Assessment. Water Resources Division, An Foras Forbartha, Dublin. 116 pp. ISBN 1 850853 060 2.

West, W. & G S. West, 1906. A comparative study of the Plankton of some Irish lakes. Proceedings of the Royal Irish Academy, 33B: 77.